SAP PRESS e-books

Print or e-book, Kindle or iPad, workplace or airplane: Choose where and how to read your SAP PRESS books! You can now get all our titles as e-books, too:

- By download and online access
- For all popular devices
- And, of course, DRM-free

Convinced? Then go to www.sap-press.com and get your e-book today.

Configuring Plant Maintenance in SAP S/4HANA®

SAP PRESS is a joint initiative of SAP and Rheinwerk Publishing. The know-how offered by SAP specialists combined with the expertise of Rheinwerk Publishing offers the reader expert books in the field. SAP PRESS features first-hand information and expert advice, and provides useful skills for professional decision-making.

SAP PRESS offers a variety of books on technical and business-related topics for the SAP user. For further information, please visit our website: *www.sap-press.com*.

Karl Liebstückel
Plant Maintenance with SAP S/4HANA: Business User Guide
2021, 665 pages, hardcover and e-book
www.sap-press.com/5180

Jawad Akhtar
Production Planning with SAP S/4HANA (2nd Edition)
2021, 1092 pages, hardcover and e-book
www.sap-press.com/5373

Namita Sachan, Aman Jain
Warehouse Management with SAP S/4HANA (4th Edition)
2024, 1001 pages, hardcover and e-book
www.sap-press.com/5886

Jawad Akhtar, Martin Murray
Materials Management with SAP S/4HANA:
Business Processes and Configuration (3rd Edition)
2024, 1018 pages, hardcover and e-book
www.sap-press.com/5835

Caetano Almeida
Material Requirements Planning with SAP S/4HANA (2nd Edition)
2024, 563 pages, hardcover and e-book
www.sap-press.com/5861

Karl Liebstückel

Configuring Plant Maintenance in SAP S/4HANA®

Editor Megan Fuerst
Acquisitions Editor Emily Nicholls
Copyeditor Doug McNair
Cover Design Graham Geary
Photo Credit Shutterstock: 2383391891/© Czajnikolandia
Layout Design Vera Brauner
Production Kelly O'Callaghan
Typesetting SatzPro, Germany
Printed and bound in the United States of America/Canada, on paper from sustainable sources

ISBN 978-1-4932-2604-7

© 2025 by Rheinwerk Publishing, Inc., Boston (MA)
2nd edition 2025

Library of Congress Cataloging-in-Publication Data
Names: Liebstuckel, Karl, author.
Title: Configuring plant maintenance in SAP S/4HANA / by Karl Liebstu?ckel.
  Description: 2nd edition. | Bonn ; Boston : Rheinwerk Publishing, 2024. |
  Includes index.
Identifiers: LCCN 2024039382 | ISBN 9781493226047 (hardcover) | ISBN
  9781493226054 (ebook)
Subjects: LCSH: SAP HANA (Electronic resource) | Plant maintenance--Data
  processing.
Classification: LCC TS192 .L54 2024 | DDC 658.2/02028553--dc23/eng/20240906
LC record available at https://lccn.loc.gov/2024039382

All rights reserved. Neither this publication nor any part of it may be copied or reproduced in any form or by any means or translated into another language, without the prior consent of Rheinwerk Publishing, 2 Heritage Drive, Suite 305, Quincy, MA 02171.

Rheinwerk Publishing makes no warranties or representations with respect to the content hereof and specifically disclaims any implied warranties of merchantability or fitness for any particular purpose. Rheinwerk Publishing assumes no responsibility for any errors that may appear in this publication.

"Rheinwerk Publishing" and the Rheinwerk Publishing logo are registered trademarks of Rheinwerk Verlag GmbH, Bonn, Germany. SAP PRESS is an imprint of Rheinwerk Verlag GmbH and Rheinwerk Publishing, Inc.

All of the screenshots and graphics reproduced in this book are subject to copyright © SAP SE, Dietmar-Hopp-Allee 16, 69190 Walldorf, Germany.

SAP, ABAP, ASAP, Concur Hipmunk, Duet, Duet Enterprise, ExpenseIt, SAP ActiveAttention, SAP Adaptive Server Enterprise, SAP Advantage Database Server, SAP ArchiveLink, SAP Ariba, SAP Business ByDesign, SAP Business Explorer (SAP BEx), SAP BusinessObjects, SAP BusinessObjects Explorer, SAP BusinessObjects Web Intelligence, SAP Business One, SAP Business Workflow, SAP BW/4HANA, SAP C/4HANA, SAP Concur, SAP Crystal Reports, SAP EarlyWatch, SAP Fieldglass, SAP Fiori, SAP Global Trade Services (SAP GTS), SAP GoingLive, SAP HANA, SAP Jam, SAP Leonardo, SAP Lumira, SAP MaxDB, SAP NetWeaver, SAP PartnerEdge, SAPPHIRE NOW, SAP PowerBuilder, SAP PowerDesigner, SAP R/2, SAP R/3, SAP Replication Server, SAP Roambi, SAP S/4HANA, SAP S/4HANA Cloud, SAP SQL Anywhere, SAP Strategic Enterprise Management (SAP SEM), SAP SuccessFactors, SAP Vora, TripIt, and Qualtrics are registered or unregistered trademarks of SAP SE, Walldorf, Germany.

All other products mentioned in this book are registered or unregistered trademarks of their respective companies.

Contents at a Glance

1	SAP Projects in Plant Maintenance	27
2	Configuring Organizational Structures	79
3	Configuring Generic Functions	109
4	Configuring the Structure of Technical Systems	169
5	Configuring the Work Order Cycle	249
6	Configuring Preventive Maintenance	415
7	Configuring Additional Business Processes	433
8	Configuring the SAP Fiori Launchpad for Plant Maintenance	581
9	Usability	623

Contents

Acknowledgments 13
About the Book 15

1 SAP Projects in Plant Maintenance   27

1.1 A Possible Process for Your Plant Maintenance Project with SAP 27
  1.1.1 Implementation Strategy 27
  1.1.2 SAP Activate Methodology 29
1.2 General Risk Factors and Success Factors in SAP Projects: An Empirical Survey 35
  1.2.1 Risk Factors 35
  1.2.2 Success Factors 38
1.3 Tips for Your Plant Maintenance Project 39
  1.3.1 Discover Phase 39
  1.3.2 Prepare Phase 48
  1.3.3 Explore Phase 65
  1.3.4 Realize Phase 74
  1.3.5 Deploy Phase 75
  1.3.6 Run Phase 77
1.4 Summary 77

2 Configuring Organizational Structures   79

2.1 General SAP Organizational Units 80
  2.1.1 Company Code 81
  2.1.2 Controlling Area 84
  2.1.3 Plant in General 87
2.2 The Plant from a Maintenance Perspective 89
2.3 Maintenance-Specific Organizational Units 94
2.4 Work Centers 97
2.5 Summary 107

Contents

3 Configuring Generic Functions — 109

3.1	Object Information	110
3.2	Status Management	113
3.3	Number Assignment	119
3.4	Warranties	124
3.5	Measuring Points and Counters	127
3.6	Permits	132
3.7	Partners	134
3.8	Documents	141
3.9	Field Selection	154
3.10	List Variants	158
3.11	Multilevel Lists	163
3.12	Change Documents	165
3.13	Summary	167

4 Configuring the Structure of Technical Systems — 169

4.1	Structuring Elements	169
4.2	Functional Locations and Reference Functional Locations	171
4.3	Equipment	182
4.4	Fleet Management	191
4.5	Links and Networks	198
4.6	Linear Asset Management	201
4.7	Materials and Assemblies	211
	4.7.1 Material Types	212
	4.7.2 Material Master Layout	217
	4.7.3 Field Selection	227
4.8	Serial Numbers	232
4.9	Bill of Materials	239
4.10	Summary	247

5 Configuring the Work Order Cycle   249

5.1 Notifications   250
5.1.1 Notification Types   252
5.1.2 Catalogs and Catalog Profiles   265
5.1.3 Other Functions   272

5.2 Orders   283
5.2.1 Order Types   287
5.2.2 Availability Check   304
5.2.3 Electronic Parts Catalogs   312
5.2.4 Material Planning   321
5.2.5 Scheduling   332
5.2.6 Capacity Planning   345
5.2.7 Printing   353
5.2.8 Estimated Costs and Costing   362
5.2.9 Order Settlement   377
5.2.10 Project Management and Investment Management   384
5.2.11 Other Functions   392

5.3 Completion Confirmations   401
5.3.1 SAP S/4HANA Asset Management Completion Confirmations   401
5.3.2 Cross-Application Time Sheet Confirmations   407

5.4 Summary   412

6 Configuring Preventive Maintenance   415

6.1 Task Lists   416
6.2 Maintenance Plans   425
6.3 Summary   431

7 Configuring Additional Business Processes   433

7.1 External Processing   434
7.1.1 External Processing as an Individual Purchase Order   435
7.1.2 External Processing with Service Specifications   440

7.2 Refurbishment   448
7.2.1 Material Master   449

	7.2.2	Refurbishment Notifications	454
	7.2.3	Refurbishment Orders	456
	7.2.4	Material Requirements Planning	461
7.3	Subcontracting		463
7.4	Calibration of Test/Measurement Equipment		468
	7.4.1	Customizing Functions for Plant Maintenance	469
	7.4.2	Customizing Functions for Quality Management	483
7.5	Pool Asset Management		500
7.6	Maintenance Event Builder		517
7.7	Shift Reports and Shift Notes		519
7.8	Checklists		529
	7.8.1	Checklist Processing (Basic Version)	530
	7.8.2	Checklist Processing (Extended Version)	538
7.9	Work Order Cycle Using the Phase Model		551
7.10	Summary		579

8 Configuring the SAP Fiori Launchpad for Plant Maintenance 581

8.1	Basics of SAP Fiori		581
	8.1.1	SAP Fiori Design	581
	8.1.2	Types of SAP Fiori Apps	582
	8.1.3	Characteristics of SAP Fiori	585
8.2	How to Configure an SAP Fiori Launchpad with SAPUI5 Apps		586
	8.2.1	Step 1: Check SAP System Data	589
	8.2.2	Step 2: Explore the SAP Fiori Apps Reference Library	590
	8.2.3	Step 3: Configure OData Services	594
	8.2.4	Step 4: Configure ICF Services	597
	8.2.5	Step 5: Create a Launchpad Catalog for the SAP Fiori Launchpad	598
	8.2.6	Step 6: Create a Business Group for the SAP Fiori Launchpad	601
	8.2.7	Step 7: Create a Role and Assign Authorizations	603
	8.2.8	Step 8: Initialize the App	604
8.3	Tabular Overview of All SAP Fiori Apps		605
8.4	How to Configure an SAP Fiori Launchpad with Non-SAPUI5 Apps		610
	8.4.1	Step 1: Check SAP System Data	612
	8.4.2	Step 2: Browse the SAP Fiori Apps Reference Library	612

	8.4.3	Step 3: Configure OData Services	614
	8.4.4	Step 4: Configure ICF Services	614
	8.4.5	Step 5: Create a Business Catalog for the SAP Fiori Launchpad	615
	8.4.6	Step 6: Create a Business Group for the SAP Fiori Launchpad	617
	8.4.7	Step 7: Create or Change a Role and Assign Authorizations	618
	8.4.8	Step 8: Initialize the App	619
8.5	Tabular Overview of All Web Dynpro Apps		620
8.6	Summary		621

9 Usability 623

9.1	Options at the User's Disposal		624
	9.1.1	General User Parameters and Default Values	624
	9.1.2	Maintenance-Specific User Parameters	627
	9.1.3	Roles, Favorites, and My Home	633
	9.1.4	List Variants and Dynamic Lists	635
	9.1.5	Personalizing Input Help	642
	9.1.6	Buttons and Key Combinations	643
	9.1.7	Table Controls	644
9.2	The IT Department's Nonprogramming Options		646
	9.2.1	Transaction Variants	646
	9.2.2	Customizing	652
	9.2.3	Action Box	656
	9.2.4	SAP Service and Asset Manager	657
	9.2.5	SAP Fiori	663
	9.2.6	GuiXT	664
	9.2.7	SAP Screen Personas	670
9.3	The IT Department's Programming Options		673
	9.3.1	Upstream Transactions	673
	9.3.2	Business Application Programming Interfaces	675
	9.3.3	Web Interface	677
	9.3.4	Customer Exits	680
	9.3.5	Business Add-Ins and Enhancement Points	683
	9.3.6	SAP Business Workflow	685
9.4	Summary		690

Appendices — 691

A	**Project Plans and Overviews**	691
B	**The Author**	709

Index 711

Acknowledgments

I would like to thank everyone who contributed to the success of this book:

- My editors, **Megan Fuerst** and **Emily Nicholls**, for their encouraging support and their constructive suggestions for improvements.
- My beloved wife, **Brigitta Liebstückel**, who has now been with me for more than 40 years, not only for her loving and moral support while I wrote this book, but also, and, in particular, for spending such a lot of time on translating many passages and proofreading the whole book.

Nightwood

About the Book

If you plan for a year, plant a seed. If for ten years, plant a tree. If for a hundred years, teach the people. When you sow a seed once, you will reap a single harvest. When you teach the people, you will reap a hundred harvests.
—KUAN CHUNG TZU, Chinese philosopher, Book of Master Kuan

You're reading the second edition of *Configuring Plant Maintenance in SAP S/4HANA*, which adopts a technical viewpoint and focuses on the knowledge of *how* to install a plant maintenance solution, for example:

- How do I set up my organizational structures?
- How should I configure the Customizing settings so the asset structures can be mapped to the users' requirements?
- How should I configure Customizing so that my business processes can be handled in the manner discussed with the users?
- How should I configure the usability tools to encourage greater user acceptance of the system?
- How should I set up an SAP Fiori launchpad?

This book provides an overview of the configuration options currently available. Using more than 35 years of experience working in SAP plant maintenance on close to 100 customer projects, I'll not only show you *how* you can adapt SAP S/4HANA to the needs of your plant maintenance department and your users, but I'll also highlight some essential dos and don'ts.

Customer examples are used to illustrate what other companies have done. Numerous practical tips are also provided, and they are equally useful whether you're a beginner or an advanced user of a previous release of the system.

To begin with, let's sort out some naming differences: Which is the correct name for the plant maintenance application of SAP?

What Is the Correct Naming Convention?

With no uniform terminology, the definition of terms for the plant maintenance application in SAP S/4HANA isn't as consistent as it sounds at first. SAP media on this subject (presentations, online documentation, [F1] help, Customizing, and roadmaps) use different names for the same item. You'll find terms and abbreviations like the following:

- SAP Asset Management
- SAP Digital Asset Management
- SAP S/4HANA Asset Management
- SAP Enterprise Asset Management (SAP EAM)
- SAP Maintenance Management
- SAP Maintenance and Service Management
- SAP Plant Maintenance (SAP PM)

Each of these terms is used synonymously. For both book authors and readers, the use of different labels for identical things is confusing; therefore, to avoid that, we settled on the most frequently used term in SAP S/4HANA publications: *SAP S/4HANA Asset Management*.

Who This Book Is For

This book always addresses *you* directly. So, who are you, and what can you expect from this book?

- If you're a *project lead* responsible for an SAP plant maintenance project, this book gives you a lot of advice about project management and the configuration and adjustment options available within the system.
- If you're a *member of a project team* that has been assigned to an SAP implementation project and are responsible for adjusting the system to the needs of the enterprise and users, this book gives you plenty of tips in relation to the configuration and adjustment options available within the system.
- If you're an *external consultant* and, as a specialist consultant, you're looking for some background information, explanations in relation to the configuration and adjustment options associated with the system, and recommendations for best practices that you can then impart to your customers, this is the book for you.
- If you're an *internal consultant* (often known as a *business consultant*) about to take over all IT-related support for the user department from the IT or organizational department, this book gives you many tips in relation to the configuration and adjustment options available within the system.
- If you're a *key user*, which means that you help your colleagues in their daily work and therefore need more background information on the system than your colleagues, this book provides a great deal of background information in relation to system behavior and how this behavior manifests itself. This book also enables you

to have more targeted and skilled discussions about the needs of your users with relevant project team members, project leads, or consultants.

- If you're a *developer* who is looking for help with programming (e.g., for interfaces or add-ons), this book is, to some extent, for you. At various points throughout the book, I'll give you information about technical implementations (e.g., customer exits, Business Application Programming Interfaces [BAPIs], and business add-ins [BAdIs]) without the implementation or the programming itself.

On the other hand, this book isn't suited for some because of what you *won't* find here:

- If you're an *end user* who is looking for user documentation in relation to the SAP system, this book isn't for you.
- If you're a *manager* and have to decide whether to implement SAP S/4HANA Asset Management, this book isn't for you because it contains only system-oriented information. In other words, it doesn't provide any specific information about whether the SAP system is suitable for you.
- If you're an *interested party* in SAP plant maintenance on a general level, this book isn't for you. Instead of adding to your basic understanding of plant maintenance or providing an overview, the many details contained in this book will most likely cause you confusion.

> **Note**
> If you're in one of the three groups for which this book isn't suited, the most relevant book for your purposes is currently *Plant Maintenance with SAP S/4HANA: Business User Guide* by Karl Liebstückel, available at *www.sap-press.com/5180*.

What This Book Can and Can't Do

You can expect this book to give you the following:

- Instructions on how to implement the SAP system, such as what to be aware of during implementation (*success factors*) and what could endanger your implementation project (*risk factors*)
- A basic understanding of the philosophy associated with configuring SAP systems, especially their options and limitations
- Steps on how to use Customizing settings to not only demonstrate how to adjust the system to the needs of your enterprise but also to highlight the limitations associated with some settings
- Recommendations on which settings to configure and which to avoid

- An introduction to technologies that you can use to adjust the system interface to the needs of your users and to identify the system's limitations (e.g., configuring the SAP Fiori launchpad)
- Many useful tips and tricks for your SAP S/4HANA Asset Management system

However, this book doesn't describe any applications or functions. It provides some programming information, but it doesn't provide any source code and isn't end-user documentation. (This usually isn't the case with SAP documentation.)

A lesson learned from experiences with past projects is that every enterprise has its own idea of how the system should be used. For example, the following statements are true:

- Every enterprise maps its technical assets differently.
- Every enterprise sets up its business processes differently.
- Every enterprise needs to connect to different systems.

Therefore, you should regard the information presented in this book as a resource for adapting the system to your enterprise's individual needs to create your own SAP S/4HANA Asset Management system.

Structure of Contents

This book is divided into nine chapters, as briefly outlined in the following list:

- **Chapter 1: SAP Projects in Plant Maintenance**
  This chapter shows you how to approach an SAP plant maintenance project in a methodical way (with SAP Activate) and what you should be aware of with this kind of project. Furthermore, you'll be presented with the findings of an empirical study in which enterprises were asked to state what they believe are the reasons for success and the risk factors in their SAP projects.
- **Chapter 2: Configuring Organizational Structures**
  Organizational structures are the basis of everything in an SAP system. Along with explaining the generic SAP organizational units, this chapter shows you which plant maintenance?specific organizational units and work centers are required for other procedures and how you should configure them.
- **Chapter 3: Configuring Generic Functions**
  The SAP system has Customizing functions that are used by several objects (e.g., field selection, status control, lists). To avoid having to describe these for every single object, this chapter describes the various configuration and adjustment options and, in each case, references which object uses which function.

Structure of Contents

- **Chapter 4: Configuring the Structure of Technical Systems**
  SAP provides various elements (e.g., equipment and functional locations) for mapping a custom asset structure. This chapter describes the settings available to you within the SAP system so that you can adjust certain aspects of structuring your technical system to your requirements.

- **Chapter 5: Configuring the Work Order Cycle**
  This chapter shows you all possible settings for a basic process that you can use to adjust the processing of maintenance tasks to your requirements. These tasks range from notification creation, order planning, and order control through to order completion. Recommendations are also provided in terms of which settings you should configure and which settings it doesn't make sense to configure.

- **Chapter 6: Configuring Preventive Maintenance**
  In this chapter, you'll learn special preventive maintenance system settings related to task lists and maintenance plans, as well as recommendations for configuring these settings.

- **Chapter 7: Configuring Additional Business Processes**
  In this chapter, we'll examine business processes that go beyond normal maintenance processing (e.g., refurbishment, checklists, subcontracting, pool asset management). The settings described in Chapter 5 will be used here, but only special, additional configuration options will be described. For example, the issue of special Customizing functions for refurbishment is addressed here as well as other things that you need to be aware of. The same applies to the other business processes mentioned here.

- **Chapter 8: Configuring the SAP Fiori Launchpad for Plant Maintenance**
  In this chapter, knowledge is imparted for the configuration of the SAP Fiori launchpad, including the three types of Fiori apps: SAPUI5 apps, Web Dynpro apps, and SAP GUI transactions. You'll see how the SAP Fiori launchpad can be the single point of entry for accessing the SAP S/4HANA system.

- **Chapter 9: Usability**
  In this chapter, you'll learn how to improve the user experience (UX) of the SAP system and therefore increase acceptance among your users. This will be examined from the following three perspectives: what can end users do themselves, what can IT do without the need for programming, and at which points can IT intervene with programming? This will be discussed without describing the actual programming itself.

- **Appendix A**
  The appendix contains additional useful information such as how to prioritize the expansion phases when implementing SAP S/4HANA Asset Management, as well as a complete list of customer exits.

How to Read This Book

To make it easier for you to use this book—in particular, Chapter 2 through Chapter 7—we've attached great importance to presenting the Customizing functions in a clear and concise manner. Because you likely won't read this book as you would a piece of fiction (i.e., from start to finish) but rather use it as a reference guide, we've provided a standard structure to explain each Customizing function. An example of the layout and content of chapter sections that explain Customizing functions follows.

Customizing Function

The heading of a section that explains a Customizing function contains the name of the Customizing function as it appears in the SAP Implementation Guide (IMG), so that you can find it there.

The first section provides a definition and general description of the function.

Prerequisites

The second section explains those prerequisites that you must fulfill so that you can configure the Customizing function. These may be other Customizing functions that you need to configure in advance, or they may be organizational prerequisites.

Customizing Path

The third section contains the exact Customizing path for the Customizing function in the IMG.

Transaction(s)

The fourth section includes one or more transactions that you can alternatively use to call many of the Customizing functions contained in the system. This is practical because transactions generally enable you to access the function much faster than using the path. Unfortunately, not every function is assigned an SAP transaction, so some transaction sections may be missing.

Settings

The fifth and last section introduces you to the various settings within the relevant Customizing function. These may be entries that you need to make or indicators that you need to set. The effects associated with such settings are described as well.

For each Customizing function, you'll also receive at least one figure as a visual aid for understanding the verbal explanations provided in the book (see a sample in Figure 1).

How to Read This Book

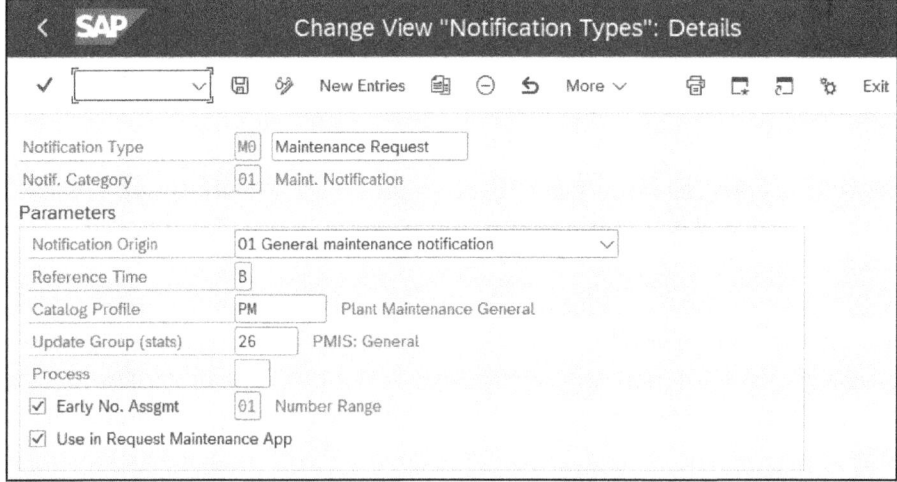

Figure 1 Sample Figure

> **Recommendations** [+]
>
> Normally, at the end of the general explanations, you'll find recommendations on which parameters to set, how to configure the Customizing function, and what to bear in mind. We'll also provide recommendations on what you should avoid. Generally, these recommendations will be contained in boxes similar to this one.

Special Icons Used in This Book

To make it easier for you to use this book, particular information is highlighted using the following special icons:

> **Warning** [!]
>
> Boxes marked with this icon provide particularly important information about the topic being discussed, or they warn you about potential sources of error.

> **Practical Tip** [+]
>
> Boxes marked with this icon give you recommendations on the Customizing settings, or they give you tips from the author's professional experience.

What's New in This Book?

You'll find that this book includes many innovations that weren't in the first edition. Parts of this second edition have been completely rewritten to cover the following subjects:

- The book contains all innovations up to on-premise SAP S/4HANA version 2023 FPS01 (e.g., all innovations that have been delivered in the business functions). There were a lot of them.
- Chapter 8, on the SAP Fiori launchpad, has been completely rewritten, and you'll find information there on how to implement the three types of Fiori apps: SAPUI5, Web Dynpro, and SAP GUI transaction apps. This chapter also describes the new configuration tools for the SAP Fiori launchpad, and it includes a complete list of available apps.
- Chapter 7, on configuring additional business processes, contains two brand-new topics: checklist processing and work order cycles using the phase model. Believe me, both involve hard work for configuration.
- The configuration of action logs has been added to Chapter 3 on configuring generic functions.
- All screenshots were newly created with the Quartz theme, SAP's newest theme for SAP GUI.
- Alterations made by SAP to menu paths and Customizing paths were taken into account.

What Is the Status of Your SAP S/4HANA System?

The range of functions of your SAP S/4HANA Asset Management system, and thus the available Customizing functions, depend crucially on which enhancement packages are installed in your system and which business functions are activated. For example, let's take a look at the **Define Category of Functional Location** Customizing function (see Chapter 4, Section 4.2). If you want to define a functional location category for linear asset management, you must activate business functions LOG_EAM_LINEAR_1 and LOG_EAM_LINEAR_2. You can check and activate business functions using Transaction SFW5 (Switch Framework – Change Business Function Status).

[!] **Warning**

Before going any further, find out the status of your SAP S/4HANA system; otherwise, you might not have some necessary functionalities at hand.

To work with functions mentioned in Chapter 2 through Chapter 9, certain business functions have to be activated. Transaction SFW5 will help you determine what kind of business functions have already been activated. If additional business functions are required, you should activate them before going any further.

In Transaction SFW5, you'll find four groups of business functions, as shown in Figure 2.

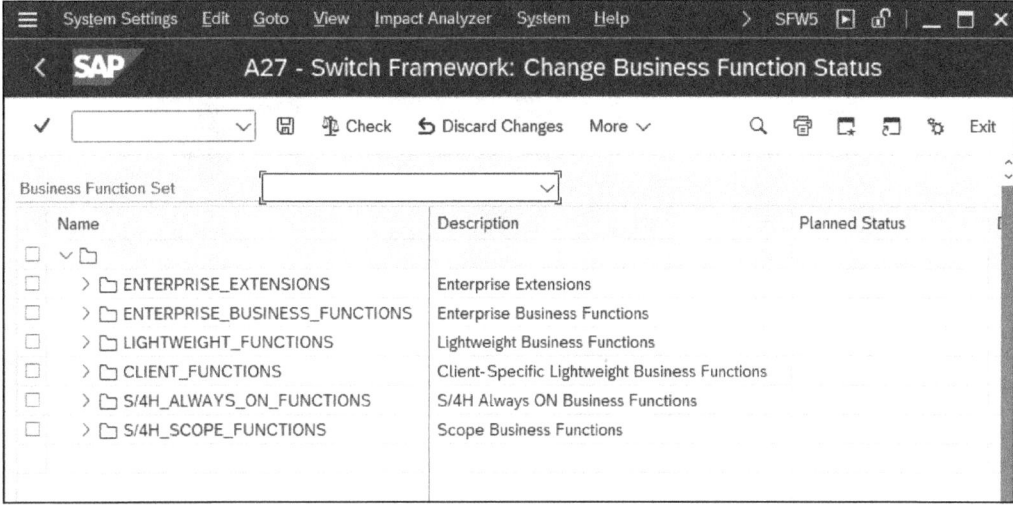

Figure 2 Business Function Groups

To use SAP S/4HANA Asset Management, the **ENTERPRISE_BUSINESS_FUNCTIONS** group and the **S/4H_ALWAYS_ON_FUNCTIONS** group are important.

In the **ENTERPRISE_BUSINESS_FUNCTIONS** group, scroll down to **LOG_EAM** (see Figure 3). You'll find information listed in columns by business function:

- **Name**
  Business function name (e.g., **LOG_EAM_CHECKLIST**) with a short **Description** (e.g., **EAM, Checklist Feature**).
- **Planned Status entries**:
  - Empty checkbox: Business function is inactive, and the customer (e.g., your IT department) could activate it by marking the indicator.
  - Selected checkbox: Business function was activated by customer (e.g., your IT department) and could be deactivated by deselecting the indicator.
  - **Business func. will remain activated**: The business function was activated by the customer (e.g., your IT department) but can't be deactivated.
- **Dependencies**
  If you see the 🗐 icon, there are dependencies to other business functions, which you can view by clicking the icon.

About the Book

- **Documentation**
  In this column, you can show business function documentation by clicking the icon.

- **Release**
  This column provides information on the enhancement package from which this business function is available (e.g., 605 = SAP ERP 6.0 Enhancement Package 5 or 107 = SAP S/4HANA 2022).

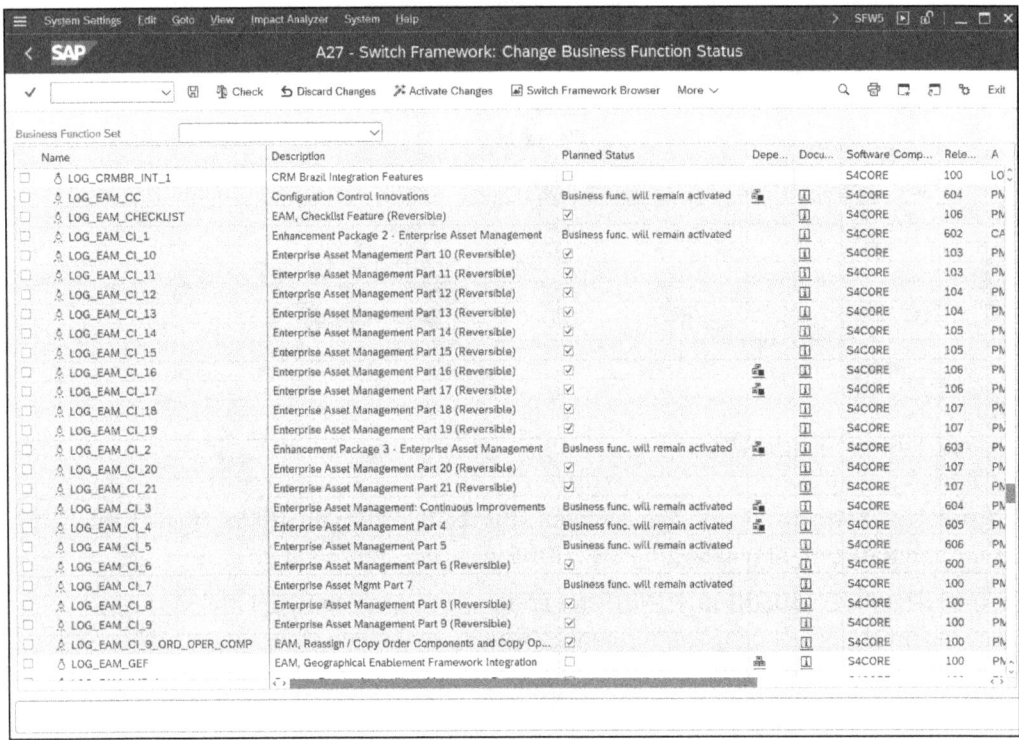

Figure 3 Business Functions: Enterprise Functions LOG_EAM

[+] **Important Tip: Activate All LOG_EAM_CI Business Functions**

Continuous improvements (CIs) are small improvements that SAP has developed to make it easier for users to work with the system and to simplify the system as a whole. Continuous improvements are free of charge and are not integrated into other SAP applications, so you can therefore activate the business functions without hesitation.

S/4H_ALWAYS_ON_FUNCTIONS (see Figure 4) have been activated by SAP for various reasons and can't be deactivated by the customer; that is, they're always available.

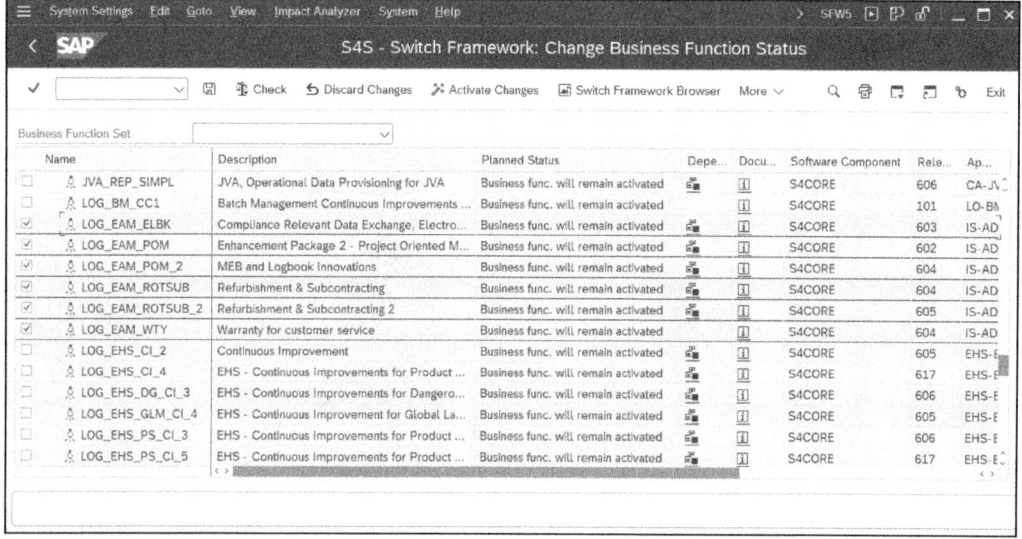

Figure 4 Business Functions: Always-On Functions LOG_EAM

Now, all that remains is to hope you'll garner both inspiration and knowledge for your work by reading and working with this book.

In the spirit of a well-known Chuang Tzu quotation, I hope that the information contained in this book falls on fertile ground and yields lasting results for you and your users.

Sincerely,
Karl Liebstückel

Chapter 1
SAP Projects in Plant Maintenance

This chapter provides a roadmap for your plant maintenance project by discussing an empirical survey of the risk factors and success factors in SAP projects. This chapter also arms you with many useful tips regarding what you need to watch out for while implementing your plant maintenance projects.

In every SAP implementation project, you lay the foundations that determine whether your users will be able to successfully use the SAP system and achieve their desired goals or whether the SAP system will be a disappointment for you, your company's top management, and your users. In an SAP implementation project, there are many things that you need to ensure are done (the *dos*)—these are your success factors. However, there are at least as many potential sources of error that you need to avoid (the *don'ts*)—these are your risk factors. This chapter will give you several useful tips to ensure that your plant maintenance project is a success.

1.1 A Possible Process for Your Plant Maintenance Project with SAP

In the basic discussion of a possible process for your plant maintenance project with SAP, you should differentiate between the implementation strategy and its accompanying methodological approach.

1.1.1 Implementation Strategy

Most enterprises implement plant maintenance in a series of stages rather than in a single step (i.e., implementing the full range of functions in all plants all at once). Therefore, during project planning, you need to make a fundamental decision about the stages in which you'll implement plant maintenance. There are two aspects of the implementation to consider:

- Functional stages
- Spatial stages

The functional stages of your plant maintenance project can be as follows:

- **Stage 1**
  Implement the structuring of technical systems. (You can omit this first stage if you can carry over asset structures from a legacy system.)
- **Stage 2**
  Implement work order processing.
- **Stage 3**
  Implement preventive maintenance.
- **Stage 4**
  Expand the system by adding potential enhancements (e.g., testing equipment processing, checking list processing, refurbishment processing, subcontracting, pool asset management, mobile solution or plant maintenance projects processing).

The spatial stages, for their part, can be as follows:

- **Stage 1**
  Implement plant maintenance in a pilot system.
- **Stage 2**
  Implement plant maintenance in one plant only.
- **Stage 3**
  Roll out plant maintenance to other plants.

This is the basis on which you can create a plan that specifies which functions are to be implemented in which operation or plant at which point in time. Once you have implemented these stages, you need to make a fundamental decision about which of the following strategies you deem the most useful for your project:

- **Horizontal strategy**
  Order processing implementation in all plants simultaneously
- **Vertical strategy**
  Full implementation in one plant, followed by a rollout to other plants
- **Combined strategy**
  Implementation of the full range of functions in one plant, followed by a rollout of order processing to all plants

Figure 1.1 shows a summary of the functional and spatial stages.

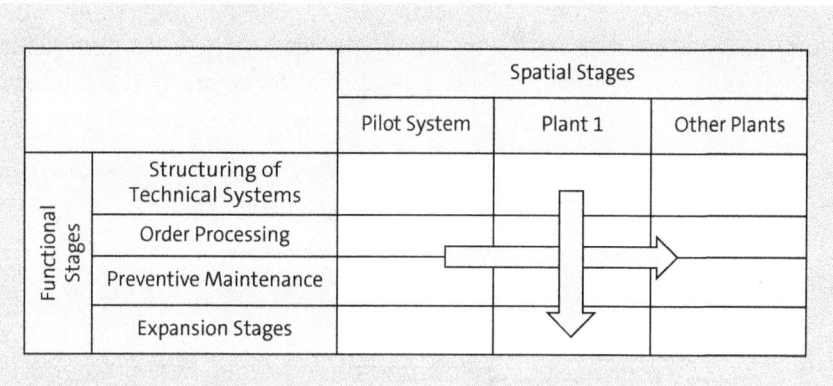

Figure 1.1 Implementation Strategy

1.1.2 SAP Activate Methodology

As is the case in all SAP projects, for plant maintenance, it's recommended that you use SAP's general methodology for implementation projects: *SAP Activate*.

SAP Activate is SAP's current methodology for implementing, upgrading, or migrating to an SAP S/4HANA landscape. SAP Activate is the successor to the ASAP methodology, but it's broader and not only includes the implementation methodology of on-premise projects such as ASAP but also offers roadmaps for cloud projects or migration projects.

SAP Activate offers a detailed roadmap to SAP S/4HANA, from scoping the project over prototyping to post go-live. SAP Activate also guides your team through integration with third-party software offerings and cloud solutions, training, project management, and other needs associated with migration.

SAP Activate is divided into the following six phases (see Figure 1.2):

1. Discover
2. Prepare
3. Explore
4. Realize
5. Deploy
6. Run

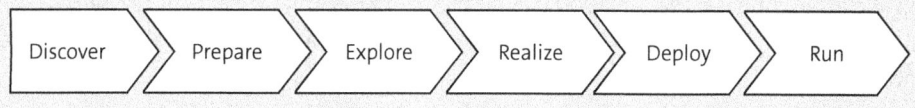

Figure 1.2 SAP Activate: Overview

1 SAP Projects in Plant Maintenance

Figure 1.3 shows the six phases on a more detailed level, as each phase has individual activities and deliverables that are linked into workstreams. There are also quality checks throughout the process and extensive resources (*accelerators*) linked to each deliverable. These accelerators provide guidance and best practices for each step.

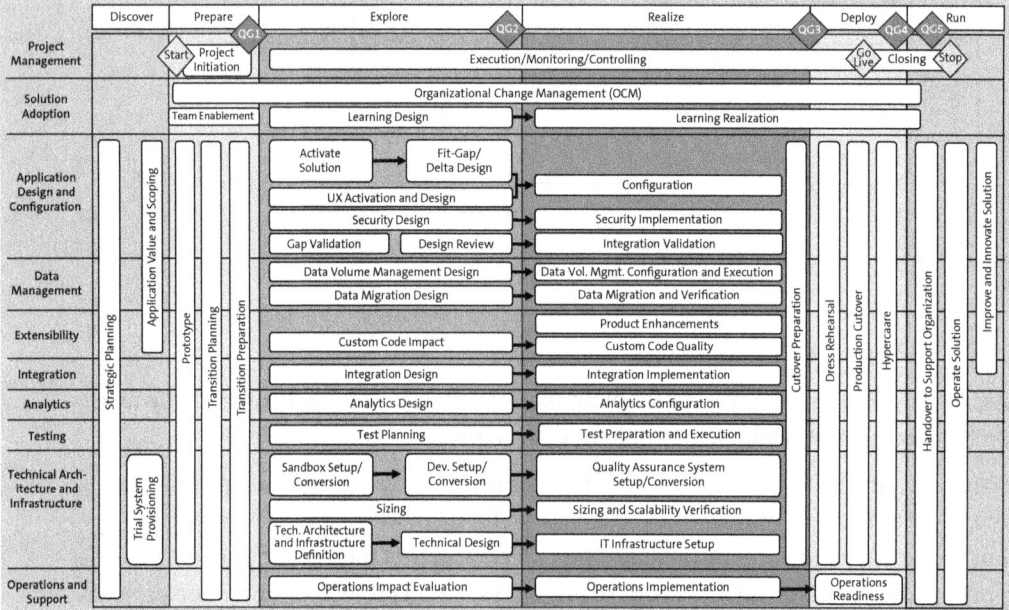

Figure 1.3 SAP Activate: Details

Phase 1: Discover

In the *discover* phase, customers become familiar with SAP S/4HANA and the implementation project. There are two important activities in this phase:

- Perform an as-is analysis and feasibility study as a starting point of the project. You'll find a detailed question list in Section 1.3.1.
- Become familiar with SAP S/4HANA, for example, by using a free two-week trial access to a cloud system. To do this, register at *https://www.sap.com/products/try-sap.html#advanced-trials*.

Phase 2: Prepare

In the second phase, *prepare*, you define the organizational aspects of the project and formulate an implementation concept. The following are the main activities in this phase:

- Define the goals of implementing an SAP system—and also define what you *don't* want to happen when you implement an SAP system.
- Define your implementation strategy.

- Define the organizational aspects of the project structure.
- Define project standards and a procedure.
- Create a preliminary project plan.
- Plan your internal marketing.
- Create a project application and get it approved.
- Decide on and adapt the documentation tools.
- Decide on and adapt the monitoring instruments.
- Define the change management process.
- Set up the system environment.
- Train the project team.
- Decide on a training strategy for key users and end users.
- Design the process map and draft the business process master list (BPML).
- Draw up a test strategy for the go-live.
- Sort through existing master data material.
- Define a data migration strategy.
- Perform an interface inventory.
- Define which project tools are to be used.

Phase 3: Explore

The third phase, *explore*, comprises all activities needed to create a detailed business blueprint. This blueprint contains a detailed to-be concept for using the SAP S/4HANA Asset Management system. During the explore phase, you also build up a prototype and perform the following activities:

- Develop a business blueprint for organizational structures.
- Develop a business blueprint for master data (in particular, the structuring of technical systems).
- Develop a business blueprint for the business processes and functions.
- Document and visualize the various business processes (e.g., using event-driven process chains, which are described in more detail in Section 1.3.2).
- Design a business blueprint for transferring legacy data, for creating new master data, or both.
- Design a business blueprint for reporting and analyses.
- Draft and establish a plan for communicating with stakeholders.
- Create a fit-gap list.
- Create a statement of work and a requirements specification for interfaces.
- Create a statement of work and a requirements specification for enhancements (a *gap list*).

- Create an archiving concept.
- Design an authorization concept.
- Create a concept for end-user training.
- Perform Customizing for prototypes.
- Build prototypes based on sample master data.
- Refine the test strategy.

Phase 4: Realize

The fourth phase, *realize*, involves implementing and mapping the business blueprint. The following are the main activities in this phase:

- Perform final Customizing.
- Adapt the layout of screens and lists.
- Develop, implement, and test the data transfer programs. Alternatively, configure one of the following SAP tools for data transfer: the Legacy System Migration Workbench (Transaction LSMW), the plant maintenance-specific tool for data transfer (Transaction IBIP), or the SAP Fiori app Migrate Your Data – Migration Cockpit.
- Enter master data (if this isn't a separate stage in the implementation).
- Implement solutions for items on the gap list.
- Develop enhancements (e.g., print programs, customer exits, BAdIs, analyses).
- Set up roles and authorizations.
- Set up the SAP Fiori launchpad.
- Create test plans and test cases.
- Perform interim and final tests on the application.
- Perform system and performance tests.
- Perform user acceptance tests.
- Configure settings for archiving and other jobs.
- Train the key users.
- Develop training courses for the end users (e.g., documentation and training examples).
- Develop user documentation.
- Plan for the transfer from the project phase to live operation.
- Develop a cutover plan.

Phase 5: Deploy

The fifth phase, *deploy*, comprises all the activities that are required shortly before and in preparation for the system go-live, the go-live itself, and activities during a short phase after go-live. These include the following:

1.1 A Possible Process for Your Plant Maintenance Project with SAP

- Train the end users.
- Transfer the legacy data to the live system.
- Perform the going live check.
- Perform the cutover plan.
- Go live.
- Provide support during live operation (i.e., support during the start-up phase).
- Provide additional training for end users.
- Hold the kickoff meeting to signal official project completion.
- Provide support for users after go-live (telephone hotline, email support, on-site presence, etc.).

Phase 6: Run

The sixth phase, *run*, mainly involves improving the operations of the SAP system and identifying any potential for optimization. These improvements include the following:

- Optimized documentation
- Optimized training
- Optimized implementation
- Optimized testing
- Optimized technical operations
- Optimized business processes
- Optimized roles and authorizations
- Optimized SAP Fiori launchpads
- Optimized system upgrades
- Optimized system maintenance

You can find all the information you need about this new methodology on SAP's Roadmap Viewer (*https://go.support.sap.com/roadmapviewer/*), where you can learn about SAP Activate in general and all the available roadmaps (see Figure 1.4).

> **Roadmap for SAP S/4HANA Asset Management**
> You should keep up to date not only with the SAP Activate methodology but also with what SAP is planning for further development of SAP S/4HANA Asset Management. You can find this information in the SAP Road Map Explorer (*https://roadmaps.sap.com/*).

Figure 1.5 shows SAP's roadmap for SAP S/4HANA Asset Management from 2024 to 2025 and further product vision.

1 SAP Projects in Plant Maintenance

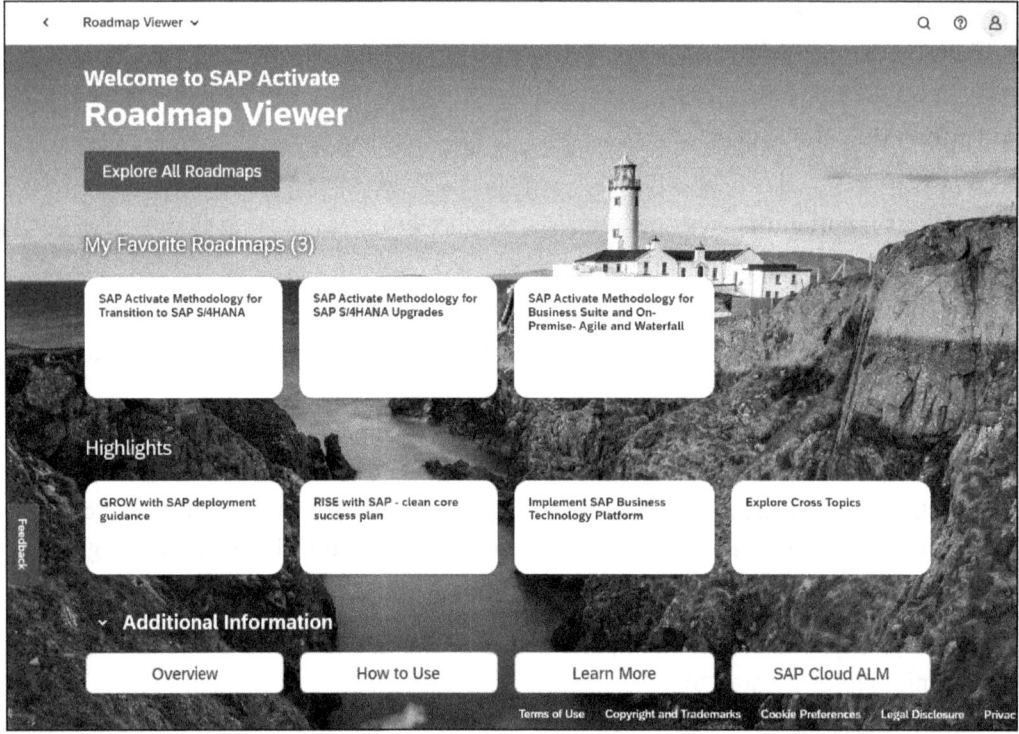

Figure 1.4 SAP Activate: Roadmap Viewer

Figure 1.5 SAP Road Map Explorer for SAP S/4HANA Asset Management

Like every project, your plant maintenance implementation project involves success factors and risk factors. It can only help your project if you benefit from the experiences of other enterprises rather than making every mistake yourself. For this reason, the results of an empirical survey are provided next. Then, based on this survey and my personal experience, this chapter will provide you with several tips for your plant maintenance project.

1.2 General Risk Factors and Success Factors in SAP Projects: An Empirical Survey

Learning from the experiences of others was the principle on which an empirical survey was conducted at the Technical University Würzburg-Schweinfurt in Germany, the content and results of which will be discussed in this section. The study was based on empirical data collected from SAP ERP customers, but the findings of the study should be applicable 1:1 to SAP S/4HANA projects.

For the survey, a catalog of 33 potential risk sources and 27 potential success factors was developed and then sent to SAP user companies. The companies were asked to rank each risk source in accordance with their own experiences as *completely true, true, not completely true,* or *not true*. To rank the success factors, a scale with the values *very important, important, neutral,* and *not important* was used. A total of 148 companies participated in the survey, and we analyzed the results.

> **Acknowledgments**
>
> I would like to thank Mr. Andreas Weber, who conducted this survey as part of his diploma thesis entitled "Risk Management in Business Software Projects: Concept and Empirical Study."

1.2.1 Risk Factors

The following potential risk factors were assigned to various categories:

- **Risk sources in the project environment**
  - R0 = Insufficient documentation of business processes
  - R02 = No experience with software implementation projects of this type and scope
  - R03 = Insufficiently high status of the SAP implementation project
  - R04 = End-user requirements not adequately considered
  - R05 = Resistance by parties involved
  - R06 = Requirements of the SAP system not prioritized
  - R07 = Change requests made continually throughout duration of project

- **Risk sources associated with the project goals**
  - R08 = Insufficient communication of goals that were set for the implementation of the SAP system
  - R09 = Hidden goals that were set for the implementation
  - R10 = Unrealistic goals
  - R11 = Unclear requirements
- **Risk sources in project management**
  - R12 = Unstructured projects procedures; milestones unknown
  - R13 = Insufficient monitoring of deadlines and costs
  - R14 = Insufficient adaptation of schedule and budget planning when change requests are made
  - R15 = Poor planning leading to retroactive change requests
  - R16 = Insufficient control of project on part of project managers
  - R17 = Insufficient inclusion of project team members in planning
  - R18 = Poor information flow in project
- **Risk sources associated with project organization**
  - R19 = Lack of clarity regarding competencies, contact persons, and responsibilities
  - R20 = Work overload; project team members not made free for project work
  - R21 = Inadequate equipment (e.g., PCs and software)
- **Risk sources associated with the project team**
  - R22 = Insufficient management of and specialist IT knowledge among project team members
  - R23 = Insufficient know-how on the part of consultants
  - R24 = Carelessness in selection of consultants
  - R25 = Lack of staff; not enough project team members
  - R26 = Poor motivation among project team members
- **Risk sources in project flow**
  - R27 = Inadequate project budget
  - R28 = Not enough time for project planning
  - R29 = Unrealistic completion date; not enough buffer time
  - R30 = Unexpected delays; deadline postponements
  - R31 = Not enough time overall
  - R32 = Problems with hardware configuration and SAP competence
  - R33 = Insufficient quality assurance

If we group together the first two categories, *completely true* and *true*, we get what is known as a *risk radar*, as shown in Figure 1.6. The farther away the points are from the center of the grid, the greater the likelihood that an implementation project will be confronted with the risk in question in the course of the project. To summarize, the following are the most important risk factors:

- In the project environment, a particularly high likelihood of occurrence was recorded for *insufficient documentation* and *change requests* during the project.
- In terms of project management, it became clear that *plans weren't adjusted* when changes occurred.
- When it came to project organization, *staff work overload* and *insufficient specialist knowledge* stood out.
- Finally, under project flow, there was a high likelihood of occurrence of *not enough time for planning*, *unexpected delays*, and *insufficient time*.

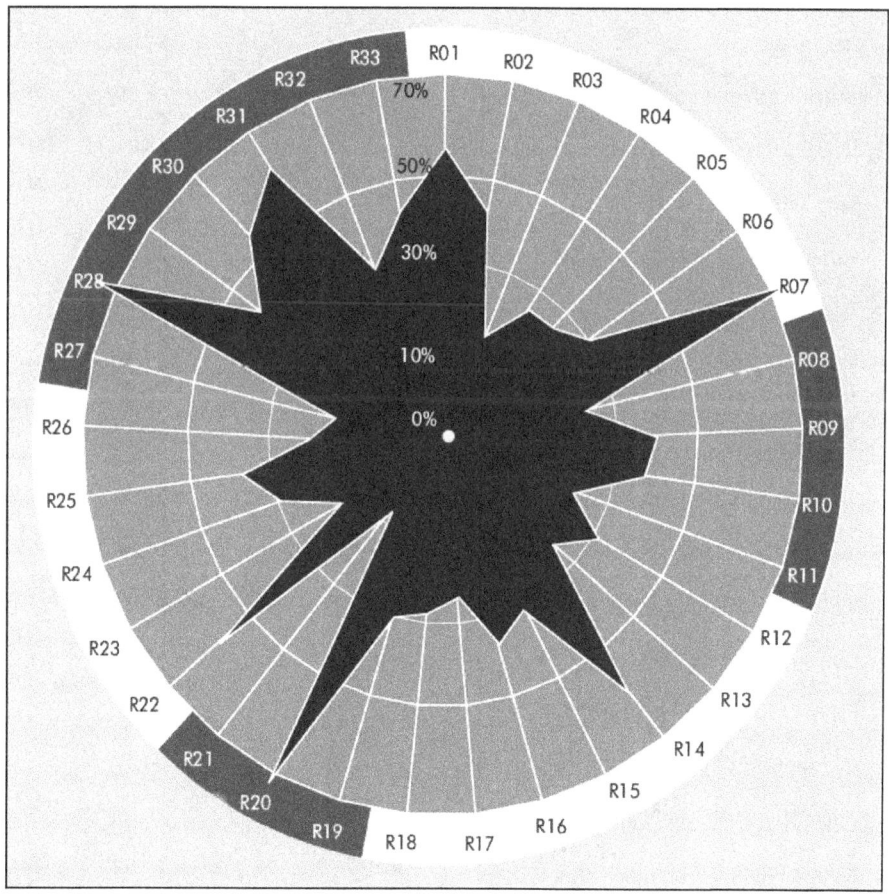

Figure 1.6 Risk Radar

1.2.2 Success Factors

Let's now look at the success factors, ranked in order of importance, as shown in Table 1.1. For example, 82.2% of respondents said that "E12 = support from top management" is very important.

Success Factor	Ranking	Very Important
E12 = Support from top management	1	82.2%
E02 = Competent project management	2	78.1%
E15 = Good collaboration	3	67.1%
E01 = Structured procedures	4	67.1%
E10 = Inclusion of all parties involved	5	65.8%
E05 = Technical competence of staff	6	64.4%
E21 = Clear goals	7	61.6%
E16 = Realistic scheduling	8	57.5%
E25 = Identification of project team members with project	9	58.9%
E19 = Early detection of problems	10	54.8%
E03 = Appropriate project organization	11	54.8%
E14 = Open information policy	12	52.1%
E26 = Appropriate reduction of daily operations tasks in project team members' workload	13	47.9%
E07 = Appropriate number of project team members	14	39.7%
E11 = Appropriate budget parameters	15	38.4%
E13 = Availability of tools	16	37.0%
E24: = Clear competence guidelines	17	45.2%
E27 = Quality assurance during project	18	39.7%
E04 = Experience from other projects	19	41.1%
E22 = High priority of implementation project	20	34.2%
E06 = Appropriate milestone planning	21	26.0%
E17 = Early training	22	26.0%
E18 = Appropriate size of work packages	23	19.4%

Table 1.1 Success Factors of SAP Projects

Success Factor	Ranking	Very Important
E23 = Small number of requirements changes	24	23.3%
E20 = Detailed project documentation	25	26.0%
E08 = Use of external consultants	26	27.4%
E09 = Small project team	27	8.2%

Table 1.1 Success Factors of SAP Projects (Cont.)

> **Weighting Factors**
> Weighting factors were used to calculate the rankings, and the values were *very important* = 4, *important* = 3, *neutral* = 2, and *not important* = 1. Therefore, the ranking doesn't correspond to the percentage value under *very important*.

According to those surveyed, the following were particularly important for the success of the project:

- Support from top management
- Competent project management
- Teamwork
- Structured procedures
- Technically competent staff

1.3 Tips for Your Plant Maintenance Project

Based on the information about the SAP Activate methodology, the results of the empirical survey, and my personal experiences with SAP S/4HANA Asset Management and other plant maintenance projects with SAP, let's consider some tips and tricks that can help you in your own projects. This section adheres to the time sequence of the SAP Activate roadmap, as described in Section 1.1.2.

1.3.1 Discover Phase

There are two important activities in this phase:

- Become familiar with SAP S/4HANA, for example, by using a free three-month trial access to a cloud system.
- Perform an as-is analysis and feasibility study as a starting point of the project.

There are different opportunities to become familiar with SAP S/4HANA, including the following:

- Start a two-week trial for SAP S/4HANA in the cloud (go to *https://www.sap.com/products/try-sap.html#advanced-trials*).
- Request a demo from SAP experts.
- Use the event finder to attend in-person or online events on SAP S/4HANA topics (go to *https://www.sap.com/events.html*).
- Attend e-learning courses that are available in SAP Learning Hub (*https://saplearninghub.plateau.com/*). The following courses are available for SAP S/4HANA Asset Management:
  - S43000: Business Processes in SAP S/4HANA Asset Management (see Figure 1.7)
  - S43100: Management of Technical Objects in SAP S/4HANA Asset Management
  - S43200: Preventive Maintenance in SAP S/4HANA Asset Management
  - S43300: Customizing in SAP S/4HANA Asset Management
  - S43400: Exploring Advanced Functions in Maintenance Processing

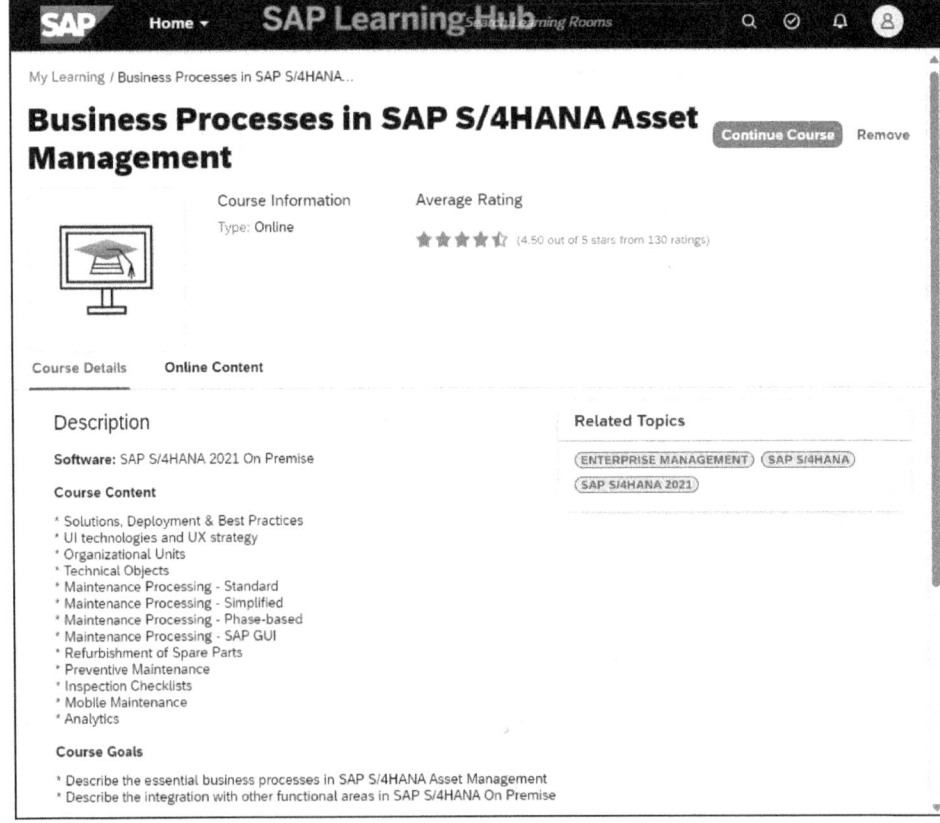

Figure 1.7 SAP Learning Hub E-Learning Course

1.3 Tips for Your Plant Maintenance Project

Perform an As-Is Analysis (Feasibility Study)

Before you do anything with your system, you need to carry out a careful as-is analysis of the framework conditions (in particular, your business processes). The time and effort you put in to perform a full and correct feasibility study will definitely be worth it.

A feasibility study should pursue several goals:

- Analyze, document, and visualize the master data, plant maintenance processes, and materials management processes used previously as well as the reporting requirements.
- Define the range of plant maintenance functions available with the SAP S/4HANA Asset Management software and assign them to potential expansion phases.
- Serve as a basis for further planning of the project details (deadlines, efforts, staff, and costs) to establish the training workshops in the explore phase and to use SAP software to draft "to-be" processes (see Figure 1.8).

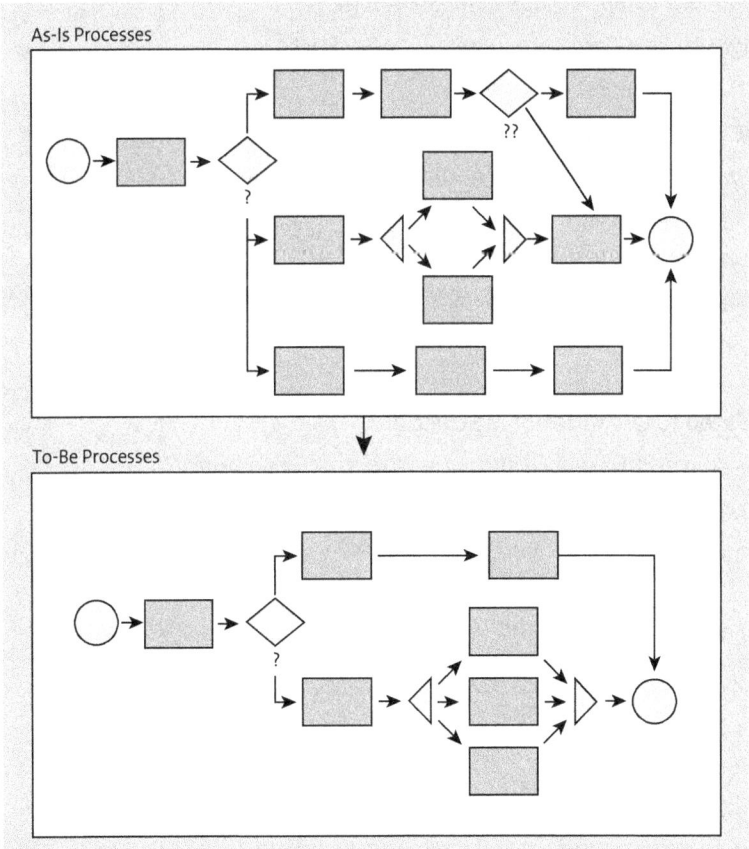

Figure 1.8 As-Is and To-Be Processes

To avoid starting from scratch, consider the following list of questions that should be discussed and answered as part of a feasibility study like this.

Key Questions Related to System Background
- *Which release level is in use? Which enhancement package is installed?*
- *Which business functions for SAP S/4HANA Asset Management (LOG_EAM, etc.) are activated?*
- *Which functionality is in use in SAP S/4HANA?*
  - *Financial accounting*
  - *Asset accounting*
  - *Controlling*
  - *Purchasing*
  - *Inventory management*
- *Are there other SAP products depending on SAP S/4HANA Asset Management? For example:*
  - *SAP Supplier Relationship Management (SAP SRM)*
  - *SAP NetWeaver Master Data Management (SAP MDM)*
  - *SAP Master Data Governance (SAP MDG)*
  - *SAP Business Warehouse (SAP BW)*
- *Are there non-SAP systems that should be getting interfaces with SAP S/4HANA Asset Management? For example:*
  - *Document management systems*
  - *Production data collection systems*
  - *Process control systems*

Key Questions Related to Organizational Structures
- *In which SAP organizational units is the enterprise structure mapped?*
  - *Controlling areas*
  - *Company codes*
  - *Plants*
  - *Storage locations*
  - *Maintenance plants*
  - *Maintenance planning plants*
  - *Purchasing organizations*
  - *Purchasing groups*

- *How are the organizational units assigned?*
  - Company codes to controlling areas (for more on this very important assignment, see Chapter 2, Section 2.1.2)
  - Plant to company codes
  - Plant to purchasing organization
  - Maintenance plant to maintenance planning plant

Key Questions Related to Technical Asset Structuring
- *What types of objects are to be managed?*
  - Buildings
  - Production assets
  - Machines
  - Fleet objects
  - Industrial trucks
  - Operating resources
  - Tools
  - Test equipment
- *Is there a distinction between fixed assets and inventories with changing installation locations?*
- *How large is the quantity structure?*

Detailed Questions Related to Technical Asset Structuring
- *Which information is to be defined for which objects?*
- *Are inventories to be kept in stock?*
- *Are several labels to be displayed in parallel for the assets?*
- *Are the inventories and assets to be classified (e.g., according to e-class or a separate classification procedure)?*
- *Are bills of materials (BOMs) to be managed for spare parts? If so:*
  - Are they single-level or multilevel BOMs?
  - Do they concern a single aggregate or category?
  - Are material numbers assigned to spare parts in the SAP system?
- *Are changes to the asset structure to be documented?*
- *Are documents (e.g., images, drawings, process instructions) to be assigned to the assets and/or inventories?*
  - Are these documents stored electronically? If so:
    - Where are they stored and how?
    - Are there master records for the documents?

- *Are warranties to be managed? If so:*
  - *Are they to be assigned to the assets and/or inventories?*
  - *Are they time-based or performance-based warranties?*
- *Are the assets and/or inventories mapped as asset master records in the SAP system?*

Key Questions Related to Plant Maintenance Processes
- *Which plant maintenance processes are to be mapped in the SAP system in the future?*
  - *Machine malfunctions*
  - *Repair*
  - *Preventive maintenance and inspection*
    - *Time-based*
    - *Performance-based*
    - *Time-based and performance-based*
    - *Condition-based*
  - *Calibration processing or inspection processing*
  - *Checklist processing*
  - *Use of external companies*
  - *Investment measures*
  - *Construction of new assets*
  - *Manufacture of spare parts*
  - *Refurbishment*
  - *Subcontracting*
  - *Shift notes*
  - *Tool and fixture construction*
- *How are the core processes currently handled?*
- *Who is involved in planning and execution, and at what point and to what extent do they become involved?*
- *What do you consider to be the weaknesses and functional deficits of the current plant maintenance processes? What will be or can be improved?*

Detailed Questions Related to Plant Maintenance Processes
- *Which workshops need to be mapped (electronics, mechanical engineering, measurement and control technology, etc.)?*
- *Are there such things as catalogs for damage codes or cause codes?*
- *Which workshop documents are required?*

- Should a mobile solution be used?
- Is there a need to perform capacity planning (available capacity, capacity load, and capacity leveling)?
- Is it necessary to perform resource planning (i.e., which employees are available to work when, for how long, and on which order)?
- Are budgets to be assigned to tasks?
- Are technical or business approval procedures to be mapped?
- What types of completion confirmations are to be mapped?
  - Time confirmations
  - Technical completion confirmations
  - Material confirmations
  - Shift notes
  - Counter readings
- How are the activities settled to the asset cost centers?
- Should shift notes be created? If so:
  - What kind of information should be recorded?
  - Should the shift notes be categorized?
- Are task lists to be managed? If so:
  - Do they concern a single object or a category?
  - What kind of information do task lists contain?
  - Are spare parts also to be defined there?
- How large is the quantity structure for task lists?
- Do you want time-based maintenance and inspection? If so:
  - Are the tasks subject to a cycle?
  - Are there different cycles within a task list? If so, what are they?
- Do you want performance-based maintenance and inspection? If so:
  - Which types of counters (operating hours, miles, pieces that are produced, etc.) are available?
  - How should the counter readings be recorded?
    - In the SAP system
    - Via mobile devices
    - By data transfer, such as from a production system
  - Are the tasks subject to a cycle?
  - Are there different cycles within a task list? If so, which are they?
  - Can you assign several counters to one maintenance plan?

- *Do you want to perform condition-based maintenance? If so:*
  - *Which types of measuring points (temperature, pressure, rotational speed, flow rate, etc.) are available?*
  - *How are the counter readings recorded?*
    - *In the SAP system*
    - *Via mobile devices*
    - *By data transfer, such as from a production system*
- *How large is the quantity structure for maintenance plans?*

Key Questions Related to Purchase and Warehouse Processes

- *Which materials management processes are to be mapped in the SAP system in the future?* For example:
  - *Procurement of direct materials (free-text purchase orders)*
  - *Procurement of spare parts in stock*
  - *Use of external companies*
    - *Individual commissioning*
    - *Outline agreements*
    - *Service specifications*
    - *Work centers for external companies*
- *How are these processes currently handled?*
- *Who is involved in the planning and execution, and at what point and to what extent do they become involved?*
- *What do you consider to be the weaknesses and functional deficits of the current materials management processes? What is to be or can be improved?*

Detailed Questions Related to Purchase and Warehouse Processes

- *Are the vendors and service providers mapped as vendor master records in the SAP system?*
- *Are outline agreements with vendors and service providers to be mapped?*
  - *If so, how is the contract release order to be triggered?*
- *Are there companies for which separate work centers are to be managed?*
  - *If so, which is the commissioning and settlement procedure for the work centers?*
- *Are the services to be created as service master data or when commissioning service specifications?*
- *Are spare parts to be mapped as material master data in the SAP system?*
- *Are spare parts to be planned in an order? If so:*
  - *Will spare parts catalogs be available on the intranet?*

- *Are automatic reservations to be triggered?*
- *Are spare parts to be subject to inventory management? If so:*
  - *What is the withdrawal process?*
  - *Are material requirements to be planned for spare parts?*
  - *Is reorder point planning to be used or something else?*
- *Are spare parts to be procured for direct consumption only?*
- *Are there vendor consignment stores? If so:*
  - *How are these managed?*
  - *Who is responsible for replenishment?*
  - *How are withdrawals settled?*
- *Are there approval procedures for the procurement processes?*
- *How large is the quantity structure for spare parts, vendors, and outline agreements?*

Key Questions Related to Reporting and Key Figures

- *Which lists and reports are currently in use, who uses them, and when are they made available?*
  - *Which of these lists and reports must be available as soon as the SAP system goes live?*
  - *Which lists could be made available later?*
- *Which key figures are currently in use, who uses them, and when are they made available?*
  - *Which of these key figures must be available as soon as the SAP system goes live?*
  - *Which key figures could be made available later?*
- *What do you consider to be the weaknesses and functional deficits of the current lists, reports, and key figures? What must be or should be improved?*

Information Acquisition

One question arises in this context: Whom do you approach to get as full and correct a picture as possible of existing processes? From my experience of carrying out as-is analyses, I recommend the following reliable information sources:

- People who are involved in executing, controlling, and monitoring business processes
- Users of identical or similar IT systems
- Customers who are communicative and often creative knowledge holders
- Business partners who work with the process (e.g., vendors)
- Subject matter experts (SMEs), who should be asked to give critical feedback
- Company management, which ultimately has to approve the processes

Another question arises here: What techniques should you use to acquire the information needed for a full and correct as-is analysis? The following techniques have proven most effective in practice:

- Establishing user workshops
- Observing the relevant employees at work
- Working with the business processes under analysis
- Taking on the role of an outsider (e.g., acting as a customer if you're the production director)
- Using questionnaires
- Holding interviews
- Brainstorming with the parties involved
- Having discussions with SMEs
- Looking through existing forms, documentation, descriptions, manuals, and other resources
- Verbally describing the organizational structure and process structure (e.g., with organigrams)

Of course, your company may have other ways of conducting information research.

Results of the Feasibility Study

The feasibility study should deliver the following results:

- Precise knowledge of which master data, plant maintenance processes, materials management processes, and reports are to be mapped in the SAP system
- Precise knowledge of which plant maintenance functions are to be implemented in the SAP system, having assigned them to three potential expansion phases: A/B/C (see Appendix A, Section A.1)
- Basic understanding to enable further planning of project details such as deadlines, efforts, staff, and costs
- A basis on which to decide which user workshops have to be established in the business blueprint during the explore phase
- A basis for using SAP software to draft to-be processes

1.3.2 Prepare Phase

In the prepare phase, you set the project goals, create a detailed schedule and a budget plan, and perform other tasks that lay the foundations of the actual content of the project, as we'll discuss in the following sections.

General Information on Project Preparation

The importance of good project preparation is often underestimated.

> **Errors in the Prepare Phase Have Serious Repercussions**
>
> A basic rule of thumb is that the earlier in the project a mistake is made, the greater the effect it will have on the project as a whole. You should therefore proceed with particular caution during the prepare phase. Leave enough time for project planning and devote all your energy to getting the most from this phase.

As shown in the empirical survey, those surveyed in the user companies believed that, in retrospect, there had been too little time for preparation in their project.

> **Schedule Sufficient Time for the Prepare Phase**
>
> After you've completed all the tasks relating to project preparation—and only then—should you start with the business blueprint.

The poorer the quality of the planning, the more likely it is that improvements will have to be made later on. Similarly, the better the project planning, the better you'll be able to handle another of the top three problems: change requests throughout the duration of the project. Experience has shown that poor advanced planning means that several retroactive improvements will have to be made later on in the process. Nonetheless, for the following reasons, you won't be able to completely avoid changes during the course of your project:

- External reasons (e.g., SAP bringing out a new release or top management deciding to restructure)
- Project-internal reasons (e.g., feedback from the reference group meaning that corrections have to be made, errors have been detected, or the concept has to be refined)

Because changes within a project are unavoidable, you should define what form these changes should take.

> **Implement a Change Management Policy**
>
> Implement a change management policy and define the following guidelines:
>
> - Which changes are permitted and which aren't
> - What types of changes you anticipate
> - Which documents are required for each change type
> - What approval procedure applies to each change type

Implementation Goal

The *goal* (i.e., what you want to achieve with this implementation) represents the starting point for each SAP implementation.

> **[+] Define the Goal of Your Plant Maintenance Project**
>
> As soon as you've defined a concrete goal, do the following:
>
> - Formulate it in writing and publish it (e.g., hang a poster in the project room, send an email to employees, make an announcement on the intranet).
> - Orient all project activities toward this goal.
> - Define what you *don't* want to happen when you implement the project (e.g., downsizing). In this way, you avoid promulgating any unfounded fears and unfulfillable hopes.

Together with your team and intended users, visualize what you want to achieve by implementing the SAP S/4HANA Asset Management system—and also visualize what to avoid. Examples of how seemingly self-evident principles can nonetheless be frequently violated are as follows:

- A goal may be formulated, but it isn't followed.
- There is a goal, but it's neither verbalized nor communicated.

The following question naturally arises: Which goals do you want to achieve by implementing an SAP S/4HANA Asset Management system in your enterprise? As a guide, the following is a short list of goals that customers often set in relation to plant maintenance with SAP S/4HANA Asset Management:

- *We want to reduce breakdowns to x%.*
- *We want to increase system availability to y%.*
- *Our customers want comprehensive documentation for our maintenance tasks so that we can continue to deliver high-quality products to them.*
- *Our vendors want comprehensive documentation of maintenance tasks so that we can continue to use their warranty measures.*
- *We want to reduce our technicians' nonpurposeful travel times and wait times by x%.*
- *We need complete asset master data.*
- *We need cost transparency.*
- *We don't want to invest any more monitoring resources in maintenance dates, and we want to allow the maintenance dates to vary by z days only.*
- *We want to make more targeted use of our employees' qualifications and availability.*
- *We have to wind down our previous system Y by a particular key date (December 31) because it's no longer being maintained. (The goal here is to wind down a legacy system.)*

- *We need an up-to-date overview of our spare parts stock.*
- *We want to reduce working capital by x%.*
- *We want to improve the delivery behavior of our vendors and service providers.*

You'll achieve the utmost success if you specify a concrete goal and work toward it. A good example is a customer in the automotive component supplier industry that wasn't satisfied with the downtime rate of its assets (6.2% at the time). Thus, *reducing downtime* was the top priority in the company's plant maintenance project. Therefore, the first phase of the implementation project was devoted to breakdown management. Breakdowns were recorded and analyzed, and countermeasures were put in place. It quickly became obvious that delayed reporting of breakdowns, especially at the ends of shifts, was causing an order bottleneck in the following shift and, consequently, unnecessarily long downtimes. To counteract this situation, floating shifts were introduced in the plant maintenance department; as a result, the downtime rate was reduced to 2.8% within six months.

Competencies and Responsibilities

As the empirical survey and personal experiences repeatedly show, *support from top management* is the most important success factor in an SAP project.

> **Secure Support from Top Management**
>
> A must for project professionals is to secure and publicize the active support of top management.

Don't settle for a simple "That's acceptable" or "Go ahead." Instead, ensure that top management plays an active role in the project. What exactly could this role consist of? For example, top management could perform the following tasks:

- Stressing the importance of the project and confirming its full support publicly at a company meeting
- Informing employees about plans for the project in a company-wide email
- Getting regular briefings on the progress of the project
- Making decisions as members of the steering committee

> **Demand the Necessary Competencies**
>
> Another must for project professionals is to ensure that the project has the necessary resources and that the project leader has the necessary competencies.

In many cases, the project is assigned to an employee who works in the line structure of a subdepartment and, as such, can at best write concepts but doesn't have the authority to make decisions about them. In this case, a major error is being committed.

The project must occupy as high a position as possible within the enterprise organization; for example, it must be at one of the following levels:

- Technical management
- Top management
- Plant management
- Area management

Make sure that the project leader can make technical- and project-related decisions (possibly subject to approval by the steering committee).

Make the Decision-Making Structure as Compact as Possible

An excessive number of decision-making instances is one of the main causes of overly long SAP projects.

You want to avoid the situation that arose within one company in the transport sector: three years into the project, the company had just about reached the stage where the structuring of technical systems had been agreed on by all decision makers.

Attach Importance to the Right Competencies

Avoid the following situation: the people with technical competence can't decide anything, and the people who have to make decisions on disciplinary grounds lack technical knowledge.

You should also note the following tip when putting together the project team.

[+]

Assemble the Right Project Team

Ensure that your plant maintenance project has all the right qualitative and quantitative components. In your project, you require the following:

- Technical competence from the user department (in mechanical engineering, electronics, measurement and control technology, building services, work scheduling, and similar areas)
- Technical competence from IT or the organization
- A knowledgeable and experienced consultant on your project team

The question of how big your project team should be isn't an easy one to answer. The staff on small plant maintenance projects in SAP sometimes consists of a single person, whereas very large projects often have dedicated teams of more than 25 people. Although this is a broad range, consider the following tip.

Have Fully Dedicated Staff on Your Team
Make sure that at least one of the project team members is 100% dedicated to the project.

In many projects, a mistaken belief is often put into practice that the same results that one 100%-dedicated project team member can achieve can also be achieved by three team members who are 50% dedicated to the project or six team members who are 30% dedicated to the project. This is a big mistake. From a project management viewpoint, the following equation applies:

1 × 100 > 3 × 50 > 6 × 30

Although this isn't mathematically correct, what it means is that an employee who can devote 100% of their time to the project is more productive than three employees at 50%, and three employees at 50% are more productive than six employees at 30%.

Why? For one thing, the larger the project team, the more time is required for consultation and communication. Furthermore, 50% can very easily turn into 30%, then 20%, then 10%, and then 5%—until finally, the team member in question has 0% availability for your project.

Although you may be thinking that this is just a theory or that it will never work here, you should ensure the following just to be safe.

Set Aside Specific Project Days
If you don't succeed in assigning employees 100% to the project, make sure that all project team members attend scheduled meetings.

For example, you can tell your team, "Every Thursday and Friday are devoted to working on this project." However, you should also ensure that your employees can work in peace on certain project days.

Obtain a Separate Project Room
Employees have too many distractions at their own desks (the telephone, answering colleagues' questions, etc.) to enable them to fully commit themselves to working on the project there. Furthermore, a project room has its own dynamic and energy, which enables everyone to concentrate solely on the project.

Closely linked to these statements is the following tip.

> **[+] Don't Overwork Your Employees**
> An employee who is already working at 120% capacity on daily tasks is incapable of working on an SAP project as well.

Despite this truth, the prevailing impression is that many companies assume that employees can work on the SAP project "alongside" their regular work.

Another issue that you should clarify at the outset is the input required for the project (e.g., to get your project budget authorized by top management).

Input Planning

Project input depends greatly on several factors that you have to consider, including the following:

- The spatial distribution of the implementation (Section 1.1.1)
- The used and targeted range and depth of functions
- Your company's internal decision-making and approval structures
- Your project competence and/or the project lead's authority to decide
- The quality of existing master data
- The number of new master data records required
- The number and complexity of the interfaces to be created
- The quantity and scope of new development work (e.g., reports, print programs, customer exits, a web user interface [UI])
- Integration with other SAP applications
- The scope of end-user training
- Whether your project concerns winding down a legacy system or a completely new implementation

With the last point, in particular, many enterprises mistakenly believe that if a legacy system is already in use, it should take even less time to implement an SAP system because, for example, the master data is already available in electronic form. Unfortunately, this isn't the case.

> **[!] Additional Effort If a Legacy System Exists**
> It takes *considerably* more effort to wind down a legacy system than to implement a new system.

There are several reasons for this:

- First, an IT system always revolves around a specific organizational structure and its individual business processes. Furthermore, it's always necessary to adjust this organizational structure when you change the IT system currently in use. This is always the case when you wind down a legacy system (e.g., replacing Maximo with SAP).

- Second, each IT system has a specific range of functions, its own terminology, and an individual layout, all of which the users have become accustomed to over the years. If you install a new IT system, everything is alien at first, looks strange, and just doesn't seem suitable. You hear concerns such as, "But the field must be on the top left of the screen, not in the middle right of the screen."
- Third, every familiarization process is difficult for all parties involved (the end users, the application consultants, IT, and the decision makers). Reservations creep in and obstacles arise, all of which have to be broken down, and this all happens repeatedly in the implementation phase.

In SAP S/4HANA Asset Management projects, project input usually fluctuates between the following values:

- Less than 50 days for small, new implementation projects with a high decision-making competence, existing master data, no interfaces, and very little additional programming
- Significantly more than 1,000 days for the winding down of legacy systems in large international projects with multilevel approval structures, manual entry of new master data, a lot of additional programming, and new connections to non-SAP systems

As an additional indicator for estimating project input, the following input distribution over the project phases may be useful:

- Discover: 5–10%
- Prepare: 5–10%
- Explore: 25–40%
- Realize: 25–40%
- Deploy: 20–30%
- Run: ongoing

For your own input planning, consider combining two procedures, as in the following tip.

Top-Down Estimate and Bottom-Up Planning

Follow these steps to carry out the procedures:

1. Carry out a *top-down estimate* (i.e., a rough input estimate based on the goals and framework conditions).
2. Break it down into the individual project phases.

3. Use the planning for individual work packages to perform bottom-up planning.
4. If the two plans are very different, examine your assumptions very closely.

Another aspect that you need to specify in the prepare phase is the type of documentation you want to create.

Documentation

One of the most important risk factors mentioned by the companies who were surveyed was insufficient documentation. Therefore, note the following important tip.

Create a Documentation Concept ("Document of Documents")

Here, define the following:

- What documents are to be created and for what purpose?
- What documentation tool will be used?
- What names will be assigned to the documents?
- Where will they be stored?
- Who is responsible for the documentation?

During the course of your project, you'll require the following documents, among others:

- **Project plan**
  A general plan that's focused on milestones and work packages, parties involved in the project, and distribution of responsibilities
- **Business case**
  A document for calculating the cost effectiveness of the project, which is usually required for budget requests
- **Monitoring documents**
  Documents that you, being the project lead, can use to monitor milestones, activities, compliance with deadlines, and so on
- **As-is analysis**
  An inventory of existing master data and business processes
- **Business process master list**
  An overview of new business processes
- **Business process procedures**
  A detailed concept of new business processes
- **Customizing**
  Documents that contain all Customizing settings

- **Gap list**
  A list of all points not covered in the standard system and for which an organizational or technical solution for the system must be found
- **WRICEF list**
  A list of all workflows, reports, interfaces, conversions, enhancements, and forms (WRICEF); that is, a list of all programming tasks, giving an overview of the solutions for the gap list
- **User roles**
  Definitions of roles, including their menus and authorizations
- **Test plan and test documents**
  Documents for planning and conducting tests as well as recording the results of such tests
- **Data transfer**
  A document that contains details about the objects that are to be transferred from the legacy system to the SAP system
- **Program development**
  Detailed programming requirements, which may also be divided into a *detailed specification* (on the business side) and a *technical specification* (on the technical side), depending on the variant
- **Training plan**
  An overview of when, where, and from whom users will receive training, as well as what topics will be covered
- **User training material**
  Material for holding the requisite training courses before the system goes live
- **User documentation**
  Reference materials
- **Cutover plan**
  Details of the transfer to a live system and a go-live
- **Feedback forms**
  Forms for problem messages
- **Sprint backlog**
  Documentation you need if you apply Scrum to your project
- **Project completion report**
  A document that is transferred to the user department involved in the go-live and that is the basis for the final presentation
- **Document of documents**
  The central document in which you describe the documents available as well as their purpose and content, if you're using a large number of document types

You may not require so many documents, you may choose to combine many of these documents into one document, or you may even need to use other documents not listed here. However, don't worry. You don't have to reinvent the wheel for the umpteenth time, as explained in the next tip.

> **SAP Templates**
>
> In SAP's Roadmap Viewer and SAP Solution Manager, SAP provides templates for all the documents listed here as well as other documents (e.g., a costing sheet for the business case or capacity calculations, presentations, checklists, report templates).

Figure 1.9 shows a list of accelerators for the SAP Activate explore phase.

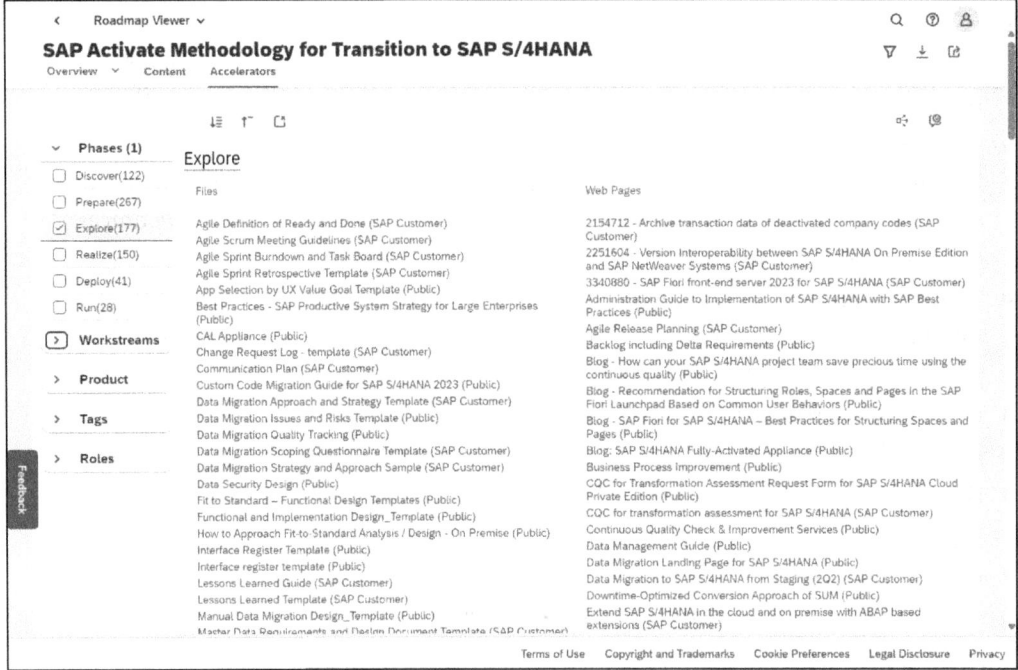

Figure 1.9 SAP Activate: Accelerators

Training

Both the empirical survey and my personal experiences have shown me that good training of project team members is a prerequisite for high-quality project work.

> **Provide Comprehensive Training for Project Team Members**
>
> Another must for project professionals is to train your project team members fully, either through attendance at SAP's standard courses or an in-house training workshop.

It isn't enough to simply give the project team members a brief introduction. If you want them to successfully perform their conceptual work, they need to have comprehensive knowledge of SAP S/4HANA Asset Management.

SAP currently offers the training courses listed in Figure 1.10.

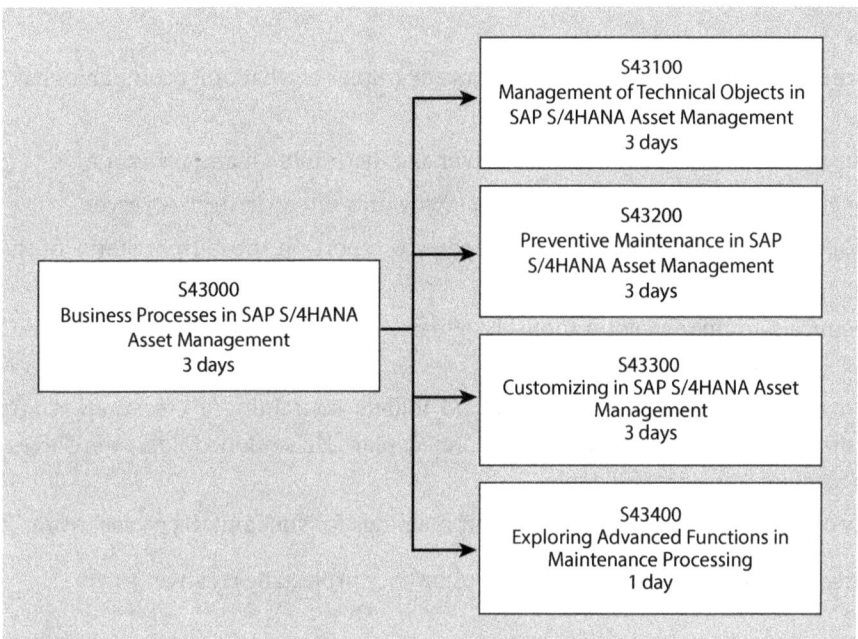

Figure 1.10 SAP Training Courses for Asset Management

> **S43000 and S43300 for Project Team Members**
>
> At a minimum, your project team members should complete courses S43000 and S43300, as well as others as required.

Another aspect of the prepare phase is marketing, which was already explained in Section 1.1.2 but will be mentioned here again for the sake of completeness.

> **Have the Courage to Leave Some Gaps**
>
> The full range of functions in the SAP system doesn't have to be (and shouldn't be) implemented all at once.

Marketing

Your project will have a greater chance of success if it has the full support of all employees, especially intended users, right from the start, and if you can create a "power base" for the project. A first step in this direction (there will be others later on) is publicity work.

> **[+] Do Something Good and Talk About It**
> Make your project known within your company (i.e., do in-house marketing).

There are many ways to do this. Here are some examples of what other companies have done:

- A transportation company designed a flyer and distributed it among its staff.
- An energy provider had a presentation playing on a screen in its main foyer.
- An automotive supplier set up a homepage to report on the current status of the project.
- A chemicals company sent a monthly newsletter to staff—in particular, intended users—and top management.
- A transportation company created cloud folders containing a presentation and description of the project as well as the project plan. These cloud folders were accessible to all interested staff people.
- Many companies convene official kickoff meetings for staff and the project team.

There are many inexpensive ways to announce your project. Be creative!

Business Process Modeling Methods

Now let's think about the external form in which the business processes will be documented and presented—that is, what business process modeling methods you'll use.

In principle, because of their superior visualization capabilities, graphical and tabular formats are better than a purely textual format. *Value chain diagrams* (VCDs, see Figure 1.11) are suitable for depicting the flow of complex processes, such as plant maintenance projects, on a general level. VCDs represent predecessor-successor relationships (including multilevel ones), such as those in work packaging and sequencing.

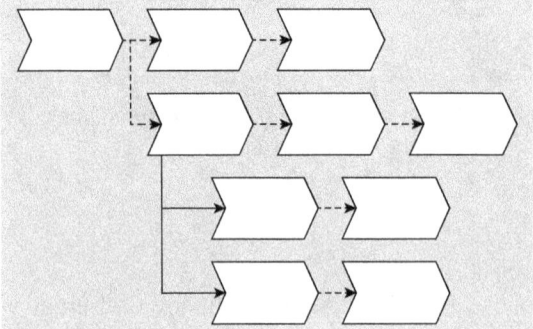

Figure 1.11 Value Chain Diagram

Event-driven process chains (EPCs) are suitable for processes of low and medium complexity. Simple and advanced versions of EPCs are available.

The simple version of an EPC represents a process flow on a more detailed level than a value chain diagram and consists of two object categories:

- Functions (e.g., print the order, check material availability)
- Events (e.g., malfunction report received, order confirmed)

Advanced EPCs are used when additional object categories have to be modeled:

- Documents (e.g., a job ticket, a confirmation slip, an object list)
- Organizational units (e.g., work scheduling, a mechanical engineering workshop)
- Process interfaces (e.g., purchase order handling, final costing)
- Information systems (e.g., Transaction IW31 in SAP S/4HANA)
- Files or databases (e.g., AFRU [order confirmation] and BANF [purchase requisition] tables)

It's up to you whether to use object categories in your business process model and, if so, which ones. Figure 1.12 shows a sample extract from such a model.

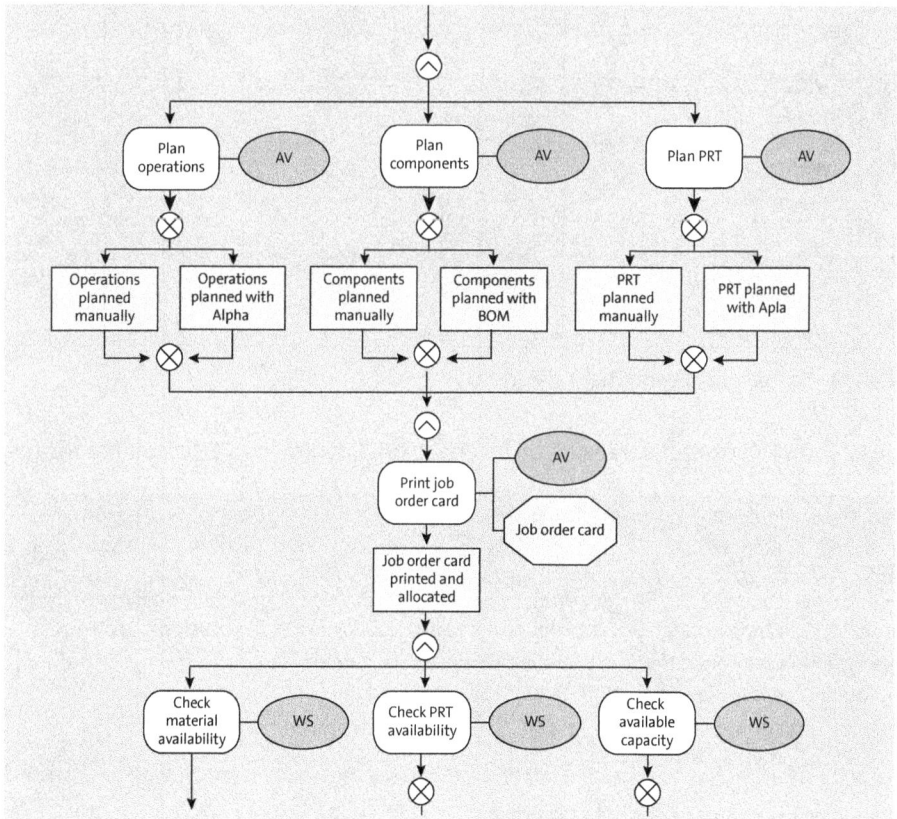

Figure 1.12 Advanced Event-Driven Process Chain

1 SAP Projects in Plant Maintenance

Transaction/event chain diagrams (ECDs) are suitable for simple processes but not for more complex processes. A central characteristic of transaction/event chain diagrams is the fact that they assign object categories to fixed columns. Transaction/ECDs are available in a purely graphical format or in a verbal tabular format. In graphical format, they have the same object categories as an advanced EPC (see Figure 1.13).

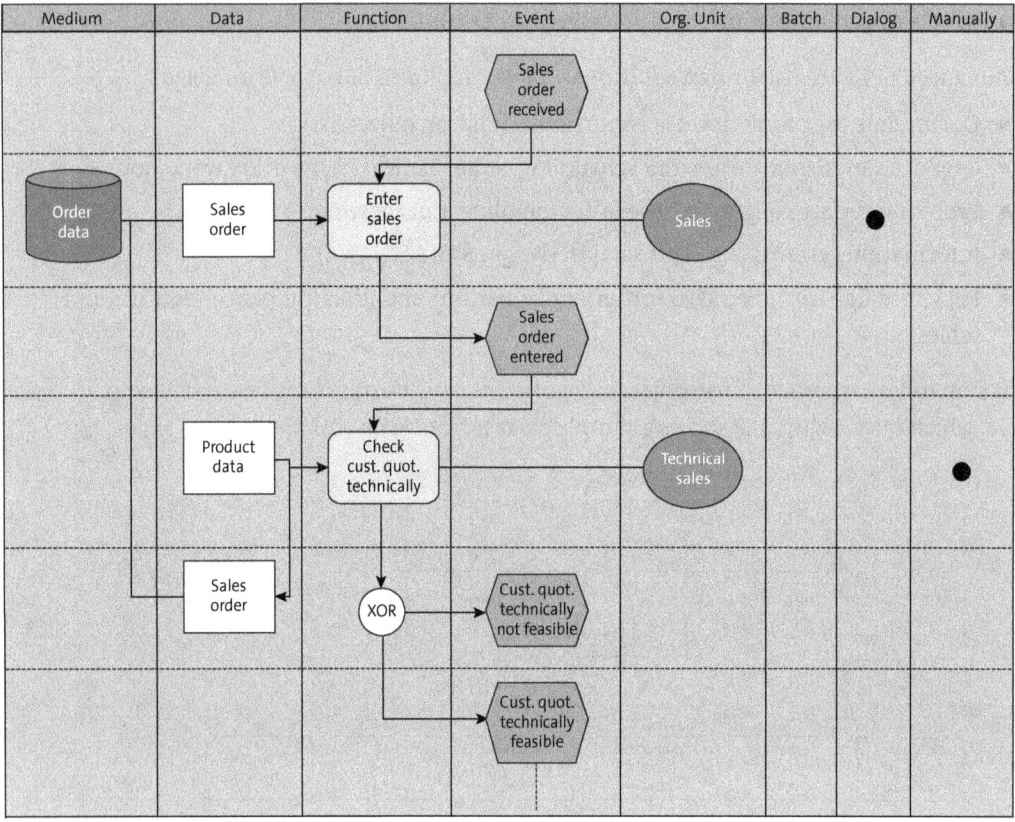

Figure 1.13 Transaction/Event Chain Diagram

Table 1.2 shows an example transaction/event chain diagram in a verbal tabular format.

No.	Event/ Result	Function/ Activity	Input (Data)	Output (Data)	Parties Involved/ Org.	Transaction Code
1.	Maintenance task completed	Confirm single entry or, in the operation list, continue with item 3.	Order number, personnel number, and time	Confirmation data record	Technician	IW41

Table 1.2 Tabular Transaction/Event Chain Diagram

1.3 Tips for Your Plant Maintenance Project

No.	Event/Result	Function/Activity	Input (Data)	Output (Data)	Parties Involved/Org.	Transaction Code
2.	Confirmation complete	Continue with item 4.		Confirmation data record		
3.	Operation list created	Select PM operations to be confirmed and branches in single record entry.	Selection data, operation list, personnel number, and time	Confirmation list	Technician	IW48
4.	Confirmations complete	Check confirmation data.	Selection data and confirmation list	Modified data records	Technician	IW47
5.	Difference detected	Make correction (cancel or follow up entry).	PM order number and correction data		Technician	IW48 or IW41

Table 1.2 Tabular Transaction/Event Chain Diagram (Cont.)

Business Process Model and Notation (BPMN) has become very popular in IT projects. Even SAP uses BPMN to visualize its so-called best practice processes. Figure 1.14 shows a part of the **4HH – Emergency Work** process.

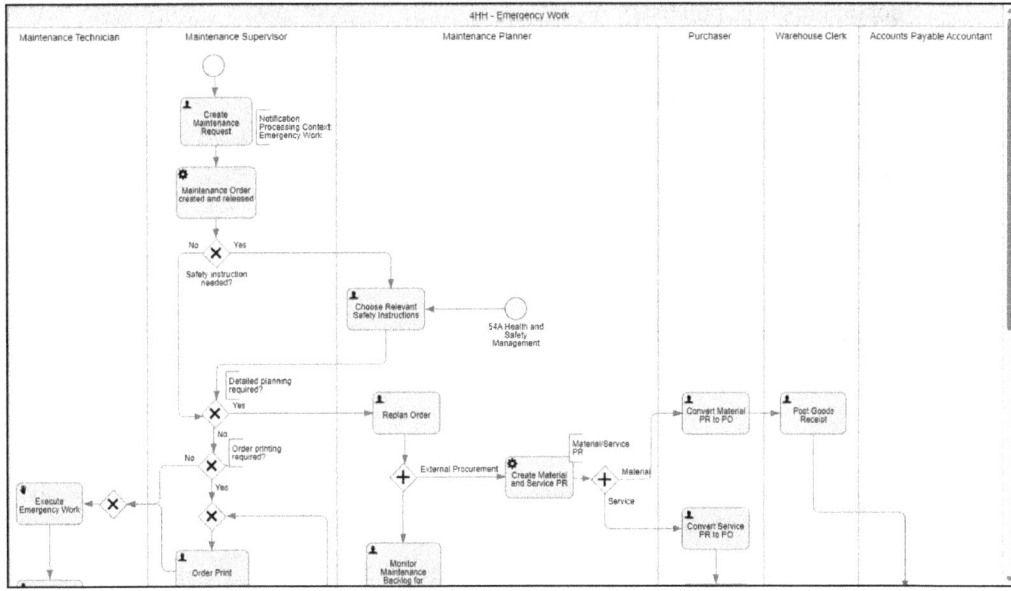

Figure 1.14 Business Process Model and Notation

BPMN uses a set of symbols that are similar to those used by EPC:

- An *activity* is a task that must be completed in a business process (e.g., recording the demand for spare parts).
- A *gateway* is a decision point, like a split/fork (e.g., in stock or not in stock), or a point at which different control flows converge, like a join/merge (e.g., courier booked and material packaged).
- An *event* is something that occurs in a business process (e.g., a message is received; a certain date is reached, such as three days after sending the purchase order; an exception situation arises).
- *Sequence flows* connect activities, gateways, and events. They represent the sequence in which activities are executed.
- A *pool* represents a participant, meaning a user, user role, or system (e.g., a buyer).
- A *lane* is a subdivision of a pool that spans the complete length of the pool (e.g., the process line of the buyer).
- A *message flow* shows two lanes or pools in a business process diagram or two elements of those exchange messages (e.g., the transfer of a requirement from the process flow of the maintenance planner to the process flow of the buyer).
- An *annotation* is a comment that can be assigned to an element of a business process.
- A *data object* represents an artifact that is processed by the business process. Data objects can be used to represent electronic objects such as documents or data records (e.g., maintenance orders) as well as physical objects such as spare parts or documents (e.g., pull lists).

[+] **Visualize the Business Processes**

Create business process models of the as-is and to-be processes via different model types:
- VCDs for an overview of complex business processes
- EPCs or BPMN for a detailed representation of complex processes or to represent less complex processes
- Transaction/ECDs in a tabular or graphical format for simple processes

System Landscape

Another error that many companies make in this phase is to make the SAP system landscape available too late.

> **Set Up a Test System Early On**
> At the very latest, a test SAP system must be available after your staff has finished their SAP training. If they don't have access to a test system at this stage, they won't be able to practice what they've learned and will quickly forget most of it. In other words, plan to set up a test system early on.

1.3.3 Explore Phase

In my book *Plant Maintenance with SAP S/4HANA: Business User Guide* (SAP PRESS, 2021), there are many tips on content that you might want to consider, especially in the explore phase of drawing up a detailed functional concept. The following is a brief list of some subject areas and common questions about them that you can ask during the explore phase:

- **Organizational units**
  What needs to be done in cross-cost accounting maintenance processing?
- **Structuring of technical systems**
  To what extent should you perform structuring of technical systems? What elements are to be used?
- **Business processes**
  How are external companies mapped? How can the layout of orders be determined?
- **Integration**
  How can equipment and assets be aligned with each other? Which material requirements planning (MRP) procedure is best for spare parts?
- **Maintenance control**
  How can active availability control be set up in budgeting? How are dynamic date calculations performed?
- **New SAP technology**
  Which SAP Fiori apps can support the business processes? Which after-event recording options are available to you?
- **Usability**
  How do you configure maintenance-specific user parameters? How can you adjust table controls to the relevant user requirements?

In the explore phase, all of these questions are reflected in your detailed concept. Next, we provide more tips that you should apply in the explore phase, especially in relation to organizational issues. The most important thing is perhaps for the employees to support the project and, in particular, the future system. You should therefore secure their support as early as possible.

Involve Your Intended Users in the Explore Phase

Only a system that is accepted by its users is a genuinely useful one. Therefore, it's important that you involve your intended users in the explore phase of the project.

Here are some suggestions for involving your users in the explore phase:

- Start a *reference group*, which is a group of selected users who will provide technical input and feedback to the project group.
- Present your interim results at regular intervals (in live presentations, in newsletters, and on your home page).
- Conduct individual surveys and get feedback from respondents.

Usability and User Acceptance

Chapter 9 of this book provides various usability improvement options. Therefore, this topic isn't discussed in detail here, but its importance is underlined in the following general tips.

Keep the System Design as Simple as Possible

Whether it's the structuring of technical systems or business process handling, user acceptance increases in direct proportion to the simplicity of the system. It's better to have 80% of the system you want with 100% user acceptance than 100% of the system you want with 20% user acceptance.

The SAP system isn't exactly known for its usability, but we'll look at the question of whether or not it really has low usability and what you can do about it in your own project. In Chapter 9, you'll find a lot of tips and tools to improve the usability of your SAP system (e.g., by using transaction variants, SAP Fiori apps, and mobile solutions).

Do Everything You Can to Make the System as User Friendly as Possible

Be particularly proactive in addressing the perceived lack of usability of the SAP system. During the implementation phase, be attentive to the concerns of your employees regarding its usability. Don't dismiss the topic; rather, accept users' fears and try to alleviate them.

The following is another tip on how you can increase user acceptance right from the start.

1.3 Tips for Your Plant Maintenance Project

> **Grab the Low-Hanging Fruit to Entice Employees**
> Use the as-is analysis to identify system aspects with which the users have problems or aspects that are perceived to be suboptimal. Offer them a solution to these specific issues that represents an improvement on the current situation. Your users will then be more prepared to support the new system.

Because these aspects are very specific to a project, here are some examples of what other companies have done:

- Created a list of information that can only be obtained following extensive research
- Created notifications in the production department directly, instead of transferring papers
- Automatically calculated maintenance dates, instead of searching through folders
- Printed the pull list directly in the warehouse for advance picking, instead of having the maintenance technicians wait a long time for the goods issue
- Sent breakdown notifications via a paging system, rather than having the worker walk back to the support desk to pick up orders
- Set up the SAP Fiori launchpad with the Maintenance Request app, for example, as an easy way to create outstanding notifications
- Automatically issued maintenance lists to production by email to avoid the worker being unable to gain access to the facility to carry out maintenance tasks because the facility was still up and running due to production being unaware of the maintenance
- Have maintenance technicians use tablets with a configured SAP Fiori launchpad, including having them use SAP Fiori apps for their daily business instead of using printouts

If you're observant, you'll notice issues like these in your own company—and if you're a little creative, you'll be able to provide your users with improvements.

Now, let's deal with an aspect that is sometimes a little neglected at the start of a project: authorizations.

Authorization Concept

The *authorization concept* controls which user in which organizational unit may execute which functions on which objects. This is a general topic, not a maintenance-specific one. As such, responsibility for the authorization concept lies with the IT department, the coordination department, or a similar department. However, you should ensure, in a reasonable amount of time, that an authorization concept is created to control the use of SAP S/4HANA Asset Management.

1 SAP Projects in Plant Maintenance

[+] **Authorization Concept: A Project within a Project**
Don't forget to design an authorization concept alongside the detailed technical concept.

Appendix A, Section A.2, contains a summary of the authorization objects that are available in SAP S/4HANA Asset Management and the organizational units, functions, and fields that are checked in each case.

This topic won't be covered in any further detail here; instead, refer to the relevant specialist literature.

[+] **Specialist Literature**
For more information, check out *Authorizations in SAP S/4HANA and SAP Fiori* by Alessandro Banzer and Alexander Sambill (SAP PRESS, 2022).

The most important settings will be explained in detail there. Examples include the following:

- How to define authorizations based on authorization objects
- How to combine several authorizations to form an authorization profile
- How to create single roles using Transaction PFCG (see Figure 1.15) and combine them to form composite roles
- How to define users with the requisite authorizations

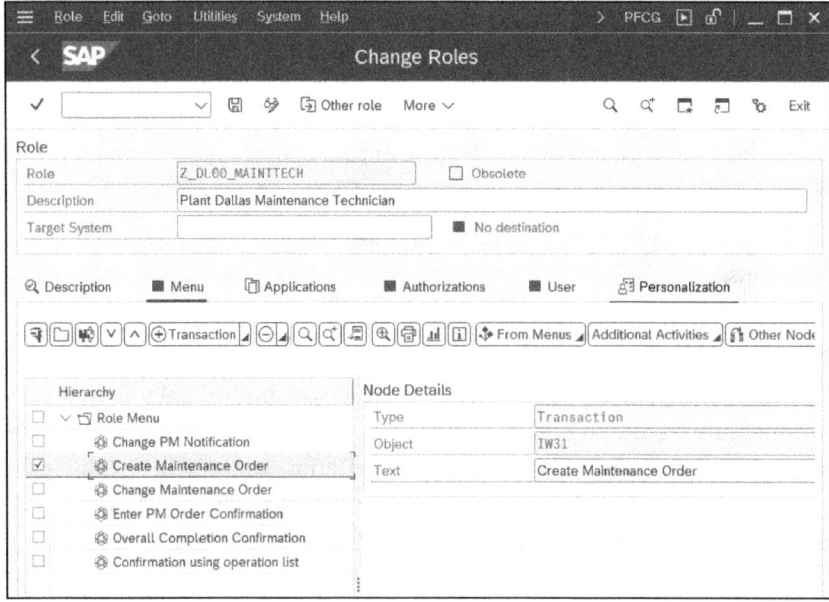

Figure 1.15 Transaction PFCG: Maintain Single Role

1.3 Tips for Your Plant Maintenance Project

Legacy Data Transfer and Master Data Maintenance

There are always two ways to transfer your master data to the SAP system: manually or automatically. If you're changing from another maintenance planning and control system to the SAP system, or if the asset data is available in another electronic format (e.g., in a CAD system), you should always try to transfer this data to the SAP system automatically.

Use SAP Tools

SAP provides the following three standard tools for transferring data from upstream or legacy systems:

- A maintenance-specific data transfer (Transaction IBIP)
- The general data transfer workbench (Transaction LSMW)
- The Migrate Your Data – Migration Cockpit SAP Fiori app

All of these tools enable the transfer of maintenance objects and their data to the SAP system.

Plant maintenance batch input, as shown in Figure 1.16, is a data transfer program tailored to the special database objects in SAP S/4HANA Asset Management.

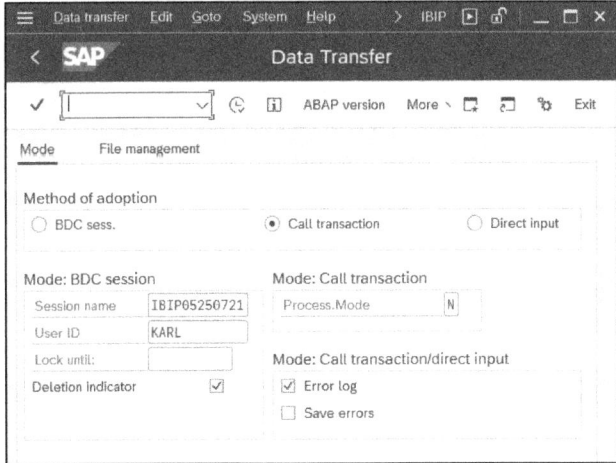

Figure 1.16 Transaction IBIP

In contrast, the Legacy System Migration Workbench (Transaction LSMW), shown in Figure 1.17, is a general data transfer workbench that can transfer not only maintenance objects but also objects from all SAP applications.

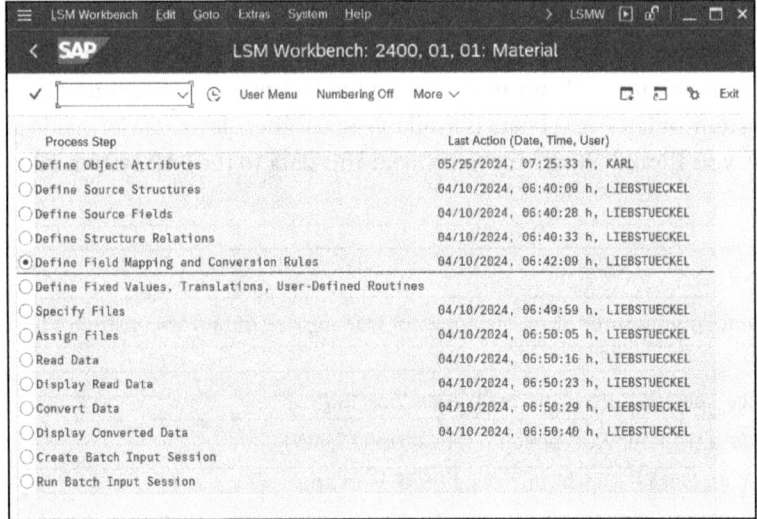

Figure 1.17 Legacy System Migration Workbench (Transaction LSMW)

The Migrate Your Data – Migration Cockpit app (see Figure 1.18) also enables the migration of data for all SAP applications.

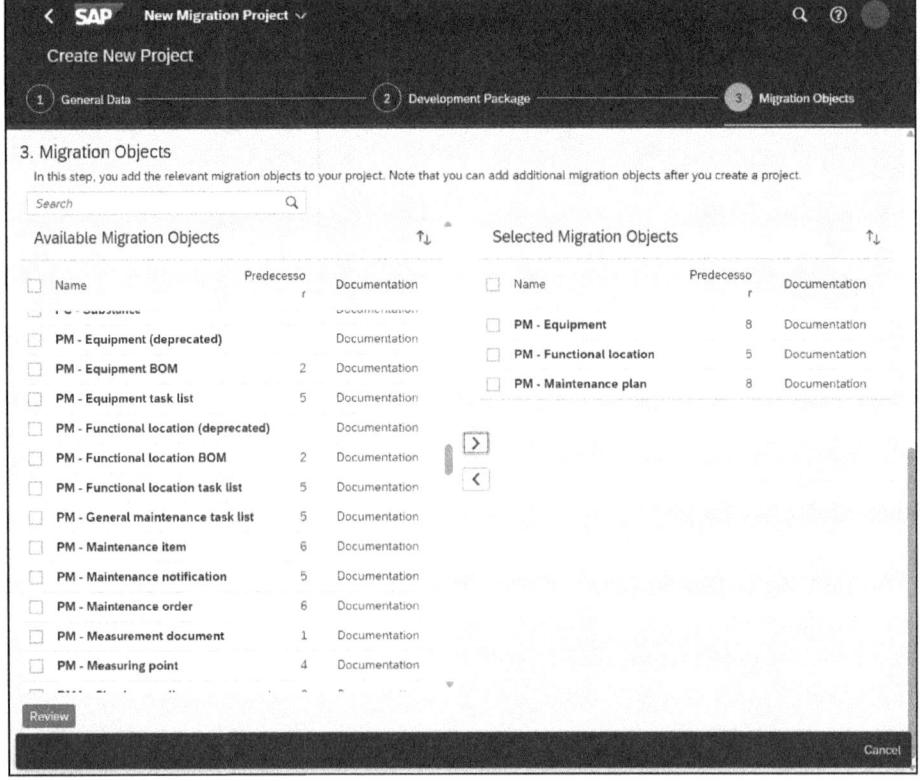

Figure 1.18 Migrate Your Data – Migration Cockpit App

The main differences among the three systems are as follows:

- **Plant maintenance batch input (Transaction IBIP)**
  Plant maintenance batch input requires a source file whose field sequence has to correspond to a predefined structure. You can find the names of these structures (e.g., IBIPEQUI for the equipment master record) in the Transaction IBIP documentation, and you can find the field structure in the Data Dictionary (Transaction SE11). No further prerequisites apply if the structure is adhered to.

- **Legacy System Migration Workbench (Transaction LSMW)**
  Transaction LSMW doesn't require any fixed structure. Instead, it uses field mapping, which assigns the fields of the source structure to the fields of the SAP object. The source structure is described on a per-project basis in the SAP system. This makes Transaction LSMW more flexible but also more complex than Transaction IBIP.

- **Migrate Your Data – Migration Cockpit app**
  Similar to Transaction IBIP, this SAP Fiori app has a fixed data structure. You can download CSV or XML templates for this purpose. These templates have to be filled in the exact field order and then must be uploaded. This makes the SAP Fiori app less flexible than the Legacy System Migration Workbench.

If you're setting up IT support for your SAP system from scratch (i.e., if you don't have any master data in electronic form), you'll have to enter the data manually.

> **Microsoft Office Tools for Data Entry**
>
> Besides the standard approach of creating data directly in the SAP system, you can use the RIACCESS program in SAP ABAP to export the SAP table structures to Microsoft Access as a local database, enter the data locally on your PC, and then transfer it to the SAP system using Transaction IBIP or Transaction LSMW.
>
> Figure 1.19 shows the initial screen of RIACCESS.

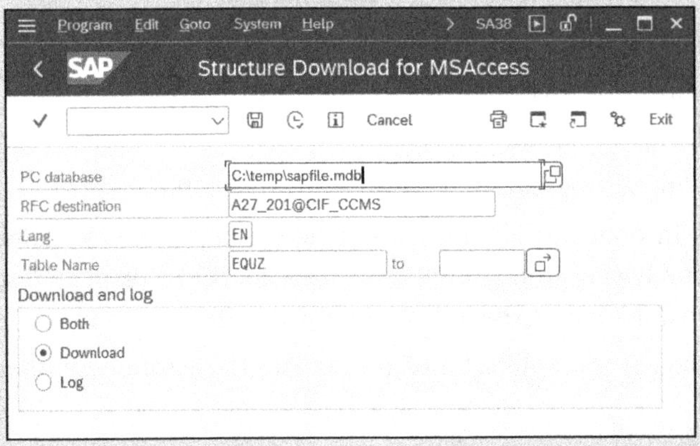

Figure 1.19 Database Structure Download to Microsoft Access

1 SAP Projects in Plant Maintenance

Regardless of what procedure you choose, you need to plan the data entry as detailed in the following tip.

> **[+] Plan the Data Entry**
>
> Create a detailed data retrieval plan. In this plan, you should specify the following points in particular:
> - Which employees are involved in the data transfer
> - What kind of templates (index cards, folders, drawings, etc.) are used
> - How the data should be formatted and prepared
> - The dates by which the data should be entered into the system
> - How the quality of the data is to be monitored
> - Who is responsible for quality control
> - How the master data is to be released
> - How long the individual activities last

Customizing for Prototypes

Finally, there is an extremely important task in the explore phase: building a prototype.

> **[+] Build a Prototype**
>
> During the explore phase, you make a large number of decisions that affect Customizing. Don't delay Customizing until the realize phase. Instead, configure all the Customizing settings in the explore phase and build a prototype in it.

The following data belongs to this prototype:
- All controlling Customizing settings for the organizational structures
- All controlling Customizing settings for the structuring of technical systems
- All controlling Customizing settings for the business processes
- All controlling Customizing settings for SAP Fiori apps
- All controlling Customizing settings for integration with other components
- Representative settings in noncontrolling Customizing settings, such as maintenance planner groups, purchasing groups, MRP controllers, locations, plant sections, and similar tables, which don't have to be fully maintained in the prototype
- Representative master data (work centers, functional location structure, equipment, task lists, etc.)

In such a prototype, you should reproduce the future live system approximately so that your users have an opportunity to see the system in advance.

1.3 Tips for Your Plant Maintenance Project

Gap List

One of the most important intermediate results at the end of the explore phase is the creation of a *gap list*. This list should include all points that are required by users but that can't be responded to in a satisfying way with Customizing or other standard tools. Programming is usually required (customer exits, Business Application Programming Interfaces [BAPIs], business add-ins [BAdIs], additional programs, etc.). Table 1.3 shows an example of what a gap list could look like.

	Reference	Request	Solution
Update lists	02_01_Work Order Cycle	Lists (e.g., IW28 notification list, IW38 order list) update automatically at a defined time interval.	Must be included in the existing lists, coding template: SAP_TIMER_DEMO
Order papers	02_01_Work Order Cycle	Specific procurement documents required to be available to the departments.	Own print programs, own layout
Order documents with documents	02_01_Work Order Cycle	Plant-specific documents required to be output during print output.	Add-on of SEAL Systems
Mail at notification creation	02_01_Work Order Cycle	Automatic mail to reporters and stakeholders when SAP message is opened by service line.	Notification workflow
Technical confirmation	02_01_Work Order Cycle	Job must not be completed if open purchase requisitions remain.	Customer exit IWO10004
Mail at appointment overtime	02_01_Work Order Cycle	Automatic mail to responsible workplace when end date is exceeded.	Order workflow
Shop floor paper	02_01_Work Order Cycle	Layout of the order paper.	Printing program and form

Table 1.3 Gap List Example

A gap list includes the following information:

- Topic
- Business blueprint source document that the topic comes from
- Request
- Acceptable solution

1.3.4 Realize Phase

In the realize phase, the plans and concepts created in the explore phase must be turned into action. To do this, you need to complete the following tasks:

- Finalize the Customizing settings. You already made some controlling Customizing settings in the explore phase so that you could build a prototype. However, in the noncontrolling Customizing settings (e.g., maintenance planner groups, purchasing groups, MRP controllers, locations, plant sections), you made representative entries only. You must now complete these settings.
- Attend to all items from the gap list.
- Implement and configure SAP Fiori launchpads (see Chapter 8).
- If necessary, set up transaction variants (see Chapter 9, Section 9.2.1).
- If necessary, implement and configure a mobile solution, such as SAP Service and Asset Manager (see Chapter 9, Section 9.2.4).
- If necessary, configure frontends with GuiXT (see Chapter 9, Section 9.2.6).
- If necessary, configure frontends with SAP Screen Personas (see Chapter 9, Section 9.2.7).
- If necessary, develop a web frontend (see Chapter 9, Section 9.3.3).
- Develop the required enhancements.
- If necessary, program customer exits or BAdIs (see Chapter 9, Section 9.3.4 and Section 9.3.5, respectively).
- If necessary, set up workflows (see Chapter 9, Section 9.3.6).
- Develop interface programs.
- Prepare the legacy data for transfer to the new system and set up the transfer tools, namely the Legacy System Migration Workbench, batch input for plant maintenance, or the Migrate Your Data – Migration Cockpit app.
- Enter the master data directly into the SAP system.
- Set up the users, roles, and authorizations.
- Develop the end-user training courses and documentation.
- Arrange for the end-user training courses to be held.

These are defined tasks that don't require any further explanation here, so let's discuss the important topic of testing. Before going live, you should test your implemented concept extensively.

Conduct Sufficient Testing

Extensive testing in the implementation phase is a prerequisite for a successful go-live. Therefore, you should schedule ample time for testing and conduct all tests in full.

There are various types of tests:

- **Functional tests**
  These are tests at the most basic level. They examine programs and transactions for errors and evaluate them. These tests focus on the program's internal functions, and their main purpose is to help you become familiar with the system in terms of your own requirements.

- **Scenario tests**
  During the Customizing process, it's necessary to test transaction chains that are dependent on one another and reflect central business processes and scenarios. These tests focus on complete business processes.

- **Integration tests**
  These are final tests that examine integration aspects by using predefined business processes or scenarios that simulate the processes in your system. These business processes already use the migrated data and are executed in the actual IT infrastructure, which comprises the SAP system, third-party software, system interfaces, and various hardware and software components.

1.3.5 Deploy Phase

The deploy phase comprises all activities that are required shortly before, during, and shortly after system go-live. These include transferring legacy data to the production system, but because the data transfer process rarely runs smoothly and in full, you'll need some time to correct and complete the data. How much time you'll need for this depends on the volume of changes required.

Allow One Day to One Week for Data Maintenance

Schedule sufficient time (one day or several days) to correct errors and complete the master data.

Another activity that you need to complete in this phase is end-user training. Although SAP training courses are recommended for project team members, this doesn't apply to end users because SAP training is generally unsuitable for end users.

Conduct In-House End-User Training

End-user training should be provided by employees in your company, ideally by the representatives of the user department in the project team. For this training, use the system environment as well as examples of the technical objects that the end users will encounter in their work later on. It's also important to provide end-user training close to the go-live date. Otherwise, the users will have forgotten much of what they learned by the time the system goes live.

Cutover Plan

At the very least, you should create a complete cutover plan by the time you conduct end-user training. A cutover plan contains all the activities that are to be carried out either directly before, during, or shortly after the go-live.

Develop a Cutover Plan

The cutover plan serves the function of a "string tied around your finger," that is, a reminder to ensure that you don't forget anything.

This plan often deals with what appear to be minor issues, but it's precisely the minor things that are easily forgotten in something big like a system go-live. This plan can contain the following items:

- Any master data that still has to be created manually because it isn't worthwhile to transfer it manually
- The settings that have to be made to the system because they can't be transported
- Any selection and display variants that need to be set
- The batch jobs that need to be triggered
- Number ranges that have to be created manually or whose transport has to be triggered manually

Go Live

Your new system then goes live on the planned key date, and you start to process the first notifications and orders there. The most important thing in this phase is to ensure that your employees support the systems.

Arrange for Project Team Members to Attend the Workshop

Organize end-user support for the first few days after the go-live. Ideally, this should be on-site user support by members of the project team. Furthermore, publish a central hotline number and an email address so that problems can be resolved quickly.

Provided that the tests were carried out in a targeted and intensive manner, you shouldn't experience any design difficulties or system problems in this phase. Experience shows that the issues that arise in this phase tend to relate to working with the software or concern detailed questions, and even these issues taper off after the initial familiarization phase.

1.3.6 Run Phase

The final phase (run) mainly involves improving the operations of SAP S/4HANA Asset Management and identifying which measures in the project phase offer potential for optimization in future projects. These measures include optimizing the following:

- Documentation
- Training
- Implementation
- Testing
- Technical operations
- Business processes
- SAP Fiori launchpads
- System upgrades
- System maintenance

This chapter concludes the project-related questions and tips. In the next chapter, we'll turn our attention to the first set of system-related questions—that is, questions related to the organizational structures that you need to set up.

1.4 Summary

As major takeaways from this chapter, you should do the following things in your plant maintenance project with SAP:

- Organize support of the top management.
- Acquire the necessary decision-making competence.
- Define clear goals for your project.
- Create your own implementation strategy.
- Follow the SAP Activate methodology.
- Take enough time to plan your project.
- Perform an as-is analysis (a feasibility study) as a starting point of the project.
- Set up the right team for this project.
- Avoid overworking your project team members.

- Take care of sufficient documentation via SAP templates.
- Involve your end users during the project.
- Do everything you can to make the system as user-friendly as possible.
- Start with low-hanging fruit.
- Build a prototype along the implementation path.
- Conduct sufficient testing and end-user training.

This chapter has provided processes and tips to help with these issues, with a focus on the SAP Activate methodology.

Chapter 2
Configuring Organizational Structures

This chapter introduces you to the indispensable elements of maintenance processing within the SAP system: the general organizational units, the maintenance-specific organizational units, and the work center.

Organizational units define your enterprise's organizational structure in general and your plant maintenance processes in particular. Consequently, they form the basis of all master data and business processes.

SAP provides a range of organizational units that you can use to map your enterprise structure:

- Some are also used in other functions in SAP S/4HANA (e.g., the company code, controlling area, or plant).
- Some are used in plant maintenance when plant maintenance has its own perspective (e.g., the plant as a maintenance planning plant and as a maintenance plant).
- Some concern plant maintenance only (e.g., the plant section or the maintenance planner group).

> **Organizational Units as Authorization Objects** [+]
>
> Each organizational unit has at least one authorization object, but usually, it has more than one. Consequently, organizational units are perfectly suited to assigning or restricting authorizations. For example, you can use the maintenance planner group to specifically assign authorizations for processing maintenance orders.

Even though it's possible to cautiously enhance the enterprise structure during live operations, deleting organizational units already in use in live systems is a very delicate matter. Changing the organizational structure involves the following tasks, among others:

- Identifying and adjusting affected Customizing tables
- Identifying and adjusting affected selection variants
- Checking custom developments because they may have been developed on the basis of certain organizational structures
- Creating new master data

2 Configuring Organizational Structures

- Checking the authorization concept
- Using data archiving to ensure that document history is retained
- Checking and correcting all potentially affected documents (e.g., you can no longer continue to process an order if the planner group has been deleted)
- Checking and correcting all potentially affected master data (e.g., you can no longer continue to process a piece of equipment if the work center has been deleted)

From a plant maintenance perspective, defining organizational structures also concerns the following areas:

- The general SAP organizational units (e.g., controlling area, company code, plant in general, storage location)
- A definition of the maintenance-specific organizational units (e.g., maintenance plant, maintenance planning plant, location, plant section)
- Customizing for the work centers in SAP S/4HANA Asset Management (e.g., mechanical engineering, electronics, measurement and control technology)

2.1 General SAP Organizational Units

Organizational units are the basis of all master data and business processes in SAP S/4HANA Asset Management. In the following sections, you'll get to know the most important organizational units from the perspective of SAP S/4HANA Asset Management and learn how you should use the associated Customizing functions. Figure 2.1 shows some typical general organizational units.

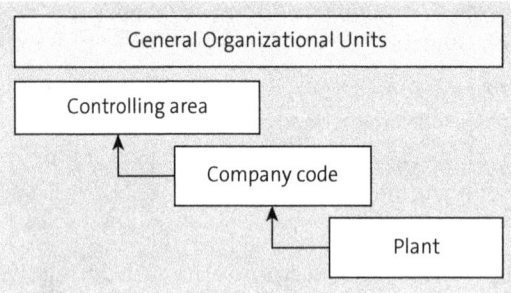

Figure 2.1 General Organizational Units

[+] **General Organizational Units Are Usually Predefined**
If you implement SAP S/4HANA Asset Management, the general organizational units in the SAP system (e.g., company code, controlling area, plant in general) are usually predefined because you defined them when you implemented other functionalities, such as controlling and materials management. Therefore, you can only influence the

design if SAP S/4HANA Asset Management is implemented from the outset, or if you define separate organizational units from a pure maintenance perspective.

2.1.1 Company Code

In SAP S/4HANA, the company code maps an independent accounting unit and is therefore the smallest organizational unit in SAP S/4HANA Finance. You therefore use a company code to map, within your corporate group (client), a company that is required to create a balance sheet, a profit and loss account, and other year-end documents.

Edit the Company Code

The company code provides the basis for many other applications, including SAP S/4HANA Asset Management. You use this Customizing function to create new company codes and to make changes to existing company codes.

Prerequisites

You must maintain the countries, languages, and currencies beforehand.

Customizing Path

Enterprise Structure • Definition • Financial Accounting • Edit, Copy, Delete, and Check Company Code

Transaction

V_T001

Settings

Choose the **Edit Company Code Data** subfunction. The system then displays a table of defined company codes (see Figure 2.2).

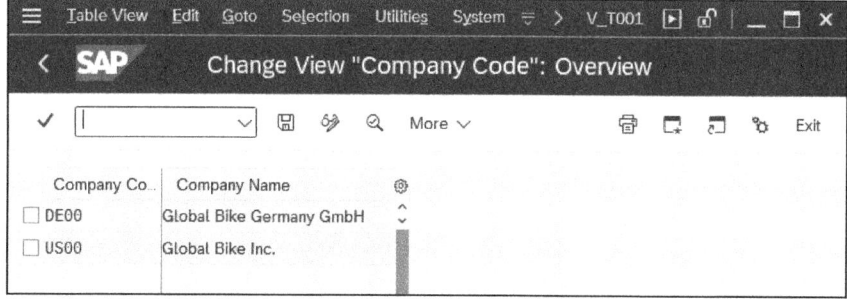

Figure 2.2 Company Code: Overview

On the **Details** screen, you maintain key parameters such as **Ctry/Reg.** and **Currency** (see Figure 2.3).

2   Configuring Organizational Structures

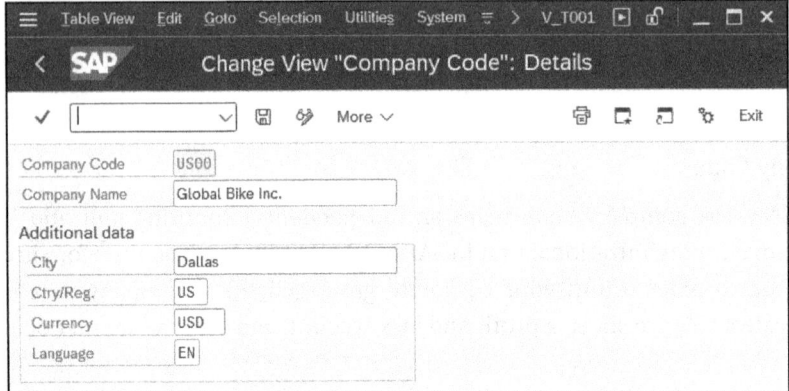

Figure 2.3 Company Code: Details

You can then use the ⊡ icon to specify the address data for the company code (see Figure 2.4).

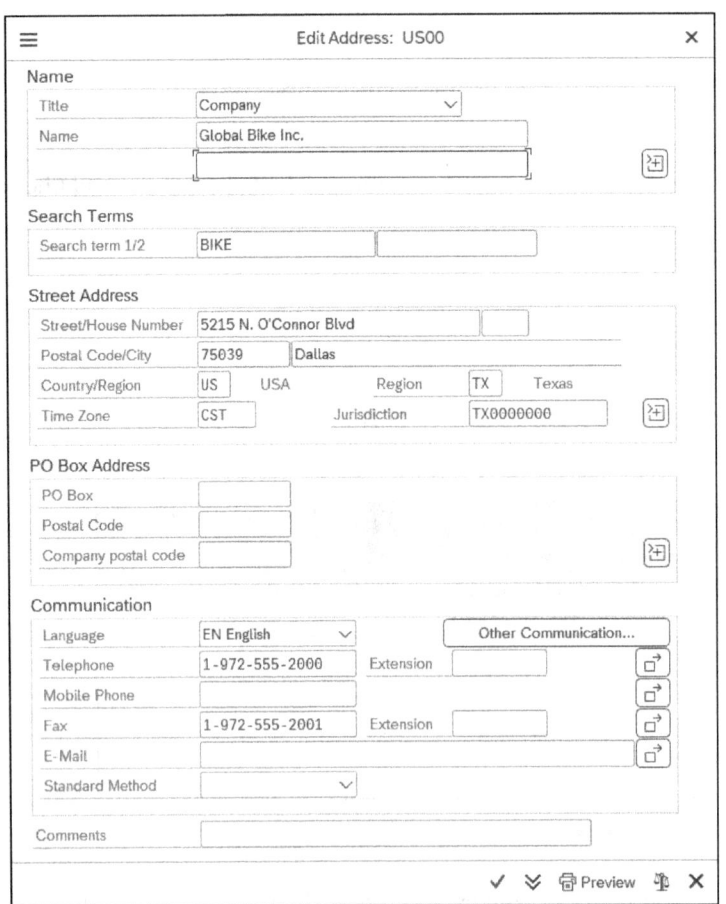

Figure 2.4 Company Code: Address Data

Global Parameters for the Company Code

You can use this Customizing function to make key specifications in relation to the company code. These include, among other things, the chart of accounts and the fiscal year variant.

Prerequisites

You must define these specifications in advance, either as master data (e.g., **Chart of Accts**) or in other Customizing functions (e.g., **Company, Credit Control Area, Fiscal Year Variant**).

Customizing Path

Financial Accounting • Financial Accounting Global Settings • Global Parameters for Company Code • Enter Global Parameters

Settings

From a plant maintenance perspective, the following fields are the most important: **Chart of Accts** and **Fiscal Year Variant** (see Figure 2.5).

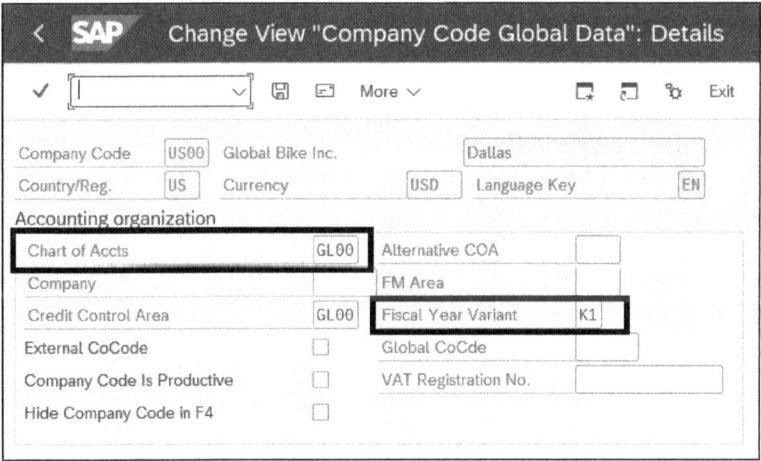

Figure 2.5 Company Code: Global Parameters

These fields represent the following:

- **Chart of Accts**

    The *chart of accounts* is a directory of all general ledger accounts. It provides the basis for all business transactions in SAP S/4HANA. For each business transaction, postings are made in the background to one or more general ledger accounts. For example, in the case of a material withdrawal to a maintenance order, at least one material stock account and a consumption account are used. A chart of accounts must be assigned to each company code, and the assigned chart of accounts is known as an *operational chart of accounts*; that is, the day-to-day postings are made to the accounts of the operational chart of accounts. The operational chart of accounts is used by SAP S/4HANA Finance.

- **Fiscal Year Variant**
  The *fiscal year variant* defines the fiscal year and the posting periods. The *fiscal year* can correspond to the calendar year, but it doesn't have to (for example, if you don't want to perform year-end closing operations during the public holidays in December). The posting periods usually contain the number of posting periods (generally twelve monthly posting periods) and the number of special periods. A *special period* is a special posting period that subdivides the last regular posting period for closing operations. A fiscal year can have a maximum of twelve posting periods and four special periods.

2.1.2 Controlling Area

The controlling area is the main organizational unit in SAP S/4HANA controlling and is the basis for all other master data (e.g., cost centers, activity types).

Maintain a Controlling Area

A *controlling area* is defined as an organizational unit within a company that is used to represent a closed system for cost accounting purposes. A controlling area may include single or multiple company codes that may use different currencies. These company codes must use the same operational chart of accounts and the same fiscal year variant.

Prerequisites
You must define the currency, chart of accounts, and fiscal year variant beforehand.

Customizing Path
Enterprise Structure • Definition • Controlling • Maintain Controlling Area

Transaction
OX06

Settings
Choose the **Maintain Controlling Area** subfunction to check or change the settings for a **Controlling Area** (see Figure 2.6).

A *company code* maps a company from an external perspective and is primarily created for tax reasons. In other words, company codes are created to optimize an enterprise's tax burden, so a company code doesn't necessarily reflect the actual set of services the company provides. In contrast, a controlling area maps a company from an internal perspective and is created for the company's actual business processes and the exchange of services among the company's various business units.

2.1 General SAP Organizational Units

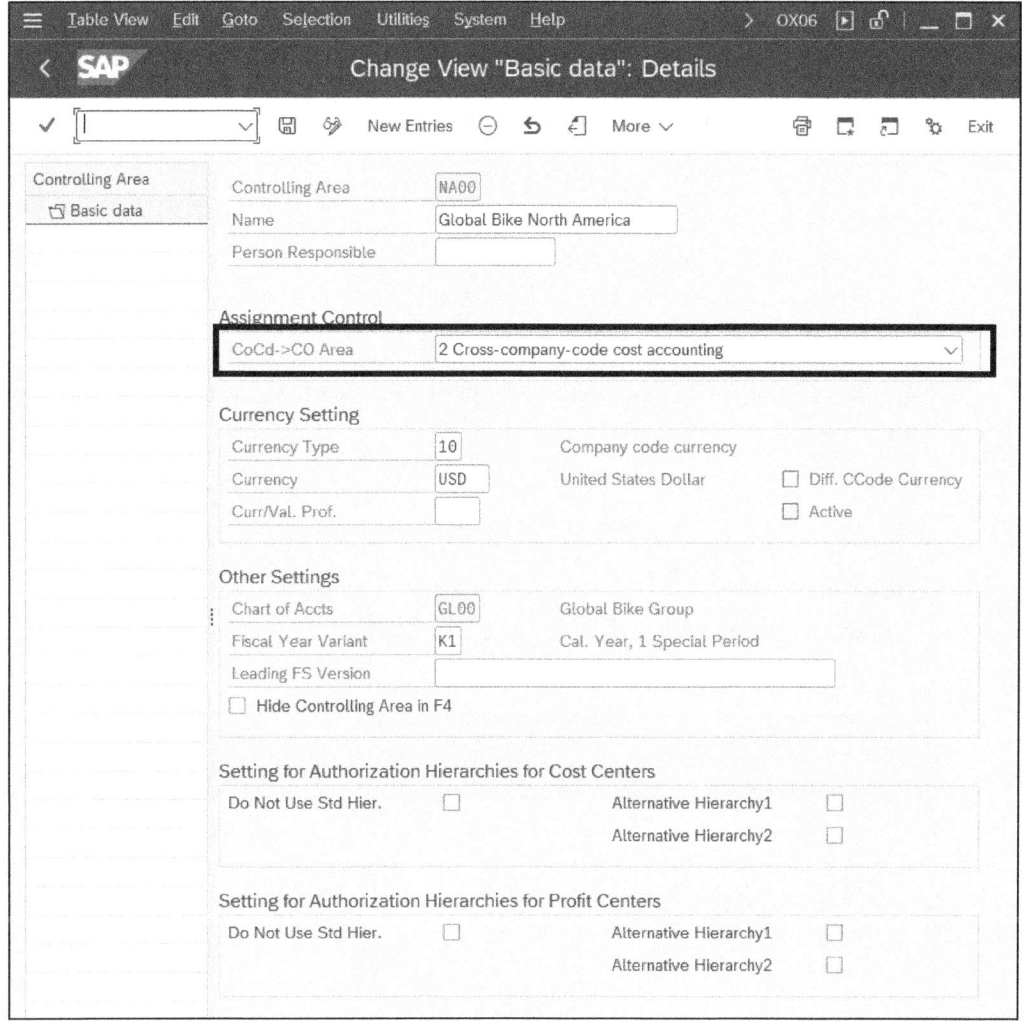

Figure 2.6 Controlling Area: Basic Data

For example, Mr. Miller, a production planner, and Mr. Lawrence, a maintenance planner, work in the same office and once belonged to the same company (company code) before the decision was made to outsource plant maintenance. Since then, Mr. Lawrence has continued to plan the repair orders and maintenance tasks for Mr. Miller's machine facilities. He still works in the same office as Mr. Miller, but he now belongs to another company (company code). The day-to-day processes haven't changed. Mr. Lawrence can continue to plan his orders in the same way as before because, from an internal perspective (controlling area), his company belongs to the same company as Mr. Miller's production, and the **2 Cross-company-code cost accounting** option is activated there.

2   Configuring Organizational Structures

Company Code: Assign a Controlling Area

You use this Customizing function to assign to a controlling area the company code that you want to perform common cost center accounting. You only have to explicitly assign company codes to the controlling area if you want to perform cross-company code cost accounting.

Prerequisites

You must define the company code and controlling area beforehand. Also, both must have the same chart of accounts and fiscal year variant.

Customizing Path

Enterprise Structure • Assignment • Controlling • Assign Company Code to Controlling Area

Transaction

OX19

Settings

As shown in Figure 2.7, you call the affected controlling area (here, "NA00") and assign the necessary company codes (here, "US00").

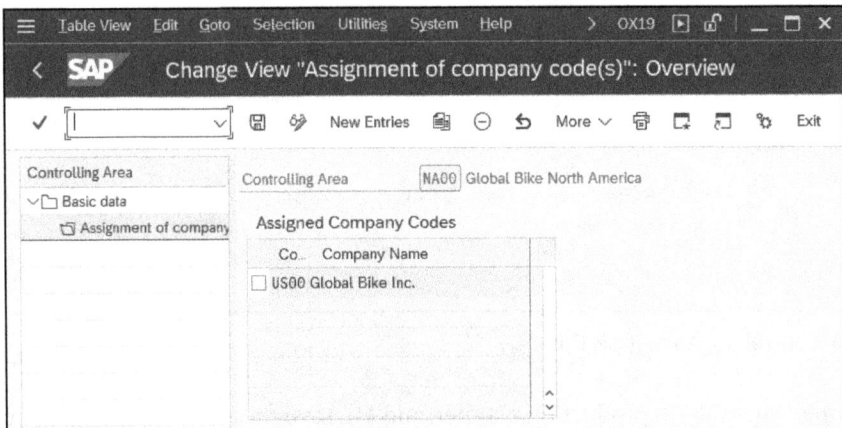

Figure 2.7 Controlling Area: Assignment of Company Codes

Due to poor experiences with other enterprise structures, I unreservedly agree with the recommendation from SAP that you only use one controlling area for the entire enterprise. In other words, you should assign all company codes to one controlling area. The advantages of this are as follows:

- A company code is permitted to be assigned to a controlling area only if both use the same fiscal year variant and chart of accounts. This is the only way you can ensure that all company codes use the same chart of accounts and thus ensure a basic level of standardization.

- You must create master data such as cost elements, cost centers, and profit centers for each controlling area. If you have only one controlling area, it reduces the time and effort needed for master data maintenance considerably.
- You can run cross-company-code analyses in controlling areas in SAP S/4HANA Finance. It's not possible to display data from different controlling areas in one report.
- Basic controlling functions—such as cross-company-code sales, transfer prices, allocations or cross-plant costing, and, of course, cross-plant and cross-company code plant maintenance—are only possible within a controlling area.

2.1.3 Plant in General

The *plant* is the central and most important organizational unit in logistics.

Define, Copy, Delete, Check Plant

You require a plant for the following areas:

- Production planning for production order processing and capacity requirements planning
- Purchasing as a purchasing or supplying plant
- Shipping as a delivering plant
- Inventory management as an organizational unit in which you manage stocks and values
- Quality management as a unit in which you edit inspection lots
- Materials management for forecasts, MRP, and product costing
- Management of all logistics master data (work centers, task lists, BOMs, and materials)

Prerequisites

You must maintain the language keys, country keys, regions, and factory calendar beforehand.

Customizing Path

Enterprise Structure • Definition • Logistics – General • Define, Copy, Delete, Check Plant

Transaction

V_T001W

Settings

Choose the **Define Plant** subfunction to access an overview of existing plants (see Figure 2.8).

2 Configuring Organizational Structures

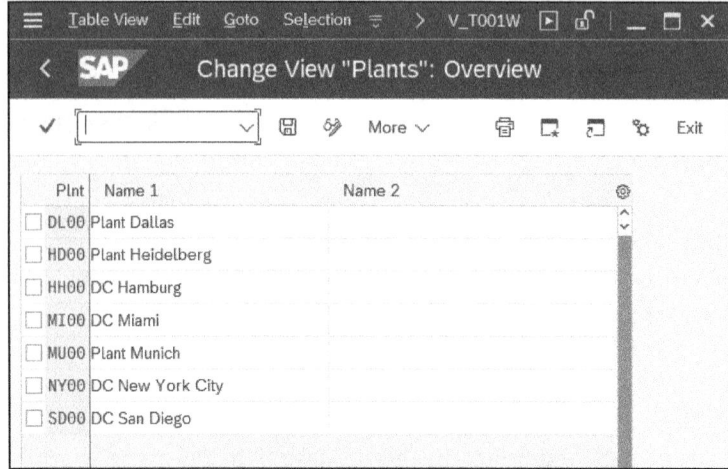

Figure 2.8 Plants: Overview

On the **Details** screen for plants, you can maintain plant addresses in particular (see Figure 2.9). These addresses are later required as object addresses for technical objects like functional location and equipment or as delivery addresses for spare parts.

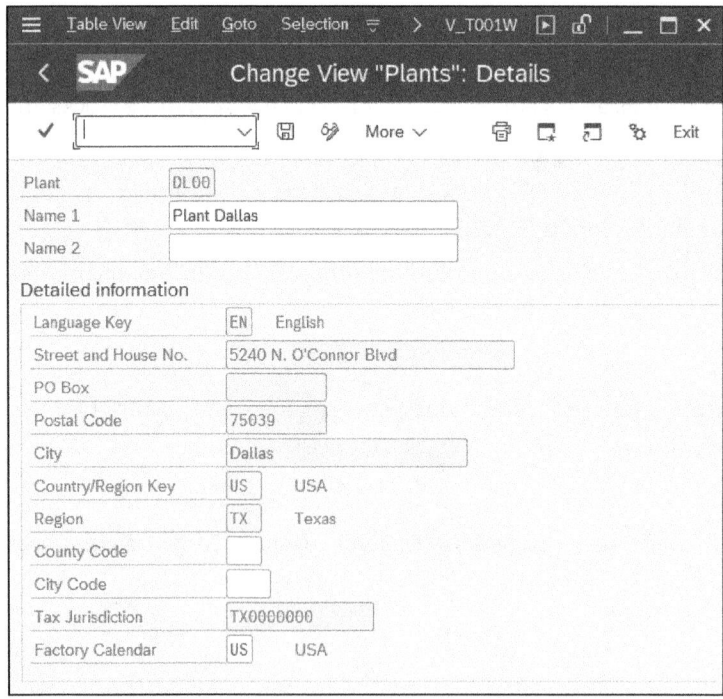

Figure 2.9 Plant: Details

Plant: Assign a Company Code

You use the Customizing function **Plant • Assign Company Code** to assign a plant to its company code. At any one time, a plant can only ever be assigned to one company code. Conversely, this means that, over time, you can assign a plant to another company code. This is necessary, for example, if you outsource your maintenance area and it becomes a separate company.

Prerequisites

You must define the plant and company code beforehand.

Customizing Path

Enterprise Structure • Assignment • Logistics – General • Assign Plant to Company Code

Transaction

OX18

Settings

As shown in Figure 2.10, you simply define a new entry when you assign the plant (e.g., **DL00**) to a company code (e.g., **US00**).

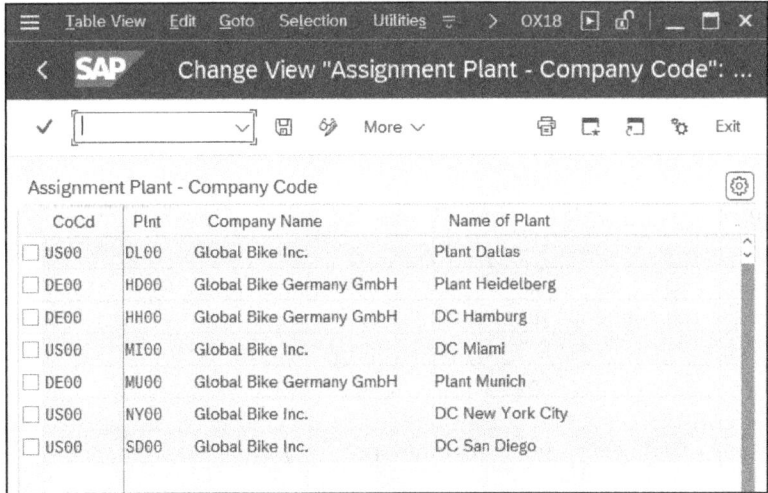

Figure 2.10 Plant: Assign Company Code

2.2 The Plant from a Maintenance Perspective

The plant is, without a doubt, the most important organizational unit for plant maintenance. It fulfills several maintenance functions as both a maintenance planning plant and as a maintenance plant. We'll look at both functions in the following sections, in addition to the topics of plant-specific maintenance and cross-plant maintenance.

2 Configuring Organizational Structures

An overview of the various organizational units in SAP S/4HANA Asset Management is provided in Figure 2.11.

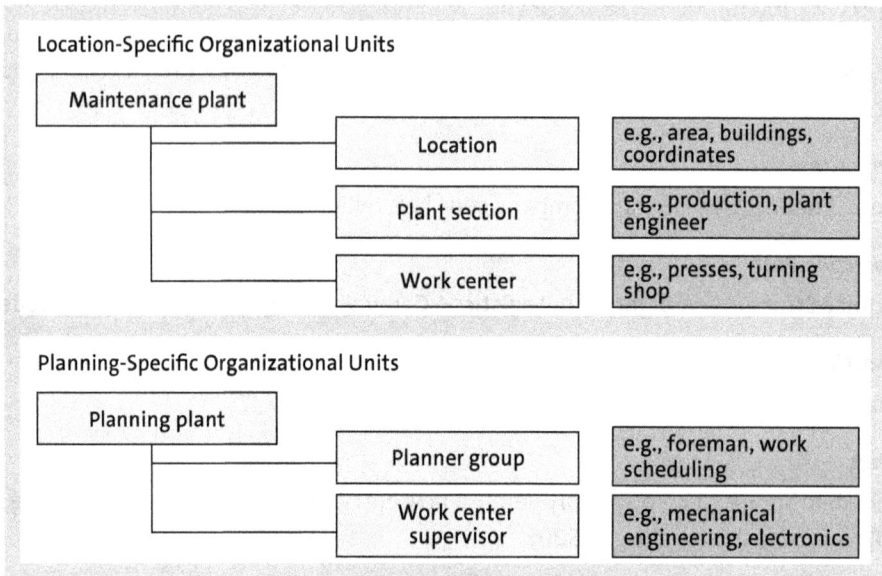

Figure 2.11 Organizational Units in SAP S/4HANA Asset Management

Maintain a Maintenance Planning Plant

A plant is responsible for planning maintenance activities. This type of plant is called a *maintenance planning plant* (or *planning plant* for short). You use the **Maintain Maintenance Planning Plant** Customizing function to convert a plant to a planning plant.

Prerequisites

You must define the maintenance planning plant as a normal logistics plant beforehand (Section 2.1.3).

Customizing Path

Enterprise Structure • **Definition** • **Plant Maintenance** • **Maintain Maintenance Planning Plant**

Settings

You obtain an overview of existing maintenance planning plants (see Figure 2.12), and you can add new plants to this overview.

2.2 The Plant from a Maintenance Perspective

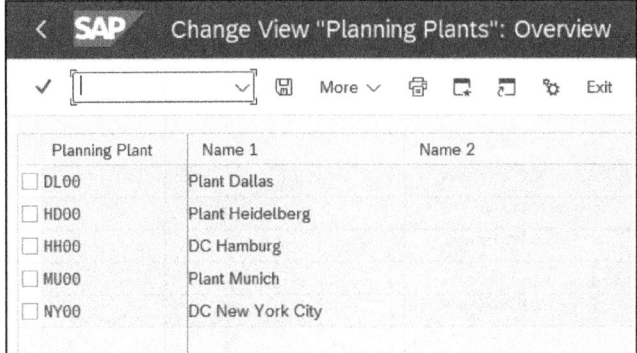

Figure 2.12 Maintenance Planning Plant: Overview

Maintenance Plant: Assign a Maintenance Planning Plant

All the technical objects to be maintained are physically present in a plant (functional location, equipment, and serial number). Here, this plant is known as a *maintenance plant*. A plant becomes a maintenance plant if you create a technical object there. Maintenance plants aren't stored in a separate Customizing table.

Every plant in which maintenance processes are to be planned or executed must have an assignment from the maintenance plant to the planning plant. You use the **Assign Maintenance Planning Plant** Customizing function to perform this assignment (see Figure 2.12).

Prerequisites

You must define both plants (the maintenance plant and the planning plant) as normal logistics plants beforehand (Section 2.1.3). If a plant isn't assigned to a planning plant (e.g., plant MI00), you can't create any technical objects there (at the maintenance plant), nor can you plan or execute maintenance tasks (at the planning plant).

Customizing Path

Enterprise Structure • Assignment • Plant Maintenance • Assign Maintenance Planning Plant to Maintenance Plant

Settings

As shown in Figure 2.13, in the **PlPl** column, you should assign a maintenance planning plant to a maintenance plant (in the **Plnt** column) if it contains assets or if you want to plan or perform maintenance work.

For business processes in SAP S/4HANA Asset Management, you need to differentiate between order planning and execution in only one plant and order planning and execution in different plants.

2 Configuring Organizational Structures

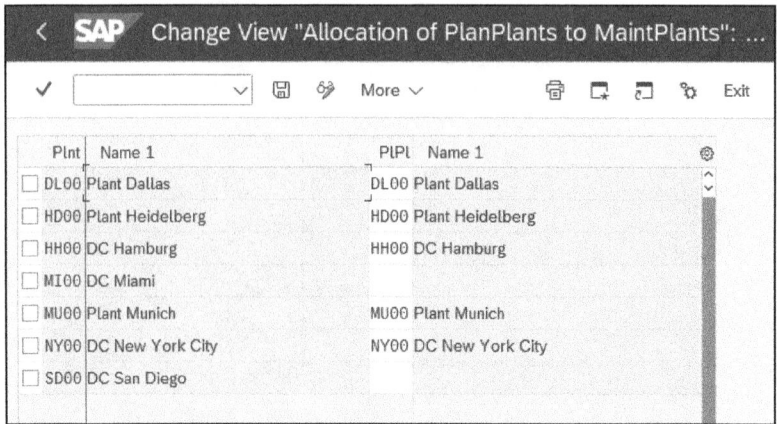

Figure 2.13 Maintenance Plant: Assign Maintenance Planning Plant

Plant-Specific Maintenance

In practice, you most frequently encounter a situation where the maintenance requirement is planned in the plant in which it originates, the orders are fulfilled by workshops in the same plant, and the spare parts are stored within the same plant. In Figure 2.14, this plant is known as **Plant 1000**. Here, the following applies: maintenance plant = planning plant = spare parts storage plant.

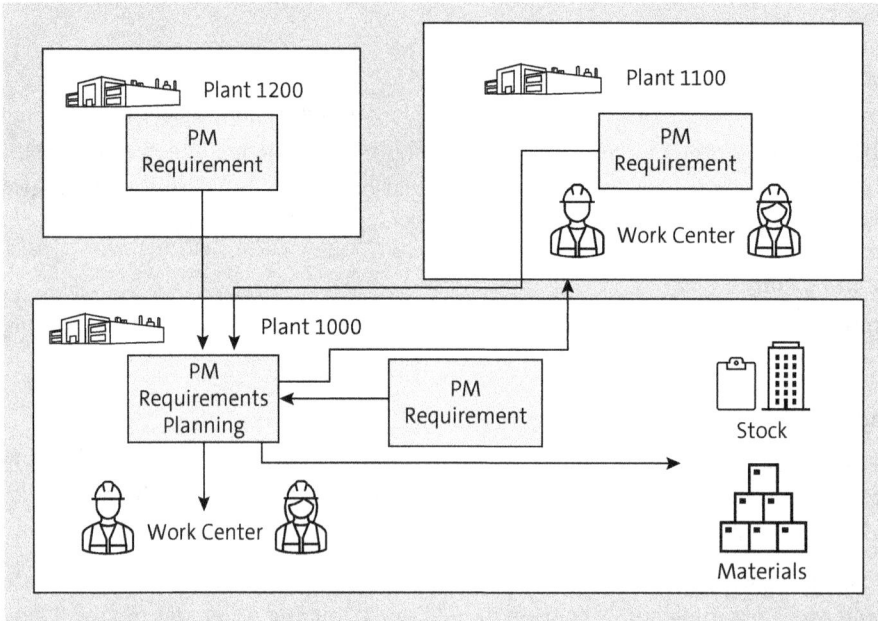

Figure 2.14 Plant and Plant Maintenance

2.2 The Plant from a Maintenance Perspective

Cross-Plant Maintenance

There are other options in addition to plant-specific maintenance:

- There is a maintenance requirement in a plant (plant **1200** in the situation depicted in Figure 2.14) because an asset is to be maintained there (i.e., in the maintenance plant), but all other functions (planning, order execution, and spare parts storage) are the responsibility of another plant (plant **1000**).
- There is a maintenance requirement in a plant (plant **1100**), and additional partial functions (order execution) are also the responsibility of this plant, but other partial functions (order planning and spare parts storage) are the responsibility of another plant (plant **1000**).

Cross-plant maintenance isn't a problem if the maintenance plant of the technical object and the plant of the executing work center are in the same company code. The same applies if the plants are in different company codes but belong to the same controlling area. This is also a standard scenario.

However, a problem occurs if the plants belong to different controlling areas. In that situation, there is no standard scenario but rather a customer-vendor relationship. Therefore, in this case, the maintenance plant (the customer) has to trigger purchase orders, and the plant of the work center (the vendor) triggers a sales order and its associated invoice. The invoice is, in turn, recorded as an incoming invoice in the maintenance plant. All in all, this is a very cumbersome procedure, but it can be simplified as described next.

> **Plants in Different Controlling Areas**
>
> If you implement cross-plant maintenance and your plants are in different controlling areas, the following approach is recommended:
>
> - In the work center plant, create a cost center for the actual maintenance plant.
> - Assign all the technical objects to the work center plant (as a maintenance plant) and to this cost center.
> - Process all maintenance orders in the work center plant.
> - Manually issue periodic invoices (e.g., monthly) from the work center plant, whereby the customer maintenance plant is debited the amount and the cost center is credited the same amount.

This procedure saves you from having to create purchase orders, sales orders, and individual invoices, as well as posting individual incoming invoices.

2.3 Maintenance-Specific Organizational Units

Additional maintenance-specific organizational units (either maintenance plant specific or planning plant specific) play an important role within a plant. Technical objects (functional locations and equipment) also contain all the maintenance plant-specific and planning plant-specific data, which is then copied to notifications and orders. This data is explained in more detail in this section.

Work centers either perform or are responsible for maintenance tasks. They reference a planning plant or a maintenance plant (Section 2.4).

Define Planner Groups

A *maintenance planner group* is responsible for planning maintenance tasks, and it also references a planning plant. You use the **Define Planner Groups** Customizing function to maintain maintenance planner groups.

Prerequisites

Because maintenance planner groups are always created for a specific maintenance planning plant, you must create the plant as both a general logistics plant and a maintenance planning plant beforehand.

Customizing Path

Plant Maintenance and Customer Service • Maintenance and Service Processing • Basic Settings • General Data • Define Planner Groups

Settings

Assign a three-digit key, a description, and, if necessary, a telephone number to each maintenance planner group (see Figure 2.15).

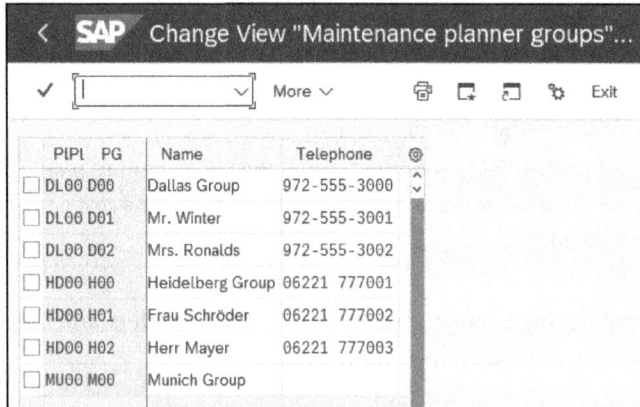

Figure 2.15 Maintenance Planner Groups

2.3 Maintenance-Specific Organizational Units

> **Recommendation for Using Maintenance Planner Groups** [+]
>
> You set up maintenance planner groups, for example, if you want to map work scheduling or individual maintenance planners known by name.

Define the Location

You use a label to indicate the physical location of a technical object, and you always define a location with reference to a maintenance plant. Furthermore, you use the **Define Location** Customizing function to maintain locations.

Prerequisites

Because a location is always created for a specific plant, you must define the maintenance plant as a normal logistics plant beforehand. You must come to an agreement on this with your colleagues from production planning and asset accounting because they will also use the location table for work centers and assets.

Customizing Path

Enterprise Structure • Definition • Logistics – General • Define Location

Settings

Assign a number (10-digit maximum) and a name to each location (see Figure 2.16).

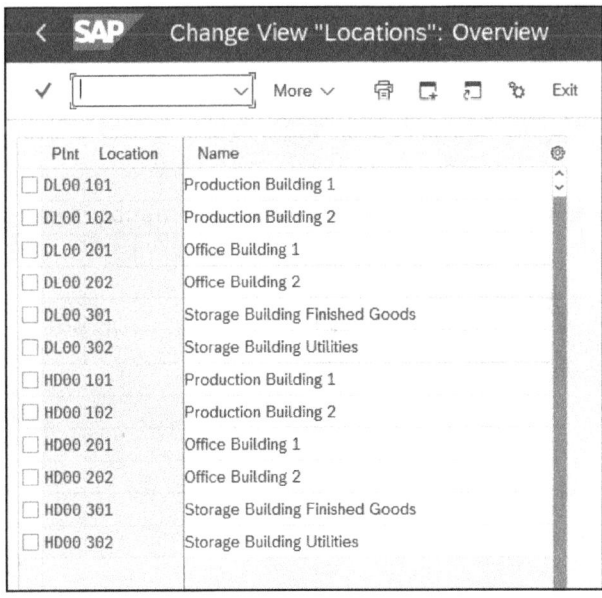

Figure 2.16 Locations: Overview

95

> **Recommendation for Assigning Names to Locations**
>
> In practice, either building numbers (e.g., F141 or WDF21) or, if they exist, plant coordinates (e.g., A01 or K15) are commonly used locations.

Define Plant Sections

The *plant section* enables you to divide a maintenance plant into production areas. The person responsible for the plant section is the contact person for all coordination between production and plant maintenance. A plant section can be assigned to each piece of equipment and each functional location, and you can also specify a plant section when processing notifications and orders.

You define the responsibilities associated with operating a (production) facility as a plant section. To this end, you use the **Define Plant Sections** Customizing function to maintain plant sections.

Prerequisites

Because plant sections are always assigned to a plant, you must define the plant as a general logistics plant beforehand.

Customizing Path

Plant Maintenance and Customer Service • Maintenance and Service Processing • Basic Settings • General Data • Define Plant Sections

Transaction

OIAB

Settings

Assign a three-digit number, a name, and, if necessary, a telephone number to each plant section (see Figure 2.17).

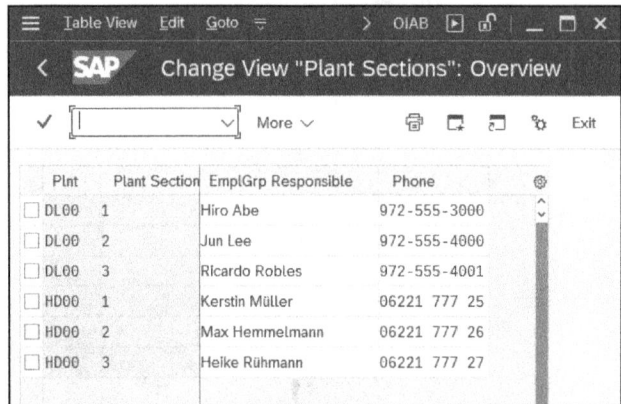

Figure 2.17 Plant Sections: Overview

> **Recommendation for the Plant Section**
> In practice, either the plant engineer responsible for the asset or the production area belonging to the asset has proven themselves as a plant section.

2.4 Work Centers

From a maintenance perspective, a *work center* represents either an individual person (e.g., Mr. John Smith, a technician) or a workshop (i.e., a group of people). In practice, the following types of workshops are commonly used:

- Mechanical
- Electrical
- Measurement and control
- Machine center
- Welding
- Paint shop
- Cleaning line
- Building services engineering

In SAP S/4HANA Asset Management, work centers are used in many locations:

- A main work center in equipment master records and functional location master records
- A main work center in a maintenance item
- A main work center in the header of a task list
- A main work center in the operations of a task list
- A main work center in the notifications
- A main work center in SAP Fiori apps
- A main work center in the order header
- A planned work center in the operations of an order
- An actual work center in time confirmations
- A main work center as a selection criterion in SAP GUI reports (e.g., IW38 Order List)
- A work center as a selection criterion in SAP GUI reports (e.g., IW47 Confirmation List)
- A main work center as a selection criterion in SAP Fiori apps (e.g., Find Orders and Notifications)

- An actual work center as a selection criterion in SAP Fiori apps (e.g., Find Confirmations)
- A planned work center as a selection criterion in SAP Fiori apps (e.g., Perform Maintenance Jobs)

The following sections cover the most important Customizing functions for creating and maintaining work centers.

Define Work Center Types and Link to the Task List Application

You use the **Define Work Center Types and Link to Task List Application** Customizing function to define the work center category. When you create a work center, you must always assign a **Work Center Category** (see Figure 2.18).

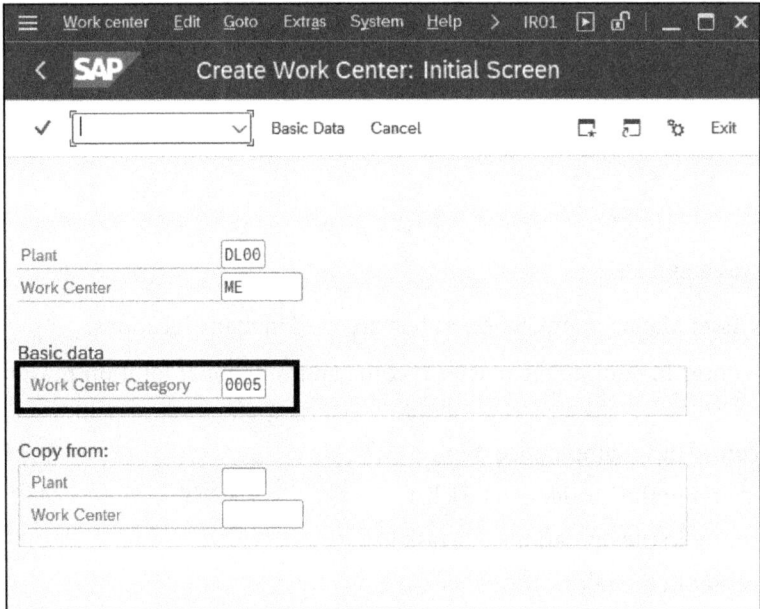

Figure 2.18 Create Work Center: Initial Screen

The work center category determines the following:

- The permitted *task list application* (e.g., SAP S/4HANA Asset Management, production planning, quality management)
- The screen sequence (e.g., **Basic Data** and **Capacities**)
- The field selection
- The use of change documents to document changes

Prerequisites
There are no prerequisites.

2.4 Work Centers

Customizing Path

Plant Maintenance and Customer Service • Maintenance Plans, Work Centers, Task Lists, and PRTs • Work Centers • General Data • Define Work Center Types and Link to Task List Application

Transaction

OIZA

Settings

SAP delivers work center category **0005 Plant maintenance** (see Figure 2.19). Usually, this work center category is sufficient. Therefore, you only need to create your own work center category if you require other control options or advanced control options.

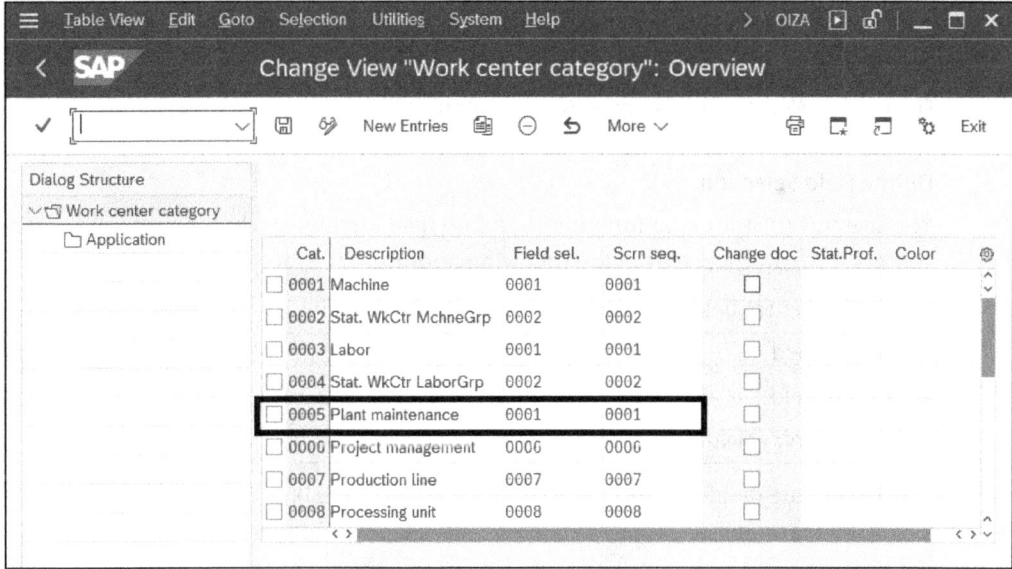

Figure 2.19 Work Center Category: Overview

To use the work center category in SAP S/4HANA Asset Management, you must assign it to a task list application (**Application**; see Figure 2.20). Make sure that your work center category is assigned to task list application I for plant maintenance. If you also want to use your work centers in other applications (e.g., in production orders as a production aid), you need to add other task list applications.

> **Work Center Category 0005 Should Suffice**
>
> Usually, work center category 0005 is sufficient for assignments to SAP S/4HANA Asset Management, and you don't have to define any other work center category.

2 Configuring Organizational Structures

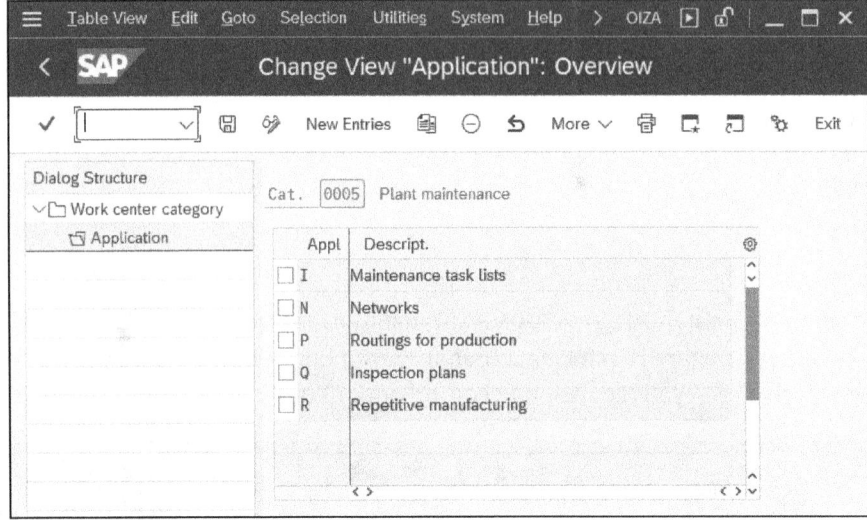

Figure 2.20 Work Center Category: Assign to Applications

Define Field Selection

You use this Customizing function to define field attributes when maintaining work centers. You can choose from the following options:

- Required entry field
- Hidden field
- Display field
- Normal entry field

These fields are divided into **Screen Groups** (see Figure 2.21).

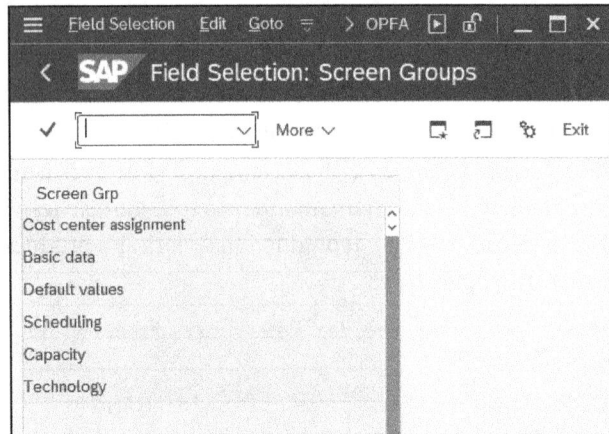

Figure 2.21 Work Center: Screen Groups

2.4 Work Centers

Prerequisites
Because I recommend that you don't configure a general field selection for all work center types but instead configure a specific field selection for each work center type, you must maintain the work center type beforehand.

Customizing Path
Plant Maintenance and Customer Service • Maintenance Plans, Work Centers, Task Lists, and PRTs • Work Centers • General Data • Define Field Selection

Transaction
OPFA

Settings
In the overview screen, click the **Influencing** button to configure the field selection on the detail screen on the basis of the relevant work center type. Figure 2.22 shows an example of the field selection for the **Basic data** screen group.

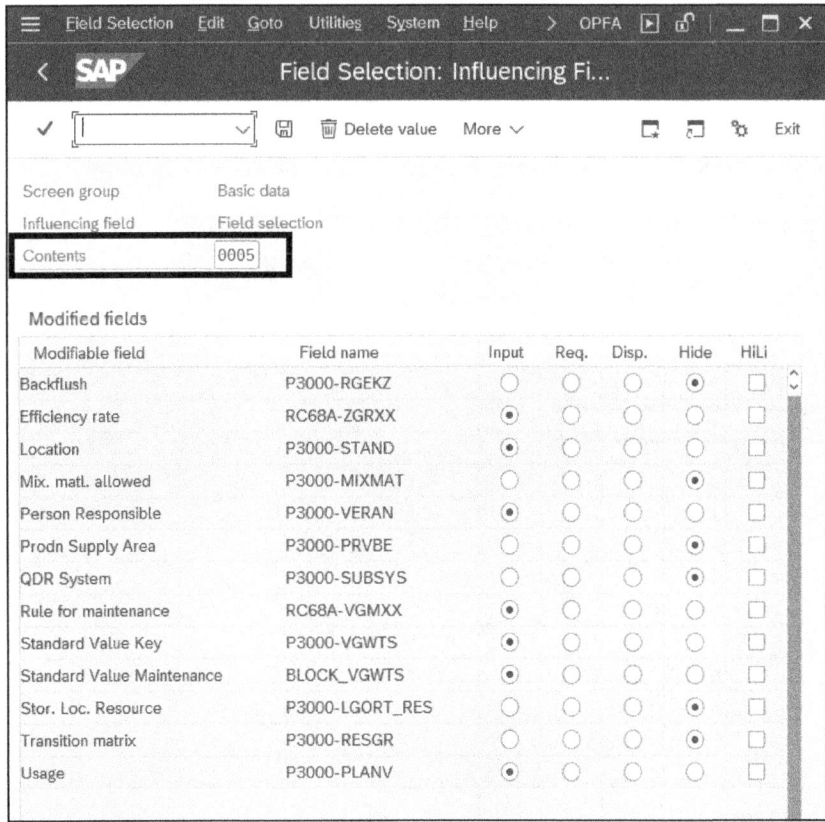

Figure 2.22 Work Center: Field Selection

2 Configuring Organizational Structures

Define Standard Value Keys

You use this Customizing function to define *standard value keys*. The standard values are plan values for determining the execution time for operations in the task list and order.

Prerequisites

There are no prerequisites.

Customizing Path

Plant Maintenance and Customer Service • Maintenance Plans, Work Centers, Task Lists, and PRTs • Work Centers • General Data • Define Standard Value Keys

Transaction

OIZ2

Settings

The task lists and orders in SAP S/4HANA Asset Management always take into account the duration of an operation for scheduling orders and the work associated with costing and capacity requirements planning. For work centers used solely by SAP S/4HANA Asset Management, use the standard value key for which no standard values are defined. SAP delivers the standard value key **SAP0** for this purpose (see Figure 2.23). Assign this key to each of your work centers in SAP S/4HANA Asset Management.

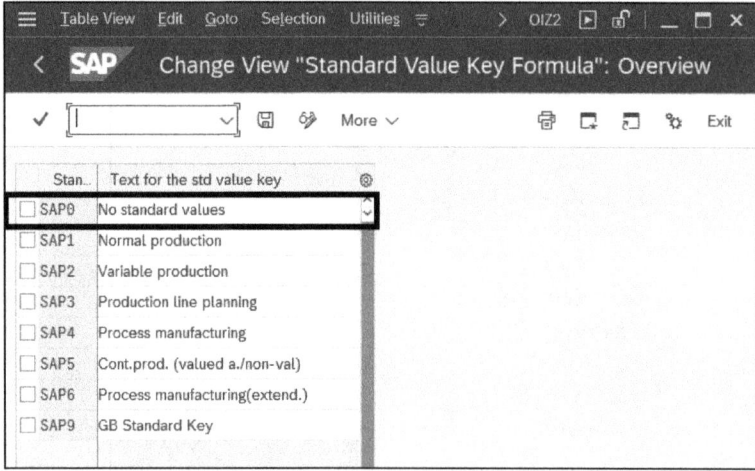

Figure 2.23 Work Center: Standard Value Key

> [+] **Assign Standard Value Key SAP0**
>
> The standard value key in the standard SAP system is sufficient. You don't have to define any other standard value keys, but make sure to assign the standard value key to each of your SAP work centers.

Define Task List Usage Keys

You use this Customizing function to specify the task lists and order categories in which a it is permitted use a work center. For example, you could differentiate between whether you want to use work centers in maintenance processing or inspection processing.

Prerequisites

There are no special prerequisites.

Customizing Path

Plant Maintenance and Customer Service • Maintenance Plans, Work Centers, Task Lists, and PRTs • Work Centers • Task List Data • Define Task List Usage Keys

Transaction

OIZD

Settings

For plant maintenance purposes, SAP delivers task list usage **004 Only maintenance task lists** and task list usage **009 All task list types** (see Figure 2.24).

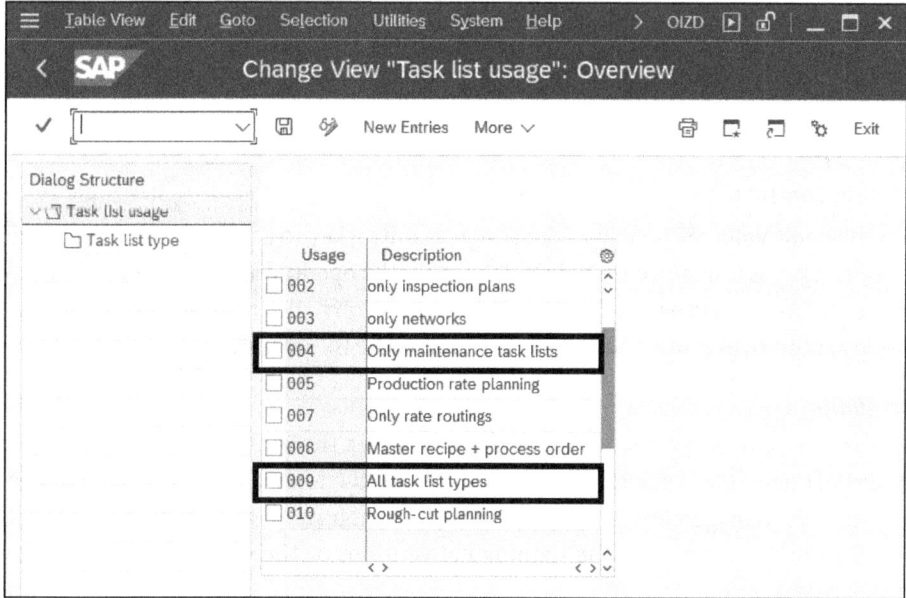

Figure 2.24 Work Center: Task List Usage

You should assign one of these task list usages to each of your work centers in SAP S/4HANA Asset Management. Specify the relevant task list types on the detail screen (e.g., **E** = equipment task lists, **A** = general maintenance task lists, **T** = functional location task lists) as shown in Figure 2.25.

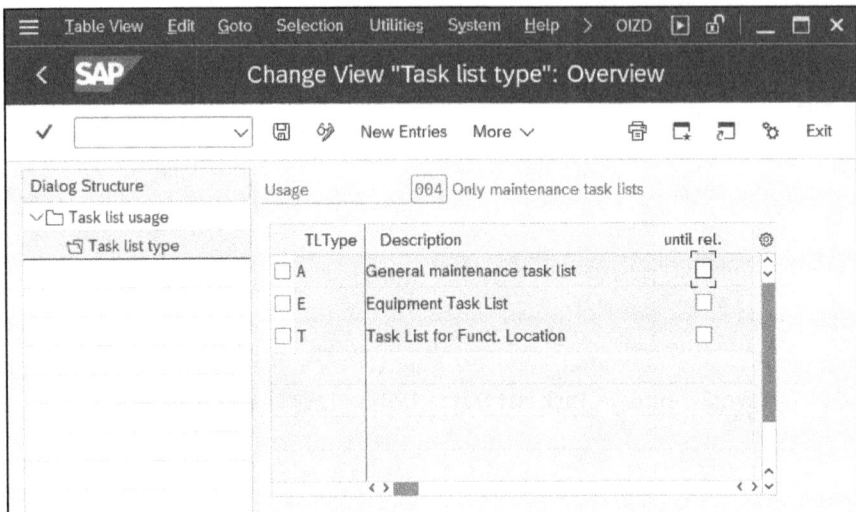

Figure 2.25 Work Center: Task List Types

> **Assign Task List Usage Key 004 or 009**
> The task list usage keys in the standard SAP system are sufficient. You don't have to define any other task list usage keys, but make sure to assign one of the standard task list usage keys to each of your work centers.

Maintain Control Keys

On the **Default Values** screen, you can assign a control key to a work center. From here, the control key is transferred as a default value to the operations associated with a task list or order. You can use the *control key* of an operation to define which business functions you want to execute or how you want to handle the operation.

Prerequisites

There are no technical prerequisites. Because, however, the control key is used not only in SAP S/4HANA Asset Management but also for production planning, quality management, project management, and customer service, you should consult with your colleagues from these areas on the naming convention for the control key (see Figure 2.26).

Customizing Path

Plant Maintenance and Customer Service • Maintenance Plans, Work Centers, Task Lists, and PRTs • Work Centers • Task List Data • Maintain Control Keys

2.4 Work Centers

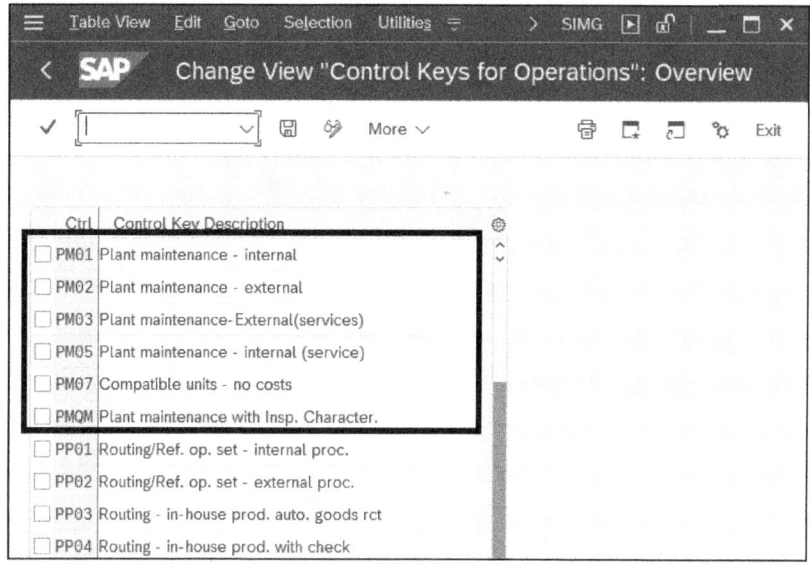

Figure 2.26 Work Center: Control Keys Overview

Settings

You use the control key to define which of the following business functions are to be executed in the order at a later stage (see Figure 2.27):

- Whether the operation is to be part of costing
- Whether the operation is to be scheduled
- Whether the operation is to generate capacity requirements
- Whether completion confirmations are expected for the operation
- Whether the operation is to be assigned to an external company
- Whether an externally processed operation is to be taken into account in scheduling
- Whether service specifications are to be set up in the operation
- Whether the operation is to be printed
- Whether time tickets or completion confirmation slips are also to be printed for the operation
- Whether inspection characteristics are to be assigned to the operation

> **Using the Control Key**
>
> You can use the control key to control, in detail, the business functions that an operation is to have (cost, print, confirm, assign externally, schedule, etc.).
>
> You require at least two control keys: a key for internal processing and a key for external processing. You can use other control keys as required.

2 Configuring Organizational Structures

> You should always define the control key in the work center as a default value so that you don't always have to manually enter it in the task list and order.

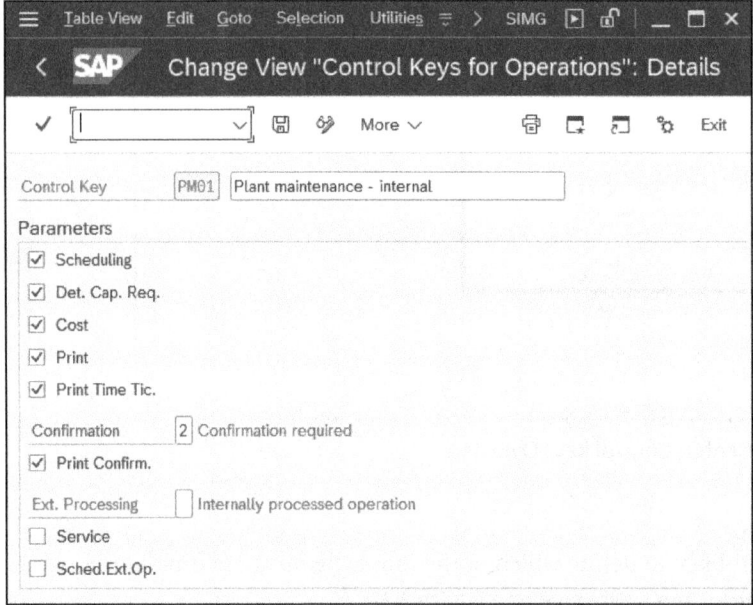

Figure 2.27 Work Center: Control Key Details

Configure the Screen Sequence for the Work Center

You use this Customizing function to define which screens are to appear when you maintain a work center, as well as the order in which they are to appear.

Prerequisites

Because the work center type determines the order in which the screens are displayed, you must define the work center type beforehand.

Customizing Path

Plant Maintenance and Customer Service • Maintenance Plans, Work Centers, Task Lists, and PRTs • Work Centers • Configure Screen Sequence for Work Center

Transaction

OIZU

Settings or Recommended Settings

Figure 2.28 shows an overview of the screens most commonly used for the work center. The following screens are available:

- Basic Data
- Defaults

- Capacities
- Scheduling
- Costing
- Technical data
- Hierarchy classification
- HR assignment
- Res.NetworkRelation

Use the **N...** column to define the order in which these screens are displayed.

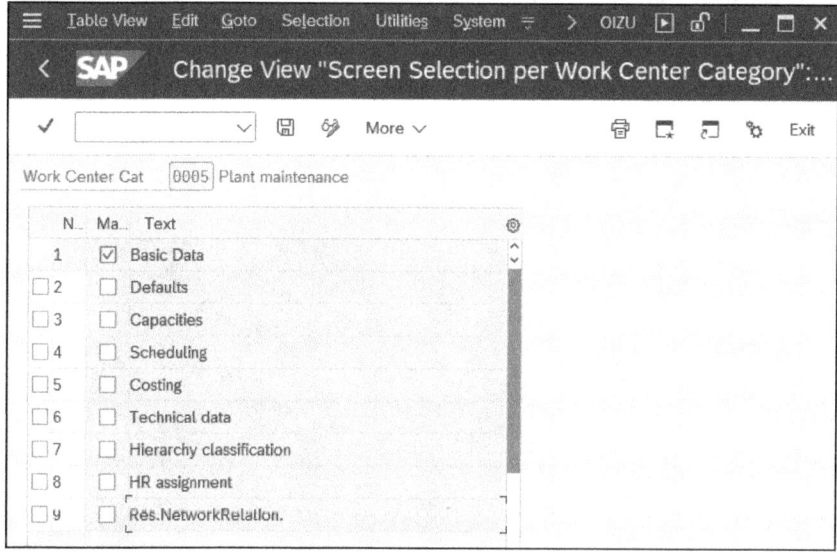

Figure 2.28 Work Center: Screen Sequence

2.5 Summary

Following are the major takeaways from this chapter for SAP S/4HANA Asset Management organizational structures:

- With organizational units, you define your enterprise's organizational structure in general and your plant maintenance structures in particular.
- Organizational structures provide a basis for all master data and business processes.
- In SAP S/4HANA Asset Management projects, the general organizational units in the SAP system (e.g., company code, controlling area, plant) are usually predefined.
- The best approach to all logistical processes is to only use one controlling area for the entire enterprise, if possible.
- The plant is the central and most important organizational unit in logistics.

- From a plant maintenance perspective, you must define maintenance planning plants and maintenance plants.
- SAP S/4HANA Asset Management supports plant-specific maintenance and cross-plant maintenance.
- From a maintenance perspective, a work center represents either an individual person or a workshop.
- In SAP S/4HANA Asset Management, work centers are used in many locations—both in SAP GUI and in the SAP Fiori apps.
- With several Customizing functions, you differentiate plant maintenance work centers from other uses (e.g., production).

This chapter has provided you with instructions for defining your organizational units, plants, and work centers.

Chapter 3
Configuring Generic Functions

This chapter introduces you to generic functions such as status management and partners, both of which are used in several locations within SAP S/4HANA Asset Management. You'll learn how to configure your tools, and you'll obtain numerous tips related to configuring Customizing settings for these functions.

In SAP S/4HANA Asset Management, some functions are used in several locations or by several objects. For example, you can use status management not only to document the condition of technical objects (functional location, equipment, and serial number) but also to manage notification and order processes. The field selection control that you use to hide fields or force entries can also be used with each SAP object.

Table 3.1 shows which generic functions can be used for which SAP object. An (X) means that the function cannot be assigned to the object link directly, only indirectly via functional location or equipment being an element of the object link.

Function	Functional Location	Equipment	Object Links	Material	Notification	Order	Task List	Maintenance Plan
Object information	X	X	(X)		X	X		
Status management	X	X	X		X	X		
Number assignment		X	(X)	X	X	X	X	X
Warranties	X	X	(X)		X	X	X	
Measuring points/counters	X	X	(X)			X	X	X
Permits	X	X	(X)			X		
Partners	X	X	(X)		X	X		
Documents	X	X	(X)	X	X	X	X	X

Table 3.1 Generic Functions: Overview

3   Configuring Generic Functions

Function	Functional Location	Equipment	Object Links	Material	Notification	Order	Task List	Maintenance Plan
Field selection	X	X		X	X	X	X	X
List variants	X	X	X	X	X	X	X	X
Multilevel lists	X	X			X	X		
Change documents	X	X	X	X	X	X	X	X

Table 3.1 Generic Functions: Overview (Cont.)

3.1   Object Information

When you receive a new notification from a requester and need to decide whether the maintenance task is to be performed or not, it's very useful to get concise information about the object by using object information.

Define Object Information Keys

You can use this Customizing function to configure a dialog box that contains concise information about the reference object, which is known as *object information* (see Figure 3.1). You can call object information in the following locations within the system:

- Equipment and serial numbers
- Functional locations
- Notifications
- Orders

Prerequisites
There are no special prerequisites.

Customizing Paths
Plant Maintenance and Customer Service • Master Data in Plant Maintenance and Customer Service • Basic Settings • Define Object Information Keys

Plant Maintenance and Customer Service • Maintenance and Service Processing • Maintenance and Service Notifications • Notification Processing • Object Information • Define Object Information Keys

Plant Maintenance and Customer Service • Maintenance and Service Processing • Maintenance and Service Orders • Object Information • Define Object Information Keys

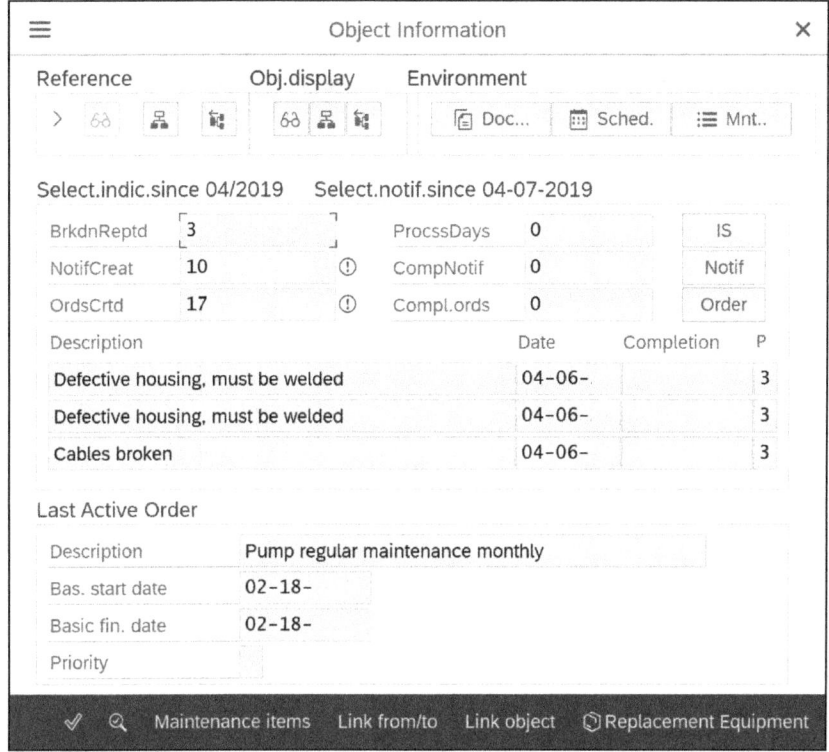

Figure 3.1 Object Information

Transaction

OIMD

Settings

You use the object information key to define which technical object data is displayed in the object information window (see Figure 3.2). The object information also contains historical data from the logistic information system as well as data from the notifications and orders processed in the system.

In the **Reference for notification and class selection** area, you define which of the following is to be displayed:

- Notifications and order information for a technical object (**Object**)
- Notifications and order information for the superior object (**SupObject**)
- Notifications and order information for the entire structure (**Structure**)

If you've selected object information for the object specified, you should check the box to set the **automatically** indicator to define whether this object information is displayed automatically (e.g., immediately after you call a notification or order). If you don't set this indicator, you can use the 🛈 icon to call the object information.

3   Configuring Generic Functions

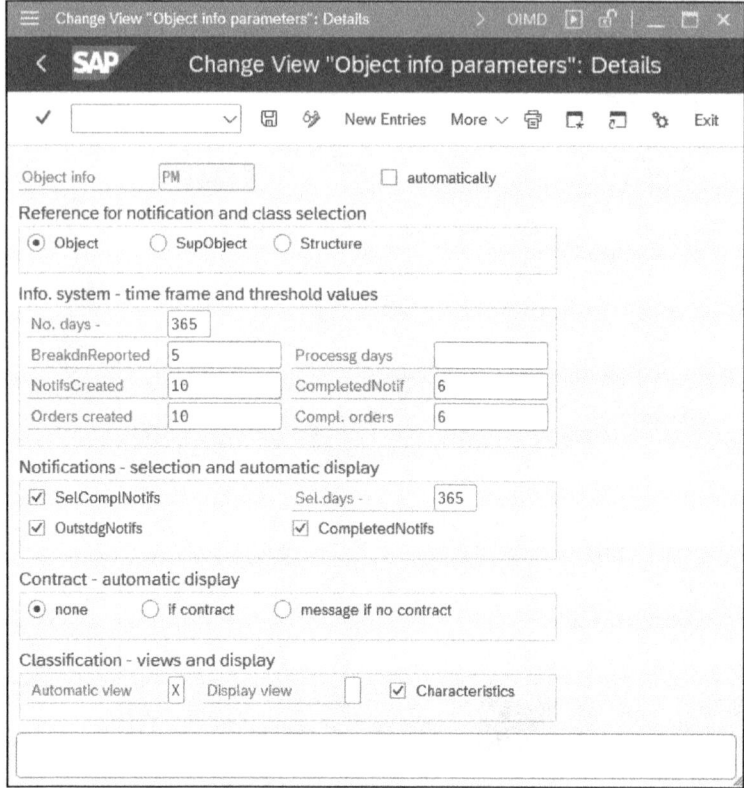

Figure 3.2 Define Object Information Key

In the **Info. system – time frame and threshold values** area, you can define the following:

- The period to be selected in the information system (highest value is 9,999 days, which is approximately 27 years)
- Threshold values that, when exceeded, are indicated by a lightning symbol in the **Object Information** window

In the **Notifications – selection and automatic display** area, you can define the following:

- The period to be selected in the notification list (highest value is 9,999 days, which is approximately 27 years)
- The notification status to be considered in the selection (outstanding and/or completed)

In the **Classification – views and display** area, you essentially define whether the characteristics of the object are to be displayed (by setting the **Characteristics** indicator).

> **Object Information Can Be Very Useful** [+]
>
> You should define object information keys in your SAP system because they will enable you to obtain a clear and concise overview of the technical object's situation in master data and in processing. However, you should only display this object information when you actually need it—so don't check the box to set the **automatically** indicator.

To assign an object information key to several objects, you use the following Customizing functions:

- Equipment
  Plant Maintenance and Customer Service • Master Data in Plant Maintenance and Customer Service • Technical Objects • Equipment • Equipment Categories • Maintain Equipment Category
- Functional location
  Plant Maintenance and Customer Service • Master Data in Plant Maintenance and Customer Service • Technical Objects • Functional Locations • Define Category of Functional Location
- Notification
  Plant Maintenance and Customer Service • Maintenance and Service Processing • Maintenance and Service Notifications • Notification Processing • Object Information • Assign Object Information Keys to Notification Types
- Order
  Plant Maintenance and Customer Service • Maintenance and Service Processing • Maintenance and Service Orders • Object Information • Assign Object Information Keys to Order Types

3.2 Status Management

Both the technical objects themselves, as well as complete notification and order processing, are associated with general SAP status management. Here, you have to distinguish between the system status and the user status.

For certain business processes, the system sets the *system statuses* both internally and automatically as part of its general status management. This happens, for example, in the equipment master **INST** when the equipment is installed at a functional location or in the order **CONF** when an order receives a final confirmation. The system status is displayed on the left-hand side of the status screen (see Figure 3.3).

3   Configuring Generic Functions

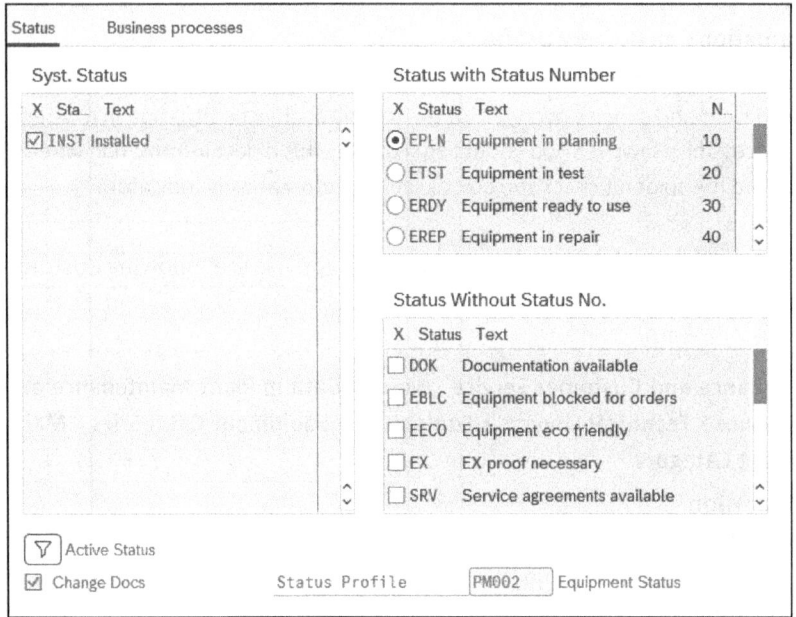

Figure 3.3 Status: System Status and User Status

Define User Status

In addition to the predefined system statuses, you can freely define *user statuses* that fulfill your requirements. The user status is set and displayed on the right-hand side of the status screen (refer to Figure 3.3). SAP distinguishes between two types of user status:

- **Status with Status Number (upper half of the screen)**
  You can select only one radio button here.
- **Status Without Status No. (lower half of the screen)**
  You can select any number of checkboxes here.

The business processes that you can execute for each of these object statuses are also defined in status management. If, for example, you set a piece of equipment to inactive, the status display informs you which business processes are still allowed (with a green traffic light), which will trigger a warning message (with a yellow traffic light), and which are forbidden (with a red traffic light) (see Figure 3.4).

You can define a user status for the following objects:

- Equipment and serial numbers
- Functional locations
- Object links
- Notifications
- Orders
- Order operations
- Work centers
- Maintenance plans

3.2 Status Management

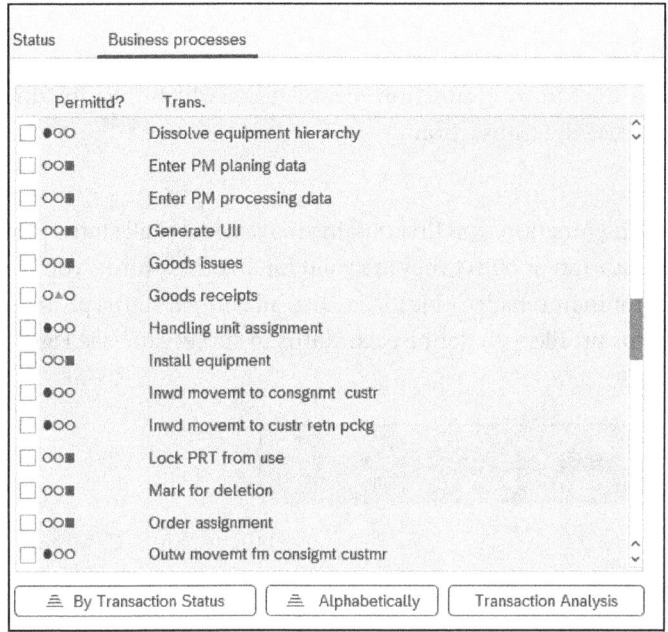

Figure 3.4 Status: Business Processes

For the sake of simplicity, we'll consider the example of a piece of equipment to illustrate status management—but be aware that status management can be used in a similar manner for all other objects.

Prerequisites
There are no special prerequisites.

Customizing Paths
The following Customizing paths are available for user status management:

- Equipment and functional location
  Plant Maintenance and Customer Service • Master Data in Plant Maintenance and Customer Service • Basic Settings • Define User Status
- Orders
  Plant Maintenance and Customer Service • Maintenance and Service Processing • Maintenance and Service Orders • General Data • User Status for Orders • Define Status Profile • Define User Status Profile for Orders
- Notifications
  Plant Maintenance and Customer Service • Maintenance and Service Processing • Maintenance and Service Notifications • Notification Processing • User Status for Notifications • Define Status Profile • Define User Status Profile for Notifications

3 Configuring Generic Functions

Transaction

OIBS

You always access the same Customizing function regardless of which Customizing path you call or whether you use the transaction.

Settings

When you call the Customizing function, you first obtain an overview of all status profiles regardless of which application or object they are valid for. In other words, you see not only the status profiles for maintenance objects but also all defined status profiles. If you now create a new status profile, you define each status in succession (see Figure 3.5).

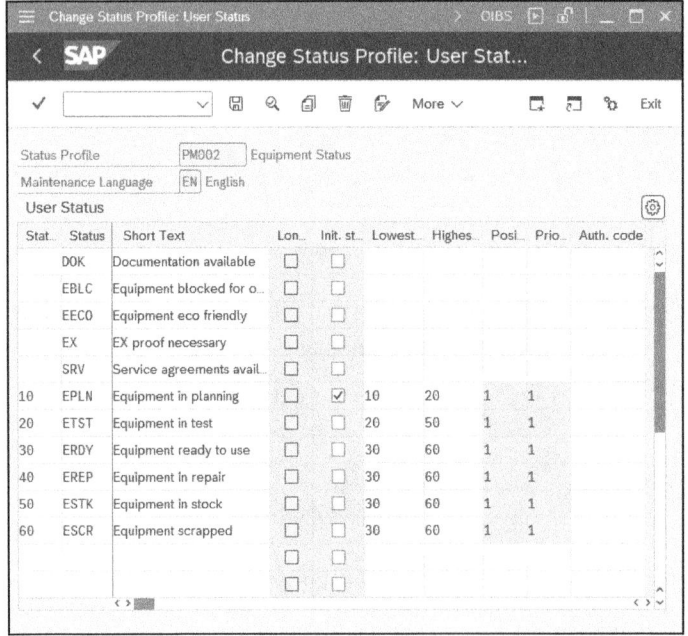

Figure 3.5 Define User Status Profile

The statuses without a status number are sorted alphabetically in the upper half of the screen, while the statuses with a status number are sorted by status number in the lower half of the screen.

Status numbers are used for the following reasons:

- Statuses with a status number are mutually exclusive. In the example shown, a piece of equipment can be ready to use (status number **30**) or in repair (status number **40**), but not both.
- You generally use the statuses with a status number to map a process (e.g., execution of a planned repair order). In the example shown, virtually the entire lifecycle of a piece of equipment (from planning and operation through scrapping) is mapped.

- One of these statuses (generally, the status with the lowest status number) is selected as the initial status and later automatically set in the object. In this case, **EPLN** is the initial status.
- If you set a specific status with a status number, you can only change to another status if it's within the upper and lower limits defined for the status number. If, in the example shown, the piece of equipment is assigned a status number of **40** (**Equipment in repair**), the next status must lie between **30** (**Equipment ready to use**) and **60** (**Equipment scrapped**). It's not possible to return to **20** (**Equipment in test**).

If you want to assign business transactions to a status, double-click the status and choose ⬜ to add new entries. For each status, you can define the business transaction control (see Figure 3.6) as follows:

- **Allowed**
  The business transaction is explicitly allowed.
- **Warning**
  The system issues a warning when you execute the business transaction.
- **Forbidd.**
  The business transaction is forbidden.

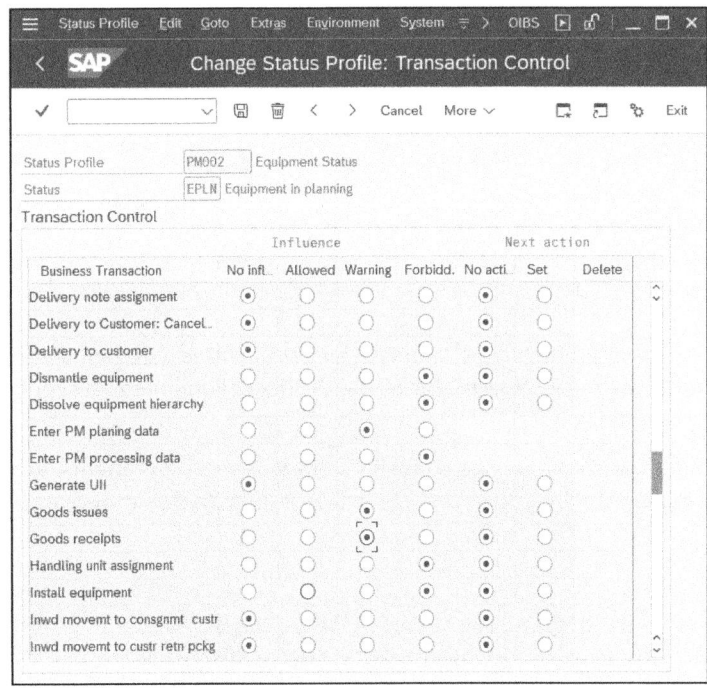

Figure 3.6 Status: Transaction Control

Finally, you must ensure that the status profile can also be used by the relevant object category. Choose the **Object Types** button to display a list of all object types associated with status management (see Figure 3.7) and then select the relevant object type. You can also permit the same status profile for several object types. For example, you can make a status profile available for both equipment and functional locations. However, you can also use a different status profile for notifications, orders, and order operations.

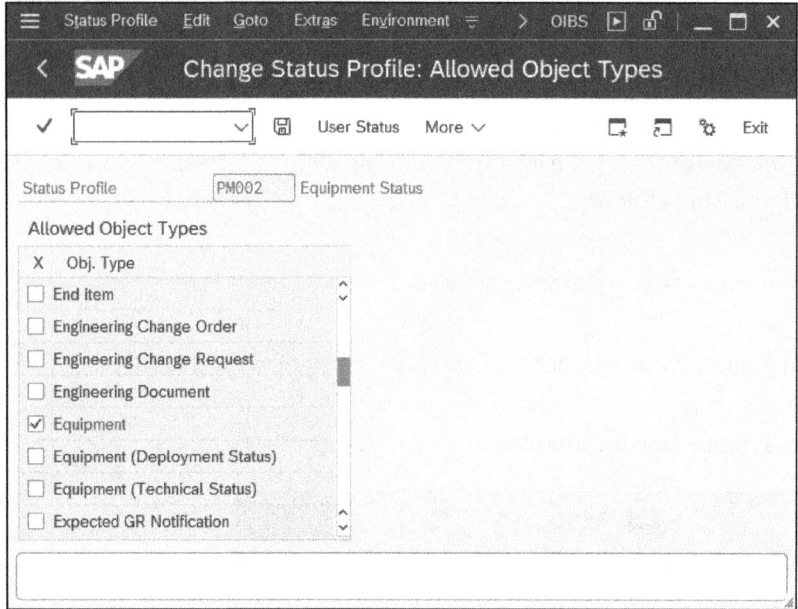

Figure 3.7 Status Profile: Allowed Object Types

You then use special object-based Customizing functions to assign a status profile to an object:

- Equipment
  Plant Maintenance and Customer Service • Master Data in Plant Maintenance and Customer Service • Basic Settings • Assign User Status Profile to Equipment Category
- Functional location
  Plant Maintenance and Customer Service • Master Data in Plant Maintenance and Customer Service • Technical Objects • Functional Locations • Define Category of Functional Location
- Object link
  Plant Maintenance and Customer Service • Master Data in Plant Maintenance and Customer Service • Technical Objects • Object Links • Define Object Types
- Notification
  Plant Maintenance and Customer Service • Maintenance and Service Processing • Maintenance and Service Notifications • Notification Processing • User Status for Notifications • Define Status Profile • Assign User Status to Notification Types

- Order and order operations
  Plant Maintenance and Customer Service · Maintenance and Service Processing · Maintenance and Service Orders · General Data · User Status for Orders · Define Status Profile · Assign User Status to Order Types
- Work centers
  Plant Maintenance and Customer Service · Maintenance Plans, Work Centers, Task Lists, and PRTs · Work Centers · General Data · Define Work Center Types and Link to Task List Application
- Maintenance plans
  Plant Maintenance and Customer Service · Maintenance Plans, Work Centers, Task Lists, and PRTs · Maintenance Plans · Set Maintenance Plan Categories

> **Make Good Use of the Status Profile Options**
>
> A purposeful status definition enables you to control in detail which business transactions are to be allowed or forbidden for which object and in which situation. It also enables you to search for objects (e.g., to display all equipment currently being planned).

3.3 Number Assignment

A number must be assigned to each data record. Apart from a few exceptions (e.g., functional locations), you adopt a similar approach to number assignment for all objects in the SAP system. You must manually assign a unique number to each object, or you must allow the system to assign a unique number. The following number assignment options are available:

- **Internal number assignment**
  The system automatically assigns consecutive numbers within a predefined number range.
- **External number assignment**
  The user assigns the numbers manually.

Define Number Ranges

You use this Customizing function to define the type of number assignment for each object. The procedure described here is used for the following maintenance objects:

- Equipment
- Serial numbers
- Object links
- Materials

- Notifications
- Shift notes
- Orders
- Measuring points and counters
- Measurement documents
- Master warranties
- Task lists
- Maintenance plans

For the sake of simplicity, we'll continue to consider the example of a piece of equipment to illustrate number assignment, and we'll adopt a similar approach to number assignment for all other objects.

Groups are always used in conjunction with number assignment. Here, several object types (e.g., equipment categories, notification types, order types) are combined to form a group, and you can then define an internal and external number range for the group. In the example shown, several equipment categories can belong to the same group and therefore have the same number range (see Figure 3.8).

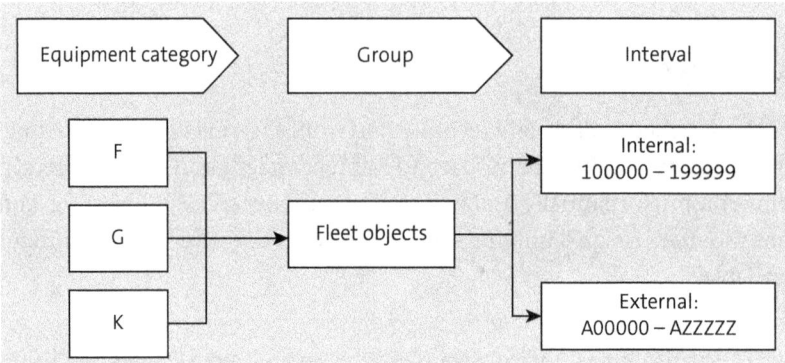

Figure 3.8 Number Assignment: Overview

Prerequisites

You must define the object types (e.g., equipment categories, functional location categories, order types, notification types) beforehand.

Customizing Paths

The following Customizing paths are available for number assignment:

- Equipment
  Plant Maintenance and Customer Service • Master Data in Plant Maintenance and Customer Service • Technical Objects • Equipment • Equipment Categories • Define Number Ranges

- Material
  Logistics – General · Material Master · Basic Settings · Material Types · Define Number Ranges for Each Material Type
- Object links
  Plant Maintenance and Customer Service · Master Data in Plant Maintenance and Customer Service · Technical Objects · Object Links · Define Number Ranges for Object Links
- Warranties
  Plant Maintenance and Customer Service · Master Data in Plant Maintenance and Customer Service · Basic Settings · Warranties · Define Number Ranges for Warranty Types
- Maintenance plans
  Plant Maintenance and Customer Service · Master Data in Plant Maintenance and Customer Service · Maintenance Plans, Work Centers, Task Lists, and PRTs · Maintenance Plans · Define Number Ranges for Maintenance Plans
- Maintenance items
  Plant Maintenance and Customer Service · Master Data in Plant Maintenance and Customer Service · Maintenance Plans, Work Centers, Task Lists, and PRTs · Maintenance Plans · Define Number Ranges for Maintenance Items
- Task lists
  Plant Maintenance and Customer Service · Master Data in Plant Maintenance and Customer Service · Maintenance Plans, Work Centers, Task Lists, and PRTs · Task Lists · Control Data
  - Define Number Ranges for General Maintenance Task Lists
  - Define Number Ranges for Equipment Task Lists
  - Define Number Ranges for Task Lists for Functional Locations
- Notifications
  Plant Maintenance and Customer Service · Master Data in Plant Maintenance and Customer Service · Maintenance and Service Processing · Maintenance and Service Notifications · Notification Creation · Notification Types · Define Number Ranges
- Shift notes
  Plant Maintenance and Customer Service · Maintenance and Service Processing · Shift Reports/Notes · Settings for Shift Notes · Define Number Ranges for Notification Type
- Orders
  Plant Maintenance and Customer Service · Master Data in Plant Maintenance and Customer Service · Maintenance and Service Processing · Maintenance and Service Orders · Functions and Settings for Order Types · Configure Number Ranges
- Measuring points
  Plant Maintenance and Customer Service · Master Data in Plant Maintenance and

Customer Service • Basic Settings • Measuring Points, Counters, and Measurement Documents • Create Number Ranges for Measuring Points
- Measuring documents
Plant Maintenance and Customer Service • Master Data in Plant Maintenance and Customer Service • Basic Settings • Measuring Points, Counters, and Measurement Documents • Create Number Ranges for Measurement Documents

Transactions

Different objects also have different transactions, as follows:

- OIEN: Equipment Number Ranges
- IW20: Notification and Shift Notes Number Ranges
- OION: Order Number Ranges
- IN20: Object Link Number Ranges
- BG20: Define Number Ranges for Warranty Types
- OIL4: General Task List Number Ranges
- OIL5: Equipment Task Lists Number Range
- OIL0: Functional Location Task Lists Number Range
- IP20: Maintenance Plan Number Assignment
- IP21: Maintenance Item Number Assignment
- IK09: Define Number Range for Measuring Points
- IK19: Define Number Range for Measuring Documents

Settings

If you want to define a number range for a group (e.g., equipment), follow these steps:

1. After you start the Customizing function, choose the **Groups** icon to bring up an overview of which groups exist and which object types (in this case, equipment categories) are assigned to which groups or which object types aren't assigned yet (see Figure 3.9).

2. Use the icon to create a new group. At the same time, the system prompts you to define one or more number ranges, namely an internal number range and/or an external number range (see Figure 3.10).

3. If you now want to assign an equipment category to the group, mark the equipment category, select the group, and choose the icon.

4. Use the **Intervals** (edit) button to make subsequent changes to the number range intervals.

5. Use the **Groups** button to subsequently change the group (e.g., the group text).

3.3 Number Assignment

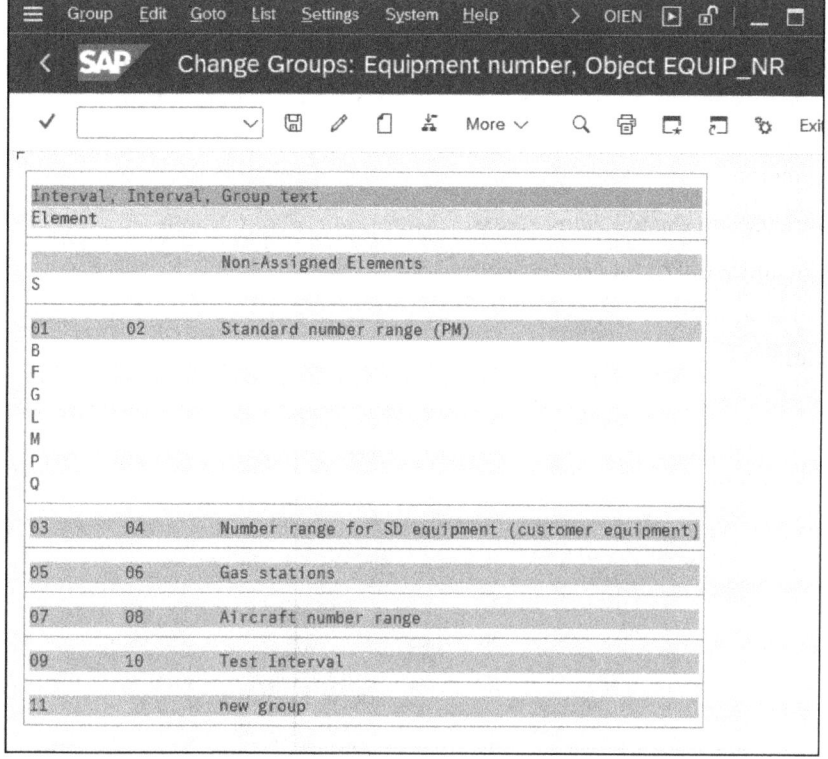

Figure 3.9 Number Assignment: Groups

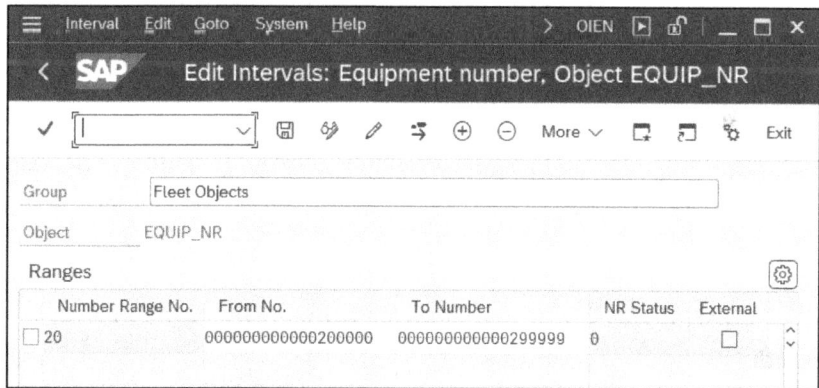

Figure 3.10 Number Assignment: Interval

> **[+] Define Short Numbers**
>
> For usability purposes, you should define numbers that are as short as possible. Your users will thank you for only having to enter a 6-digit number (e.g., 800123) rather than a 12-digit number (e.g., 800121232343).

3.4 Warranties

A *warranty* is a commitment from a manufacturer, vendor, or salesperson to a customer that services are guaranteed for a particular period, without the customer being billed. A warranty always refers to a technical object (a functional location, equipment, or serial number).

Warranties are used in the following areas:

- Equipment and serial numbers (see Figure 3.11)
- Functional locations
- Notifications
- Orders

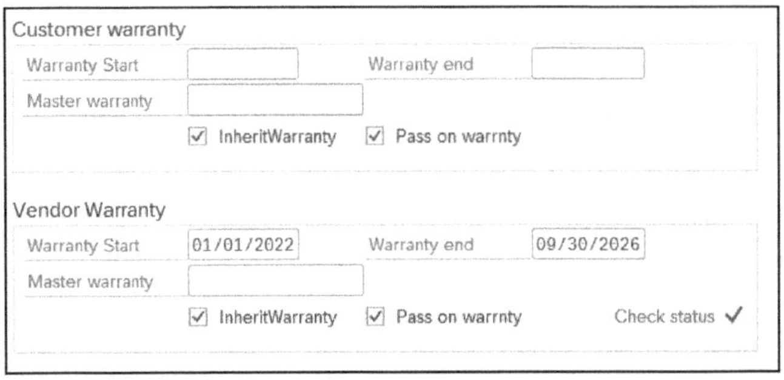

Customer warranty			
Warranty Start		Warranty end	
Master warranty			
	☑ InheritWarranty	☑ Pass on warrnty	
Vendor Warranty			
Warranty Start	01/01/2022	Warranty end	09/30/2026
Master warranty			
	☑ InheritWarranty	☑ Pass on warrnty	Check status ✓

Figure 3.11 Warranty in Equipment Master Record

If you want to assign warranties to technical objects and then use these warranties in maintenance processing, the following Customizing functions are available to you.

Check Warranty Categories

You use this Customizing function to define whether you're the guarantor or the warrantee.

Prerequisites

There are no special prerequisites.

Customizing Path

Plant Maintenance and Customer Service • Master Data in Plant Maintenance and Customer Service • Basic Settings • Warranties • Check Warranty Categories

Transaction

GM01

Settings

In the standard SAP delivery, a distinction is made between the following warranty categories (see Figure 3.12):

- **Warranty category I (inbound)**
  This warranty category exists if you're the warrantee.
- **Warranty category O (outbound)**
  This warranty category exists if you're the guarantor.

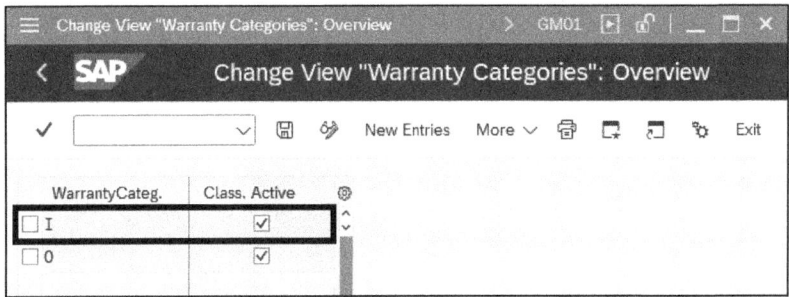

Figure 3.12 Warranty Categories: Overview

> **Delete Customer Warranties**
> In plant maintenance, you are always the warrantee, and for this, you'll need warranty category I only. Therefore, you can delete warranty category O because there won't be any occurrences requiring you to manage warranties granted to customers.

Define Warranty Types

You use this Customizing function to define whether you want to manage one or more warranty types for a warranty category. For example, for warranty category I (inbound), you can define one warranty type for a manufacturer warranty and another warranty type for a vendor warranty.

Prerequisites

You must define the warranty categories beforehand. If you want to assign a partner determination procedure and status profile, you must also define them beforehand.

Customizing Path

Plant Maintenance and Customer Service • Master Data in Plant Maintenance and Customer Service • Basic Settings • Warranties • Define Warranty Types

Settings

If you want to distinguish between vendor warranties and manufacturer warranties, you can define two warranty types for this purpose (see Figure 3.13).

For each warranty type, define the following:

- Do you want the system to display a dialog box when you open a notification?
- Do you want the system to display a dialog box when you open an order?

- Do you want the system to display a dialog box when you open an invoice?
- Do you want to assign a status profile?
- Do you want to assign a partner determination procedure?

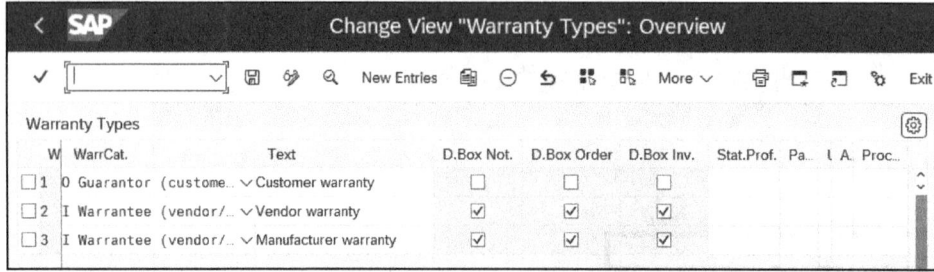

Figure 3.13 Warranty Types: Overview

Vendor Warranties and Manufacturer Warranties

As a precaution, you should always create two warranty types for the inbound warranty category:

- One for the manufacturer warranty
- One for the vendor warranty

Define Warranty Counters

You use this Customizing function to define counters for use in warranty processing. However, this Customizing function is only necessary if you work with master warranties or performance-based warranties. You don't require counters if you enter only basic warranty data (start and end dates) directly in the technical object (equipment or functional location).

Prerequisites

You must define the counters as characteristics in the classification system beforehand.

Customizing Path

Plant Maintenance and Customer Service • Master Data in Plant Maintenance and Customer Service • Basic Settings • Warranties • Define Warranty Counters

Transaction

GM04

Settings

When you create a master warranty, the system automatically proposes the counters marked as **Default**. Furthermore, if you want to measure time, you should always specify a counter that you've defined as a time-dependent counter (see Figure 3.14).

3.5 Measuring Points and Counters

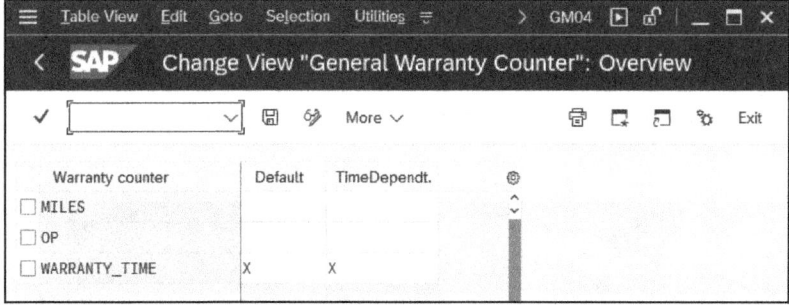

Figure 3.14 Warranty Counters: Overview

3.5 Measuring Points and Counters

Measuring points mark those locations at which the current condition of an asset is described (e.g., temperature, number of revolutions, pressure, level of contamination, viscosity). At individual measuring points, you can specify target values as well as upper and/or lower limits. Measuring points are a basic prerequisite for condition-based maintenance (see Figure 3.15).

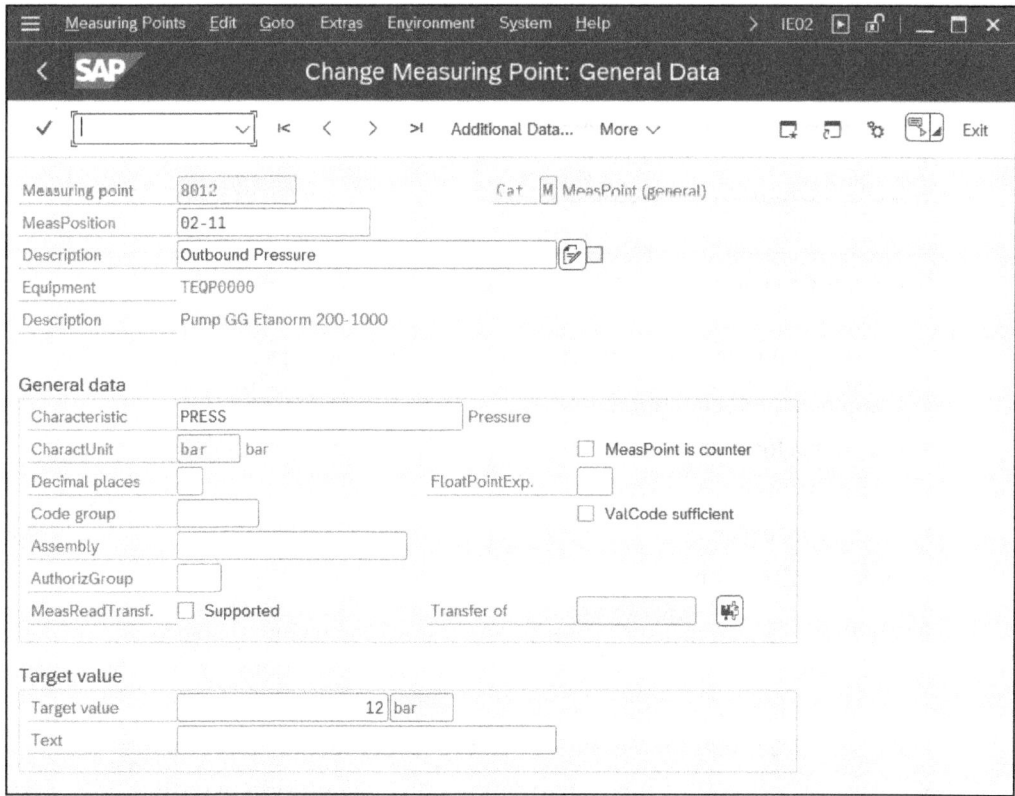

Figure 3.15 Measuring Point

3   Configuring Generic Functions

Counters are the tools you can use in the SAP system to represent wear and tear on an object, consumption of an object, or the reduction of an object's useful life (e.g., an odometer, an operation hours counter, numbers of pieces, output in tons). Having counters is a basic prerequisite for performance-based plant maintenance (see Figure 3.16).

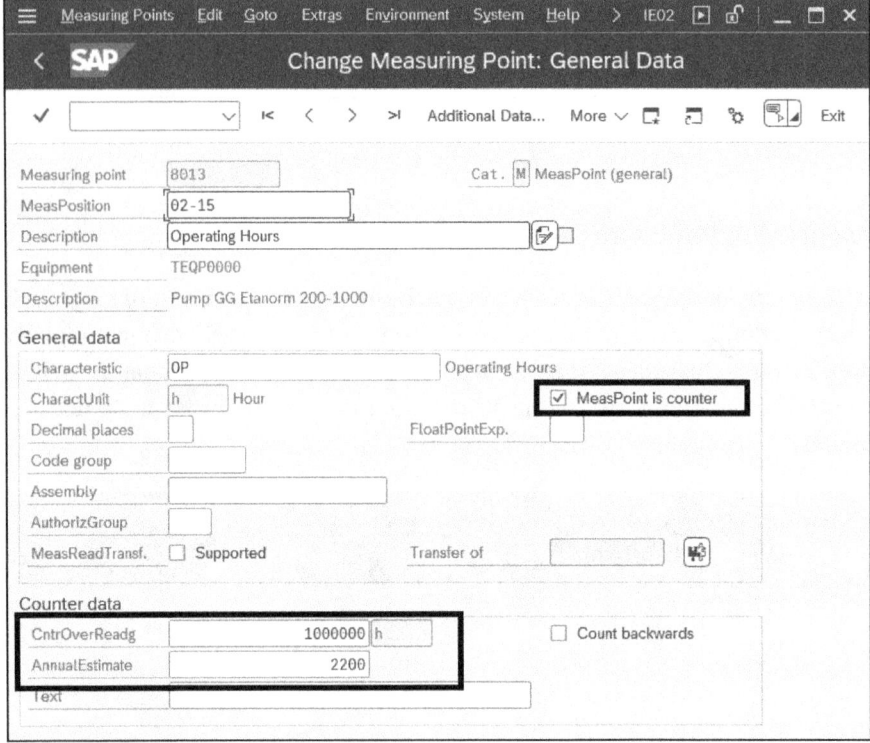

Figure 3.16 Counter

Measuring points and counters are used in the following objects within the system:

- Equipment and serial numbers
- Functional locations
- Task lists
- Maintenance plans
- Order confirmations

If you want to assign measuring points and counters to technical objects and then use them in maintenance processing, several Customizing functions are available to you.

Make System Settings for Measuring Points and Measurement Documents

You use this Customizing function to define a single global system setting: the generation of interval documents.

What are interval documents? If you use the measurement document transfer, difference measurement documents are transferred to the subordinate pieces of equipment when you enter a measurement document for a functional location. In the example shown in Figure 3.17, twelve measurement documents are generated for the pieces of equipment installed at functional location K1-M each time you enter a measurement document for the functional location.

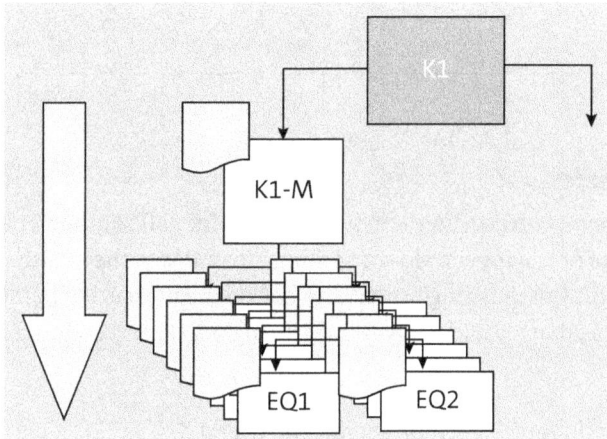

Figure 3.17 Measurement Document Transfer

Therefore, in the case of larger assets, the system may easily generate several hundred measurement documents each time you enter one measurement document. If you don't want this to happen, you can activate interval documents. Then, a single interval document is created (and updated repeatedly) for the length of time that the piece of equipment is installed at the functional location. Therefore, the *interval document* represents a recurring document that is repeatedly overwritten with the latest data. A new interval document is only generated if the installation location of the piece of equipment changes.

Prerequisites
There are no special prerequisites.

Customizing Path
Plant Maintenance and Customer Service • Master Data in Plant Maintenance and Customer Service • Basic Settings • Measuring Points, Counters, and Measurement Documents • Make System Settings for Measuring Points and Measurement Documents

Settings
Check the box to set the **Transfer generates interval documents** indicator (see Figure 3.18).

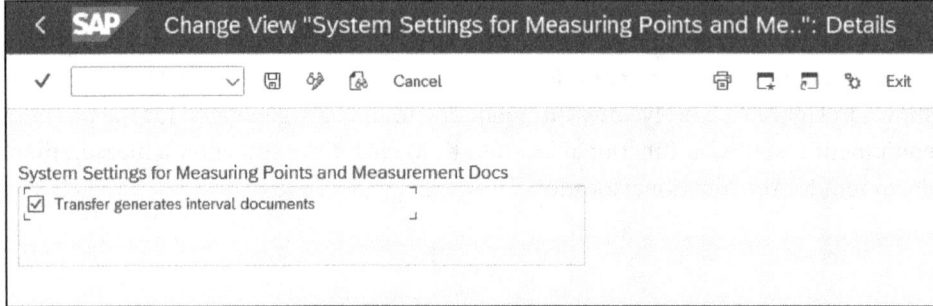

Figure 3.18 Interval Documents

Define Measuring Point Categories

You must assign a *measuring point category* to each measuring point and counter. The measuring point category, in turn, assigns certain controlling attributes to the measuring point or counter. You use this Customizing function to define the measuring point categories and generic attributes that should apply to measuring points and counters.

Prerequisites

If you want to assign catalogs, you must maintain them beforehand.

Customizing Path

Plant Maintenance and Customer Service • Master Data in Plant Maintenance and Customer Service • Basic Settings • Measuring Points, Counters, and Measurement Documents • Define Measuring Point Categories

Settings

You can use the measuring point category to assign the following attributes to a measuring point or counter (see Figure 3.19):

- **MeasPosUniqnss**
  This determines whether you want the measurement position to be *unique*. You use the measurement position to define the installation position of the measuring point associated with the measuring point object (for example, measurement position 1 [measurement of inside temperature] or measurement position 2 [measurement of outside temperature]). You can use the uniqueness check for measurement positions to prevent a measurement position from being defined several times for each object. The following settings are available to you: **0** = no check, **1** = check at object level, and **2** = check at client level.

- **Catalog type**
  This is used when entering the measurement reading to record the inspection results (e.g., **C** = damage codes).

- **MeasRge message**
  This determines the system response that occurs when the measurement range limits are exceeded: blank = no message, **W** = warning message, or **E** = error message.
- **TolPeriod (sec)**
  This specifies which tolerance period is permitted when entering future measurement readings if, in the event of a data transfer from an external system, the external server has a different system time than the SAP system. The tolerance period is specified in seconds.
- **Linear Asset**
  This determines whether you can enter linear data for a measuring point.
- **Business Event**
  You can use this checkbox to enable business events for a measuring point and a measurement document.
- **Create Notifications**
  This allows you to create condition-based maintenance notifications for a measuring point category.

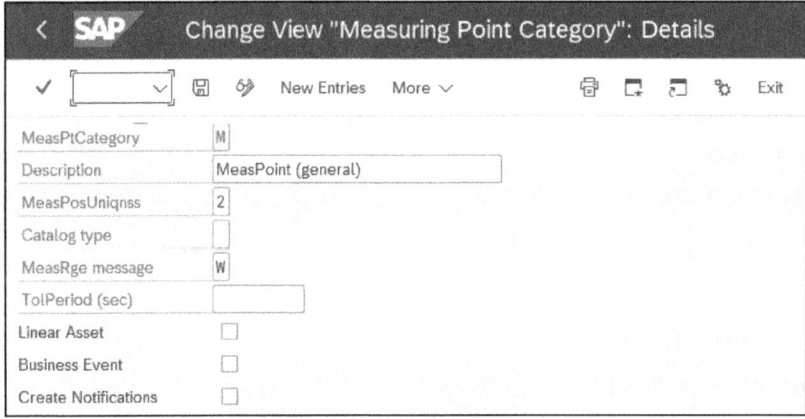

Figure 3.19 Measurement Point Category Details

Other Customizing functions are available for controlling the measuring points and counters. However, they have already been described previously as generic functions:

- **Create Number Ranges for Measuring Points**
  Number assignment is explained in detail in Section 3.3. Because measuring point numbers are only assigned internally, you should create only one number range, namely 01.
- **Create Number Ranges for Measurement Documents**
  Number assignment is explained in detail in Section 3.3. Because measurement document numbers are only assigned internally, you should create only one number range, namely 01.

- **Define Field Selection for Measuring Points and Measurement Documents**
  Field selection is explained in detail in Section 3.9.
- **Set List Editing for Measuring Point Lists**
  For more information about the technique used to design lists, see Section 3.10.
- **Set List Editing for Measurement Document Lists**
  For more information about the technique used to design lists, see Section 3.10.

3.6 Permits

Certain regulations or conditions must be considered when using some technical objects or performing maintenance work on them. You can define such regulations for technical objects as permits, and these permits must be granted for the orders during maintenance processing. Important permits in plant maintenance include the following:

- Fire permits
- Notifications of environmental protection
- Welding permits
- Drivers' licenses
- Fire protection permits
- Vat access permits
- Activation authorizations
- Technical inspection certificates
- Explosion protection zones

Permits are used in the following locations:

- Equipment and serial numbers
- Functional locations
- Orders

To define permits, use Transaction IPMD and then assign equipment (for example) via **Goto • Permits** (see Figure 3.20).

When you assign permits for technical objects, you also define whether an order is to be placed, must be placed, or must not be placed when an order is released (in the **OR** column) or when an order is technically complete (in the **OC** column).

Furthermore, you can define whether the relevant approval is to be printed on the order paper (in the **Print** column) and whether the permit is to be transferred to the processing data (e.g., for an order or notification in the **Proposal** column).

The Customizing functions available to you for use with permits are described next.

3.6 Permits

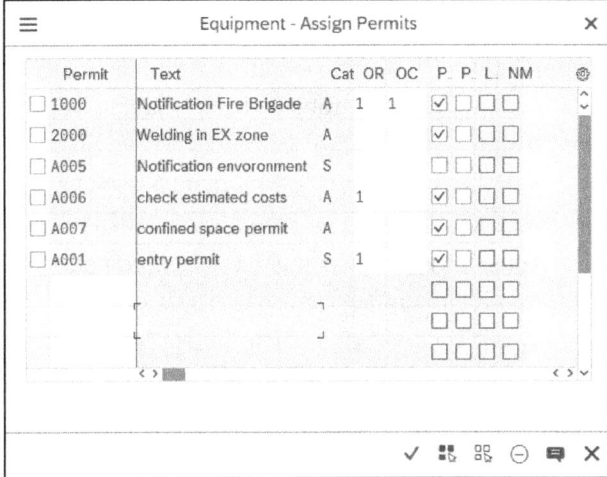

Figure 3.20 Equipment with Permits

Define Permit Categories

You use this Customizing function to define permit categories, and you can use the permit category to group permits together. To facilitate this, you must assign each permit to a permit category, but the permit categories aren't used for controlling. Rather, they are used only for grouping and selection purposes.

Prerequisites

There are no specific prerequisites.

Customizing Path

Plant Maintenance and Customer Service • Master Data in Plant Maintenance and Customer Service • Basic Settings • Permits • Define Permit Categories

Settings

Define the categories according to their individual requirements (see Figure 3.21).

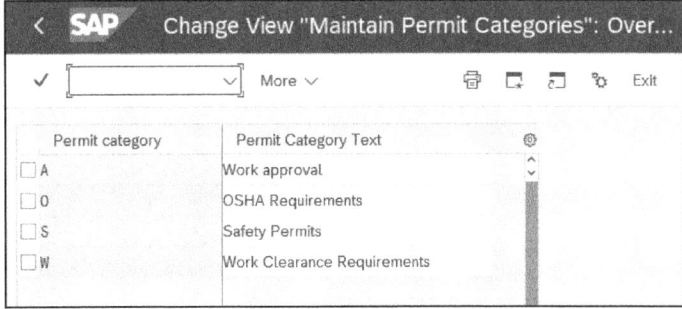

Figure 3.21 Permit Categories: Overview

133

3 Configuring Generic Functions

3.7 Partners

By default, the SAP system knows only a few organizational units that you can assign to a technical object or order (e.g., maintenance planner group and responsible work center).

By defining *partners*, you can expand these competencies and responsibilities considerably and specify them in greater detail. A partner can be an individual or legal entity, and it's either an internal organizational unit (e.g., department, cost center, person) or an external organizational unit (e.g., vendor, manufacturer, service provider).

Partners are used in the following locations:

- Equipment and serial numbers
- Functional locations
- Notifications
- Orders

Figure 3.22 shows the following procedure for assigning partners to technical objects or processing objects.

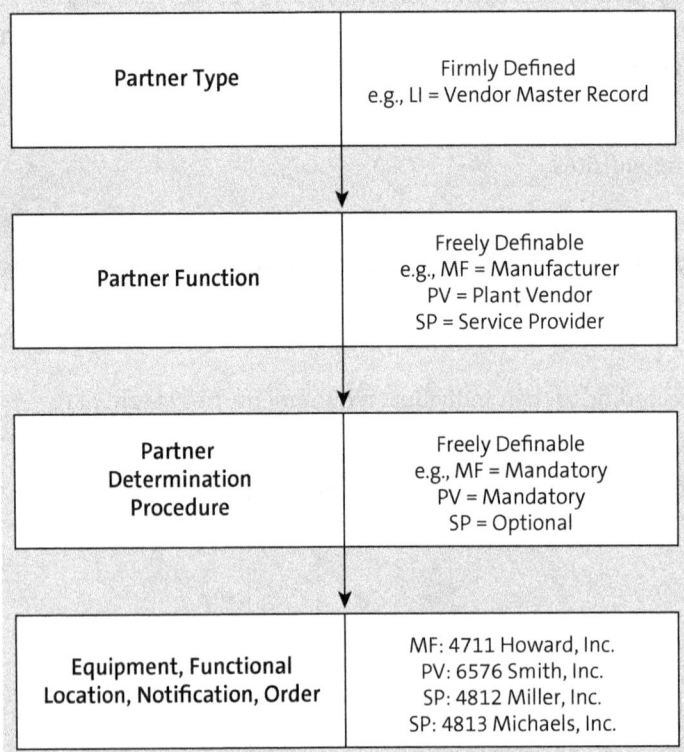

Figure 3.22 Partner: Overview

- **Partner type**
  The *partner type* is predefined by SAP and always contains a database table (customer, contact person, vendor, user, personnel number, organizational unit, and position).
- **Partner function**
  In Customizing (**Define Partner Determination Procedure and Partner Function**), you can freely define *partner functions* with reference to a partner type. For example, you can define partner functions such as the manufacturer, plant vendor, and service provider, and you can refer all functions to the vendor database table.
- **Partner determination procedure**
  You can freely define a *partner determination procedure*, which is a grouping of partner functions that specifies which partner functions are permitted or must always be specified. For example, you can determine that the manufacturer and vendor must always be specified for a piece of equipment but that the service provider is an optional specification.

Define Partner Determination Procedure and Partner Function

This Customizing function is to be used to define partner functions. You use it to define the partner type to which a partner function is to belong, and you can also use it to define the partner determination procedures and to assign the corresponding partner functions to a partner determination procedure.

Prerequisites

There are no specific prerequisites.

Customizing Paths

The following Customizing paths are available for partner functions:

- **Equipment and functional location**
  Plant Maintenance and Customer Service • Master Data in Plant Maintenance and Customer Service • Basic Settings • Partners • Define Partner Determination Procedure and Partner Function
- **Notifications**
  Plant Maintenance and Customer Service • Maintenance and Service Processing • Maintenance and Service Notifications • Notification Creation • Partners • Define Partner Determination Procedure and Partner Function • Define Partner Determination Procedure
- **Orders**
  Plant Maintenance and Customer Service • Maintenance and Service Processing • Maintenance and Service Orders • Partners • Define Partner Determination Procedure and Partner Function • Define Partner Determination Procedure

3 Configuring Generic Functions

Transaction

VOP2

Settings

Choose **Plant Maintenance** and then click **Change Partner**. You use the **Partner Functions** function (see Figure 3.23) to define the partner functions that you require in technical objects or processing objects.

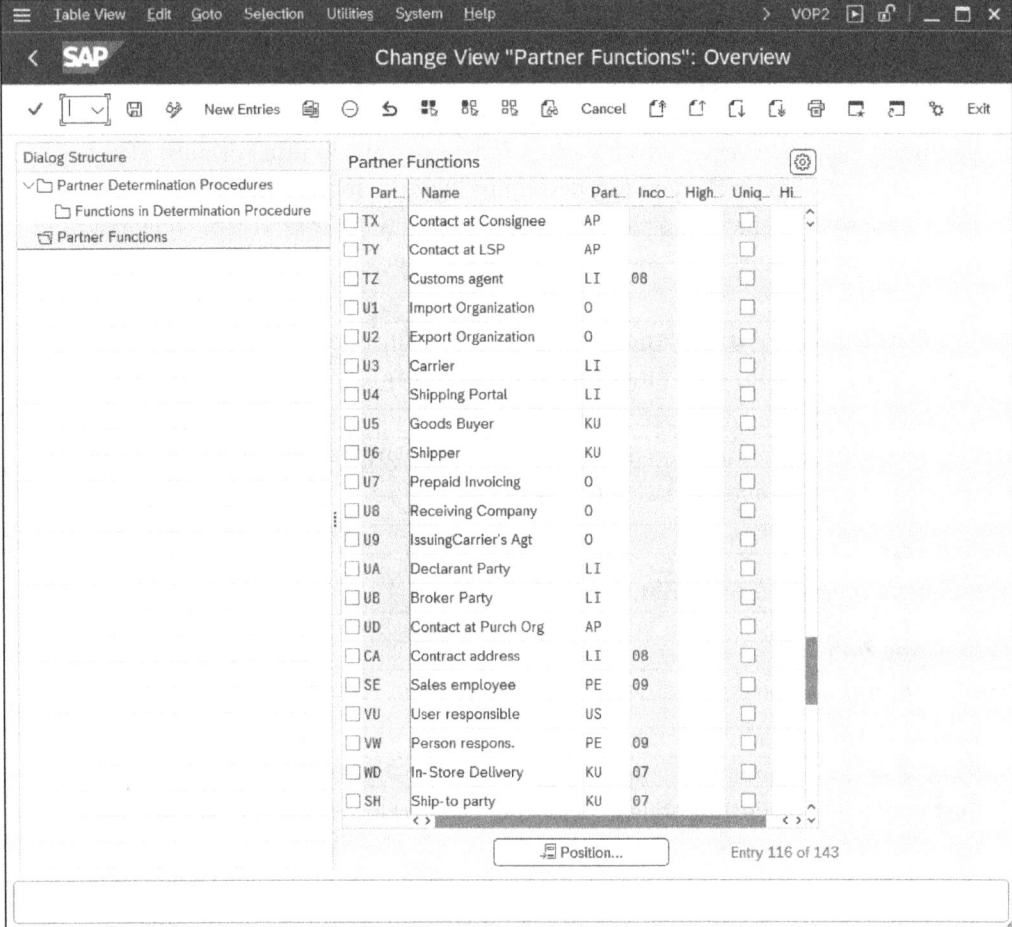

Figure 3.23 Partner Functions: Overview

> [!] **Take Care When Editing Partner Functions**
>
> You use the **Define Partner Functions** Customizing function to maintain not only partner functions in SAP S/4HANA Asset Management but also all partner functions used in the SAP system (e.g., in quality management or customer service).

3.7 Partners

You assign each partner function to a partner type. From an SAP S/4HANA Asset Management perspective, the following partner types are relevant:

- **A**: Work center
- **PE**: Person
- **US**: SAP system user
- **LI**: Vendor
- **S**: Position
- **O**: Organizational unit
- **MA**: Mailing address
- **WK**: Plant

You use the **Partner Determination Procedures** function to maintain the partner determination procedures that you require, and you use the **Functions in Determination Procedure** function to assign the partner functions required in each case. You also define the following attributes here (the four indicators to the right of **Name in** Figure 3.24):

- **No change possible**
  Whether it will be possible to change the partner retroactively
- **Partner mandatory**
  Whether the partner function is to be mandatory
- **Unique**
  Whether the partner function is unique (i.e., if the partner function isn't permitted to be assigned to the same object repeatedly)
- **Appointments**
  Whether an appointment is to be entered in the employee's calendar

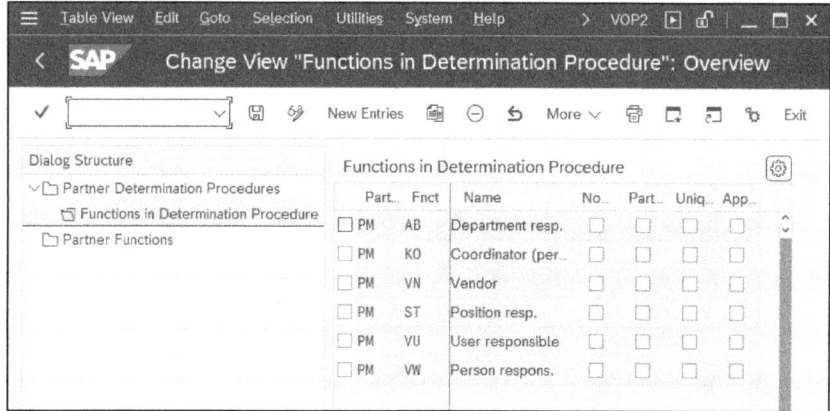

Figure 3.24 Partner Determination Procedure

Assign Partner Determination Procedure to Plant Maintenance Objects

Although not a specific Customizing function, you need to assign the partner determination procedure to plant maintenance objects. This includes the equipment, functional location, notification, and order.

Prerequisites

The partner determination procedure has to be set up.

Customizing Paths

The following Customizing paths are available for assigning the partner determination procedure:

- Equipment
  Plant Maintenance and Customer Service • Master Data in Plant Maintenance and Customer Service • Technical Objects • Equipment • Equipment Categories • Assign Partner Determination Procedure to Equipment Category
- Functional location
  Plant Maintenance and Customer Service • Master Data in Plant Maintenance and Customer Service • Technical Objects • Functional Locations • Define Category of Functional Location
- Notifications
  Plant Maintenance and Customer Service • Maintenance and Service Processing • Maintenance and Service Notifications • Notification Creation • Partners • Define Partner Determination Procedure and Partner Function • Assign Partner Determination Procedure to Notification Type
- Orders
  Plant Maintenance and Customer Service • Maintenance and Service Processing • Maintenance and Service Orders • Partners • Define Partner Determination Procedure and Partner Function • Assign Partner Determination Procedure to Order

Transactions

The following transactions are available for assigning the partner determination procedure:

- OIEV: Used for equipment
- S_ALR_87000811_2: Used for notifications
- OIOM: Used for orders

Settings

Figure 3.25 shows the assignment of the partner determination procedure (**PartDet Prc.**) to the equipment category, as an example.

3.7 Partners

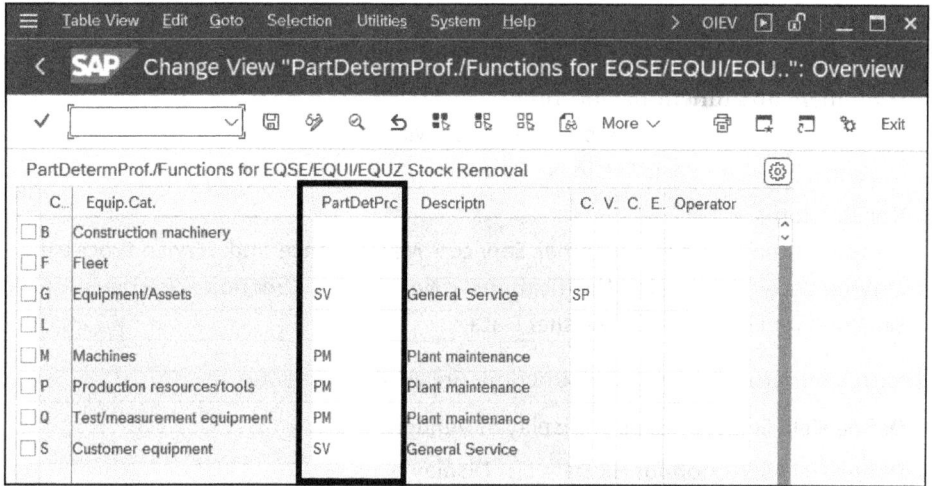

Figure 3.25 Assigning Partner Determination Procedure to Equipment Category

Now, it's possible to assign partners to the objects. For example, Figure 3.26 shows you the partners of an equipment object.

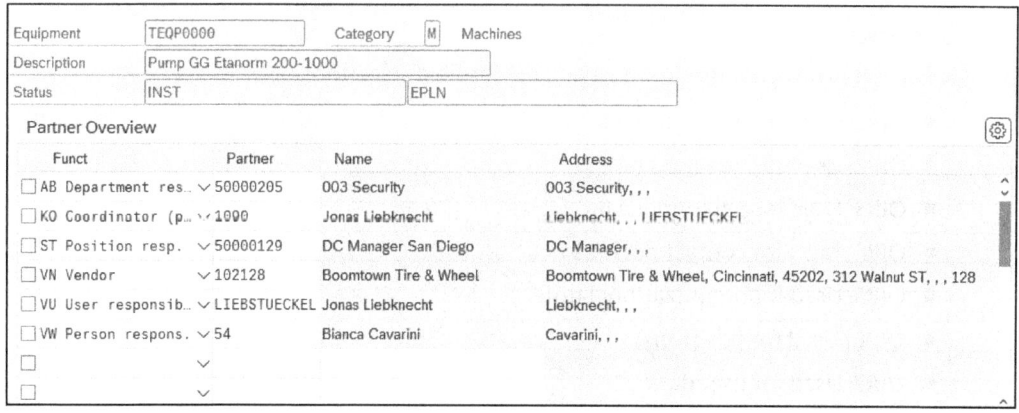

Figure 3.26 Equipment with Partners

Customizing Functions for the Field Selection for Partners

You use these Customizing functions to define which fields are to be displayed for which partner types at their place of use (e.g., in the equipment master or order).

Prerequisites

You must define partner functions for each partner type beforehand.

139

3 Configuring Generic Functions

Customizing Paths

The following Customizing paths are available for partner field selection:

- Equipment and functional location
  Plant Maintenance and Customer Service • Master Data in Plant Maintenance and Customer Service • Basic Settings • Partners

- Notification
  Plant Maintenance and Customer Service • Maintenance and Service Processing • Maintenance and Service Notifications • Notification Creation • Partners • Field Selection for List Display of Partner Data

Then, the following Customizing functions are available to you:

- Define Field Selection for List Display of Vendor Data
- Define Field Selection for HR Data List Display
- Define Field Selection for Organizational Unit List Display
- Define Field Selection for Position List Display
- Define Field Selection for List Display of User Data
- Define Field Selection for Address Data List Display

Transactions

The following transactions are available for field selection:

- OIR: Used for customer data
- OIR2: Used for vendor data
- OIR3: Used for personnel data
- OIR4: Used for contact persons
- OIR5: Used for organizational units
- OIR6: Used for positions
- OIR7: Used for user data
- OIR8: Used for address data

Settings

Figure 3.27 shows the settings for the vendor field selection, but these are representative of all other settings for field selections. Here, you define which fields are to be invisible as well as the display positions for those fields that you want to display.

3.8 Documents

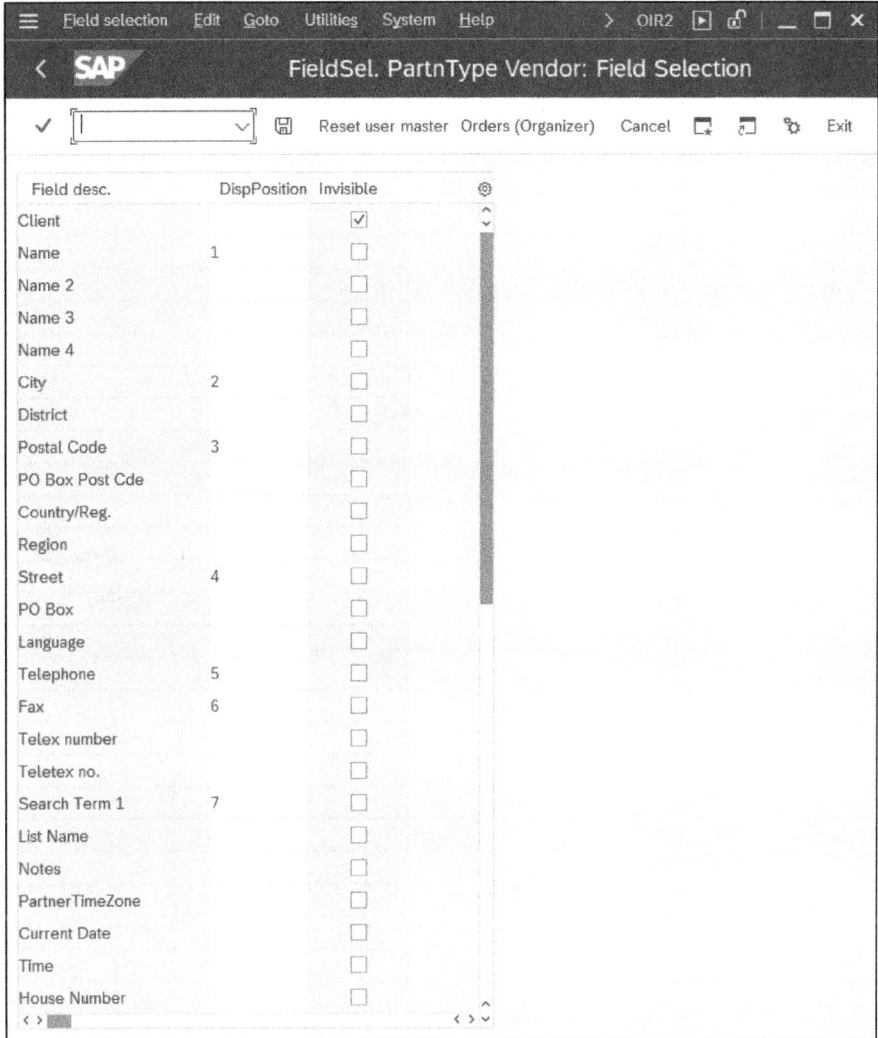

Figure 3.27 Partner Field Selection

3.8 Documents

Many companies want to link the technical objects and/or processing objects with documents such as design drawings, work instructions, checklists, images, inspection instructions, exploded drawings, measuring and control technology profiles, 3D models, and so on.

In the SAP system, you may manage your documents as document master records using Transactions CV01N to CV04N (Document Create, Change, Display, and Find, respectively) or you can create the document master record directly within plant maintenance objects like equipment or orders. Within these document master records, there's a link to the original documents (e.g., a link to the storage location of the CAD file or a link to the file server for Microsoft Office documents) whereby the link is retained or the original documents are uploaded to an SAP database or an external document management system.

In SAP S/4HANA Asset Management, documents are used in the following objects:

- Equipment and serial numbers
- Functional locations
- BOMs
- Notifications
- Orders
- Operations
- Task lists
- Task list operations
- Maintenance plans
- Object links
- Materials
- Measuring points
- Shift notices

Figure 3.28 shows the **Linked Documents** screen group of an equipment master.

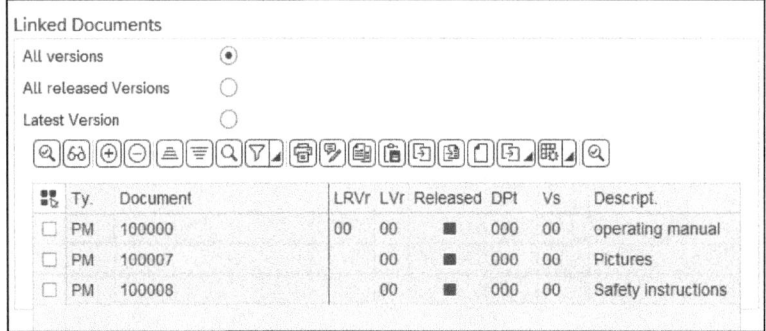

Figure 3.28 Linked Documents Screen Group

The next section doesn't describe the Customizing required for the full scope of functions in the SAP document management system because, for pure plant maintenance purposes, it's generally sufficient to define a reduced scenario and configure the necessary Customizing. This scenario looks like this:

- To differentiate this scenario from other application areas, you want this scenario to have its own *document type*.
- Furthermore, you want to be able to use it for *plant maintenance objects* only.
- To prevent other enterprise areas from encroaching on this scenario, you want to create a separate *secure storage area* in the SAP system and store the original documents there.
- All other functions in the SAP document management system (e.g., status profiles and versioning) don't play any role here.

We'll discuss how to map this scenario in Customizing next.

Define Document Types

You use this Customizing function to configure the basic settings for a document type (e.g., number assignment, object links).

Prerequisites

It is vitally important to define the number ranges (external and internal) beforehand.

To assign documents to orders, task lists, and operations, you must activate business function LOG_EAM_CI_6.

To assign documents to maintenance plans, you must activate business function LOG_EAM_CI_13.

To assign documents to shift notices, you must activate business function LOG_PP_SRN_CONF.

Customizing Path

Cross-Application Components • Document Management • Control Data • Define Document Types

Transaction

DC10

Settings

Figure 3.29 shows the basic settings for a document type, and the following list describes how to configure these settings from an SAP S/4HANA Asset Management perspective:

- **Use KPro**
  This indicator controls the use of Knowledge Provider to store original files in defined storage systems (known as *content repositories*). If you define a new document type, you should use only Knowledge Provider to control file storage. Don't check the box to set this indicator retroactively for a document type. If original files are already maintained for a document type and Knowledge Provider wasn't used previously to control file storage, the system can no longer manage these original files.

- **Version Assgmt**
  If you check the box to set this indicator, the system assigns the new version number automatically. Because version numbers are rarely used in plant maintenance, version 00 is automatically assigned for the document.

- **Archiving Authorization**
  This indicator controls whether the original files can be archived.

3 Configuring Generic Functions

- **Change Docs**
  Check the box to set this indicator if you want to use change documents to log changes to the document info record.

- **Internal Number Range/External Number Range**
  You use the number range controls to assign the external and internal number ranges.

- **File Size**
  The specifications in relation to the maximum file size are relevant only if you don't use Knowledge Provider to store the files.

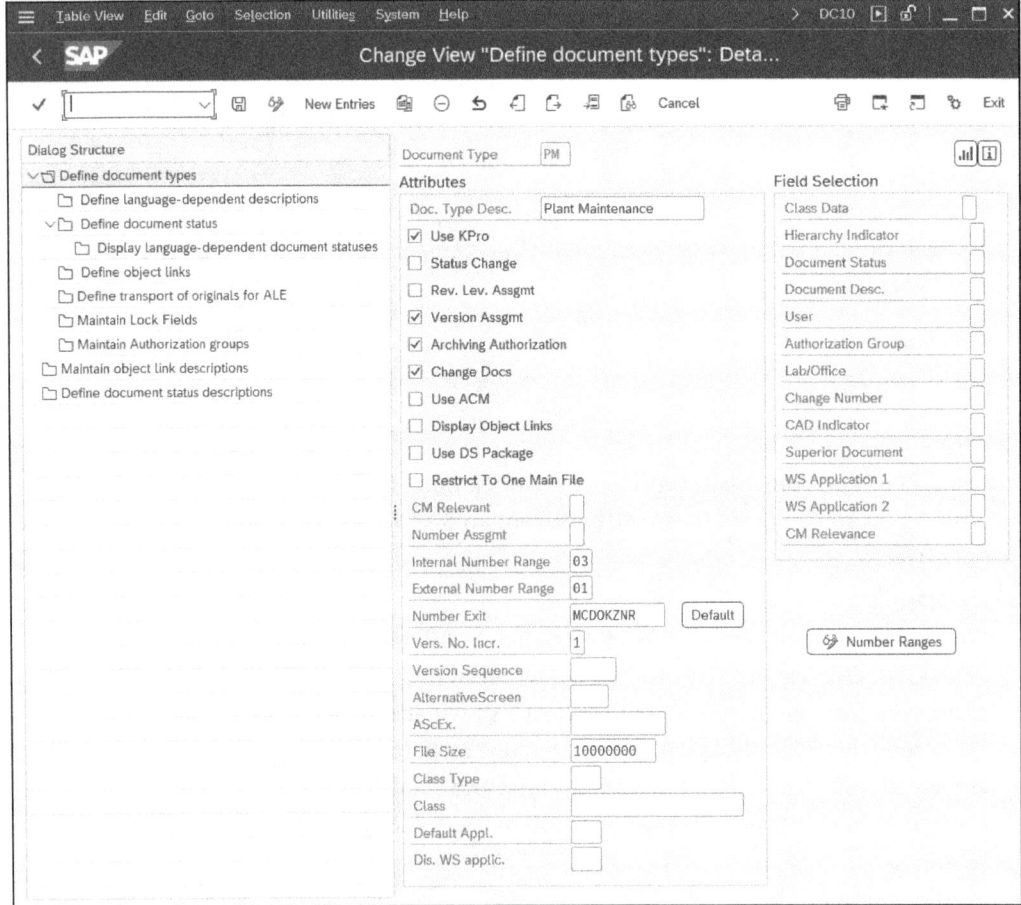

Figure 3.29 Document Type: Basic Settings

You use the **Define document status** subfunction (see Figure 3.30) to define the status to be used. In SAP S/4HANA Asset Management, you generally don't need to use a multilevel status profile such as those used in product development. Instead, you only need to use a status that has activated the **Release Flag** indicator so that the document can be used in other business processes.

3.8 Documents

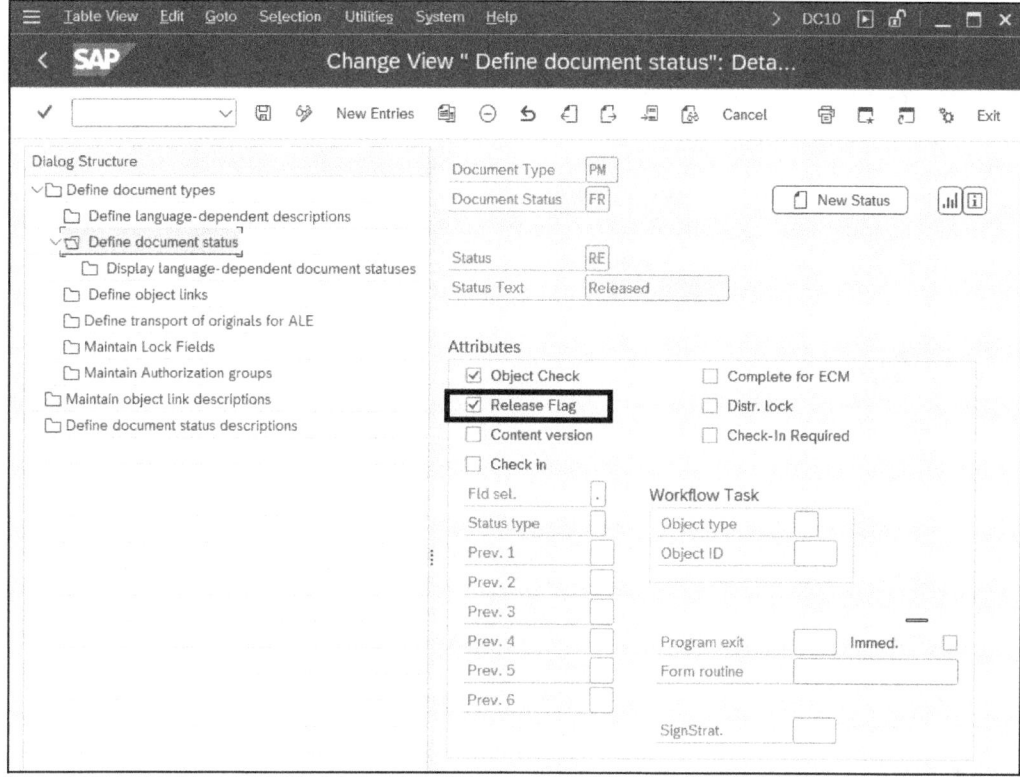

Figure 3.30 Document Type: Status

You use the **Define object links** subfunction (see Figure 3.31) to define the SAP objects to which documents of this type can be assigned. From an SAP S/4HANA Asset Management perspective, this refers to the following objects:

- Equipment Master (EQUI)
- Functional Location (IFLOT)
- Reference Location (IRLOT)
- Object Link (INET)
- Material Master (MARA)
- Plant Material (MARC)
- Measuring Points (IMPTT)
- Maintenance Notific. (PMQMEL)
- Appropriation Req. (IMAV)
- Order Header (PMAUFK)
- Order Operation (PMAFVC)
- BOM Header (STKO_DOC)

3 Configuring Generic Functions

- BOM Item (STPO_DOC)
- Task List Header (PMPLKO)
- Task List Operation (PMPLPO)
- Maintenance Plan (MPLA)

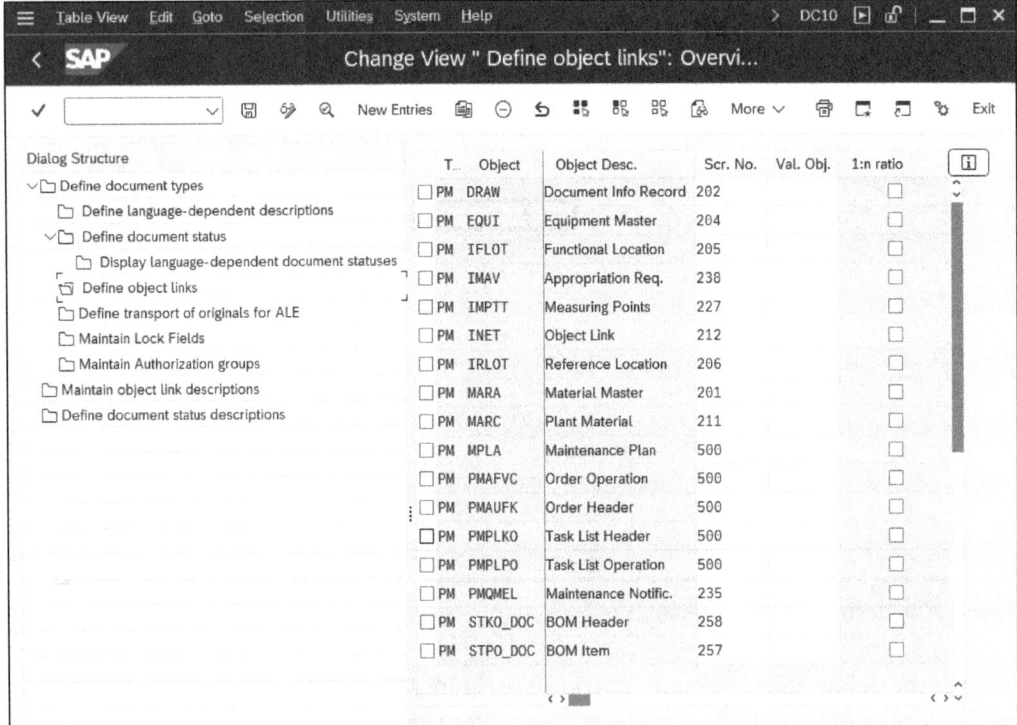

Figure 3.31 Document Type: Object Links Overview

On the object link detail screen (see Figure 3.32), you find more important information.

The **Create Document** field enables you to determine how a new document is created from the master data processing of an object (e.g., an order):

- If you type "2" in the **Create Document** field, you can create a document info record only by calling Transaction CV01N (Create Document). You can link the document info record to the object (e.g., order) only after creating the document.

- If you type "1" in the **Create Document** field, you define a plain creation process: link one original application file without processing the document info record. Select the original application file in question from the file manager and the document info record for this original application is created automatically (by internal number assignment).

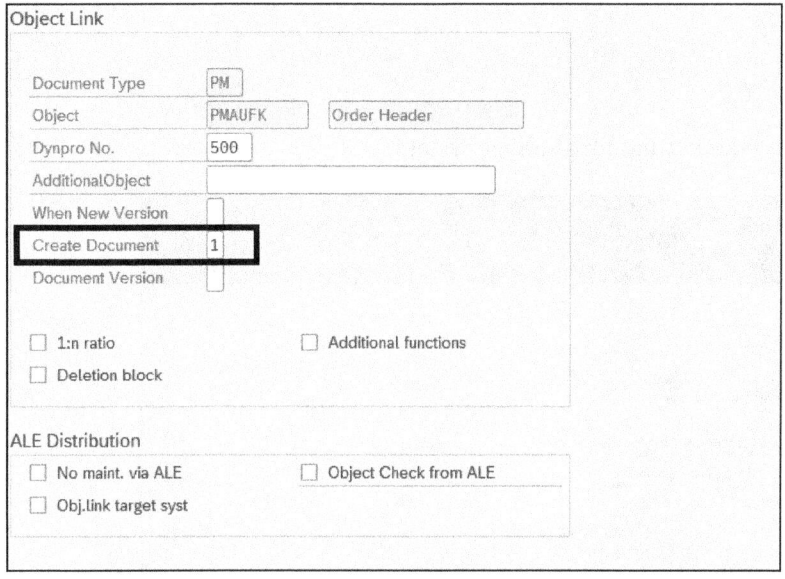

Figure 3.32 Document Type: Object Link Detail

Set Create Document to "1"

For some SAP S/4HANA Asset Management objects, such as notifications, orders, task lists, and maintenance plans, it may be helpful to type "1" in the **Create Document** field to create the document master record automatically. It's much easier and less time-consuming than a manual procedure, which is necessary if you type "2" in the field.

Don't check the box to set the **1:n ratio** indicator. If you do, it means that you can only assign a single document of this document type to the object. Generally, this doesn't make sense.

Maintain Storage System

You use this Customizing function to define the storage location (known as the content repository) for the original files. The system uses these storage systems for documents whose document type allows for storage via Knowledge Provider.

Prerequisites

If you want to use your own database tables to physically store documents, you must define these beforehand in Transaction SE11. Furthermore, their structure must correspond to the structure of the SDOKCONT1 database table.

Customizing Path

Cross-Application Components • Document Management • General Data • Settings for Storage Systems • Maintain Storage System

3 Configuring Generic Functions

Transaction

OACO

Settings

Figure 3.33 shows the settings for a storage system.

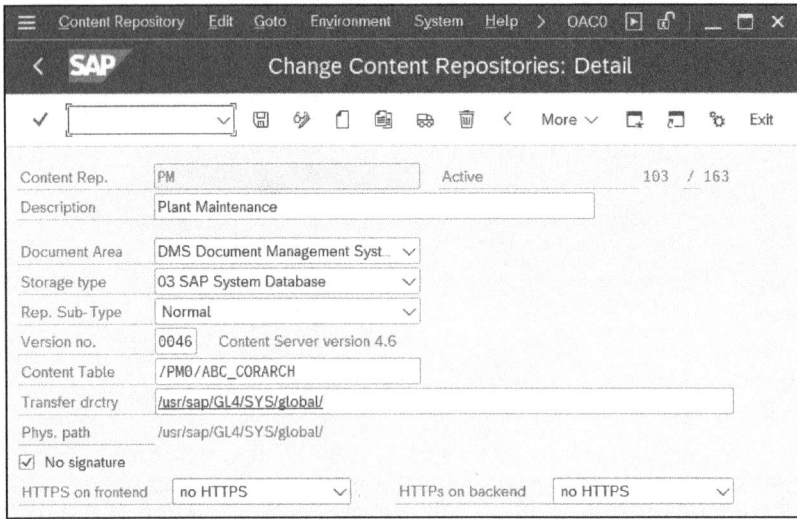

Figure 3.33 Content Repository Details

The following describes how to configure these settings from a plant maintenance perspective:

- The database of the SAP S/4HANA system should be the **Storage type**.
- You should always select the latest **Version no.** (which is currently **0046**) for the database interface.
- The **Content Table** is the database table in which the documents are stored. You can define this table as a client-specific or cross-client table. You must define your own database tables in Transaction SE11 beforehand.

Maintain Storage Category

You use this Customizing function to set up one or more logical views (storage categories) of a storage system (content repository). Having several views of the same storage system enables you to store the original files in a highly structured manner.

Prerequisites

You must define the content repository (storage system) beforehand.

Customizing Path

Cross-Application Components • Document Management • General Data • Settings for Storage Systems • Maintain Storage Category

Transaction

OACT

Settings

Figure 3.34 shows you how to assign a category to a content repository.

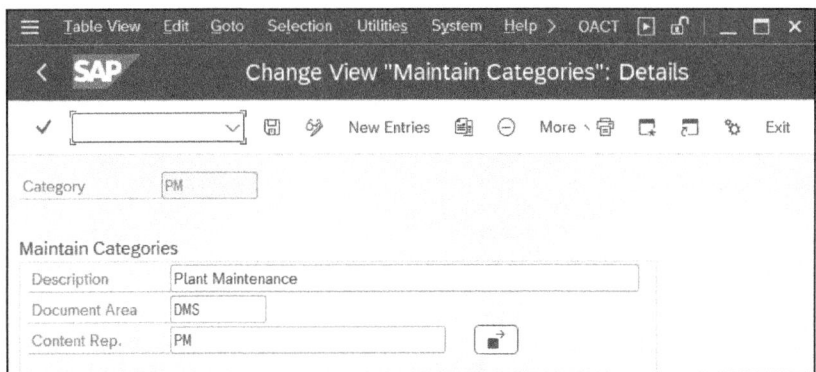

Figure 3.34 Storage Category Details

Define Workstation Application

For all file types that you want to use in SAP S/4HANA Asset Management (e.g., PDF, JPG, DWG, PNG DOC), you must use this Customizing function to specify the application to which they belong as well as how you want to control these file types.

Prerequisites

You must install the necessary programs for starting the relevant file types beforehand.

Customizing Path

Cross-Application Components • Document Management • General Data • Define Workstation Application

Settings

Figure 3.35 shows an overview table of all applications and associated file types. The standard SAP delivery already contains many applications and file types, which you should check for accuracy and completeness. If the necessary file types don't exist, you must add the corresponding new entries to this table.

As an example, Figure 3.36 shows the detail screen for pictures. Here, the following is important:

- **File format**
  Be sure to assign *all* potential file formats (e.g., DOC application with the file formats JPG, JPEG, PNG, GIF).

- **Start Authorization**
  Check the box to select this indicator so that the application starts immediately after

3 Configuring Generic Functions

you double-click the file (e.g., a JPEG picture in the equipment master) and is permitted to display the document.

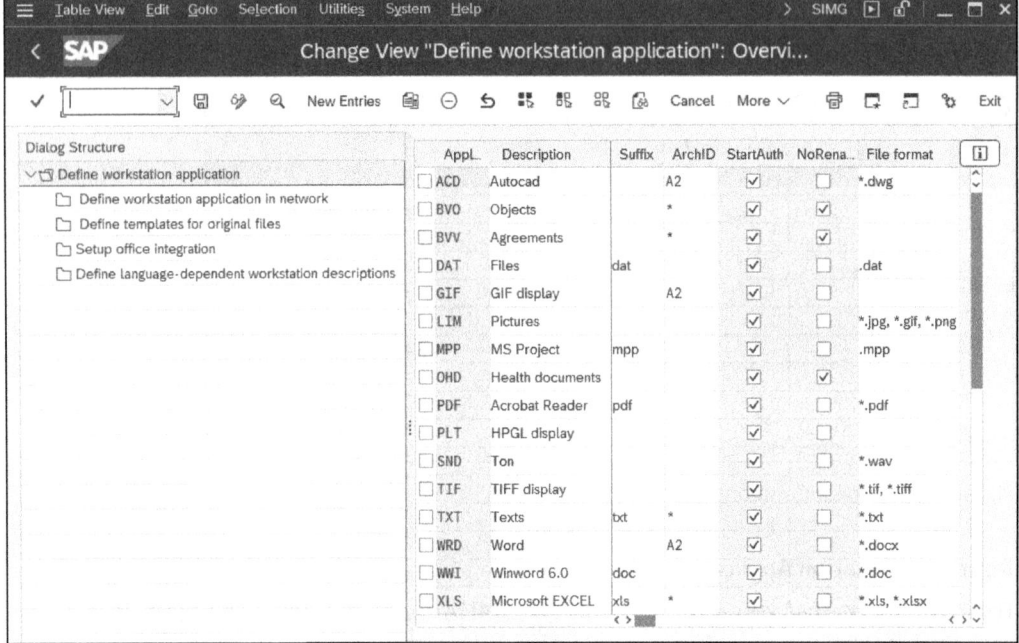

Figure 3.35 Workstation Applications: Overview

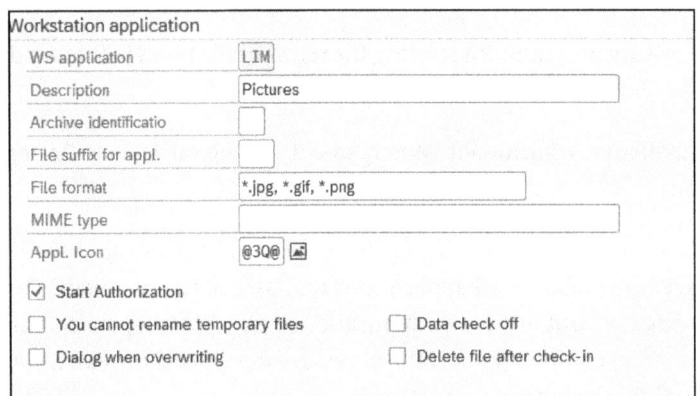

Figure 3.36 Workstation Applications: Detail

You use the **Define workstation application in network** subfunction to define which program is to start when a certain file type is called. In Figure 3.37, this is shown for picture files.

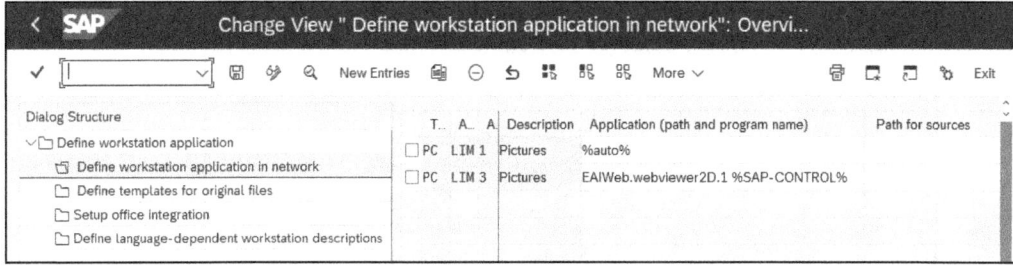

Figure 3.37 Workstation Applications in Network

Note the following regarding this subfunction:

- It always makes sense to type "%auto%" in the **Application (path and program name)** column as the first alternative so that the workstation then automatically uses the general Multipurpose Internet Mail Extensions (MIME) technology to search for the correct program.
- As a second alternative, enter a specific path for the program that you want to start.

For certain purposes, SAP also provides integrated viewers that can be called using a predefined string. Examples include the following:

- 2D CAD drawings via EAIWeb.webviewer2D.1 %SAP-CONTROL%
- 3D CAD drawings via EAIWeb.webviewer3D.1 %SAP-CONTROL%
- Right-hemisphere files via %VIEWER-CONTROL% %SAPPROVIS%

Define Profiles

In this work step, you define profiles that combine default values and settings for processing original application files.

The background information in a profile is standard information that is often required in similar constellations for processing original application files. This makes it easier to handle and to start documents.

It is best to assign only one role or several users to a profile.

Prerequisites

If you want to assign roles, you must define roles beforehand.

Customizing Path

Cross-Application Components • Document Management • General Data • Define Profiles

Settings

Figure 3.38 shows an overview table of all profiles.

3 Configuring Generic Functions

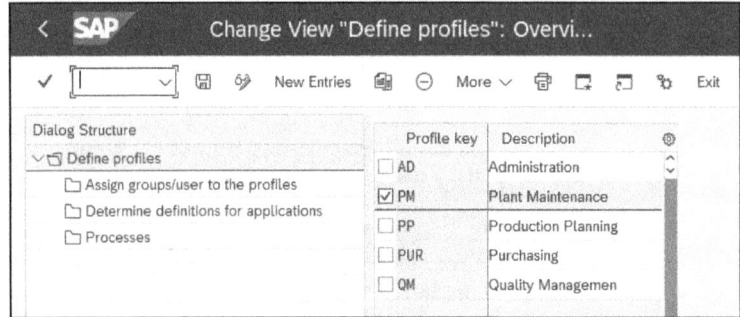

Figure 3.38 Profiles: Overview

Next, you carry out the **Assign groups/user to the profiles** step. You can assign a profile to a role as well as to several users (see Figure 3.39).

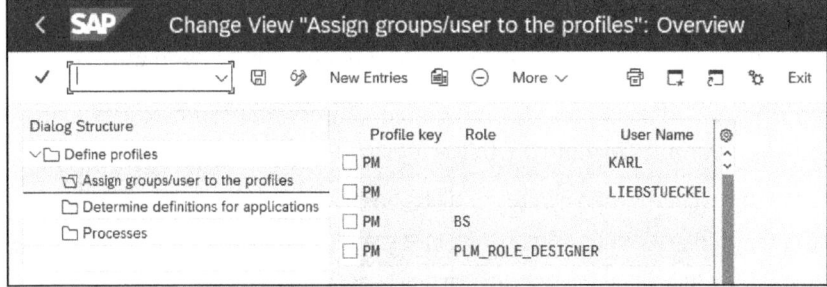

Figure 3.39 Profile: Assign Roles and Users

Select a profile and carry out the **Determine definitions for applications** step (see Figure 3.40). Enter the following default values for the application. These are the values you want when processing an original application file:

- The workstation application that you want to automatically start when printing or displaying a document
- The **Storage category** for storing the original application file using Knowledge Provider

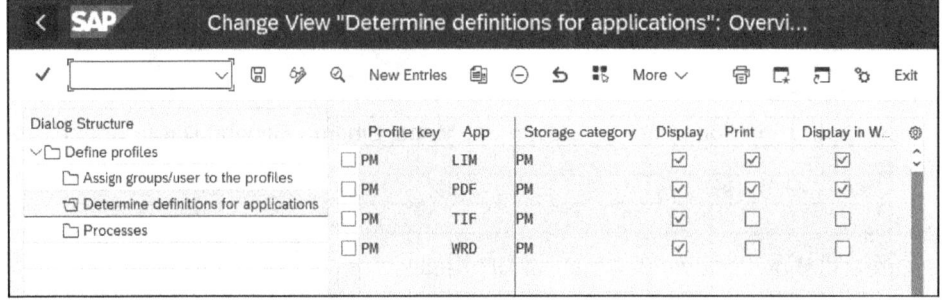

Figure 3.40 Profile: Determine Definitions for Applications

3.8 Documents

Activate Functions for Enterprise Asset Management

To be able to assign documents for specific usages, it's necessary to activate additional functions for SAP S/4HANA Asset Management.

Prerequisites

You must activate business functions LOG_EAM_CI_6 for document assignments in maintenance orders and LOG_EAM_SIMPLICITY_2 for document assignment on the web UI.

Customizing Path

Plant Maintenance and Customer Service • System Enhancements and Data Transfer • Activate Functions for Enterprise Asset Management

Settings

For document usages, several features must be activated, as follows (see Figure 3.41):

- **DOC_BOV Document Assignment in Basic Order View**
  This function enables you to display and assign relevant documents when processing maintenance orders in the basic order view.

- **DOC_TL_GUI Document Assignment to Task List in SAP GUI Transactions**
  This function enables you to display and assign relevant documents at the header and operation levels when processing task lists in SAP GUI transactions for SAP S/4HANA Asset Management.

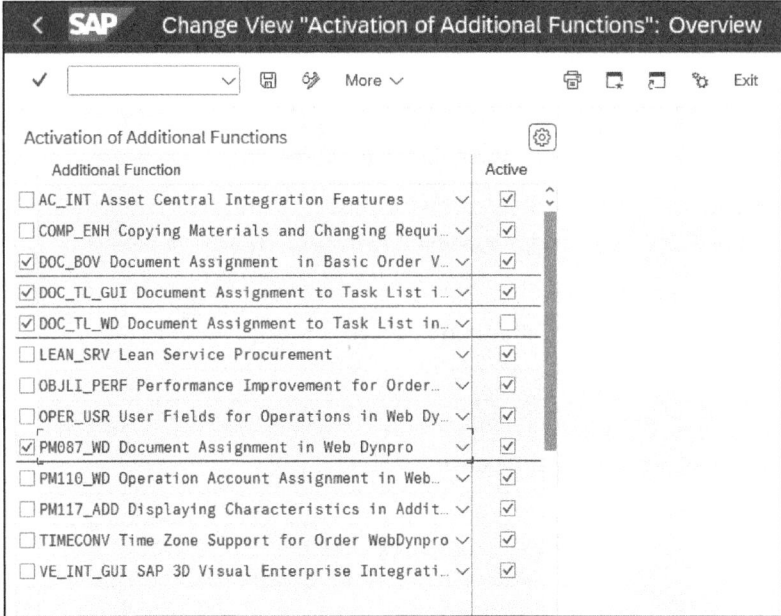

Figure 3.41 Additional Functions

- **DOC_TL_WD Document Assignment to Task List in Web Dynpro**
  This function enables you to display and assign relevant documents at the header and operation levels when processing task lists using the web UI for SAP S/4HANA Asset Management.

- **PM087_WD Document Assignment in Web Dynpro**
  This function enables you to display, create, and assign relevant documents when processing maintenance orders on the web UI for SAP S/4HANA Asset Management. You can create new documents or search for existing documents and assign them at the order level or operation level. You can navigate to display document info records and open the original file of each document as long as it's stored on a central document server. You can also display documents that aren't directly assigned to the maintenance order but are assigned to a dependent object (e.g., an assembly). This list of documents can be used for searching and sorting documents as well as defining and managing filters.

3.9 Field Selection

You use field selections to define attributes for each object type. For example, you can determine the following:

- Which field must be maintained
- Whether the field is merely a display field
- Whether it's a normal entry field
- Whether the field is to be hidden

Furthermore, you can and should base the field selection on *influencing fields*.

The field selection technique is available to you for almost all SAP objects. From an SAP S/4HANA Asset Management perspective, the following objects are available:

- Equipment and serial numbers
- Functional locations and reference functional locations
- Materials
- Notifications
- Orders, order operations, and components
- Completion confirmations
- Measuring points and measurement documents
- Task lists
- Work centers

3.9 Field Selection

Customizing Functions for Field Selection

For each of the preceding objects, the system has separate Customizing functions for controlling the field selection. However, the technique is always the same.

Prerequisites

If you want to base the field selection on influencing fields (e.g., the field selection for a functional location is based on the *functional location category* or the field selection for an order is based on the *order type*), you must define these influencing factors beforehand.

Customizing Paths

The following Customizing paths are available for field selection:

- Functional location
  Plant Maintenance and Customer Service • Master Data in Plant Maintenance and Customer Service • Technical Objects • Functional Locations • Define Field Selection for Functional Locations
- Equipment
  Plant Maintenance and Customer Service • Master Data in Plant Maintenance and Customer Service • Technical Objects • Equipment • Define Field Selection for the Equipment Master Record
- Measurement points and measurement documents
  Plant Maintenance and Customer Service • Master Data in Plant Maintenance and Customer Service • Basic Settings • Measuring Points, Counters and Measurement Documents • Define Field Selection for Measuring Points and Measurement Documents
- Material
  Logistics – General • Material Master • Field Selection • Assign Fields to Field Selection Groups, Maintain Field Selection for Data Screens, Define Industry Sectors and Industry-Sector-Specific Field Selection, and Define Plant-Specific Field Selection and Plant-Specific Screen Selection
- Task lists
  Plant Maintenance and Customer Service • Master Data in Plant Maintenance and Customer Service • Maintenance Plans, Work Centers, Task Lists, and PRTs • Task Lists • Control Data • Define Field Selection
- Work centers
  Plant Maintenance and Customer Service • Master Data in Plant Maintenance and Customer Service • Maintenance Plans, Work Centers, Task Lists, and PRTs • Work Centers • General Data • Define Field Selection
- Notifications
  Plant Maintenance and Customer Service • Maintenance and Service Processing • Maintenance and Service Notifications • Notification Creation • Notification Types • Set Field Selection for Notifications

3 Configuring Generic Functions

- Orders
  Plant Maintenance and Customer Service • Maintenance and Service Processing • Maintenance and Service Orders • Define Field Selection for Order Header Data (PM)
- Order operations
  Plant Maintenance and Customer Service • Maintenance and Service Processing • Maintenance and Service Orders • Define Field Selection for Order Operation (PM and CS)
- Order components
  Plant Maintenance and Customer Service • Maintenance and Service Processing • Maintenance and Service Orders • Define Field Selection for Components (PM and CS)
- Completion confirmation
  Plant Maintenance and Customer Service • Maintenance and Service Processing • Completion Confirmations • Set Field Selection for Completion Confirmation

Transactions

The following transactions are available for field selection:

- OIAE: Used for functional location-specific data
- OIAF: Used for fields that are common to both the functional location and the piece of equipment
- OIAD: Used for equipment-specific data
- OMSR, OMS9, OMS3, OMSA: Used for material
- OP5A through OP5H: Used for task lists
- OPFA: Used for work centers
- OIAL: Used for notifications
- OIAN and OIAZ: Used for order header data
- OIOPD and OIOPL: Used for order operations
- OICMPD and OICMPL: Used for components
- OIZN: Used for completion confirmations

Settings

Next, we'll discuss this technique using the example of field selection for functional locations.

You use the **Field Selection for Special Functional Location Fields** subfunction to define the field selection for fields that concern the functional location only.

You use the **Field Selection for Functional Location (Common Equipment/Location Fields)** subfunction to define those fields that are common to both the functional location and the piece of equipment.

3.9 Field Selection

After you select the function and choose **Influencing**, you access an overview of all influencing fields. In the case of the functional location, the following fields are suitable:

- Activity category (**Add**, **Change**, and/or **Display**)
- Object type
- Functional location category

In particular, the functional location category is frequently selected so that the field selection is based on this category. You then have the following options for each field (see Figure 3.42):

- **Input**
  The field is a normal-entry field.
- **Req.**
  The field is a required-entry field.
- **Disp.**
  The field is a display-only field.
- **Hide**
  The field is hidden.
- **HiLi**
  The field is highlighted (in a different color).

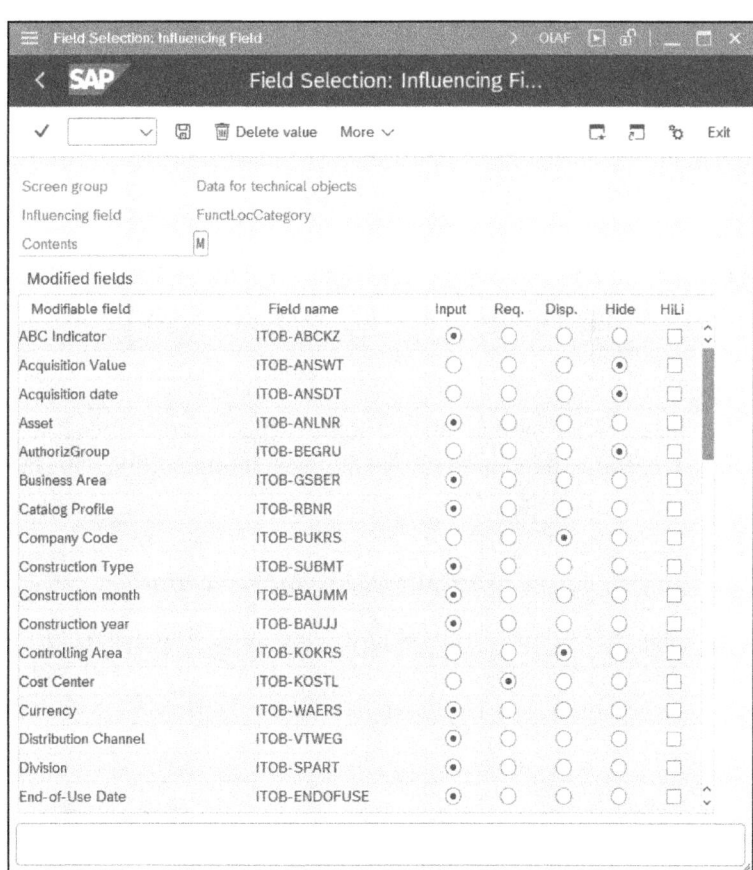

Figure 3.42 Field Selection for Functional Location

If you want to base the field selection on several influencing fields (e.g., on the functional location category and object type), then two influencing factors apply to a field. SAP has clearly defined priority rules for such cases (see Figure 3.43).

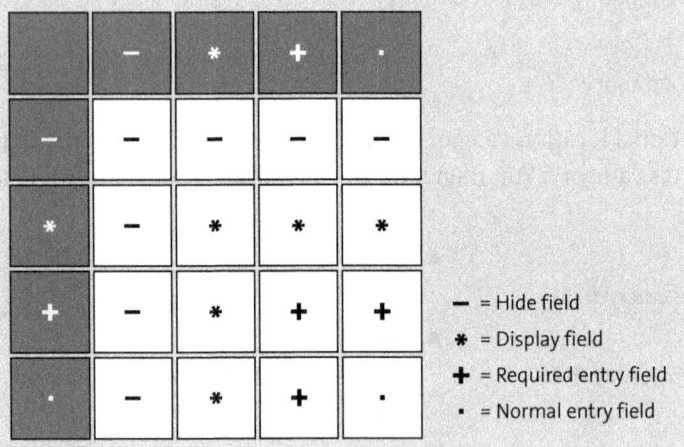

Figure 3.43 Field Selection Priority Rules

The priority rules are explained here:

- **Priority 1: Hide**
  If one of the influencing factors defines a field as hidden, then it's hidden regardless of the field selection control for other influencing fields.
- **Priority 2: Display**
  If it's not hidden and an influencing factor defines it as a display field, it becomes a display field, even if other influencing factors define it as a required-entry or normal-entry field.
- **Priority 3: Required**
  If it's not a hidden field or display field and an influencing factor defines it as a required-entry field, it becomes a required-entry field, even if other influencing factors define it as a normal-entry field.
- **Priority 4: Normal**
  If no influencing factor defines it elsewhere, this field remains a normal-entry field.

Highlighting fields in color isn't subject to any priority rules. If an influencing factor defines a field as highlighted, it's flagged with another color.

3.10 List Variants

In SAP S/4HANA Asset Management, many lists are available to you. SAP provides lists for the following objects, accessed via the transaction codes in parentheses:

- Equipment (Transactions IE05 and IH08)
- Fleet objects (Transactions IE36 and IE37)
- Serial numbers (Transactions IQ08 and IQ09)
- Functional locations (Transactions IL05 and IH06)
- Reference functional locations (Transactions IL15 and IH07)
- Object links for functional locations (Transactions IN15 and IN16)
- Object links for equipment (Transactions IN18 and IN19)
- Materials (Transaction IH09)
- Notifications (Transactions IW28 and IW29)
- Notification items (Transactions IW68 and IW69)
- Activities (Transactions IW64 and IW65)
- Tasks (Transactions IW66 and IW67)
- Orders (Transactions IW38 and IW39)
- Material availability lists (Transaction IW38A)
- Order operations (Transactions IW37 and IW49)
- Order and operations (Transaction IW37N)
- Completion confirmations (Transaction IW47)
- Measuring points (Transactions IK07 and IK08)
- Measurement documents (Transactions IK17 and IK18)
- Maintenance plans (Transactions IP15 and IP16)
- Maintenance items (Transactions IP17 and IP18)
- Task lists (Transactions IA08 and IA09)
- Task lists and operations (Transactions IA38 and IA39)
- Shift notes (Transactions SHN4, ISHN4, and SHN5)
- Shift reports (Transactions SHR4 and ISHR4)

In the case of all lists, many Customizing functions for selecting and determining the layout of the lists are available to you. Each of them is described next.

Customizing Functions for Configuring Lists

You can use these Customizing functions to define the following defaults for all users:

- Determine all potential selection fields.
- Set specific values for the selection fields.
- Perform the field selection for the list display.

These settings apply to all users who haven't created their own user-specific variants or don't use other general variants.

3 Configuring Generic Functions

Prerequisites

If you want to define a selection or display variant that refers to other Customizing functions (e.g., functional location category), you must define it beforehand.

To create list variants for task lists and operations, you must activate business function LOG_EAM_CI_14.

To create list variants for shift notes and shift reports, you must activate business function LOG_PP_SRN_CONF.

Customizing Paths

The following Customizing paths are available for list variants:

- Functional locations
  Plant Maintenance and Customer Service • Master Data in Plant Maintenance and Customer Service • Technical Objects • Functional Locations • Set List Editing for Functional Locations

- Equipment
  Plant Maintenance and Customer Service • Master Data in Plant Maintenance and Customer Service • Technical Objects • Equipment • Set List Editing for Equipment

- Serial numbers
  Plant Maintenance and Customer Service • Master Data in Plant Maintenance and Customer Service • Technical Objects • Serial Number Management • Set List Editing for Serial Numbers

- Maintenance plans
  Plant Maintenance and Customer Service • Maintenance Plans, Work Centers, Task Lists, and PRTs • Maintenance Plans •
  - Set List Editing for Maintenance Plans
  - Set List Editing for Maintenance Items
  - Set List Editing for Maintenance Item Dates

- Task lists
  Plant Maintenance and Customer Service • Maintenance Plans, Work Centers, Task Lists, and PRTs • Task Lists • Set List Editing for Task Lists

- Notifications
  Plant Maintenance and Customer Service • Maintenance and Service Processing • Maintenance and Service Notifications • Notification Processing • Define List Variants

- Orders
  Plant Maintenance and Customer Service • Maintenance and Service Processing • Maintenance and Service Orders • List Editing • Define List Variants

- Completion confirmations
  Plant Maintenance and Customer Service • Maintenance and Service Processing • Completion Confirmations • Set List Editing for Completion Confirmations

3.10 List Variants

Transactions

The following transactions are available for list editing:

- OIW6: Used for functional location lists
- OIYC: Used for equipment lists
- OIYH: Used for serial number lists
- OIWW: Used for maintenance plan lists
- OIWY: Used for maintenance item lists
- OIW5: Used for maintenance dates lists
- OIWO: Used for list of task lists
- OIWI: Used for notification lists
- OIWL: Used for order lists
- OIWU: Used for operation lists
- OIW3: Used for lists of order confirmations

Settings

Similarly, list editing, which is introduced here, applies to all objects. For the sake of simplicity, we'll use the example of a functional location to illustrate the list editing technique.

When you execute the Customizing function, you obtain the following three subfunctions:

- List for selection screen: Display mode
- List for selection screen: Change mode
- Maintain list for field selection

You should first maintain the selection screen. Then, depending on your selection, you'll obtain a list of all potential selection fields. You can then choose the **Attributes** button to access the detail screen. Here, you can decide how to proceed for each field (see Figure 3.44).

The settings on this screen have the following meanings:

- **Protect field**
  If selected, the relevant selection criterion is write protected at runtime (e.g., if a user starts the variant).

- **Hide field**
  If selected, the relevant selection criterion is hidden when a user starts a variant or changes the values.

- **Hide field 'TO'**
  If selected, the same applies here as in **Hide field**, and then it's not possible to specify an interval.

3 Configuring Generic Functions

- **Required field**
  If selected, a value must be entered for the relevant selection criterion when the user starts a variant.

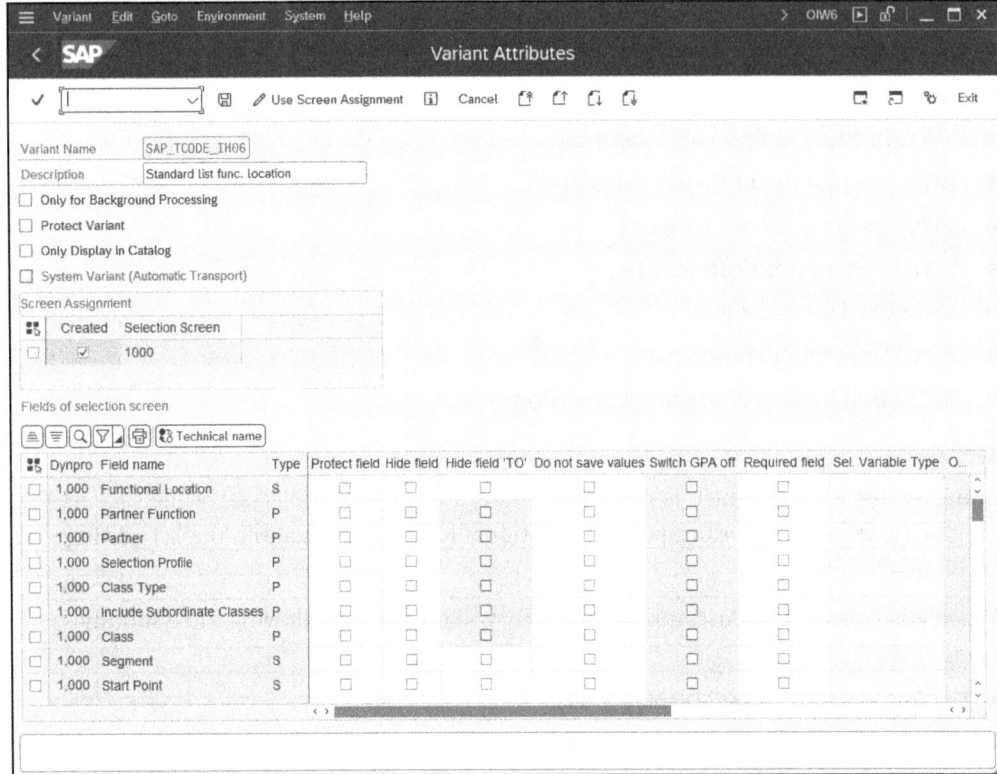

Figure 3.44 Selection Variant for Functional Location

You use the **Maintain field selection list** subfunction to determine the layout of the standard list. You have to execute the list and then choose 📋 to access a dialog box in which you can configure the following settings (see Figure 3.45):

- **Displayed Columns**
  Which fields should be displayed?
- **Sort Order**
  According to which criteria is the list to be sorted?
- **Filter**
  Is a filter to be used?
- **View**
  Do you want to output an SAP list or a Microsoft Excel table?
- **Display**
  For example, do you want a striped pattern, or do you want to optimize the column width?

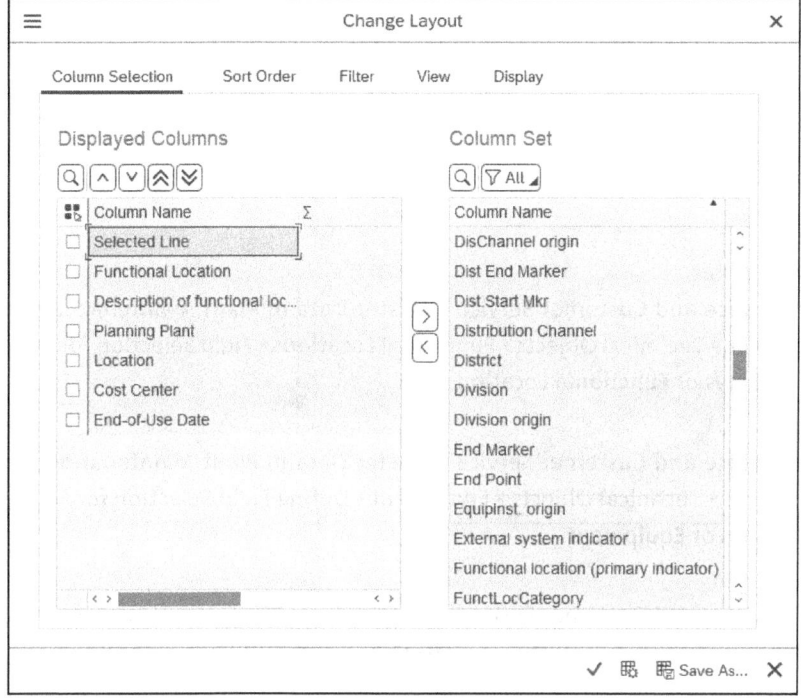

Figure 3.45 List Variant for Functional Location

3.11 Multilevel Lists

From an SAP S/4HANA Asset Management perspective, multilevel lists are available for the following transactions:

- IE07: Used for equipment
- IL07: Used for functional locations
- IH18: Used for reference functional locations
- IW30: Used for notifications
- IW40: Used for orders
- IA10: Used for task lists

In the case of a list object (e.g., a functional location), multilevel lists are used to display other objects (e.g., orders, notifications, documents) for a functional location.

Customizing Functions for the Field Selection for Multilevel List Displays

You use this Customizing function to adjust the defaults for multilevel lists. Here, you define which fields are to be displayed for which objects. These settings apply to all users who haven't created their own user-specific variants or don't use other general variants.

The multilevel list technique is available to you for all the preceding SAP objects. For the sake of simplicity, we'll use the example of a functional location to illustrate the multilevel list technique.

Prerequisites

There are no specific prerequisites.

Customizing Paths

- Functional location
  Plant Maintenance and Customer Service • Master Data in Plant Maintenance and Customer Service • Technical Objects • Functional Locations • Field Selection for Multilevel List Displays of Functional Location

- Equipment
  Plant Maintenance and Customer Service • Master Data in Plant Maintenance and Customer Service • Technical Objects • Equipment • Define Field Selection for Multilevel List Displays of Equipment

- Task lists
  Plant Maintenance and Customer Service • Maintenance Plans, Work Centers, Task Lists, and PRTs • Task Lists • Presetting for List Display of Multilevel Task Lists

- Notifications
  Plant Maintenance and Customer Service • Maintenance and Service Processing • Maintenance and Service Notifications • Notification Processing • Define Field Selection for Multilevel List Displays of Notifications

- Orders
  Plant Maintenance and Customer Service • Maintenance and Service Processing • Maintenance and Service Orders • Field Selection for Multilevel List Displays of Orders

The following subfunctions are available for the individual objects:

- Define Field Selection for Functional Location Fields
- Define Field Selection for Fields of the Reference Functional Location
- Define Field Selection for Equipment Usage Data Fields
- Define Field Selection for Equipment Master Data Fields
- Define Field Selection for Partner Data Fields
- Define Field Selection for Notification Data Fields
- Define Field Selection for Order Data Fields
- Define Field Selection for Class Data Fields
- Define Field Selection for Characteristic Data Fields
- Define Field Selection for Document Management Fields
- Define Field Selection for Object Link Data Fields

- Define Field Selection for Measuring Point and Counter Fields
- Define Field Selection for Measurement Document and Counter Reading Fields
- Define Field Selection for Permit Fields

Settings

Figure 3.46 shows the settings for the order segment in the multilevel list for functional locations. However, the following are representative of all other objects:

- DispPosition
  You define, as a presetting, the position at which the field is to appear. By default, fields without an item number aren't displayed. When designing their individual layouts, users can change or delete the item number as well as enter an item number for other fields so that they are displayed.
- Invisible
  You define whether the field is to be invisible for user-specific list designs. If a field is invisible, a user can't select it for their variant.

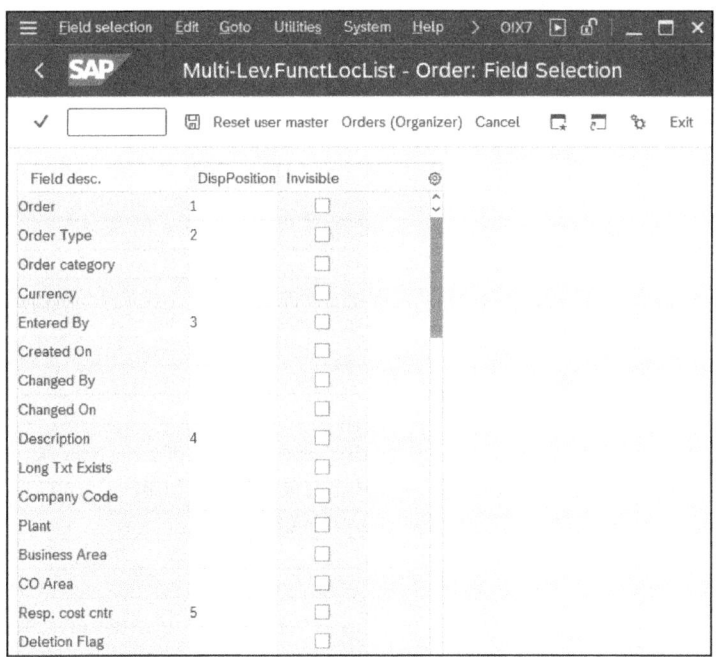

Figure 3.46 Order Segment for Multilevel Lists

3.12 Change Documents

For many SAP objects, changes are documented in so-called *change documents*, which show who made which changes and when. In SAP S/4HANA Asset Management, change documents are created for the following objects:

3   Configuring Generic Functions

- Equipment
- Functional locations
- Reference functional locations
- Notifications
- Orders
- Task lists
- Maintenance plans

In most cases, the change documents are presented as an action log. Figure 3.47 shows the **Action Log** of a piece of equipment.

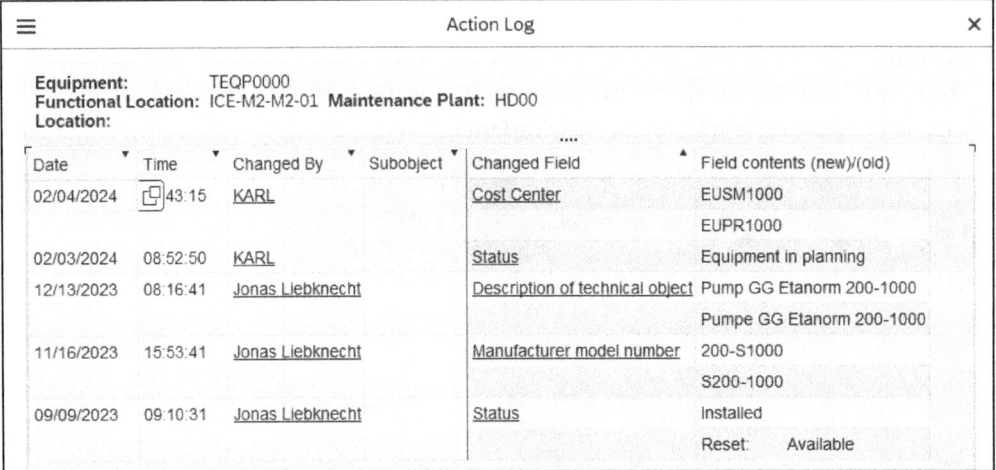

Figure 3.47 Equipment Action Log

Customizing Functions for Change Documents

For some objects (such as for notifications), change documents are automatically created. For other objects (such as orders, functional locations, and equipment), the creation of change documents must be activated using a customizing function.

Prerequisites

To generate change documents for maintenance plan scheduling, you must activate business function LOG_EAM_CI_4.

Customizing Paths

There are several customizing functions for activating change documents:

- **Functional location**
  Plant Maintenance and Customer Service • Master Data in Plant Maintenance and Customer Service • Technical Objects • Functional Locations • Define Category of Functional Location

- Equipment
  Plant Maintenance and Customer Service • Master Data in Plant Maintenance and Customer Service • Technical Objects • Equipment • Equipment Categories • Maintain Equipment Category
- Work orders
  Plant Maintenance and Customer Service • Maintenance and Service Processing • Maintenance and Service Orders • Functions and Settings for Order Types • Define Change Documents, Collective Purchase Requisition, MRP Relevance
- Maintenance plans
  Plant Maintenance and Customer Service • Master Data in Plant Maintenance and Customer Service • Maintenance Plans, Work Centers, Task Lists, and PRTs • Maintenance Plans • General Data • Set Maintenance Plan Categories
- Work centers
  Plant Maintenance and Customer Service • Master Data in Plant Maintenance and Customer Service • Maintenance Plans, Work Centers, Task Lists, and PRTs • Work Centers • General Data • Define Work Center Types and Link to Task List Application

Settings

Figure 3.48 shows the settings how to activate change documents for pieces of equipment using the **C** indicator (change document) per equipment category.

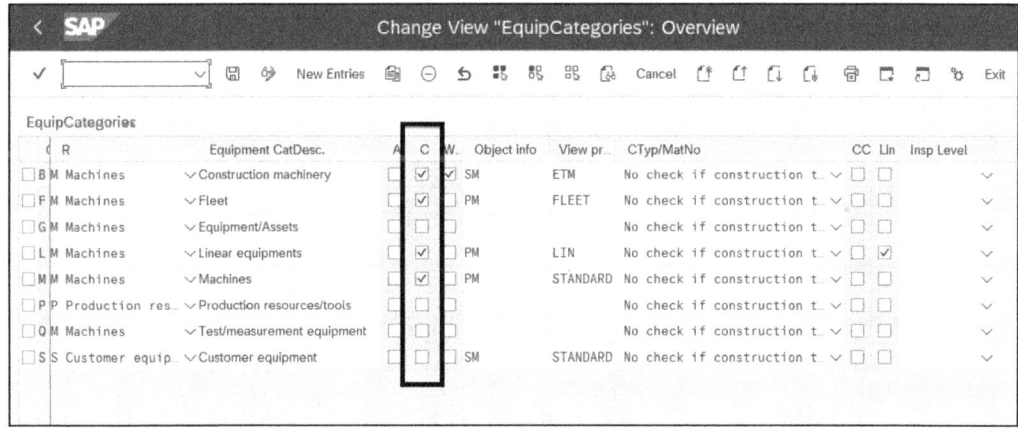

Figure 3.48 Change Documents for Equipment Category

3.13 Summary

In SAP S/4HANA Asset Management, the following functions are used in several locations or for several objects, so they have a generic character:

- Object information
- Status management

- Number assignment
- Warranties
- Measuring points and counters
- Permits
- Partners
- Documents
- Field selection
- List variants
- Multilevel lists
- Change documents

Each of these functions (e.g., multilevel lists) is available for at least four of the SAP S/4HANA Asset Management objects, while some of the functions (e.g., list variants and documents) are available for all SAP S/4HANA Asset Management objects.

In this chapter, you learned how to configure these generic functions mostly based on only one object as an example, but this example is applicable to all the other objects.

Chapter 4
Configuring the Structure of Technical Systems

This chapter introduces you to the structuring elements of SAP S/4HANA Asset Management. You'll learn how to configure these elements, and you'll find tips that deal with configuring the Customizing settings for asset structures.

A requirements-oriented approach to the structuring of technical systems is the basis for being able to use SAP S/4HANA Asset Management to map and subsequently process business processes in plant maintenance. The following issues are discussed in this chapter:

- Structuring elements to use
- Functional locations
- Reference functional locations
- Equipment
- Fleet management
- Links and networks
- Linear asset management
- Materials and assemblies
- Serial numbers
- Bills of materials (BOMs)

4.1 Structuring Elements

SAP S/4HANA Asset Management provides you with a broad range of potential *structuring elements*, including functional locations, reference functional locations, equipment, object links, linear assets, serial numbers, maintenance assemblies, materials, and different types of BOMs.

Functional locations represent a complex, generally multilevel asset structure, and you must create each element of the asset structure as a functional location. Therefore, functional locations are used to establish a vertical asset structure. Functional locations also usually represent immovable, functional units, and examples of functional locations include process plants in the chemical and pharmaceutical industries, power plants, production lines, buildings, pipeline systems, infrastructure, and computer networks.

Reference functional locations are solely templates for generating real functional location structures or for subsequently passing on data to real functional locations.

Equipment represents movable, individual aggregates (inventories). Examples include machines, pumps, engines, production resources/tools (PRTs), fleet objects (cars, trucks, forklifts, and industrial trucks), and IT inventories (PCs, printers, monitors, notebooks, and projectors). Equipment categories that are moved infrequently (e.g., pumps) are intended to be installed in functional locations, while other equipment categories (e.g., fleet objects) aren't installed in functional locations due to their constant movement.

You establish *object links* between different technical objects (e.g., pieces of equipment or functional locations). Such links exist, for example, among individual production units, production plants and supply systems, and supply systems and disposal systems. You use object links to form an *object network*, and in this way, your assets can have a horizontal structure. Object links can't be the subject of business processes (e.g., malfunction reports). Rather, they are used for information and visualization purposes only.

Linear assets (linear asset management) are technical systems with a linear infrastructure whose properties and conditions may change from stage to stage (dynamic segmentation). Examples of linear assets include pipelines, road networks, railway networks, overhead cables, pipes, and so on. You can create linear assets as technical objects (e.g., functional locations or pieces of equipment) and store linear data there.

Unlike a piece of equipment, a *material* doesn't represent an individual item but rather an object category (e.g., the *pump normal 400–100* category or the *three-phase normal engine SM/I, 220/380 V, 50 Hz, 0.18 kW* category). A material includes a specific number of objects of the corresponding category. You require materials for spare parts and for pieces of equipment and maintenance assemblies that are suitable for storage.

A *maintenance assembly* serves to provide a functional location or piece of equipment with a deeper structure. For example, a forklift could comprise the following maintenance assemblies: lift type, chassis, brake system, and drive assembly. You can also assign a maintenance assembly to a notification, order, or maintenance plan to specify the location of any damage that occurs.

You create *serial numbers* for a material number. Here, you can specify as many serial numbers as you want for one material number. A serial number is an individual item and corresponds to a piece of equipment, and the serial number function enables you to place the equipment in storage. In terms of usage, a material serial number corresponds to a piece of equipment.

An *equipment BOM* generally comprises a list of spare parts, and it's assigned directly to a piece of equipment. This means that only this piece of equipment can use this BOM.

A *BOM for a functional location* generally comprises a list of spare parts, and it's assigned directly to a functional location. This means that only this functional location can use this BOM.

A *material BOM* also comprises a list of spare parts, but you can make the material BOM available for any number of pieces of equipment or functional locations by means of an *indirect assignment*.

In the following sections, you'll learn about the main settings for the individual technical objects.

4.2 Functional Locations and Reference Functional Locations

Functional locations represent a complex, generally multilevel, immovable asset structure. Therefore, functional locations are used to establish a vertical asset structure.

Reference functional locations are solely templates for generating real functional location structures or for subsequently passing on data to real functional locations. Reference functional locations can't be the subject of business processes (e.g., malfunction reports). Rather, they are used as a reference template.

Practical examples of functional locations include the following:

- Processing plants in the chemical, pharmaceutical, and food industries
- Power plants such as coal-fired, hydroelectric, and nuclear power plants
- Production lines in discrete manufacturing
- Complex machines such as automated systems and flexible production cells
- Real estate
- Power systems, gas systems, water systems, and heating systems
- Pipeline systems
- Infrastructure such as roads, public places, tracks, tunnels, and bridges
- Computer networks
- Complex fleet objects such as locomotives, high-speed trains, and tractors
- Aircraft

You should create functional locations in the following cases:

- You want to manage individual data from a maintenance perspective (technical data such as performance data or organizational data such as the work center).
- You want to create notifications, orders, or maintenance plans.
- You want to manage an asset pool.

4   Configuring the Structure of Technical Systems

- You have documentation obligations, so providing evidence of performance of maintenance tasks is mandatory.
- You want to collate and analyze technical data (such as causes of damage, measurement readings, or counter readings).
- You want to verify costs.
- Different perspectives among technical systems are required (for example, one perspective for electrical engineering and one perspective for measurement and control).

Figure 4.1 provides an overview of the Customizing functions for functional locations.

Figure 4.1 Functional Locations: Customizing Functions Overview

The most important of these Customizing functions are covered in the following sections.

Create a Structure Indicator for Reference Locations/Functional Locations

You always assign an external number to a functional location and reference a functional location in accordance with the *structure indicator*. You manually assign this functional location number according to the specifications made using the structure indicator.

Prerequisites
There are no special prerequisites.

4.2 Functional Locations and Reference Functional Locations

Customizing Path

Plant Maintenance and Customer Service • Master Data in Plant Maintenance and Customer Service • Technical Objects • Functional Locations • Create Structure Indicator for Reference Locations/Functional Locations

Transaction

OIPK

Settings

Figure 4.2 provides an overview of existing structure indicators. If you create a new structure indicator, you define the name and description of the structure indicator in the Customizing table.

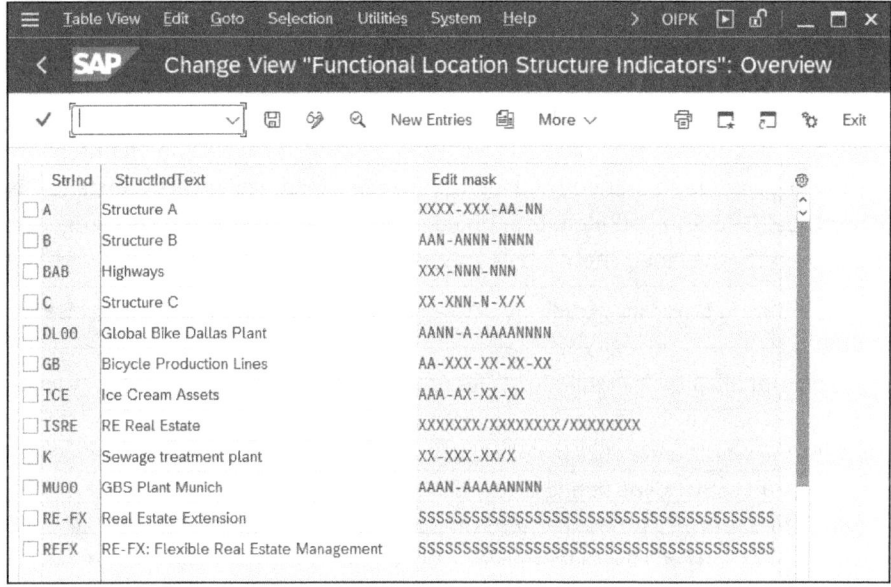

Figure 4.2 Functional Locations: Structure Indicators Overview

On the detail screen, you configure the following additional settings (see Figure 4.3):

- Length of the functional location number
- Number of levels for the asset structure
- Number of characters for the individual level numbers
- Rule used to establish the functional location number:
  - A: Alphabetic characters
  - N: Numeric characters
  - X: Alphanumeric characters
  - S: Special characters and alphanumeric characters

4 Configuring the Structure of Technical Systems

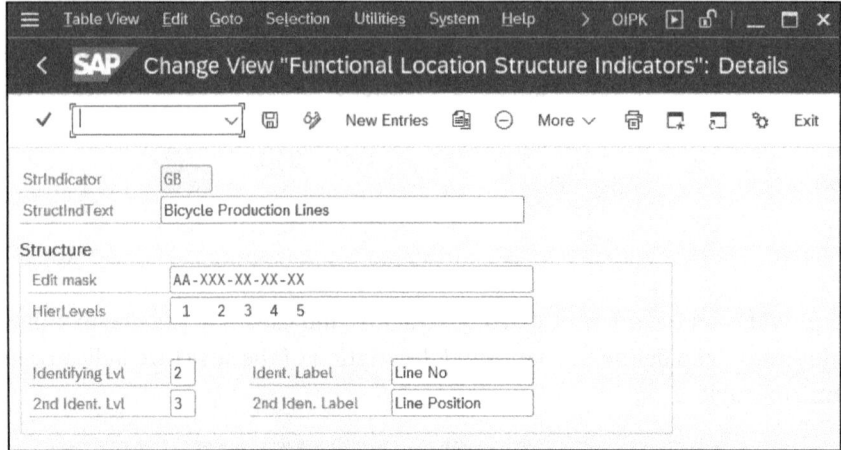

Figure 4.3 Functional Locations: Structure Indicators Detail Screen

Finally, you define the number and the label, the latter of which identifies the level in the structure of a functional location. The label will be displayed in maintenance objects and documents so that you can identify where a technical object is installed. In the **Ident. Label** field and **2nd Iden. Label** field, you can provide descriptive language-dependent text for the label. These labels are then displayed on the user interface.

> [+] **Example**
>
> The structure of the **Bicycle Production Lines** functional location has four hierarchy levels. At the second level, you have the line number, such as assembly line 1. At the third level, there are work centers like assembly station brakes used in different functional locations. In this activity, you define the second level as the identifying level, and when you search for a brakes station, the resulting list also shows the second level of the structure so that you can easily identify in which functional location brake stations are installed.

Set View Profiles for Technical Objects

You can use this Customizing function to define the layout of screens associated with technical objects (e.g., equipment, functional locations). You determine the following:

- Number of tabs
- Which screen groups to position on which tab and where
- Names of tabs
- Whether an icon is to be assigned to a tab

Prerequisites

There are no special prerequisites.

4.2 Functional Locations and Reference Functional Locations

Customizing Path

Plant Maintenance and Customer Service • Master Data in Plant Maintenance and Customer Service • Technical Objects • General Data • Set View Profiles for Technical Objects

Settings

First, define a name for the view profile. Then, specify whether you want it to be valid for equipment or functional locations (see Figure 4.4).

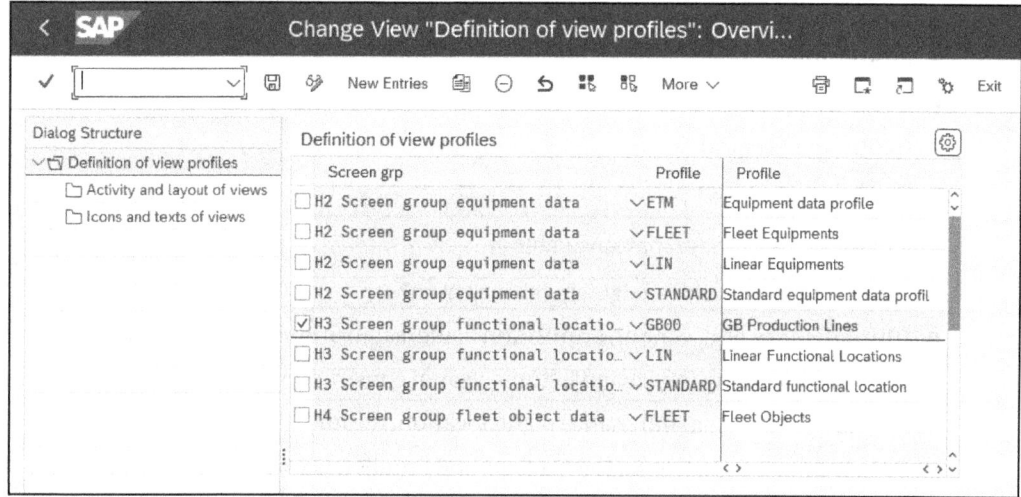

Figure 4.4 Functional Locations: View Profiles Overview

Next, use the **Activity and layout of views** subfunction to assign the relevant screen areas to the tabs (see Figure 4.5).

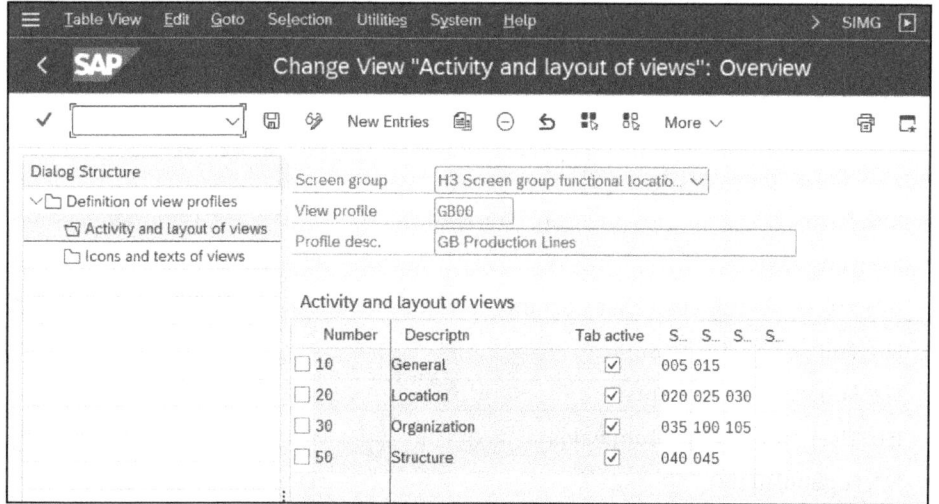

Figure 4.5 Functional Locations: View Profiles Screen Layout

The following tabs are available:

- 0010: General
- 0020: Location
- 0030: Organization
- 0050: Structure
- 0060: Additional data 1
- 0080: Additional data 2
- 0085: Additional data 3
- 0090: Other
- 0300: Linear data
- 0400: Asset central
- 0500: Spatial editor

The following screen areas are available:

- 005: **General** (e.g., size, weight, inventory number)
- 010: **Reference** (e.g., vendor, acquisition value, acquisition date)
- 015: **Manufacturing** (e.g., manufacturer, year of construction, month of construction)
- 020: **Location** (e.g., maintenance plant, location, room)
- 025: **Address** (e.g., zip code, town, telephone number, fax number)
- 030: **Account assignment** (e.g., cost center, asset, company code)
- 035: **Responsibilities** (e.g., planner group, responsible work center)
- 040: **Structuring** (e.g., superior functional location, position)
- 045: **Equipment** (e.g., installed equipment)
- 095: Full-screen standard class
- 096: Standard class with 20 characteristics
- 097: Subscreen standard class
- 098: Classification
- 099: Class characteristics
- 100: **Partner** (e.g., partner function, name, address)
- 105: Long text
- 110: **Warranty** (e.g., customer warranty, vendor warranty)
- 120: Full-screen documents
- 122: Subscreen documents
- 150: User data with custom fields
- 141: Logbook data
- 300: Linear data for functional locations

4.2 Functional Locations and Reference Functional Locations

- 350: Linear data for characteristics
- 380 to 388: Asset central subscreens
- 500: Geospatial editor

Finally, you can use the **Icons and texts of views** subfunction to define your own titles for the tabs and to assign an icon (see Figure 4.6). You can choose from more than 1,000 predefined icons.

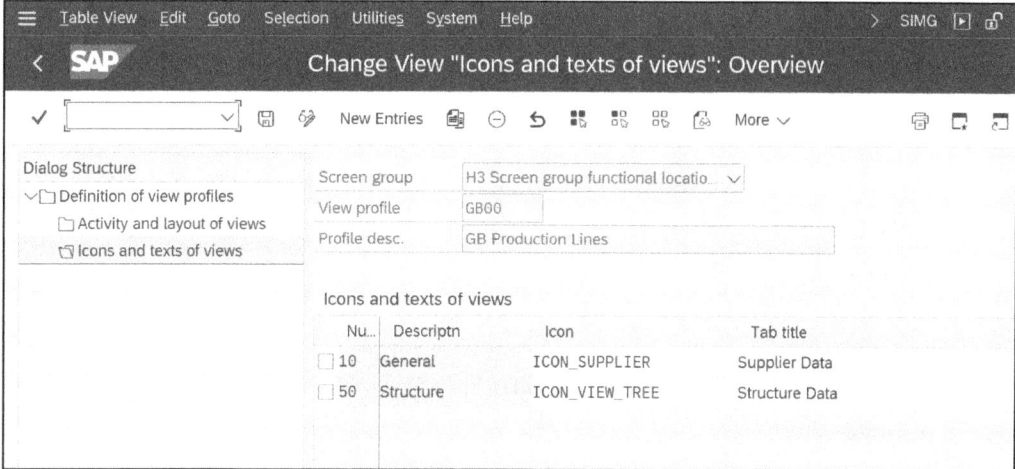

Figure 4.6 Functional Locations: View Profile with Icons and Texts

Define Category of Functional Location

You can use this Customizing function to define different functional location categories.

Prerequisites

If you want to assign a status profile, partner determination procedure, measuring point category, and object information key, you must define them beforehand.

If you want to define a functional location category for linear asset management, you must activate business functions LOG_EAM_LINEAR_1 and LOG_EAM_LINEAR_2.

Customizing Path

Plant Maintenance and Customer Service • Master Data in Plant Maintenance and Customer Service • Technical Objects • Functional Locations • Define Category of Functional Location

Settings

Figure 4.7 provides an overview of the functional location categories. Here, you can define the name and description of each category.

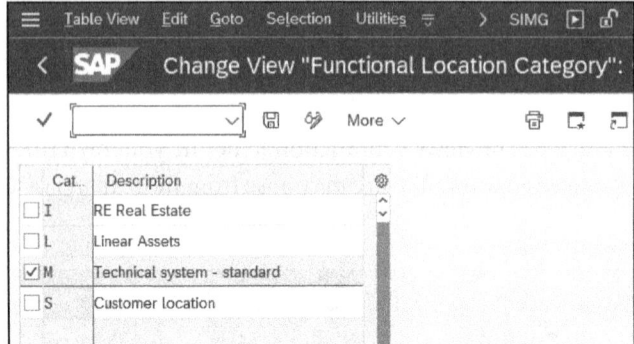

Figure 4.7 Functional Location: Category Overview

On the detail screen, you can determine the following information for each category (see Figure 4.8):

- **ChangeDocuments**
  Change documents generated when changing data.
- **CustObject**
  The category concerns customer assets.
- **Linear Asset**
  Linear data to enter for the functional locations.
- **Inspection Level**
  Allows you to specify the level of an object for inspection purposes. Inspections can be executed on different levels of an asset, and while scheduling an inspection for an asset, you can select any of these objects and all the objects underneath will be included.
- **Status Profile**
  The status profile to be assigned.
- **PartnDet.Proc.**
  The partner determination procedure to apply.
- **MeasPtCategory**
  The measuring point category to use.
- **View profile**
  The layout of the master record.
- **Change Docs During Creation**
  Change documents generated when creating a new functional location.
- **Object info**
  Information to be displayed as object information.

You can also base the field selection on the functional location category.

4.2 Functional Locations and Reference Functional Locations

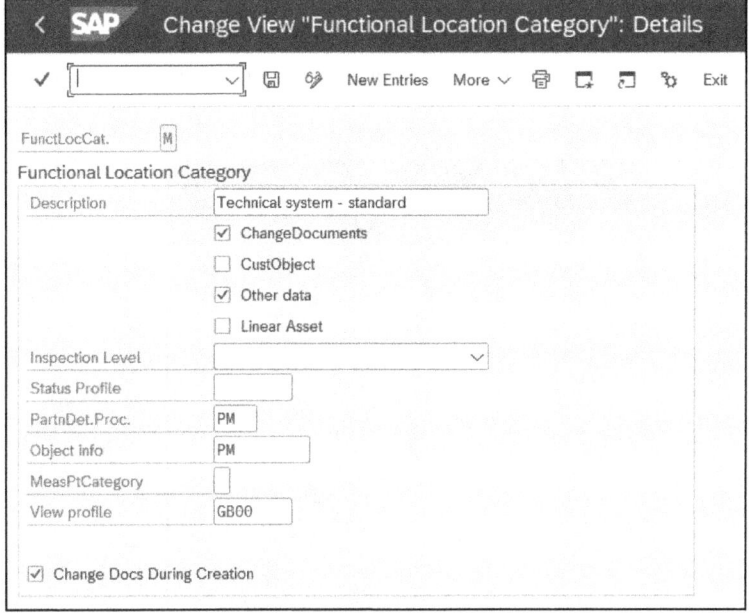

Figure 4.8 Functional Location: Category Details

Define Structural Display for Functional Locations

You can use this Customizing function to define the layout of the structural display for functional locations in Transaction IH01 (Functional Location Structure).

The fields you select are used for the field selection. If a user doesn't define their own field selection, the system displays the fields that you've activated.

Prerequisites
There are no special prerequisites.

Customizing Path
Plant Maintenance and Customer Service • Master Data in Plant Maintenance and Customer Service • Technical Objects • Functional Locations • Define Structural Display for Functional Locations

Transaction
OIWP

Settings
You can use numbering to define the order in which the fields are displayed in the list. In other words, the field assigned the number 1 is the first field displayed in the list, the field assigned the number 2 is the second field displayed in the list, and so on (see Figure 4.9).

Each user can design their user-related list based on the field list that you've defined. Here, however, the user can only work with those fields that are visible. In other words, an individual user can't activate fields marked as **Invisible** for their own user-specific field selection.

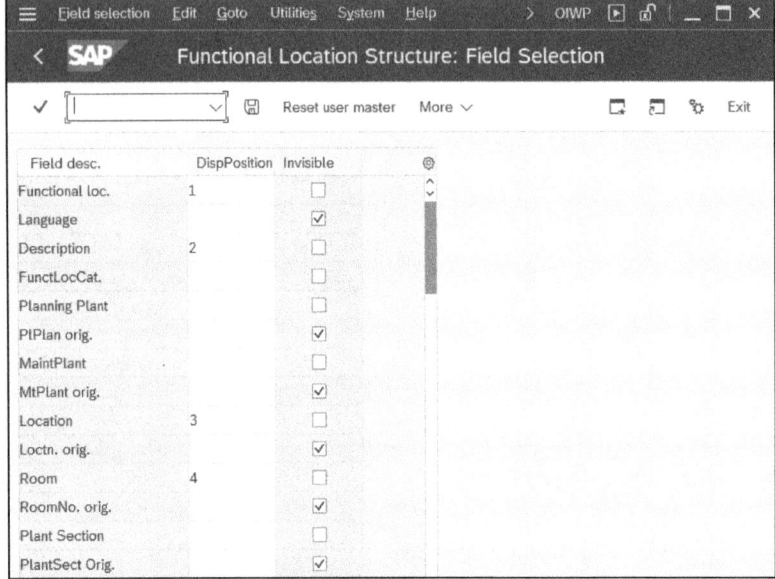

Figure 4.9 Functional Location: Structure Field Selection List

Alternative Labeling of Functional Locations

You should activate alternative labeling if you want to do any of the following later:

- Assign several numbers to a functional location structure or to each individual location for different perspectives (e.g., own naming conventions versus customer naming conventions)
- Create primary and secondary functional location structures (e.g., a primary structure for a power supply and a secondary structure for building services)
- Change the number assigned to a functional location (e.g., if you've assigned an incorrect number or you're assigning parts of an asset to another asset)

Prerequisites

There are no special prerequisites for simply activating alternative labeling. If, however, you want to use several labeling systems, you require the corresponding structure indicators (see the start of Section 4.2, where we discuss the structure indicator).

Customizing Path

Plant Maintenance and Customer Service • Master Data in Plant Maintenance and Customer Service • Technical Objects • Functional Locations • Alternative Labeling of Functional Locations • Define Labeling Systems for Functional Locations

4.2 Functional Locations and Reference Functional Locations

Transactions

There are two transactions used for labeling functional locations:

- OIPU, used for activating alternative labeling
- OIPV, used for defining labeling systems for functional locations

Settings

To activate alternative labeling, set the **Alt. label active** (activate alternative labeling) indicator (see Figure 4.10).

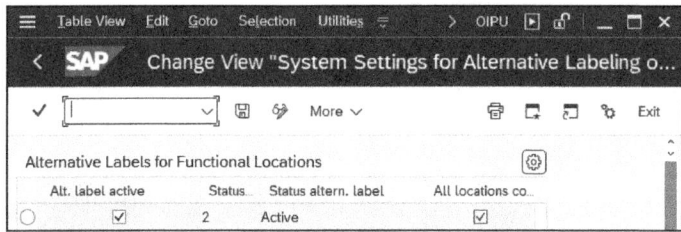

Figure 4.10 Functional Locations: Alternative Labeling Activation

The following consequences are associated with activating alternative labeling:

- You can change the number of the functional location.
- Then, the following two database tables for managing the functional locations are activated: IFLOS (Labels for Functional Locations) and IFLOT (Root Table for Functional Locations).
- Table IFLOS contains the internal number of the functional location TPLNR and a reference to the number STRNO, which can be displayed on the interface and may change over time. Figure 4.11 depicts such a scenario: during processing, the functional location number ?0100000000000000006 was changed from **ICE-FR-03** to **ICE-FR-04**.
- Table IFLOT contains the actual information about the functional location (maintenance planning plant, maintenance planner group, description, etc.).
- If you activate alternative labeling retroactively (i.e., if you've already entered functional locations), you must run report RI_IFLOT2IFLOS. The program generates an entry in table IFLOS for each entry in table IFLOT. Each time you enter a new functional location, the system automatically creates two entries—one in table IFLOS and one in table IFLOT.
- You can set up several labeling systems for functional locations.

As shown in Figure 4.12, you use the **Define Labeling Systems for Functional Locations** Customizing function to define alternative labeling systems (e.g., or if you want to differentiate between internal structures and customer structures, or if you want to define parallel views—a mechanical view and an electrical view—for the functional locations).

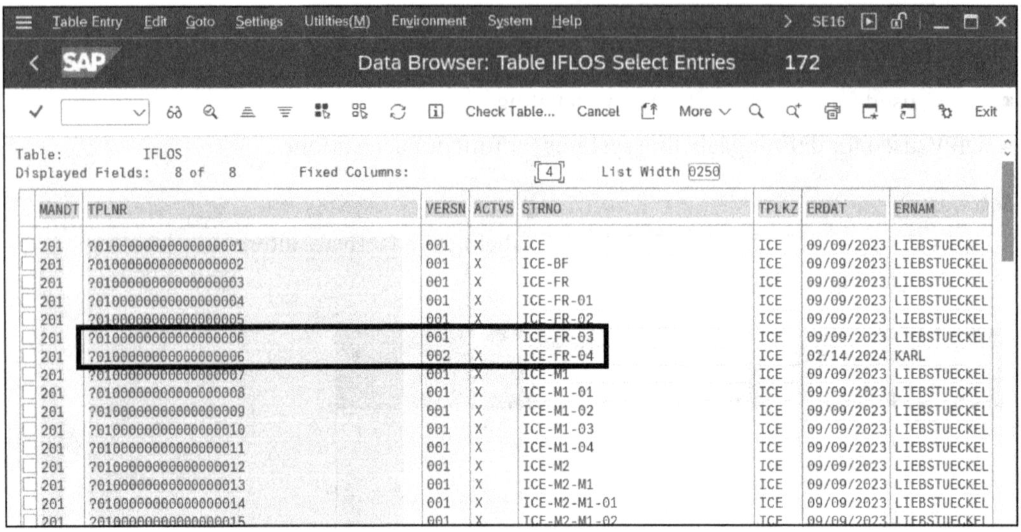

4 Configuring the Structure of Technical Systems

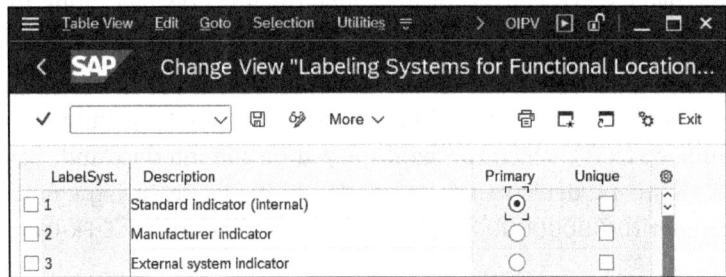

Figure 4.11 Table IFLOS

You can also define whether you want to check the uniqueness of one labeling system against all other labeling systems. A uniqueness check is always performed within a labeling system.

Figure 4.12 Functional Location: Alternative Labeling Systems

One of the labeling systems must be selected as the main labeling system. This should be the system that is used by most users.

4.3 Equipment

A piece of *equipment* represents a movable, individual aggregate (inventory). Examples of equipment include the following:

- Machines
- Means of production (pumps and engines)
- Production resources/tools (PRTs)

- Test equipment (scales and gauges)
- Fleet objects (cars, trucks, forklift trucks, and industrial trucks)
- IT inventories (PCs, printers, monitors, notebooks, and projectors)
- Robots

You create a piece of equipment in the following scenarios:

- You install the devices in functional locations and want to provide evidence of a usage history.
- You want to store the devices.
- You want to manage individual data (technical data such as performance data or organizational data such as the work center) from a maintenance perspective.
- You want to open notifications, orders, or maintenance plans.
- You have documentation commitments and must provide evidence of maintenance tasks performed.
- You want to collate and analyze technical data such as causes of damage, measurement readings, or counter readings.
- You want to verify costs.
- You want to refurbish a device.
- You want to subcontract an external supplier.
- You manage devices in a pool and want to support the borrowing process.
- You want or need to perform calibrations or checks.

Figure 4.13 provides an overview of the Customizing functions for equipment.

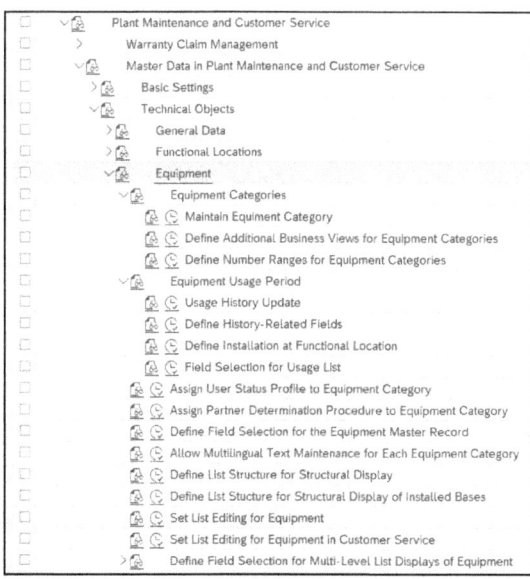

Figure 4.13 Equipment: Customizing Functions Overview

Now, let's discuss the most important of these Customizing functions (which will be covered in the following sections). The equipment category is at the heart of Customizing for equipment (see Figure 4.14).

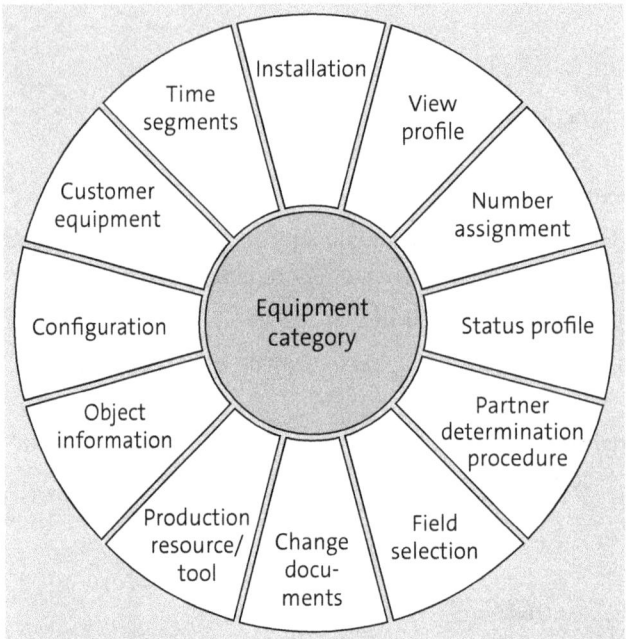

Figure 4.14 Equipment Category

You can use the *equipment category* to define the following:

- The number assignment or which equipment numbers can be assigned
- Whether change documents are written
- Which information is displayed as object information
- The layout of the equipment master (view profile) and which fields are to be maintained (field selection)
- Whether the equipment category concerns a linear object
- Which user status profile is used
- Whether the piece of equipment is a PRT
- Which partner determination procedure is used
- Whether the equipment category concerns a piece of customer equipment
- Whether the piece of equipment can be configured
- Whether time segments are written for changes and, if so, for which fields
- Whether the piece of equipment is permitted to be installed on functional locations

4.3 Equipment

Maintain Equipment Category

You use this Customizing function to define equipment categories. This enables you to control which basic properties you want a corresponding equipment master record to have. You can use control properties to define both the number and the type of equipment categories you require.

Prerequisites

To assign equipment categories, you must define the relevant object information keys and view profile beforehand.

If you want to define an equipment category for linear asset management, you must activate business functions LOG_EAM_LINEAR_1 and LOG_EAM_LINEAR_2.

Customizing Path

Plant Maintenance and Customer Service · Master Data in Plant Maintenance and Customer Service · Technical Objects · Equipment · Equipment Categories · Maintain Equipment Category

Settings

The following settings are available to you in the columns shown in Figure 4.15:

- **R**

  You assign a reference category that is used to define history-relevant fields in an equipment master record. The following three reference categories are available: **M Machines**, **S Customer equipment**, and **P Production Resources/Tools**. **M Machines** is the default reference category.

- **A**

  In the case of external assignment, you can prohibit alphanumeric numbers.

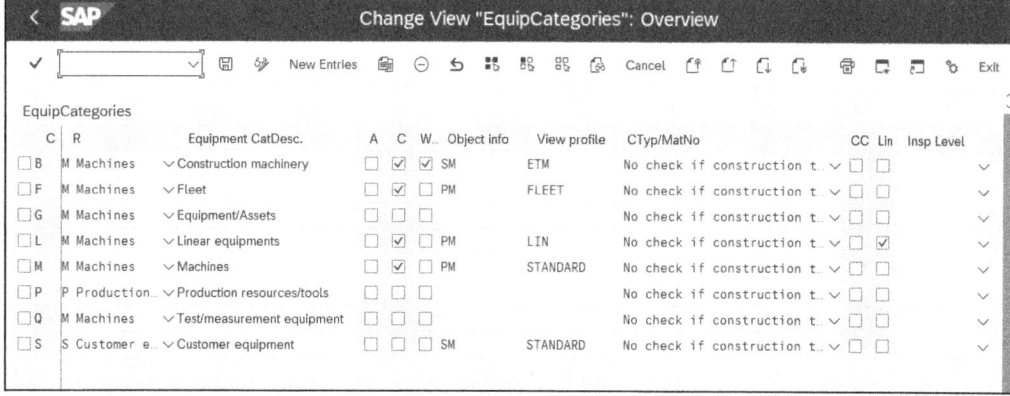

Figure 4.15 Equipment Categories: Overview

4 Configuring the Structure of Technical Systems

- **C**
  You can activate change documents for verifying master data changes.
- **W...**
  You can trigger certain workflow events during master data maintenance.
- **Object info**
  You assign the **Object info** key.
- **View profile**
  You assign the view profile.
- **CTyp/MatNo**
  You decide whether you want to check the synchronization between the material number and the construction type. If you do, you define whether you want the system to issue a warning message or error message if there is an inconsistency.
- **CC**
  You define whether a change document is to be generated when creating a piece of equipment.
- **Lin**
  You determine whether you can enter linear data for a piece of equipment.
- **Insp Level**
  This allows you to specify the level of an object for inspection purposes. Inspections can be executed on different levels of an asset, and while scheduling an inspection for an asset, you can select any of these objects. All the objects underneath will be included.

You can also base the equipment category on the field selection. (For more information, see Chapter 3, Section 3.9, which provides information about the field selection.)

Define Additional Business Views for Equipment Categories

In this work step, you can define additional views for each equipment category. The views represent additional tabs within the equipment master record.

Prerequisites

You must define the equipment categories beforehand.

Customizing Path

Plant Maintenance and Customer Service • Master Data in Plant Maintenance and Customer Service • Technical Objects • Equipment • Equipment Categories • Define Additional Business Views for Equipment Categories

Settings

You can activate or deactivate the following views in the columns shown in Figure 4.16:

4.3 Equipment

- **Production Res/Tools**
  Activate the PRT view if you want to perform calibration of the test and measurement equipment business process. For more information, see Chapter 7, Section 7.4.
- **S**
  Activate the **S** (sales data) view if you want to maintain fields such as **Sales organization**, **Division**, **Sold-to party**, and **Ship-to party**.
- **C**
  Activate the **C** (equipment configuration) view if you want to assign a configurable material to the piece of equipment.
- **SD**
  Activate the **SD** (serial data) view if you want to store serial data in addition to the normal equipment data (e.g., the material number, serial number, stock type, storage location, stock batch) or if you want to generate a serial number history.
- **Other data**
  Activate the **Other data** view to see an additional, freely configurable tab.
- **CC**
  Activate the **CC** (configuration control) view.

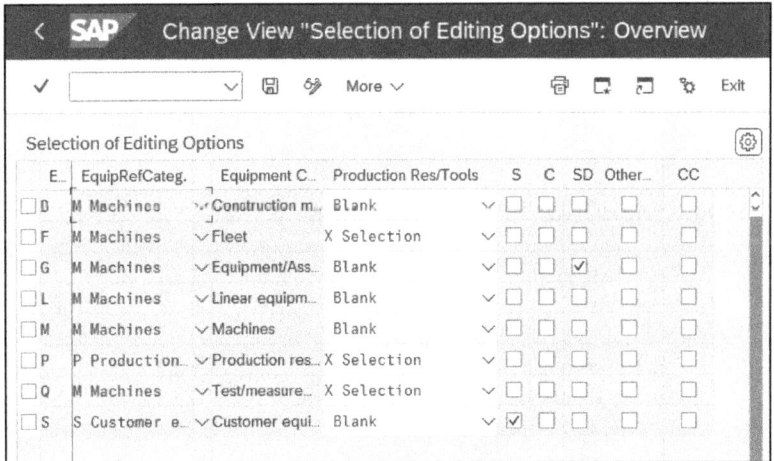

Figure 4.16 Equipment Category: Additional Business Views

When you activate a view, the assigned data fields and screens are automatically offered during maintenance. If the views are deactivated, you can maintain the data fields and screens, but you must explicitly call them using a function key.

Usage History Update

In this work step, you define whether a usage history is to be created for each equipment category. In contrast to the change documentation associated with change documents, the system not only stores the field contents of the fields both before and after the change but also retains the time-dependent data fields for the entire equipment master record when the change is made. However, you can't base the usage history on all fields. Rather, you can base it only on the time-dependent data. The former includes, among others, the description, class, manufacturer, or warranty data. The latter includes, among others, the functional location, cost center, responsible work center, or maintenance planner group. In the **Define History-Related Fields** Customizing function, you define those fields for which you want the usage history to be updated for each reference category.

Prerequisites

You must define the equipment categories beforehand.

Customizing Path

Plant Maintenance and Customer Service • Master Data in Plant Maintenance and Customer Service • Technical Objects • Equipment • Equipment Usage Period • Usage History Update

Settings

Select the **Time seg.** checkbox for those equipment categories for which you want the usage history to be updated (see Figure 4.17).

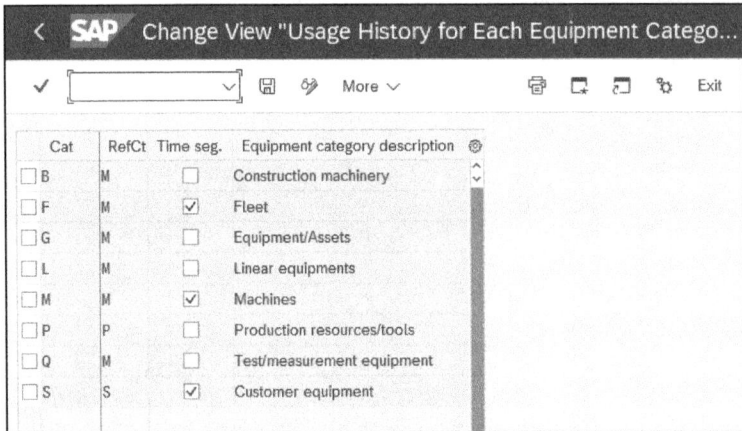

Figure 4.17 Equipment Category: Usage History

Define History-Related Fields

In this work step, you define the reference category for which you want the usage history to be updated when you change fields on the basis of the **Usage History Update** Customizing function. Note that the usage history is always updated when you change the fields in the **Location** and **Structuring** screen areas if you've activated the update. These include the location, maintenance plant, functional location, or superior piece of equipment, for example.

4.3 Equipment

Prerequisites

If you want to perform an update, you must activate the update for each equipment category beforehand.

Customizing Path

Plant Maintenance and Customer Service • Master Data in Plant Maintenance and Customer Service • Technical Objects • Equipment • Equipment Usage Period • Define History-Related Fields

Transaction

OIEZ

Settings

For each reference category, select those fields for which you want the usage history to be updated (see Figure 4.18).

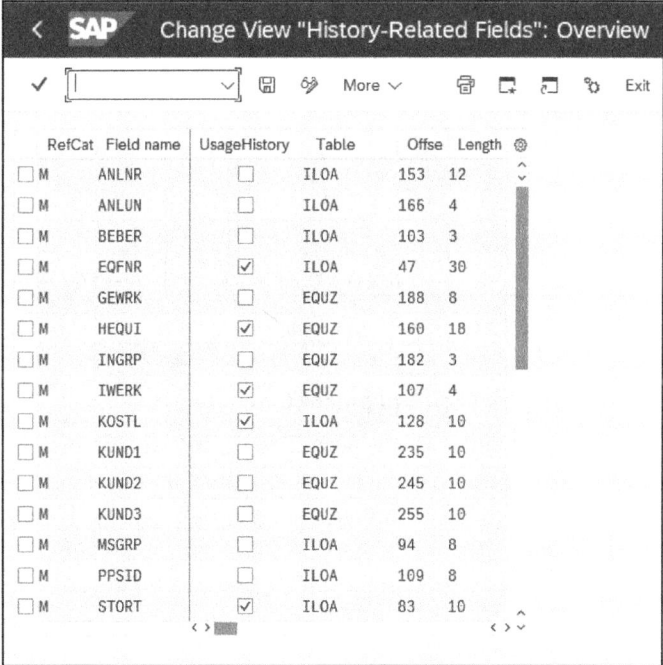

Figure 4.18 Equipment Reference Category: History-Related Fields

Define Installation at Functional Location

In this work step, you define whether to permit an installation at a functional location for each equipment category.

4 Configuring the Structure of Technical Systems

Prerequisites
You must define the equipment categories beforehand.

Customizing Path
Plant Maintenance and Customer Service • Master Data in Plant Maintenance and Customer Service • Technical Objects • Equipment • Equipment Usage Period • Define Installation at Functional Location

Settings
Select the equipment categories for which you want to permit an installation at functional locations (see Figure 4.19).

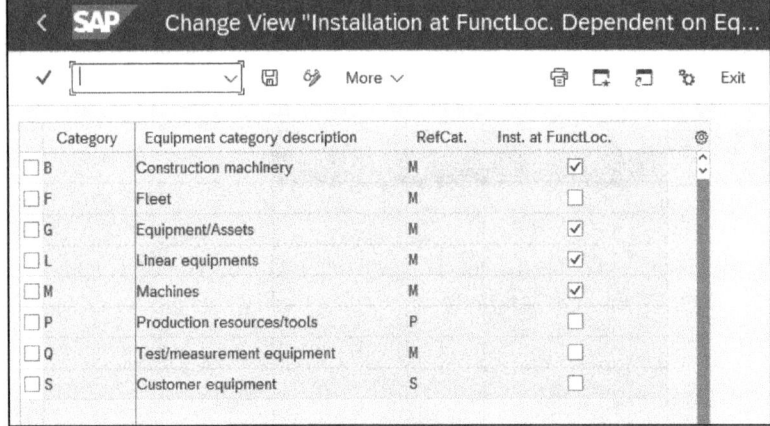

Figure 4.19 Equipment Category: Installation at Functional Locations

Allow Multilingual Text Maintenance for Each Equipment Category
In this work step, you define whether you want to permit multilingual text maintenance for each equipment category. This is necessary, for example, if your company is part of a group that operates internationally and stores its equipment on a cross-country and therefore cross-lingual basis.

Prerequisites
You must define the equipment categories beforehand.

Customizing Path
Plant Maintenance and Customer Service • Master Data in Plant Maintenance and Customer Service • Technical Objects • Equipment • Equipment Usage Period • Allow Multilingual Text Maintenance for Each Equipment Category

Settings
Select the equipment categories for which you want to permit multilingual text maintenance (see Figure 4.20).

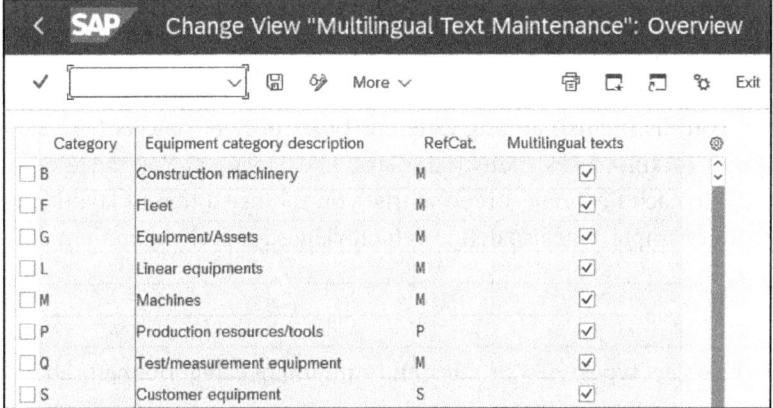

Figure 4.20 Equipment Category: Multilingual Texts

4.4 Fleet Management

In addition to the general Customizing settings, some special Customizing functions are available for managing equipment and/or functional locations as fleet objects. For example, you can configure the following:

- Fleet object-specific view profiles
- Consumable types
- Engine types
- Usage types
- Costing and account assignment of fuel consumption
- Gas stations

Figure 4.21 provides an overview of the Customizing functions for fleet objects.

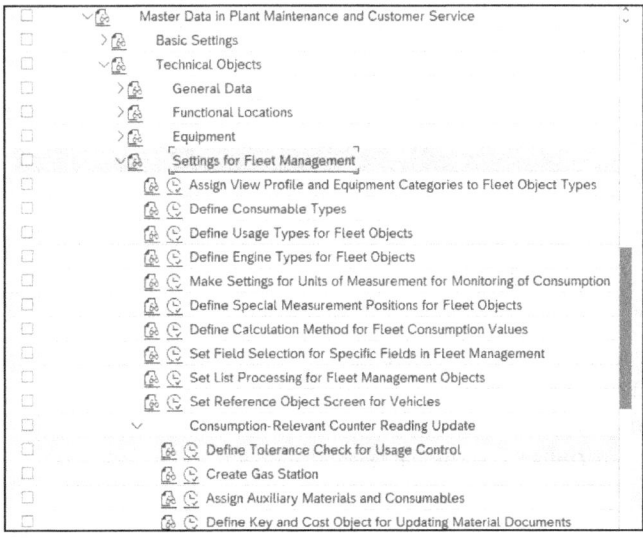

Figure 4.21 Fleet Management: Customizing Functions Overview

The most important of these Customizing functions are explained next.

Assign View Profile and Equipment Categories to Fleet Object Types

In this work step, you distinguish among different types of fleet objects (e.g., cars, trucks, mobile cranes, forklift trucks, industrial trucks, wheel loaders). You can assign a custom view profile to each fleet object type so that you can use different layouts for forklifts and cars, for example. You also define default values for each equipment category.

Prerequisites

You must define the object types, view profiles, and equipment categories beforehand.

Customizing Path

Plant Maintenance and Customer Service • Master Data in Plant Maintenance and Customer Service • Technical Objects • Settings for Fleet Management • Assign View Profile and Equipment Categories to Fleet Object Types

Settings

In the table, you should add object types that represent the fleet objects and assign the relevant view profile and permitted equipment categories to each fleet object type (see Figure 4.22).

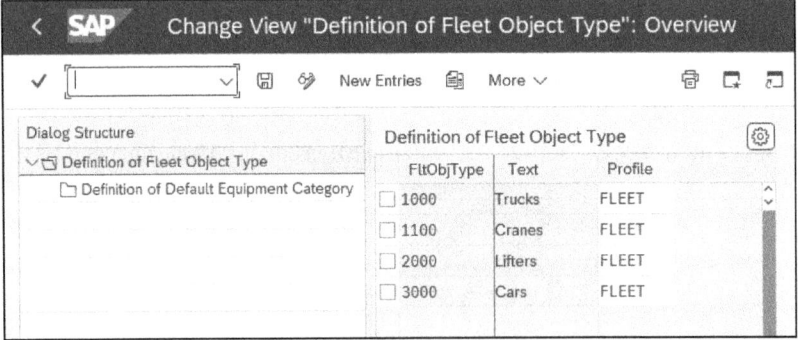

Figure 4.22 Fleet Objects: Types

> [!] **Fleet Object Tabs Are Also Displayed**
>
> The tabs for the view profile specified here are displayed in addition to the tabs for the view profile for the equipment category. If, for example, the equipment category has view profile **A** with tabs **010** (**General Data**), **020** (**Location Data**), and **030** (**Organizational Data**), while the fleet object type has view profile **B** with tabs **130** (**Fleet Object ID/Dimensions**) and **140** (**Fleet Object – Technical Data**), then all five tabs—010, 020, 030, 130, and 140—are displayed in the equipment master record.

4.4 Fleet Management

Define Consumable Types

You can use this Customizing function to define consumable types for fleet objects, and you can assign up to three consumable types to each fleet object.

Prerequisites

There are no special prerequisites.

Customizing Path

Plant Maintenance and Customer Service • Master Data in Plant Maintenance and Customer Service • Technical Objects • Settings for Fleet Management • Define Consumable Types

Settings

In the table, add the consumable types that are used to run your fleet objects and/or are required as a lubricant (see Figure 4.23).

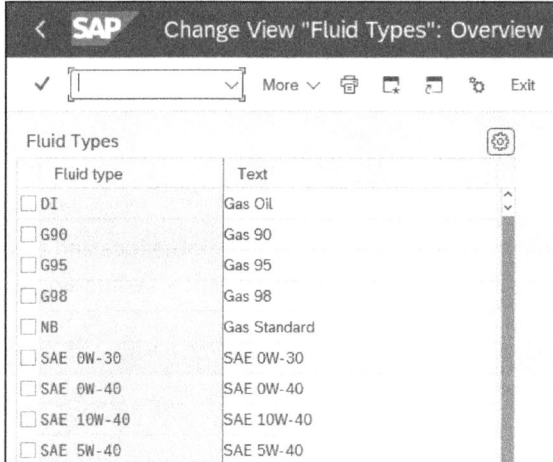

Figure 4.23 Fleet Objects: Fluid Types

Define Usage Types for Fleet Objects

You can use this Customizing function to define different usage types for fleet objects. For example, you can define whether the fleet objects are used primarily for business or transportation purposes.

Prerequisites

There are no special prerequisites.

Customizing Path

Plant Maintenance and Customer Service • Master Data in Plant Maintenance and Customer Service • Technical Objects • Settings for Fleet Management • Define Usage Types for Fleet Objects

4 Configuring the Structure of Technical Systems

Settings

In the table, add the usage types that you require for your fleet objects (see Figure 4.24).

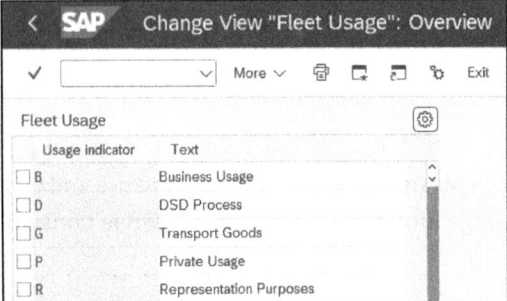

Figure 4.24 Fleet Objects: Fleet Usage

Define Engine Types for Fleet Objects

You can use this Customizing function to define different engine types for fleet objects (e.g., whether the fleet objects run on diesel, gas, or electricity).

Prerequisites

There are no special prerequisites.

Customizing Path

Plant Maintenance and Customer Service • Master Data in Plant Maintenance and Customer Service • Technical Objects • Settings for Fleet Management • Define Engine Types for Fleet Objects

Settings

In the table, add the engine types that you require for the different types of fleet objects in your fleet (see Figure 4.25).

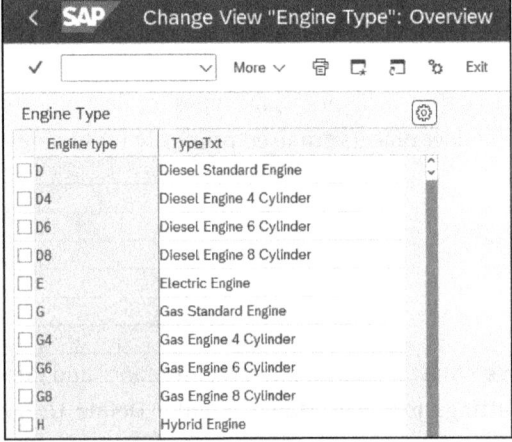

Figure 4.25 Fleet Objects: Engine Type

Make Settings for Units of Measurement for Monitoring of Consumption

You can use this Customizing function to configure settings for determining the fuel consumption of your fleet objects. Here, fuel consumption is gauged using either a time-related calculation (e.g., liters per hour) or a distance-related calculation (e.g., miles per gallon).

Prerequisites

You must define the units of measurement beforehand (by using the **Check Units of Measurement** Customizing function).

Customizing Path

Plant Maintenance and Customer Service • Master Data in Plant Maintenance and Customer Service • Technical Objects • Settings for Fleet Management • Make Settings for Units of Measurement for Monitoring of Consumption

Settings

In the table, add the consumption units that you want to use to measure your fuel consumption (see Figure 4.26).

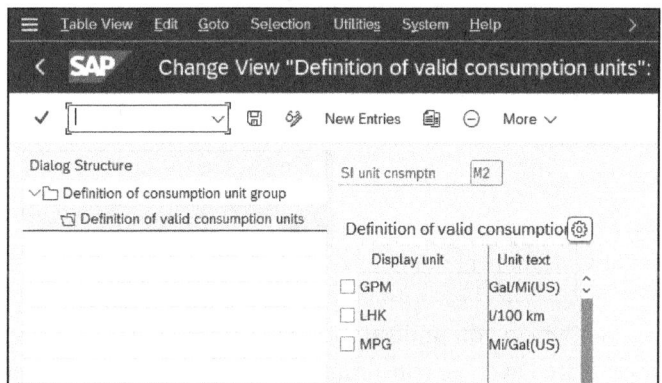

Figure 4.26 Fleet Objects: Consumption Units

You then use the **Definition of valid consumption units** subfunction to define the corresponding units of measurement (e.g., liters per 100 km or miles per gallon for the volume consumption per distance).

> **Note**
> You can define any number of units of measurement for each consumption unit group.

Define Special Measurement Positions for Fleet Objects

You can use this Customizing function to create special measurement positions that are filled with special fleet management functions. The measurement positions are used to perform fleet object-specific calculations (e.g., consumption) and are used as a basis for analyses by the plant maintenance information system (PMIS) via Transaction MCIZ (PMIS: Vehicle Consumption Analysis).

Prerequisites

You must define the units of measurement beforehand (by using the **Check Units of Measurement** Customizing function).

Customizing Path

Plant Maintenance and Customer Service • Master Data in Plant Maintenance and Customer Service • Technical Objects • Settings for Fleet Management • Define Special Measurement Positions for Fleet Objects

Settings

In the table, add the types of measurement positions according to which you want to calculate your fuel consumption in PMIS (see Figure 4.27).

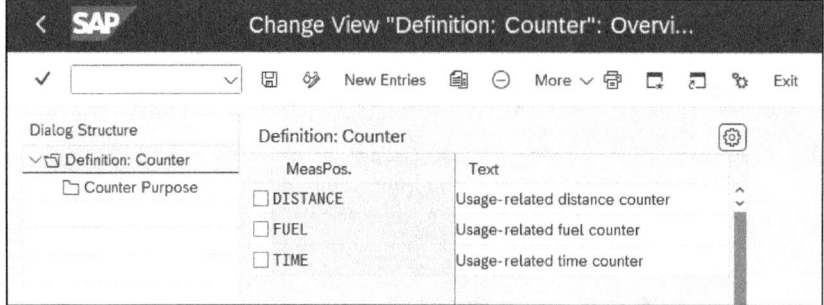

Figure 4.27 Fleet Objects: Special Measurement Positions

Consider the following example to help you understand the calculation method. You want to calculate the distance-related average consumption for a truck (e.g., miles per gallon diesel), so you create two counters in the piece of equipment (see Figure 4.28):

- You require a counter that enables you to measure distance. For this counter, you enter, for example, the counter definition "DISTANCE" in the **MeasPos.** field, which, in turn, was defined in the **Counter Purpose** subfunction as **1 = Usage-related distance counter**.

- You require a second counter that enables you to measure consumption. For this counter, you enter, for example, the counter definition "FUEL" in the **MeasPos.** field, which, in turn, was defined in the **Counter Purpose** subfunction as **3 = Usage-related fuel counter**.

In your daily work, you'll now use the first counter with the measurement position **DISTANCE** to enter the distance data and the second counter **FUEL** to enter the consumption data.

4.4 Fleet Management

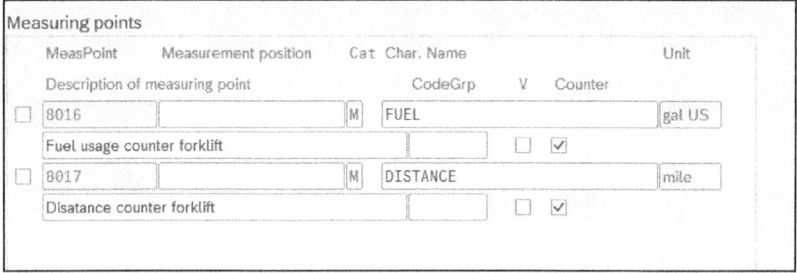

Figure 4.28 Fleet Objects: Counters

PMIS can then use this information to calculate the average consumption. You define the method for calculating average consumption in the **Define Calculation Method for Fleet Consumption Values** Customizing function.

Define Calculation Method for Fleet Consumption Values

You can use this Customizing function to define the methods for calculating the consumption values of fleet objects in PMIS (Transaction MCIZ [PMIS: Vehicle Consumption Analysis]).

Prerequisites

You must define the units of measurement (by using the **Check Units of Measurement** Customizing function) and the measurement positions beforehand.

Customizing Path

Plant Maintenance and Customer Service · Master Data in Plant Maintenance and Customer Service · Technical Objects · Settings for Fleet Management · Define Calculation Method for Fleet Consumption Values

Settings

You need to define the following on the screen shown in Figure 4.29:

- **Unit text**
  Define the unit that is to be displayed (e.g., **l/100 km, Mi/Gal**).

- **Fuel counter**
  To calculate consumption, type in the fleet object counters you want to use in the consumption calculation (here, "FUEL").

- **Primary counter**
  Type in the specific *primary counter* (here, "DISTANCE"), which is the fleet object counter that you use as a basis for measuring the distance or operation time in relation to the consumption measured for the fleet object.

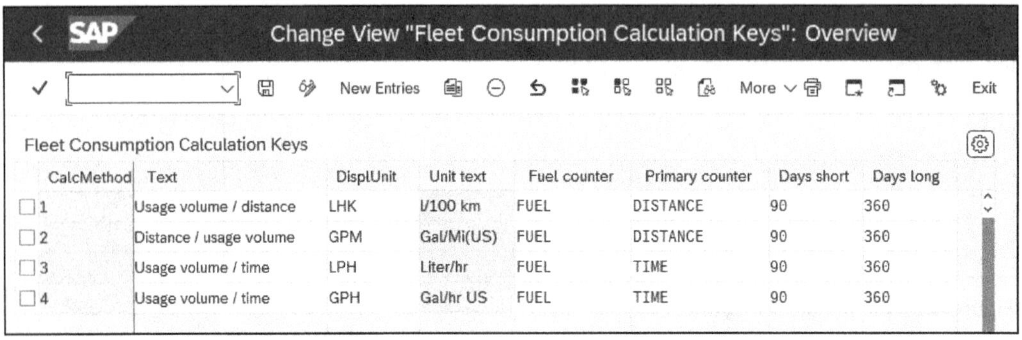
Change View "Fleet Consumption Calculation Keys": Overview
New Entries
More
Exit
Fleet Consumption Calculation Keys
CalcMethod Text DisplUnit Unit text Fuel counter Primary counter Days short Days long
1 Usage volume / distance LHK l/100 km FUEL DISTANCE 90 360
2 Distance / usage volume GPM Gal/Mi(US) FUEL DISTANCE 90 360
3 Usage volume / time LPH Liter/hr FUEL TIME 90 360
4 Usage volume / time GPH Gal/hr US FUEL TIME 90 360

Figure 4.29 Fleet Objects: Calculation Keys

Although they aren't discussed in detail here, the following are some additional Customizing functions related to fleet object consumption:

- **Define Tolerance Check for Usage Control**
  Define whether (and, if so, to what extent) the consumption entered should differ from the short-term or long-term average consumption before the system issues warning or error messages.
- **Create Gas Station**
  Create internal gas stations with reference to the plant and storage location. When a consumption value is entered, the goods issue posting is automatically made from the relevant storage location.
- **Assign Auxiliary Materials and Consumables**
  Define the storage location in which a particular material number is to be posted for a particular consumable type.
- **Define Key and Cost Object for Updating Material Documents**
  Define the account assignment for the automatic issue posting (e.g., the cost center for a piece of equipment).

4.5 Links and Networks

SAP S/4HANA Asset Management provides functions that enable you to map links between or among different technical objects or systems (equipment or functional locations). For example, you can establish the following links:

- Among production units
- Between production plants and supply systems
- Between supply systems and disposal systems

Such object links form an *object network* that can give your assets a horizontal structure. Figure 4.30 provides an overview of the Customizing functions for object links.

4.5 Links and Networks

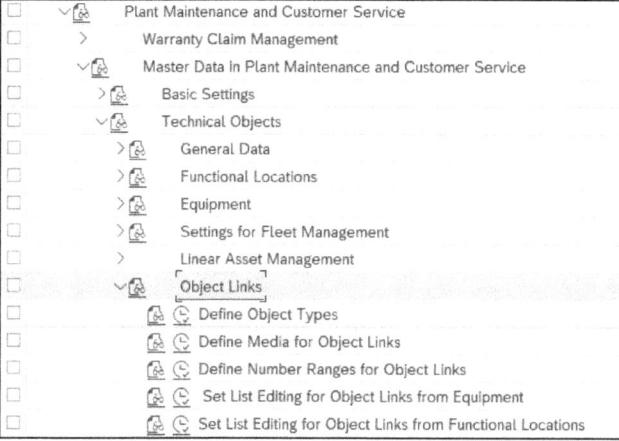

Figure 4.30 Object Links: Customizing Functions Overview

The most important of these Customizing functions are explained in the following sections.

Define Object Types

You use this Customizing function to define object types for object links (similar to defining equipment categories for equipment or categories for functional locations).

Prerequisites

If you want to assign a status profile, you must define it beforehand.

Customizing Path

Plant Maintenance and Customer Service • Master Data in Plant Maintenance and Customer Service • Technical Objects • Object Links • Define Object Types

Settings

You can define the following on the screen shown in Figure 4.31.

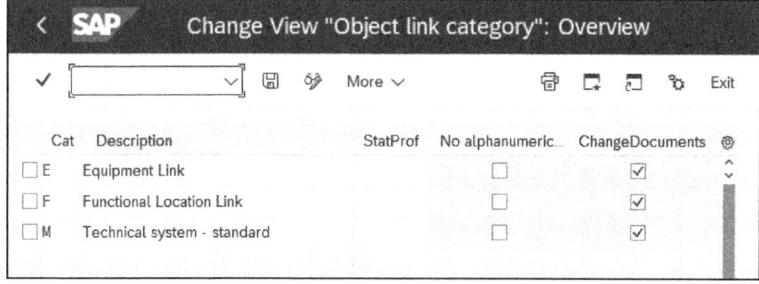

Figure 4.31 Object Links: Category Overview

199

4 Configuring the Structure of Technical Systems

- **StatProf**
  Assign a status profile.
- **No alphanumeric nos**
  Define whether alphanumeric characters are permitted.
- **ChangeDocuments**
  Define whether change documents are generated.

Define Media for Object Links

You use this Customizing function to define the media permitted for object links (e.g., water, gas, or granulate).

Prerequisites

If you want to assign a material number or color, you must define these beforehand.

Customizing Path

Plant Maintenance and Customer Service • Master Data in Plant Maintenance and Customer Service • Technical Objects • Object Links • Define Media for Object Links

Transaction

OINM

Settings

You can define the following on the screen shown in Figure 4.32:

- **Medium**
  Define whether the medium is managed under a material number.
- **PN-Relev.**
  Define whether the medium is relevant for production.

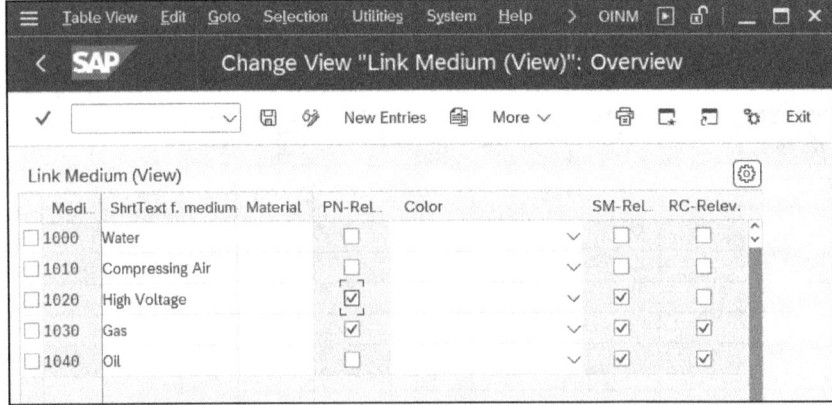

Figure 4.32 Object Links: Medium

- **Color**
  Choose which color is assigned to the medium.
- **SM-Relev.**
  Check the box to set this indicator to select media that you want to be available when you create a scale method. For example, the **Oil** medium is relevant for viscosity scaling.
- **RC-Relev.**
  This indicator is used to specify the media that are relevant for reclassification.

4.6 Linear Asset Management

Linear assets are technical systems with a linear infrastructure whose properties and conditions may change from stage to stage (in a process called *dynamic segmentation*). Examples of linear assets include the following:

- Pipelines
- Road networks
- Railway networks
- Overhead cables

As you can tell from these examples, this function is of particular interest to the oil and gas industry, utility companies (electricity, gas, and water), road maintenance companies, public authorities, and rail operators.

If you want to maintain linear assets, you must activate business functions LOG_EAM_LINEAR_1 and LOG_EAM_LINEAR_2.

To use linear asset management to its full potential, you must create the following conditions:

- In Customizing for the classification system, you use the **Maintain Object Types and Class Types** Customizing function to create an organizational area (e.g., the organizational area **L** for linear objects).
- To enable you to enter linear data for measuring points and measurement documents, you use the **Define Measuring Point Categories** Customizing function to define a measuring point category, and you also set the **Linear Asset** indicator.
- To define a view profile for functional locations and a view profile for equipment, you use the **Set View Profiles for Technical Objects** Customizing function. You also assign the field groups for linear data there (refer to the **Set View Profiles for Technical Objects** Customizing function in Figure 4.4 and Figure 4.5).
- To define linear data for functional locations, you use the **Define Category of Functional Location** Customizing function to define a functional location category. You

assign the view profile there and set the **Linear Asset** indicator (refer to the **Define Category of Functional Location** Customizing function and Figure 4.8).

- To define linear data for equipment, you use the **Maintain Equipment Category** Customizing function to define an equipment category. In particular, you assign the view profile there and set the **Linear Asset** indicator (refer to the **Maintain Equipment Category** Customizing function and Figure 4.15).

Additional Customizing functions are available within linear asset management, an overview of which is provided in Figure 4.33.

Figure 4.33 Linear Asset Management: Customizing Functions Overview

The most important of these Customizing functions are covered in the following sections.

Define Offset Types

You use this Customizing function to define the necessary offset types for linear objects. Offset data is always required if, when defining a linear asset structure, you integrate objects that have a certain offset in relation to the actual linear object (e.g., if, for the linear object **Road**, you want to define a signaling device with a horizontal offset of 2 yards from the edge of the road and a vertical offset of 350 yards).

4.6 Linear Asset Management

Prerequisites

You must define the units of measurement used for the offset type beforehand (by using the **Check Units of Measurement** Customizing function).

Customizing Path

Plant Maintenance and Customer Service • Master Data in Plant Maintenance and Customer Service • Technical Objects • Linear Asset Management • Define Offset Types

Settings

You can define the following on the screen shown in Figure 4.34:

- **OTC**
  A two-digit code.

- **Offset Type Desc.**
  A description of the offset type.

- **UoM**
  The unit of measurement for the offset type

- **Default Offset**
  Whether the offset type is to be proposed when you define linear objects

- **Documentation Object**
  The name of the text module

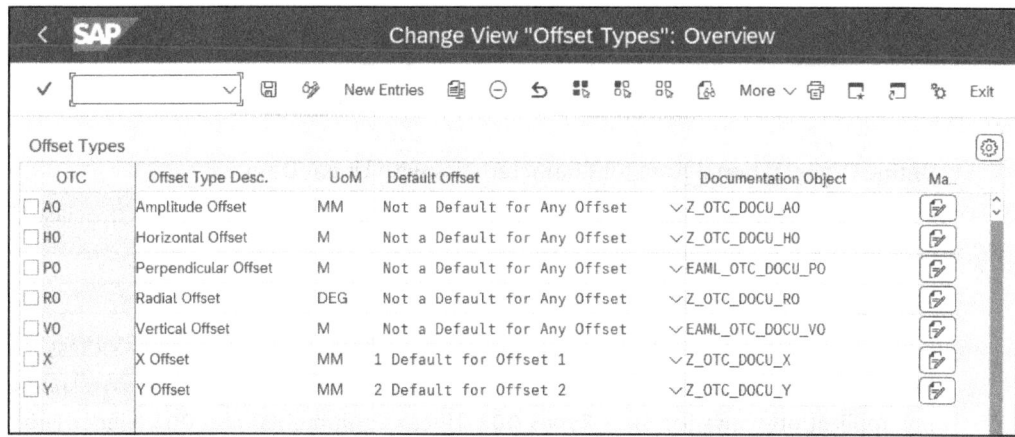

Figure 4.34 Linear Asset Management Offset Types

Define Linear Reference Pattern Type

You use this Customizing function to define the necessary linear reference pattern types for linear objects.

Linear reference patterns are created so that you can assign a descriptive reference point to real technical objects or documents (orders, notifications, etc.). You can then specify these patterns in a notification; for example, you can say, "The damage

4 Configuring the Structure of Technical Systems

occurred ten miles after the vehicle stopped at the Kennebunk rest area and service station." You assign a linear reference pattern type to each linear reference pattern.

Prerequisites
There are no specific prerequisites.

Customizing Path
Plant Maintenance and Customer Service • Master Data in Plant Maintenance and Customer Service • Technical Objects • Linear Asset Management • Define LRP Type

Settings
You create an ID (with a maximum of four characters) and a description for the linear reference pattern type (see Figure 4.35).

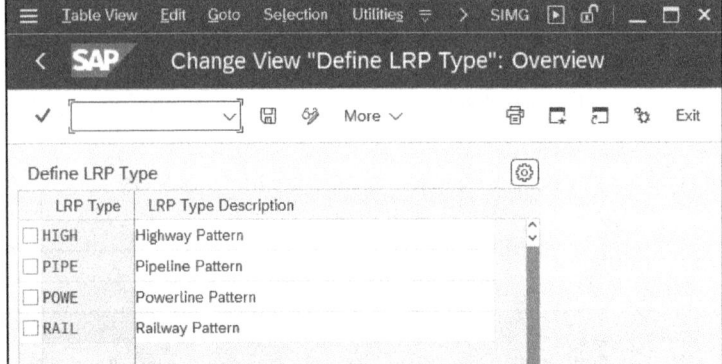

Figure 4.35 Linear Asset Management: Linear Reference Pattern Type

Define Organizational Area for Characteristics with Linear Data

You use this Customizing function to define which view you want to use in the classification system for the linear objects.

Prerequisites
You must use the **Cross-Application Components • Classification System • Classes • Maintain Object Types and Class Types** Customizing function to define the corresponding application views for **Class Types 002 (Linear Equipments)** and **003 (Linear Functional Locations)**. See Figure 4.36.

Customizing Path
Plant Maintenance and Customer Service • Master Data in Plant Maintenance and Customer Service • Technical Objects • Linear Asset Management • Define Organizational Area for Characteristics with Linear Data

4.6 Linear Asset Management

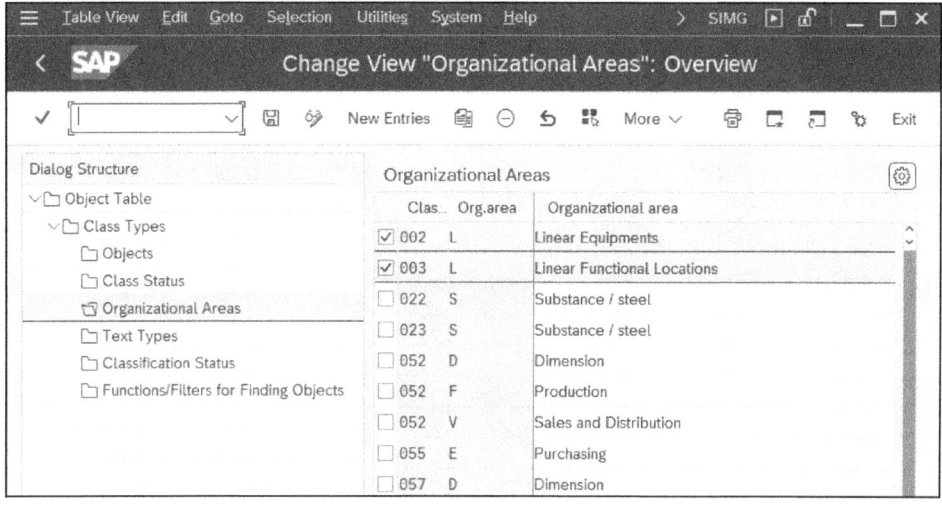

Figure 4.36 Classification: Organizational Areas for Linear Assets

Settings

You assign the corresponding view to class types **002** (**Linear Equipments**) and **003** (**Linear Functional Locations**) (see Figure 4.37). Other class types and objects aren't possible.

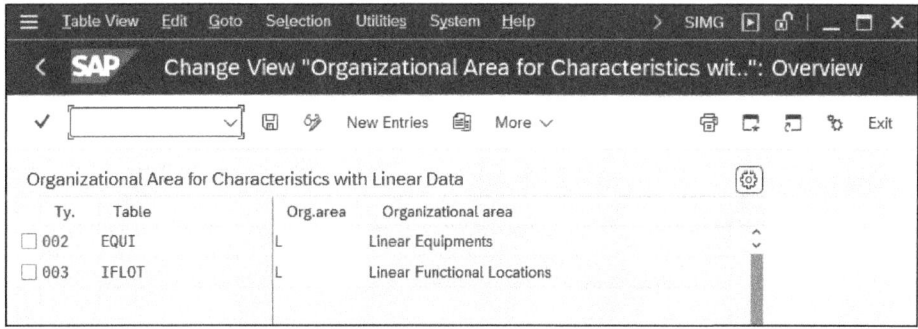

Figure 4.37 Linear Asset Management: Organizational Area

Object Networks

When business function LOG_EAM_LINEAR_2 is activated, you can map your linear assets as object networks that comprise several object links between functional locations or pieces of equipment. You can enter linear data at the network or object level, and you can display the object network in a graphical diagram. You can also use network attribute values and properties to define additional information about your object networks (Transaction IN21). The following single-level and multilevel list reports are available to you for analyzing your object network data: Transaction IN24 and Transaction IN25. Furthermore, you can use some Customizing functions that enable you to adjust this function to your own requirements.

4 Configuring the Structure of Technical Systems

Define Network Types

You can use this Customizing activity to create network types. Then, in Transaction IN21 (Create Object Network), you select one of these network types for the object network.

To enter network attribute values and any changes to these values, you use the **Assign Network Attribute Categories to Network Types** Customizing function to assign your network attribute categories to the network types that you've created. You can assign one network attribute category to several network types.

Prerequisites

There are no special prerequisites.

Customizing Path

Plant Maintenance and Customer Service • Master Data in Plant Maintenance and Customer Service • Technical Objects • Linear Asset Management • Define Additional Data for Object Links and Networks • Define Network Types

Settings

Create the necessary (four-character) **Network Type** with a corresponding **Network Type Description** (see Figure 4.38). You can also determine whether change documents are to be generated when you change and/or create a network in the last two columns.

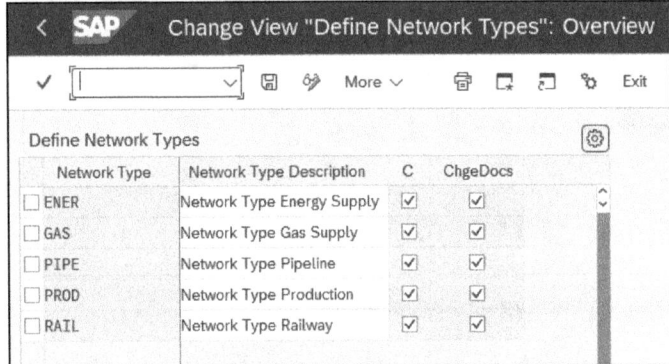

Figure 4.38 Linear Asset Management: Network Types

Define Network Group

You can use this Customizing function to define network groups. If you create object networks, you can assign each object network to a network group. The network group is then used as a selection criterion in the list displays.

Prerequisites

There are no special prerequisites.

4.6 Linear Asset Management

Customizing Path

Plant Maintenance and Customer Service • Master Data in Plant Maintenance and Customer Service • Technical Objects • Linear Asset Management • Define Additional Data for Object Links and Networks • Define Network Group

Settings

Create the necessary (10-character) network groups (**NetGroup**) with a corresponding **Network Group Description** (see Figure 4.39). Because the object networks don't contain a field for assignment to a plant, it makes sense to use a network group for this purpose. However, you can also set up network groups for all other usage purposes.

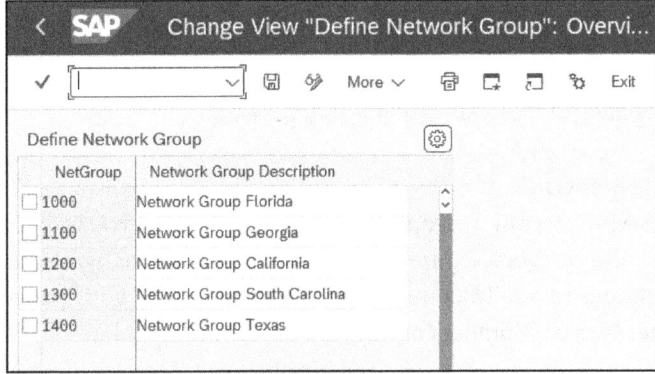

Figure 4.39 Linear Asset Management Network Groups

Define Network Attribute Categories

You can use this Customizing function to define network attribute categories. You can then define network attributes and assign them to predefined categories.

Prerequisites

There are no special prerequisites.

Customizing Path

Plant Maintenance and Customer Service • Master Data in Plant Maintenance and Customer Service • Technical Objects • Linear Asset Management • Define Settings for Network Attributes • Define Network Attribute Categories

Settings

This Customizing function is composed of several subfunctions, as follows:

- **Define Network Attribute Categories**
  This subfunction defines the necessary (12-character) network groups with a corresponding description (see Figure 4.40). Here, you can specify whether this category and its attributes are linear assets by selecting the **Lin. Cat.** checkbox. If you check this box, the **End Point** field for the lower-level attributes becomes ready for input. If

you don't, the system assumes that this category and its attributes aren't linear, and the **End Point** field isn't ready for input.

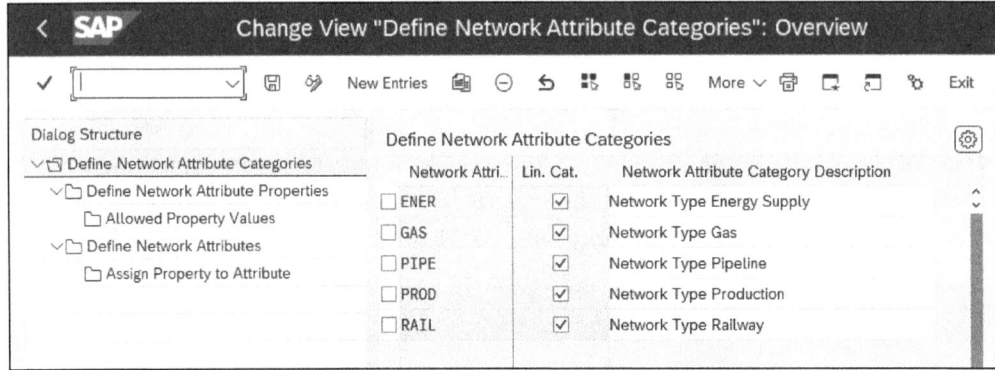

Figure 4.40 Linear Asset Management: Network Attribute Categories

- **Define Network Attribute Properties**
  This subfunction defines which technical properties or other properties you want to assign to a network attribute category. Figure 4.41 shows that you can assign track, sleeper, or line type attributes to a railway infrastructure. In each case, you specify the format (e.g., **Character Format, Number Format**) and the field length.

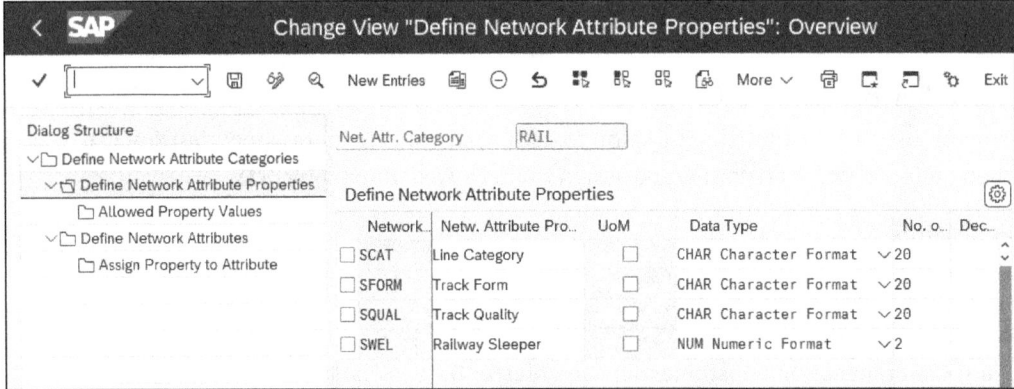

Figure 4.41 Linear Asset Management: Network Attribute Properties

- **Allowed Property Values**
  This subfunction defines the values permitted for each network attribute. Figure 4.42 shows, for example, the attributes permitted for the **Line Category** property (e.g., regional route or long-distance route, passenger route, freight route, mixed route). If you don't define a value table, you can freely define the relevant property in the master record for the object network.

4.6 Linear Asset Management

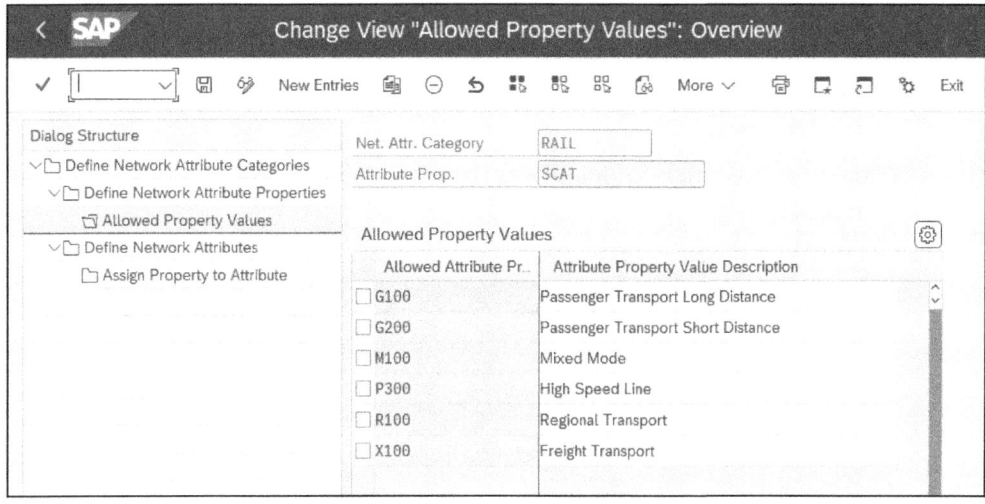

Figure 4.42 Linear Asset Management: Network Property Values

Assign Network Attribute Categories to Network Types

You use this Customizing function to define which network attribute categories are used for which network type.

Prerequisites

You must define the network types and network attribute categories beforehand.

Customizing Path

Plant Maintenance and Customer Service • **Master Data in Plant Maintenance and Customer Service** • **Technical Objects** • **Linear Asset Management** • **Define Settings for Network Attributes** • **Assign Network Attribute Categories to Network Types**

Settings

You assign the network attribute category to the network type. Figure 4.43 shows a 1:1 assignment, but you can also assign several network attribute categories to a network type and the same network attribute category to different network types. In other words, an N:M assignment is also possible.

Customizing Functions for List Displays

Linear asset management has a range of Customizing functions for field selections:

- **Define Field Selection for Linear Data Fields in Multilevel Lists**, Transaction OIUXA (i.e., for Transactions IE07, IL07, IW30, and IW40)
- **Define Field Selection for Linear Data Fields for Char. in Multilevel Lists**, Transaction OIUXB (i.e., for Transactions IE07 and IL07)

- **Define Field Selection for Linear Data Fields in Structure Lists, Transaction OIUXC** (i.e., for Transactions IH01 and IH03)
- **Define Field Selection for Multilevel List Displays of Funct. Loc. Network** (i.e., for Transaction IN24)
- **Define Field Selection for Multilevel List Displays of Equipment Network** (i.e., for Transaction IN25)

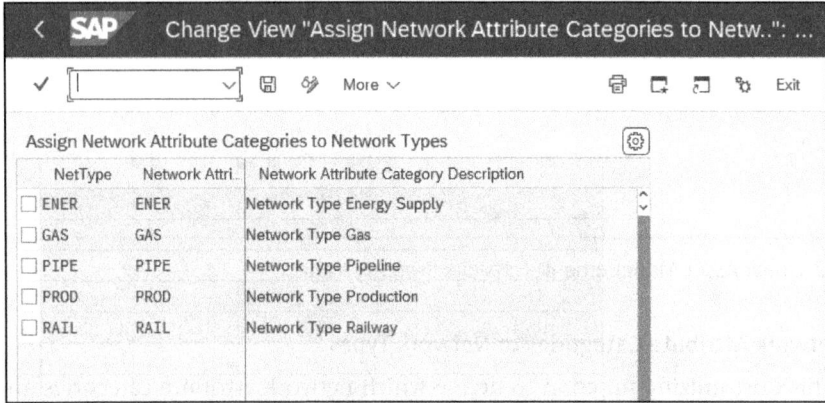

Figure 4.43 Linear Asset Management: Assign Network Types and Categories

The last two functions in the preceding list have the following subfunctions:

- **Define Field Selection for Object Network Fields,** Transactions OIUXG and OIUXO
- **Define Field Selection for Network Attribute Values,** Transactions OIUXH and OIUXP
- **Define Field Selection for Object Link Data Fields,** Transactions OIUXI and OIUXQ
- **Define Field Selection for Functional Location Fields,** Transaction OIUXJ
- **Define Field Selection for Equipment Fields,** Transaction OIUXR
- **Define Field Selection for Notification Data Fields,** Transactions OIUXK and OIUXS
- **Define Field Selection for Order Data Fields,** Transactions OIUXL and OIUXT
- **Define Field Selection for Measuring Point and Counter Fields,** Transactions OIUXM and OIUXU
- **Define Field Selection for Measurement Document and Counter Reading Fields,** Transactions OIUXN and OIUXV

How to make these field definitions was described in Chapter 3, Section 3.11. Figure 4.44 shows, for example, which data fields are selected for linear objects in the structural display for functional locations (Transaction IH01).

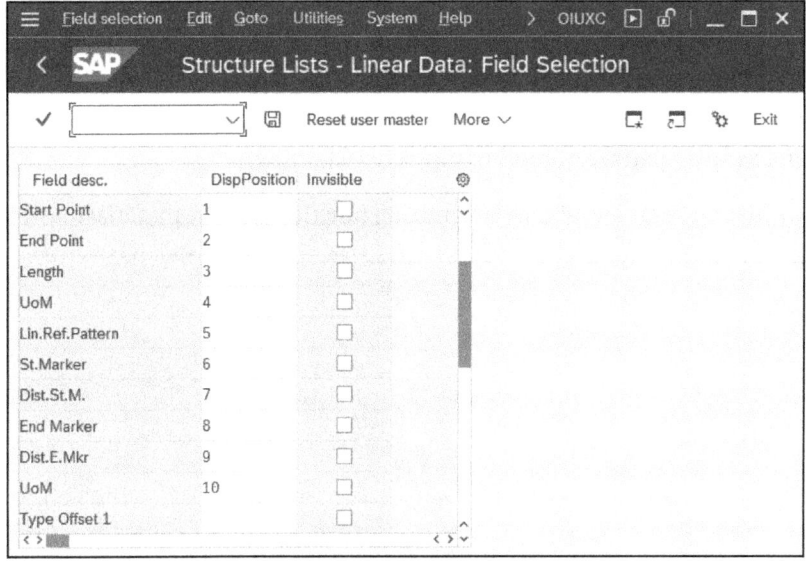

Figure 4.44 Linear Asset Management: Structure List with Linear Data

4.7 Materials and Assemblies

The *material master* contains information about materials that a company designs, procures, produces, stores, and sells. It integrates data from various areas within a company (e.g., purchasing, accounting). Unlike a piece of equipment or a functional location, a material master record doesn't describe just one thing, but rather similar items. In other words, a material master record generally represents several similar spare parts, assemblies, raw materials, and so on.

From a maintenance perspective, material master records are used for the following purposes:

- Some material master records are purchased as spare parts and then stored.
- Some material master records are purchased as equipment and then placed in storage as material/serial numbers.
- Some material master records are construction types that solely encompass a group of similar pieces of equipment or functional locations. This makes it possible to execute shared functions for the entire group (e.g., management of shared BOMs or general maintenance task lists).
- Some material master records serve as maintenance assemblies for substructures of pieces of equipment or functional locations.

4 Configuring the Structure of Technical Systems

- Some material master records fulfill the same function as spare parts. The only difference is that they aren't placed in storage; instead, they are procured each time as nonstock items (e.g., because they are too expensive, too large, or rarely used).
- Some material master records are used as operating resources for maintenance tasks, but not as spare parts (e.g., tools or protective clothing).

Figure 4.45 provides an overview of the Customizing functions for the material master.

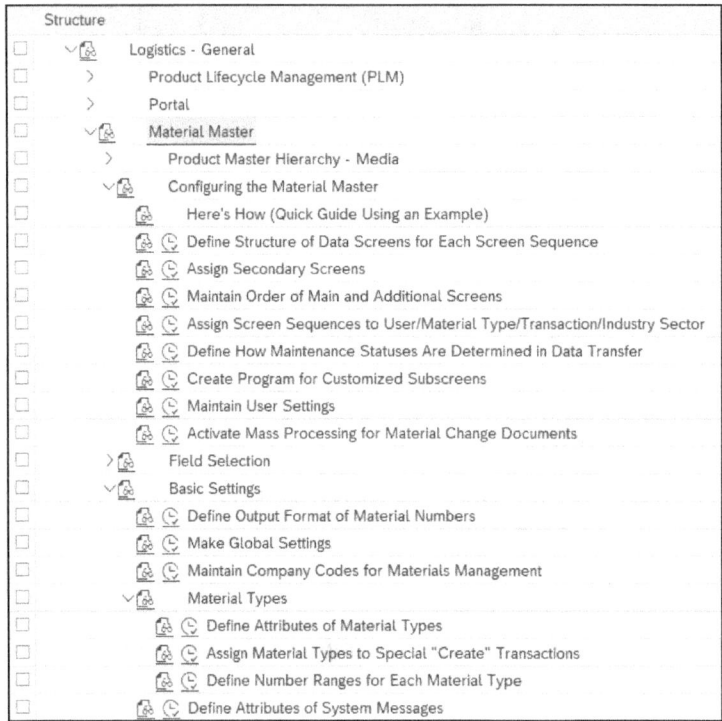

Figure 4.45 Material Master: Customizing Functions Overview

The most important of these Customizing functions for material types, the material master, and field selection are covered in the following sections.

4.7.1 Material Types

You group materials that have the same basic characteristics by assigning them to one common material type. This enables you to manage different materials in a uniform manner, according to your company's requirements. An overview of potential material types is provided in Figure 4.46.

From a maintenance perspective, the following material types are of practical relevance (shown with dark blue background in Figure 4.46):

- **Maintenance assemblies**
  Maintenance assemblies aren't standalone objects but are rather logical elements that divide the technical objects in plant maintenance into clearly defined units. A forklift, for example, can be a technical object, while the lifting plant, gear shift, chassis, and so on, can be the associated maintenance assemblies.

- **Nonstock materials**
  Nonstock materials aren't held in storage but are procured as required and used immediately. This may be the case, for example, because the parts are too expensive, very rarely required, or difficult to store (too large, too heavy, or too bulky). The master record of nonstock materials comprises purchasing data only.

- **Operating resources**
  Operating resources are required for maintenance tasks (e.g., tools and devices, test equipment, protective clothing, safety devices). Unlike expendable supplies, they don't perish but wear out with use over time.

- **Expendable supplies**
  Expendable supplies aren't part of the finished product but are required for the production process (e.g., lubricant, power, grease, oil).

- **Spare parts**
  Spare parts are used to replace defective parts. They can be purchased and placed in storage.

- **Auxiliary materials**
  Auxiliary materials form part of a finished product but are insignificant and barely seen in the finished product (e.g., screws, adhesive, welded joints).

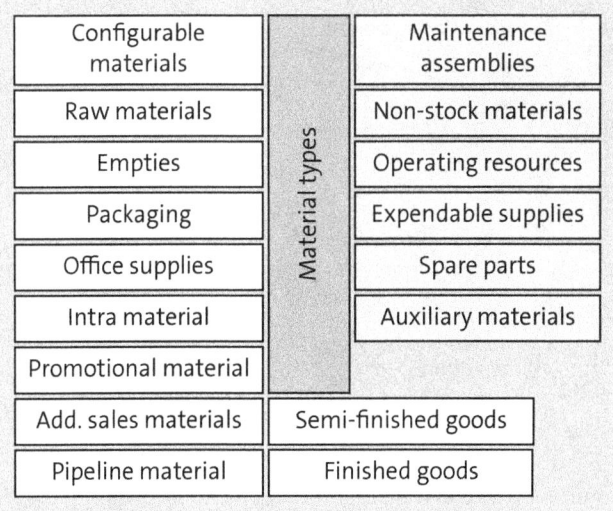

Figure 4.46 Material Master: Material Types

In Customizing, the material type is the main control element for the material master. You define the following for the material type (see Figure 4.47):

- Views that comprise the material master record (purchasing, storage, MRP, accounting, etc.)
- Whether the material number can be assigned internally or externally
- The number range interval for the material number
- Which screens appear and in which sequence
- Which screens are main screens and which are secondary screens
- Which fields are to be maintained (field selection)
- Whether quantity and value changes are updated
- How the material prices are to be defined (standard price, moving average price, etc.)
- Whether the material is to be produced in house or procured externally
- Accounts to which material consumption and material stock are posted if a material is placed in storage or removed from storage
- Authorization check for access to the material master
- Which statuses can be assigned to the material master

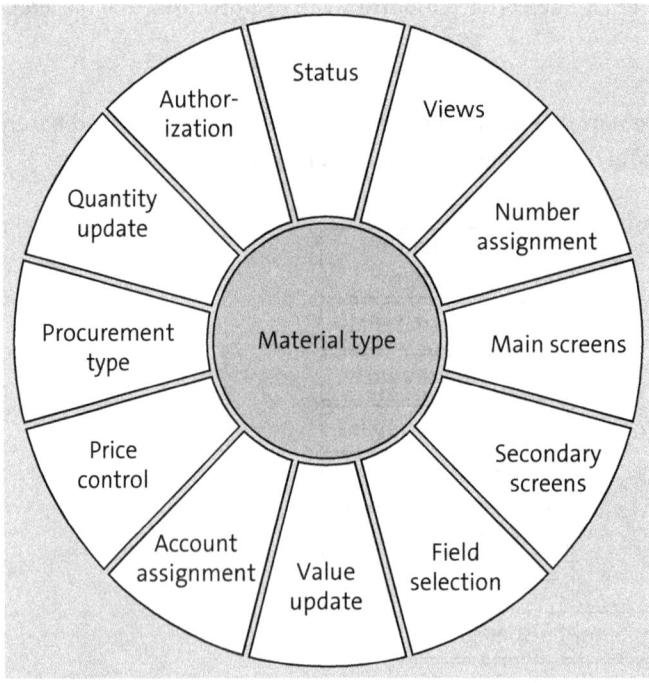

Figure 4.47 Material Type Customizing Functions

Define Attributes of Material Types

You can use this Customizing function to define the basic definitions for a material type (e.g., the user departments, the reference to the screen sequence, the statuses permitted).

Prerequisites

You must have configured the following Customizing settings:

- The field reference for the field selection
- The screen reference for the screen sequence
- The account category reference for automatic account assignment
- The valuation areas for the quantity and value update

If configured beforehand, you can also use the following Customizing settings:

- Authorization group
- Cross-plant material status

Customizing Path

Logistics – General • Material Master • Basic Settings • Material Types • Define Attributes of Material Types

Settings

Figure 4.48 shows an example of a material type for spare parts. You define the following settings for a material type:

- Field reference
  This setting establishes a link to the Customizing functions for field selections. Together with other influencing factors, these functions control the fields to be maintained for the material type.

- SRef: material type
  This setting establishes a link to the Customizing functions for the screen control (**Configuring the Material Master**). Together with other influencing factors, these functions control the screens (main and secondary) for the material type.

- X-Plant Mat.Status
  This setting defines whether an initial material status is to be proposed for the materials of this material type. In this way, you can control, for example, that materials are first locked for certain uses (e.g., purchasing, BOMs) and then need to be explicitly released.

- User departments
  This setting defines which user departments are permitted to maintain material masters. In the case of maintenance materials, these are usually purchasing, MRP, storage, accounting, and classification, as well as plant stock and warehouse stock. You therefore also indirectly determine the screen sequence.

4   Configuring the Structure of Technical Systems

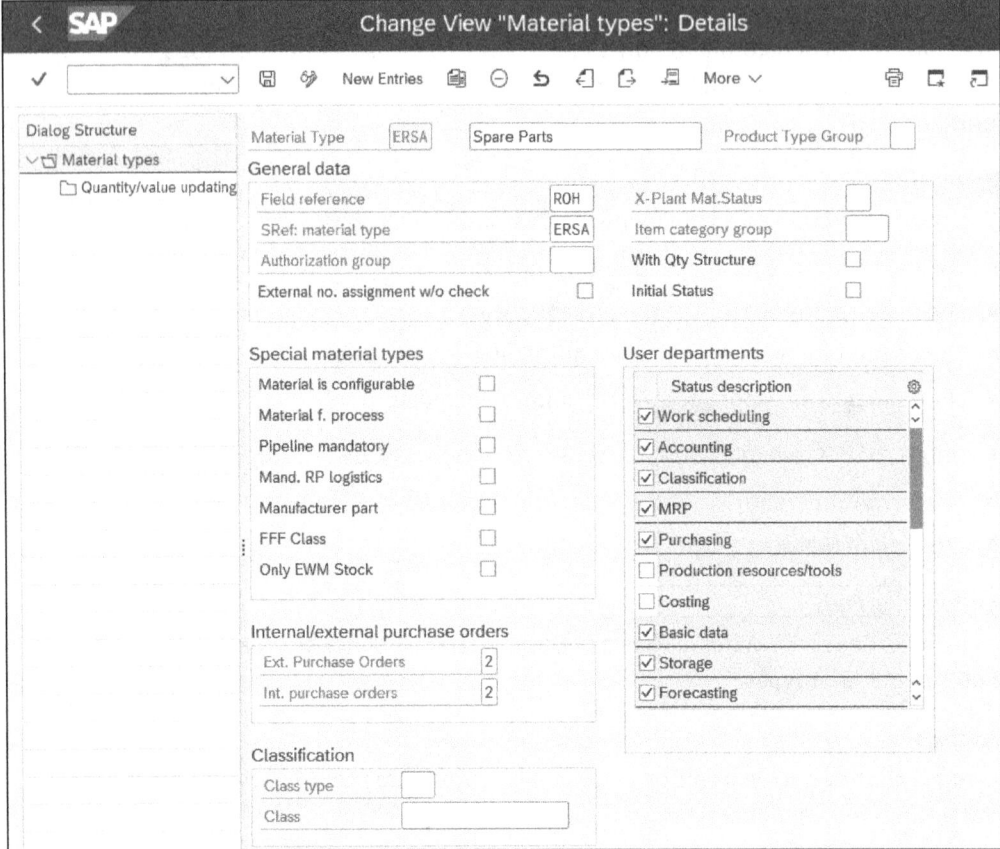

Figure 4.48 Material Type: Basic Settings

- **Internal/external purchase orders**
  This setting defines the procurement type—internal and/or external—for materials of this material type. You control whether the procurement type is prohibited (**0**), permitted but with a warning message issued (**1**), or always permitted (**2**). Maintenance materials are primarily procured externally.

- **Price control**
  This setting defines which price control procedure—standard price or moving average price—is to be proposed for materials of this material type. In the case of maintenance materials, both procedures are used.

  If you don't want to allow this default value to be changed in the material master, check the **Price Ctrl.Mandatory** box.

- **Acct cat. reference**
  This setting defines (together with other factors like the company code, chart of accounts, or business process) the automatic account assignment for materials of

this material type. Consult with your colleagues from accounting or ask them which account category reference is to be used for maintenance materials.

- **Quantity/value updating**
  This setting defines whether updates are to be handled in the same manner in all valuation areas (which usually correspond to the plants) or handled differently in each valuation area. In the case of maintenance materials, you generally base the quantity and price update on the valuation area and therefore on the plant.

If you want to control quantity and value updating on the basis of the plant, use the **Quantity/value updating** subfunctions to define which update is to occur for which plant (see Figure 4.49).

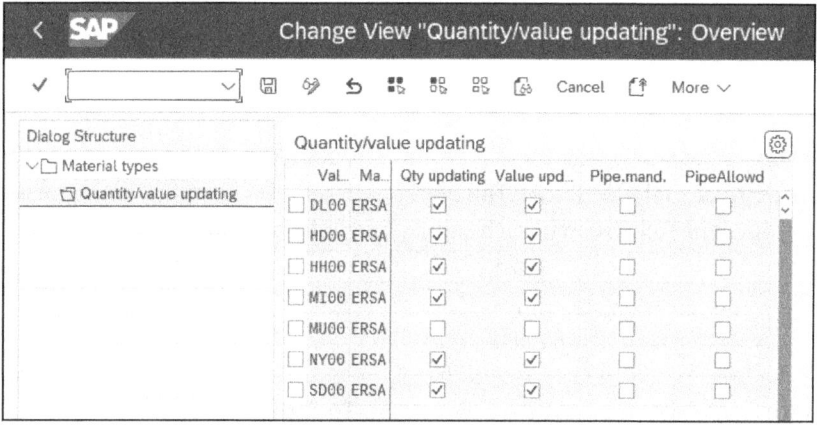

Figure 4.49 Material Type: Quantity and Value Update

4.7.2 Material Master Layout

You can use the following Customizing activities to define your own material master layout. You can base your layout on the following influencing factors (see Figure 4.50):

- Transaction
- Material type
- User
- Industry sector

The influencing factors determine the *screen sequence*, and for each screen sequence, you define the following:

- Number and type of main screens
- Number and type of subscreens (screen groups) for each main screen
- Number and type of secondary screens
- Order of main screens and secondary screens

4 Configuring the Structure of Technical Systems

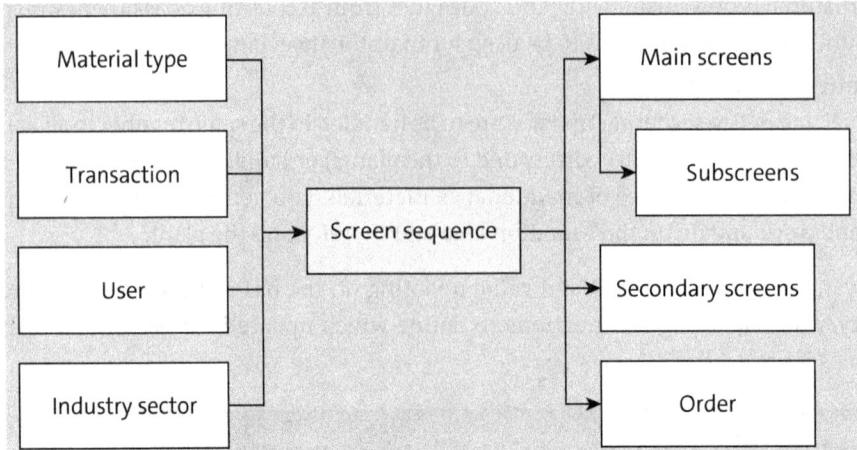

Figure 4.50 Material Master: Screen Sequence

The following example shows you the screen content definition for a typical material type: spare parts (material type ERSA). The material master will contain the following main screens only (one full screen in each case) in the following order:

- Basic Data
- Classification
- Documents
- Purchasing
- MRP
- Plant/Storage Data
- Accounting

Each of the main screens will contain a maximum of three screen groups, which provide typical maintenance information.

The following secondary screens are required in the following order:

- Multilingual Texts
- Units of Measure
- Basic Data Text

Now, let's look at the most important Customizing activities for defining and maintaining screens in the material master.

Define Structure of Data Screens for Each Screen Sequence

You use this Customizing function to specify a screen sequence and to assign the necessary main screens and their subscreens.

4.7 Materials and Assemblies

Prerequisites
There are no specific prerequisites.

Customizing Path
Logistics – General • Material Master • Configuring the Material Master • Define Structure of Data Screens for Each Screen Sequence

Transaction
OMT3B

Settings
The easiest way to generate a new screen sequence is to copy and adjust (i.e., reduce) screen sequence **01**, which is delivered in the standard SAP system. For our example, we copied screen sequence **ZE** for **Spare Parts** (see Figure 4.51).

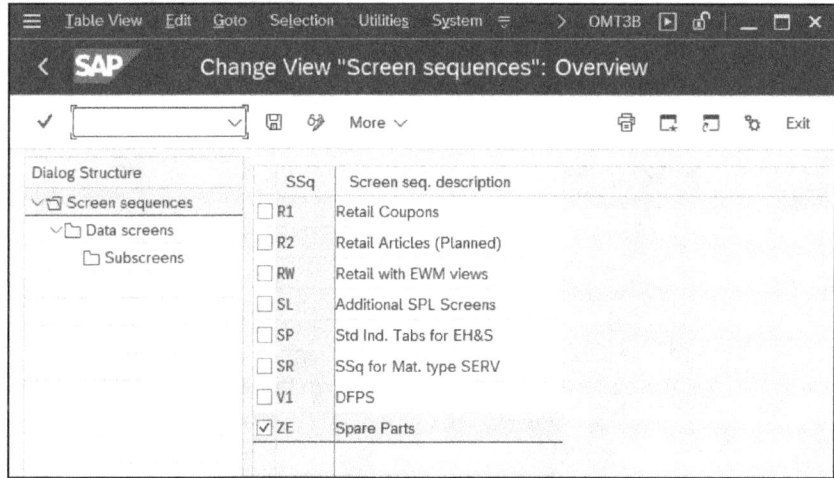

Figure 4.51 Material Master: Screen Sequences Overview

You should then use the **Data screens** subfunction (see Figure 4.52) to delete those data screens that you copied from screen sequence **01** but that aren't required for plant maintenance.

On this screen, you enter following information:

- **Screen description**
  Choose your own screen description for each screen.

- **T**
  Use the screen type to define whether the screen is a main screen (**1**) or a secondary screen (**2**).

- **SCon**
  Use the container screen to define the screen structure. In addition, you can press the F4 key to choose from approximately 50 different variants. For example, you

4 Configuring the Structure of Technical Systems

can select whether the screen is a general container screen and, if so, with how many modules (any number between 6 and 20 modules), or whether the screen is a special screen (e.g., for a description or classification).

- **Maint. status**
  Define which maintenance status is to be set if you maintain this screen. In Figure 4.52, you can obtain the maintenance status associated with the relevant screen (e.g., **E** for purchasing, **B** for accounting).

- **GUI status**
  Refine the screen types for main screens and secondary screens. The standard screen type for main screens is **DATE**, while the standard screen type for secondary screens is **ADD0001**.

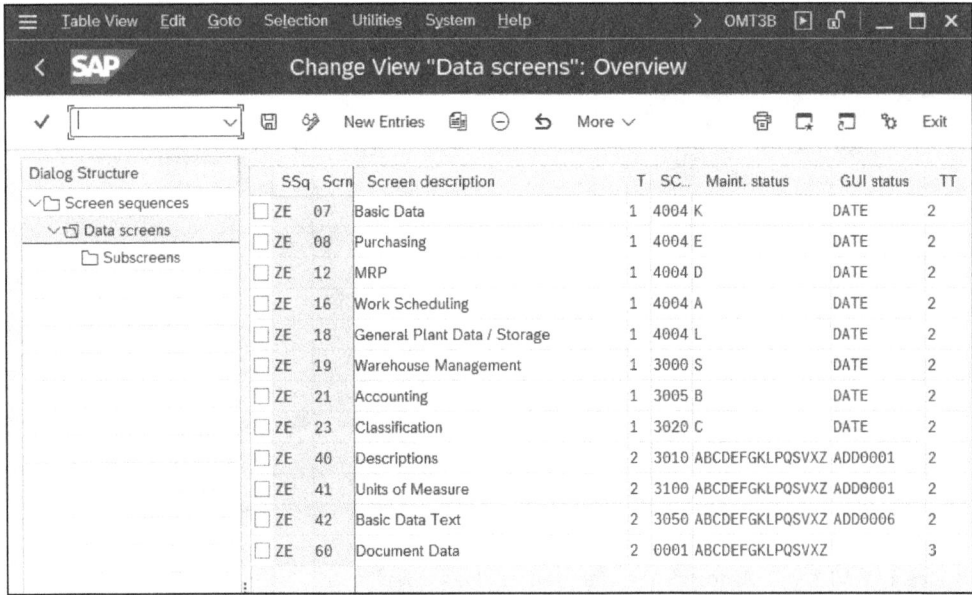

Figure 4.52 Material Master: Data Screens Overview

You use the **Subscreens** subfunction to determine the screen areas required for each screen. Here, you can choose from a selection of almost 300 screen areas for the various views. The following is a representative selection of screen areas for the views in our example.

Figure 4.53 shows a collection of screen areas for **Basic Data**:

- 1002: Material Numbers
- 2001: General Data
- 2007: Dimensions and EAN
- 2003: Document without Document Master Record

- 2004: Document with Document Master Record
- 2002: Other Data

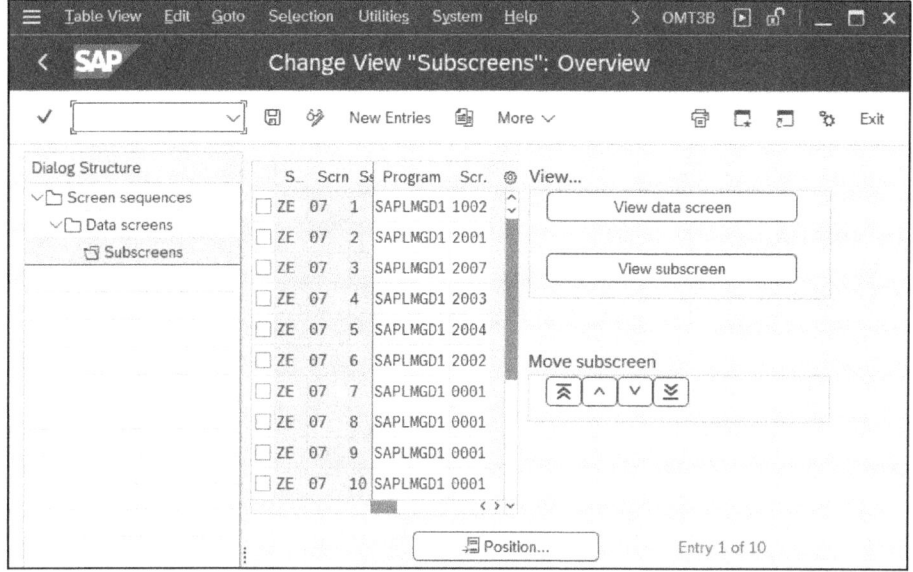

Figure 4.53 Material Master: General Data Subscreens

> **[!] Important Notes on the Definition**
>
> Screen area **0001** represents an empty placeholder. If, for example, you select a container screen with 15 screen areas (as is the case here) but you require only five, you must enter "0001" in the remaining screen areas.
>
> You can click the **View subscreen** button to preview the layout of the screen areas to be selected. You can click the **View data screen** button to display the finished data screen.

Figure 4.54 shows a representative collection of screen areas for **Purchasing**:

- 1001: Material Numbers with Plant
- 2301: Purchasing Data – General Data
- 2302: Purchasing Data – Purchasing Values
- 2303: Purchasing Data – Other Data

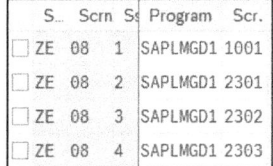

Figure 4.54 Material Master: Purchasing Subscreens

4 Configuring the Structure of Technical Systems

Figure 4.55 shows a representative collection of data for **MRP**:

- 1001: Material Numbers with Plant
- 2482: MRP Procedure
- 2483: Lot-Size Data
- 2484: Procurement
- 2485: Scheduling
- 2493: Availability Check

S...	Scrn	S	Program	Scr.
☐ ZE	12	1	SAPLMGD1	1001
☐ ZE	12	2	SAPLMGD1	2482
☐ ZE	12	3	SAPLMGD1	2483
☐ ZE	12	4	SAPLMGD1	2484
☐ ZE	12	5	SAPLMGD1	2485
☐ ZE	12	6	SAPLMGD1	2493

Figure 4.55 Material Master: MRP Subscreens

Figure 4.56 shows the most important screen areas for **Plant/Storage Data**:

- 1005: Material Numbers with Plant and Storage Location
- 2701: Storage Data – General Data
- 2702: Storage Data – Shelf Life Data

S...	Scrn	S	Program	Scr.
☐ ZE	18	1	SAPLMGD1	1005
☐ ZE	18	2	SAPLMGD1	2701
☐ ZE	18	3	SAPLMGD1	2702
☐ ZE	18	4	SAPLMGD1	2703
☐ ZE	18	5	SAPLMGD1	5801

Figure 4.56 Material Master: Plant/Storage Subscreens

Figure 4.57 shows the most important screen areas for **Accounting**:

- 1008: Material Numbers with Plant and Valuation Type
- 2801: Valuation Data for Current Period – General Data
- 2802: Valuation Data for Current Period – Current Valuation

S...	Scrn	S	Program	Scr.
☐ ZE	21	1	SAPLMGD1	1008
☐ ZE	21	2	SAPLMGD1	2801
☐ ZE	21	3	SAPLMGD1	2802

Figure 4.57 Material Master: Accounting Subscreens

4.7 Materials and Assemblies

Assign Secondary Screens

You use this Customizing function to define whether a secondary screen is to be displayed as an additional screen if you choose → **Additional Data** to call it or whether it's to be displayed as a secondary screen when you use the relevant button to call it directly.

Prerequisites

You must use the **Define Structure of Data Screens for Each Screen Sequence** Customizing function to define the basic layout of the screen sequence, and you must assign type 2 screens (secondary screen). If no type 2 screens exist, you can't assign any secondary screens.

Customizing Path

Logistics – General • Material Master • Configuring the Material Master • Assign Secondary Screens

Transaction

OMT3Z

Settings

If you want to display the screen as an additional screen, don't enter a function code in the **FCode** field (see Figure 4.58). The system then automatically assigns the next available number to the additional screen.

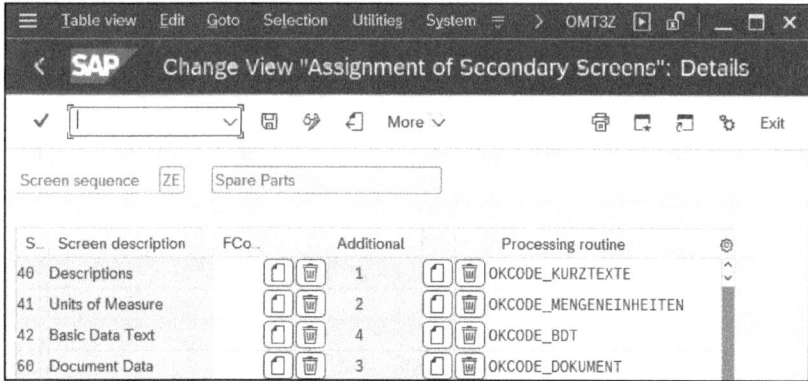

Figure 4.58 Material Master: Secondary Screens

If you want the screen to be displayed as a secondary screen, enter the relevant function code in the **FCode** column (e.g., "PB26" for **Basic Data Text**, "PB13" for **Material Forecast**). Here, you can choose from a total of 29 function codes.

When you use additional data or function codes to call screens, you must assign a suitable **Processing routine**. Here, you can choose from a total of 38 predefined processing routines.

4 Configuring the Structure of Technical Systems

Maintain Order of Main and Additional Screens

You can use this Customizing function to maintain the order in which the *main screens* are displayed when you maintain material master records in the **View Selection** dialog box as well as the order in which the main screens and *additional screens* are displayed as tabs.

Prerequisites

You must use the **Define Structure of Data Screens for Each Screen Sequence** Customizing function to define the basic layout of the screen sequence, and you must use the **Assign Secondary Screens** Customizing function to define the layout of the secondary screens.

Customizing Path

Logistics – General • Material Master • Configuring the Material Master • Maintain Order of Main and Additional Screens

Transaction

OMT3R

Settings

You define the order in which the main screens and secondary screens are displayed by assigning a status number to each main screen and additional screen (see Figure 4.59). Here, it's best to use numbers in intervals of 10 so that you have the option to insert additional screens later. The screens with the lowest numbers are displayed first.

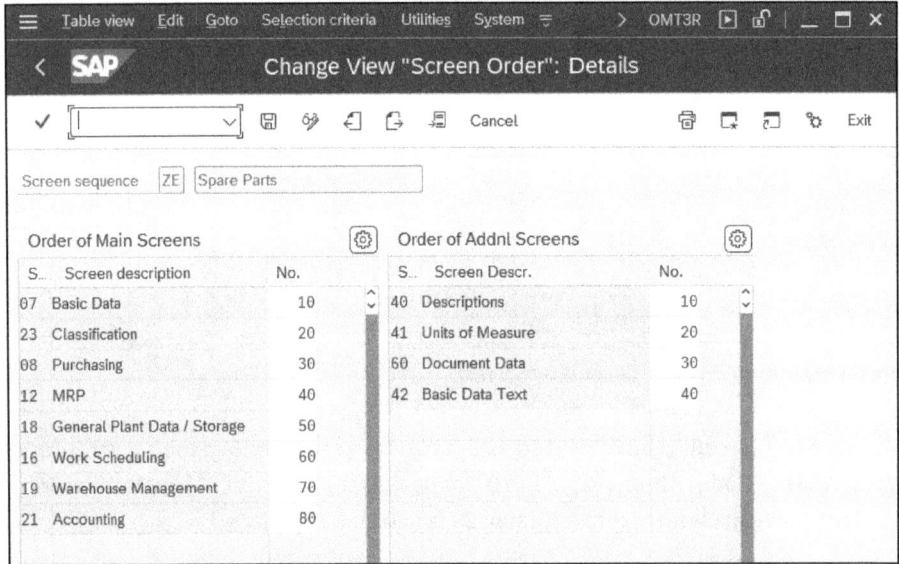

Figure 4.59 Material Master: Screen Order

Assign Screen Sequences to User/Material Type/Transaction/Industry Sector

You use this Customizing function to assign a screen sequence to a combination of one or more transactions, users, material types, or industry sectors. The system automatically displays this screen sequence if the relevant users choose this transaction, material type, or industry sector.

Prerequisites

You must use the **Define Structure of Data Screens for Each Screen Sequence** Customizing function to define the basic layout of the screen sequence, and you must use the **Assign Secondary Screens** Customizing function to define the layout of the secondary screens.

Customizing Path

Logistics – General • Material Master • Configuring the Material Master • Assign Screen Sequences to User/Material Type/Transaction/Industry Sector

Transaction

OMT3E

Settings

First, you use the **User screen reference** subfunctions to form groups. For example, you can form groups of transactions, users, material types, and industry sectors, and then use the **Screen sequence control** main function to assign the most suitable screen layout to the material master.

For this example, the **Material type screen reference** subfunction is used to define a separate material type reference **ERSA** for the material type **ERSA**. As you can see in Figure 4.60, you can group several material types together to form a material type screen reference. You can also combine or individualize users, transactions, and industry sectors in the same way.

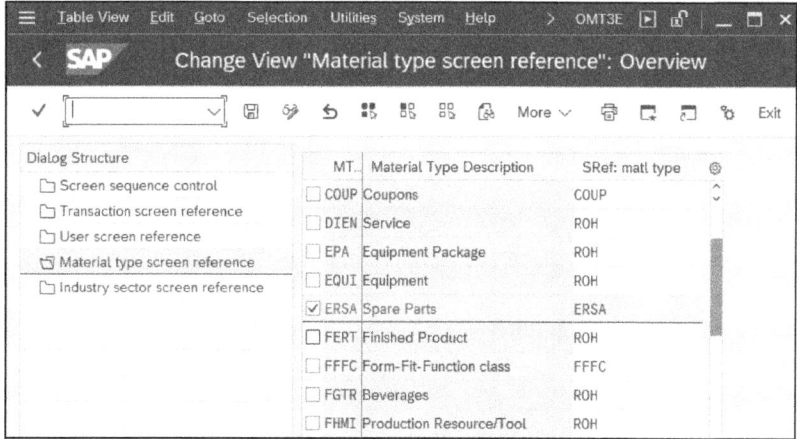

Figure 4.60 Material Master: Material Type Screen Reference

4 Configuring the Structure of Technical Systems

You use the **Screen sequence control** main function to assign a screen sequence on the basis of the screen references. It isn't necessary to have created a screen reference for each object. Rather, you can replace each object with a wild card (*) so that the combination doesn't refer to a particular screen reference. However, SAP has defined some unique priorities here, as shown in Table 4.1. In this table, X typically denotes that an entry exists for this value (e.g., material type = ERSA), while * denotes that no entry exists for this value.

Priority	Transaction	User	Material Type	Industry Sector
1	X	X	X	X
2	X	X	*	*
3	X	*	X	X
4	X	*	X	*
5	X	*	*	*

Table 4.1 Priorities for Controlling Screen Sequences

If these priorities are applied to the example, it means the following:

- Because the user isn't assigned to a user group, the **Transaction/Material Type** combination is used so that the screen sequence **ZE (Spare Parts)** is displayed when you maintain a material master (see Figure 4.61).
- If, however, a user from the user group **BO** needs to maintain a material master of the material type **ERSA**, screen sequence **01** is assigned because the **Transaction/User** entry is given higher priority than the **Transaction/Material Type** entry.

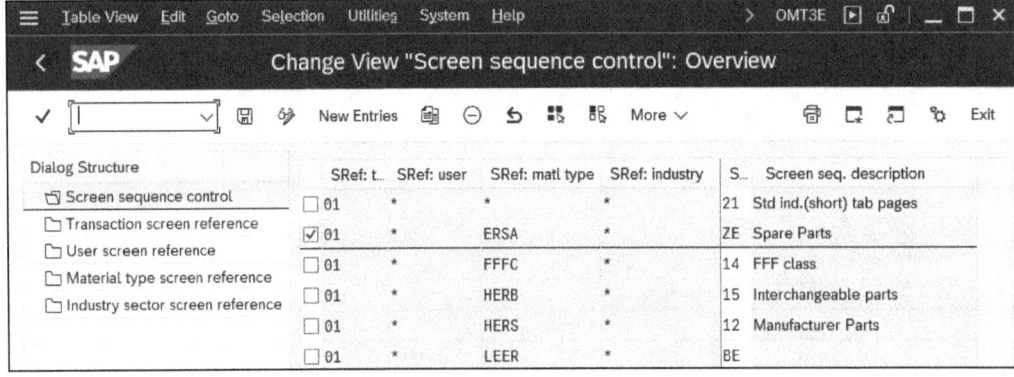

Figure 4.61 Material Master: Screen Sequence Control

4.7.3 Field Selection

The field selection technique, introduced in Chapter 3, Section 3.9, can be used in conjunction with equipment or orders, for example. You can also use the field selection option with the material master. However, the field selection technique deployed here is completely different from the technique described previously. Therefore, the next section discusses how you can configure the field selection for use with the material master.

In the case of the material master, the field selection depends on three factors (see Figure 4.62):

- Material type
- Industry sector
- Plant

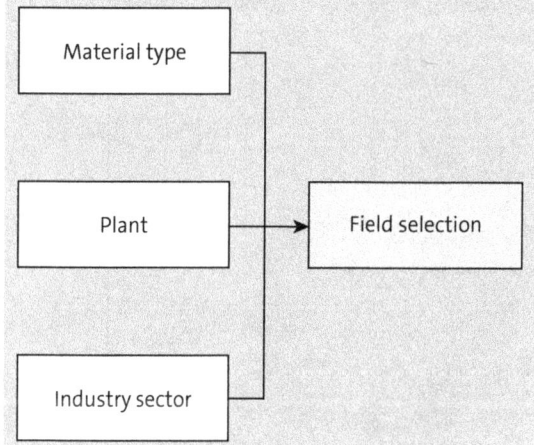

Figure 4.62 Material Master: Field Selection

You've used the **Define Attributes of Material Types** Customizing function to define a **Field reference** for the material type (refer to the **Define Attributes of Material Types** Customizing function). Let's now look at the most important Customizing functions in this area.

Assign Fields to Field Selection Groups

You use this Customizing function to assign the individual fields of the material master to a field selection group. You can assign each field to only one field selection group.

Prerequisites

There are no specific prerequisites.

4 Configuring the Structure of Technical Systems

Customizing Path

Logistics – General • Material Master • Field Selection • Assign Fields to Field Selection Groups

Transaction

OMSR

Settings

SAP has reserved certain areas, groups 111–120 and 211–240, as available to you for your own field selections. However, it's not necessary for you to form your own groups. You can simply retain the standard SAP assignment and base your field selections on that assignment. Figure 4.63 shows the various field selection groups available.

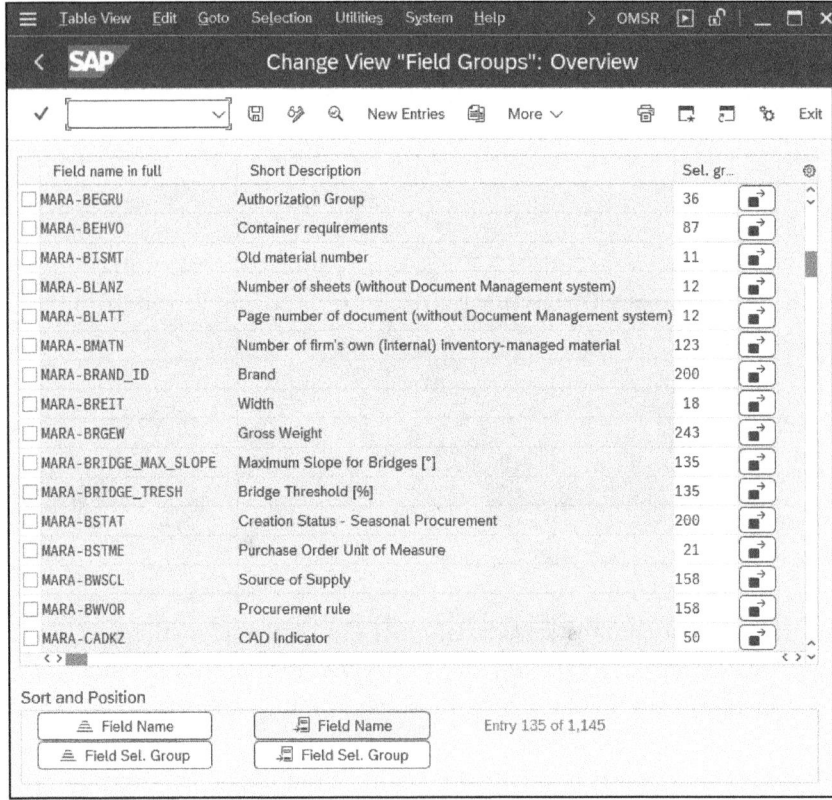

Figure 4.63 Material Master: Field Groups

Define Industry Sectors and Industry Sector-Specific Field Selection

You use this Customizing function to define your industry sectors and to determine the field reference that your industry sectors will use to influence the field selection.

Prerequisites

There are no special prerequisites.

4.7 Materials and Assemblies

Customizing Path

Logistics – General • Material Master • Field Selection • Define Industry Sectors and Industry Sector-Specific Field Selection

Transaction

OMS3

Settings

You assign a field reference to the industry sectors that will influence the field selection. Here, you can assign a custom field reference to each industry sector, or you can combine several industry sectors (see Figure 4.64).

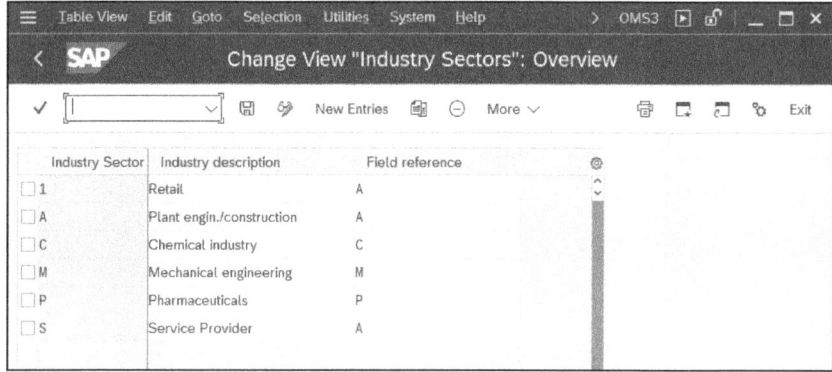

Figure 4.64 Material Master: Industry-Specific Field Selection

Define Plant-Specific Field Selection and Plant-Specific Screen Selection

You use this Customizing function to define the field reference that your plants will use to influence the field selection.

Prerequisites

You must define the plants beforehand.

Customizing Path

Logistics – General • Material Master • Field Selection • Define Plant-Specific Field Selection and Plant-Specific Screen Selection

Transaction

OMSA

Settings

You assign a field reference to the plants that will influence the field selection. Here, you can assign a custom field reference to each plant, or you can combine several plants (see Figure 4.65).

4 Configuring the Structure of Technical Systems

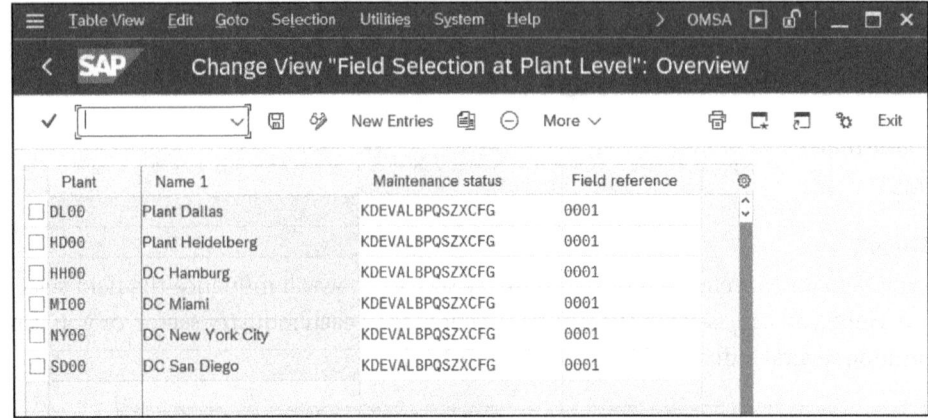

Figure 4.65 Material Master Plant-Specific Field Selection

Maintain Field Selection for Data Screens

You use this Customizing function to define a particular field reference and its influence on the field selection:

- Display field
- Required entry field
- Normal entry field
- Hide

Prerequisites

You must define the plants beforehand.

Customizing Path

Logistics – General • Material Master • Field Selection • Maintain Field Selection for Data Screens

Transaction

OMS9

Settings

Several influencing factors can now affect the field selection (see Figure 4.66), as follows:

- The field selection of the material type **ERSA** can define a field as a required entry field.
- On the other hand, the field selection of plant **DL00** can define the field as a normal entry field.
- At the same time, the field selection of industry sector **M** wants to hide this field.

4.7 Materials and Assemblies

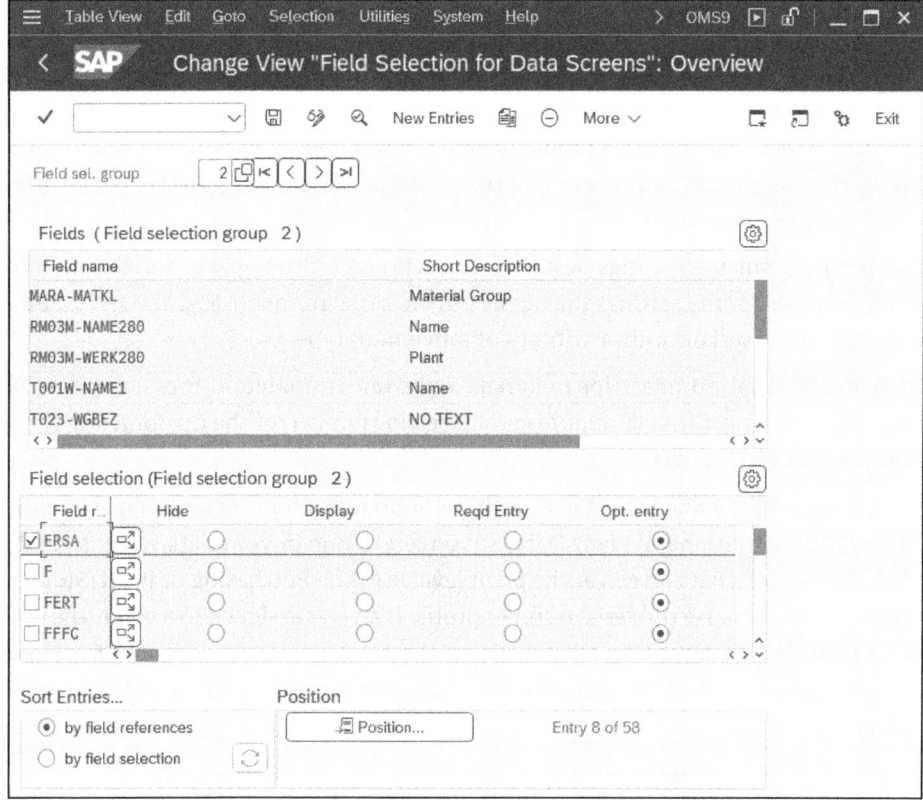

Figure 4.66 Material Master: Field Selection

If several different influencing factors affect the field selection, the priority rules described earlier also apply here (see Chapter 3, Section 3.9):

- **Priority 1: Hide**
  If one of the influencing factors defines a field as hidden, it's hidden regardless of the field selection control for other influencing fields.

- **Priority 2: Display**
  If it's not hidden, and if an influencing factor defines it as a display field, it becomes a display field, even if other influencing factors define it as a required entry or normal entry field.

- **Priority 3: Required entry**
  If it's not a hidden field or a display field, and if an influencing factor defines it as a required entry field, it becomes a required entry field, even if other influencing factors define it as a normal entry field.

- **Priority 4: Normal entry field**
  If no influencing factor defines it elsewhere, this field remains a normal entry field.

4.8 Serial Numbers

You create serial numbers for a material number, and you can specify as many serial numbers as you want for one material number. A *serial number* is an individual item that corresponds to a piece of equipment, and the serial number function enables you to place the equipment in storage. In terms of usage, a material serial number corresponds to a piece of equipment.

All of the Customizing settings described in Section 4.3 also apply to serial numbers, but there are also some settings that apply only to serial numbers (e.g., the serial number profile or the serialization attributes of movement types).

If you create serial numbers for materials and want to maintain these as pieces of equipment, you must first configure certain settings in two of the Customizing functions described previously:

- In **Configuring the Material Master** with the **Subscreens** subfunction, you assign the **SAPLMGD1 5801 (General Plant Data)** subscreen to one of your data screens. To do this, you select a data screen at the plant level (e.g., the **Purchasing** or **Plant/Storage Data** screen) because the serial number profile is always assigned to a plant. The system then displays a subscreen in the material master (as shown in Figure 4.67). Here, you can define serial data such as the serial number profile and the serial number level.

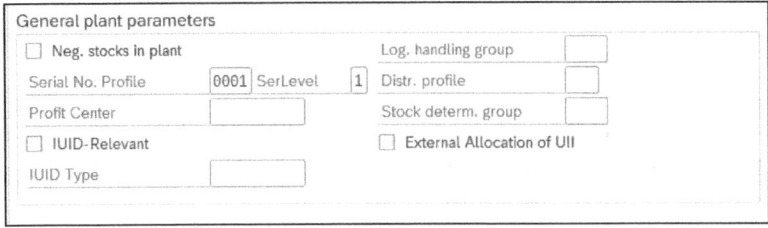

Figure 4.67 Material Master: Subscreen Serial Numbers

- In **Define Additional Business Views for Equipment Categories**, you must check the box to set the **SD** (serial data; refer to Figure 4.16) indicator for the equipment category to which you want to assign the serial number. Then, the system displays the serial data subscreens in the equipment master (see Figure 4.68). Here, you can assign the relevant serial data (e.g., the **Serial Number**, **Plant**, **StorageLocation**, and **Stock batch**) to the piece of equipment.

Figure 4.69 provides an overview of the Customizing functions for serial numbers.

The most important of these Customizing functions are covered in the following sections.

4.8 Serial Numbers

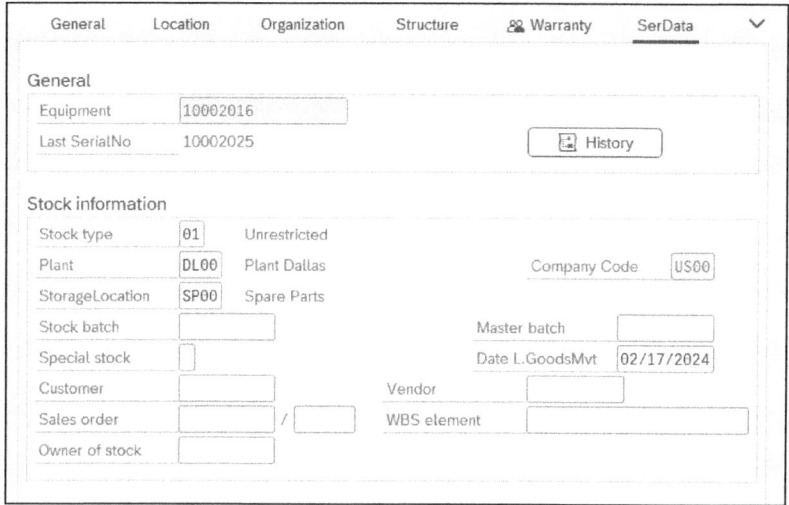

Figure 4.68 Equipment Master with Serial Data

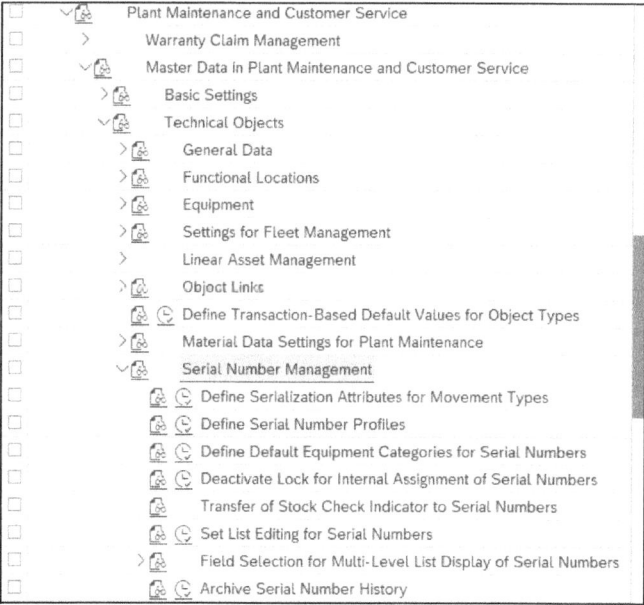

Figure 4.69 Serial Number Management Customizing Functions Overview

Define Serial Number Profiles

You use this Customizing function to define serial number profiles. In the material master, you must assign a serial number profile at the plant level to each *material* that you want to serialize. This means that for each plant, you can assign a custom serial number profile to a material. A material may require serial numbers in one plant but not in another.

4 Configuring the Structure of Technical Systems

Prerequisites
There are no special prerequisites.

Customizing Path
Plant Maintenance and Customer Service • Master Data in Plant Maintenance and Customer Service • Technical Objects • Serial Number Management • Define Serial Number Profiles

Transaction
OIS2

Settings
On the **Overview** screen for serial number profiles, you configure the following settings in addition to the profile name and profile description (see Figure 4.70):

- **ExistReq.**
  Check the box to set this indicator if you want the serial number entered for a business process (e.g., for a goods receipt) to have been created as a master record beforehand. This is a good cautionary step that prevents the system from being able to automatically create serial numbers for a goods receipt, for example.

- **Category**
  Enter the equipment category that you want the system to assign when automatically creating serial numbers for a business process, such as receipt of goods.

- **StkCk**
  Use this stock check indicator to define whether you want the system to perform a stock check when assigning serial numbers. If you do, you also define the type of message (a warning message or an error message) that you want the system to issue if any stock inconsistencies arise in relation to inventory management.

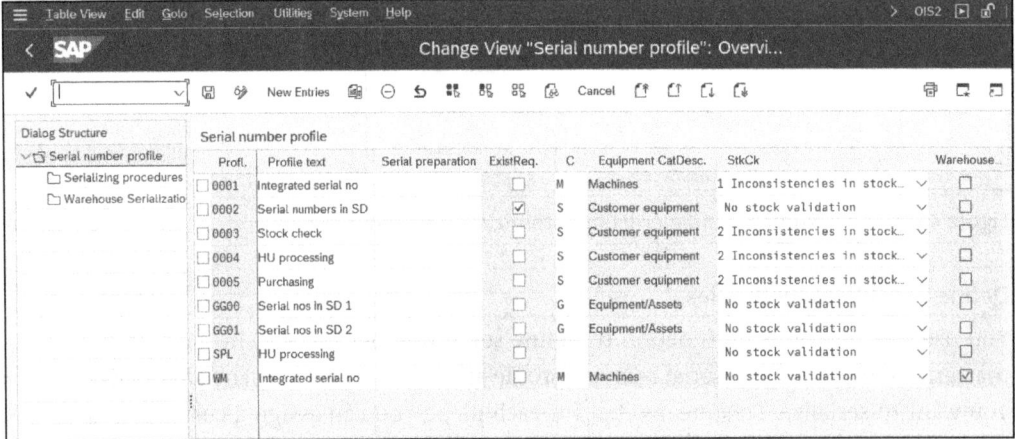

Figure 4.70 Serial Number Management: Serial Number Profile Overview

234

- **Warehouse**
  Via data transmission, the serial number profile is harmonized with SAP Extended Warehouse Management (SAP EWM).

Then, on the detail screen, you define the **Serializing procedures** that are permitted for this serial number profile (see Figure 4.71).

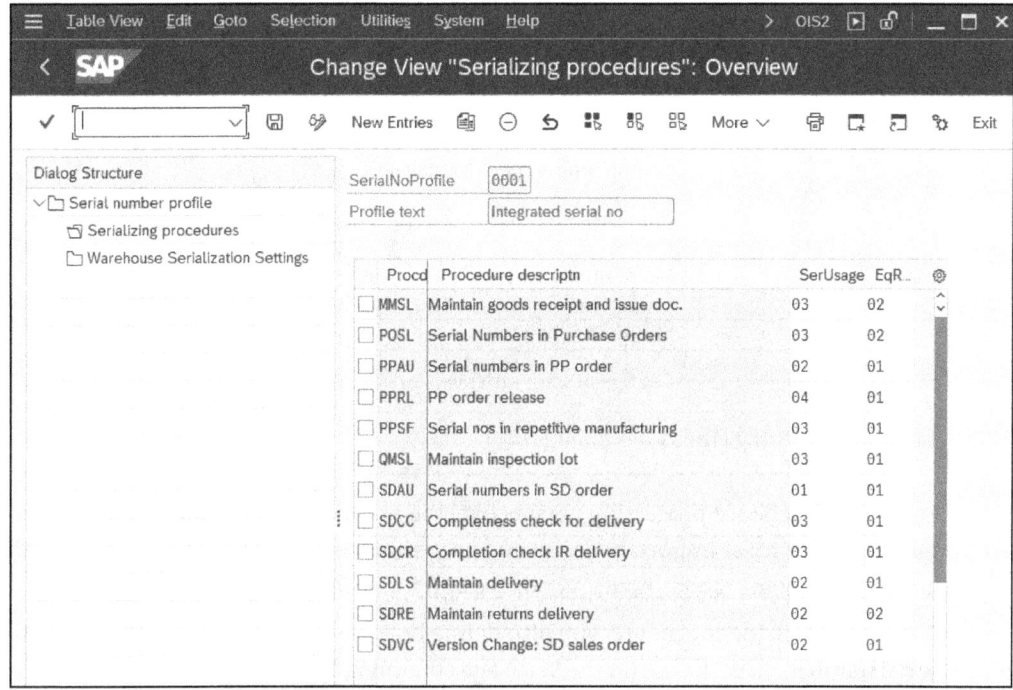

Figure 4.71 Serial Number Management: Serializing Procedures

Table 4.2 lists the serializing procedures available.

Serializing Procedure	Description
ADVR	Material inspection in advanced returns management
HUSL	Handling unit
JITC	Assign serial number to customer just-in-time (JIT) call
MMSL	Maintain goods receipt and issue document
PIAU	Serial numbers in process orders
PIRL	Release process order
POSL	Serial numbers in purchase orders

Table 4.2 Serializing Procedures

4 Configuring the Structure of Technical Systems

Serializing Procedure	Description
PPAU	Serial numbers in production planning order
PPRL	Production planning order release
PPSF	Serial numbers in repetitive manufacturing
PRSL	Serial numbers in purchase requisitions
QMSL	Maintain inspection lot
SDAU	Serial numbers in sales order
SDCC	Completeness check for delivery
SDCR	Completeness check for returns delivery
SDLS	Maintain delivery
SDRE	Returns delivery
SDVC	Change in sales order version

Table 4.2 Serializing Procedures (Cont.)

From a maintenance processing perspective, the serializing procedure **MMSL** is, without doubt, the most important serializing procedure because it enables you to permit goods movements (goods issues, goods receipt) for serial numbers.

For each serializing procedure, you also define how you want the system to deal with serial numbers in each case (in the **SerUsage** column in Figure 4.71):

- 01
  No serial numbers are assigned.
- 02
  Serial numbers can be assigned.
- 03
  Serial numbers must be assigned.
- 04
  Serial numbers are automatically assigned.

If serial numbers are automatically assigned to a procedure, the dialog box in which you enter serial numbers isn't displayed. Instead, the system automatically assigns the next available number in numeric order.

You use the **EqReq** (equipment required) column to control whether a piece of equipment must be created for the relevant serializing procedure. If you choose indicator **01** (default: without equipment), you can still decide at a later stage to create an equipment master record when assigning serial numbers in the relevant dialog box. If you choose indicator **02** (always with equipment), it is binding, meaning you can't undo it.

4.8 Serial Numbers

If you want to work with serial numbers in inbound or outbound deliveries in SAP EWM, you must define a serial number profile in which you determine whether the serial number requirement should exist at the following levels (see Figure 4.72):

- At the document level
- At the warehouse level
- At the stock level

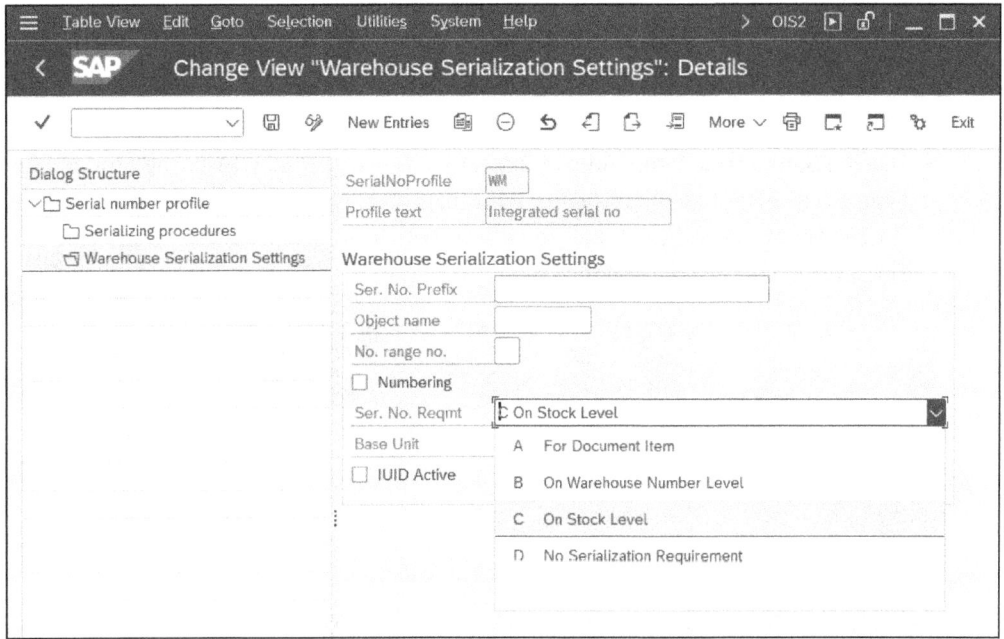

Figure 4.72 Serial Number Management: Serial Numbers in SAP EWM

Define Serialization Attributes for Movement Types

You can use this Customizing function to define serialization attributes for individual *movement types* or *movement type groups*. You only need to maintain this setting if the assignment of serialization attributes at the *serializing procedure level* is too broad and you want to differentiate it further.

At present, only the serializing procedure **MMSL** is supported.

Prerequisites
You must maintain the serial number profiles beforehand.

Customizing Path
Plant Maintenance and Customer Service • Master Data in Plant Maintenance and Customer Service • Technical Objects • Serial Number Management • Define Serialization Attributes for Movement Types

4   Configuring the Structure of Technical Systems

Transactions

S_ALR_87000310_1 and S_ALR_87000310_2

Settings

For example, in the serial number profile **0001** for the serializing procedure **MMSL**, you can set **SerUsage** (serial number usage) to **02**. In other words, you can specify a serial number for goods receipts, goods issues, and stock transfers, but it's not mandatory for you to do so.

If, however, you now want to make it mandatory to enter serial numbers for a goods issue from a purchase order or a goods issue from a production order (and only there), you can achieve this by configuring the following settings:

- Use the **Define Flow Type Groups** subfunction (see Figure 4.73) to define a movement type group (here, **PM01**) for the serial number profile **0001** and the serializing procedure **MMSL**. Here, set **SerUsage** (serial number usage) to **03**, which will indicate that serial numbers must be assigned.

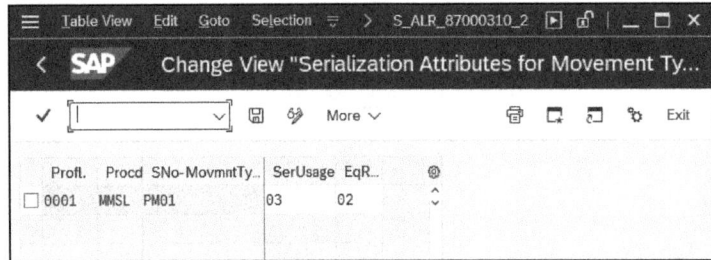

Figure 4.73 Serial Number Management: Attributes for Movement Type Groups

- You use the **Assign Movement Type Groups to Movement Types** subfunction (see Figure 4.74) to assign your movement type group **PM01** to movement types **101** (**Goods Issue from Purchase Order**) and **101** (**Goods Issue from Production Order**), both without consumption.

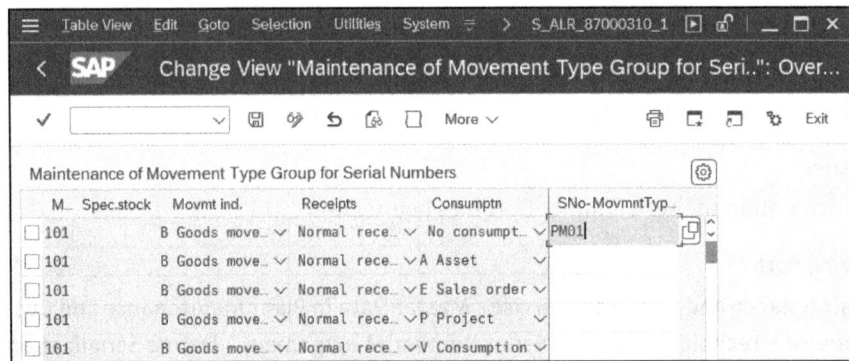

Figure 4.74 Serial Number Management: Assigning Movement Type Groups to Movement Types

4.9 Bill of Materials

A *bill of materials* (BOM) is essentially a complete, formally structured list of all components that belong to a product or assembly. The components can be parts or assemblies that can or can't be kept in storage, and the assemblies, in turn, can have BOMs that describe them in greater detail. This gives rise to a multilevel BOM structure.

In plant maintenance, you can primarily use BOMs for the following two purposes:

- **Structure description**
  This is the act of describing the structure of a technical object or material. You can use the BOM to pinpoint the location of damage that has occurred or the location where maintenance tasks are performed on a technical object.

- **Spare parts assignment**
  This is the assignment of spare parts for a technical object or material.

From a maintenance perspective, there are three different types of BOMs:

- An equipment BOM for a special piece of equipment
- A functional location BOM for a special functional location
- A material BOM for a construction type and therefore for a group of functional locations or pieces of equipment

The SAP Reference IMG has many Customizing functions for managing BOMs (see Figure 4.75) because they are used not only in plant maintenance but also in production planning, costing, quality management, and other areas.

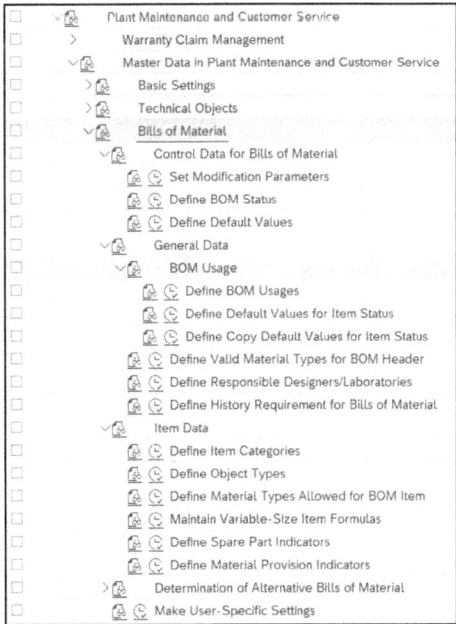

Figure 4.75 BOM: Customizing Functions Overview

4   Configuring the Structure of Technical Systems

However, many of the Customizing functions that relate to BOMs aren't required for plant maintenance. Consequently, this section only introduces those Customizing functions that are of interest from a maintenance perspective and that therefore should be examined.

Define BOM Status

You use this Customizing function to define the BOM status that you require. The BOM status in the BOM header enables you to control whether the BOM is still locked or released for use. If it's released, you should specify the purpose for which it has been released.

Prerequisites

There are no special prerequisites.

Customizing Path

Plant Maintenance and Customer Service • Master Data in Plant Maintenance and Customer Service • Bills of Material • Control Data for Bills of Material • Define BOM Status

Transaction

OICH

Settings

Figure 4.76 shows the standard SAP delivery. If you don't want to release a BOM yet, check the box to set status **2** (**Inactive**) in the BOM header. Otherwise, check the box to set status **1** (**Active**).

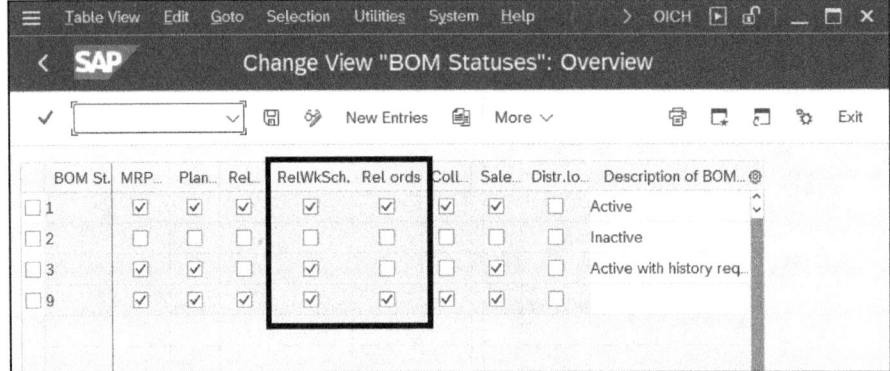

Figure 4.76 BOM: Statuses Overview

If you define your own statuses, take note of the functions that play a role in plant maintenance:

- RelWkSch.
  Released for the task list

4.9 Bill of Materials

- **Rel ords**
  Released for the order

The remaining functions aren't used in plant maintenance.

Define BOM Usages

You use this Customizing function to define which user departments (e.g., production, plant maintenance, construction) use BOMs of a certain *usage*.

Prerequisites
There are no special prerequisites.

Customizing Path
Plant Maintenance and Customer Service • Master Data in Plant Maintenance and Customer Service • Bills of Material • General Data • BOM Usage • Define BOM Usages

Transaction
OICD

Settings
Figure 4.77 shows the standard SAP delivery. SAP delivers usage **4** (**Plant Maintenance**) for plant maintenance purposes.

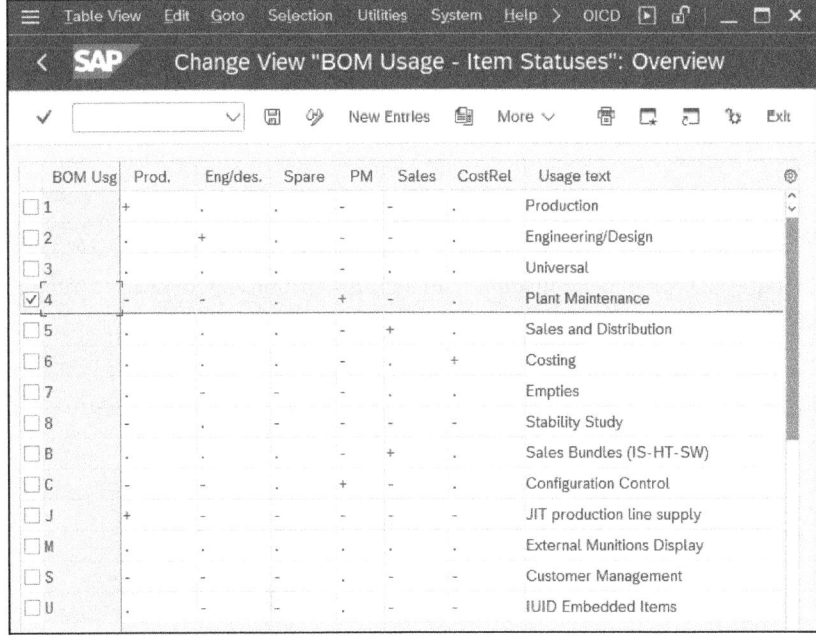

Figure 4.77 BOM: Usage

Generally, this usage is sufficient. We've only ever encountered one customer who, in addition to usage 4, set up usage 1 for information BOMs, which, for their company, contained alternative spare parts for the material number in the BOM header. In other words, the BOM items list all materials that can be used as an alternative to the spare part in the BOM header.

If you define your own BOM usages, make sure that you select the **PM** (relevant for plant maintenance) column.

Define Valid Material Types for the BOM Header

You require this Customizing function only if you want to create BOMs for construction types. If you want to work solely with equipment BOMs or BOMs for a functional location, this Customizing function doesn't play any role.

You use this Customizing function to define the material types that you want to create for the BOMs for construction types.

Prerequisites

You must define the necessary material types beforehand (see the **Define Attributes of Material Types** Customizing function).

Customizing Path

Plant Maintenance and Customer Service • Master Data in Plant Maintenance and Customer Service • Bills of Material • General Data • Define Valid Material Types for BOM Header

Transaction

OICG

Settings

If you don't want to make any restrictions whatsoever, a fully masked entry (BOM usage = * and material type in BOM header = * in Figure 4.78) suffices. In other words, all material types can be used for all BOM usages.

If you want to restrict all BOMs apart from maintenance BOMs, which you don't want to restrict, mask only usage 4 (BOM usage = 4 and material type in BOM header = *). In other words, all material types can be used for maintenance BOMs.

Otherwise, list exactly those material types that are permitted to be used for BOMs for construction types in plant maintenance (e.g., **ERSA**, **FERT**, **IBAU** in Figure 4.78).

Define Item Categories

In the BOM, you must assign an item category to each item. You use this Customizing function to define the item categories and their properties, which you require in your BOMs.

4.9 Bill of Materials

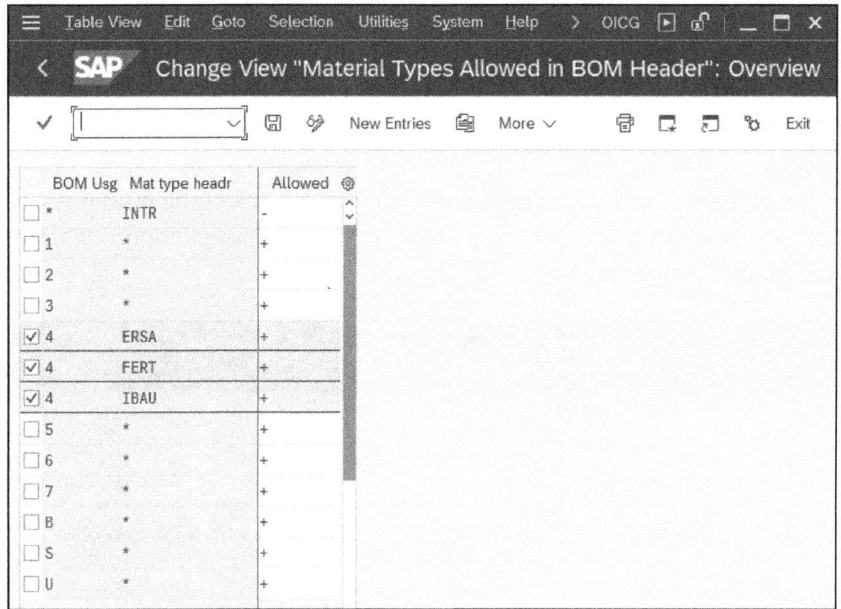

Figure 4.78 BOM: Material Types for BOM Header

Prerequisites

There are no special prerequisites.

Customizing Path

Plant Maintenance and Customer Service · Master Data in Plant Maintenance and Customer Service · Bills of Material · Item Data · Define Item Categories

Transaction

OICK

Settings

Figure 4.79 shows the standard SAP delivery, which generally more than suffices here. From a maintenance perspective, the following item categories play a role:

- **L: Stock item**
  If you later assign a material with item category **L** to an order, it triggers a reservation.
- **N: Non-stock item**
  If you later assign a material with item category **N** to an order, it triggers a purchase requisition.
- **I: PM structure element**
  You use these items solely to define a substructure. Neither a reservation nor a purchase requisition is triggered.

243

4 Configuring the Structure of Technical Systems

- **D: Document item**
  You use these items as an alternative to direct document assignment to an object or to assign additional documents to an object. Neither a reservation nor a purchase requisition is triggered.

- **T: Text item**
  You can use text items to set headings above certain sections of the BOM (e.g., the text item "The electrical spare parts start here."). Neither a reservation nor a purchase requisition is triggered.

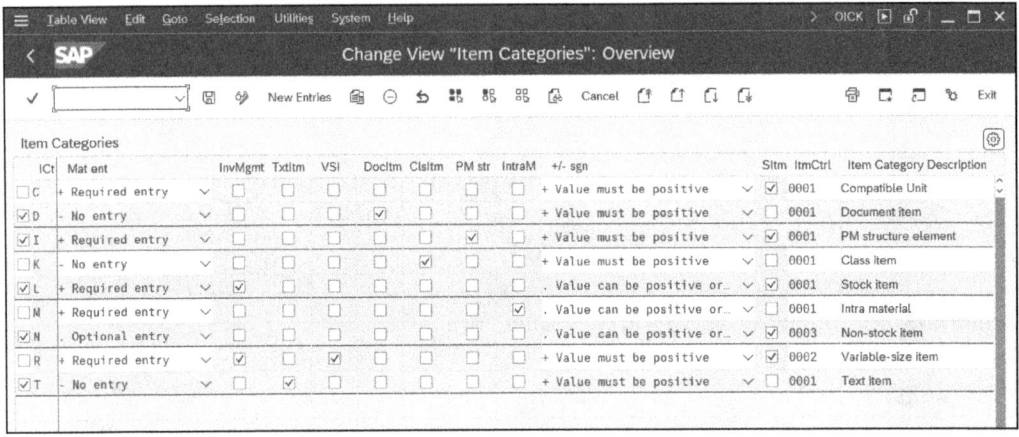

Figure 4.79 BOM: Item Categories

Define Material Types Allowed for BOM Items

You use this Customizing function to define which *material types* are provided in conjunction with the *BOM usage* and the material type in the BOM header for creating BOM items.

Prerequisites

You must maintain the material types in the BOM item, the material types permitted in the BOM header, and the BOM usage beforehand.

Customizing Path

Plant Maintenance and Customer Service • Master Data in Plant Maintenance and Customer Service • Bills of Material • Item Data • Define Material Types Allowed for BOM Item

Transaction

OICP

Settings

As with the material types for the BOM header, you have various options here (see Figure 4.80):

4.9 Bill of Materials

- If you don't want to impose any restrictions whatsoever, create a fully masked entry (i.e., usage = *, material type in BOM header = *, and material type in BOM item = *). Specific entries are necessary only if you want to exclude certain material types.
- If you don't want to impose any restrictions for plant maintenance, mask only the material types (i.e., BOM usage = **4**, material type in BOM header = *, and material type in BOM item = *). Specific entries are necessary only if you want to exclude certain material types.
- If you want to make a restriction for only one particular material type for plant maintenance, mask only the material type at the item level (i.e., BOM usage = **4**, material type in BOM header = **ERSA**, and material type in BOM item = *). Specific entries are necessary only if you want to exclude certain material types.
- If you want to define precisely which material types at the item level are to be permitted for which item types at the header level in plant maintenance, create specific entries accordingly (e.g., BOM usage = **4**, material type in BOM header = **IBAU**, material type in BOM item = **ERSA**).

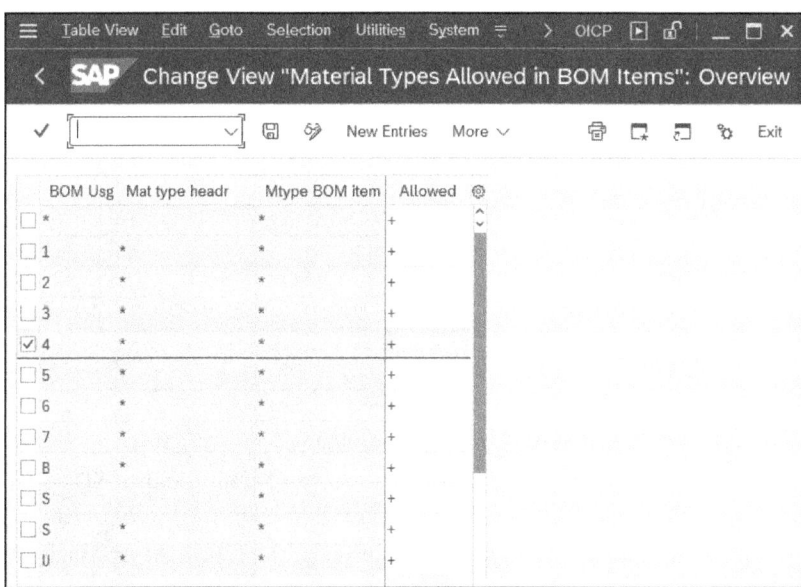

Figure 4.80 BOM: Material Types for BOM Items

Define Priorities for BOM Usage

You use this Customizing function to define your priorities for the automatic selection of a BOM usage.

Prerequisites

You must maintain the BOM usages beforehand.

Customizing Path

Plant Maintenance and Customer Service • Master Data in Plant Maintenance and Customer Service • Bills of Material • Determination of Alternative Bills of Material • Define Priorities for BOM Usage

Transaction

OICJ

Settings

In SAP S/4HANA Asset Management, you require only one priority for BOM usage **4**, which you can then process further using the **Define Selection Criteria for Alternative Determination** Customizing function (see Figure 4.81).

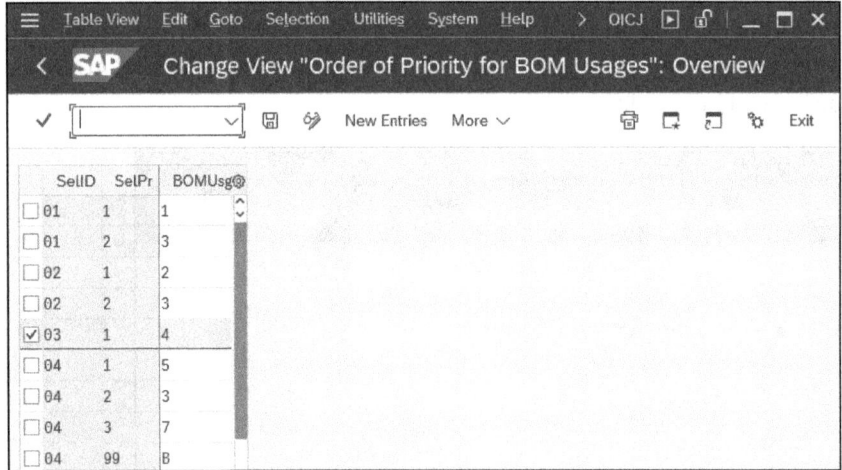

Figure 4.81 BOM: Priority of BOM Usages

Define Selection Criteria for Alternative Determination

You use this Customizing function to define, in the case of maintenance BOMs, an application (e.g., **INST**) that accesses the selection ID for determining *alternative BOMs* according to priorities (e.g., **03**). This application is used in the following transactions for selecting BOMs:

- Transaction IH01 (Structural Display for Functional Locations)
- Transaction IH03 (Structural Display for Equipment)
- Transaction IH04 (Structural Display for Equipment Bill of Materials)
- Transaction IH05 (Structural Display for Material BOM)
- Transaction IH12 (Structural Display for Bill of Materials for Functional Location)

Prerequisites

You must maintain the selection ID for determining alternative BOMs according to priorities beforehand.

Customizing Path

Plant Maintenance and Customer Service • Master Data in Plant Maintenance and Customer Service • Bills of Material • Determination of Alternative Bills of Material • PM-Specific Selection Criteria for Alternative Determination • Define Selection Criteria for Alternative Determination

Transaction

OICQ

Settings

Define an application (e.g., **INST**) in which you assign the selection ID for determining alternative BOMs according to priorities (e.g., **03**) (see Figure 4.82).

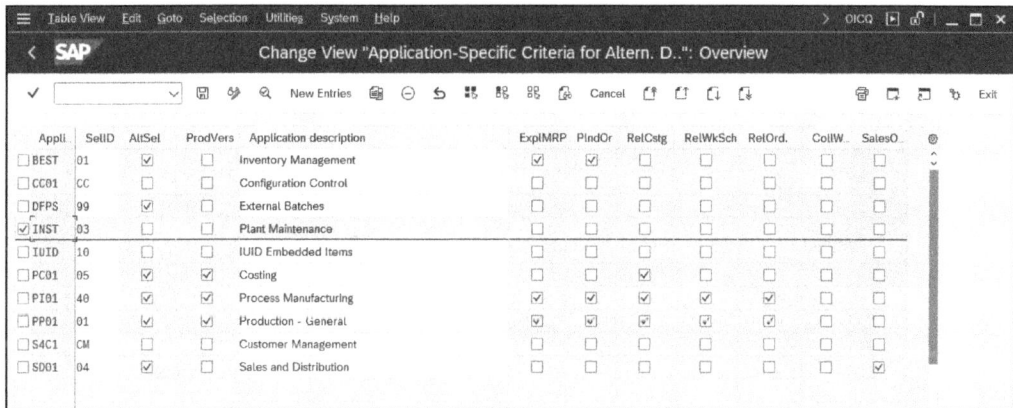

Figure 4.82 BOM: Alternative Determination

4.10 Summary

SAP S/4HANA Asset Management provides a broad range of potential structuring elements, among which, functional location and equipment are the major ones. In this chapter, you were shown their most important Customizing functions. The key takeaways from this chapter are as follows:

- The most important functional location Customizing function is maintaining the structure indicator, which determines length and rules for the functional location number, number of levels, and labeling procedure.
- Another important functional location Customizing function is defining the functional location category, which influences screen layout and status profile.

- Also important is the activation of alternative labeling to get the opportunity to rename functional locations.
- The most important equipment Customizing function belongs to equipment category, which influences, for example, number ranges, screen layout, status profile, and the ability to be installed at functional locations.

In this chapter, you also learned about the Customizing functions of other structuring elements:

- Fleet objects (e.g., assigning special fleet screens or defining consumable and engine types)
- Links and networks (e.g., defining types and media)
- Linear asset management (e.g., defining linear pattern types or network attributes)
- Material (e.g., defining material types or configuring a material master layout)
- Serial numbers (e.g., defining a serial number profile and assigning serializing procedures)
- BOM (e.g., defining BOM usages or valid material types for BOM headers and items)

Chapter 5
Configuring the Work Order Cycle

This chapter introduces you to the Customizing functions that will enable you to adjust the plant maintenance work order cycle to your requirements. You'll learn how to configure your tools, and you'll get numerous tips related to configuring Customizing settings for these functions.

This chapter explains the core business process of plant maintenance, that is, the work order cycle. After seeing many companies from the inside, it's clear to me that each company has its own ideas about what business processes in plant maintenance should be like and how they must be mapped in the SAP system. Your first thoughts must be how you can map your day-to-day activities in SAP S/4HANA Asset Management and how SAP S/4HANA Asset Management will support you in accomplishing your tasks. No book on earth can do this work for you, but this chapter will actively support you in your work.

To give you as complete a picture as possible of the Customizing functions available, this chapter focuses on the *scheduled repair* business process because this very complex maintenance process contains most of the Customizing functions you need. Figure 5.1 shows the individual steps within this business process.

The following describes these business process steps:

1. **Create notification**
   First, you enter the notification in relation to specific damage or any other request (e.g., a request for a modification). The following information is important here: the technical object, a description of what has happened and what repairs are necessary, and the date by which the work is to be completed.

2. **Plan order**
   The order is created and planned from the notification. Typical planning tasks include creating operations, reserving spare parts, appointing external companies, and planning operating times. The account assignment rules are also defined in this step.

3. **Schedule order**
   The order is transferred to scheduling, where you check the corresponding availability (in particular, material availability), provide the required capacities, and print out the shop papers.

4. **Execute order**
   This processing phase involves the withdrawal of the spare parts from the warehouse and the actual processing of the order.

5. **Confirm order**
   After you complete the tasks, the required actual times are confirmed in completion. Technical completion confirmations, regarding how the damage was handled and the condition of the technical object, are entered here.

6. **Complete order**
   From a technical perspective, the order is now complete. The maintenance planner administers the technical completion, whereas the business completion has to be settled by controllers. All the information will be updated in the order history.

Step	Content	Typical Role	Integration
1. Create Notification	Problem, Technical Object, Priority, Date	Employee	PM
2. Plan Order	Operations, Materials, Externals, Work, PRT	Maintenance Planner	PM, MM, FI, CO, HR
3. Schedule Order	Check Availabilities, Permits, Release, Print	Maintenance Planner	PM, MM, FI, CO, HR
4. Execute Order	Goods Issues Planned and Unplanned	Maintenance Technician	PM, MM, FI, CO
5. Confirm Order	Time and Technical Confirmations	Maintenance Technician	PM, CO, HR
6. Complete Order	Technical Completion, Settlement, Business Completion	Maintenance Planner Controller	PM, CO
History			

Figure 5.1 Work Order Cycle

This process requires three business objects in the SAP system—notifications, orders, and completion confirmations—which are described separately in the following sections.

5.1 Notifications

In maintenance processing, you use *notifications* in exceptional operational situations to perform the following activities:

- Describing the technical emergency associated with an object
- Requesting a required task in the maintenance department
- Documenting the work performed

Consequently, maintenance tasks are documented in notifications so that they can be analyzed in the long term. Figure 5.2 shows the structure and content of a maintenance notification.

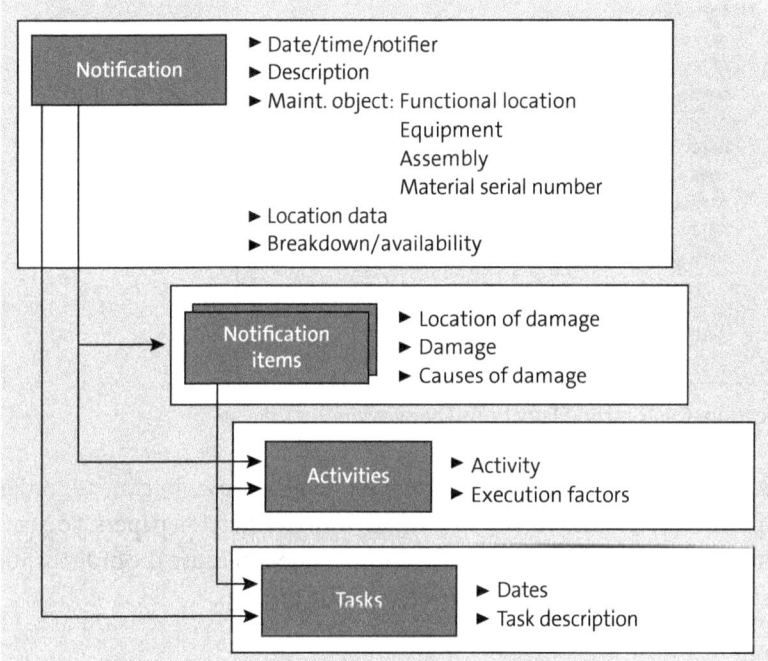

Figure 5.2 Notification: Structure and Content

The Customizing functions that you can use to adjust notifications to your requirements are described in this subsection. A complete overview of the Customizing functions for notifications is provided in Figure 5.3.

The following Customizing functions and their subfunctions also apply to notifications, but they've already been described in Chapter 3:

- **Define Number Ranges**
- **Set Field Selection for Notifications**
- **Define Partner Determination Procedure and Partner Function**
- **Define Object Information Keys**
- **User Status for Notifications**
- **Define List Variants**
- **Define Field Selection for Multilevel List Displays for Notifications**

5 Configuring the Work Order Cycle

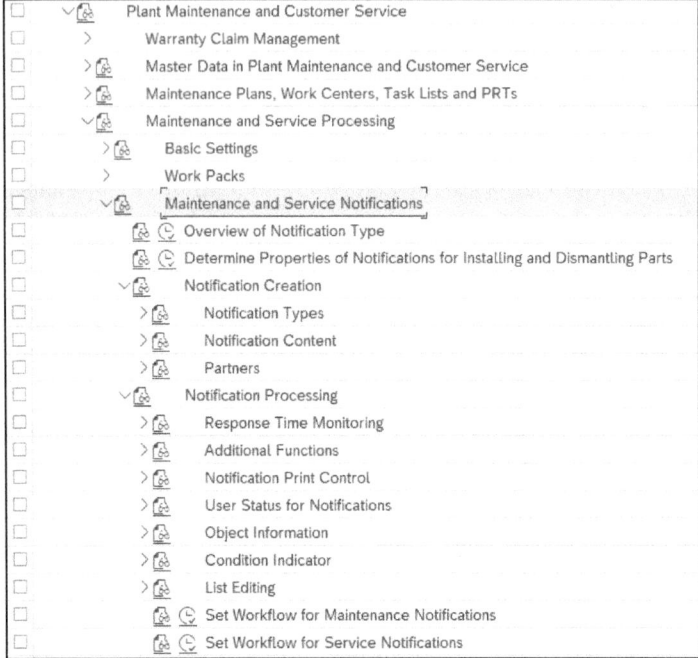

Figure 5.3 Notification: Customizing Functions Overview

The print procedure for notifications isn't described here because, in general, order papers (not notifications) are printed. If you want to print notification papers, you can refer to the information provided in relation to printing orders because the order paper printing settings are the same as the notification printing settings.

5.1.1 Notification Types

In Customizing, the notification type is the main control element for notification processing, and you define the following for the notification type (see Figure 5.4):

- Screen layout design
- The number to be assigned to the notification
- Long text control
- Whether or not the notification type may be used in the Request Maintenance app
- The order type to which the notification is to be assigned
- Field selection
- Which catalogs to use
- Which priorities must be available with the notification
- Which prioritization procedure should apply to this notification type
- The partners to assign

- Information to appear as object information
- The user status to assign to a notification
- Whether to use a class to include additional information in the notification
- The date to use as the reference time in the history

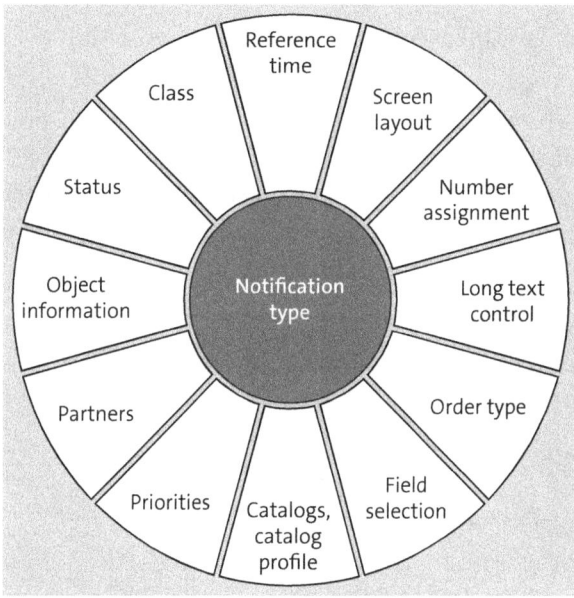

Figure 5.4 Notification Type: Overview

The first Customizing function you need to execute is **Define Notification Types**.

Define Notification Types

You use this Customizing function to define the notification types that you require. The functions described in Figure 5.4 should be taken into consideration when defining the number of notification types you need as well as which particular notification types you want to use. If, for example, you require different data entry layouts, or if you want to assign different number ranges for internal reasons, you require different notification types for these purposes.

> **Tip**
> In practice, it has proven beneficial to use between one and five notification types.

Prerequisites

There are no technical prerequisites, but you must consult with your colleagues from quality management, service management, and claim management on the naming convention for notification types because they also use notifications.

Customizing Path
Plant Maintenance and Customer Service • Maintenance and Service Processing • Maintenance and Service Notifications • Notification Creation • Notification Types • Define Notification Types

Settings
Specifically, you can or should configure the following settings (see Figure 5.5):

- **Notification Type**
  The notification type has a two-character ID. Because notifications are used not only in SAP S/4HANA Asset Management but also in quality management, customer service processing, and claims management, and because they can be created as *generic notifications*, you must consult with your colleagues on the naming convention for maintenance notification types (e.g., always use Mx or Ix).

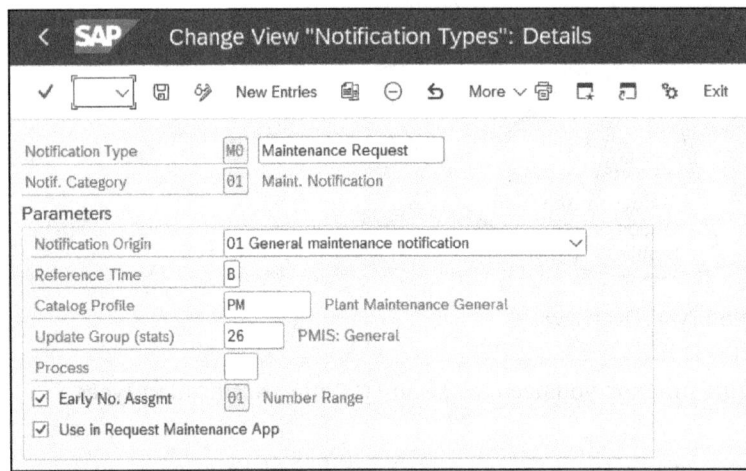

Figure 5.5 Notification Type: Details

- **Notif. Category**
  The notification category determines the application in which this notification type is used. Always assign **01 Maint. Notification** to those notification types that you want to use in SAP S/4HANA Asset Management. Other notification categories include **02 Quality Notification**, **03 Service Notification**, **04 Claim**, and **05 General Notification**.

- **Notification Origin**
  This setting determines the nature of the notification. Therefore, specify whether it concerns a **General maintenance notification (01)**, a **Malfunction report (02)**, or an **Activity report (03)**. Other origin types (e.g., **Q3 = Problem message, 05 = Activity report**) don't play any role in SAP S/4HANA Asset Management. The notification origin is used in the logistics information system (LIS) for statistical analyses. Otherwise, it doesn't have any other effect.

- **Reference Time**
  This setting specifies the proposed date and time by which the notification is to be completed:
  - **A**: Start of breakdown
  - **B**: Notification receipt
  - **C**: End of breakdown
  - **D**: Completion of notification

  This date then controls the chronological order of the notifications in the history.

- **Catalog Profile**
  The catalog profile entered in the notification type is used if you haven't assigned a catalog profile to the piece of equipment or functional location. For more information, refer to the Customizing functions for catalogs and catalog profiles.

- **Update Group (stats)**
  This is a control element within the update definition, which different business processes in different applications can use to influence the statistics update in the LIS in a uniform manner. SAP delivers update group 26 for plant maintenance, so if you haven't set up your own information structures in the LIS, or if you don't require your own update rules, always enter "26" here.

- **Early No. Assgmt.**
  Checking the box to set this early number assignment indicator specifies that the system assigns the notification number immediately after you call this transaction and not when you first save the notification. This is always recommended if, for example, you receive notifications via a central hotline. Then, the hotline representative can immediately quote the notification number to the person who has reported the issue.

- **Use in Request Maintenance App**
  If you want to create a notification with this kind of notification type, an indicator must be set. For this purpose, you can choose a notification type from the input help. The input help only lists the notification types for which you've checked the box. Note that you'll have to check the box for at least one notification type. Otherwise, it's not possible to create notifications in the Request Maintenance app.

Set Screen Templates for the Notification Type

You use this Customizing function to define the screen template layout. You define the following areas:

- The number of tabs
- The title to assign to each tab
- Whether to identify the tabs by a special icon

5 Configuring the Work Order Cycle

- The field groups (subscreens) to appear on each tab
- Layout changes based on the type of activity (creating, changing, and displaying)

Prerequisites

You must define the notification types beforehand.

Customizing Path

Plant Maintenance and Customer Service • **Maintenance and Service Processing** • **Maintenance and Service Notifications** • **Notification Creation** • **Notification Types** • **Set Screen Templates for the Notification Type**

Settings

As an example, the following two tabs have been defined for notification type **M0**: **Notification 1** and **Additional data 1** (see Figure 5.6).

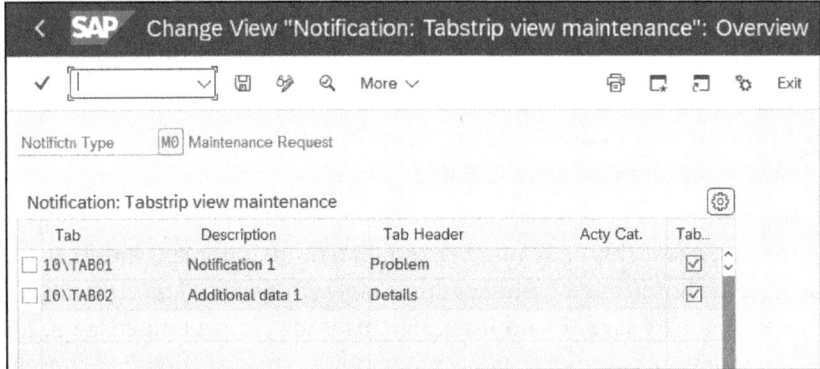

Figure 5.6 Notification Type: Template Overview

On the detail screen, configure several settings as they are shown in Figure 5.7.

The most important setting here is the definition of the **Tab**, and the following predefined tabs are available:

- **10\TAB00: Simplified notification**
- **10\TAB01*: Notification 1**
- **10\TAB02*: Additional data 1**
- **10\TAB03*: Additional data 2**
- **10\TAB04: System availability**
- **10\TAB05: Malfunction data**
- **10\TAB06: Location**
- **10\TAB07: Dates**
- **10\TAB08: Maintenance plan**
- **10\TAB09*: Additional data 3**

- 10\TAB10: Item
- 10\TAB11: Tasks
- 10\TAB12: Activities
- 10\TAB13: Item 2
- 10\TAB14: All tasks
- 10\TAB15: All activities
- 10\TAB16: All causes
- 10\TAB17*: Additional data 4
- 10\TAB18: Form
- 10\TAB19*: Customer data 1
- 10\TAB20*: Customer data 2
- 10\TAB21*: Customer data 3
- 10\TAB23: Pool Asset management
- 10\TAB26: Asset central

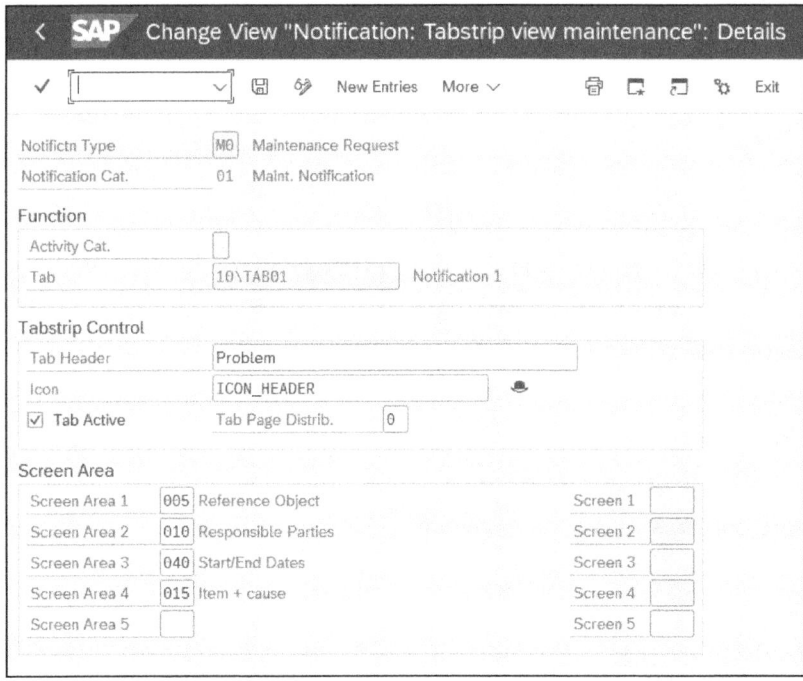

Figure 5.7 Notification Type: Template Detail

You can then define the following subtabs for the **10\TAB10 (Item)** tab:

- 20\TAB01: Item overview
- 20\TAB02: Item causes

5 Configuring the Work Order Cycle

- 20\TAB03: Item tasks
- 20\TAB04: Item activities

Because the tabs not marked with * are predefined by SAP, there are no further settings for these tabs.

For those tabs marked with *, you can define the screen areas yourself (as shown in Figure 5.8, which is the second tab for notification type **M0**). The following screen areas are available:

- 005: Reference object
- 010: Responsible parties
- 015: Item + cause
- 020: System availability
- 030: Subject and long text
- 032: Subject and long text (Windows, 32-bit)
- 033: Subject and long text (Web GUI)
- 035: Malfunction data
- 039: Reference documents/inspection lot
- 040: Start/end dates
- 045: Activities
- 050: Tasks
- 055: Causes
- 060: Notification and object address
- 065: Partner overview (as a table control)
- 070: Installation hierarchy
- 080: Warranty
- 085: Notification overview
- 090: Customer subscreen (one screen)
- 091: Customer subscreen (several subscreens)
- 092: DMS links
- 130: Shift note
- 131: Shift note attachments
- 192: DMS large screen
- 850: Pool asset management
- 860: Asset central indicators

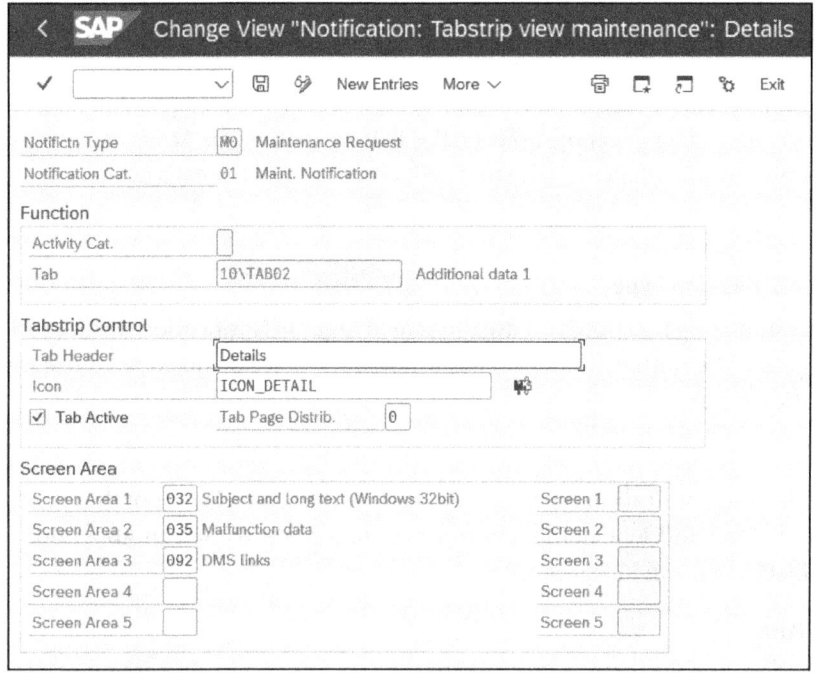

Figure 5.8 Notification Type: Template Details

You still have the following additional settings for each tab:

- **Activity Cat.**
  You can define the layout based on the activity category, which means you can configure the screen layout so that a different layout appears when you change data compared to when you add data. You might need this option, for example, when you want to provide a production employee with a screen that is as basic as possible for entering a notification. However, if a maintenance employee subsequently calls the same notification at a later stage, they should be able to update the notification with the additional required information. Consequently, if the same notification is called in change mode, it may then contain additional tabs and subscreens.

- **Tab Header**
  You can define the tab header yourself; you don't need to adhere to the header defined by SAP.

- **Icon**
  If you want a tab to receive particular attention, you can add an icon to the header. More than 1,000 predefined icons are available for selection.

One to Three Tabs

In practice, the following principle should always apply: "As much as necessary, but as little as possible." In this spirit, one to three tabs usually suffices.

> In addition, use the **Activity Cat.** option to have different layouts for adding and changing data.
>
> It's very important to note that if you want the system to display a particular tab when you process a notification, you must check the box to set the **Tab Active** indicator under **Tabstrip Control**. Otherwise, the tab is defined in the layout, but it's not shown.

Overview of Notification Type

Within this Customizing function, you use the **Screen Areas in Notification Header** subfunction to define the notification header and reference objects you want to use for each notification type.

Prerequisites

You must define the notification types themselves and the screen layout of these notification types beforehand. One of the tabs must contain the entry **Screen Area 1 005 = Reference Object**, as shown in Figure 5.7.

Customizing Path

Plant Maintenance and Customer Service • Maintenance and Service Processing • Maintenance and Service Notifications • Overview of Notification Type. Once there, choose the **Screen Areas in Notification Header** subfunction.

Settings

Figure 5.9 shows the settings for this subfunction. In **Screen Type Hdr** (screen type header), always enter "H100" (**Header maintenance notification**) unless you've defined your own headers. The **Header maintenance notification** contains information such as the notification type, notification number, short text, status, and assigned order number.

For **ScrnType Object** (screen type object), the following settings are available:

- O100: Functional location + equipment + assembly
- O110: Equipment + assembly
- O120: Functional location (30) + equipment + assembly
- O130: Serial number + material number + device data
- O140: Without a reference object
- O150: Equipment only
- O160: Functional location only
- O170: Equipment + serial no. + material No.
- O180: Functional location 1:1 + equipment + assembly
- O190: Material sample

5.1 Notifications

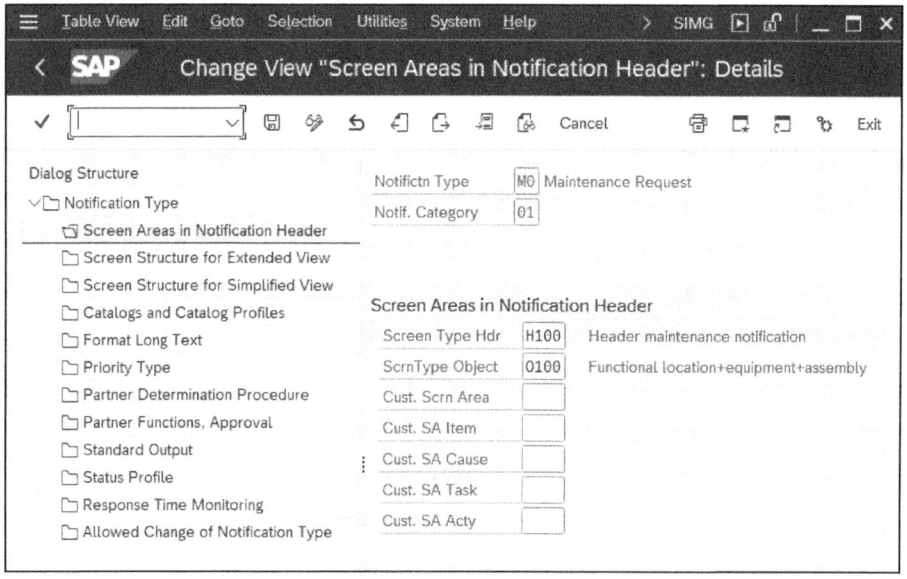

Figure 5.9 Notification Type: Screen Areas in Notification Header

> **Screen Types H100 and O100** [+]
>
> The screen type header (**Screen Type Hdr**) is always **H100**. For the screen type object (**ScrnType Object**), you should enter the screen type that you use most often for this notification type. Most users use **O100**.
>
> If, however, you require a different screen type in a notification, you can choose **Extras** • **Settings** • **Reference Object View** within the notification to change the screen type.

Define Long Text Control for Notification Types

You use this Customizing function to define how you want to handle long texts within a notification.

Prerequisites

You must define the notification types beforehand.

Customizing Path

Plant Maintenance and Customer Service • Maintenance and Service Processing • Maintenance and Service Notifications • Notification Creation • Notification Types • Define Long Text Control for Notification Types

Settings

The following settings are available to you (see Figure 5.10):

261

5 Configuring the Work Order Cycle

- **Log line**
  If you check the box to set this indicator, the system automatically inserts a line above the long text entered that contains information about the name of the person who entered the long text and the date and time when the entry was made.

- **No text change**
  If you check the box to set this indicator, you won't be able to change long texts retroactively.

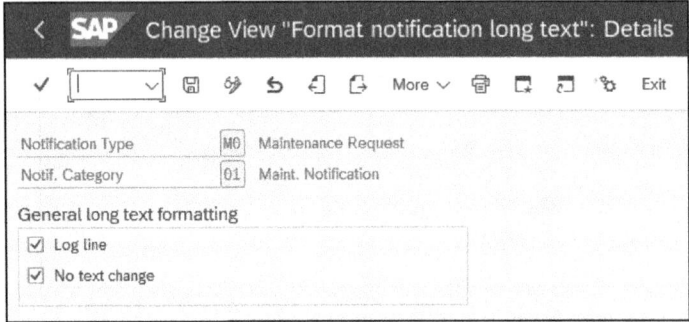

Figure 5.10 Notification Type: Long Text Formatting

Log Line and No Text Change Functions
You should activate the **Log line** function so that you know who entered which texts and when. To prevent any manipulation of the texts, you should also prohibit retroactive changes to the texts by checking the **No text change** box. Unfortunately, these options aren't available elsewhere in the SAP system.

Define Transaction Start Values

You use this Customizing function to define whether default values for calling the entry and maintenance transactions for the notification (in particular, Transactions IW21, IW24, IW25, and IW26) are to be defined.

Prerequisites

You must define the notification types beforehand.

Customizing Path

Plant Maintenance and Customer Service • Maintenance and Service Processing • Maintenance and Service Notifications • Notification Creation • Notification Types • Define Transaction Start Values

Settings

In the **Notification Type** field, enter the notification type that you want the system to propose when you call the transaction (see Figure 5.11).

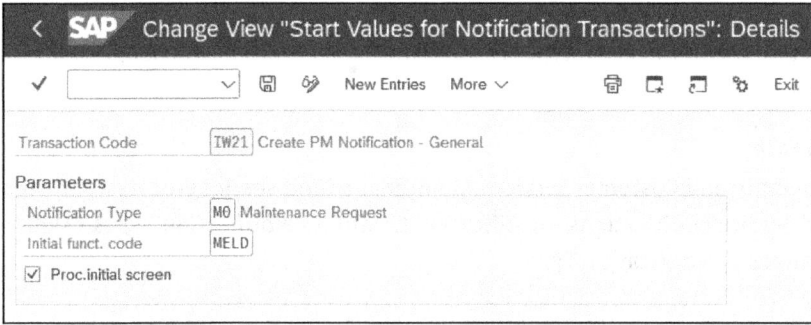

Figure 5.11 Notification Type: Transaction Start Values

If you have only one notification type, you don't need to check the box to set the **Proc.initial screen** (process initial screen) indicator, which signifies that you don't access an initial screen. Instead, you immediately access the entry screens for the notification, and the notification type is already set there.

If you have several notification types, you should check the box to set the **Proc.initial screen** indicator so that you access the initial screen for the notification. Although the notification type is also already set there as a default value, you can change it.

> **Processing the Initial Screen** [+]
> You should check the box to set the indicator for processing the initial screen. Otherwise, the system will always jump immediately back to the SAP Easy Access menu.

Define Allowed Changes of Notification Type

You use this Customizing function to define which changes to the notification type are permitted.

> **Extended Change Options** [+]
> You can change the notification type in *simple transactions* (Transactions IQS21 and IQS22), as well as in Transactions IW21 through IW26, if you activate the **Extended change of notification type** parameter.

You can change the notification type under the following conditions:

- The change was expressly permitted in this table.
- The notification still has the **Open** status.

5 Configuring the Work Order Cycle

Prerequisites

You must define the notification types beforehand. For the extended change of notification type, you must activate business function LOG_EAM_CI_7.

Customizing Path

Plant Maintenance and Customer Service • **Maintenance and Service Processing** • **Maintenance and Service Notifications** • **Notification Creation** • **Notification Types** • **Define Allowed Changes of Notification Type**

Settings

Enter all permitted changes for **Original Notification Type** and **Target Notification Type** (see Figure 5.12).

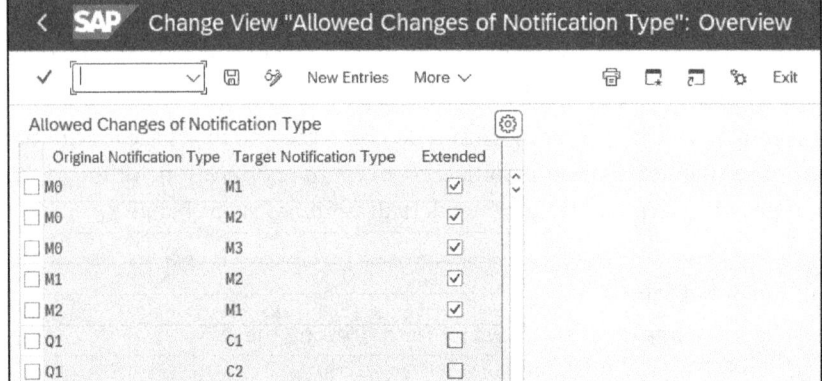

Figure 5.12 Notification Type: Allowed Changes

Assign Notification Types to Order Types

You use this Customizing function to define which order type is to be proposed for which notification type, if an order is generated from a notification.

Prerequisites

You must define the notification types and order types beforehand.

Customizing Path

Plant Maintenance and Customer Service • **Maintenance and Service Processing** • **Maintenance and Service Notifications** • **Notification Creation** • **Notification Types** • **Assign Notification Types to Order Types**

Settings

Enter which order type is to be proposed for each notification type (see Figure 5.13).

5.1 Notifications

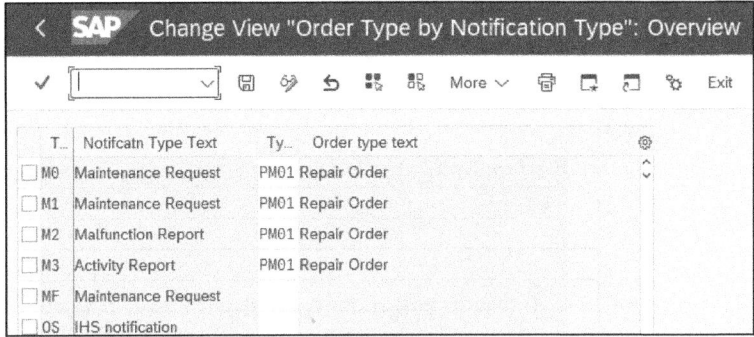

Figure 5.13 Notification Type: Assign Order Type

5.1.2 Catalogs and Catalog Profiles

In addition to organizational information (e.g., dates, responsibilities, cost centers), you can store technical information about problems, breakdowns, damage, causes and solutions, or damage repair in a notification. This information is part of the notification and is included in the history. A special feature of this information (when compared with all other information) is the fact that you can formalize this information in catalogs and therefore make it available for analysis.

Damage or *cause of damage* catalogs are generally used in SAP S/4HANA Asset Management (see Figure 5.14). Each catalog has the following three-tier structure: catalog → code groups → codes.

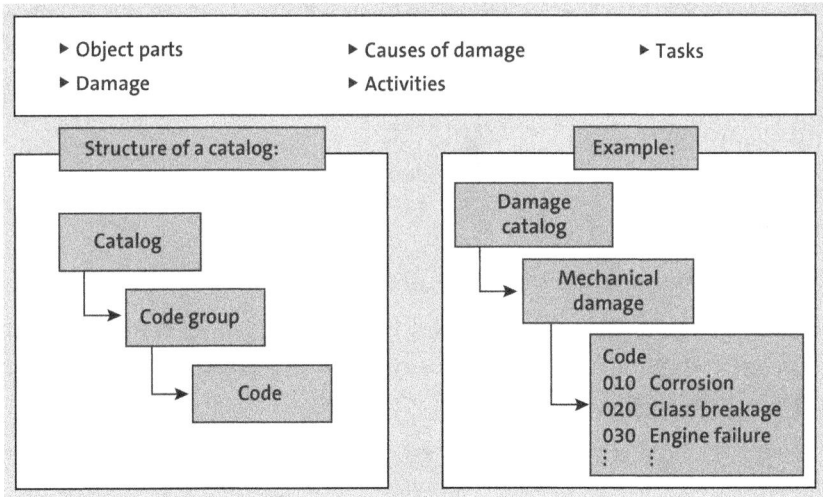

Figure 5.14 Notification: Catalog Structure

A code exists for each set of findings. Furthermore, the codes, in turn, are grouped together to form code groups according to characteristics.

265

Next, let's discuss the Customizing functions for creating catalogs and assigning notification types.

Maintain Catalogs

You use this Customizing function to maintain the catalogs, code groups, and codes that you want to use.

Prerequisites

There are no special prerequisites. However, you must consult with your colleagues from quality management and customer service on the naming convention for catalogs because they also use catalogs.

Customizing Path

Plant Maintenance and Customer Service · Maintenance and Service Processing · Maintenance and Service Notifications · Notification Creation · Notification Content · Maintain Catalogs

Transactions

S_ALR_87001010_1 and QS41

Settings

Use the **Define Catalogs** subfunction to define the first hierarchy level (i.e., the number and description of catalog types that you want to use in SAP S/4HANA Asset Management). Figure 5.15 shows the type and naming convention for the catalog types that are generally used in SAP S/4HANA Asset Management, as follows:

- 2: Tasks
- 5: Causes
- A: Activities (PM)
- B: Object Parts
- C: Overview of Damage

Of course, your own naming conventions may differ from these.

Use the **Edit Catalogs** subfunction to define the second hierarchy level. Here, define the categorization into code groups for each catalog type. This then raises the key question of how to categorize catalog types.

Categorization Criteria for Catalogs

User companies frequently use the following criteria for categorizing code groups in catalogs:

- **Functional criteria**
  Mechanical damage, electrical causes of damage, hydraulic object parts, and so on

5.1 Notifications

- **Object-related criteria**
  Damage to engines, causes of damage to pumps, object parts for forklifts, and so on

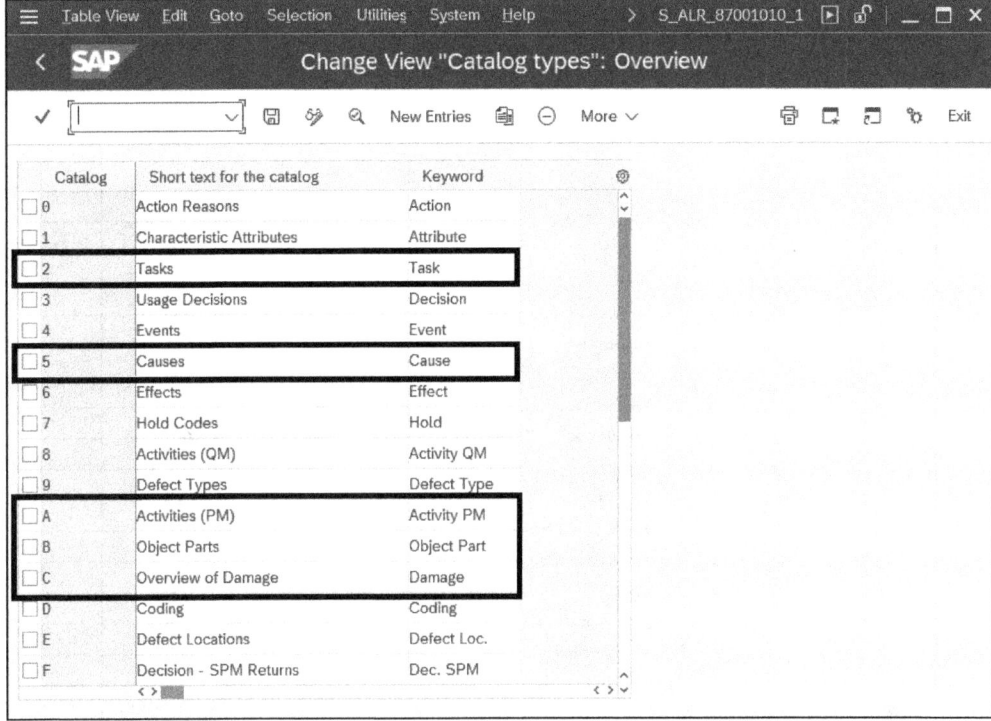

Figure 5.15 Notification: Catalog Types

Figure 5.16 uses the example of **Catalog C** = **Overview of Damage** to show one possible categorization. If you want to use the code group in notifications, you must set the **Status of Code Group** to **2 Released**.

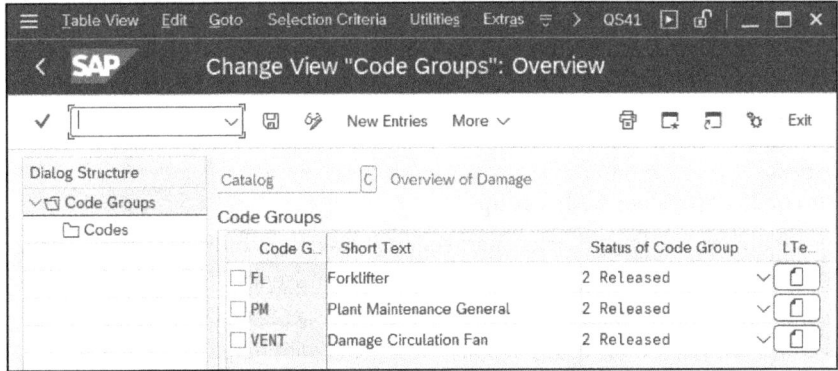

Figure 5.16 Notification: Code Groups Overview

Then, in the third hierarchy level, define the actual codes for each code group. Figure 5.17 shows, for example, the code groups defined for **Catalog C = Overview of Damage** and **Code Group FL = Forklifter**.

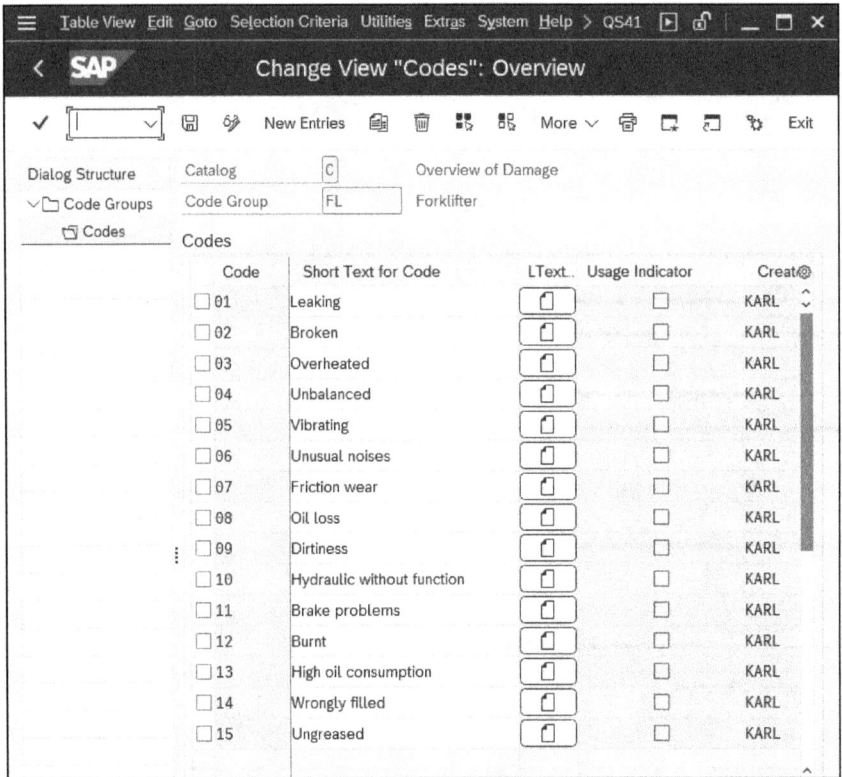

Figure 5.17 Notification: Codes Overview

Codes already used can't be erased, and codes that are no longer used or necessary should be deactivated. To deactivate codes, business function LOG_EAM_QM_CODE_DEACT must be activated. In this business function, codes can be deactivated retrospectively. Deactivated codes are no longer available in the input help and can no longer be assigned manually either.

Activating business function LOG_EAM_CI_12, you can sort out the codes by text (not only by code number).

[+] **Maximum of 25 Entries per Code Group**
Define clear damage, cause, and object part codes. A maximum of 25 codes should be available to the user or it will be too time-consuming for users to find the code they need and data quality and system acceptance will suffer as a result. Once again, the "as much as is necessary, but as little as possible" principle applies here.

Define Catalog Profile

In this work step, you define the *catalog profiles* that you can assign to the technical objects (functional location and equipment) and the notification type:

- Define the catalogs and code groups from the individual catalogs that are to be grouped together. You can then choose one of the code groups contained in the catalog by specifying a generic key for the code groups. You can also make multiple assignments for each catalog type (see Figure 5.18).
- Define the class in class type 015 to be proposed when classifying notification items. You can also include additional nonstandard fields in the notification.
- Define the type of message the system will issue if the catalog profile isn't complied with.

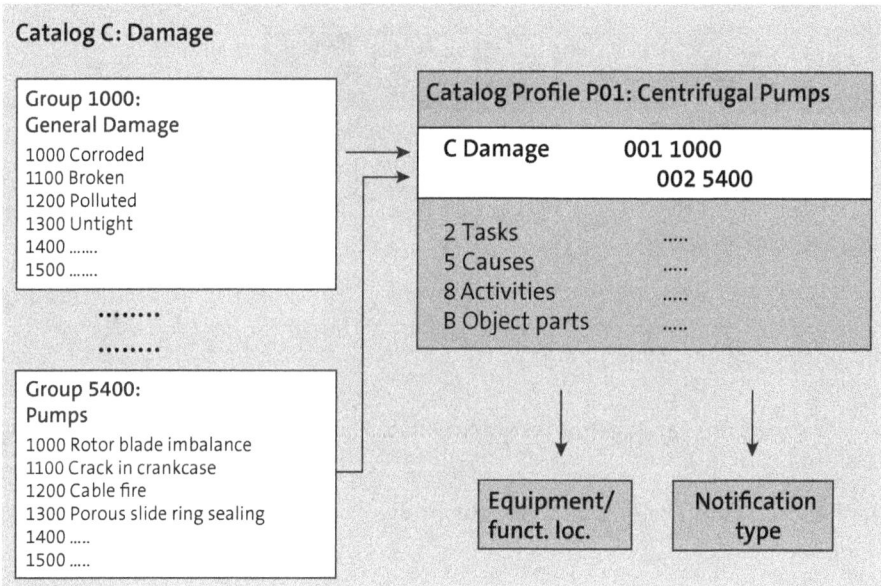

Figure 5.18 Notification: Catalog Profile Overview

Prerequisites

You must define the catalog types, code groups, and codes beforehand. If you also want to assign a class, you must use Transaction CT04 (Characteristics) to define the necessary characteristics and Transaction CL02 (Classes) to create a class of class type 015.

Customizing Path

Plant Maintenance and Customer Service • Maintenance and Service Processing • Maintenance and Service Notifications • Notification Creation • Notification Content • Define Catalog Profile

5 Configuring the Work Order Cycle

Transaction

OQN6

Settings

Figure 5.19 shows, for example, how you can use the **Catalogs/Code Groups** subfunction to assign code groups to a catalog profile:

- You can fully qualify the assignment, for example, with **Catalog C** and **Cd.Grp Pat FL**.
- You can also partly qualify the assignment. For example, the entry **Catalog 5** and **Cd.Grp Pat PM*** ensures that, in the notification, all groups whose group name starts with **P*** are proposed for the damage codes.
- You can assign several groups from the same catalog type. For example, **FL** assigns object-specific entries, while **PM*** assigns generic entries such as general damage.

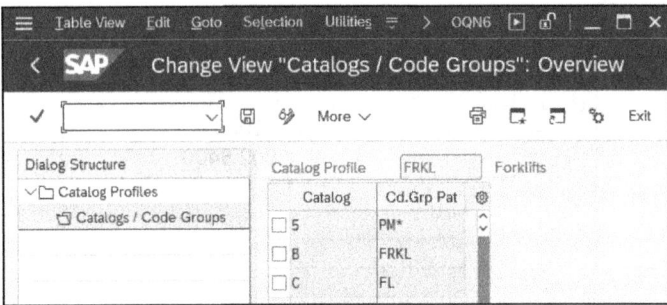

Figure 5.19 Notification: Catalog Profile

Figure 5.20 shows the **Catalog Profiles** subfunction. Here, you can define the following for each catalog profile:

- Whether you want to assign a class to the profile and, if so, which class you want to assign
- Whether a message is to be output when you change the catalog profile in a notification and, if so, what type of message: I = **Information**, W = **Warning**, or E = **Error** (the last of which means the catalog profile can't be changed later)

You can now enter the catalog profile in the following places:

- Equipment master (**Catalog Profile** field)
- Master record for the functional location (**Catalog Profile** field)
- Notification type (see the **Change Catalogs and Catalog Profile for Notification Type** Customizing function)

[+] **Priority Rule**

The following priority rule applies to the catalog profile: equipment → functional location → notification type.

5.1 Notifications

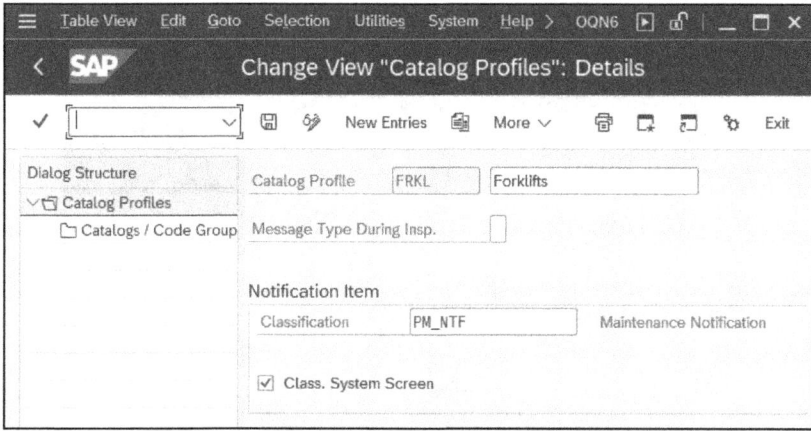

Figure 5.20 Notification Catalog Profile Detail

In other words, a catalog profile is determined in accordance with the following rules:

- When you determine the catalog profile in the notification, the system checks whether a catalog profile is already entered in the equipment master record. If it is, the profile is copied to the notification.
- If it isn't, the system checks whether a catalog profile is already entered in the master record for the functional location. If it is, the profile is copied to the notification.
- If it isn't, the system checks whether a catalog profile is already entered in the notification type. If it is, the profile is copied to the notification.
- If it isn't, the system doesn't propose any catalog profile in the notification. However, you can choose **Extras** • **Settings** • **Catalog Profile** • **Selection** at any time to assign a catalog profile to a notification.

Change Catalogs and Catalog Profile for Notification Type

You use this Customizing function to assign the relevant catalogs to the notification types.

Prerequisites

You must define the catalog types, catalog profiles, and notification types beforehand.

Customizing Path

Plant Maintenance and Customer Service • Maintenance and Service Processing • Maintenance and Service Notifications • Notification Creation • Notification Content • Change Catalogs and Catalog Profile for Notification Type

Settings

For **Problems** (damage), **Causes**, **Tasks**, **Activities**, and **Object Parts**, assign the catalog types that you want to use with this notification type (see Figure 5.21). The **Coding** entry

271

isn't used in SAP S/4HANA Asset Management. You can therefore access different catalog types for different notification types.

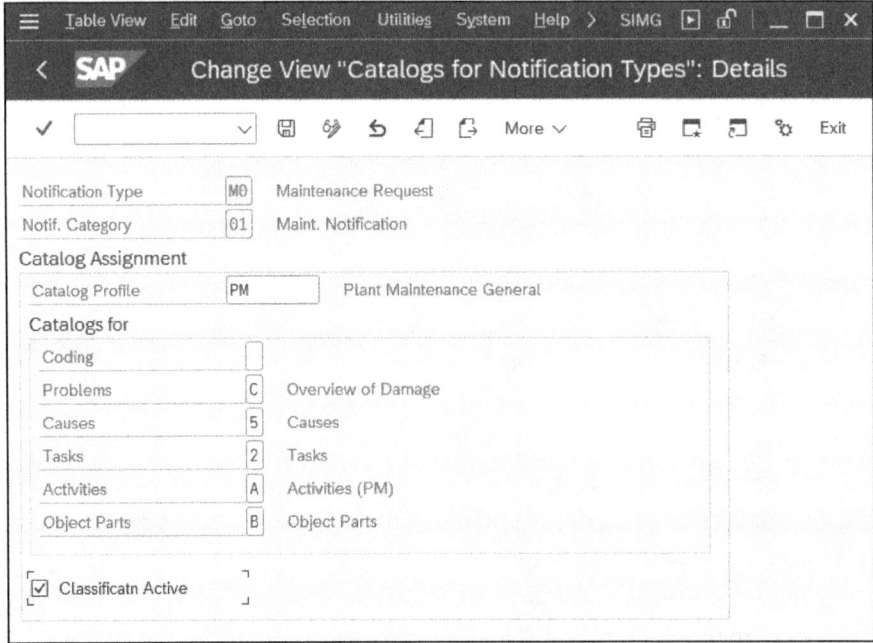

Figure 5.21 Notification Type: Assigning Catalogs

> [!] **Assign Catalog Types to Notification Types**
> You must enter the relevant catalogs for the notification types so that the system can display the corresponding fields in the notification. For example, if you don't assign a catalog type to a notification type in the case of catalogs for **Problems**, the notification won't contain any fields in relation to damage.

You can also assign a **Catalog Profile** to the notification type, and the system will then always propose this catalog profile in the notification if neither the piece of equipment nor the functional location has a catalog profile.

5.1.3 Other Functions

Of the many other Customizing functions available for the notification, we'll now cover the following:

- Priorities
- Response time monitoring

5.1 Notifications

- Action boxes
- System conditions
- Operational effects

Define Priorities

You use this Customizing function to define the priority types and associated priorities and to assign a notification type.

Prerequisites

You must define the notification types beforehand, and you must also consult with your colleagues from quality management, customer service, and claims management on the naming convention for priority types because they also use priorities.

Customizing Path

Plant Maintenance and Customer Service • Maintenance and Service Processing • Maintenance and Service Notifications • Notification Processing • Response Time Monitoring • Define Priorities

Transactions

S_ALR_87000248_2, S_ALR_87001148_1, and S_ALR_87001148_3

Settings

Use the **Define Priority Types** subfunction to determine a priority type. A *priority type* contains the actual priorities and is assigned to a notification type (see Figure 5.22).

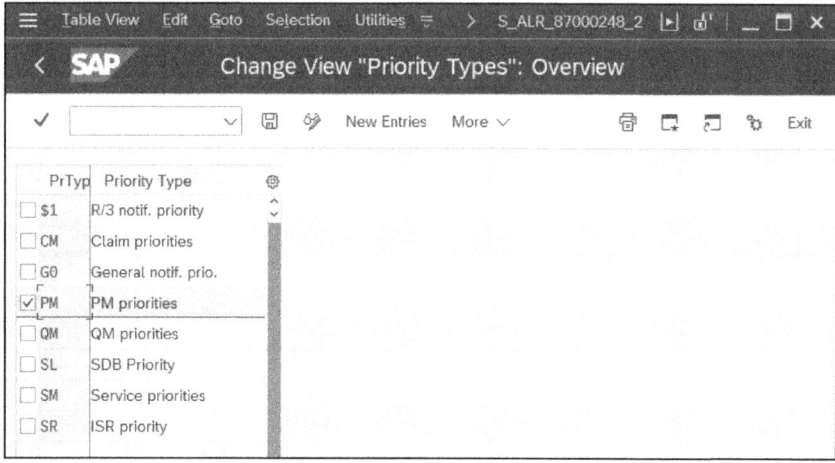

Figure 5.22 Notification: Priority Types

> **One Priority Type for All Notification Types**
> In most cases, one priority type should suffice for all notification types.

5 Configuring the Work Order Cycle

If, however, you want to handle the notification types differently, you'll need several notification types. This opens the possibility of defining multitiered priorities for your different notification types (e.g., a three-tier priority type for a malfunction report type and a five-tier priority type for normal maintenance requests).

Use the **Define Priorities for Each Priority Type** subfunction to define the relevant priorities for each priority type. In addition to the priority code and associated text, define the relative start and relative end (in their respective units of measurement) for each priority (see Figure 5.23):

- **Rel.Start**
  The system uses the relative start and its unit of measurement to calculate the *required start* for the notification, starting from the time when the notification was entered. In the example shown, the required start for the notification always corresponds to the date on which the notification was entered because no priorities have entered a relative start. The scheduling basis is the start date unit (**SDUn.**).

- **Rel. End**
  The system uses the relative end and its unit of measurement to calculate the *required end* for the notification, starting from the time when the notification was entered. In the example shown, priority **3** generates the following required end for the notification: entry date + 14 days. The scheduling basis is the end date unit (**EDUn.**).

- **Duration for Final Due Date**
  This defines the duration for calculating the final due date by which a maintenance activity must be completed.

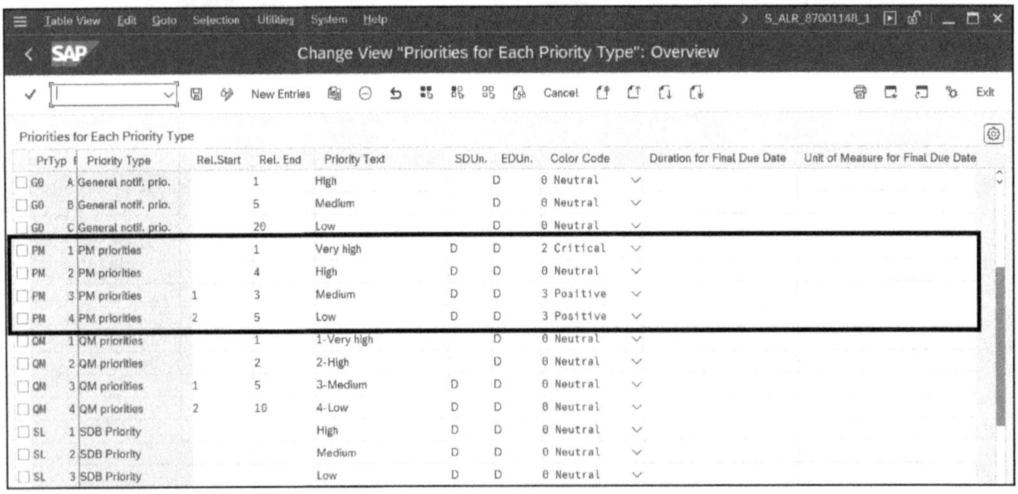

PrTyp	P	Priority Type	Rel.Start	Rel. End	Priority Text	SDUn.	EDUn.	Color Code	Duration for Final Due Date	Unit of Measure for Final Due Date
G0	A	General notif. prio.		1	High		D	0 Neutral		
G0	B	General notif. prio.		5	Medium		D	0 Neutral		
G0	C	General notif. prio.		20	Low		D	0 Neutral		
PM	1	PM priorities		1	Very high	D	D	2 Critical		
PM	2	PM priorities		4	High	D	D	0 Neutral		
PM	3	PM priorities	1	3	Medium	D	D	3 Positive		
PM	4	PM priorities	2	5	Low	D	D	3 Positive		
QM	1	QM priorities		1	1-Very high		D	0 Neutral		
QM	2	QM priorities		2	2-High		D	0 Neutral		
QM	3	QM priorities	1	5	3-Medium	D	D	0 Neutral		
QM	4	QM priorities	2	10	4-Low	D	D	0 Neutral		
SL	1	SDB Priority			High	D	D	0 Neutral		
SL	2	SDB Priority			Medium	D	D	0 Neutral		
SL	3	SDB Priority			Low	D	D	0 Neutral		

Figure 5.23 Notification: Priorities

- **Unit of Measure for Final Due Date**
  This defines the unit of measure for the duration that helps in calculating the final due date by which a maintenance activity must be completed.

Use the **Assign Priority Types to Notification Types** subfunction to define which priority type you want to use with which notification type (see Figure 5.24). Generally, you have just one priority type, and this priority type is assigned to all notification types.

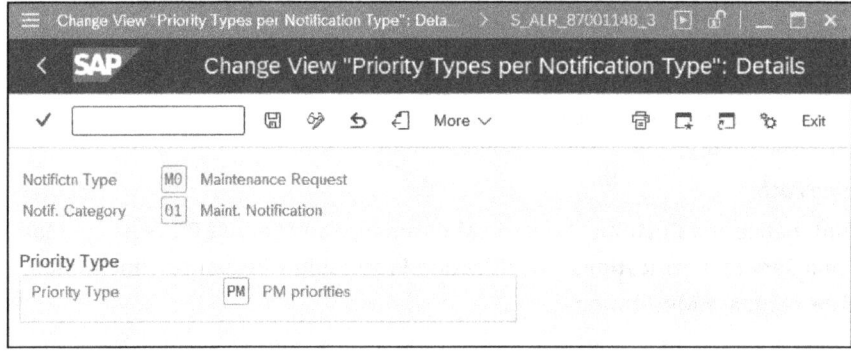

Figure 5.24 Notification Type: Priority Type

[+] **Always Assign a Priority Type (Even a Priority Type without Priorities)**

If you don't assign a priority type to a notification type and then attempt to enter a notification, the system will issue a warning message in a dialog box, indicating that the notification type doesn't have a priority type. To prevent this, you should assign a priority type to each notification type.

If, in general, you work without priorities or you don't want to have priorities for certain notification types, you should define a blank priority type (i.e., a priority type that doesn't contain any priorities) to circumvent this irritating warning message.

Define Response Monitoring

You use this Customizing function to define response monitoring for notifications. To do this, you determine *response profiles* and *service profiles* and then assign these two profiles to the notification types (see Figure 5.25).

Prerequisites

You must define the priority types and the *tasks* catalog type and its corresponding code groups and codes beforehand.

You must consult with your colleagues from quality management, customer service, and claims management on the naming convention for service profiles and response profiles because they also use response monitoring.

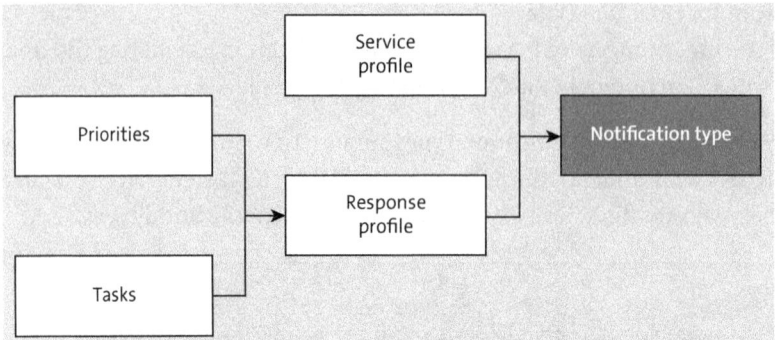

Figure 5.25 Notification Type: Service and Response Profile

Customizing Path

Plant Maintenance and Customer Service • Maintenance and Service Processing • Maintenance and Service Notifications • Notification Processing • Response Time Monitoring • Define Response Monitoring

Transactions

V_T355E, OIMF, and S_ALR_87000921_1

Settings

Use the **Response profile** subfunction to define a name and description for response profiles. Use the **Response Times** subfunction to define the following:

- Priority (e.g., **1**)
- Task (generally, catalog type **2** = **Tasks**)
- Time frame (calculated from the required start for the notification)
- Responsible person (e.g., the **Person responsible** partner function)

Figure 5.26 shows the following examples:

- The *notify plant engineer* task (catalog type **2**, code group **PM1**, and code **1**) in the first line is to be completed by Mr. Moore (partner function **VW** = person responsible with personnel number **54**), no later than 15 minutes after the required start for the notification.
- The *create order* task (catalog type **2**, code group **PM1**, and code **2**) in the second line is to be completed by Ms. Henry (partner function **VW** = person responsible with personnel number **1089**), no later than one hour after the required start for the notification.

[+] **Generate Tasks Automatically**
If you've defined a response profile with tasks, you can choose **Edit** • **Tasks** • **Determine Automatically** to generate the corresponding tasks in the notification.

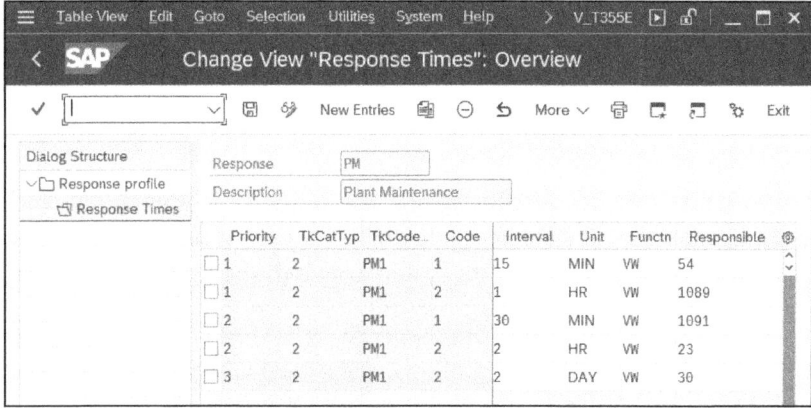

Figure 5.26 Notification: Response Profile

Use the **Define Service Profile** subfunction to define the times (within individual weekdays) during which a notification can be processed (see Figure 5.27). You can also schedule several service time intervals for each weekday and assign a separate response profile to each interval.

Figure 5.27 Notification: Service Profile

Use the **Response Time Monitoring for Notification Type** subfunction to assign a service profile and response profile to a notification type (see Figure 5.28).

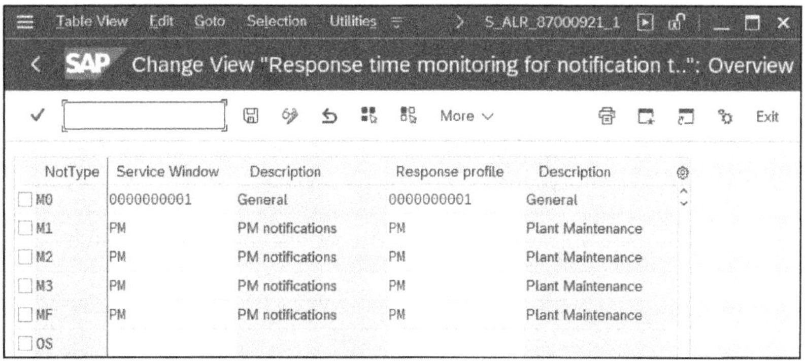

Figure 5.28 Notification: Response Time Monitoring

If you've already assigned a response profile to a service profile, it overrides the response profile for the notification type.

Define Action Box

You use this Customizing function to define functions that you can use to generate *tasks* or *activities* (known as *follow-up actions*) for a notification. They are defined directly for a notification type and are available when you process notifications at the header level. When you call the function, the system copies the defined task or activity to the notification.

Prerequisites

You must define the notification types and requisite tasks and activities beforehand.

Customizing Path

Plant Maintenance and Customer Service • Maintenance and Service Processing • Maintenance and Service Notifications • Notification Processing • Additional Functions • Define Action Box

Transactions

S_ALR_87000995_1

Settings

Use the **Define Follow-Up Functions (Generated Tasks/Activities)** subfunction and then **Activities** to define which follow-up activities you want to assign to the notification type (see Figure 5.29).

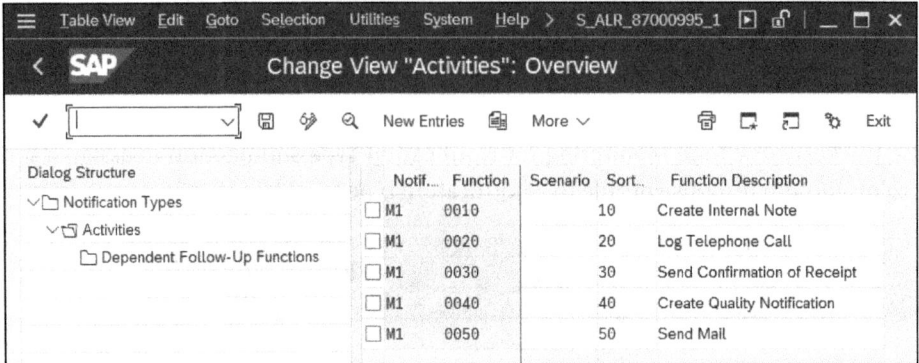

Figure 5.29 Notification: Action Box Overview

On the detail screen, you can define detailed information about each activity (see Figure 5.30).

5.1 Notifications

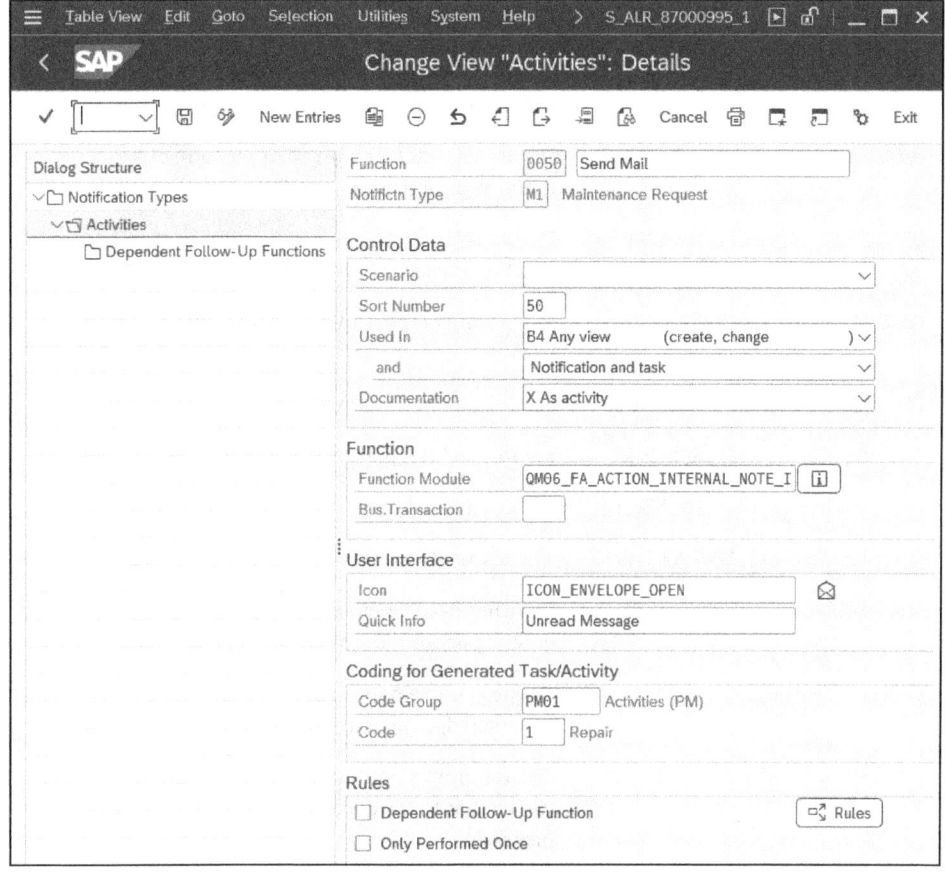

Figure 5.30 Notification: Activity Detail Screen

> **The Action Box Contains Certain Function Modules** [+]
>
> The QM06_FA* function modules document an activity, whereas the QM06_FM* function modules document a task.

The most important piece of information here is, without doubt, the **Function Module** selected. You can use either custom-programmed function modules or function modules delivered by SAP. Table 5.1 lists the function modules.

Function Module	Description
QM06_FA_ACTION_INTERNAL_NOTE_1	Documents a note in the form of a notification activity as a table control
QM06_FA_ACTION_INTERNAL_NOTE_2	Documents a note in the form of a notification activity as an OCX window

Table 5.1 Function Modules

Function Module	Description
QM06_FA_ACTION_TELEPHONCALL_1	Documents a telephone call in the form of a notification activity as a table control
QM06_FA_ACTION_TELEPHONCALL_2	Documents a telephone call in the form of a notification activity as an OCX window
QM06_FA_ACTION_TELEPHONCALL_3	Documents a notification activity when the SAPphone function is used to commence a telephone call
QM06_FM_PARTNER_ACTION	Starts a query to an internal or external notification partner
QM06_FM_TASK_CLAIM_SEND_PAPER	Sends claim information
QM06_FM_TASK_CREATE_8DREPORT	Creates an 8D report from a notification
QM06_FM_TASK_CREATE_QM_NOTIFIC	Creates a quality notification from a notification
QM06_FM_TASK_CRT_8DREPORT_PDF	Creates an 8D report as a PDF from a notification
QM06_FM_TASK_GOODS_MOVEMENT	Branches from a notification to the function for entering goods movements
QM06_FM_TASK_GOODS_MOVEMENT_2	Defines, from a notification, the quantities that are to be repaired, scrapped, and returned
QM06_FM_TASK_REQ_8DREPORT_PDF	Creates, from a notification, an 8D report as a PDF that can be sent to the person responsible for further processing
QM06_FM_TASK_REQUEST_8DREPORT	Creates, from a notification, an 8D report with print formatting, which can be sent to the person responsible for further processing
QM06_FM_TASK_Q_LEVEL_RESET	Resets the quality level
QM06_FM_TASK_Q_LEVEL_TIGHTEN	Tightens the quality level
QM06_FM_TASK_REQ_8DREPORT_PDF	Requests an 8D report as a PDF
QM06_FM_TASK_REQUEST_8DREPORT	Requests an 8D report
QM06_FM_TASK_RMA_ORDER_CREATE	Creates a repair order from a notification
QM06_FM_TASK_SALES_ORDER_STOCK	Transfers, from a notification, the usage decision for reclaimed goods to the corresponding repair order
QM06_FM_TASK_SEND_PAPER	Creates and sends different documents from a notification
QM06_FM_TASK_SEND_PAPER_ATTACH	Creates and sends different documents from a notification

Table 5.1 Function Modules (Cont.)

Function Module	Description
QM06_FM_TASK_SEND_PAPER_PDF	Creates and sends different documents from a notification
QISD_CREATE_DOCUMENT_FLOW	Updates the main document flow
QISD_DISPLAY_ATTACHED_LINKS	Performs the follow-up activity for displaying assigned symptoms/solutions
QISD_DISPLAY_SOLUTION	Displays details for a solution
QISD_SOLUTION_DATABASE	Performs the follow-up activity for calling the solution database

Table 5.1 Function Modules (Cont.)

In addition, you can specify the following:

- Define the **Function** description.
- Use the **Sort Number** to define the order in which the activities are to appear in the action box.
- Use **Used In** to define whether the activity is to appear for every view or only certain views (add, change, or display) and define whether the activity is to be available in notification processing, task processing, or both.
- For **Documentation**, define whether the activities performed are to be documented and, if so, whether they are to be defined in the notification as tasks or activities.
- Assign an **Icon** to the activity. Here, you can choose from more than a thousand icons.
- Position the cursor on the activity to see the **Quick Info** displayed.
- Use **Code Group** and **Code** to define the activity or task that is to be used to document the activity in the notification. If you select a QM06_FA* function module, it implies an activity; if you select a QM06_FM* function module, it implies a task.

Define System Conditions

In this work step, you define the possible statuses of your system (i.e., functional location and equipment) that can be used in notification processing.

Prerequisites
There are no specific prerequisites.

Customizing Path
Plant Maintenance and Customer Service · Maintenance and Service Processing · Maintenance and Service Notifications · Notification Processing · Condition Indicator · Define System Conditions

5 Configuring the Work Order Cycle

Settings

As an example, you may define the system statuses as fully operational, in limited operation, or out of order (see Figure 5.31).

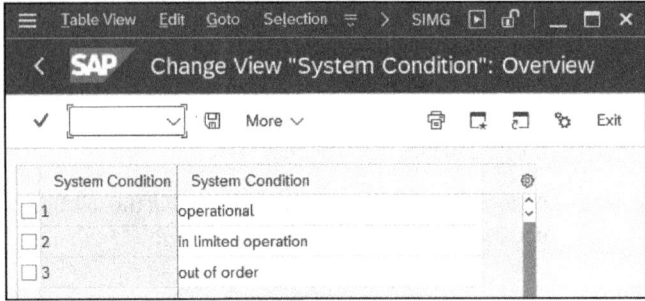

Figure 5.31 Notification: System Conditions

Define Operational Effects

In this step, you define the company-internal effects malfunctions have on production. The settings can be used as selection and evaluation criteria.

The operational effects are part of the SAP Fiori apps Request Maintenance and Create Maintenance Request.

Prerequisites

There are no specific prerequisites.

Customizing Path

Plant Maintenance and Customer Service · Maintenance and Service Processing · Maintenance and Service Notifications · Notification Processing · Condition Indicator · Define Operational Effects

Settings

As an example, you could define operational effects for no effect, limited production, or production breakdown (see Figure 5.32).

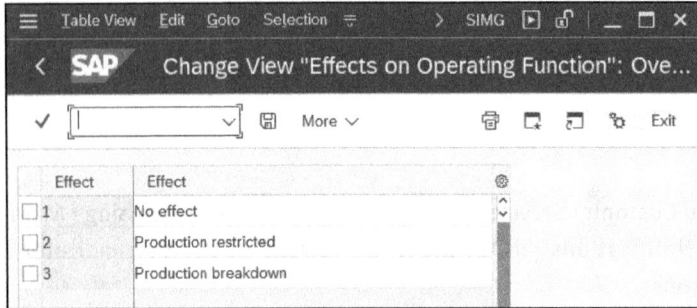

Figure 5.32 Notification: Operational Effects

5.2 Orders

Notifications are primarily used for documentation purposes and to request maintenance tasks. In contrast, *orders* are used by maintenance planners and technicians as tools for planning and carrying out maintenance work. The characteristics of an order are as follows:

- An order is used to plan and carry out a maintenance task.
- An order primarily contains processing information.
- An order is a highly integrative object with many connections to applications such as warehouse management, purchasing, and controlling.

Figure 5.33 shows the structure and content of an order, as described in the following list:

- **Order header**
  The data in the order header applies to the entire order and is used to identify and manage the order.
- **Object list**
  If the order concerns several objects (e.g., in the case of an inspection), you can enter the objects in an object list. The *object list* contains all objects for which the order is executed (functional locations, equipment, assemblies, and notifications).
- **Operation**
  You can use *operations* to describe the work that is to be performed when implementing an order. Operations are carried out either by your own employees or by external companies.
- **Material list**
  The *material list* contains spare parts that are required and consumed when an order is being executed. This is either stock material for which a reservation is subsequently generated or nonstock material for which a purchase requisition is created.
- **PRTs**
  To execute an order, you require PRTs (e.g., tools, protective clothing, and hand pallet trucks). However, unlike materials, PRTs aren't consumed.
- **Settlement rule**
  In the *settlement rule*, you specify the cost object (e.g., the cost center) to which the costs must be charged. The settlement rule applies to the entire order. Alternatively, you can assign different account assignments to the operations.
- **Cost data**
  Cost data informs you about how high the estimated, planned, and actual costs are in the value categories for this order and which cost elements are relevant for the order. Furthermore, it specifies which key figures in the plant maintenance information system (PMIS) are updated using the value categories and how these key figures

are updated with the actual costs of the order. You obtain cost information for individual operations and as the total for the complete order.

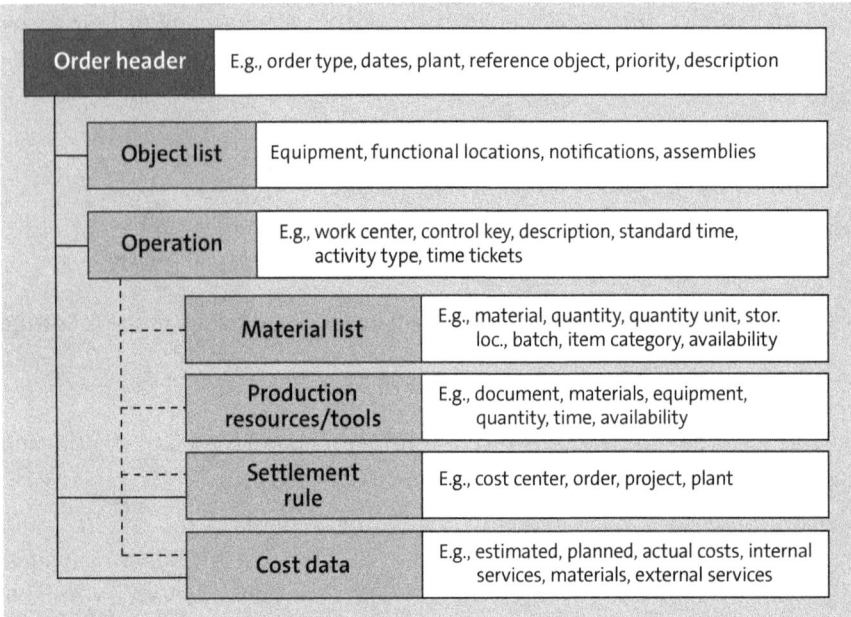

Figure 5.33 Order: Structure and Content

There are six ways in which you can create an order (see Figure 5.34):

- **Automatically generating an order from a maintenance item**
  You define a maintenance plan, which automatically creates an order at regular intervals based on the information stored there (e.g., the reference object, order type, and task list).
- **Creating an order from a notification**
  You receive an individual notification from a requester (e.g., from production), and you then create an order from this notification.
- **Grouping several notifications into one order**
  You receive several notifications that are to be processed within a single order (e.g., several malfunction reports that refer to the same asset), and you have the option to create an order from the notification list.
- **Creating an order directly without a notification**
  You create an order directly (i.e., without a notification).
- **Creating an order directly with a notification**
  Similar to the previous option, you create an order directly, but you supplement the order with information such as damage and cause of damage or breakdown.
- **Creating an order directly with a retroactive notification**
  You create an order without a notification but, when completing the order, you

enter technical confirmation data in a notification in addition to time confirmation data.

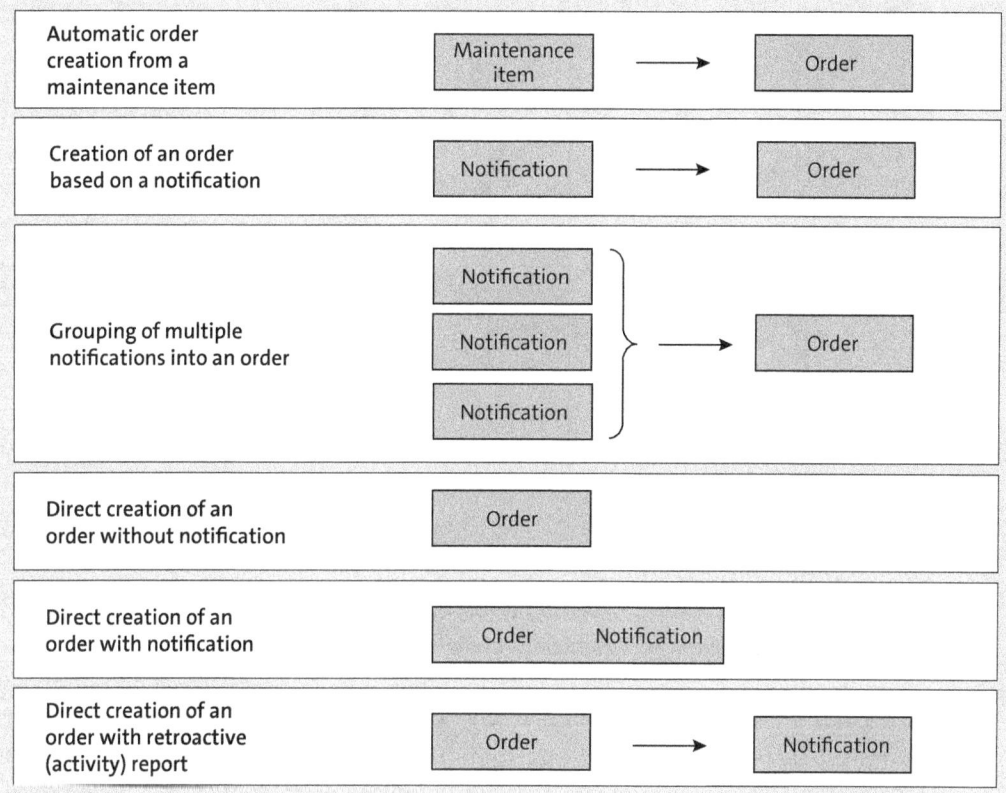

Figure 5.34 Creating an Order

Next, I'll describe the Customizing functions that you can use to adjust orders to your requirements. An overview of the Customizing functions for orders is provided in Figure 5.35.

The following Customizing functions and their subfunctions are also relevant for orders, but they've already been described in Chapter 3, so they are simply listed here:

- Number ranges
- Field selection
- Partner determination procedure and partner function
- Object information key
- User status
- List variants
- Field selection for multilevel list displays

5 Configuring the Work Order Cycle

Figure 5.35 Order: Customizing Functions Overview

To provide a clearer overview of the many Customizing functions for orders, this section is subdivided into the following sections, which cover all Customizing functions for each topic:

- Section 5.2.1 contains all basic functions that you need to configure an order type.
- Section 5.2.2 shows you all Customizing functions for availability checks.
- Section 5.2.3 shows you all Customizing functions for electronic parts catalogs.
- Section 5.2.4 shows you all remaining Customizing functions for material planning, such as material documents.
- Section 5.2.5 contains all Customizing functions for scheduling.
- Section 5.2.6 contains all Customizing functions for capacity planning.
- Section 5.2.7 shows you all Customizing functions for print control in relation to shop papers and messages.
- Section 5.2.8 contains all Customizing functions for cost estimates and costing.
- Section 5.2.9 contains all Customizing functions for order settlement.
- Section 5.2.10 contains all Customizing functions regarding project-based maintenance.
- Section 5.2.11 introduces all other Customizing functions that don't relate to the other sections but are nevertheless important.

5.2.1 Order Types

In Customizing, the *order type* is the main control element for order processing. Unlike with the notification type, where the plant plays almost no role whatsoever, it's necessary, in the case of the Customizing functions for orders, to distinguish between the following:

- Whether the Customizing function for the order type is generally applicable (i.e., across all plants)
- Whether the Customizing function for the order type refers to only one plant

For each order type without specific reference to a plant (i.e., when the settings apply to all plants), you define, among other things, the following in Customizing (see Figure 5.36):

- Which maintenance activity types are permitted for the order type
- Which number is to be assigned to an order
- How you want to settle an order
- How the order is to be integrated with the notification
- Which fields you want to enter
- Which shop papers are to be printed
- Which priorities you want to use in the order

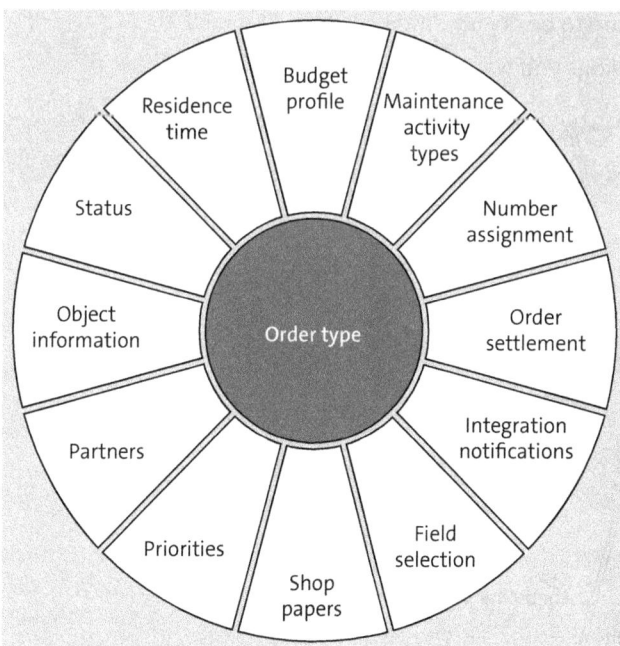

Figure 5.36 Order Type: Customizing Function in General

- Which partners you want to assign to an order
- Which object information key is to be used
- How long you want to keep an order in the system
- What effect budgets are to have
- Which status profile is to be used

For each order type that is to apply to only one particular plant (i.e., each order type with a plant reference), you define, among other things, the following in Customizing (see Figure 5.37):

- How the screen layout will appear
- How the address data for the delivery address is to be accessed
- Which values are to be proposed for procurement transactions
- How you want to access task lists
- How you want to schedule orders
- How availability checks are to be performed
- How you want to conduct preliminary and final costings for orders
- Whether you want to show the costs at the header level or the operation level
- For which changes the system is to create change documents
- Whether you want the order to contain linear data
- Which goods movements are to be documented
- Which electronic parts catalogs you want to integrate

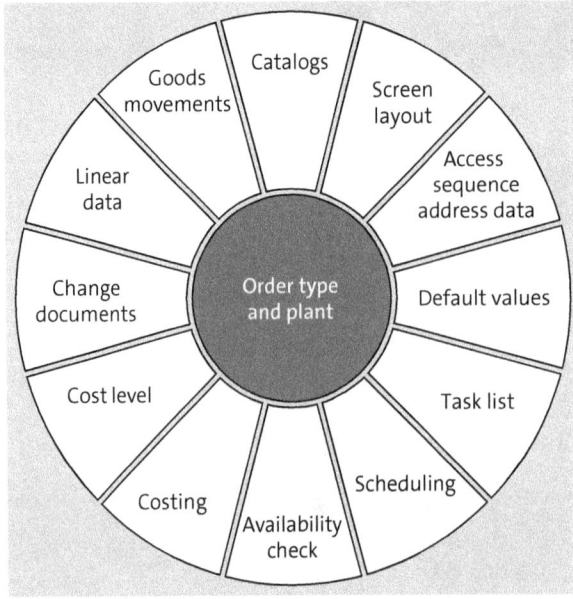

Figure 5.37 Customizing Functions per Plant

5.2 Orders

The following sections explain the basic Customizing functions for configuring an order type.

Configure Order Types

You use this Customizing function to define the order type and some basic parameters.

Prerequisites

Because order types are used not only in SAP S/4HANA Asset Management but also in other user departments and their corresponding areas in SAP S/4HANA (see Figure 5.38), you must consult with your colleagues from the other user departments (production planning, quality management, project management, controlling, customer service, and materials management) on the naming convention and number ranges for order types.

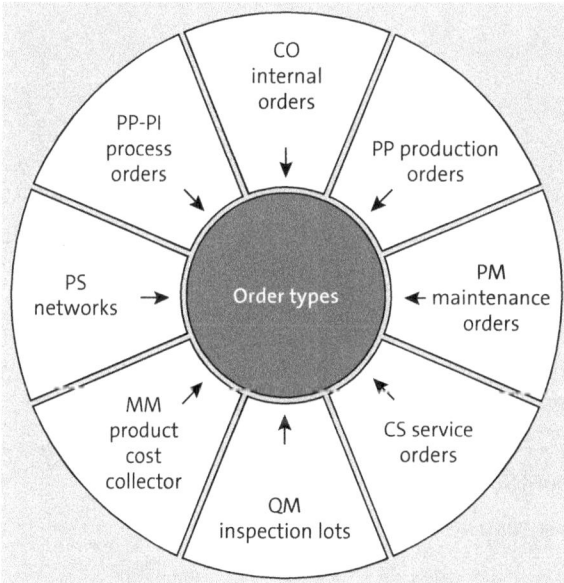

Figure 5.38 Order Categories and Order Types

These different applications result in the use of *order categories*. Each order type must be assigned to an order category, which, in turn, is predefined by SAP. The following order categories exist:

- 01: Internal order (Controlling)
- 02: Accrual calculation order (Controlling)
- 03: Model order (Controlling)
- 04: CO production order
- 05: Product cost collector

5 Configuring the Work Order Cycle

- 06: QM order
- 10: PP production order
- 20: Network
- 30: Service order and maintenance order
- 40: Process order
- 50: Inspection lot
- 60: Personnel order
- 70: Shipping deadline

Customizing Path

Plant Maintenance and Customer Service • Maintenance and Service Processing • Maintenance and Service Orders • Functions and Settings for Order Types • Configure Order Types

Transaction

OIOA

Prerequisites

You must define a settlement profile and a budget profile beforehand. If you want to assign the order type to a phase model, you must activate business feature **EAM_PHASE_MODEL_PROCESSING** (see Chapter 7, Section 7.9).

Settings

When defining which order types you want to configure, you should be guided by the Customizing functions and Customizing options that are explained next.

> **Three to Seven Order Types Should Suffice**
>
> You'll generally require only the following order types:
> - Repair
> - Maintenance and inspection
> - Investment
> - Refurbishment (if used)
> - Calibration of test equipment (if used)
> - Pool asset management (if used)
> - External processing (optional)
>
> You can use maintenance activity types for further refinements and subdivisions.

You can use this Customizing function to configure the following settings (see Figure 5.39):

5.2 Orders

- **Order category**
  In this Customizing function, the system automatically assigns the **30 Maintenance order**.

- **Order Type**
  Specify the naming convention and description.

- **Open item management**
  Use this option to define whether you want to use the planned costs associated with the order as a basis for creating *order commitments*.

- **Settlement Profile**
  Use this option to specify how the order is to be settled (for more information, see Section 5.2.9).

- **Budget Profile**
  Use this option to specify whether a budget availability check is to be performed and, if so, how (for more information, see Section 5.2.8).

- **Residence Time1**
  Use this option to specify, in months, the time gap between the user manually setting the deletion flag and archiving automatically setting the deletion indicator.

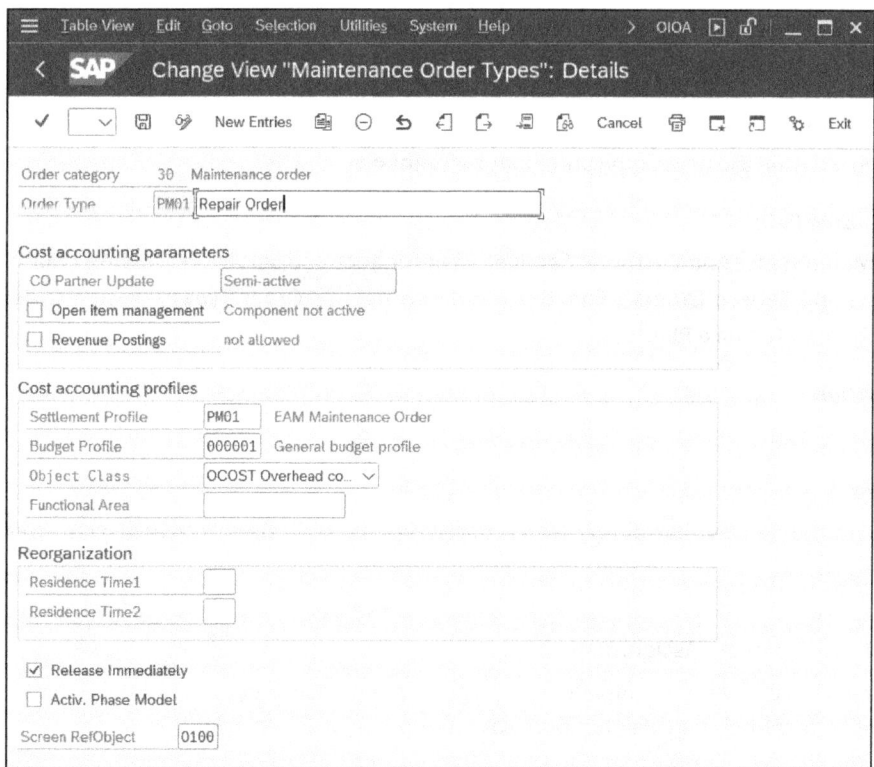

Figure 5.39 Order: Configure Order Type

- **Residence Time2**
  Use this option to specify, in months, the time gap between setting the deletion indicator and deleting the order.

- **Release Immediately**
  This indicator only takes effect if the order is automatically retrieved from a maintenance plan or is generated from a notification. The effect of this setting is that, in these two scenarios, the system immediately sets the **REL** (released) status and not the **CRTD** (created) status in the order.

- **Activ. Phase Model**
  This indicates that maintenance orders of this order type are processed along a specific phase model. If this checkbox is selected, all orders of this order type are processed with additional screening, approval, dispatching, and scheduling steps, which are also reflected in additional system statuses (see Chapter 7, Section 7.9).

- **Screen RefObject**
  Use this field to control which reference object can be entered in the order header (e.g., **O100 = functional location, piece of equipment, and assembly**).

Assign Order Types to Maintenance Plants

If you want to use an order type in a planning plant, you must use this Customizing function to make the order type available for the plant.

Prerequisites

You must define the order type and plant beforehand.

Customizing Path

Plant Maintenance and Customer Service • Maintenance and Service Processing • Maintenance and Service Orders • Functions and Settings for Order Types • Assign Order Types to Maintenance Plants

Transaction

OIOD

Settings

Generate the entries that define which order type is to be used in which plant (see Figure 5.40).

> **Ensure a Complete Assignment**
> This list must have the complete order type to maintenance planning plant assignments. Otherwise, you won't be able to use the order type in the plant.

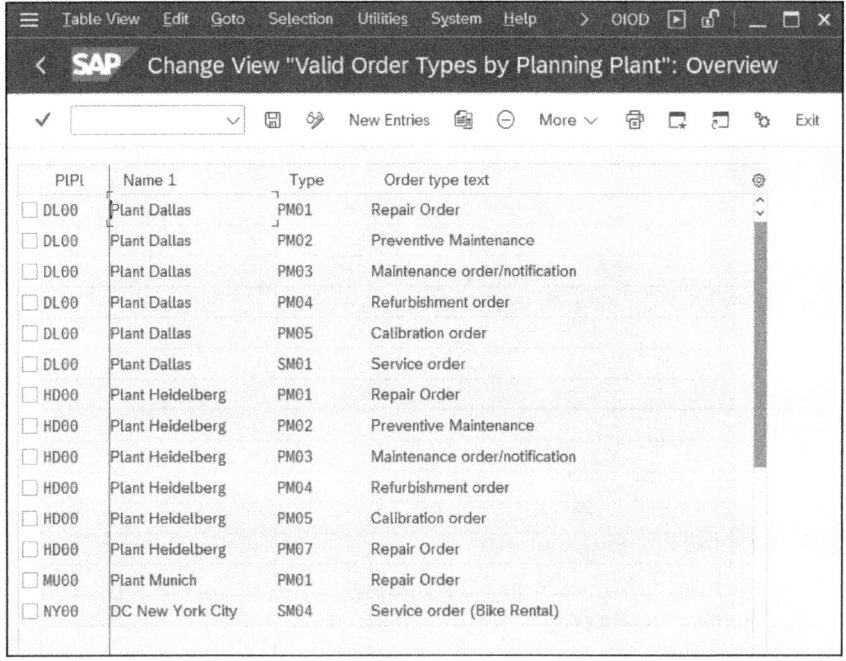

Figure 5.40 Order Type: Assign to Maintenance Planning Plant

Define a Default Value for the Planning Indicator for Each Order Type

You use this Customizing function to define, for each order type, a default value for the order planning indicator (see Figure 5.41). The order planning indicator is solely used to select and perform statistical analyses in the LIS. It doesn't have any controlling effect whatsoever, and it can be changed within an order.

Prerequisites
You must define the order types beforehand.

Customizing Path
Plant Maintenance and Customer Service • Maintenance and Service Processing • Maintenance and Service Orders • Functions and Settings for Order Types • Define Default Value for Planning Indicator for Each Order Type

Transaction
OIOS

Settings
The LIS contains key figures for the number of *immediate orders*, *planned orders*, and *unplanned orders*.

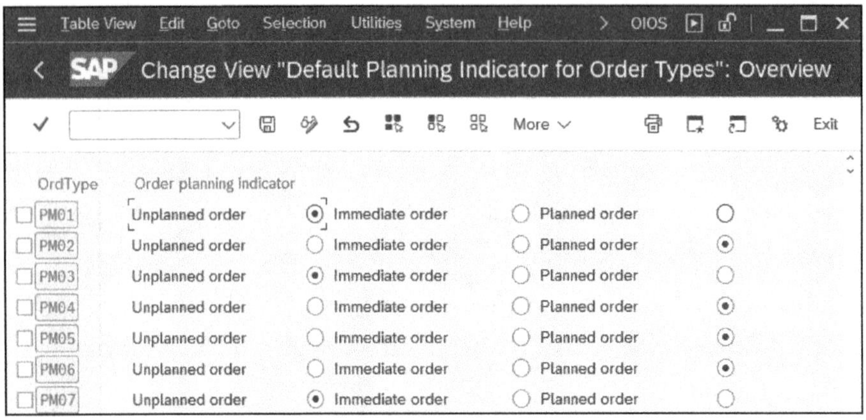

5 Configuring the Work Order Cycle

Figure 5.41 Order Type: Planning Indicator

[+] **No Fixed Rule for the Planning Indicator**

You're completely free to choose which order type you want to assign to which planning indicator. For example, you can do the following:

- Assign the **Planned order** planning indicator to all order types associated with maintenance and inspection.
- Assign the **Immediate order** planning indicator to all order types associated with repairing a breakdown.
- Assign the **Unplanned order** planning indicator to all other order types.

Define Notification and Order Integration

You use this Customizing function to define the following:

- Whether you want to maintain notification and order data on one screen
- Whether you want the system to automatically create a notification when you enter an order
- Whether the system is to automatically copy the long text of the notification header to the order header if an order is generated from a notification
- Whether and to what extent you'll permit the assignment of operations to object list entries in an order

Prerequisites

You must define the order types and notification types beforehand.

For enhanced object list assignment, you must activate business function LOG_EAM_CI_4.

5.2 Orders

Customizing Path

Plant Maintenance and Customer Service • Maintenance and Service Processing • Maintenance and Service Orders • Functions and Settings for Order Types • Define Notification and Order Integration

Settings

The following settings are available for each order type (see Figure 5.42):

- **Notif.Type**

  In this column, specify the notification type for the automatically generated notification.

- **Notifctn**

  If you check the box to set the notification indicator, you can maintain order and notification data on one screen. Furthermore, if you create an order directly, the system automatically creates the notification in the background.

- **Long Text**

  If you check the box to set this indicator, then when the system generates an order, it copies the long text of the notification header to the long text of the order. However, subsequent changes to the notification long text are no longer copied.

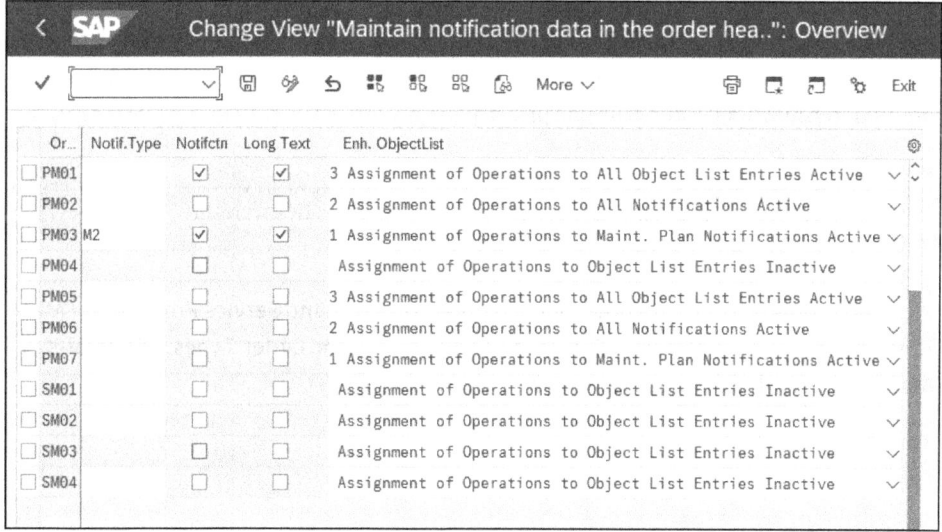

Figure 5.42 Order Type: Notification Integration

- **Enh. ObjectList**

  Use the entry in the enhanced object list column to control the automatic assignment of operations to object list entries (e.g., notification, piece of equipment, functional location) if a notification is assigned to an order. The following settings are available:

- <Blank>: **Assignment of Operations to Object List Entries Inactive**
  The enhanced functions for the object list are inactive, meaning the operations of the order can't be assigned to entries in the object list.
- 1: **Assignment of Operations to Maint. Plan Notifications Active**
  When you assign a notification to an order, the system can automatically assign the operations solely to those notifications in the object list that were generated from maintenance plans. You can display the operations assigned to a notification.
- 2: **Assignment of Operations to All Notifications Active**
  When you assign a notification to an order, the system can automatically assign the operations to all notifications in the object list (i.e., to notifications created manually and to maintenance plan notifications). You can display the operations assigned to a notification.
- 3: **Assignment of Operations to All Object List Entries Active**
  The system can automatically assign the operations to any object list entry (e.g., to notifications, equipment, or functional locations).

Maintain Default Values for Control Keys for Order Types

Chapter 2, Section 2.4 explained both the functionality and the function of the control key. You use this Customizing function to define which control key is to be proposed for which order type if a control key isn't defined in the work center. If a control key is defined in the work center, this has priority.

Prerequisites

You must define the plants, order types, and control keys beforehand.

Customizing Path

Plant Maintenance and Customer Service • Maintenance and Service Processing • Maintenance and Service Orders • Functions and Settings for Order Types • Control Key • Maintain Default Values for Control Keys for Order Types

Transaction

OIO6

Settings

Define which control key is to be proposed for which order type and in which plant (see Figure 5.43).

5.2 Orders

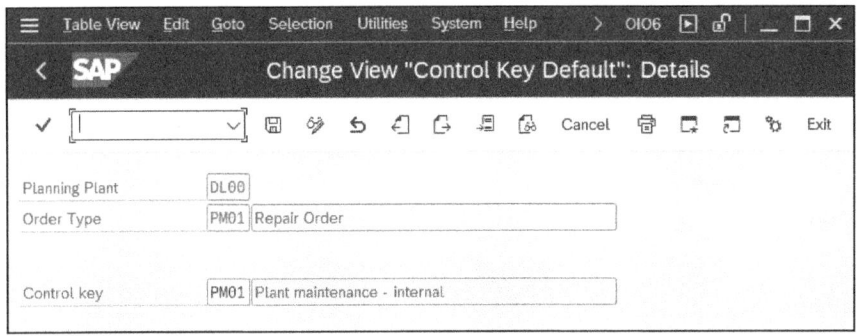

Figure 5.43 Order Type: Control Key

Maintenance Activity Types

You use these Customizing functions to define the maintenance activity types and to assign them to the order types. The *maintenance activity types* are used to refine the order types further and are important selection criteria for order lists. You can do the following with the maintenance activity types:

- Categorize the costs in the LIS. In other words, you can use maintenance activity types to summarize the maintenance costs.
- Indicate whether a maintenance activity type concerns a warranty case or an insurance case.

Prerequisites

You must define the order types beforehand.

If you activate business function LOG_EAM_CI_8, you may assign maintenance activity types to notification headers (subscreen 040 start/end dates) and maintenance plans as well as to orders.

Customizing Paths

The following Customizing paths are available for activity types:

- Defining
  Plant Maintenance and Customer Service • Maintenance and Service Processing • Maintenance and Service Orders • Functions and Settings for Order Types • Maintenance Activity Type • Define Maintenance Activity Types
- Assigning
  Plant Maintenance and Customer Service • Maintenance and Service Processing • Maintenance and Service Orders • Functions and Settings for Order Types • Maintenance Activity Type • Assign Valid Maintenance Activity Types to Maintenance Order Types

5 Configuring the Work Order Cycle

- **Default values**
  Plant Maintenance and Customer Service • Maintenance and Service Processing • Maintenance and Service Orders • Functions and Settings for Order Types • Maintenance Activity Type • Default Values for Maintenance Activity Type for Each Order Type

Transactions

The following transactions are available for activity types:

- OIO5 (Valid Maintenance Activity Types by Order Type)
- OIO4 (Default Maintenance Activity Type by Order Type)

Settings

Figure 5.44 lists the maintenance activity types that are commonly set up in projects.

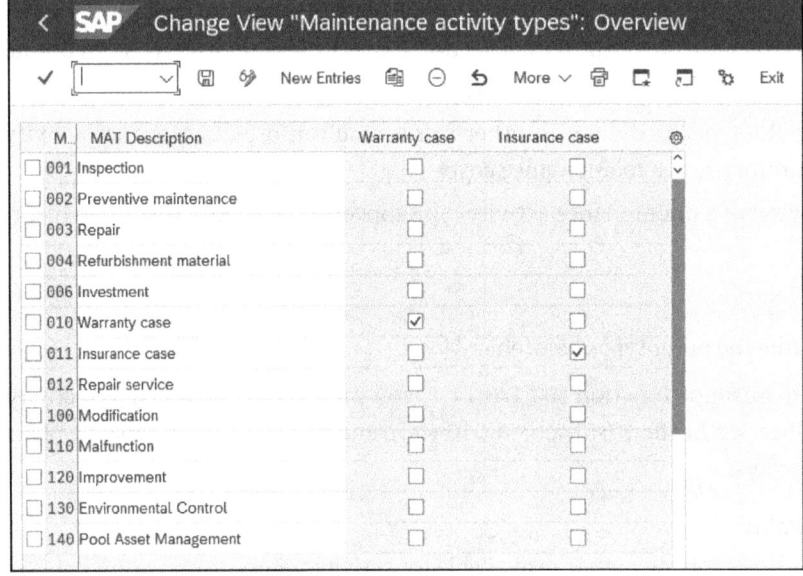

Figure 5.44 Order: Maintenance Activity Types

> **Four to Ten Maintenance Activity Types Should Suffice**
>
> You should always map the following four maintenance activity types in accordance with international standard ISO 55000:
>
> - Repair
> - Inspection
> - Preventive maintenance
> - Improvement

If you use these processes, you also need the following:
- **Refurbishment**
- **Pool Asset Management**
- **Calibration**

Use additional maintenance activity types if necessary.

Then, use the **Assign Valid Maintenance Activity Types to Maintenance Order Types** Customizing function to define which maintenance activity types are valid for which order types (see Figure 5.45). You will then be unable to select any other maintenance activity types for the respective order type.

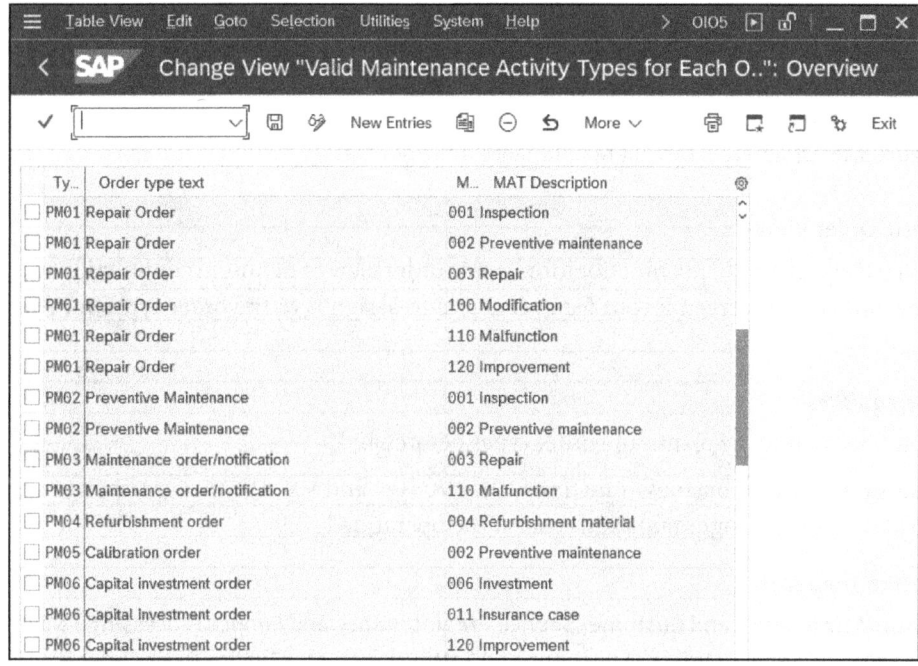

Figure 5.45 Order Type: Valid Maintenance Activity Types

Finally, use the **Default Values for Maintenance Activity Type for Each Order Type** Customizing function to define which maintenance activity type is to be proposed for which order type (see Figure 5.46). Here, you can propose only one maintenance activity type, which you must have also previously assigned to the order type.

5 Configuring the Work Order Cycle

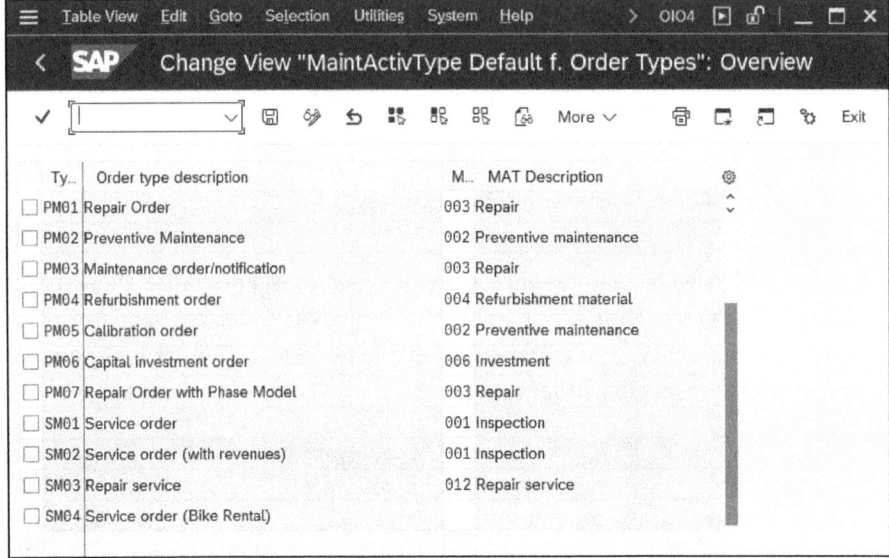

Figure 5.46 Order Type: Default Maintenance Activity Types

Basic Order Views

You use the Customizing functions for a basic order view to define, in conjunction with *view profiles*, the screen layout for orders and to assign it to the order types for each plant.

Prerequisites

You must define the plants and order types beforehand.

You must activate business functions LOG_EAM_SIMP and LOG_EAM_CI_6 (to change and display account assignments and costs at the operation level).

Customizing Path

Plant Maintenance and Customer Service • **Maintenance and Service Processing** • **Maintenance and Service Orders** • **Functions and Settings for Order Types** • **Basic Order View** • **Define View Profiles** and **Assign View Profiles to Order Types**

Settings

First, use the **Define View Profiles** Customizing function to define a name and description for the view profile (see Figure 5.47).

Use the **Tabs** subfunction to determine the tabs that the view profile is to contain (see Figure 5.48). When defining tabs, note the following:

- The tab number is used as a key for the order in which the tabs are displayed.
- You can assign a maximum of eight screen areas to a tab. The order in which the screen areas are assigned also determines the order in which they are displayed on the tab.

5.2 Orders

- Some screen areas occupy the entire tab (full screen). You must always enter these screen areas in the field for screen area 1.
- You can also assign an icon to each tab. Here, you can choose from more than a thousand icons.

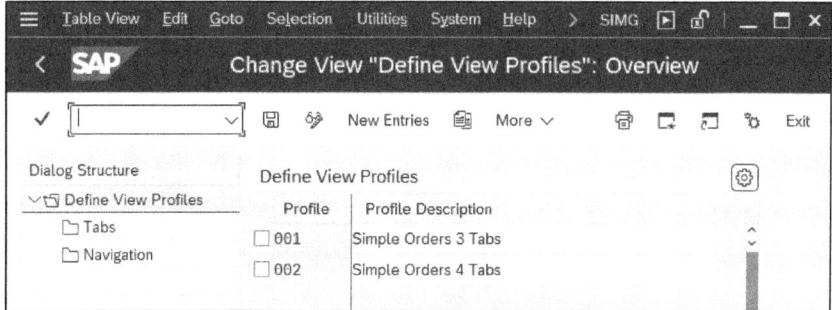

Figure 5.47 Order: View Profiles Overview

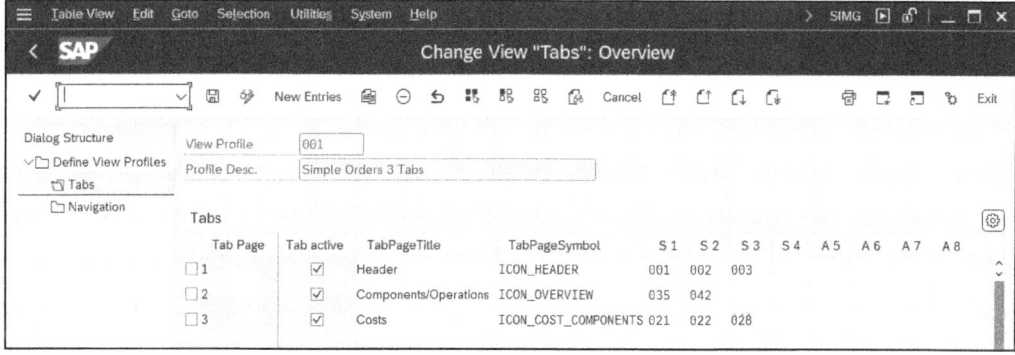

Figure 5.48 Order: View Profile Details

Table 5.2 lists the available screen areas.

ID	Screen Area
001	Responsibilities
002	Reference object
003	Dates (overview)
004	Dates (enhanced)
005	Header long text as control (23 lines)
006	Header long text as control (10 lines)
007	Operation details (with tabs)

Table 5.2 Screen Areas

ID	Screen Area
008	Operation details (selected fields)
010	Partner overview
011	Partner overview (small)
012	Addresses
013	Service data
014	Notification data
015	Linear data
021	Cost overview
022	Cost detail
023	System condition and planning indicator
024	Organization
025	Parameters, administrative data, and reservation/purchase requisition
026	Location data
027	Sales and distribution data for location
028	Account assignment
033	Operations (full screen)
034	Operations (as a subscreen, 15 lines)
035	Operations (as a subscreen, 6 lines)
040	Components (full screen)
041	Components (as a subscreen, 15 lines)
042	Components (as a subscreen, 6 lines)
050	Object list
090	Customer enhancement
091	Customer enhancement (full screen)

Table 5.2 Screen Areas (Cont.)

[+] **Activate the Tab**

Don't forget to activate the tab by checking the box to set the **Tab active** indicator. Otherwise, the tab is defined, but it's not displayed.

5.2 Orders

Use the **Navigation** subfunction to define which button will bring you to which of your defined tabs (see Figure 5.49).

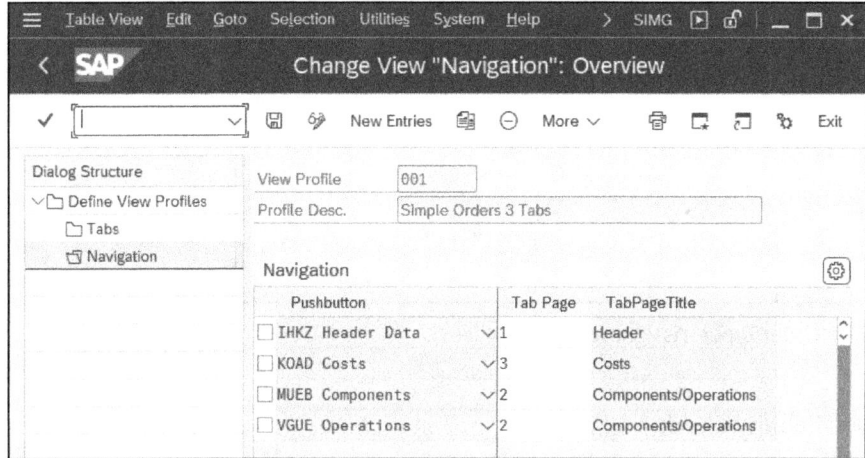

Figure 5.49 Order: View Profile Navigation

Table 5.3 lists all available buttons.

ID	Button
IHKD	Additional Data
IHKZ	Header Data
IHPL	Planning Data
KOAD	Costs
KOBK	Settlement Rule
MUEB	Components
VGUE	Operations
VWPA	Control Data

Table 5.3 Buttons

Finally, use the **Assign View Profiles to Order Types** Customizing function to define which view profile is to be used for which order type and in which plant (see Figure 5.50).

> **View Profiles for Each Order Type and Plant**
> View profiles allow different plants to use their own screen layout for the same order type.

5 Configuring the Work Order Cycle

Type	PlPl	Activity Category	Profile	Profile Description
PM01	DL00		001	Simple Orders 3 Tabs
PM02	DL00		002	Simple Orders 4 Tabs
PM03	DL00		001	Simple Orders 3 Tabs
PM03	HD00		001	Simple Orders 3 Tabs
PM05	DL00		002	Simple Orders 4 Tabs
PM05	HD00		002	Simple Orders 4 Tabs

Figure 5.50 Order Types: Assigning View Profiles

5.2.2 Availability Check

The system always performs a dynamic *availability check*, which checks whether enough of the required material is available in the plant on the requirements date (i.e., when the order is being executed). Figure 5.51 illustrates how the current warehouse stock, based on the current date and considering safety stocks, can change up to the requirements date because of stock movements (planned goods receipts, planned goods issues, and planned stock transfers). The *available-to-promise* (ATP) quantity is determined on the requirements date.

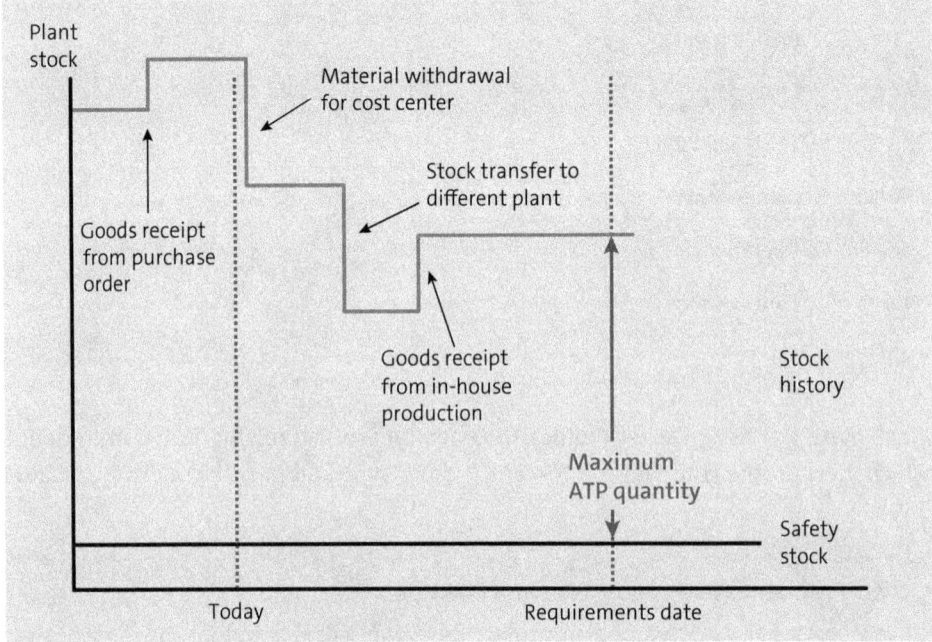

Figure 5.51 Order: Material Availability Check

Figure 5.52 shows the relationship between the Customizing functions and the availability check as described here: You define the *checking group* in Customizing and then assign it to the material master in the MRP view. Next, you define the *checking rule* in Customizing and assign it to an order type for each plant. Finally, based on the checking group and checking rule, you define the *scope of check* in Customizing.

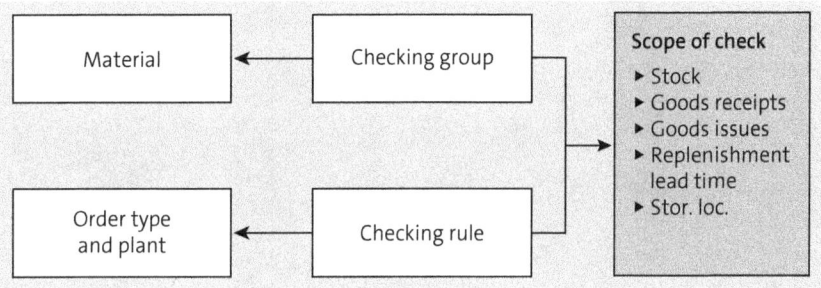

Figure 5.52 Material Availability Check Overview

Let's now look at the most important Customizing activities for availability checks.

Define Checking Rules

You use this Customizing function to define checking rules.

Prerequisites

There are no specific prerequisites.

Customizing Path

Plant Maintenance and Customer Service • Maintenance and Service Processing • Maintenance and Service Orders • Functions and Settings for Order Types • Availability Check for Material, PRTs, and Capacities • Define Checking Rules

Settings

Figure 5.53 provides an overview of the checking rules. You should define a name and description for each checking rule.

> **Separate Checking Rules for Plant Maintenance** [+]
> Create a checking rule that you'll use later for maintenance materials only. Initially, the checking rule doesn't have any other function.

5 Configuring the Work Order Cycle

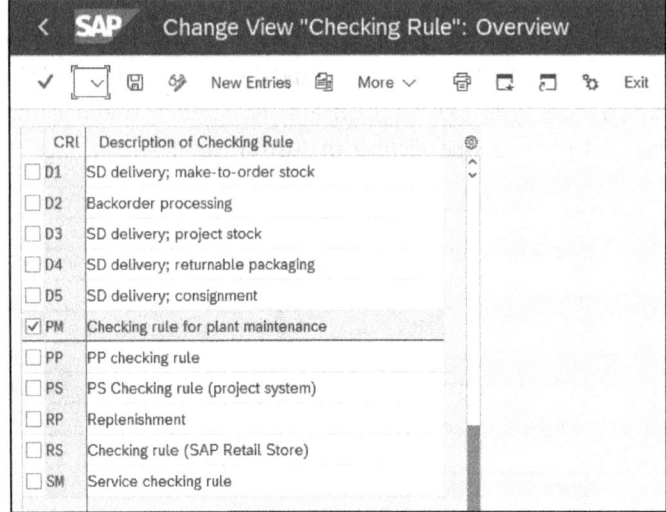

Figure 5.53 Material Availability Check: Checking Rules

Define Inspection Control

You use this Customizing function to define, for each order type, plant, and operation, which checking rule you want to use for the availability check.

Prerequisites

You must define the plants, order types, and checking rules beforehand.

Customizing Path

Plant Maintenance and Customer Service • Maintenance and Service Processing • Maintenance and Service Orders • Functions and Settings for Order Types • Availability Check for Material, PRTs, and Capacities • Define Inspection Control

Transaction

OIOI

Settings

As you can see from Figure 5.54, there are two entries for each order type and plant:

- **Availability transaction 1**
  This refers to created orders, meaning this material availability check is performed as long as the order has the **CRTD** (created) status.

- **Availability transaction 2**
  This applies at the time when an order is released or, if an availability check is performed, after the order has been released.

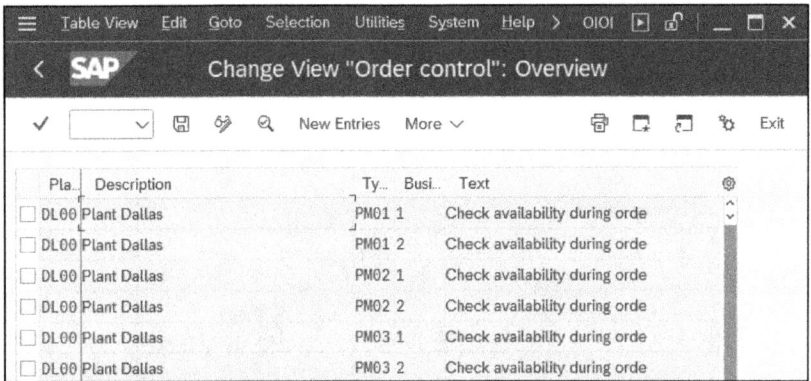

Figure 5.54 Material Availability Check: Inspection Control Overview

The following settings are available for each entry (see Figure 5.55):

- **No check**
  You can check the box for this indicator to deactivate the material availability check, making all other entries have no effect.

- **Status check**
  This indicator is available for availability transaction **2** only. You can check it to determine that the system will only ascertain whether the **MSPT** (material shortage) *system status* will be active when it performs a check before releasing an order. You can also check the box for this indicator to prevent the system from triggering a new *availability check* each time you save the order, thus reducing the system load.

- **Check when saving**
  If you set this indicator, the system automatically performs the material availability check when you save the order after having made all your material-relevant changes to the order (e.g., new components, changed date, or new quantity).

- **Checking Rule**
  For this field, you can enter the checking rule that you defined for SAP S/4HANA Asset Management.

- **Component Check Type**
  This field is always **ATP check** (i.e., a dynamic availability check on the requirements date for the component).

- **Release material**
  This indicator is available for availability transaction **2** only and has the following options:
  - 1: User decides on release if parts are missing
  - 2: Release permitted despite missing parts
  - 3: No release if parts are missing

5 Configuring the Work Order Cycle

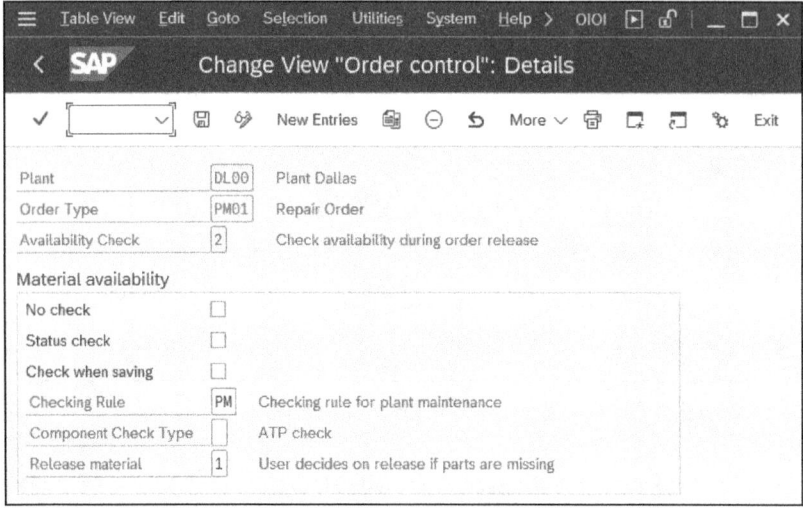

Figure 5.55 Material Availability Check: Inspection Control Details

[+] **Always Let the User Decide**

You should always set the **Release material** option to **1** so that the planner is alerted to the fact that there is a material shortfall. Only then can the planner, and not the system, decide how to deal with such a situation.

Define Checking Group

You use this Customizing function to define checking groups (see Figure 5.56).

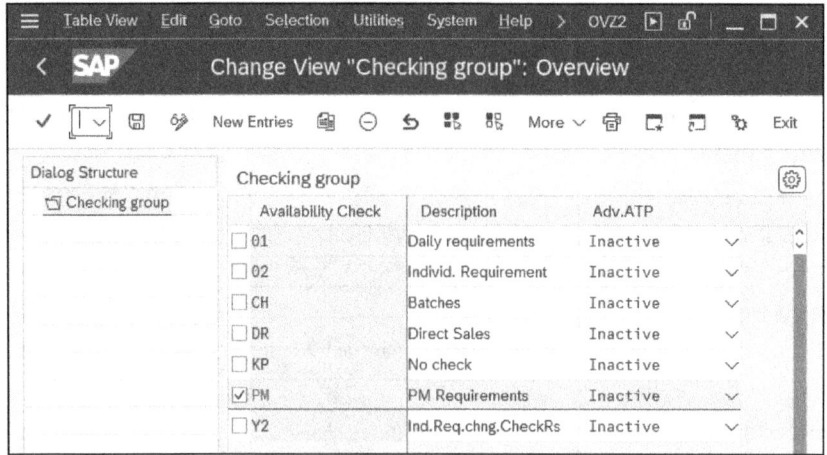

Figure 5.56 Material Availability Check: Checking Group

308

Prerequisites
There are no specific prerequisites.

Customizing Path
Production • Shop Floor Control • Operations • Availability Check • Define Checking Group and Define Basic Availability Check Group Behavior

> **The Checking Group Is Not on the Menu for Plant Maintenance** [!]
> The Customizing function for defining checking groups isn't located on the **Plant Maintenance** menu; look for it on the **Production** menu instead.

Transaction
OVZ2

Settings
For the **Checking group**, you have the following option: You can activate the advanced availability check (**Adv.ATP** column). If you choose **Active** in the **Adv.ATP** dropdown list, the system uses the advanced product availability check instead of the standard product availability check. The advanced product availability check is optimized for large orders with a great number of components and mass processing. Besides other restrictions, you need your own licenses and your own ATP server for this functionality.

In the Customizing function **Define Basic Availability Check Group Behavior**, you have the following options (see Figure 5.57):

- **No PAC**

  Checking the box for the **No PAC** (no product availability check) indicator switches off the product availability check in accordance with the ATP logic for the checking group concerned.

- **Accumulation**

  With these new ATP checking procedures, you have the following setting options when creating/changing a sales order:
  - Cumulation of the confirmed quantities when creating and changing
  - Cumulation of the requirement quantity when creating and no cumulation when changing
  - Cumulation of the requirement quantity when creating and cumulation of the confirmed quantity when changing
  - No cumulation (old checking logic)

- **Relevant for Check Against Planning**

  This new indicator specifies in which cases the availability of a material must be

5 Configuring the Work Order Cycle

checked against planning (planned independent requirements) within the framework of a component availability check:

- No check.
- Always check.
- Check only if dummy.

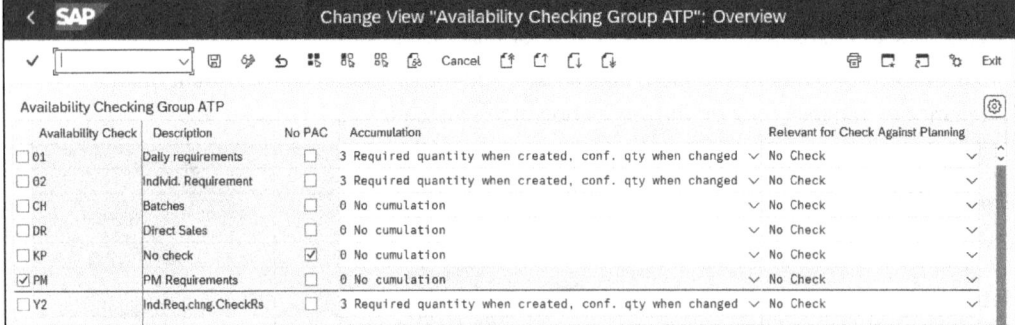

Figure 5.57 Material Availability Check: Define Basic Availability Check Group Behavior

> **[+] Check Each Individual Requirement**
>
> From a maintenance perspective, it doesn't make sense to cumulate requirements. Therefore, use the **Accumulation** indicator **0 = No cumulation** to check the material requirements individually. The plant maintenance requirements are not relevant for planning (**No Check**).

You enter the checking group in the material master (see Figure 5.58). This field belongs to the **Availability Check** screen group, which is generally displayed in the **MRP** view.

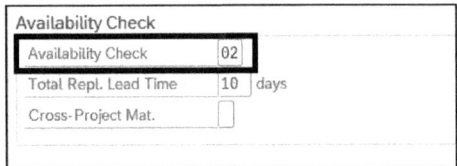

Figure 5.58 Material Master: Checking Group

Define Scope of Check

You use this Customizing function to control how you want to perform the availability check for each checking group and checking rule (i.e., which stock types and which goods receipts/goods issues you want the system to check).

Prerequisites

You must define the checking group and checking rule beforehand.

Customizing Path

Plant Maintenance and Customer Service • Maintenance and Service Processing • Maintenance and Service Orders • Functions and Settings for Order Types • Availability Check for Material, PRTs, and Capacities • Define Scope of Check

Transaction

OPJJ

Settings

You can make the following settings on the screen shown in Figure 5.59:

- Stocks

  This area contains the *stock types* to be included in the calculation. Therefore, if you want the material currently located in the quality inspection stock to be considered, check the box to set the **With Quality Inspection Stock** indicator.

- Future Supply

  This area contains the movement types that cause a planned goods receipt. Therefore, if you want to consider purchase orders whose goods receipt is to be expected up to the requirements date of the material in the order (as a result of the delivery date), set the **With Purchase Orders** option.

- Requirements

  This area contains the movement types that cause a planned goods issue. Therefore, if you want to take into account reservations for which the issuance of goods is to be expected up to the requirements date of the material in the order, check the box to set the **With Reservations** indicator (i.e., reservations that were manually created using Transaction MB21) and **With Dependent Reservations** (i.e., reservations that were created as a result of another order).

- Without Replenishment Lead Time

  This option specifies whether or not the system considers the *replenishment lead time* during the availability check. Materials whose requirements date lies within the replenishment lead time but for which there is no stock are confirmed after the replenishment lead time. In the case of materials whose delivery date lies outside the replenishment lead time, the system assumes that a complete confirmation is possible. If you want the system to take the replenishment lead time into account, you must leave this field blank in the standard system.

- Without Storage Location Check

  You can check the box to set this indicator if, for this scope of checking, you want to deactivate the availability check at the storage location level, meaning the check will be performed at the plant level (that is, across all storage locations).

5 Configuring the Work Order Cycle

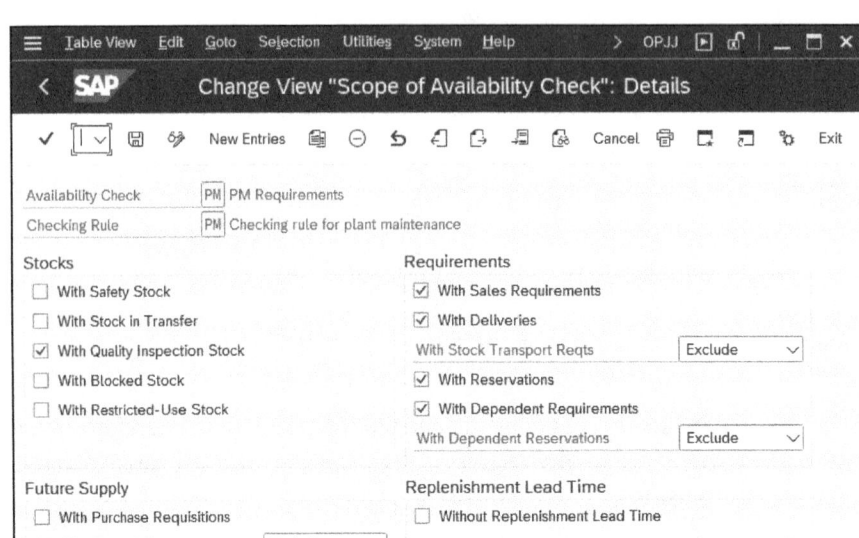

Figure 5.59 Material Availability Check: Scope of Check

[+] **Settings for the Scope of Check**

If you want to perform a reliable material availability check, follow these steps:

1. Deactivate all stock types.
2. Activate all requirements that result in a planned goods issue (e.g., reservations).
3. Deactivate all requirements that result in a planned goods receipt (e.g., purchase requisitions).
4. Deactivate the **Without Replenishment Lead Time** and **Without Storage Location Check** indicators.

5.2.3 Electronic Parts Catalogs

You can connect electronic parts catalogs to SAP S/4HANA Asset Management and thereby provide support for the procurement of spare parts. If you connect more than one catalog, the system always navigates directly to the catalog that you select where you select the parts you require. You then use the **Add** function to place them in your shopping basket and use the **Checkout** function to transfer them to your SAP system (see Figure 5.60).

5.2 Orders

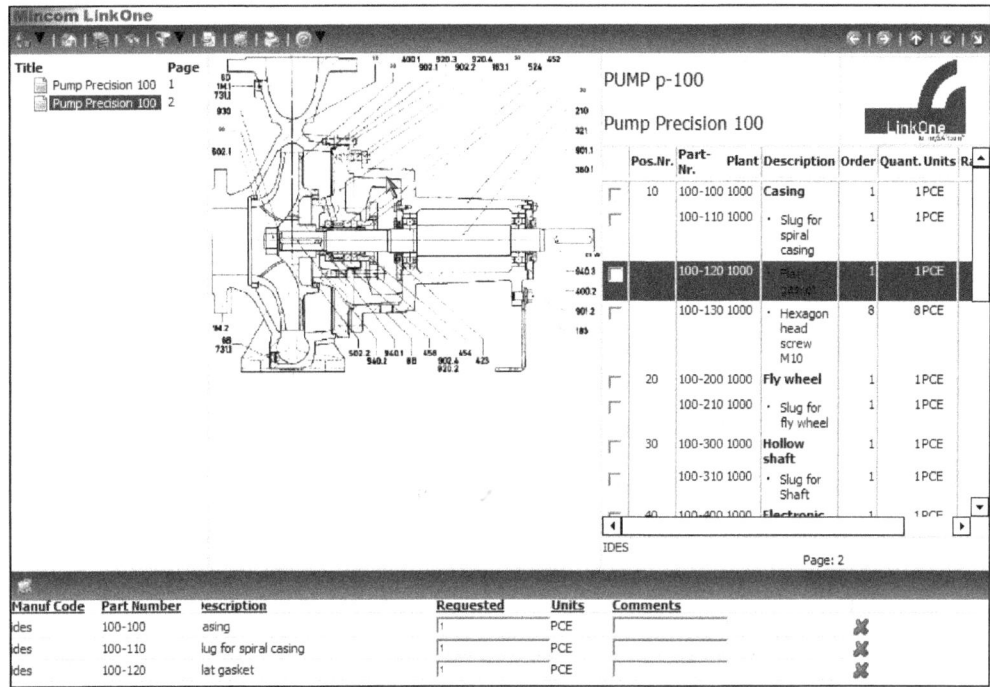

Figure 5.60 Order: Electronic Parts Catalog

> **[!] Activate Enterprise Extension EA-PLM**
>
> If you want to use the catalog connection, you must first use the Switch Framework (Transaction SFW5) to check whether enterprise extension EA-PLM is part of S/4H_ALWAYS_ON_FUNCTIONS. Otherwise, you must activate enterprise extension EA-PLM.

The **Interface for Procurement Using Catalogs (OCI)** Customizing function won't appear in the menu until you activate enterprise extension EA-PLM.

The following sections describe how to configure the necessary settings to make electronic parts catalogs available for material planning within an order.

Define Catalogs

You use this Customizing function to define all catalogs and their call structure.

Prerequisites
You need to have activated enterprise extension EA-PLM.

Customizing Path
Plant Maintenance and Customer Service • Maintenance and Service Processing • Maintenance and Service Orders • Interface for Procurement Using Catalogs (OCI) • Define Catalogs

5 Configuring the Work Order Cycle

Settings

First, define the **Catalog ID** and **Catalog name** (see Figure 5.61).

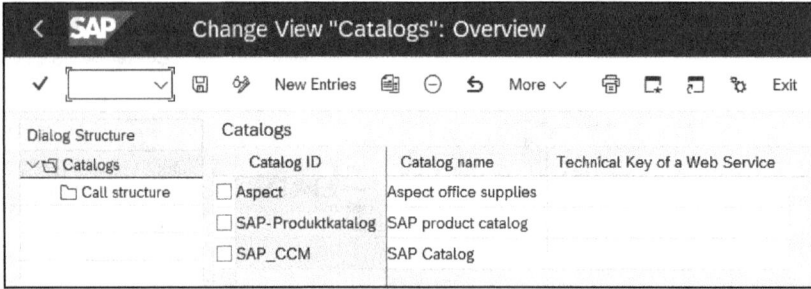

Figure 5.61 Order: Electronic Parts Catalog Overview

If you want to call a catalog from material planning within an order, you must define its internal call structure (i.e., the requisite data for addressing the catalog).

[!] **Note**

Because the internal call structure of a spare parts catalog is always catalog specific, ask the catalog provider for the catalog's internal call structure.

This generally comprises information such as the following (see Figure 5.62).

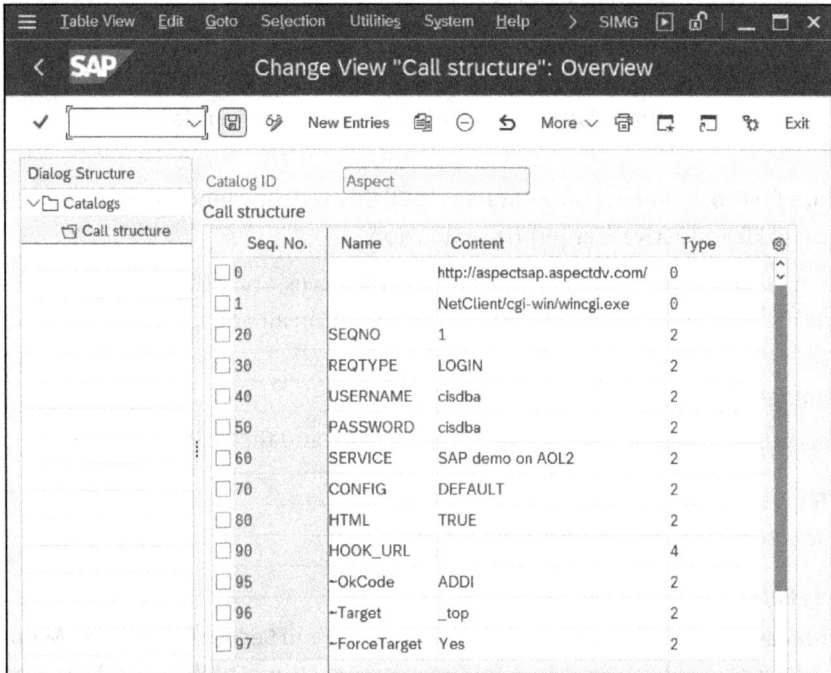

Figure 5.62 Order: Electronic Parts Catalog Call Structure

- URL
- Internal username
- Password
- Language

Assign Catalog to Order Type

You use this Customizing function to define which catalogs are to be used in which plants and for which order types. Several catalogs are permitted for each order type.

Prerequisites

You need an activated EA-PLM enterprise extension, and you must define the catalogs beforehand, including call structure, order types, and plants.

Customizing Path

Plant Maintenance and Customer Service • Maintenance and Service Processing • Maintenance and Service Orders • Interface for Procurement Using Catalogs (OCI) • Assign Catalog to Order Type

Settings

Define which catalogs are to be used for each order type and plant. As you can see in Figure 5.63, this concerns an *n:m* relationship, which means you can assign several catalogs to each order type for each plant, and you can use the same catalog in several order type/plant combinations.

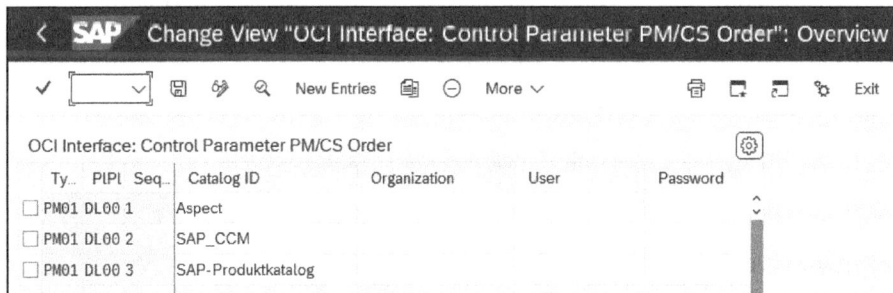

Figure 5.63 Order Types: Assigning Electronic Parts Catalogs

Convert HTML Fields to SAP Fields

You use this Customizing function to define how the HTML fields from the catalog are assigned to the fields in SAP S/4HANA Asset Management.

Prerequisites

You need an activated EA-PLM enterprise extension.

Customizing Path

Plant Maintenance and Customer Service • Maintenance and Service Processing • Maintenance and Service Orders • Interface for Procurement Using Catalogs (OCI) • Convert HTML Fields to SAP Fields

Settings

Table 5.4 shows the HTML fields and an explanation of how they are defined for the Open Catalog Interface (OCI).

HTML Field	Explanation
NEW_ITEM-MATNR	Material number
NEW_ITEM-VENDORMAT	Material number for the vendor
NEW_ITEM-VENDOR	Vendor number
NEW_ITEM-DESCRIPTION	Description of the item to be ordered
NEW_ITEM-MATGROUP	Material group
NEW_ITEM-CONTRACT	Contract number
NEW_ITEM-CONTRACT_ITEM	Contract item
NEW_ITEM-CURRENCY	Currency, which is required when a price is returned (otherwise, optional); must have the standard ISO code for currencies
NEW_ITEM-PRICEUNIT	Number of units that must be ordered at a certain price
NEW_ITEM-PRICE	Price according to the catalog
NEW_ITEM-QUANTITY	Required amount
NEW_ITEM-UNIT	Unit of measure (in standard ISO code)
NEW_ITEM-MANUFACTCODE	Manufacturer number
NEW_ITEM-MANUFACTMAT	Product number of manufacturer
NEW_ITEM-LEADTIME	Delivery time in days
NEW_ITEM-SERVICE	Required if services are to also be copied from the catalog
NEW_ITEM-CUST_FIELD1	Customer-specific field
NEW_ITEM-CUST_FIELD2	Customer-specific field
NEW_ITEM-CUST_FIELD3	Customer-specific field

Table 5.4 HTML Fields for OCI

HTML Field	Explanation
NEW_ITEM-CUST_FIELD4	Customer-specific field
NEW_ITEM-CUST_FIELD5	Customer-specific field

Table 5.4 HTML Fields for OCI (Cont.)

Table 5.5 shows which fields are copied to the target fields of the component in the order.

SAP Target Field	SAP Receiver Structure Field	HTML Field
RESBD-MATNR	RIHFCOM_XL-MATNR	NEW_ITEM-MATNR
RESBD-IDNLF	RIHFCOM_XL-IDNLF	NEW_ITEM-VENDORMAT
RESBD-POSTP	RIHFCOM_XL-POSTP	N/A
RESBD-POTX1	RIHFCOM_XL-KTEXT	NEW_ITEM-DESCRIPTION
RESBD-WERKS	RIHFCOM_XL-WERKS	N/A
RESBD-BDMNG	RIHFCOM_XL-MENGE	NEW_ITEM-QUANTITY
RESBD-MENGE	RIHFCOM_XL-MENGE	NEW_ITEM-QUANTITY
RESBD-ERFMG	RIHFCOM_XL-MENGE	NEW_ITEM-QUANTITY
RESBD-ERFME	RIHFCOM_XL-MEINH	NEW_ITEM-UNIT
RESBD-EINHEIT	RIHFCOM_XL-MEINH	NEW_ITEM-UNIT
RESBD-LIFNR	RIHFCOM_XL-FLIEF	NEW_ITEM-VENDOR
RESBD-INFNR	RIHFCOM_XL-INFNR	N/A
RESBD-KONNR	RIHFCOM_XL-CONTRACT	NEW_ITEM-CONTRACT
RESBD-KTPNR	RIHFCOM_XL-CONTRACT_ITEM	NEW_ITEM-CONTRACT_ITEM
RESBD-MFRPN	RIHFCOM_XL-MFRPN	NEW_ITEM-MANUFACTMAT
RESBD-MFRNR	RIHFCOM_XL-MFRNR	NEW_ITEM-MANUFACTCODE
RESBD-LIFZT	RIHFCOM_XL-LEADTIME	NEW_ITEM-LEADTIME
RESBD-MATKL	RIHFCOM_XL-MATKL	NEW_ITEM-MATGROUP
RESBD-EKGRP	RIHFCOM_XL-EKGRP	N/A

Table 5.5 SAP and HTML Fields

5 Configuring the Work Order Cycle

SAP Target Field	SAP Receiver Structure Field	HTML Field
RESBD-EKORG	RIHFCOM_XL-EKORG	N/A
RESBD-GPREIS	RIHFCOM_XL-PREIS	NEW_ITEM-PRICE
RESBD-PEINH	RIHFCOM_XL-PEINH	NEW_ITEM-PRICEUNIT
RESBD-SAKNR	RIHFCOM_XL-SAKTO	N/A
RESBD-CATALOGID	RIHFCOM_XL-CATALOGID	N/A
RESBD-CATALOG_PROD_ID	RIHFCOM_XL-CATALOG_PROD_ID	N/A

Table 5.5 SAP and HTML Fields (Cont.)

Figure 5.64 shows an example of such an assignment.

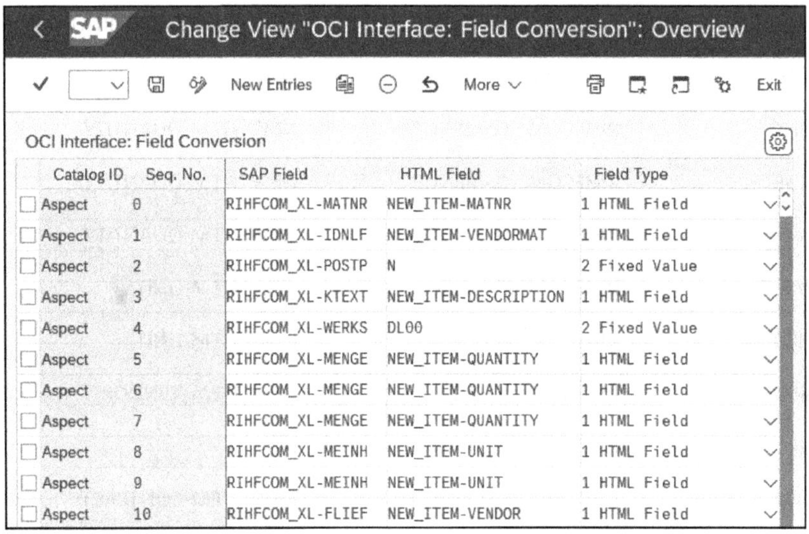

Figure 5.64 Order: Electronic Parts Catalog Fields

Convert HTML Field Values

You use this Customizing function to define which standard SAP values are used to overwrite the values from the catalog.

Prerequisites

You must have an activated `EA-PLM` enterprise extension and must define the units of measurement beforehand.

5.2 Orders

Customizing Path

Plant Maintenance and Customer Service • **Maintenance and Service Processing** • **Maintenance and Service Orders** • **Interface for Procurement Using Catalogs (OCI)** • **Convert HTML Field Values**

Settings

Make corresponding entries in terms of which catalog values are to be overwritten with fixed SAP values. If, for example, this concerns an American catalog, the unit of measure **EA** or the currency **$** is possibly used in the catalog. However, you can overwrite these with the SAP values **ST** and **USD** (see Figure 5.65).

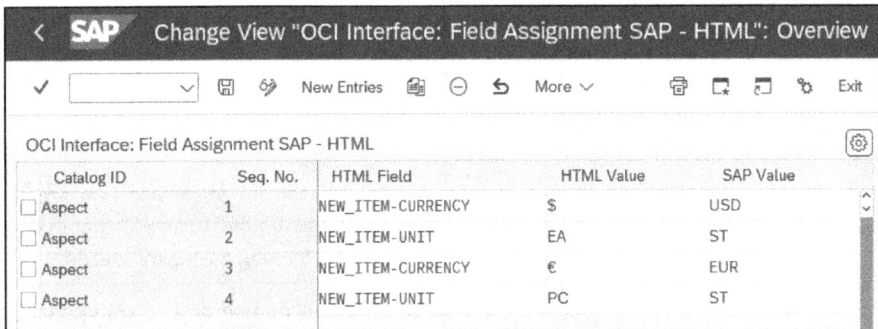

Figure 5.65 Order: Electronic Parts Catalog Field Assignment

Define Conversion Modules

You can use this Customizing function to define conversion modules for converting the data from the catalog into valid SAP values.

Prerequisites

You must have an activated EA-PLM enterprise extension and must create and program the conversion modules beforehand.

Customizing Path

Plant Maintenance and Customer Service • **Maintenance and Service Processing** • **Maintenance and Service Orders** • **Interface for Procurement Using Catalogs (OCI)** • **Define Conversion Modules**

Settings

Enter which function module is to be processed for which catalog and in which order (see Figure 5.66).

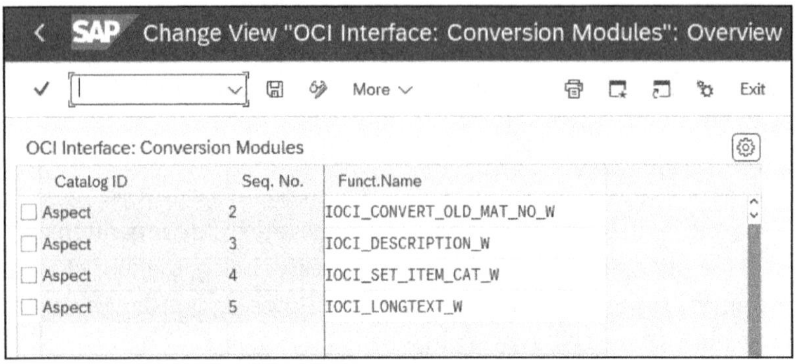

5   Configuring the Work Order Cycle

Figure 5.66 Order: Electronic Parts Catalog Conversion Modules

Table 5.6 shows the conversion modules delivered with the interface in the standard SAP system.

Name	Explanation
IOCI_CONVERT_INFO_REC_W	Converts the catalog material number to an SAP material number in accordance with the purchasing info record
IOCI_CONVERT_MPM_W	Converts the catalog material number to an SAP material number in accordance with the manufacturer number
IOCI_CONVERT_OLD_MAT_NO_W	Converts to an existing material number using the **Old Material Number** field
IOCI_DESCRIPTION_W	Copies the catalog material number and description to the SAP field for the short text
IOCI_SET_ITEM_CAT_W	Searches for the item category in accordance with the Customizing settings for the material type
IOCI_LONGTEXT_W	Copies the catalog long text to the material long text or the service long text
IOCI_EXAMPLE_W	Copies the template for customer-specific conversion modules

Table 5.6 OCI Conversion Modules

However, you can program your own conversion modules at any time, and you can use the IOCI_EXAMPLE_W function module as a template for your own conversion modules.

To use the supplied conversion modules, follow these steps:

1. Call the **Convert HTML Fields to SAP Fields** Customizing function.
2. In the **SAP Field** column, specify the fields listed in Table 5.7.

3. Fill in the columns with the catalog fields you require. Table 5.7 shows which field is used in which conversion module.

Conversion Module	SAP Field 1	SAP Field 2
IOCI_CONVERT_INFO_REC_W	VENDOR	VENDORMAT
IOCI_CONVERT_MPM_W	VENDOR	VENDORMAT
IOCI_CONVERT_OLD_MAT_NO_W	MATERIAL	N/A
IOCI_DESCRIPTION_W	MATERIAL	DESCRIPTION
IOCI_LONGTEXT_W	LONGTEXT	N/A

Table 5.7 Fields in Conversion Modules

> **[+] Use Function Modules Purposefully**
>
> To receive the right data at the right location, use the function modules in a purposeful manner. For example, you should always process the IOCI_SET_ITEM_CAT_W function module because it determines whether a reservation or purchase requisition is triggered.

5.2.4 Material Planning

In addition to the preceding groups (availability check and electronic parts catalogs), there are still other Customizing functions that you can use to influence or support material planning within an order, as described in the following sections.

Define Default Values for Component Item Categories

If you plan a component in an order, you must assign an item category. The most frequently used item categories in plant maintenance are **L** (stock material) and **N** (non-stock material).

You can use this Customizing function to define which item category is to be proposed for a component for each plant and material type. If you don't assign a default value, you must enter the item category manually.

Prerequisites

You must define the plants, material types, and item categories beforehand.

Customizing Path

Plant Maintenance and Customer Service • Maintenance and Service Processing • Maintenance and Service Orders • General Data • Define Default Values for Component Item Categories

Settings

If, for example, you want to trigger a reservation for spare parts (material type **ERSA**), assign item category **L** (see Figure 5.67). If you want to generate a purchase requisition for all nonstock materials (material type **NLAG**), assign item category **N**.

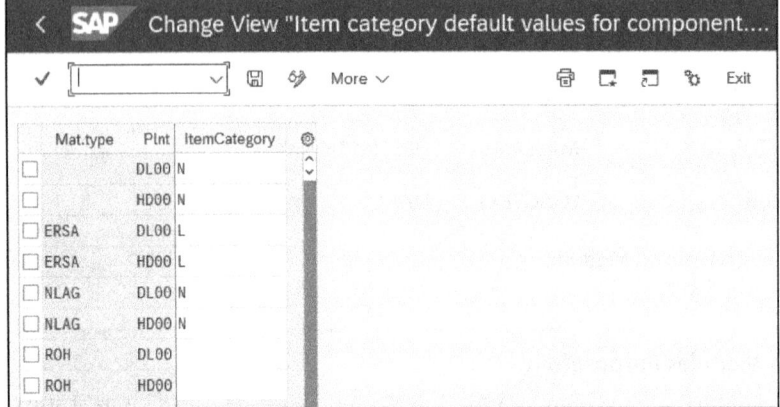

Figure 5.67 Order: Default Values for Item Category

> [+] **Blank Entry for Components without a Material Number**
>
> You should always include an entry without a material type (blank value for the material type). This entry is always used if you want to procure a component without a material number within the order. You should assign the item category for nonstock components to this entry. Usually, this is category **N**.

Determine Movement Types for Material Reservations

During material planning, reservations are created for components that can be kept in stock. Such *reservations* receive a movement type for the material withdrawal. You use this Customizing function to define which movement types are to be entered.

Prerequisites

You must define the movement types beforehand.

Customizing Path

Plant Maintenance and Customer Service • Maintenance and Service Processing • Maintenance and Service Orders • General Data • Define Movement Types for Material Reservations

Settings

In this table, the entry **IW01** (**Appl. Development R/3 Plant Maintenance Order Processing**) is crucial (see Figure 5.68).

5.2 Orders

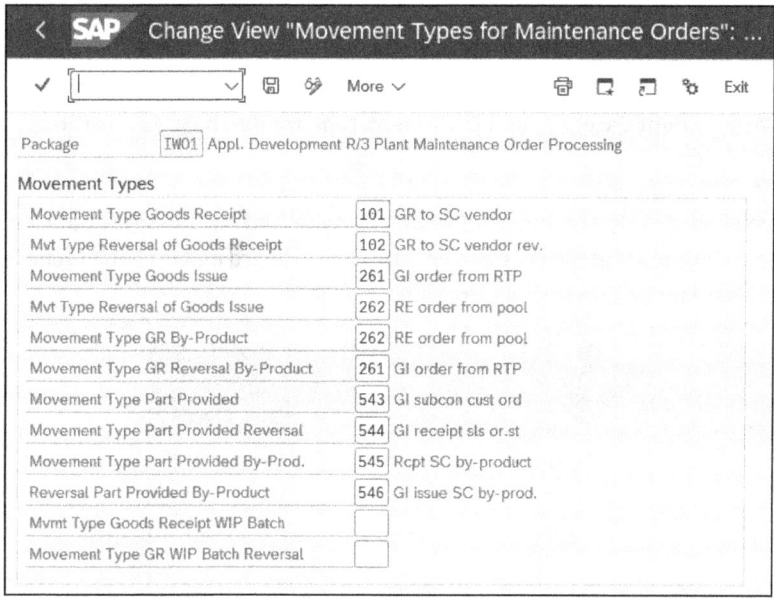

Figure 5.68 Order: Movement Types

For maintenance orders, only the entries **Movement Type Goods Issue** and **Mvt Type Reversal of Goods Issue** are used with standard movement types **261** and **262**, respectively.

For refurbishment orders, you also require the entries **Movement Type Goods Receipt** and **Mvt Type Reversal of Goods Receipt** with standard movement types **101** and **102**, respectively.

> **No Change to Movement Types Is Necessary**
>
> Generally, you can use the standard entries delivered by SAP. You only need to adjust them if you've defined your own movement types.
>
> Note that the entry applies to all orders. Unfortunately, it's not possible to differentiate according to order type or plant.

Define Account Assignment Category and Document Type for Purchase Requisitions

You use this Customizing function to define the *account assignment category* and the document type for purchase requisitions created from orders.

Prerequisites

You must define the document type and account assignment category beforehand.

5 Configuring the Work Order Cycle

Customizing Path

Plant Maintenance and Customer Service • Maintenance and Service Processing • Maintenance and Service Orders • Functions and Settings for Order Types • Procurement • Define Account Assignment Category and Document Type for Purchase Requisitions

Settings

For maintenance orders, only the entries **Document type** and **Acct.Assgmt.Cat.Gen.** (general account assignment category) are used with the standard entries **NB** (**Purchase Requisition**) and **F** (**Order**), as shown in Figure 5.69.

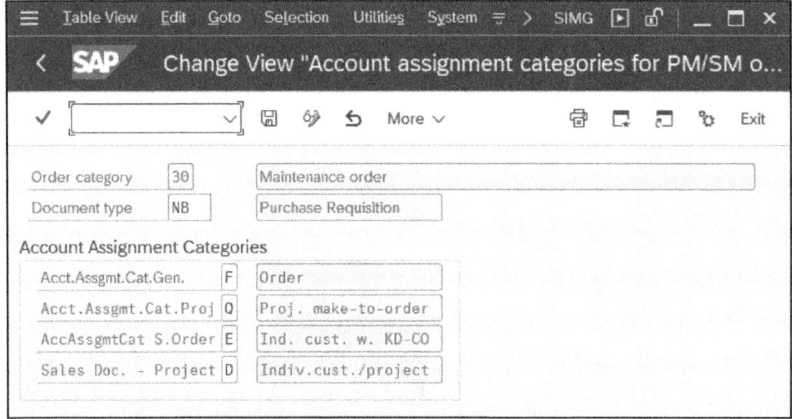

Figure 5.69 Order: Account Assignment Category

No Change to Document Type and Account Assignment Category Is Necessary

Generally, you can use the standard entries delivered by SAP. You only need to adjust them if you've defined your own document types and account assignment categories.

Note that the entry applies to all orders. Unfortunately, it's not possible to differentiate according to order type or plant.

Create Default Value Profiles for External Procurement

You use this Customizing function to define profiles for the external procurement of services and materials. In the profile, you can define default values for generating purchase requisitions from orders. In each profile, you can define default values for the following:

- Cost element
- Purchasing organization
- Purchasing group
- Material group

For external procurement, it's always necessary to differentiate between the external procurement of materials and the external procurement of services.

5.2 Orders

Prerequisites

You must define the requisite values for the cost element, purchasing organization, purchasing group, and material group beforehand.

Customizing Path

Plant Maintenance and Customer Service • Maintenance and Service Processing • Maintenance and Service Orders • Functions and Settings for Order Types • Procurement • Create Default Value Profiles for External Procurement

Settings

Make all the entries that are to be copied to purchase requisitions later. Figure 5.70 shows some examples.

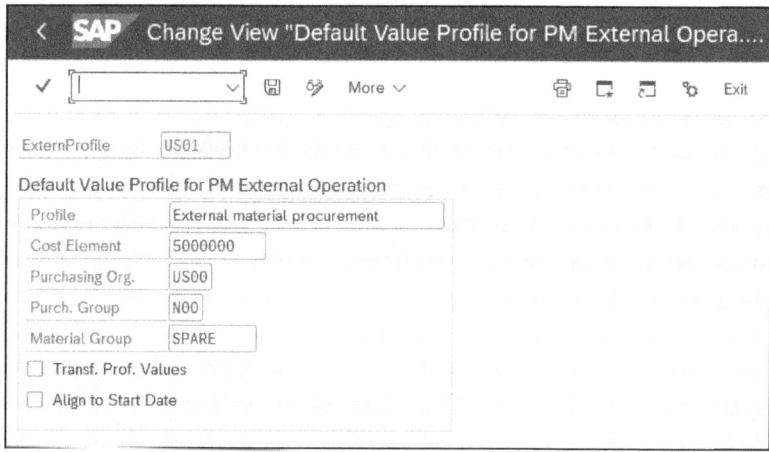

Figure 5.70 Orders: Value Profile for External Procurement

> **One Profile for Material and One Profile for Services** [+]
>
> To be able to differentiate later between whether you want to order material from a supplier or commission a contractor with a service, you have to define a profile for the procurement of material and a profile for the purchase of services.

Then, use the **Default Values for Task List Data and Profile Assignments** Customizing function (Section 5.2.11) to make the relevant assignment for each order type and plant.

Define Access Sequence for Determining Address Data

You can assign an address other than the plant address to the purchase requisitions generated from the orders. You can also assign an address to maintenance objects (i.e., a piece of equipment, functional location, notification, or order).

325

5 Configuring the Work Order Cycle

You use this Customizing function to define, for each plant and order type, the sequence in which the delivery address of nonstock components is to be determined from the addresses of maintenance objects (see Figure 5.71).

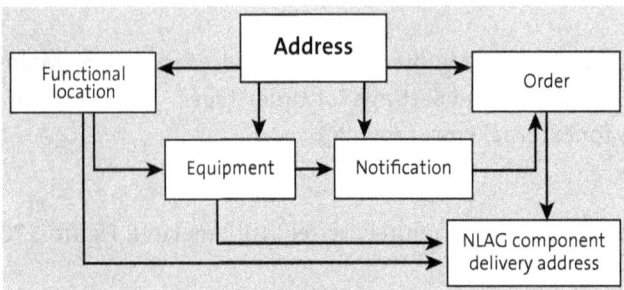

Figure 5.71 Order: Delivery Address Overview

Prerequisites

You must define the plant and order type beforehand.

Customizing Path

Plant Maintenance and Customer Service • Maintenance and Service Processing • Maintenance and Service Orders • Functions and Settings for Order Types • Define Access Sequence for Determining Address Data

Settings

For each order type and plant, define an access sequence with up to five priorities. The following objects are available to you:

- 1: Order
- 2: Functional location
- 3: Equipment
- 4: Sold-to party (order partner)
- 5: Ship-to party (order partner)

In fields **A1** to **A5**, define the sequence in which the system is to access the relevant addresses (see Figure 5.72).

Access Sequences Make Sense if the Delivery Address Differs from Plant Address

An access sequence always makes sense if the delivery address for the components differs from the plant address (e.g., if objects such as the power supply or the infrastructure measures lie outside the plant premises).

If you operate typical on-site maintenance, you generally don't require access sequences.

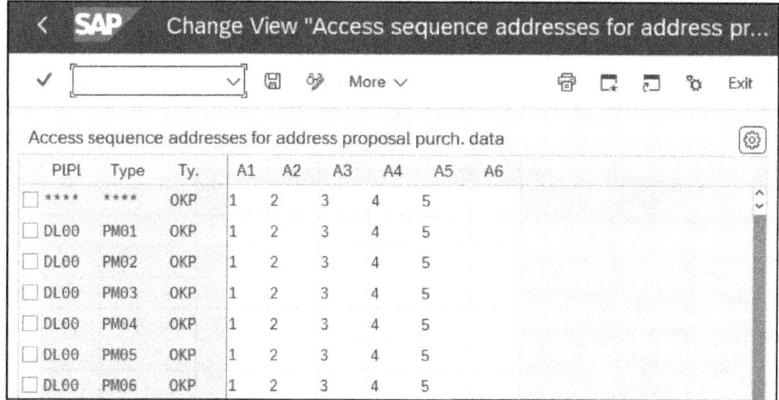

Figure 5.72 Order: Determining Delivery Address

Define Collective Purchase Requisition and MRP Relevance

You use this Customizing function to define, for each order type and plant, various order parameters (including parameters that you can use to control material planning).

Prerequisites

You must define the plants and order types beforehand.

Customizing Path

Plant Maintenance and Customer Service • Maintenance and Service Processing • Maintenance and Service Orders • Functions and Settings for Order Types • Procurement • Define Collective Purchase Requisition and MRP Relevance

Settings

Figure 5.73 shows the fields within this table, which you can use to influence material planning:

- CollReqstn (collective purchase requisition)
  If you check the box to set this indicator, all the purchase requisition items that result from an order number are grouped together under a single purchase requisition number. If you don't set this indicator, the system generates a separate purchase requisition number for each purchase requisition item.

- Res/PurRq (reservation/purchase requisition)
  Use this indicator to define the point in time when purchase requisitions or reservations are to become MRP relevant. The following options are available on the drop-down menu:
  - 1: Never
  - 2: From Release
  - 3: Immediately

5 Configuring the Work Order Cycle

- **Net Order Price**
  Check the box to set this indicator to specify that the net price from the purchase requisition must be copied unchanged to the purchase order and can't be changed there.

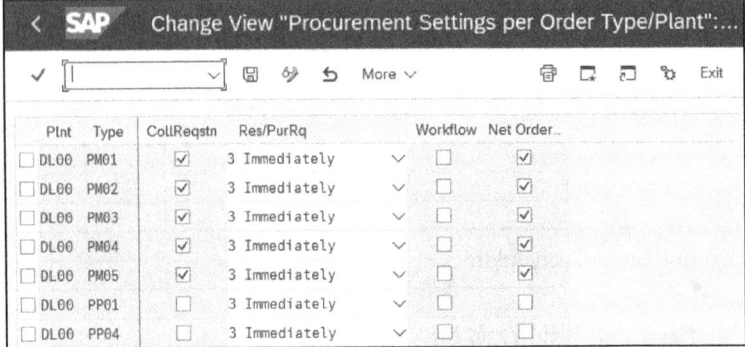

Figure 5.73 Order Type: Material Planning Parameters

> **Notes on the Res/PurRq Indicator**
> Some notes on the **Res/PurRq** indicator are as follows:
> - Unfortunately, it's not possible to differentiate between reservations and purchase requisitions. Instead, these are always activated simultaneously. For example, you can't control the fact that purchase requisitions are generated immediately but that reservations don't become MRP relevant until they are released.
> - You can't prevent the system from generating reservations. If you plan a stock material, the system always generates a reservation document even if the indicator is set to **1 (Never)** or **2 (From Release)**. The sole purpose of these indicators is to ensure that the reservation never becomes MRP relevant or only becomes MRP relevant after orders have been released.

Define Documentation for Goods Movements for the Order

You use this Customizing function to define which materials movements are to be recorded for each order type and plant. You can then view these in the orders (either in the document flow or in the documented goods movements).

Prerequisites

You must define the order type and plant beforehand.

Customizing Path

Plant Maintenance and Customer Service • Maintenance and Service Processing • Maintenance and Service Orders • Functions and Settings for Order Types • Goods Movements for Order • Define Documentation for Goods Movements for the Order

328

5.2 Orders

Settings

You can document the following goods movements for the maintenance orders separately using these indicators (see Figure 5.74):

- **PurchOrder**
  Goods receipts for the purchase order
- **PlGoodsIss**
  Planned goods issues (goods issues for a reservation)
- **UnplGdsIss**
  Unplanned goods issues (goods issues without a reservation)
- **GR refrbshmnt**
  Goods receipts for refurbishment orders

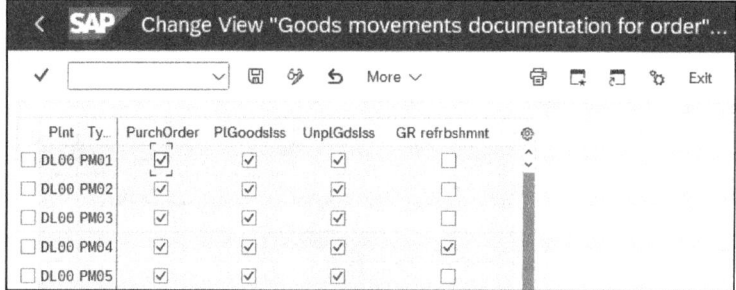

Figure 5.74 Order Type: Goods Movement Documentation

Notes on Documenting Goods Movements

Some notes on documenting goods movements are as follows:

- You may be inclined to document all types of materials movements. However, this increases the number of documents used.
- Regardless of these settings, the materials movements are always documented in materials management, and you can use the materials management transactions (e.g., Transaction MB51 [Material Document List]) to display these documents.
- Activate only the materials movements that are also to be documented in the order.
- If you change the settings retroactively, you must start report RIAUFM00, which re-creates the documents for the materials movements in maintenance orders.

Texts for Purchase Requisitions

Previously, you could only add a single long text for the nonstock item, which then was transferred to the purchase requisition as item text. The LOG_EAM_CI_12 business function brings in another innovation here: if you activate it, up to four texts are available.

5 Configuring the Work Order Cycle

This is flexible, so that you can use the Customizing function to define the type of text for purchase requisitions, as follows:

- Which types of text you want to use
- Which text type of the nonstock item should be copied to which text type of the purchase requisition

Figure 5.75 shows you possible text types (e.g., note, remark, comment), which you can then call up individually using a button.

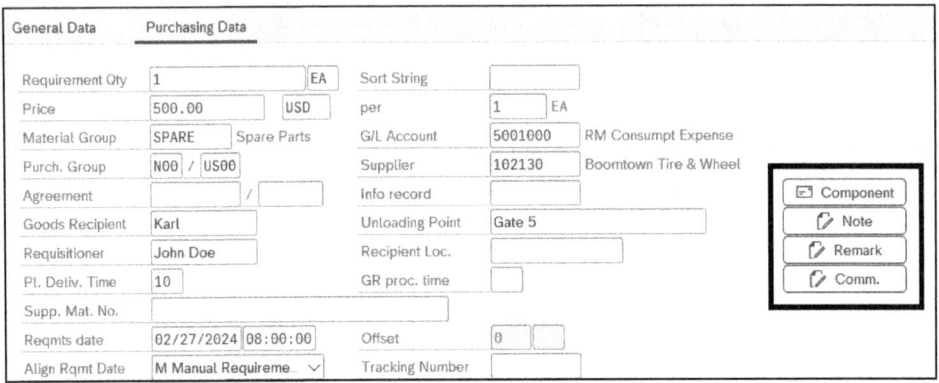

Figure 5.75 Nonstock Material or External Services: Text Types

Prerequisites

You must activate business function LOG_EAM_CI_12.

Customizing Path

The following Customizing paths are available for purchase requisitions:

- **Material Management • Purchasing • Purchase Requisition • Texts for Purchase Requisitions • Define Text Types**
- **Material Management • Purchasing • Purchase Requisition • Texts for Purchase Requisitions • Define Copying Rules**
- **Plant Maintenance and Customer Service • Maintenance and Service Processing • Maintenance and Service Orders • General Data • Specify Labels for Additional Operation and Component Texts**

Settings

With the **Define Text Types** Customizing function, you define target text types for purchase requisitions (e.g., **01** = **Item Text**, **02** = **Item Note**, **03** = **Delivery Text**). Figure 5.76 shows you a list of potential target text types.

Next, you have to link each source text type with a target text type (e.g., source text purchase requisition item note with target text type item note). If you want to transfer texts from an SAP S/4HANA Asset Management order to a purchase requisition, you must link the order operation and/or order component with a purchase requisition target text type (see Figure 5.77).

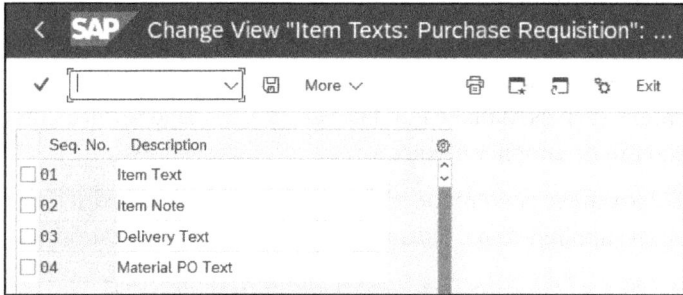

Figure 5.76 Purchase Requisition: Text Types

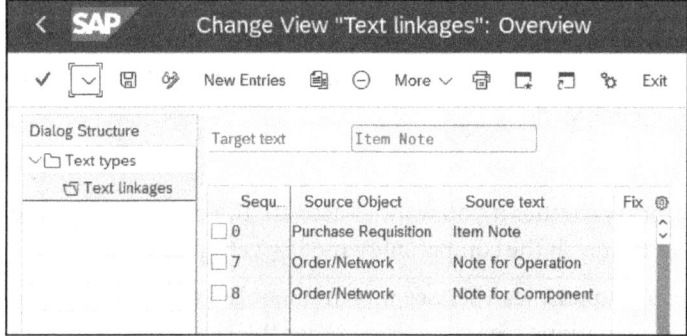

Figure 5.77 Purchase Requisition: Text Linkages

With the **Specify Labels for Additional Operation and Component Texts** Customizing function, you define text labels that should appear on the user interface within SAP S/4HANA Asset Management for order components and order operations (see Figure 5.78).

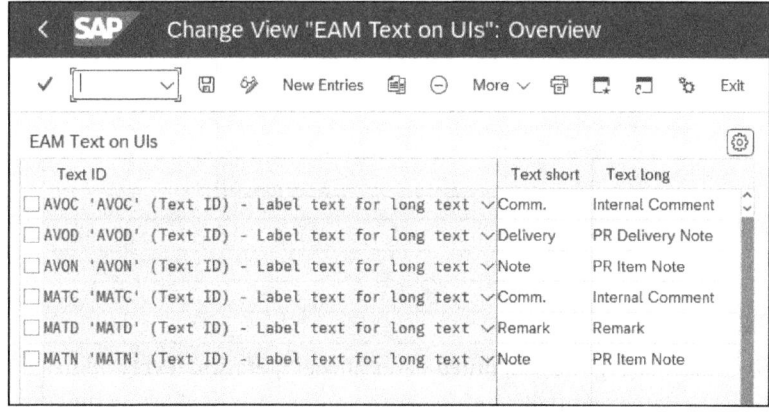

Figure 5.78 Order: Text Labels

5 Configuring the Work Order Cycle

Activate Enhanced Procurement Mode

You use this Customizing function to define whether or not to activate the enhanced procurement mode for each order type and plant.

The enhanced procurement mode enables the procurement of *lean services*, which are a type of service that can be purchased in the same way as materials using the SAP Fiori apps for service procurement such as Manage Purchase Orders, Manage Service Entry Sheets – Lean Services, and Create Supplier Invoice.

Lean services are intended for the procurement of small or highly specialized services to be carried out by external suppliers as part of an internal or external operation.

Prerequisites

You must define the plants and order types beforehand.

Customizing Path

Plant Maintenance and Customer Service • Maintenance and Service Processing • Maintenance and Service Orders • Functions and Settings for Order Types • Procurement • Activate Enhanced Procurement Mode

Settings

You define plant and order type either if you want to activate the enhanced procurement mode or if you want to stay in the compatibility mode (see Figure 5.79).

Compatibility Mode doesn't support lean service procurement but allows you to procure externally processed operations regardless of whether they are defined as a service by their control key (e.g. **PM02** and **PM03**). The system default setting is the **Compatibility Mode**.

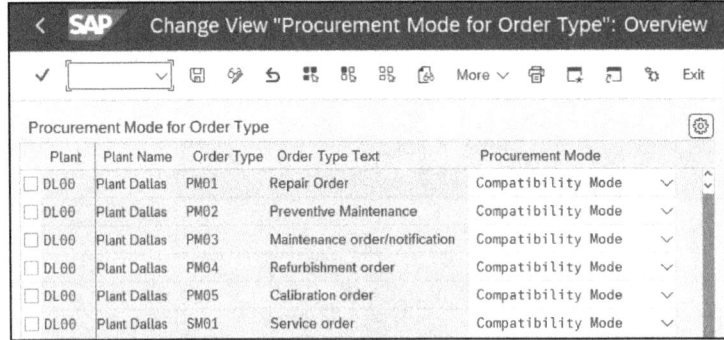

Figure 5.79 Order: Procurement Mode

5.2.5 Scheduling

Scheduling involves calculating the scheduled dates at the operation level and header level based on the basic dates manually defined in the order and considering the durations at the operation level.

> **Scheduling Doesn't Always Make Sense**
> Scheduling in the SAP system only makes sense if you've saved standard times. If you don't have these, you shouldn't activate scheduling initially. You can ensure this by using a control key for which the **Scheduling** option isn't activated.

In Chapter 2, Section 2.4, you learned how to define control keys (see also the Maintain Control Keys subsection and Figure 2.27 in that chapter).

There are many ways to perform scheduling:

- Lead-time scheduling is executed either as forward scheduling or backward scheduling.
- Both forward and backward scheduling are executed for network scheduling, while taking relationships into account. The basic start date is used as the basis for calculating the earliest scheduled dates, while the basic finish date is used as the basis for calculating the latest scheduled dates. The difference between the earliest date and the latest date gives rise to a *buffer*, which is shown not only for each operation but also at the header level.

To perform scheduling, you must fulfill the following prerequisites:

- The relevant duration must be entered in all operations.
- A control key must be assigned to the operation, and the **Scheduling** option must be activated for this control key.
- A formula must be assigned for the *duration of internal processing* to the work center. This must point to the **DAUNO** field (that is, the duration from the operation). In the standard system, this is formula **SAP004**.
- If you want to perform lead-time scheduling, the **Set Scheduling Parameters** Customizing function must first be used to assign the scheduling type (forwards or backwards) to the order type in the plant.

Figure 5.80 shows you the relationships among scheduling, capacity requirements, and costing, and the procedures associated with each, as follows:

- **Formula parameters**
  In Customizing, you define *formula parameters* for the **DAUNO** and **ARBEI** fields. (In the standard delivery, these are formula parameters **SAP_10** and **SAP_07**.)
- **Formula definitions**
  In Customizing, you create *formula definitions* for the formula parameters. (In the standard delivery, these are formulas **SAP004** and **SAP008**.)
- **Work center**
  In the *work center*, you enter the formula for duration in the **Duration of Internal Processing** field on the **Scheduling** tab. You enter the formula for work in the **Other**

5 Configuring the Work Order Cycle

Formula field on the **Capacities** tab and in the **Formula Key for Costing** field on the **Costing** tab.

- **Order**
  You can then schedule the *order* based on the duration. Furthermore, you can generate capacity requirements based on the work and perform a costing.

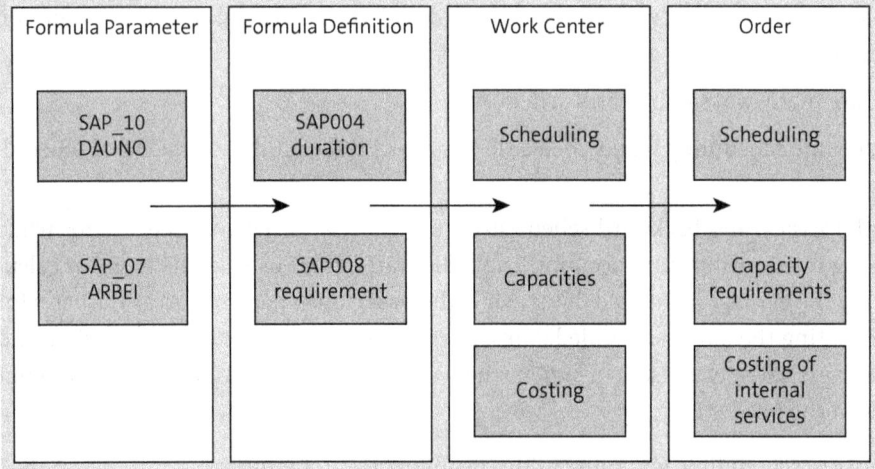

Figure 5.80 Relationships Among Scheduling, Capacities, and Costing

[+] **Scheduling Is Only Partly Available in the Plant Maintenance Menu**
For the most part, the Customizing functions for configuring scheduling aren't available in the **Plant Maintenance** menu. Therefore, use the Customizing menu for **Production** or simply call the relevant transactions directly.

Define Formula Parameters

You use this Customizing function to define the parameters that you'll later use as a basis for scheduling, costing, and capacity requirements determination. From a maintenance perspective, you require the **DAUNO** (duration) and **ARBEI** (work) fields for this purpose.

Prerequisites
There are no special prerequisites.

Customizing Path
Production • Shop Floor Control • Operations • Scheduling • Define Formula Parameters

Transaction
OP7B

334

5.2 Orders

Settings

In the standard delivery, the formula parameters **SAP_07** for **ARBEI** and **SAP_10** for **DAUNO** are available in Customizing for this purpose (see Figure 5.81).

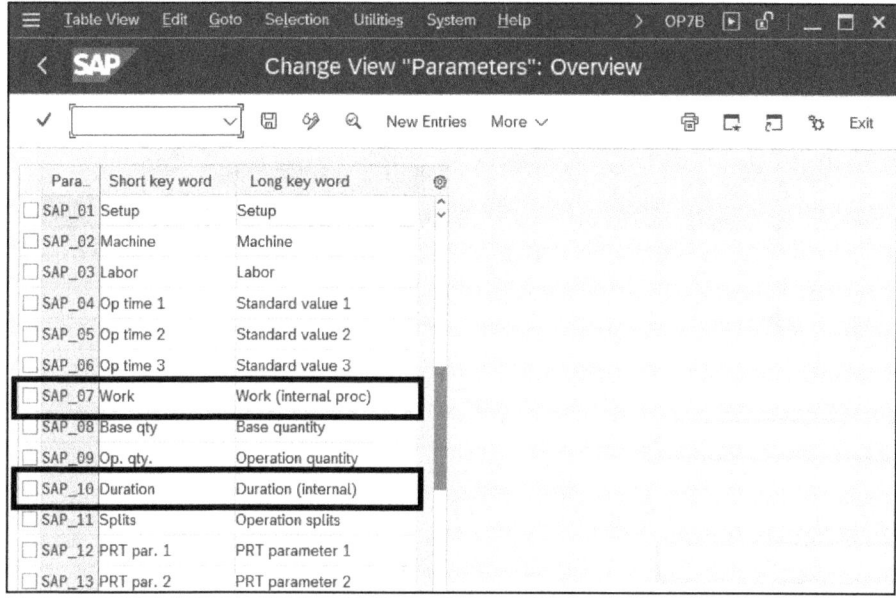

Figure 5.81 Order: Formula Parameters

> **Use the Formula Parameters in the Standard SAP System** [+]
>
> Generally, you can use the formula parameters delivered in the standard SAP system without having to make any adjustments to them. You only require other formula parameters if you want to define your own calculations.

Configure Formula Definition

You use this Customizing function to define the formulas used to calculate results on the basis of formula parameters that contain mathematical functions for the following:

- Capacity planning
- Scheduling
- Costing

Prerequisites

You must define the formula parameters beforehand.

Customizing Path

Production • Shop Floor Control • Operations • Scheduling • Define Formulas

335

5 Configuring the Work Order Cycle

Transaction

OPK3

Settings

In the standard delivery, the formula definition **SAP004** for the duration and the formula definition **SAP008** for the requirement are defined for plant maintenance (see Figure 5.82).

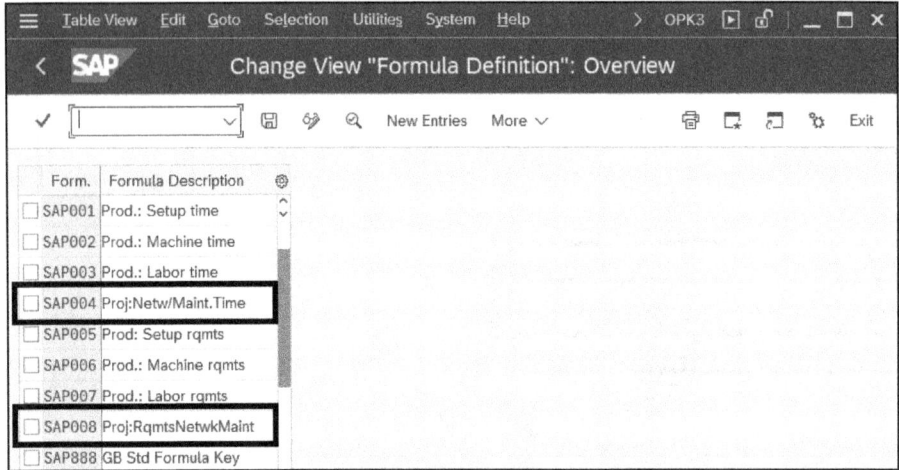

Figure 5.82 Order: Formula Definition Overview

The **SAP004** formula key (see Figure 5.83), which calculates the resulting lead time, is based on the **SAP_10** formula parameter and therefore on the **DAUNO** (duration) field. It's important that you check the box to set the **Allowed for Scheduling** indicator here.

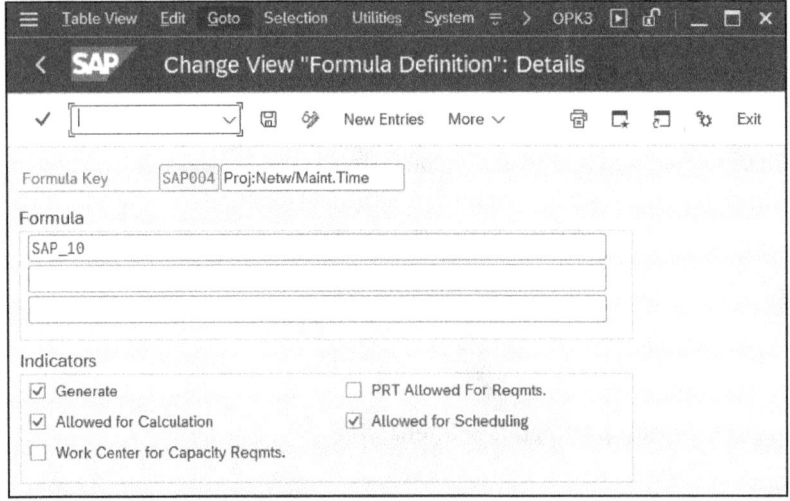

Figure 5.83 Order: Formula Definition for Scheduling

The **SAP008** formula key (see Figure 5.84) is based on the **SAP_07** formula parameter and therefore on the **ARBEI** (work) field, and it can calculate capacity requirements and costs for internal services. It's important that you check the boxes to set the **Allowed for Calculation** and **Work Center for Capacity Reqmts.** indicators here.

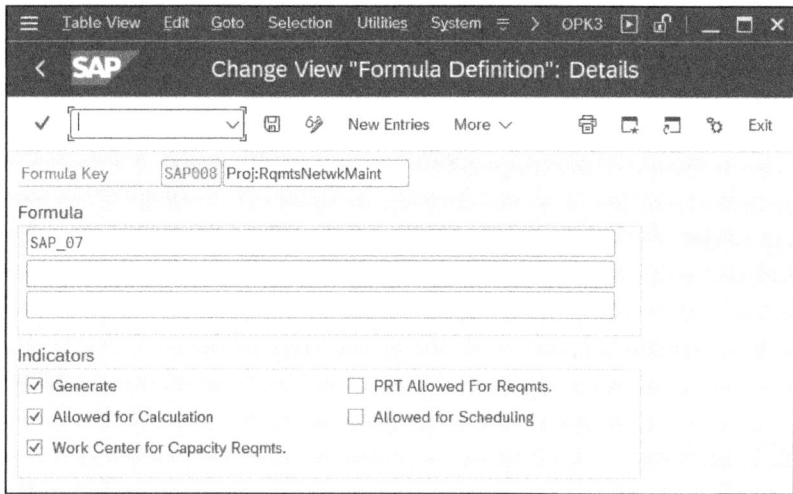

Figure 5.84 Order: Formula Definition for Capacity Requirements and Costing

> **Use the Formula Definitions in the Standard SAP System** [+]
>
> Generally, you can use the standard SAP definitions for formula definitions to schedule orders, calculate costs, and generate capacity requirements. You only need to change the standard SAP definitions if you want to perform your own calculations.

Specify Scheduling Type

You use this Customizing function to define the scheduling types that you want to use. In the standard delivery, SAP supports the following scheduling types:

- **Forwards**
  The system uses the basic start date and adds the duration of operations to calculate the *earliest scheduled start and end dates* at both the header level and the operation level.

- **Forwards in time**
  This is the same as **Forwards**, except that you specify the required start time in addition to the basic start date.

- **Backwards**
  The system uses the basic finish date and subtracts the duration of operations to calculate the *latest scheduled start and end dates* at both the header level and the operation level.

- **Backwards in time**
  This is the same as **Backwards**, except that you specify the required end time in addition to the basic finish date.
- **Current date**
  This is the same as **Forwards in time**, except that the basic start date and time are the current date and the current time.
- **Only capacity requirements**
  The scheduled dates are equated with the basic dates, and the work is distributed evenly. For example, say you have an order with a basic start date of November 1 and a basic finish date of November 10. For the operation, you've scheduled forty hours of work, so as a result, from November 1 through November 10, a capacity requirement of four hours is allocated to each day.
- **Don't schedule**
  The system avoids scheduling and generating capacity requirements.

Prerequisites

There are no special prerequisites.

Customizing Path

Production • Shop Floor Control • Operations • Scheduling • Specify Scheduling Type

Transaction

OPJN

Settings

Figure 5.85 shows the table of scheduling types delivered by SAP.

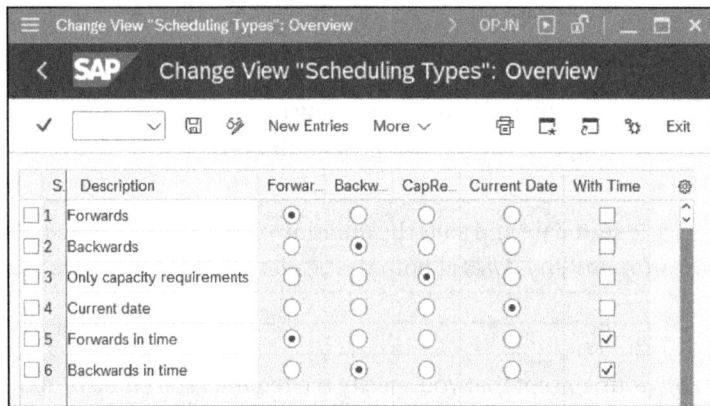

Figure 5.85 Order: Scheduling Types

> **All Scheduling Types Available** [+]
>
> Generally, you can use the standard SAP definitions for scheduling types. You only need to change the standard SAP definitions if you want to exclude scheduling types or change their descriptions.

Set Scheduling Parameters

You use this Customizing function to define, for each order type and plant, which scheduling procedure is to be used.

> **Production Scheduler = *** [+]
>
> When you assign the scheduling parameters to the order type and plant, the production scheduler must always be * because plant maintenance doesn't use the production scheduler. Only production uses the production scheduler.

Prerequisites

You must define the scheduling type, order type, and plant beforehand.

Customizing Path

Plant Maintenance and Customer Service • Maintenance and Service Processing • Maintenance and Service Orders • Scheduling • Set Scheduling Parameters

Transaction

OPU7

Settings

This Customizing function contains the following settings (see Figure 5.86):

- **Adjust Dates**
  Use this option to specify whether, as a result of scheduling, the basic dates are to be automatically adjusted to the scheduled dates and whether the requirements dates for the material components are to be based on the basic start date or the respective operations dates.

- **Scheduling Type**
  Use this indicator to specify the preferred scheduling type for the plant and order type. This scheduling type is then copied to the order as a default value, but it can be changed there.

- **Automatic Scheduling**
  Check the box for this indicator to ensure that an order is automatically scheduled when you choose **Save**.

5 Configuring the Work Order Cycle

- **Automatic log**
  Check the box for this indicator to ensure that a scheduling log is automatically displayed after scheduling.

- **Scheduling with breaks**
  Check the box for this indicator to determine that break times are to be taken into careful consideration. If you set this indicator, a date calculated by the system can no longer occur during a break time.

- **Shift Order**
  Check the box for this indicator to specify that, in the case of partially confirmed operations, existing actual dates aren't considered when scheduling is repeated.

- **Latest Staging Date**
  Check the box for this indicator to specify that the requirements date for the material components is to be based on the latest date. If you don't check it, the requirements date will be based on the earliest date. This indicator only plays a role in network scheduling.

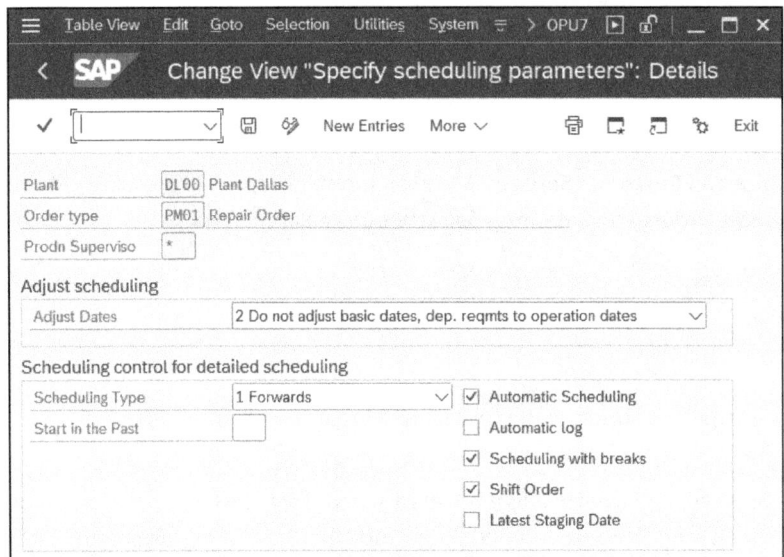

Figure 5.86 Order Type: Scheduling Parameters

> [+] **Scheduling Recommendations**
>
> Some recommendations for scheduling are as follows:
>
> - Set the **Adjust Dates** field to **Do not adjust basic dates**. If you don't, then during scheduling, the basic dates that you preassigned manually are overwritten by the scheduled dates, which means they are lost and can't be restored.
>
> - In plant maintenance, **Forward scheduling** is the normal scenario for lead-time scheduling.

5.2 Orders

- If you haven't deactivated scheduling, you should check the box to set the **Automatic Scheduling** indicator.
- To ensure that the system doesn't display the scheduling results in irritating popup logs each time you save the order, uncheck the box to deactivate the **Automatic log** indicator.
- To obtain as up-to-date scheduling data as possible, check the box to set the **Scheduling with breaks** indicator.
- To ensure that scheduling is adjusted to the actual situation, check the box to set the **Shift Order** indicator.
- Because the earliest materials requirements dates should always be used in the order, don't check the box to set the **Latest Staging Date** indicator.

Activate Default Value for Current Date as Basic Date

You use this Customizing function to specify, for each plant, whether the current date is to be proposed as the basic date.

Prerequisites

You must define the plants beforehand.

Customizing Path

Plant Maintenance and Customer Service • Maintenance and Service Processing • Maintenance and Service Orders • General Data • Activate Default Value for Current Date as Basic Date

Settings

Select the plants for which the current date is to be proposed as the basic date (see Figure 5.87).

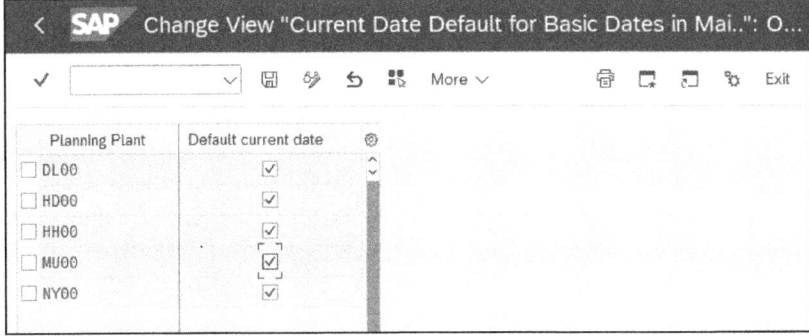

Figure 5.87 Order: Default Current Date

Define Default Values for Units for Operation

You use this Customizing function to define, for each plant, which units of measurement are to be proposed for the **Work** and **Duration** fields.

Prerequisites

You must define the plants and units of measurement beforehand.

Customizing Path

Plant Maintenance and Customer Service • Maintenance and Service Processing • Maintenance and Service Orders • General Data • Define Default Values for Units for Operation

Transaction

OIO9

Settings

Define which units of measurement are to be proposed for the **Work** and **Duration** fields for each plant (see Figure 5.88). If, however, the work center for the operation contains different units of measurement as default values, the default values for the plant are overwritten. If necessary, you can always overwrite those default values manually.

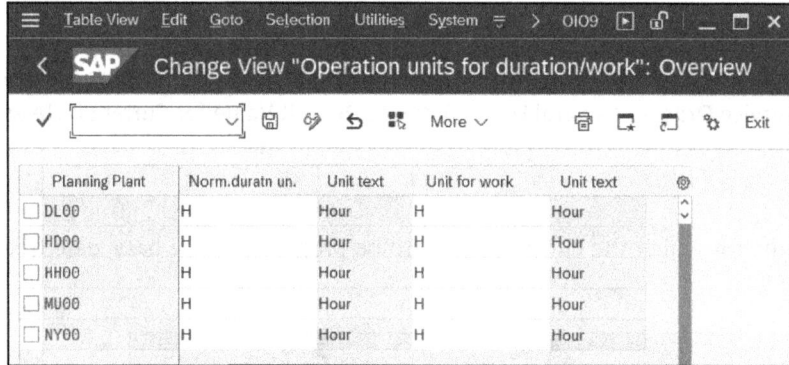

Figure 5.88 Order: Default Units of Measurement

Bar Chart: Define Graphic Profiles

You use this Customizing function to define a graphic profile that you can use as a basis for displaying a bar chart in Transaction I_GRAPH_MONITOR. This bar chart shows you which technician processed which orders and on which dates. You call the graphic profile immediately after you start the transaction.

Prerequisites

There are no special prerequisites.

5.2 Orders

Customizing Path

ABAP Platform • Application Server • Frontend Services • Bar Chart • Define Graphic Profiles

Settings

The graphic profile defines the complete graphical layout. It comprises the following parts (see Figure 5.89):

- **Chart group**
  The chart types made available by the application are grouped together in this group, where you can define the structure and content of the graphic (i.e., the title, areas, headings, columns, etc.), the background color for the areas, and similar items.

- **Element Group**
  The graphical element types made available by the application are grouped together in this group, where you can define, for example, the type, size, and color of the symbols and lines used.

- **Color group**
  The color categories made available by the application are grouped together in this group, where you can define the color categories that are subsequently used in the graphical element group.

- **Form group**
  The form types made available by the application are grouped together in this group, where you can define the form types that are subsequently used in the graphical element group.

- **Options profile**
  The data related to the window and the processing modes are specified in this profile, where you can define, for example, the window sizes, time interval, scroll bar, options available, and other properties of the graphic.

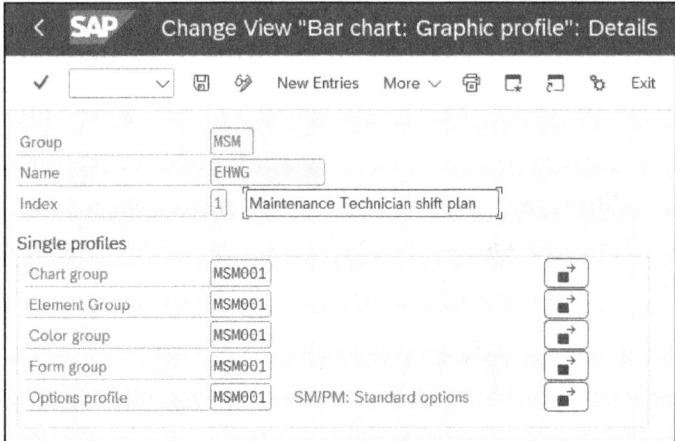

Figure 5.89 Order Bar Chart Graphic Profile

> **Use Bar Chart Graphic Profiles from the Standard SAP System**
> You don't have to define bar chart graphic profiles in the first phase of the implementation project. Instead, you can use one of the profiles delivered in the standard SAP system (e.g., **MSM**), and then, you can create your own profiles only if the standard SAP profiles prove to be unsuitable. The best way to do the latter is to copy a standard SAP profile and adjust it accordingly.

Network: Define Graphic Profiles

You use this Customizing function to define a graphic profile that you can use as a basis for displaying the network in Transaction IW32 (under **Goto** · **Graphic** · **Network Structure**) or a bar chart of the order in Transaction IW32 (under **Goto** · **Graphic** · **Gantt Chart**).

Prerequisites

There are no special prerequisites.

Customizing Path

ABAP Platform · Application Server · Frontend Services · Network/Hierarchy · Define Graphic Profiles

Settings

A graphic profile defines the complete graphical layout and comprises the following parts (see Figure 5.90):

- **Node group**
  The node types made available by the application are grouped together in this group, where you can define colors, shapes, lines, and sizes of nodes (operations) in the network.

- **Link group**
  The link types made available by the application are grouped together in this group, where you can define colors, shapes, lines, and the thickness of the links in the network.

- **Frame group**
  The frame types made available by the application are grouped together in this group, where you can define colors and shapes as well as the width and thickness of the frames in the network.

- **Options profile**
  Data related to the window, processing modes, and colors is included in this profile, where you can define, for example, the window sizes, time interval, area sequence, options available, and other properties of the graphic.

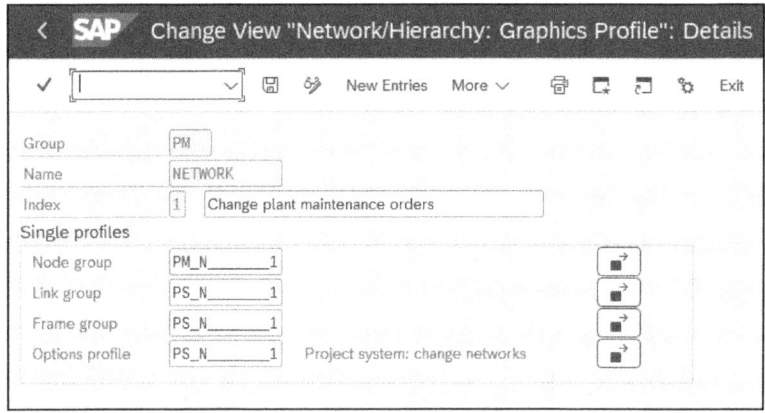

Figure 5.90 Order: Network Graphic Profile

Don't call the graphic profile for the network directly within an order. Instead, use the **Create Default Value Profiles for General Order Data** and **Default Values for Task List Data and Profile Assignments** Customizing functions to assign the graphic profile to each plant and order type.

Use Network Graphic Profiles from the Standard SAP System
You don't have to define network graphic profiles in the first phase of the implementation project. Instead, you can use one of the profiles delivered in the standard SAP system (e.g., **PM NETWORK**).

Then, you can create your own profiles only if the standard SAP profiles prove to be unsuitable. The best way to do this is to copy a standard SAP profile and adjust it accordingly.

5.2.6 Capacity Planning

In *capacity planning*, you compare the available capacity to the capacity requirement. You maintain the available capacity in the work center, and the capacity requirement is determined by the standard times associated with the operations in the order (the **Work** field). However, capacity planning doesn't always make sense.

First, Check Whether Capacity Planning Makes Sense
Essentially, capacity planning only makes sense if you have a sufficiently large order volume and reasonably accurate standard times. Unlike in production, these prerequisites are frequently not fulfilled in plant maintenance. Either there are no exact specifications in relation to the standard times, or you have many unplanned orders.

> If these prerequisites aren't fulfilled, you shouldn't activate capacity planning. You can ensure that it doesn't get activated by using a control key for which the **Det. Cap. Req.** (determine capacity requirements) option isn't activated.

If you perform capacity planning, it is always composed of three stages (see Figure 5.91).

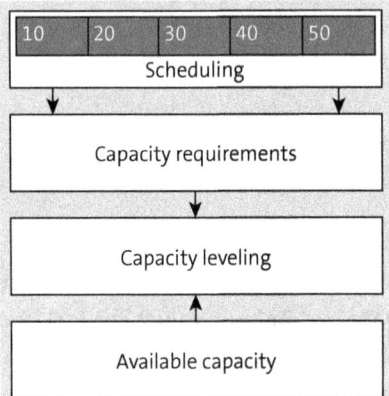

Figure 5.91 Capacity Planning Overview

You define the *available capacity* on the **Capacities** tab within the work center (either as the standard available capacity or with shift models).

Capacity requirements are created based on orders (more specifically, on specifications in the operations in the **Work** field). To generate capacity requirements, you must fulfill the following prerequisites:

- In the work center, you must enter a formula for calculating the capacity requirement for internal processing. This formula must point to the **ARBEI** field (that is, the work from the operation). In the standard system, this is the formula **SAP008** (see also the **Define Formula Parameters** and **Define Formulas** Customizing functions).
- In the case of operations, you must use a control key for which the **Capacity Requirements** function is activated.
- You must schedule the order (Section 5.2.5).

For *capacity leveling*, you have the following options:

- You can perform a *capacity evaluation* (Transaction CM01, used for work center capacity load), which is a clear comparison of the available capacity and the capacity requirement, and thus ensure capacity leveling (e.g., by shifting the orders).
- Alternatively, you can use the planning table (Transaction CM33, used for planning graphical tables, or Transaction CM34, used for planning tabular tables) to perform order scheduling on a capacity basis.

5.2 Orders

The next section describes the most important Customizing settings for capacity planning in SAP S/4HANA Asset Management.

> **Capacity Planning Is Not in the Plant Maintenance Menu** [+]
>
> The Customizing functions for configuring capacity planning aren't available in the **Plant Maintenance** menu. Therefore, you should use the Customizing menu for **Production** or simply call the relevant transactions directly.

Define Capacity Category

This Customizing function is the starting point for capacity planning. Here, you define the capacity types that you want to assign to a work center and use as a basis for capacity planning.

Prerequisites

There are no special prerequisites.

Customizing Path

Production • Capacity Requirements Planning • Master Data • Capacity Data • Define Capacity Category

Settings

From a maintenance perspective, labor capacity is the most important capacity category. Therefore, it's important to use a capacity category for which the **CapCat. Per** indicator is set. In the standard delivery, this is capacity category **002** (**Person**; see Figure 5.92).

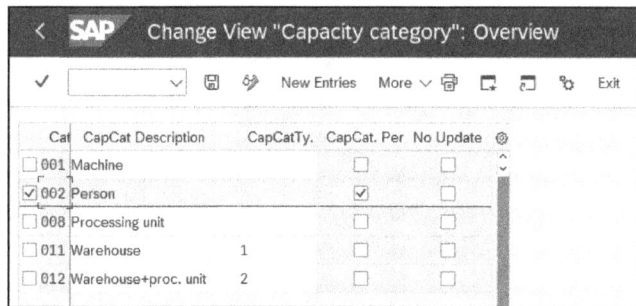

Figure 5.92 Capacity Planning: Capacity Category

> **Use Capacity Category 002** [+]
>
> You can also use the standard SAP capacity category for plant maintenance (**002**). You only need to define your own capacity category if you want to differentiate yourself from production. Ensure that you check the box to set the **CapCat. Per** indicator.

5 Configuring the Work Order Cycle

Define Shift Sequences

If you want to define only the standard available capacity for the work center, you don't have to configure any further Customizing settings for the available capacity.

If, however, you want to define different available capacities and shifts, you must use this Customizing function to define the shift models that you'll use in SAP S/4HANA Asset Management.

Prerequisites

There are no special prerequisites.

Customizing Path

Production • Capacity Requirements Planning • Operations • Available Capacity • Define Shift Sequences

Transaction

OP4A

Settings

Define shift sequences in the three consecutive, interrelated subfunctions of **Work break schedules**, **Shift definitions**, and **Shift sequences**.

Use the **Work break schedules** subfunction to define parameters that describe the individual break schedules in more detail (see Figure 5.93):

- Start
- End
- After
- Break duration

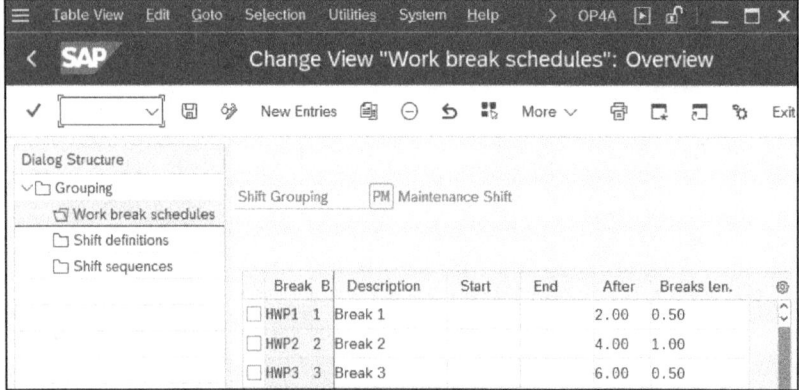

Figure 5.93 Capacity Planning: Work Break Schedules

You can enter these as either fixed times or relative to the start of the shift (e.g., after 2 hours).

Use the **Shift definitions** subfunction to define cross-work center parameters that describe the individual shifts in more detail (see Figure 5.94). Examples include the following:

- Start time
- End time
- Break times
- Validity period of a shift

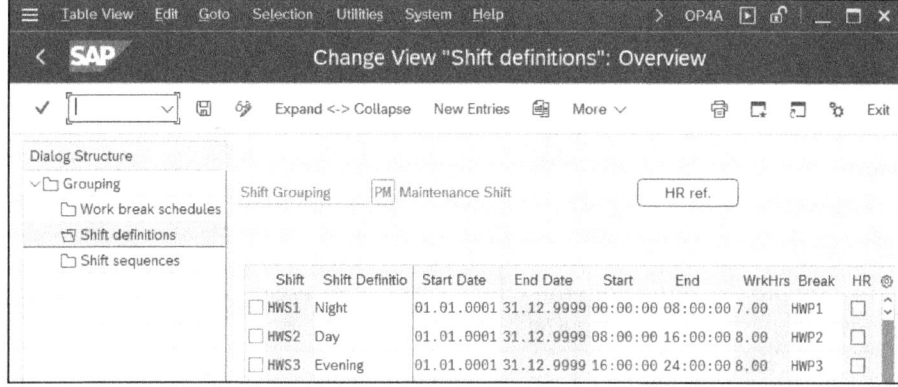

Figure 5.94 Capacity Planning: Shift Definition

Here, incorporate one of the work break schedules that were defined in the work break schedules (in the **Break** field).

Use the **Shift sequences** subfunction to define a series of shifts per weekday (see Figure 5.95). The shift sequence is therefore a grouping of *shift definitions*.

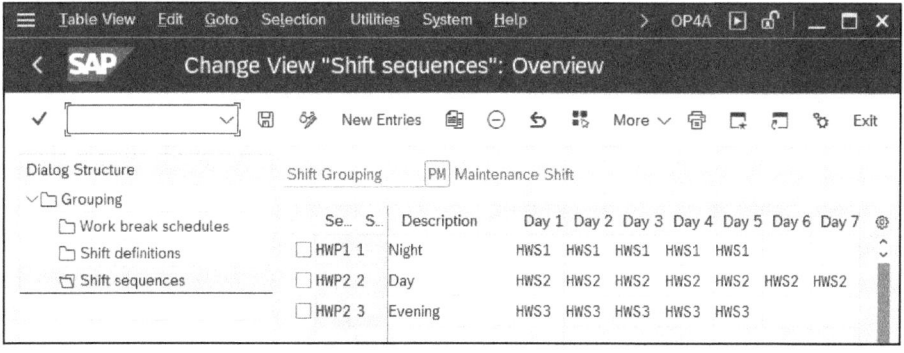

Figure 5.95 Capacity Planning: Shift Sequences

In the work center, you can assign the shift sequence to a capacity category as available capacity. In conjunction with the number of persons assigned to the capacity category, the SAP system uses the predefined values from the shift definition to determine the available capacity for a work center.

5 Configuring the Work Order Cycle

Define Overall Profiles in Capacity Evaluation

For capacity evaluation (Transaction CM01, used for work center capacity load), you use this Customizing function to define the selection and layout.

Prerequisites

There are no special prerequisites.

Customizing Path

Production • Capacity Requirements Planning • Evaluation • Profiles • Define Overall Profiles

Transaction

OPA6

Settings

An overall profile is composed of the following single profiles (see Figure 5.96):

- **Sel. profile**
  In this subprofile, you define the units (work center types, order categories, and capacity categories) and periods (start date, end date, and unit) for which you want to perform capacity planning.

- **Option profile**
  In this subprofile, you define (among other things) whether you want to cumulate the capacity evaluation, whether you want a hierarchical analysis, or which period type is to be the underlying basis here.

- **List profile**
  In this subprofile, you define the level of detail for the capacity evaluation and the interface with Microsoft Excel.

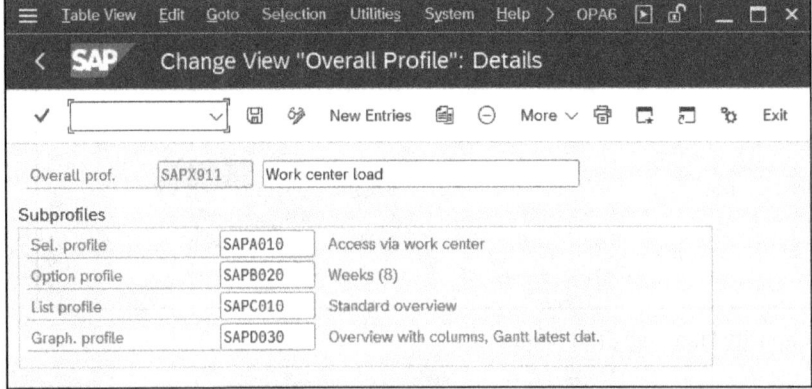

Figure 5.96 Capacity Planning: Evaluation Profiles

- **Graph. profile**
  In this subprofile, you define the analysis display (columns, bars, and lines) and the sort order for the capacity details.

If you start Transaction CM01, you can choose **Planning • Profiles** to select an overall profile and each of the subprofiles separately.

> **Capacity Evaluation Profiles from the Standard SAP System** [+]
>
> You don't have to define capacity evaluation profiles in the first phase of the implementation project. Instead, you can use one of the profiles delivered in the standard SAP system (e.g., **SAPX911**).
>
> Then, you should only create your own profiles if the standard SAP profiles prove to be unsuitable. The best way to do this is to copy a standard SAP profile and adjust it accordingly.

Define Overall Profile in Capacity Leveling

For capacity leveling (Transaction CM33 and Transaction CM34), you use this Customizing function to define the selection, control, and layout for the planning table in which you'll perform capacity leveling.

Prerequisites

There are no special prerequisites.

Customizing Path

Production • Capacity Requirements Planning • Capacity Leveling and Extended Evaluation • Define Overall Profile

Transaction

OPDO

Settings

An overall profile contains the following subprofiles (see Figure 5.97):

- **Selection profile**
  In this subprofile, you define the primary selection criteria (generally, the plant and work center) for accessing the transaction. You can also specify additional filtering criteria here (e.g., maintenance planner group, functional location, and plant section).
- **Control profile**
  In this subprofile, you define whether you want to use the graphical or tabular planning table.

5 Configuring the Work Order Cycle

- **Time profile**
  In this subprofile, you define which past or future periods are to appear in the planning table.

- **Evaluation Profile**
  In this subprofile, you define the unit of measurement and whether you want to perform capacity cumulation over this period.

- **Strategy prof.**
  In this subprofile, you define which functions are to be available when scheduling orders (e.g., operation date check) as well as the sequence in which scheduling is to take place (e.g., latest start date).

- **Period profile**
  In this subprofile, you define whether capacities are to be planned on a daily, weekly, or monthly basis.

- **Planning tab.profile**
  In this subprofile (the profile of the graphical planning table), you define the layout of the graphical planning table (e.g., screen areas, time axis, and graphical elements).

- **Prof. pln.tab. (tab)**
  In this subprofile (the profile of the tabular planning table), you define the layout of the tabular planning table (e.g., columns, rows, and totaling).

- **List profile**
  In this subprofile, you define which information is to be available in the detail list and how this list is to be sorted.

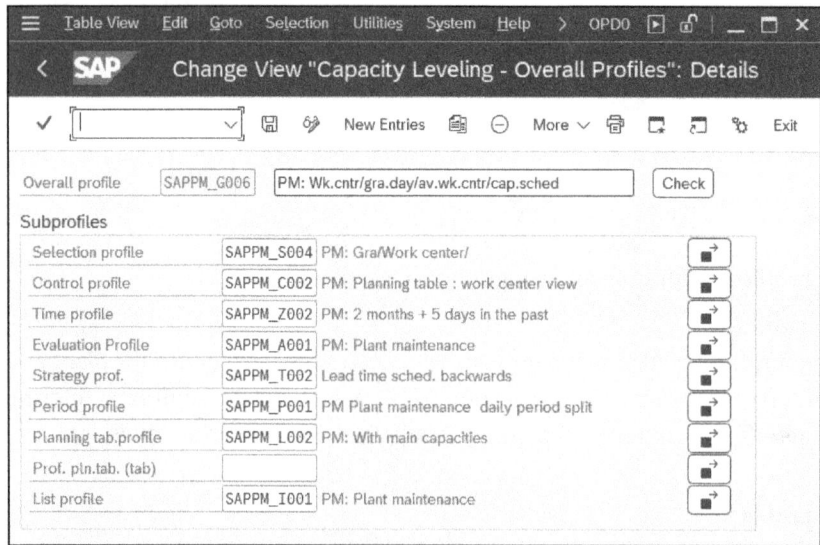

Figure 5.97 Capacity Planning: Capacity Leveling Profiles

5.2 Orders

> **Use Capacity Leveling Profiles from the Standard SAP System**
>
> You don't have to define capacity leveling profiles in the first phase of the implementation project. Instead, use one of the profiles delivered in the standard SAP system (e.g., **SAPPM_G006**). Then, only create your own profiles if the standard SAP profiles prove to be unsuitable. The best way to do this is to copy a standard SAP profile and adjust it accordingly.

5.2.7 Printing

The SAP system has many design options in relation to printing orders and notifications. Generally, only the order is printed when notifications and orders are used, so I'll just concentrate on order printing here.

> **Freedom of Choice with Shop Paper Printing**
>
> You have considerable freedom of choice in relation to the following:
>
> - How many shop papers to print
> - Which shop papers to print
> - The titles to assign to shop papers
> - The layout of the shop papers
> - Which shop paper is to be output and on which output medium

For example, you could print the following documents as shop papers (see Figure 5.98):

- An operation control ticket for the maintenance supervisor
- A job ticket for the part to be repaired
- Time tickets and completion confirmation slips for the technician
- A material pick list for the warehouse
- Material issue slips for the technician
- An object list for inspections

The following output media are available:

- Local printers
- Network printers
- Fax machines
- Email
- PC downloads

The following Customizing functions are available for controlling the output of order documents.

5 Configuring the Work Order Cycle

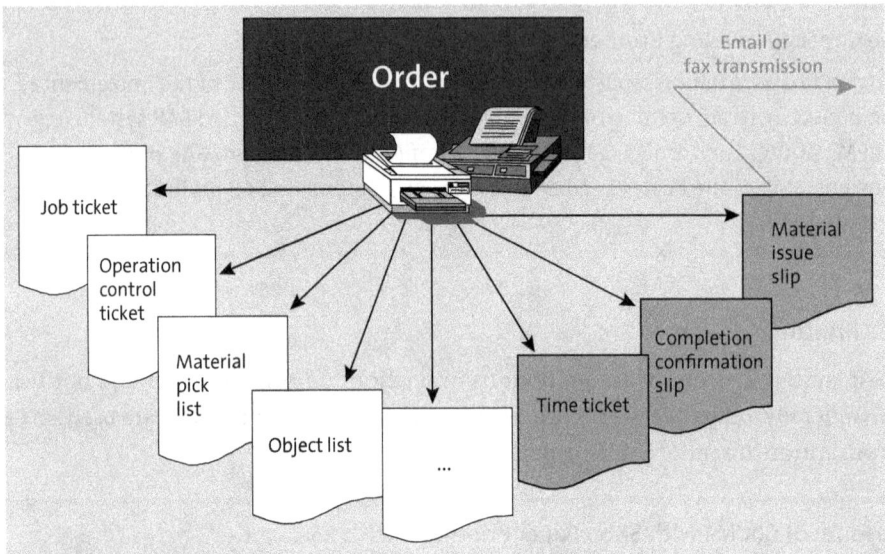

Figure 5.98 Order: Shop Papers

Define Shop Papers, Forms, and Output Programs

You can use this Customizing function to define your shop papers and their print layout. You can also determine which shop paper is to be printed for which order type.

Prerequisites

You must program the print report and define the layout in an SAPscript form beforehand. You must also define the order type.

If you want to print papers before an order is released or after an order is technically completed, you must activate business function LOG_EAM_CI_7.

Customizing Path

Plant Maintenance and Customer Service • Maintenance and Service Processing • Maintenance and Service Orders • Print Control • Define Shop Papers, Forms, and Output Programs

Transactions

The following transactions are available for defining and assigning shop papers:

- OIDF (PM Shop Papers for Orders)
- OIDG (PM Shop Papers by Order Type)

Settings

This Customizing function has two subfunctions: **Define Shop Papers** and **Define Shop Papers for Order Type**. Use the **Define Shop Papers** subfunction (see Figure 5.99) to define the following:

354

5.2 Orders

- **Shop paper**
  A four-digit number (maximum) for the shop paper.

- **Program name**
  The report name used for the print control.

- **ABAP form**
  The form routine used for the print control.

- **Form**
  The SAPscript form used as a basis for designing the pages. The form contains different elements that are used to control the layout of the individual pages and to show design information for the texts to be output on the individual pages.

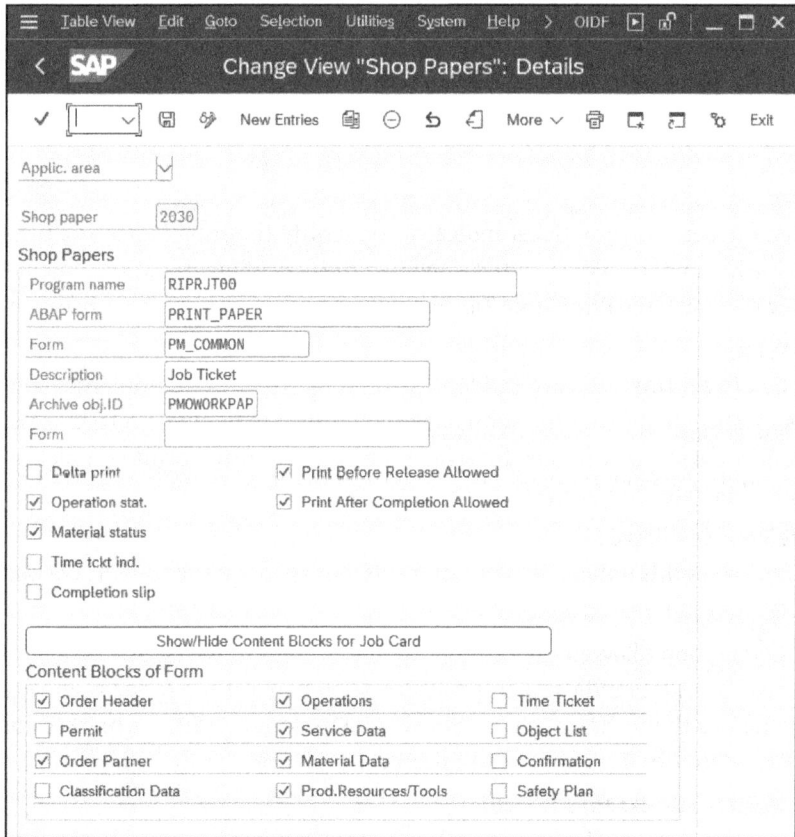

Figure 5.99 Order: Define Shop Paper

You also have the following control options:

- **Delta print**
  Check the box to set this indicator to make delta print the default print setting for this shop paper. Delta print means that only those operations and components that don't yet have the status **Printed** are printed.

- **Operation stat.**
  Check the box to set this indicator for operation-based shop papers (e.g., time tickets and completion confirmation slips). This causes the system to set the **Printed** status when a shop paper is printed, meaning the system sets the **Printed** status as soon as you print a shop paper for which the **Operation stat.** indicator has been set. If you don't check the box to set this indicator for any of your shop papers, the operation can never obtain the **Printed** status, meaning that in the case of a delta print, the system will print the shop papers for all operations.

- **Material status**
  Check the box to set this indicator for component-based shop papers (material issue slips). This causes the system to set the **Printed** status when a shop paper is printed, meaning the system sets the **Printed** status as soon as you print a shop paper for which the **Material status** indicator has been set. If you don't set this indicator for any of your shop papers, the component can never obtain the **Printed** status, meaning that in the case of a delta print, the system will print the shop papers for all components.

- **Time tckt ind.**
  Check the box to set this time ticket indicator to identify the shop paper as a time ticket. This means that it will be printed only if, in the case of an operation, you use a control key for which the **Print Time Tic.** (print time tickets) indicator has been set.

- **Completion slip**
  Check this box to set this indicator to identify the shop paper as a completion confirmation slip. This means that the shop paper will be printed only if, in the case of an operation, you use a control key for which the **Print Confirm.** (print completion confirmation) indicator has been set.

- **Print Before Release Allowed**
  Check the box to set this indicator to preprint this shop paper if the order is created (status **CRTD**) but not released (status **REL**).

- **Print After Completion Allowed**
  Check the box to set this indicator to print this shop paper if the order is already completed and has the technically completed (**TECO**), closed (**CLSD**), or marked for deletion (**DLFL**) status, or if the deletion indicator (**DLT**) is set.

- **Show/Hide Content Blocks for Job Card**
  Use this button to specify whether the related content block is output on the job card. For each form, you can select the content blocks that are output when printing a job card and define the content blocks individually. The resulting document contains all the content blocks that are active in the selected forms.

Use the **Define Shop Papers for Order Type** subfunction to define which shop paper is to be printed for which order type (in the **Do...** column for document type; see Figure 5.100). You can use masking here whereby the masked entries apply to all order types that aren't explicitly specified.

5.2 Orders

Use the **Selection** column to specify which assigned shop papers are to be proposed for printing. If you suppress the print dialog box, the system prints exactly these shop papers.

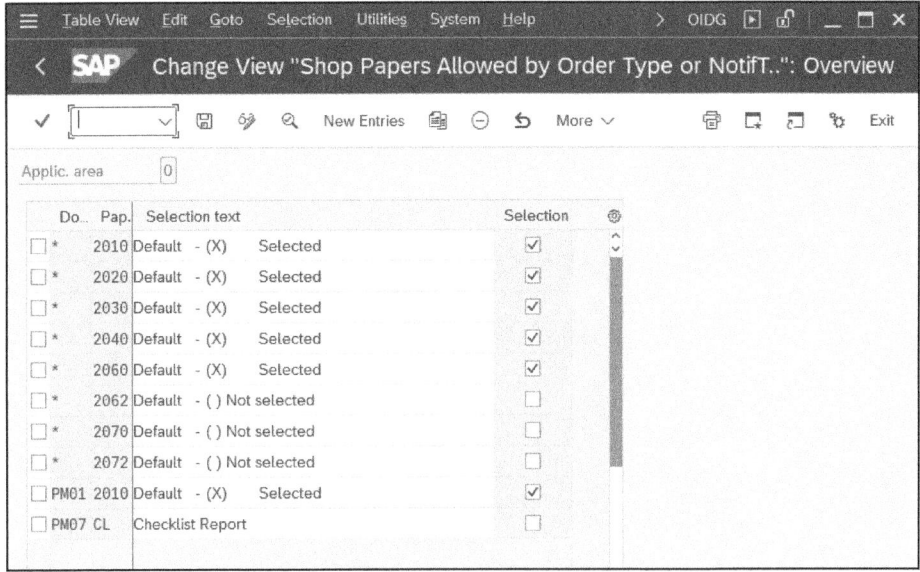

Figure 5.100 Order Type: Shop Papers

Define Printer

You use this Customizing function to control which shop paper is to be printed on which printer. User-specific print control and print diversion functions are available here.

Prerequisites

You must define the user master records, shop papers, plants, maintenance planner groups, and printers beforehand.

If you want to print papers before an order is released or after an order is technically completed, you must activate business function LOG_EAM_CI_7.

Customizing Path

Plant Maintenance and Customer Service • Maintenance and Service Processing • Maintenance and Service Orders • Print Control • Define Printer

Transaction

The following transaction codes are available for print control:

- Transaction OIDH for user-specific print control
- Transaction OIDI for defining fields for the print diversion
- Transaction OIDJ for defining the field contents for the print diversion

Settings

Use the **User-Specific Print Control** subfunction to define which user prints which shop paper and on which printer (see Figure 5.101).

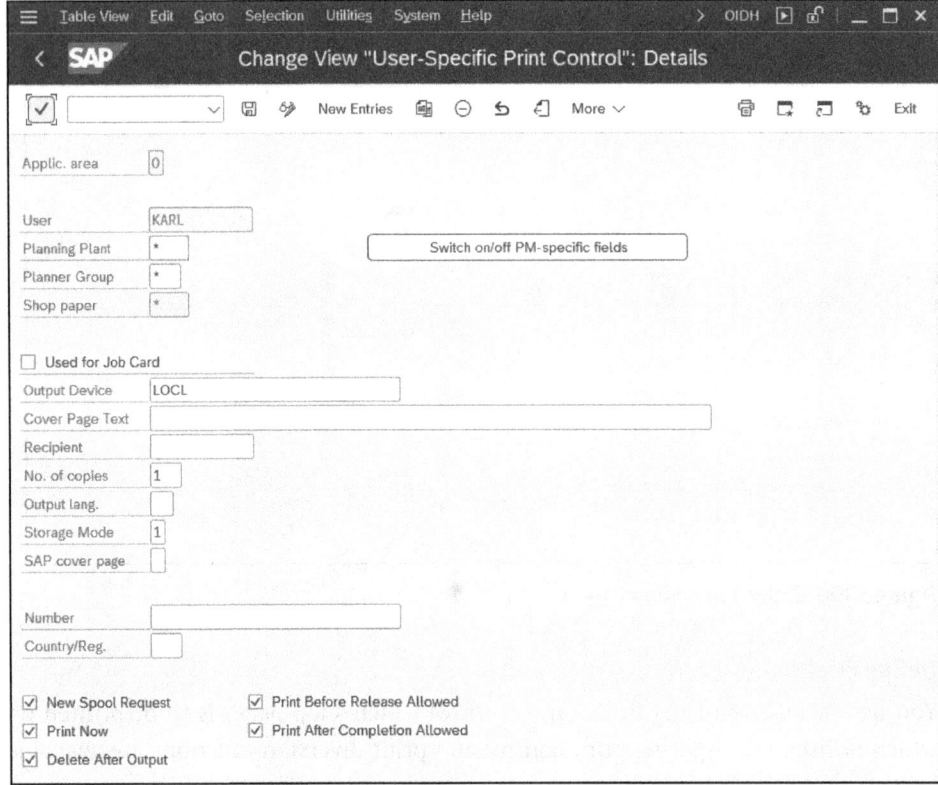

Figure 5.101 Order: User-Specific Print Control

You can define the plant and the maintenance planner group as additional influencing factors. However, you can also use *masking*, whereby the respective shop paper is always printed on the same printer, regardless of the plant and/or maintenance planner group. You can choose one of the following printers:

- The user's local Windows printer (printer **LOCL**)
- A network printer
- Email, in which a PDF attachment is generated instead of a printout

Checking the box to set the **Print Before Release Allowed** indicator allows users to preprint shop papers if an order is created (status **CRTD**) but not released (status **REL**).

Checking the box to set the **Print After Completion Allowed** indicator allows users to print out shop papers if an order is already completed and has the technically completed (**TECO**), closed (**CLSD**), or marked for deletion (**DLFL**) status, or if the deletion indicator (**DLT**) is set.

You can use the **Define Print Diversion** subfunction to activate a *print diversion*, in which the shop papers aren't printed using the standard printer defined in the **User-Specific Print Control** subfunction. Instead, the print output is diverted to another printer. There are two types of print diversion:

- **Print diversion of shop papers to the work center printer**
  The work center printer is used for completion confirmation tickets and time tickets. In this case, you define the printer in the work center (Transaction IR02, **Default Values** tab, **Shop Paper Printer** field).

- **Print diversion via Customizing**
  When determining a printer, print diversion (if activated) always has the highest priority, meaning it has priority over the user-specific selection and the work center printer.

Depending on which shop paper it concerns, print diversion via Customizing occurs at the header, operation, or component level. Table 5.8 provides an overview of which shop paper is diverted via which level or levels (H = header, O = operation, M = material components, and W = work center printer).

Shop Paper	Description	H	O	M	W
2010	Operation control ticket	X			
2020	Material pick list	X			
2030	Job ticket	X			
2040	Material issue slip	X	X	X	
2060	Time ticket		X	X	X
2062	Time ticket for split records		X	X	X
2070	Completion confirmation slip		X	X	X
2072	Completion confirmation slip for split records		X	X	X

Table 5.8 Orders: Assignment of Print Diversion

Accordingly, the fields from the following tables are available for activating the print diversion (see Figure 5.102):

- Table CAUFVD: Orders (header)
- Table AFVGD: Order operations
- Table RESBD: Material components
- Table RIPWO: Technical objects
- Table ILOA: Technical objects

If you've activated print diversion, you can use the **Print Diversion According to Field Contents** subfunction to define special printers and print parameters for a specific shop paper based on particular field contents (see Figure 5.103). At any given time, only

5 Configuring the Work Order Cycle

one table entry can be active for each shop paper. Otherwise, the uniqueness of the printer assignment can't be ensured.

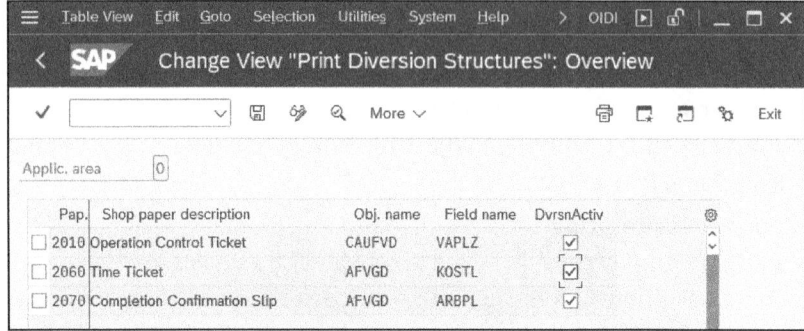

Figure 5.102 Order: Activate Print Diversion

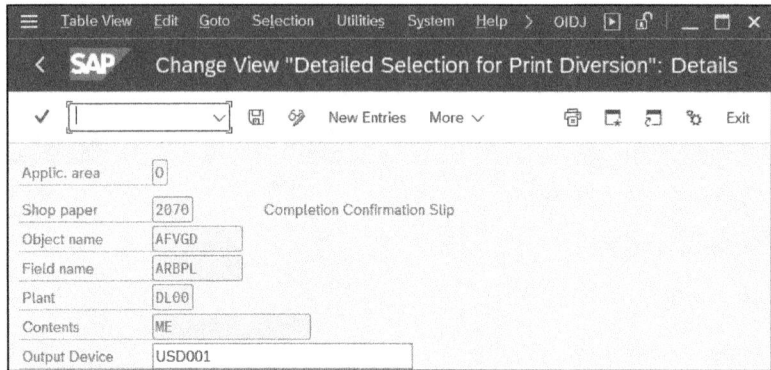

Figure 5.103 Order: Print Diversion Control

For each table entry, you can define any number of records that differ in terms of their field content or field name (e.g., different printers for the mechanical engineering work center and the electronics work center).

You can define further information for the following scenarios:

- To output the shop paper to a fax number (with a country key) instead of a printer
- To output the shop paper several times (with a number of printouts)
- To output the shop paper immediately
- To use a spool to control shop paper printing

Send Shop Papers by Email

A special feature here is that you can also send your shop papers by email (e.g., to a smartphone) rather than printing them out. Because this is becoming more and more common, the instructions are provided here.

To send shop papers by email, follow these steps:

1. Set up automatic forwarding in the workplace using Transaction SBWP (**Settings • Office Settings • Automatic Forwarding**). Here, enter the email address to which you want to send emails.

2. Use Transaction SCOT to set up a send job for INT (**Business Communication Administration • Settings • SMTP Connection • Outbound Message • Settings**). Here, define the frequency with which emails are to be sent from the SAP system to the external email address (e.g., every five minutes).

3. Use Transaction SPAD to set up a printer called EMAIL, then assign the device type **PDF1** to this printer. You must also set the access method for the host spool to **M** (email to recipient/owner).

4. Use the **User-Specific Print Control** or **Define Print Diversion** subfunction to assign the **EMAIL** printer as an output device.

If you then print an order, the SAP system sends emails with a PDF attachment (see Figure 5.104) instead of outputting the document to a physical printer. You can then decide whether you want to output the PDF as a shop paper.

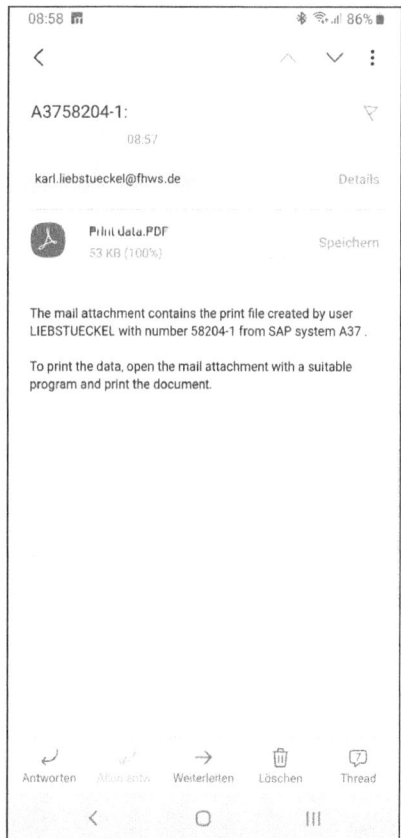
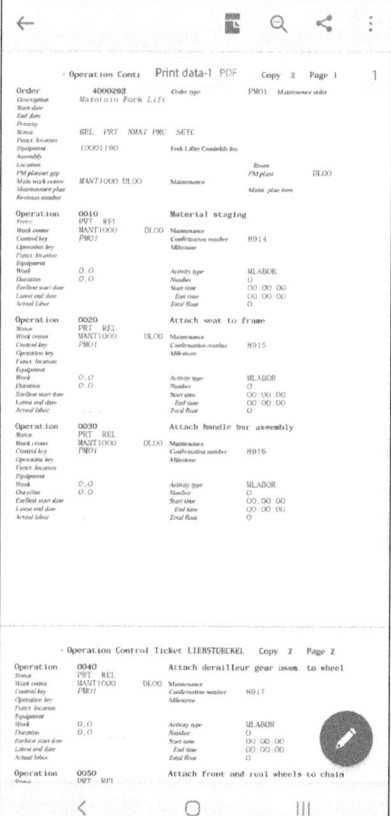

Figure 5.104 Order: Print by Email

5 Configuring the Work Order Cycle

Define Status Control for Shop Papers

You can use this Customizing function to specify, in the case of maintenance orders, whether the timeliness of printed shop papers should be tracked.

If an order or operation is changed after printing the shop papers, the system displays the corresponding status in the order. Furthermore, this option means that, in this case, the status of the order can't be technically completed before the shop papers are reprinted.

Prerequisites

You must define the plants and order types beforehand.

Customizing Path

Plant Maintenance and Customer Service • Maintenance and Service Processing • Maintenance and Service Orders • Functions and Settings for Order Types • Define Status Control for Shop Papers

Settings

Check the box to activate status control for each plant and order type (see Figure 5.105).

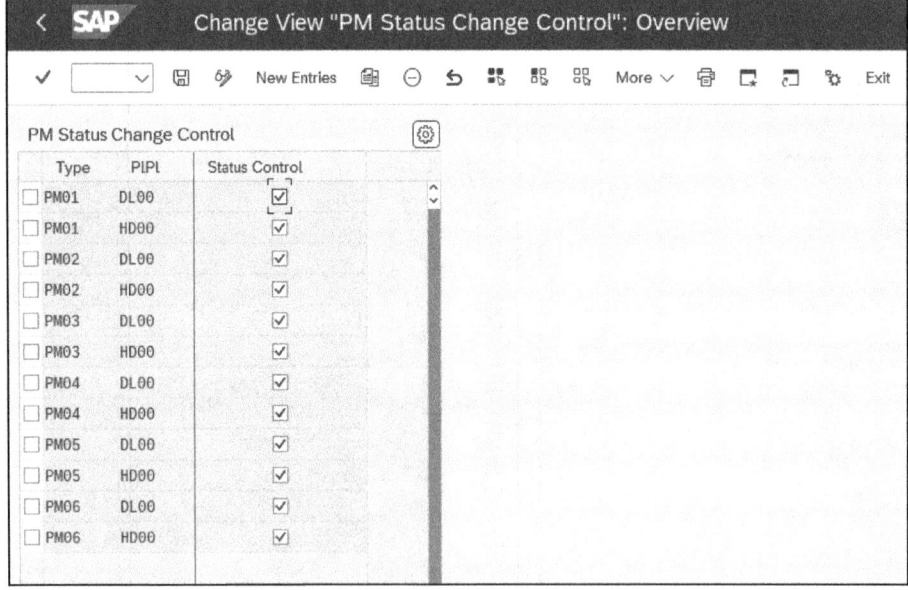

Figure 5.105 Order: Print Status Control

5.2.8 Estimated Costs and Costing

In *costing*, costs are automatically calculated based on the cost rates defined in the system and the amounts specified in resource planning. Consequently, the planned costs (and also the actual costs subsequently accrued) aren't planned manually.

You can also enter *estimated costs*, which aren't calculated. Instead, they are specified manually because they are solely based on a maintenance planner's empirical values and not on the planned resources. There are three different ways to enter estimated costs:

- At the operation level, differentiated for each value category. The total for all operations is then shown at the header level for each value category and as a total.
- At the header level, differentiated for each value category. The total is then created automatically.
- As the total for the entire order.

The planning of resources leads to the accrual of planned costs on the order. Here, there are two different ways to display the costs:

- **According to cost elements**
 The cost display according to cost elements shows all cost elements included in the costing. This is more like a controlling-oriented view because it displays a comparison of planned and actual costs.

- **According to value categories**
 The cost display according to value categories groups several cost elements into *value categories*. In general, this is a clearer layout for maintenance purposes because it displays a comparison of estimated, planned, and actual costs.

The costs of an order are calculated at two *summarization levels*. You can also display the costs at these two levels:

- **Operation level**
 The cost display at the operation level shows you all costs included in the relevant operation, either based on the cost elements as a planned/actual comparison or on the basis of the value categories as a comparison of the estimated/planned and actual costs. Orders whose costs are displayed at the operation level obtain the **ACAS** (activity account assignment) status, meaning you can use this status to select the orders in question.

- **Header level**
 The cost display at the header level summarizes all costs at the operation level and shows you all costs included in the order, either on the basis of the cost elements as a planned/actual comparison or on the basis of the value categories as a comparison of the estimated/planned and actual costs. The costs at the header level are always displayed regardless of whether you've activated costs at the operation level.

To assign estimated costs within an order and to display costs at the value category level, you must configure the following Customizing functions.

5 Configuring the Work Order Cycle

Maintain Value Categories

You use this Customizing function to define, for each controlling area, all the value categories you want to manage within plant maintenance (**Component** entry "PM").

Prerequisites

Because value categories are always defined for each controlling area, you must define the controlling areas beforehand.

Customizing Path

Plant Maintenance and Customer Service • Maintenance and Service Processing • Basic Settings • Settings for Display of Costs • Maintain Value Categories

Transaction

OIK1

Settings

In addition to a number and description for the value category, define whether the category concerns costs or revenues (see Figure 5.106). Internal maintenance orders always fall into the **Costs** category, while the **Revenues** category only comes into play if you render services for a third party and want to use maintenance orders to process these services. Furthermore, specify whether you want to maintain other values (e.g., hours) in addition to costs (in the **UM** column).

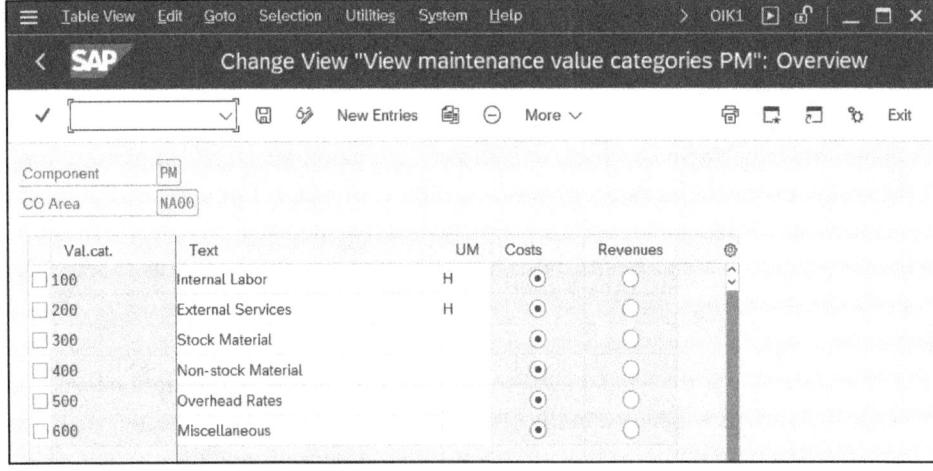

Figure 5.106 Order Value Categories

> [+] **Five Value Categories Are Usually Required**
>
> In plant maintenance, you generally require the following value categories:
>
> - Internal Labor
> - External Services

- Stock Material
- Non-stock Material
- Overhead Rates
- Miscellaneous

If necessary, you can add other value categories.

Assign Cost Elements to Value Categories

You use this Customizing function to control, for each controlling area, which cost elements are to be displayed in which value categories.

Prerequisites
You must define controlling areas, value categories, and cost elements beforehand.

Customizing Path
Plant Maintenance and Customer Service • Maintenance and Service Processing • Basic Settings • Settings for Display of Costs • Assign Cost Elements to Value Categories

Transaction
OIK2

Settings
When assigning cost elements to value categories, you have the following options (see Figure 5.107):

- One or more individual cost elements
- One or more intervals
- One or more cost element groups

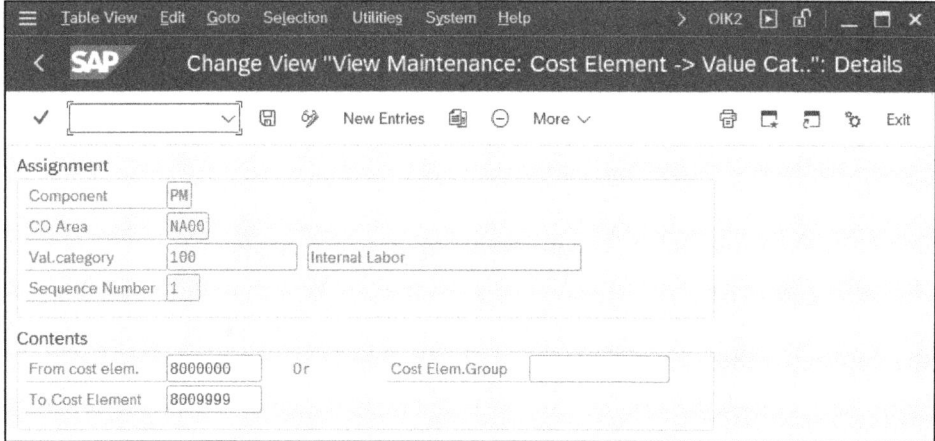

Figure 5.107 Order: Assign Cost Elements to Value Categories

5 Configuring the Work Order Cycle

> [!] **Take Care When Making Retroactive Changes to Value Categories**
> If you subsequently change the assignment of cost elements to value categories, you must use program RIPMCO00 to recreate the *PMCO* and *PMCOQT* files. You must also use report RIPMS001 to rebuild the information structures in PMIS.

Define Default Values for Value Categories

You use this Customizing function to select those value categories that are defined using the **Maintain Value Categories** Customizing function and are to appear in the order for entering estimated costs as default values.

Prerequisites

You must define the value categories beforehand.

Customizing Path

Plant Maintenance and Customer Service • Maintenance and Service Processing • Basic Settings • Settings for Display of Costs • Define Default Values for Value Categories

Settings

Define the default value categories (see Figure 5.108).

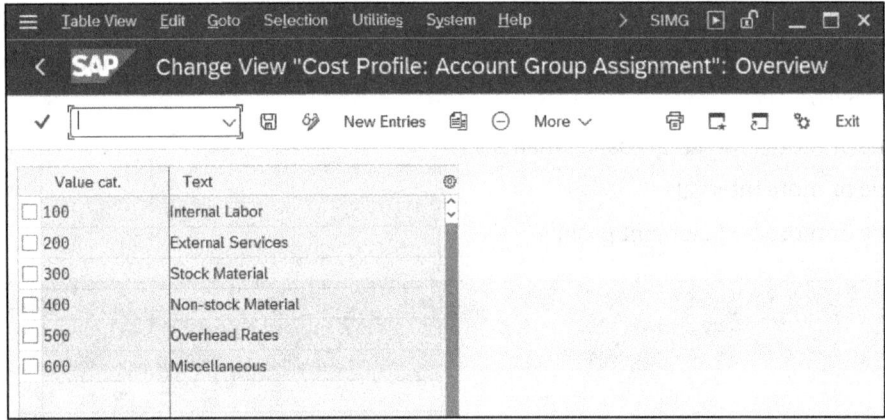

Figure 5.108 Order Default Value Categories

Check Consistency of Value Category Assignment

In this activity, you check the correct assignment of cost elements to value categories. When you call the report, you specify the controlling area for which you want to run the consistency check.

Once the consistency check has run, the system issues a list containing information about the assignment of value categories for each controlling area.

Prerequisites

You must define the value categories, the controlling area, and the assignment of cost elements to value categories beforehand.

Customizing Path

Plant Maintenance and Customer Service • Maintenance and Service Processing • Basic Settings • Settings for Display of Costs • Check Consistency of Value Category Assignment

Transaction

OIVC

Settings

Run the **Check Consistency of Value Category Assignment** Customizing function and the system displays the following results (see Figure 5.109):

- ○○■

  Everything is okay.

- ○▲○

  Warning messages are issued, for example, if cost elements haven't been assigned.

- ●○○

  Errors have occurred, for example, if cost elements have been assigned several times.

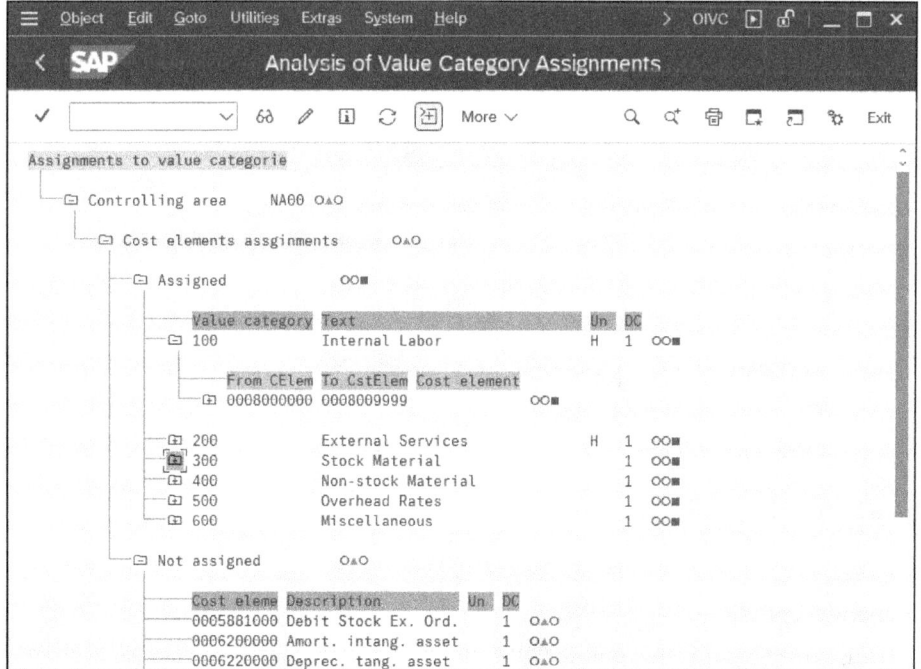

Figure 5.109 Order: Analysis of Value Category Assignments

5 Configuring the Work Order Cycle

[+] **Check Consistency**
You can use the **Check Consistency of Value Category Assignment** Customizing function to check whether errors occur when you assign cost elements to value categories.

Now, the costs associated with the orders should be calculated. Figure 5.110 shows the relationship between the Customizing functions so that you can calculate the costs associated with an order (i.e., the planned costs and actual costs):

- **Costing sheet**
  Define the cost elements that the costing is to take into consideration. Furthermore, determine which charge lines are to be used to calculate the costs (e.g., overhead rates).

- **Valuation variant**
  Use this variant to control which cost rates are to be used to calculate the costs (e.g., material price control).

- **Costing type**
  Although not important in plant maintenance, the costing type is significant in applications such as production planning or controlling. Together with the valuation variant, its sole purpose is to define the costing type of a costing variant.

- **Costing parameters per order type, plant**
  Use this to assign the costing variant for both the preliminary costing and the actual costing for each order type and plant.

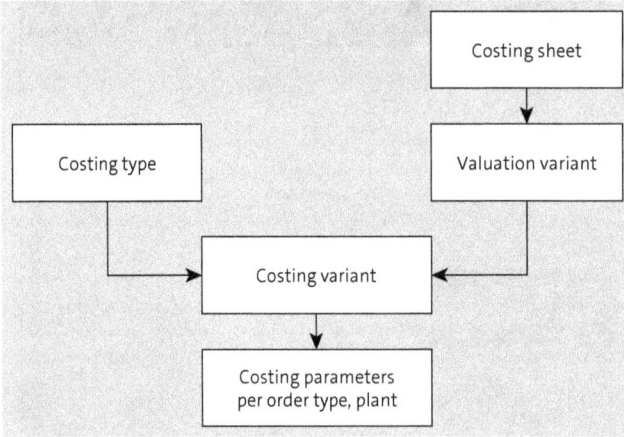

Figure 5.110 Order Costing Overview

[+] **Separate Costing**
Because costing is used not only in SAP S/4HANA Asset Management but also in other applications in SAP S/4HANA, such as project processing, production planning,

368

5.2 Orders

customer service, and controlling, you should always define a separate costing for maintenance processing to make it more flexible and to differentiate it from other types of costing.

Maintain Costing Sheet

You use this Customizing function to define the costing sheet, and you then use the costing sheet to determine the costing structure and content.

Prerequisites

You must define the cost elements beforehand. If you work with overhead rates, you must also define the objects to be credited (e.g., the warehouse cost center) beforehand.

Customizing Path

Plant Maintenance and Customer Service • Maintenance and Service Processing • Maintenance and Service Orders • Functions and Settings for Order Types • Costing Data for Maintenance and Service Orders • Maintain Costing Sheet

Settings

Use the **Costing Sheets** subfunction to define a name and description for your costing sheet (see Figure 5.111).

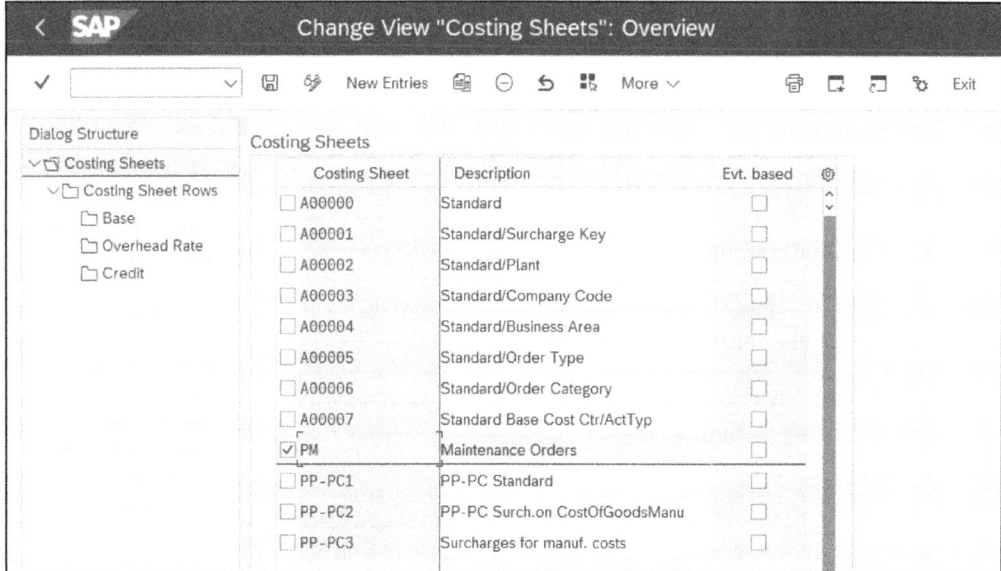

Figure 5.111 Order: Costing Sheets Overview

Use the **Costing Sheet Rows** subfunction to define the costing elements and structure. The following elements should be contained here (see Figure 5.112):

5   Configuring the Work Order Cycle

- Internal service
- Overhead rate for internal service
- Material
- Overhead rate for material

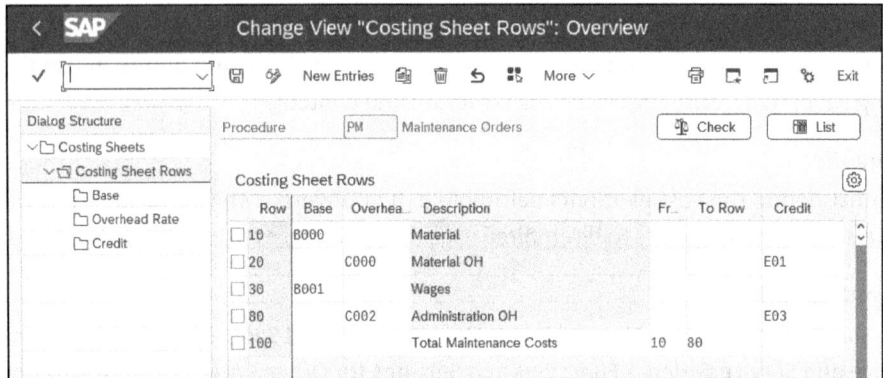

Figure 5.112 Order: Costing Sheet Rows

[+] **Take Care When Assigning Rows**

In the **From Row** and **To Row** columns, ensure the following:
- The subtotals and totals are calculated correctly.
- The overhead rows refer to the correct base rows.

Use the **Base** subfunction to define the assignment of cost elements as a basis for calculating overheads (see Figure 5.113). The following options are associated with assigning cost elements:

- Individual cost elements
- Cost element intervals
- Cost element groups

Use the **Overhead Rate** subfunction to define your overhead rates with the following information (see Figure 5.114):

- **Valid From/To**
  This can be used to manage different overhead rates in different validity periods.
- **Ovrhd Type**
  This is the overhead type for which the overhead rate is to apply (1 = standard cost estimate, 2 = actual costing, and 3 = commitment costing).
- **Percentage/Unit**
  This is the amount and unit of measurement for the overhead rate (e.g., 20% or five dollars).

5.2 Orders

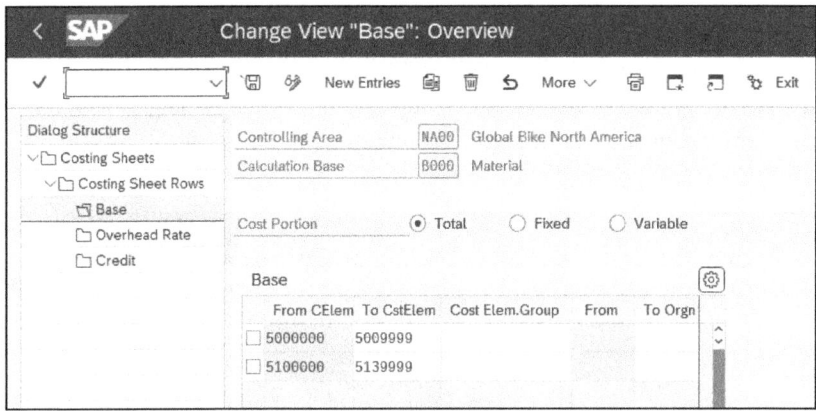

Figure 5.113 Order: Costing Sheet Baseline

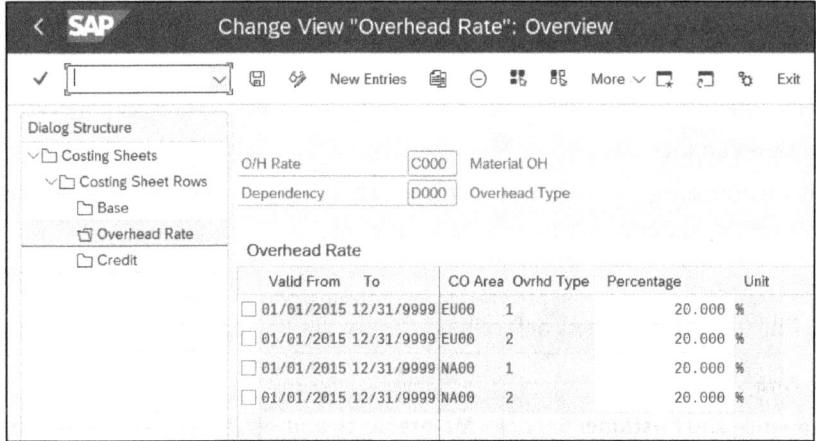

Figure 5.114 Order: Costing Sheet Overhead Rate

If you work with overhead rates, you *must* assign an account assignment object that is to be credited with the values from the overhead calculation. Use the **Credit** subfunction to define this account assignment object (see Figure 5.115). For example, when you define an overhead rate for material postings, you award percentage or absolute overhead costs to each goods receipt or goods issue, and you then specify the warehouse cost center as a credit object to which the corresponding amounts are credited.

Recommendations for Overhead Costs

For overhead costs, the following is true:
- You should always use the same overhead rates for all overhead types, especially planned costing and actual costing. At some point, all other overhead rates will cease to make sense.

- You can specify percentage or absolute overhead rates. For example, you can define a cost of five dollars for each material withdrawal.
- You must specify an account assignment object that is credited (e.g., the warehouse cost center, which is credited with the material overhead costs).

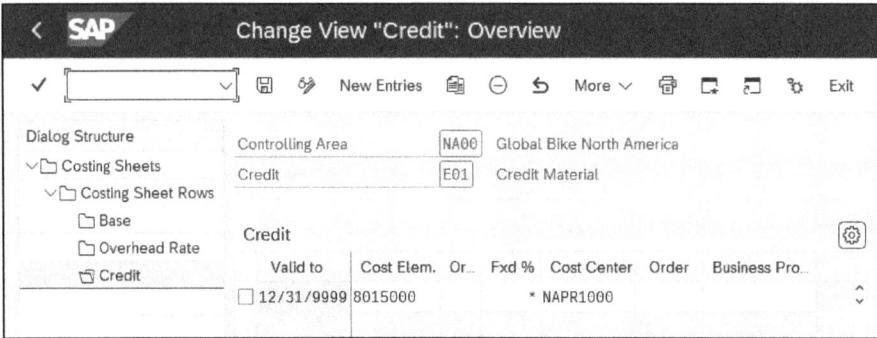

Figure 5.115 Order: Costing Sheet Overhead Credits

Define Valuation Variants

You use this Customizing function to control which valuation methods (e.g., which price controls for the material) are to be used for the costing.

Prerequisites

You must define the costing sheets beforehand.

Customizing Path

Plant Maintenance and Customer Service • Maintenance and Service Processing • Maintenance and Service Orders • Functions and Settings for Order Types • Costing Data for Maintenance and Service Orders • Define Valuation Variants

Transaction

OKP8

Settings

On the **Material Val.** (material valuation) tab, define the priority sequence for determining the material price (see Figure 5.116).

Recommendation

The best choice is **7 Valuation Price According to Price Control in Mat. Master**.

On the **ActivityTypes/Processes** tab, define the priority sequence for determining the prices when calculating the cost of the internal service (see Figure 5.117).

5.2 Orders

> **Recommendation**
> The best choice is **1 Plan Price for the Period**.

On the **Ext. Processing** tab, define the priority sequence for determining the prices for external processing (see Figure 5.118).

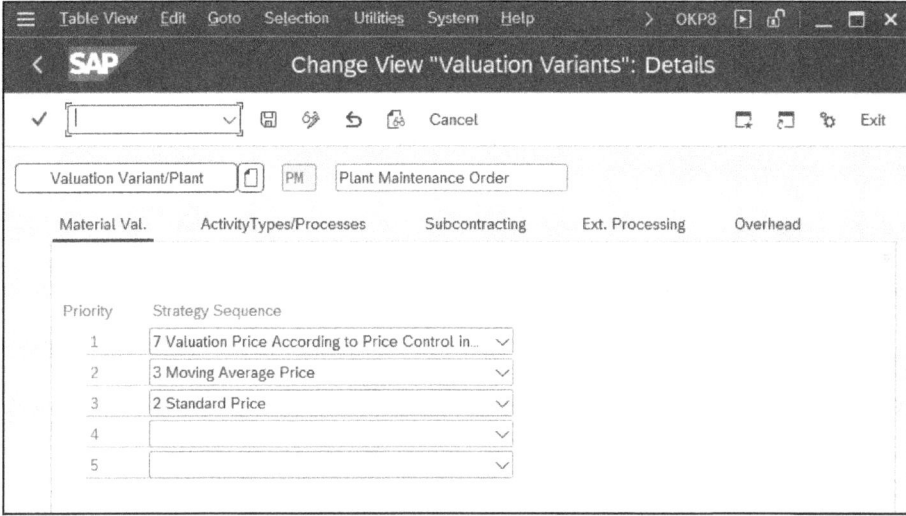

Figure 5.116 Order: Costing Material Valuation

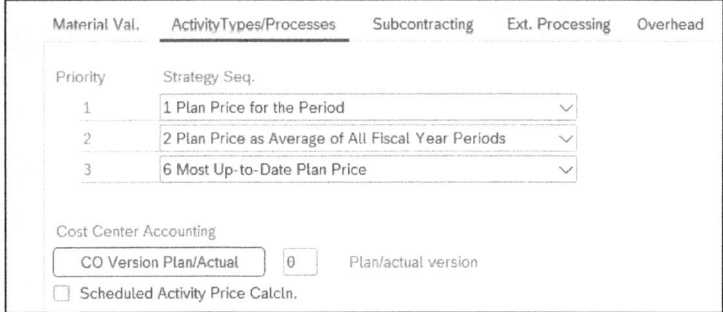

Figure 5.117 Order: Costing Internal Activity Valuation

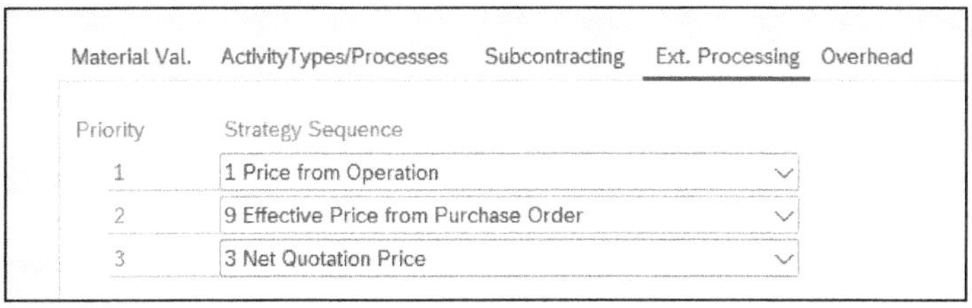

Figure 5.118 Order: Costing External Processing Valuation

5 Configuring the Work Order Cycle

> **[+] Recommendation**
>
> The best choices are **1 Price from Operation** for the planned costing and **9 Effective Price from Purchase Order** for the actual costing. If you require two different valuation approaches, define two different valuation variants.

The information on the **Subcontracting** tab will be required during the *subcontracting* business process (see Chapter 7, Section 7.3) and will therefore be discussed there.

On the **Overhead** tab, assign the costing sheet (see Figure 5.119).

Figure 5.119 Order: Costing Overhead Valuation

Maintain Costing Variants

You use this Customizing function to group the costing type and the valuation variant together to form a costing variant.

Prerequisites

You must define the costing type and valuation variant beforehand. The costing type doesn't have any controlling effect in SAP S/4HANA Asset Management (unlike in production planning, controlling, and project management) and therefore isn't discussed further here. The only thing to ensure here is that a separate costing type (e.g., **05** = **Maintenance Order**) is defined here for SAP S/4HANA Asset Management.

Customizing Path

Plant Maintenance and Customer Service • Maintenance and Service Processing • Maintenance and Service Orders • Functions and Settings for Order Types • Costing Data for Maintenance and Service Orders • Maintain Costing Variants

Transaction

OKP6

Settings

Define a name and description for the costing variants, then assign the costing type and valuation variant (see Figure 5.120).

> **[+] One or Two Costing Variants**
>
> If you have two different valuation variants (e.g., one for planned costing and one for actual costing), define two different costing variants here.

374

5.2 Orders

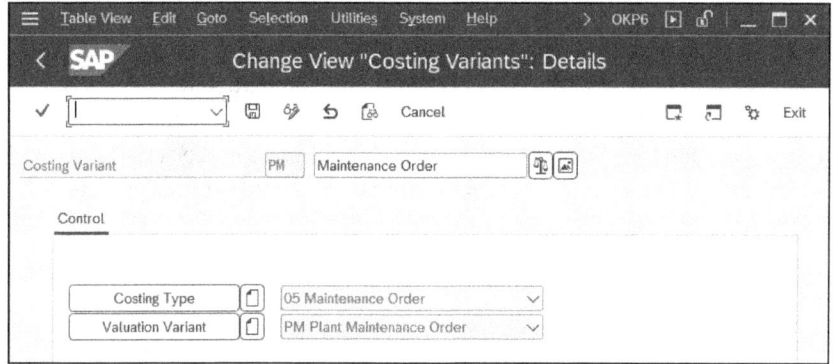

Figure 5.120 Order: Costing Variant

Assign Costing Parameters and Results Analysis Keys

You use this Customizing function to assign, for each order type and plant, the costing variants that you want to use for the planned costing and actual costing.

Prerequisites

You must define the plants, order types, and costing variants beforehand.

Customizing Path

Plant Maintenance and Customer Service • Maintenance and Service Processing • Maintenance and Service Orders • Functions and Settings for Order Types • Costing Data for Maintenance and Service Orders • Assign Costing Parameters and Results Analysis Keys

Transaction

OIOF

Settings

For **Planned var. cal.** or **Actual var. calc.**, enter the costing variants that you want to use for the planned costing or actual costing, respectively (see Figure 5.121).

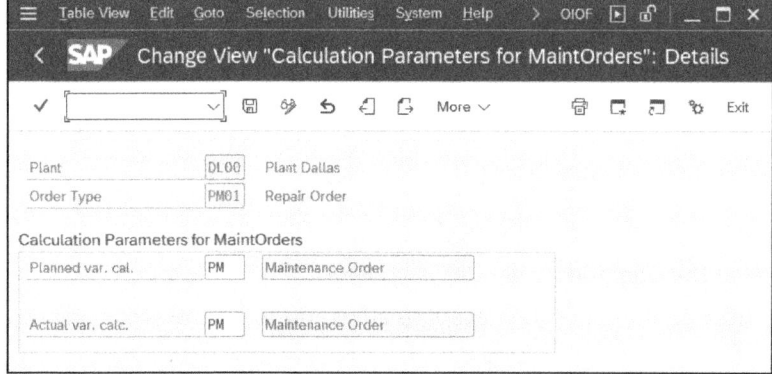

Figure 5.121 Order: Assign Costing Parameters

375

5 Configuring the Work Order Cycle

[+] **Planned Costing and Actual Costing: The Same or Different?**

You have the option of using two different costing variants to structure planned costing and actual costing differently. For example, in planned costing, you can use the price from the operation to valuate external processing, whereas in the actual costing, you can use the effective price from the purchase order to valuate external processing.

Define Cost Settings

You use this Customizing function to define, for each order type and plant, whether you want to manage costs at the operation level or the header level within the order.

Prerequisites

You must define the plants and order types beforehand.

If you want to manage costs on the operation level, you must activate business functions LOG_EAM_OLC and LOG_EAM_OLC_2.

Customizing Path

Plant Maintenance and Customer Service • Maintenance and Service Processing • Maintenance and Service Orders • Functions and Settings for Order Types • Costs at Operation Level • Define Cost Settings

Settings

When you have plants and order types for which you want to activate costs at the operation level, check the box to set the indicator in column **A** (see Figure 5.122).

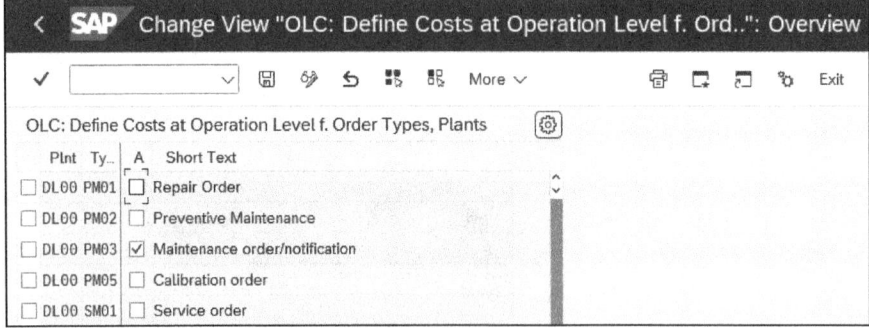

Figure 5.122 Order: Costs at Operation Level

[+] **Effects of Costs at the Operation Level**

If you activate the costs at the operation level, this affects the entire cost processing of the order:

- The costs associated with the planned costing and actual costing are calculated at the operation level, and the totals are displayed at the header level.
- The estimated costs are assigned at the operation level and totaled at the header level.
- The order is settled with reference to the operation, meaning that each operation obtains a separate settlement rule.
- All documents (e.g., purchase requisitions, purchase orders, material withdrawals, invoice receipts) must be assigned to the order number *and* operation number.
- If you activate the costs at the operation level retroactively, you must use report RIPMCO01 or Transaction OLI5N to rebuild the tables with the order costs PMCO and PMCO_OP.
- The same applies if you undo the activation retroactively.

To enable the system to determine the costs based on your resource planning, you must fulfill the following Customizing setting.

> **[+] Missing Order Costs**
>
> If no costs are displayed in your order or the only costs displayed are for certain value categories, then check all the following customizing settings and other prerequisites. You'll most certainly find the performance hindrance there and be able to rectify it.
>
> - Use the **Costing** function to assign a control key to the operations that are to be included in the costing.
> - Assign **Work** to the operations that are to be included in the costing.
> - For the cost center of the work center, use Transaction KP26 to define a fixed price (the **Fixed Price** option) and/or variable price (the **Variable Price** option) for the version you use, the fiscal year, the cost center, and the activity type.
> - In the work center, enter the standard value key "SAP0" on the **Basic Data** tab.
> - In the work center, maintain the following data on the **Costing** tab: **Cost Center**, **Activity Type**, and **Formula Key**.
> - The **Formula Key** must point to the **ARBEI** field (the work from the operation). In the standard system, this is the formula **SAP008**.
> - Transfer prices (either standard prices or moving average prices) must be defined for the materials whose costs are to be calculated.
> - Enter a value for nonstock materials.
> - Enter a value for external processing.

5.2.9 Order Settlement

Figure 5.123 shows the relationships among the Customizing functions for order settlement, as follows:

- In the allocation structure, you define how you want to convert the debit cost elements into credit cost elements.
- In the settlement profile, you define the allocation structure, the permitted account assignment objects, and other settlement parameters.
- You then assign the settlement profile to the order type.

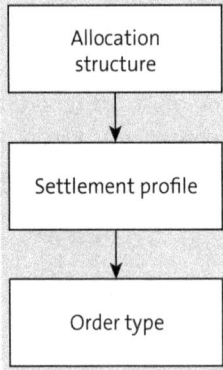

Figure 5.123 Order Settlement Overview

[+] **Separate Order Settlement**
Because order settlement is used not only in SAP S/4HANA Asset Management but also in other areas of SAP S/4HANA, such as project management, production planning, customer service, and controlling, you should always define a separate order settlement for maintenance orders to make it more flexible and to differentiate it from other types of order settlement.

The following sections cover the most important Customizing activities for order settlement.

Maintain Allocation Structures

You use this Customizing function to define how the cost elements used to debit the orders are to be converted into credit cost elements.

Prerequisites

You must define the cost elements and, if necessary, the cost element groups, beforehand.

Customizing Path

Plant Maintenance and Customer Service • Maintenance and Service Processing • Basic Settings • General Order Settlement • Maintain Allocation Structures

5.2 Orders

Settings

Use the **Maintain Allocation Structures** function to define a name and description for the allocation structure that you want to use in plant maintenance (see Figure 5.124). You can have more than one allocation structure.

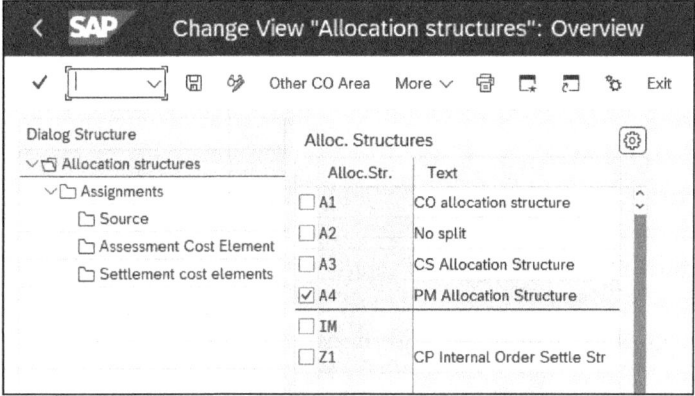

Figure 5.124 Order: Allocation Structures Overview

Use the **Assignments** subfunction to define the groupings (not to be confused with cost element groups) and therefore the level of detail for the order settlement (see Figure 5.125). Common assignments include the following:

- Internal services
- External processing
- Stock material
- External material
- Overhead costs
- Settlement costs (if orders have been debited with order settlements that are to be allocated further)
- Other costs

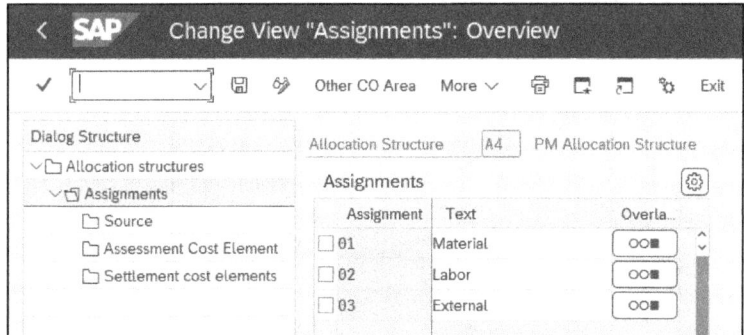

Figure 5.125 Order: Allocation Structure Assignment

5 Configuring the Work Order Cycle

Use the **Source** subfunction to define, for each assignment and controlling area, the cost elements or cost element groups that need to be settled (see Figure 5.126).

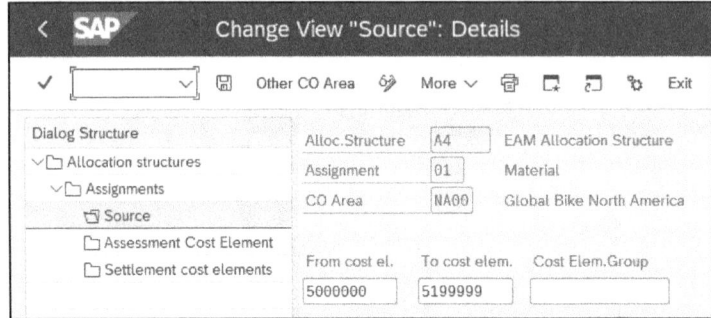

Figure 5.126 Order: Allocation Structure Assignment Source

The allocation structure must fulfill the following criteria:

- **Completeness**
  All cost elements under which debits accrue must be represented in the relevant allocation structure.

- **Uniqueness**
  Each cost element under which debits accrue must occur only once within an allocation structure. In other words, within an allocation structure, only one settlement cost element is permitted to be assigned to a source.

Use the **Settlement cost elements** subfunction to define the settlement procedure for each assignment, controlling area, and receiver category (e.g., the **CTR** cost center or the **ORD** order, as shown in Figure 5.127). Here, you have two options:

- **Assign the debit cost element groups to a settlement cost element**
  In this case, the settlement cost element must have cost element category **21** (**Internal Settlement**).

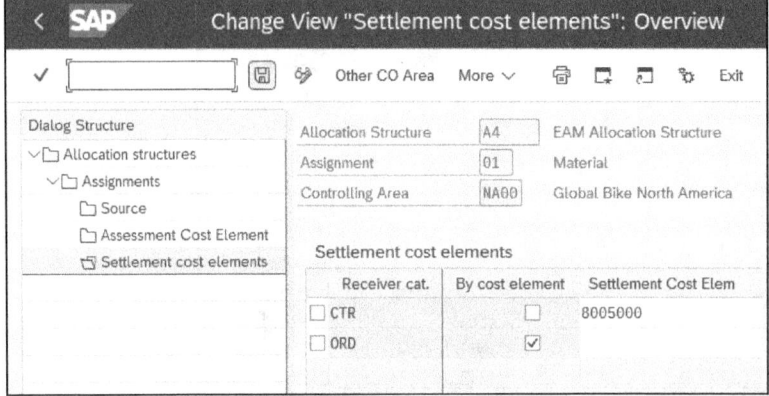

Figure 5.127 Order: Allocation Structure Settlement

- **Perform a cost element-based settlement**
  In this case, the debit cost element is also used as the settlement cost element, and all cost element categories are permitted as the settlement cost element.

Maintain Settlement Profiles

You use this Customizing function to define the allocation structure, the permitted account assignment objects, and other settlement parameters.

Prerequisites

You must define the allocation structures and document types for the settlement documents beforehand.

Customizing Path

Plant Maintenance and Customer Service • Maintenance and Service Processing • Basic Settings • General Order Settlement • Maintain Settlement Profiles

Settings

You must define the following on the screen shown in Figure 5.128:

- **Actual Costs/Cost of Sales**
  Use the radio buttons in this area to specify whether your maintenance orders are to be settled. Generally, you specify that the orders are to be settled in full. Only in exceptional cases (e.g., if you don't use controlling) are orders not settled.

- **Valid Receivers**
  In this field group, define the account assignment objects that you want to permit as settlement receivers for maintenance orders. The cost center, order, work breakdown structure (WBS) element, and network are the most common receivers for costs from maintenance orders.

  You can then enter one of these valid receivers as a **Default Object Type**, which is then used to automatically create the settlement rule. For example, if you specify **Cost Center** under **Valid Receivers** and you specify **CTR** as the **Default Object Type**, the cost center from the functional location or piece of equipment is automatically entered in the settlement rule for the order when the order is released or technically completed.

- **%-Settlement**
  If you check the box to set this indicator, you can define, within the settlement rule, distribution rules governing the settlement of costs according to a percentage settlement.

- **Amount Settlement**
  If you check the box to set this indicator, you can define, within the settlement rule, distribution rules governing the settlement of costs according to fixed amounts.

5 Configuring the Work Order Cycle

- **Max.No.Dist.Rls**

  Use this field to determine the rule for the maximum number of distributions permitted.

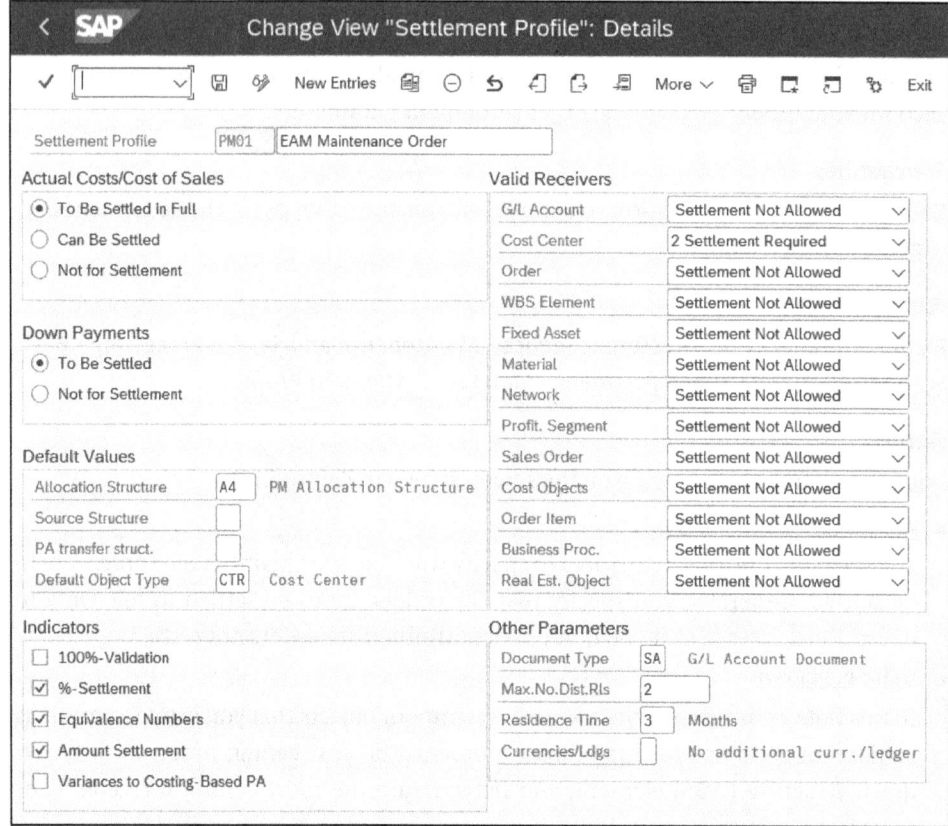

Figure 5.128 Order: Settlement Profile

> **Recommendations for the Settlement Profile**
>
> Following are some recommendations for the settlement profile:
>
> - Settle the maintenance orders in full.
> - Permit at least one cost center and one order as **Valid Receivers**.
> - Set the most common receiver as the **Default Object Type**. This is generally the cost center.
> - Check the box to activate **%-Settlement** so that you can distribute the entire amount to be settled across several receivers.
> - Check the box to activate **Amount Settlement** so that you implement the settlement according to fixed prices, which is a very popular form of settlement in plant maintenance. Amount settlement is also very much in keeping with the sold-to party/contractor principle.

Settlement Rule: Define Time and Creation of Distribution Rule

You use this Customizing function to control the point in time (release or technical completion) when the settlement rule is to be created and to control how you want to handle an order hierarchy in terms of its settlement rule.

Prerequisites

You must define the order types beforehand.

Customizing Path

Plant Maintenance and Customer Service • Maintenance and Service Processing • Maintenance and Service Orders • Functions and Settings for Order Types • Settlement Rule: Define Time and Creation of Distribution Rule

Settings

Use the **Distrib. rule sub-order** option to define how you want to handle the settlement rule for suborders. You have the following five options (see Figure 5.129):

- 1: Copy the distribution rule(s) for the main order (highest superior order to all other orders)
- 2: Copy the distribution rule(s) for the superior order (e.g., order 1 to suborder 1)
- 3: Main order will be settlement receiver (all order will be settled to highest superior order)
- 4: Superior order will be settlement receiver (e.g., suborder 2 will be settled to order 1)
- <Blank>: No special handling of suborders (each order will get its own settlement rule)

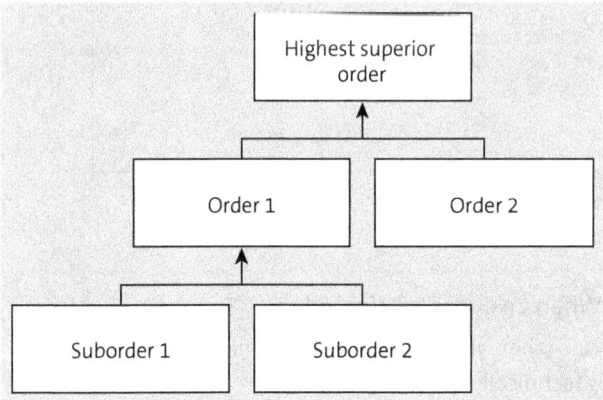

Figure 5.129 Order Hierarchy Settlement

For settings 1–4, the default values from the settlement profile are overridden. If you leave the entry blank, the settings from the settlement profile apply.

Use the **Settlement rule time** option to define when and how the settlement rule is to be created. You have the following three options (see Figure 5.130):

- **1: Mandatory for release**
  On the basis of the settlement profile, the system tries to automatically create the settlement rule when the order is released. If it's unsuccessful, you must maintain the settlement rule manually.

- **2: Distribution rule for release**
  On the basis of the settlement profile, the system tries to automatically create the settlement rule when the order is released. If it's unsuccessful, the order is released without a settlement rule.

- **<Blank>: Mandatory for completion**
  On the basis of the settlement profile, the system tries to automatically create the settlement rule when the order is technically completed. If it's unsuccessful, you must maintain the settlement rule manually.

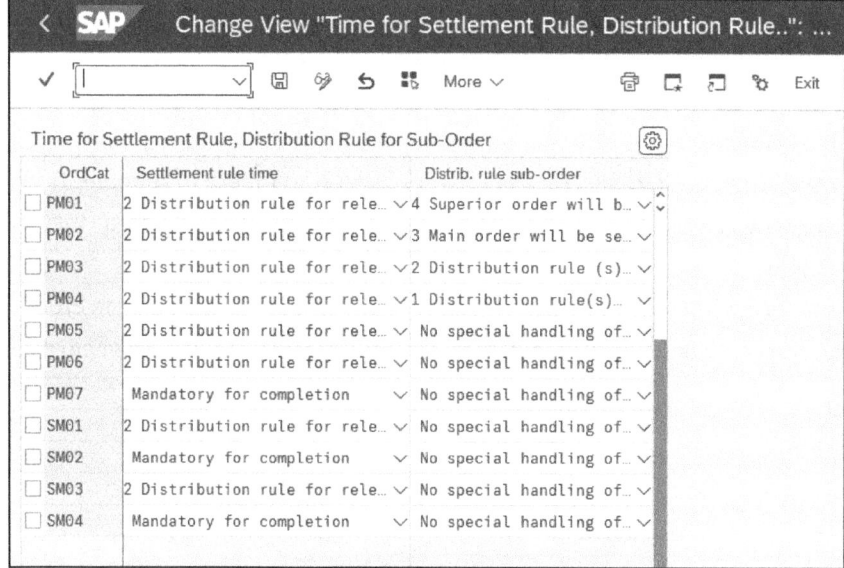

Figure 5.130 Order: Time for Settlement Rule and Order Hierarchy

[!] **Create the Settlement Rule When an Order Is Released**

Always create the settlement rule when an order is released. Otherwise, you can't settle the order before the order is technically completed.

5.2.10 Project Management and Investment Management

You can assign SAP S/4HANA Asset Management orders to project system objects (the project, WBS element, and network) and to investment management objects (the investment program and investment program position). For these assignments to work, you need some Customizing settings.

384

First, you must decide, for each order type, whether you want to assign project system objects or investment management objects to the SAP S/4HANA Asset Management order (see Figure 5.131). It's not possible to assign both simultaneously.

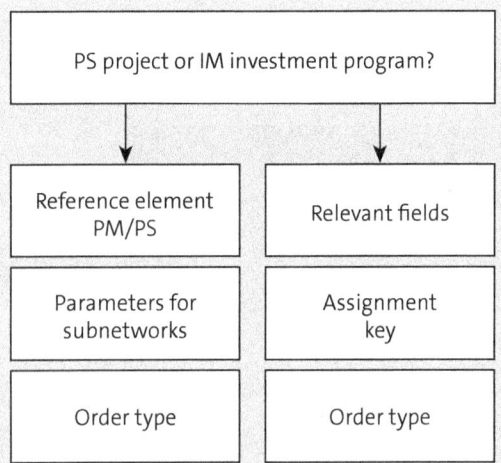

Figure 5.131 Order: Project System or Investment Management

Then, depending on whether you want to assign project system objects or investment management objects, additional settings are necessary:

- If you want to assign project system objects, use the *PM/PS reference element* to determine the parameters for an automatic assignment, and use the parameters for subnetworks to determine the permitted project system network type/maintenance order type combinations.
- If you want to assign investment management objects, determine the relevant field for the assignment and then use an *assignment key* to assign the objects to a maintenance order type.

Define Transfer of Project or Investment Program

You use this Customizing function to control, for each object type, whether project numbers maintained in the master data of maintenance objects *or* investment programs are to be automatically copied to the maintenance order. It isn't possible to copy both to an SAP S/4HANA Asset Management order simultaneously.

Prerequisites

You must define the order types beforehand.

Customizing Path

Plant Maintenance and Customer Service • Maintenance and Service Processing • Maintenance and Service Orders • Functions and Settings for Order Types • Define Transfer of Project or Investment Program

Settings

If you want to copy project numbers from the piece of equipment or functional location to the maintenance order, enter "X" in the **Copy assignment** column (see Figure 5.132).

If you want to copy investment programs from the piece of equipment or functional location to the maintenance order, enter "1" in the **Copy assignment** column.

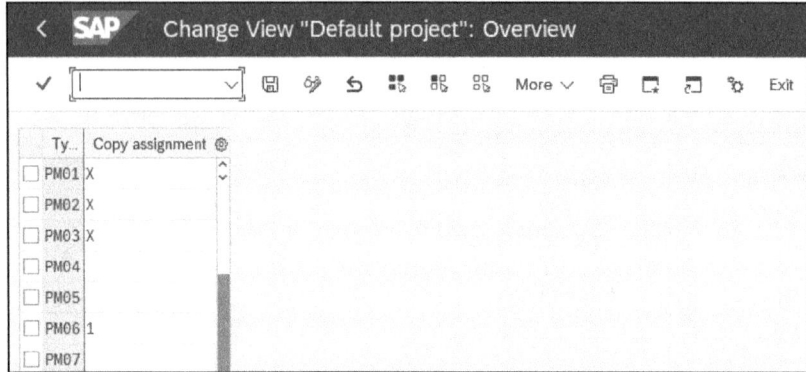

Figure 5.132 Order: Assign Project or Investment Program

Define Relevant Fields for Assignment of IM Program

You can use this Customizing function to define, for each assignment key, the fields that are to be used to automatically determine an *investment program*. This concerns fields in the maintenance order or fields from the master data of the reference object (e.g., the functional location or piece of equipment) assigned to the order.

Prerequisites

There are no specific prerequisites.

Customizing Path

Plant Maintenance and Customer Service • Maintenance and Service Processing • Maintenance and Service Orders • Functions and Settings for Order Types • Define Relevant Fields for Assignment of IM Program

Settings

You can use the following fields to control the automatic assignment of the investment program (see Figure 5.133):

- **Fiscal yr**
  Fiscal year.
- **Comp. code**
  Company code.

- **Bus. area**
  Business area.
- **CO area**
  Controlling area.
- **Plnt**
  Plant. Choose from **1** = maintenance plant, **2** = planning plant, or **<Blank>** = no effect.
- **CCtr**
  Cost center. Choose from **1** = cost center of reference object, **2** = cost center of main work center, or **<Blank>** = no effect.
- **Profit ctr**
  Profit center.
- **Plant sec.**
  Plant section.
- **FL**
  Functional location. Choose from **1** = functional location of order, **2** = top functional location of order, or **<Blank>** = no effect.

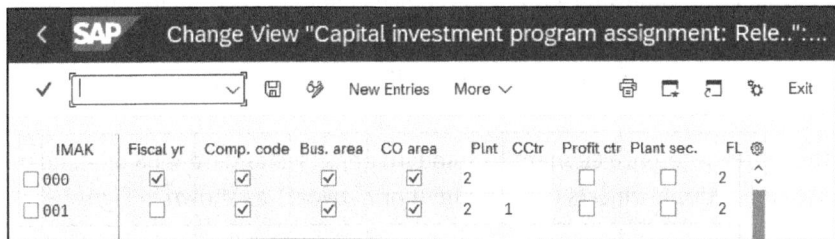

Figure 5.133 Order: Assign Investment Program

Assign IM Assignment Key to Order Types

You can use this Customizing function to define, for each order type, an assignment key that can be used to automatically find a specific investment program.

Prerequisites
You must define the order types and assignment keys beforehand.

Customizing Path
Plant Maintenance and Customer Service • Maintenance and Service Processing • Maintenance and Service Orders • Functions and Settings for Order Types • Assign IM Assignment Key to Order Types

Settings
Enter the corresponding assignment key for the order types for which you want the system to automatically propose investment programs (see Figure 5.134).

5 Configuring the Work Order Cycle

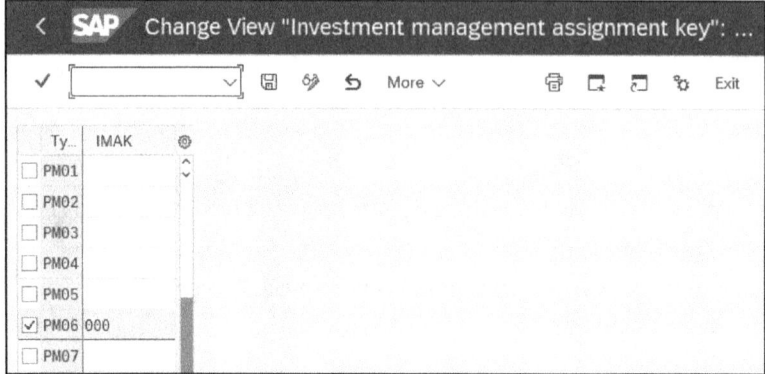

Figure 5.134 Order: Assign IM Assignment Key

Define Field Values for PM/PS Reference Element

You use this Customizing function to define the fields that you want to use for the automatic assignment of maintenance orders to WBS elements. To ensure that this automatic assignment works, you must fulfill the following prerequisites:

- Use the **Define Field Values for PM/PS Reference Element** Customizing function to maintain fields that will later establish a direct connection.
- Assign the **PM/PS** reference element to a task list (see Figure 5.135) or maintenance item.
- Assign the **PM/PS** reference element to standard objects (standard WBS or standard network) or operational objects (WBS element or network), as shown in Figure 5.135.

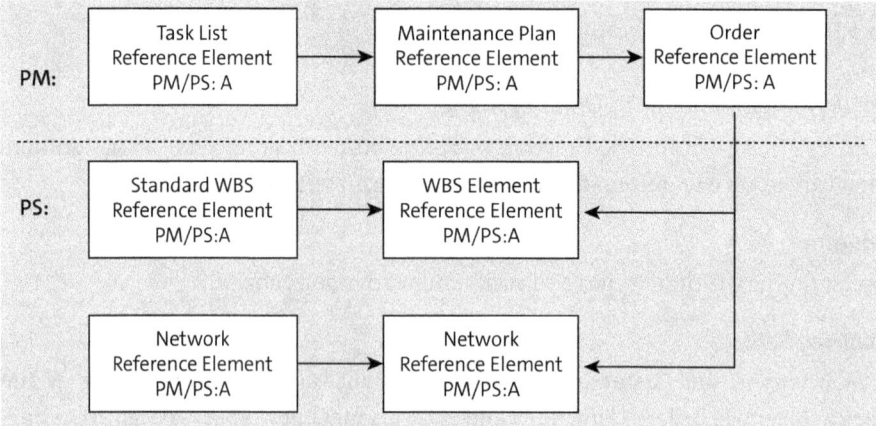

Figure 5.135 Order: PM/PS Reference Element

You can use Transaction ADPMPS to perform an automatic or semiautomatic assignment of maintenance orders to WBS elements or networks.

5.2 Orders

Prerequisites

There are no specific prerequisites.

Customizing Path

Plant Maintenance and Customer Service • Maintenance and Service Processing • Define Field Values for PM/PS Reference Element

Settings

Define the **PM/PS Reference Element** as a grouping for subsequent maintenance tasks (see Figure 5.136).

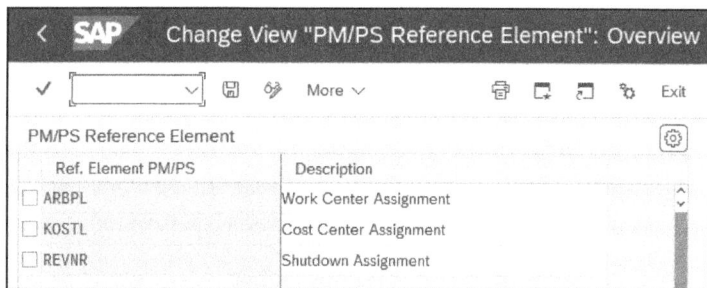

Figure 5.136 Order: Examples of PM/PS Reference Elements

Define Parameters for Subnetworks

You use this Customizing function to define possible combinations for the assignment of maintenance order types to network types. From a project processing perspective, maintenance orders are a special kind of *subnetwork*.

Prerequisites

You must define the order types and network types beforehand.

Customizing Path

Project System • Structures • Operative Structures • Network • Define Parameters for Subnetworks

Transaction

OPTP

Settings

In this Customizing function, you define each SAP S/4HANA order type that you want to assign to a project systems network type. In addition, you define which control key and dates are to be copied from the network to the maintenance order (see Figure 5.137).

5 Configuring the Work Order Cycle

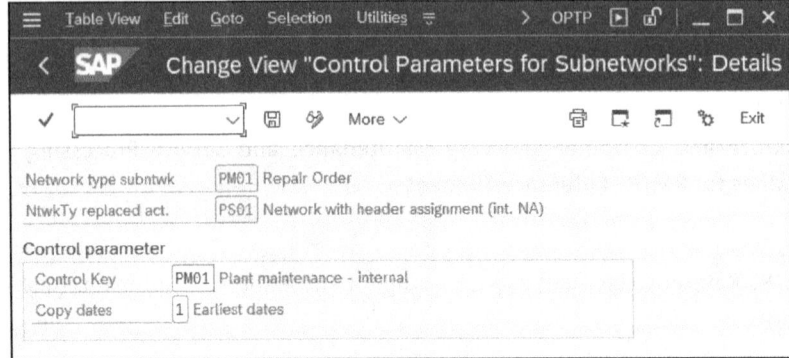

Figure 5.137 Order: Control Parameters for Subnetworks

Define Investment Profiles

You use this Customizing function to control, in the case of investment measures, the integration into asset accounting. You control whether an asset under construction (AuC) is to be created automatically and, if so, to which asset class it is to belong.

Prerequisites

You must define the asset classes beforehand.

Customizing Path

Plant Maintenance and Customer Service • Maintenance and Service Processing • Maintenance and Service Orders • Functions and Settings for Order Types • Order Types and Investment Management • Define Investment Profiles

Transaction

OITA

Settings

Figure 5.138 shows **Investment Profile 000001**, which is an investment profile template provided by SAP. You can use this template as a basis for defining your own investment profiles, as follows:

- **Manage AuC**
  Check the box to set this indicator to define the parallel management of an AuC. If you don't set this indicator, parallel management of the associated investment measures isn't possible in asset accounting.

- **AuC per source structure/assignmt**
  Check the box to set this indicator if you want to manage more than one AuC for each investment measure.

- **Inv.meas. ast.class (asset under construction)**
  Check the box to use this option to determine the asset class for your automatically created AuC.

- **Settlement**
  In this screen area, use the radio buttons to determine whether investment measures are to be managed on a line item or summary basis. **Summary settlement** means that only flat-rate distribution rules can be entered for the entire debit or for specific cost elements. In contrast with **Line item settlement and list of origins**, the **Summary settlement** has the advantage of better performance. However, when you use **Summary settlement**, the system can't display the origins of the line items.

- **Type of distribution rules**
  In this area, you define how rules are to be entered for the depreciation forecast (distribution according to percentages, amounts, equivalence numbers, etc.).

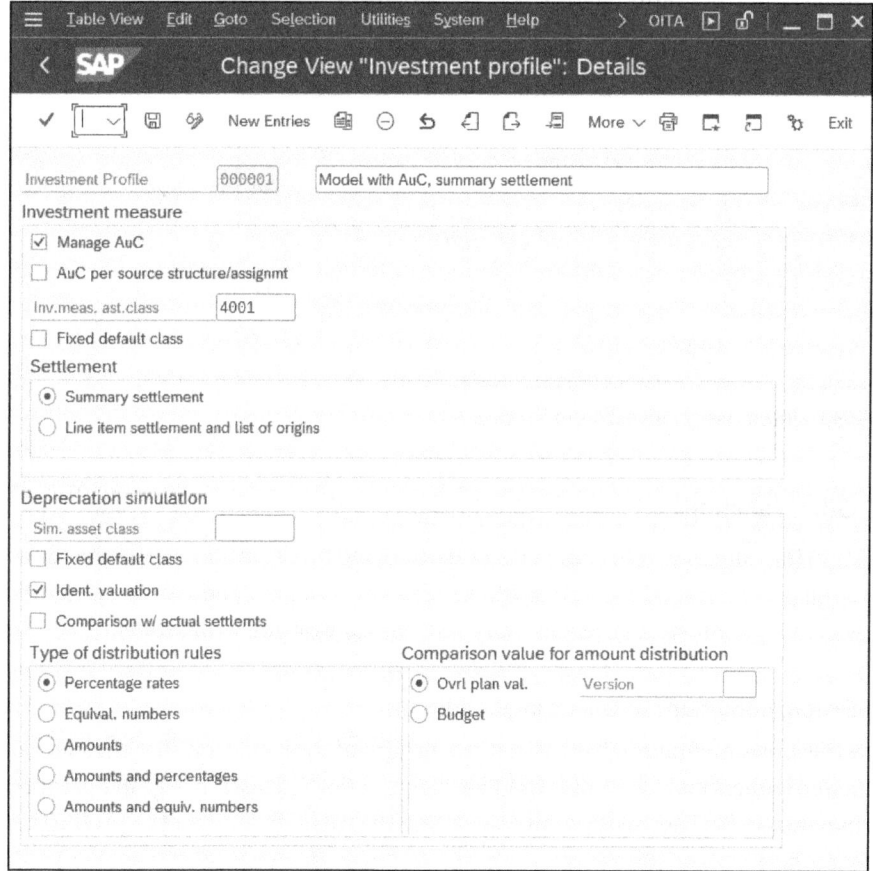

Figure 5.138 Order: Investment Profile

Indicate Order Types for Investment Measures

You use this Customizing function to define which of your order types is available for investments and which investment profile is to be used.

5 Configuring the Work Order Cycle

Prerequisites

You must define the order types and investment profiles beforehand.

Customizing Path

Plant Maintenance and Customer Service • Maintenance and Service Processing • Maintenance and Service Orders • Functions and Settings for Order Types • Order Types and Investment Management • Indicate Order Types for Investment Measures

Settings

If you want to manage investment orders, activate an order type by checking the box to set the relevant indicator in the **Invest. order** column. Then, assign a suitable investment profile in the **InvProfile** column (see Figure 5.139).

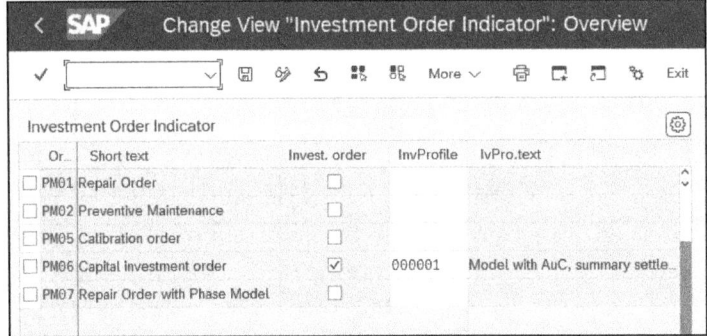

Figure 5.139 Order: Assign Investment Profile

5.2.11 Other Functions

This section introduces you to some other Customizing functions for orders that don't relate to the other subsections in this chapter but nevertheless need to be configured in the system to complete all the necessary settings for working with orders.

Create System Conditions or Operating Conditions

You use this Customizing function to define the operating condition that the piece of equipment or functional location is to have during order processing (e.g., **in operation** or **not in operation**). If necessary, you also define the capacity load on the production work center here.

Prerequisites

There are no specific prerequisites.

Customizing Path

Plant Maintenance and Customer Service • Maintenance and Service Processing • Maintenance and Service Orders • General Data • Create System Conditions or Operating Conditions

Settings

When integrating SAP S/4HANA Asset Management with production planning, you can define a production work center (in the **Work Center** field) in the reference objects (a functional location and/or a piece of equipment). If your maintenance tasks are to influence the capacities in the production work center, you must define an operating condition by selecting the relevant indicator in the **Reservation by PM** column (see Figure 5.140). The maintenance order then generates a capacity load for the production work center, but this doesn't affect the capacity load on the maintenance work center.

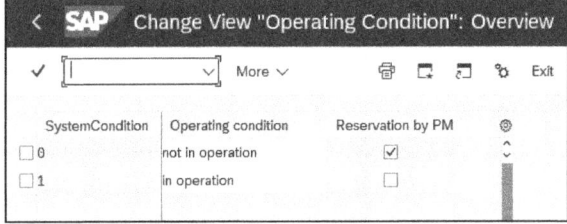

Figure 5.140 Order: Operating Conditions

Define PRT Control Keys

When you process orders, you can assign PRTs at the operation level. PRTs such as protective clothing, hand pallet trucks, drawings, or tools are needed to process the order, but unlike components, they aren't consumed. For each assignment, you must specify a *PRT control key*. You use this Customizing function to define PRT control keys.

> **PRT Control Keys versus Control Keys**
>
> Don't confuse PRT control keys with the control keys that you assign to a work center, to an operation in a task list, or to an operation in an order. PRT control keys have nothing to do with these control keys.

Prerequisites

There are no specific prerequisites.

Customizing Path

Plant Maintenance and Customer Service • Maintenance and Service Processing • Maintenance and Service Orders • Production Resource/Tool Assignments • Define PRT Control Keys

Settings

You can use PRT control keys to specify the following (see Figure 5.141):

- Schedule
- Calculate

5 Configuring the Work Order Cycle

- **Confirm**
- **Expand** (additional information when printing)
- **Print**

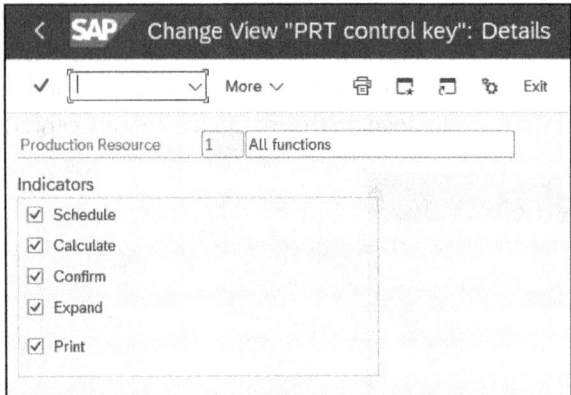

Figure 5.141 Order: PRT Control Key

> **[+] Only the Print Indicator Is Relevant**
>
> At present, only the **Print** indicator is relevant for maintenance orders. You also control whether the PRTs are to be output during shop paper printing, but none of the other functions have been programmed for maintenance orders.

Message Control

You use this Customizing function to define the way a system message concerning orders is output (e.g., in an error message or a warning).

Prerequisites

There are no specific prerequisites.

Customizing Path

Plant Maintenance and Customer Service • Maintenance and Service Processing • Maintenance and Service Orders • Message Control

Settings

Define how each message is to be output. The following options are available (see Figure 5.142):

- **<Blank>**: No message
- **W:** Warning
- **E:** Error message
- **S:** Result message

394

- **P:** Dialog box
- **I:** Information
- **X:** Program termination (a little joke among developers)

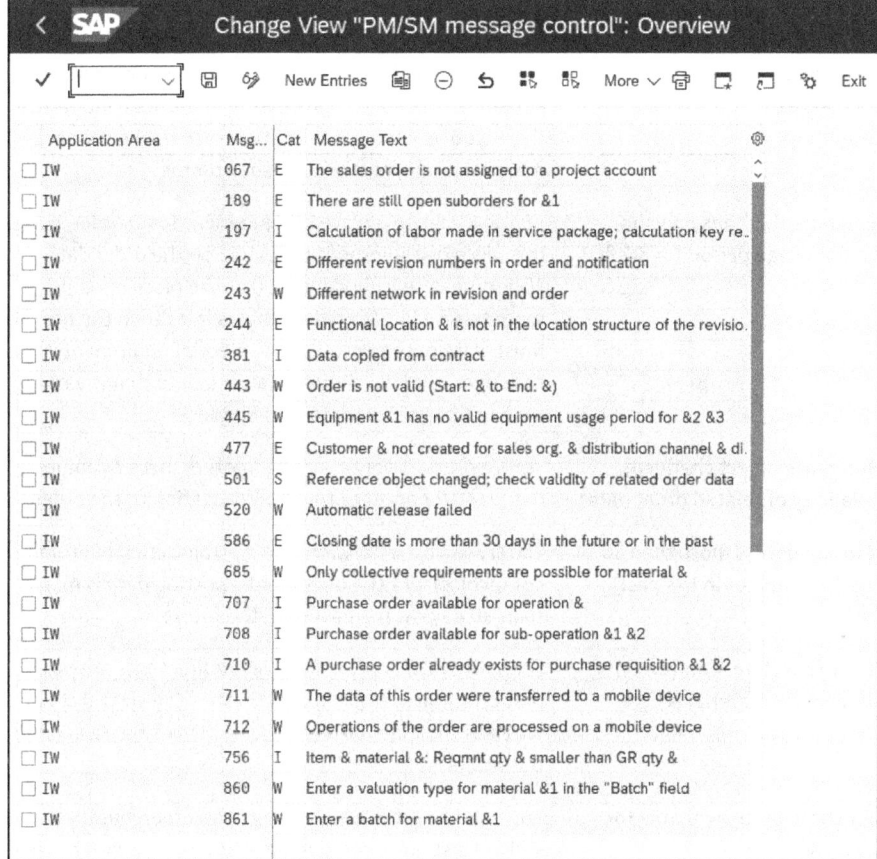

Figure 5.142 Order: Message Control

Table 5.9 lists the most important areas in which system messages occur.

System Message	Description
189 (There are still open suborders for &1)	You want to technically complete a superior order, but at least one suborder has yet to be technically completed.
242 (Different revision numbers in order and notification)	You've assigned different revision numbers to a notification and an order.
243 (Different network in revision and order)	You've assigned different networks to a revision and an order.

Table 5.9 System Messages

System Message	Description
244 (Functional location & is not in the location structure of the revision)	You've assigned a functional location number to an order, but the functional location number isn't in the functional location structure that you've assigned to a revision number.
381 (Data copied from contract)	You've assigned a contract number to a nonstock component or an external processing operation, and from there, you've copied information such as the vendor, material group, and price to the order.
445 (Equipment &1 has no valid equipment usage period for &2 &3)	You've assigned a piece of equipment to an order, but this piece of equipment isn't valid on the order date. Either the date isn't valid yet (i.e., the piece of equipment has a valid-from date that is later than the basic finish date of the order) or the piece of equipment is no longer valid (i.e., it has a valid-to date that is earlier than the basic start date of the order).
501 (Reference object changed; check validity of related order data)	You've retroactively assigned another piece of equipment and/or another functional location to the order.
586 (Closing date is more than 30 days in the future or in the past)	You try to post a document (e.g., Transaction FB60 vendor invoice), but the document's posting date is more than 30 days in the past or in the future.
685 (Only collective requirements are possible for material &)	For material &, only collective requirements are possible according to the material master. You should check whether the individual requirements are necessary for this material.
707 (Purchase order available for operation &)	You want to set a deletion flag in an order, but there is still at least one open purchase order (i.e., a purchase order for which there is no goods receipt or invoice receipt).
708 (Purchase order available for suboperation &1 &2)	You want to set a deletion flag in a suborder, but there is still at least one open purchase order (i.e., a purchase order for which there is no goods receipt or invoice receipt).
711 (The data of this order were transferred to a mobile device)	You try to change order header data of an order that has been transferred to a mobile device.
712 (Operations of the order are processed on a mobile device)	You try to change order operation data of an order that has been transferred to a mobile device.

Table 5.9 System Messages (Cont.)

5.2 Orders

System Message	Description
860 (Enter a valuation type for material &1 in the "Batch" field)	This concerns a refurbishment order (see Chapter 7, Section 7.2). You've assigned a component to an order, the component is subject to a separate valuation, but you haven't specified a valuation type.
861 (Enter a batch for material &1)	You're currently planning a component that is subject to a batch management requirement, but you haven't entered a batch.

Table 5.9 System Messages (Cont.)

Define Proposed Reference Time for Technical Completion

You use this Customizing function to define which date is to be proposed as the date for the technical completion of an order. This is crucial when sorting orders into chronological order in the history (e.g., sorting orders by the months to which they are assigned in PMIS).

Prerequisites

You must define the order types beforehand.

Customizing Path

Plant Maintenance and Customer Service • Maintenance and Service Processing • Maintenance and Service Orders • Functions and Settings for Order Types • Define Proposed Reference Time for Technical Completion

Settings

Define, for each order type, which date is to be proposed as the date for the technical completion of the order (see Figure 5.143).

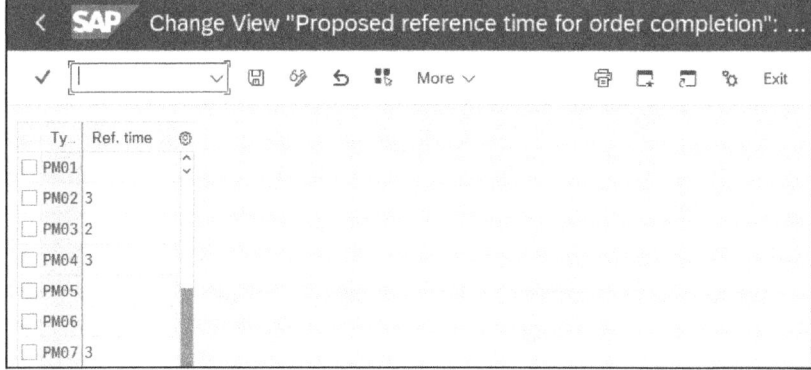

Figure 5.143 Order: Reference Time for Technical Completion

5 Configuring the Work Order Cycle

Here, you have the following options:

- <Blank>: Current date
- 1: Entry date
- 2: Basic start date
- 3: Basic finish date

Regardless of the date proposed, you can overwrite the technical completion date.

Create Default Value Profiles for General Order Data

You use this Customizing function to define certain default values that you assign, together with other default values in the **Default Values for Task List Data and Profile Assignments** Customizing function, to the order type and plant. Figure 5.144 shows the relationships between these default values.

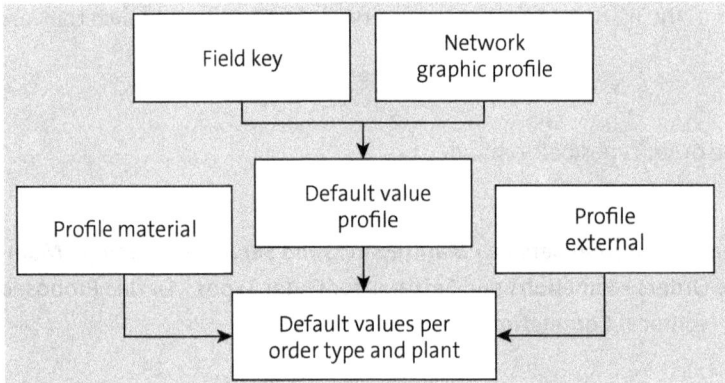

Figure 5.144 Order: Relationships among Default Values

Prerequisites

You must define the field key and network graphic profiles beforehand.

Customizing Path

Plant Maintenance and Customer Service • Maintenance and Service Processing • Maintenance and Service Orders • Functions and Settings for Order Types • Create Default Value Profiles for General Order Data

Settings

You can define the following default values in a default value profile (see Figure 5.145):

- **Field key**
  Use this if you need additional nonstandard fields in the operation.

5.2 Orders

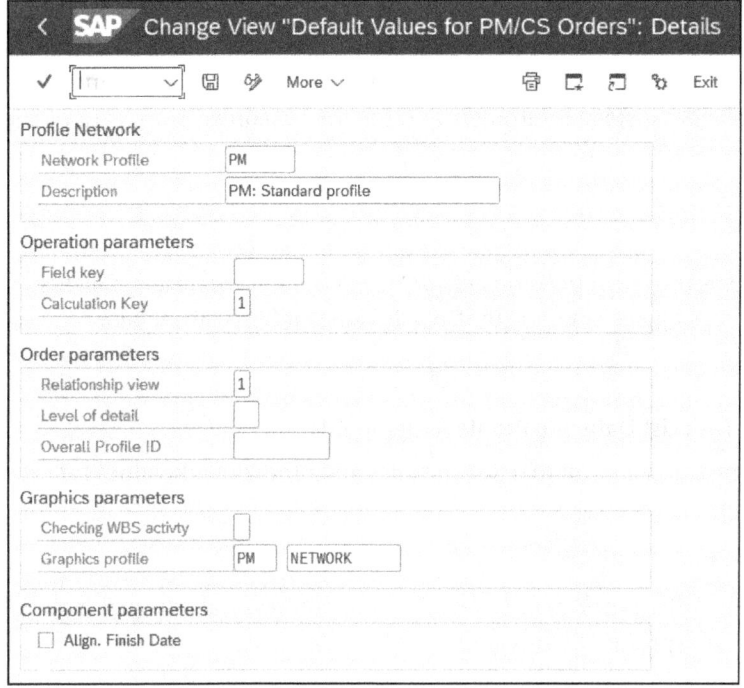

Figure 5.145 Order: Default Value Profile

- **Calculation Key**
  Use this to control the relationship between the **Work** and **Duration** fields. Here, you have the following options:
    - 0: Maintain manually
    - 1: Calculate duration
    - 2: Calculate work
    - 3: Calculate number of capacities
- **Relationship view**
  In this field, you define the view from which the relationships in the operations are to be displayed. Here, you have the following options:
    - 1: Predecessor
    - 2: Successor
    - 3: Mixed
    - 4: As created
- **Graphics profile**
  Use this to define the complete layout of the network graphic. Detailed explanations are provided in Section 5.2.5.

399

- **Align. Finish Date**
  If you check the box to set this indicator, you also need to specify which components to stage for the scheduled finish date of the operation. If you don't set this indicator, the components will be staged automatically for the scheduled start date of the operation.

> [+] **Note**
> Don't be confused by the use of the term *network profile* in this Customizing function. It is a term from project processing, and the **Default Value Profile** is meant here.

Default Values for Task List Data and Profile Assignments

You use this Customizing function to define profiles and other default values for each order type and plant.

Prerequisites

You must define the order types, plants, default value profiles, and external material and external processing profiles beforehand.

Customizing Path

Plant Maintenance and Customer Service • Maintenance and Service Processing • Maintenance and Service Orders • Functions and Settings for Order Types • Default Values for Task List Data and Profile Assignments

Settings

Make the following assignments for each order type and plant (see Figure 5.146):

- **ExternProfile**
  This assigns default values for the cost element, purchasing organization, material group, and purchasing group for external processing (Section 5.2.4).
- **Mat.Profile**
  This assigns default values for the cost element, purchasing organization, material group, and purchasing group for external material (Section 5.2.4).
- **MaintProfile**
  This assigns the default value profiles.
- **Act/op UoM**
  This defines the default value for the unit of measure for the operation.

In addition to these default values, you can use this Customizing function to define these aspects of how task lists are to be transferred to orders:

- Whether a dialog box for selecting operations is to appear
- Whether a dialog box for selecting work centers is to appear

- Whether the operations selected are to be renumbered consecutively in intervals of ten operations
- Whether a task list can be fully integrated only once
- Whether the sequence of operations is the same as in the task list

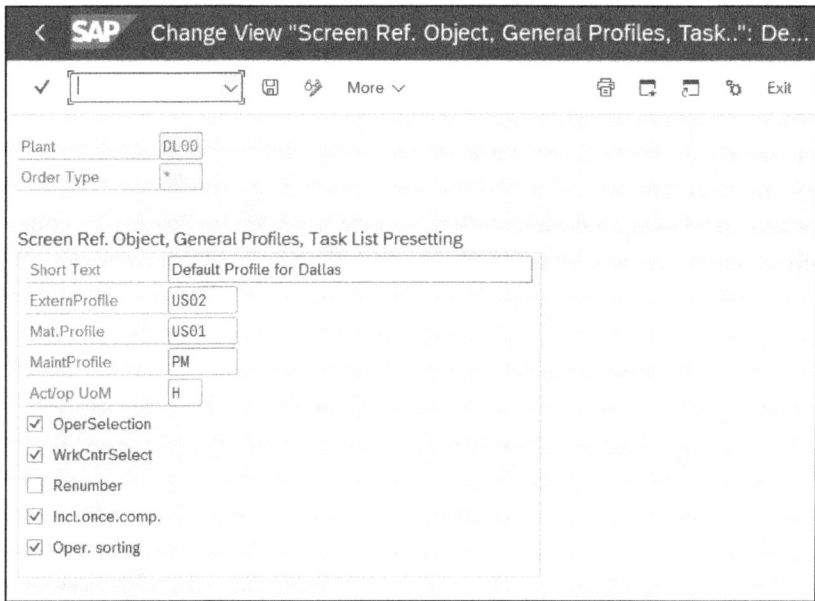

Figure 5.146 Order: Default Values

Note

These default values and settings are overwritten by user-specific default values, which the user can configure in the order under **Extras • Settings • Default Values**.

5.3 Completion Confirmations

The Customizing functions available to you for confirmation reporting are the maintenance transactions (Transaction IW41 and on) and the Customizing functions for the cross-application time sheet (CATS).

5.3.1 SAP S/4HANA Asset Management Completion Confirmations

The following Customizing functions are available to you if you want to use completion confirmations in conjunction with the maintenance Transactions IW41 and so on.

Define Control Parameters for Completion Confirmations

You use this Customizing function to define, for each order type and plant, the control parameters for confirmation reporting in SAP S/4HANA Asset Management. This is a basic prerequisite for all other functions, regardless of whether you use maintenance transactions, CATS transactions, or other transactions to execute your completion confirmations.

Prerequisites

You must define the order types and plants beforehand.

Customizing Path

Plant Maintenance and Customer Service • Maintenance and Service Processing • Completion Confirmations • Define Control Parameters for Completion Confirmations

Transaction

OIOR

Settings

The following control options are available for each order type and plant (see Figure 5.147):

- **Final Confirmation**
  Check the box to set this indicator if you want the system to propose the final confirmation indicator for the completion confirmation.

- **Clear Open Reservs.**
  Check the box to set this indicator if you want the system to propose the **Clear Open Reservs.** indicator for the completion confirmation. When you set this indicator, open reservations (i.e., reservations that haven't been withdrawn) are deleted with the final confirmation.

- **Propose Dates**
  Check the box to set this indicator if you want the system to propose the operation dates (scheduled start date and scheduled finish date) as completion confirmation dates.

- **Propose Activities**
  Check the box to set this indicator if you want the system to propose the work as actual work or, in the case of existing completion confirmations, the remaining work as actual work.

- **Calc. performance**
  Check the box to set this indicator to determine whether the system is to calculate the actual work if a change is made to the dependent fields (e.g., the **Final Confirmation** field or the **Remaining Work** field).

5.3 Completion Confirmations

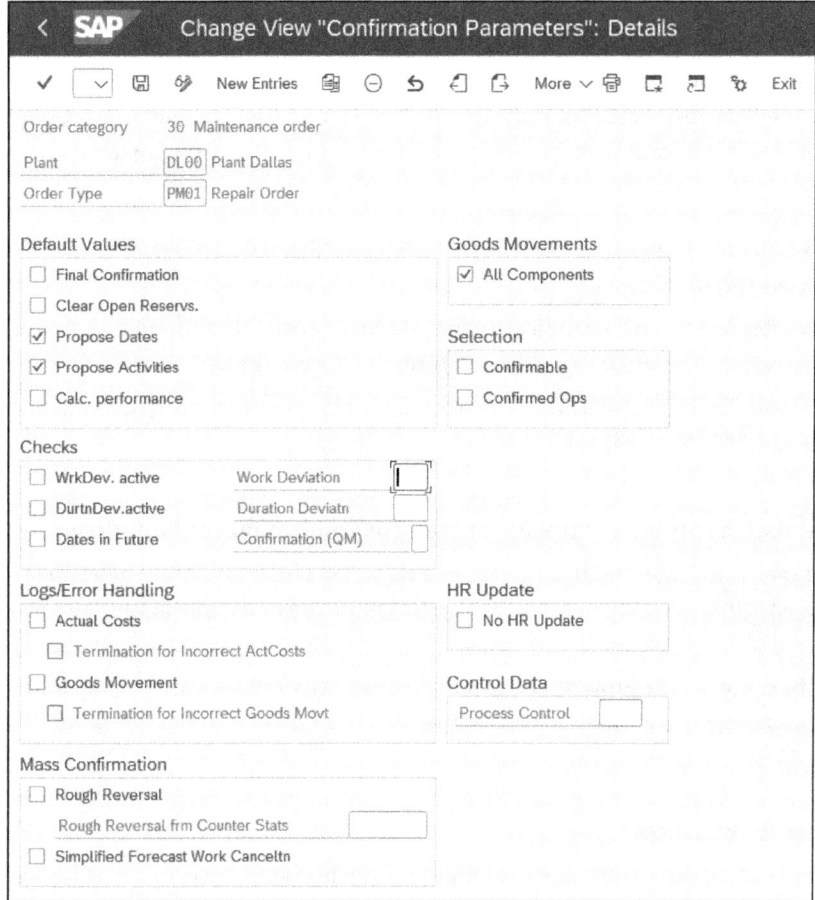

Figure 5.147 Completion Confirmation: Parameters

- **WrkDev. active**
  Check the box to set this indicator if you want to activate the deviation check for work. Then, in the **Work Deviation** field, specify the percentage that this deviation must exceed before the system issues a warning message.

- **DurtnDev.active**
  Check the box to set this indicator if you want to activate the deviation check for the duration. Then, in the **Duration Deviatn** field, specify the percentage that this deviation must exceed before the system issues a warning message.

- **Dates in Future**
  Check the box to set this indicator if you want the system to permit the use of dates in the future (e.g., the posting date, the end of execution time). For example, this is necessary if you enter a time recording on October 31 but you want to add this time to the month of November (the posting date).

403

5 Configuring the Work Order Cycle

- **Logs/Error Handling**
  In this screen area, specify that the system will automatically display a log if the actual cost calculation or material posting is incorrect. Check the box to set the **Termination for Incorrect ActCosts** indicator or the **Termination for Incorrect Goods Movt** indicator if you want to ensure that the time confirmation isn't posted if errors occur.

Use the **All Components** indicator to decide which components are displayed in the backflushing overview:

- If you check the box to set this indicator, the system displays all components regardless of whether the backflushing indicator is set for the component in the order.
- If you don't set this indicator, the system displays only those components for which the backflushing indicator is set.

> **Recommendations for the Control Parameters for Completion Confirmations**
> Make sure that you define the control parameters for all plants and order types that are to be used. Otherwise, you may not be able to confirm an order type in the plant in question.
>
> Check the boxes to set the **Propose Dates** and **Propose Activities** indicators and possibly the **All Components** indicator. Leave all other indicators and default values blank.

Define Causes for Variances

You use this Customizing function to define, for each plant, reasons for variances between the actual times and planned times. You can then use these reasons in your analyses.

Prerequisites
You must define plants beforehand.

Customizing Path
Plant Maintenance and Customer Service • Maintenance and Service Processing • Completion Confirmations • Define Causes for Variances

Settings
For each plant, define the variance reasons that you want to track later (see Figure 5.148).

5.3 Completion Confirmations

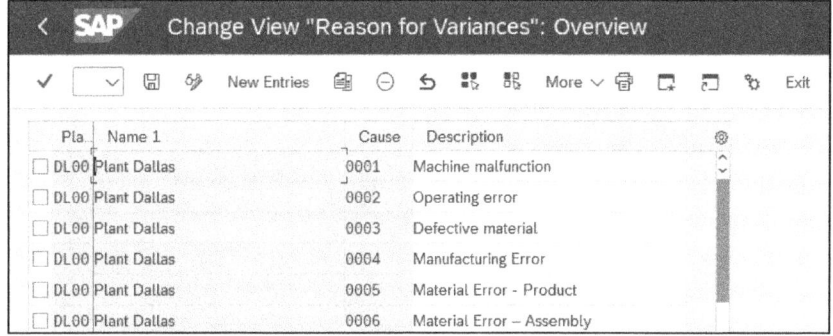

Figure 5.148 Completion Confirmation: Causes for Variances

Set Screen Templates for Completion Confirmation

You can use this Customizing function to compile the individual screen areas for the overall completion confirmation (Transaction IW42 [Overall Completion Confirmation]). You can define different profiles here, and each user can later select the most appropriate profile in their user settings for the overall completion confirmation (in Transaction IW42 under **Extras • Settings**).

Prerequisites

There are no specific prerequisites.

Customizing Path

Plant Maintenance and Customer Service • Maintenance and Service Processing • Completion Confirmations • Set Screen Templates for Completion Confirmation

Settings

Each profile contains five rows that, in turn, contain one or two screen areas (see Figure 5.149).

IW42 – Overall Completion Confirmation	
Subscreen 1	Subscreen 2
Subscreen 3	Subscreen 4
Subscreen 5	Subscreen 6
Subscreen 7	Subscreen 8
Subscreen 9	Subscreen 10

Figure 5.149 Completion Confirmation: Overall Confirmation Structure

The following screen areas are available (see Figure 5.150):

- Times
- Services
- Goods movements
- Notification items
- Causes of damage
- Meas./cntr readings
- Tasks
- Activities
- Items

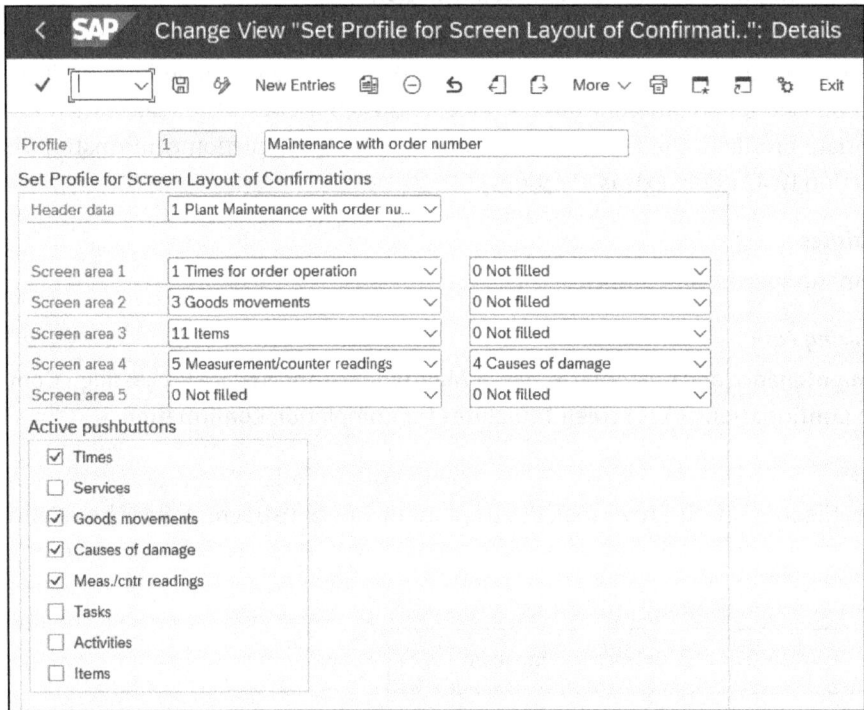

Figure 5.150 Completion Confirmation: Overall Confirmation Content

If, in a particular row, you fill in only the left screen area, then the screen area in the Transaction IW42 entry screen spans the entire width of the screen. If you fill in the left and right screen areas, then the two screen areas are displayed side by side.

In addition to the layout of the screen areas, you can specify which buttons are to appear (e.g., **Times** or **Goods movements**). Then, in the overall completion confirmation, use the pushbuttons to switch to the corresponding full screen. You then no longer have six entry rows but rather an entire overview screen.

> **Define Separate Layouts for the Overall Completion Confirmation** [+]
>
> Seize this opportunity to provide your users with suitable profiles for their work environment. The "as much as necessary, but as little as possible" principle from earlier in the chapter applies here as well.

5.3.2 Cross-Application Time Sheet Confirmations

CATS is an application that you can use to enter several person-related actual times. This application is not only available in SAP S/4HANA Asset Management but also in other SAP S/4HANA application areas, such as human resources and production planning.

The following sections discuss the most important Customizing functions for use in plant maintenance.

Set Up Data Entry Profiles

This Customizing function is a prerequisite for using CATS. In the data entry profiles, you define, for example, whether a separate release and approval procedure is required for all actual times entered, the number of periods for which times can be entered simultaneously, and whether you want to work with worklists.

Prerequisites

There are no specific prerequisites.

Customizing Path

Cross-Application Components • Time Sheet • Specific Settings for CATS Classic • Set Up Data Entry Profiles

Settings

Because the data entry profile is a long screen with many settings, I've divided it into two parts. Figure 5.151 shows the upper part of the data entry profile.

The following settings are the most important settings from an SAP S/4HANA Asset Management perspective:

- **Profile changeable**
  Check the box to set this indicator if you want the user to be able to change the profile settings (with restrictions; for example, displaying or hiding a totals column) while entering the relevant data.

- **With totals line**
  Check the box to set this indicator if you want the system to display a daily total of all times entered.

- **Release on saving**
  Check the box to set this indicator if you want the system to release all entered times when the user chooses **Save**. If you don't set this indicator, the user must release all entered times separately.

- **Period type**
  Type a number in this field to control the period view used with the profile (e.g., **2** = week and **4** = month).

- **Periods**
  Enter the number of periods to be entered.

- **Enter for several personnel nos**
  Check the box to set this indicator if you want to enter this data for several persons and specify how these persons are to be selected (e.g., according to cost center or organizational unit).

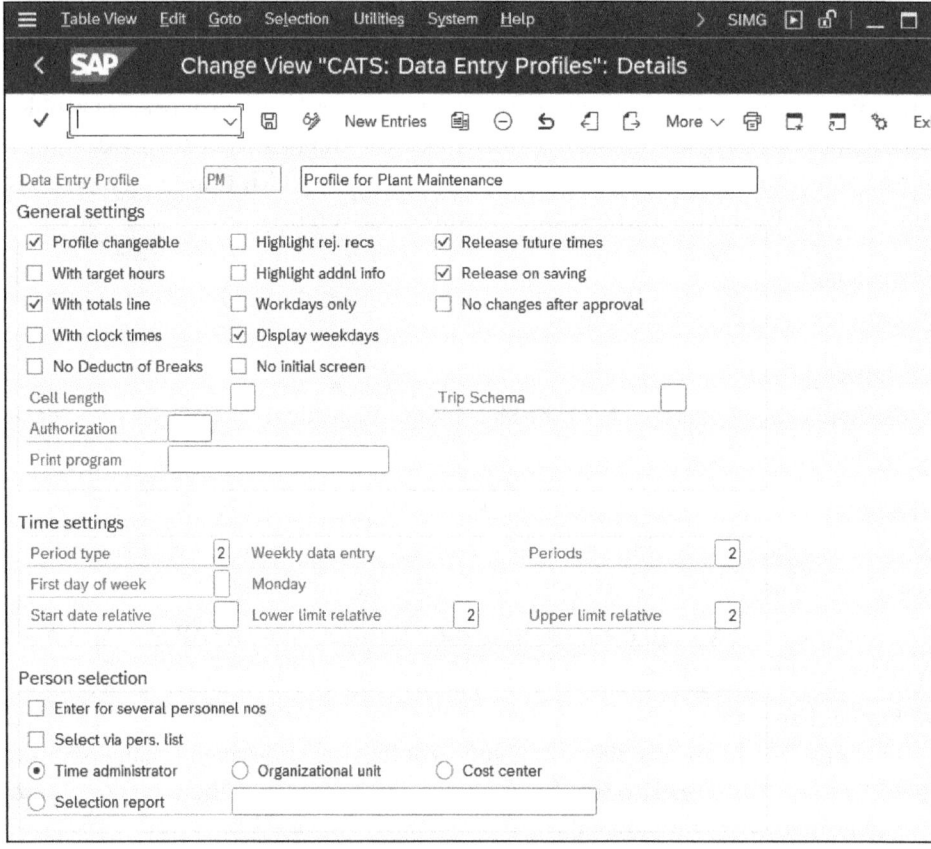

Figure 5.151 CATS: Data Entry Profile Part 1

5.3 Completion Confirmations

Figure 5.152 shows the lower part of the data entry profile. Here, you can make the following settings:

- **Approval**
  In this screen area, specify whether all entered and released times are to be subject to an approval procedure before they are transferred to SAP S/4HANA Asset Management. (Because this section covers a standard scenario in which such times aren't subject to an additional approval procedure, the approval procedure isn't discussed here.)

- **Cost accounting variant**
  In this area, specify how personnel costs are to be handled. In SAP S/4HANA Asset Management, the standard scenario is that the personnel costs are assigned to the master cost center.

- **Default values**
  In this area, define the default values that the system is to automatically preassign when entering times. From an SAP S/4HANA Asset Management perspective, the following default values make sense: **Controlling area**, **Activity type**, and **Att./absence type**.

- **Worklist**
  In this area, check the box to set the **With worklist** indicator if you want to work with a worklist. I recommend that you don't work with a worklist because it makes the entry of actual times very unclear. If, however, you want to work with a worklist, you must define what type of worklist is to be created and how (e.g., based on the operations for the work center).

Recommendations for the CATS Data Entry Profile

Following are some recommendations for the data entry profile in CATS:

- Allow the user to change the profile while entering the relevant data.
- Allow the system to release all entered times when the user chooses **Save**.
- Display a totals row for each day.
- Enter the times on a weekly basis.
- Don't enter the times for more than one person simultaneously.
- Transfer the times without an approval procedure.
- Check the boxes to set the default values for **Controlling area** and **Activity type**.
- Enter the times without a worklist.

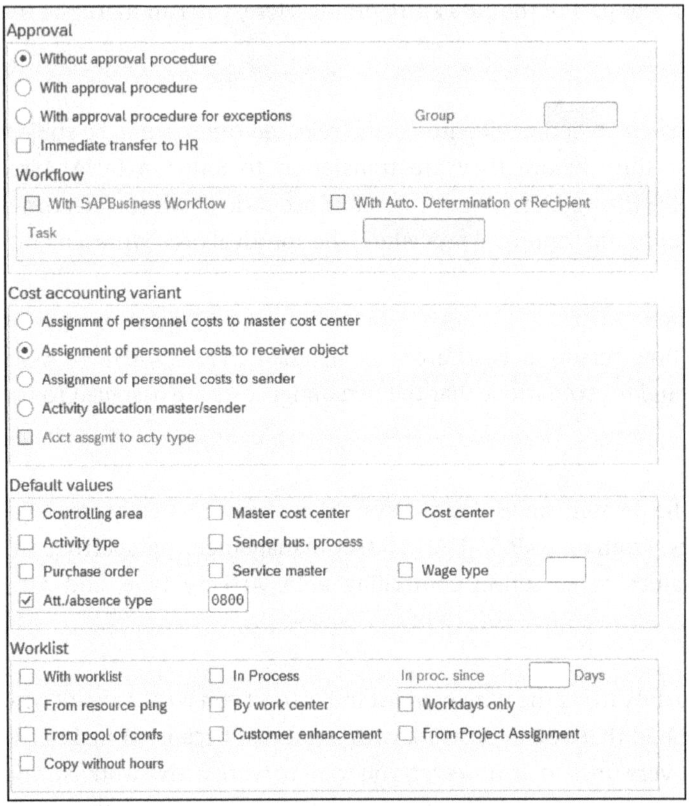

5 Configuring the Work Order Cycle

Figure 5.152 CATS: Data Entry Profile Part 2

Define Field Selection

The Customizing functions for field selection were explained in Chapter 3, Section 3.9 as generic functions. However, because the field selection is particularly important for CATS, let's discuss some of its special features here.

Prerequisites
You must set up the data entry profiles beforehand.

Customizing Path
Cross-Application Components · Time Sheet · Settings for All User Interfaces · Time Recording · Choose Fields

Transaction
CAC2

Settings or Recommended Settings
Set up the field selection for the **Data entry section** screen group, the **Data Entry Profile** influencing value, and the data entry profile that you've set up for plant maintenance (**PM** in our example; see Figure 5.153).

5.3 Completion Confirmations

Hide Fields

Because the data entry section has a large number of fields that aren't required from an SAP S/4HANA Asset Management perspective, it makes sense to hide all the fields that you don't require and to leave the remaining fields ready for input so that the data entry section remains clear.

From a maintenance perspective, you require only the following fields (all others can and should be hidden):

- Work Center: CATSD-ARBPL
- Receiver Order: CATSD-RAUFNR
- Final Confirmation: CATSDB-AUERU
- Activity Type: CATSD-LSTAR
- Total: CATSD-SUMDAYS
- Partial confirmation: CATSDB-ERUZU
- Activity: CATSD-VORNR
- Plant: CATSD-WERKS

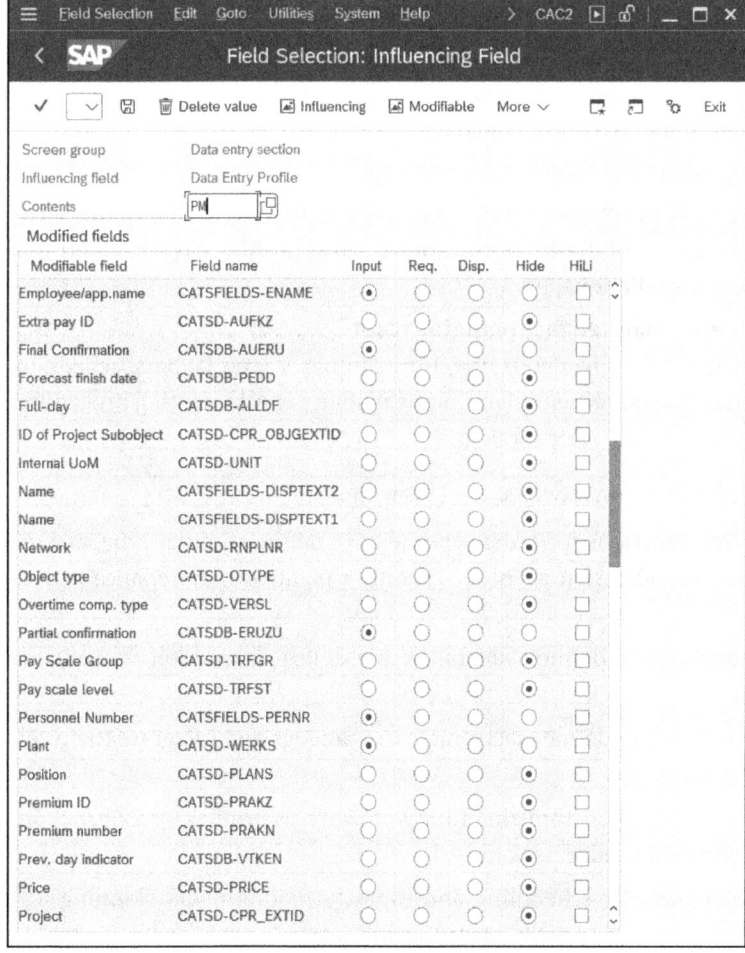

Figure 5.153 CATS: Field Selection

5.4 Summary

In this chapter you were introduced to the Customizing functions that will enable you to adjust the plant maintenance work order cycle in SAP S/4HANA Asset Management to your requirements. This process requires three business objects: notification, order, and completion confirmation.

The notification sections covered the following:

- The notification type shapes the screen layout, number assignment, long text control, status profile, object information, partner determination, priorities, or order type.
- Notification Customizing enables you to maintain catalogs by defining, for example, damage codes and damage cause. It also enables you to assign them via catalog profile to equipment or functional locations.
- There are other notification Customizing functions such as defining priorities, response monitoring, and action boxes.

In order Customizing functions, a distinction must be drawn between Customizing functions that belong to order types in general and those that belong to order types defined for each maintenance planning plant:

- **Order types in general**
  These Customizing functions determine, for example, number assignments, settlement procedures, integration with notifications, status profiles, partner determination, priorities, or shop floor papers.
- **Order types for each maintenance planning plant**
  These Customizing functions determine, for example, screen layouts, scheduling procedures, capacity requirement planning, availability checks, costing procedures, or integration of electronic parts catalogs.

The order Customizing sections covered the following:

- Basic Customizing functions for order types (e.g., to define an order type, assign a planning indicator, and assign an order type to a plant or maintenance activity types)
- Customizing functions to define availability checks (e.g., inspection controls and scope of check)
- Customizing functions to define electronic parts catalogs and integrate them into orders
- Several material planning Customizing functions, such as profiles for external procurement or determining address data
- Customizing functions for scheduling and capacity requirements planning (e.g., scheduling parameters and capacity planning profiles)

- Customizing functions for printing procedures (e.g., how to define shop papers, assign shop papers to users, and configure emailing instead of printing)
- Customizing functions for order cost functions: estimated costs (e.g., how to define value categories, how to assign cost elements), costing (e.g., how to define a costing sheet, how to assign it to an order type), and order settlement (e.g., how to set up a settlement profile with allocation structures)
- Customizing functions to integrate orders with projects or investment programs and how to set up several profiles for order types

The confirmation section covered the following Customizing functions:

- **Within SAP S/4HANA Asset Management**
  For example, how to define control parameters and how to set up screen layout profiles for overall confirmations
- **Within CATS**
  For example, how to define a data entry profile and how to assign a fitting field selection

Chapter 6
Configuring Preventive Maintenance

In this chapter, you'll learn about the Customizing functions for preventive maintenance. Here, the focus is on those options available within the SAP system for adjusting task lists and maintenance plans to your requirements.

Preventive maintenance is characterized by the fact that you can plan for required resources (work centers, materials, external companies, etc.) in terms of content and scheduling. This type of business process is associated with the following scenario: deadline monitoring (program RISTRA20H) runs automatically as a batch job and ensures that the maintenance call objects (e.g., the orders) are automatically generated on the due date. Figure 6.1 shows SAP's objects for preventive maintenance.

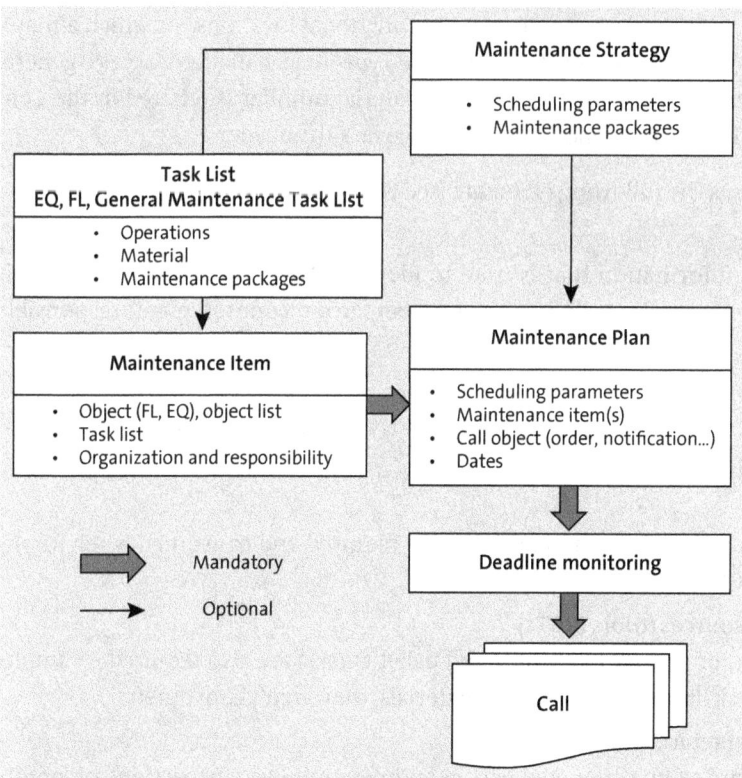

Figure 6.1 Preventive Maintenance Overview

6.1 Task Lists

A *task list* essentially describes activities (operations) and contains materials that are required during activity processing. From a maintenance perspective, there are three different types of task lists:

- **Equipment task list**
  You create an *equipment task list* for exactly one piece of equipment if you want to map its specific features. However, you can only use this task list in connection with this one piece of equipment.

- **Functional location task list**
  You create a *functional location task list* for exactly one functional location if you want to map its specific features. As with equipment task lists, you can only use this functional location task list in connection with this one functional location.

- **General maintenance task list**
  A *general maintenance task list* isn't object specific, which means that it isn't assigned to any particular piece of equipment or functional location. In fact, you can indirectly make a general maintenance task list available for several pieces of equipment and/or functional locations. To do this, use the **Construction Type** field in the **Structuring** screen group in the master record for the piece of equipment or functional location. All pieces of equipment and functional locations for which a material number is entered in the **Construction Type** field can then access general maintenance task lists for which the same material number is entered in the **Construction Type** field in the general maintenance task list header.

A task list comprises the following elements (see Figure 6.2):

- **Header data**
  Header data is information that is used to identify and manage the task list. This data applies to the entire task list (e.g., number, group counter, plant, responsible work center).

- **Operations**
  Operations are used to describe the work that is to be performed when implementing the task list.

- **Material list**
  The *material list* contains spare parts that are required and consumed when implementing the task list.

- **Production resources/tools (PRTs)**
  PRTs (e.g., tools, protective clothing, hand pallet trucks) are also required for implementing the task list. However, unlike materials, they aren't consumed.

- **Inspection characteristics**
  If inspections are to be conducted within an operation (e.g., inspections of length, weight, and function), you can define them as *inspection characteristics*.

- **Maintenance packages**
  If the task list is used in a strategy maintenance plan, you use *maintenance packages* to control the frequency with which the maintenance work is performed—either on a time-dependent basis (e.g., once every 3 months) or on a performance-dependent basis (e.g., once every 1,200 operating hours).

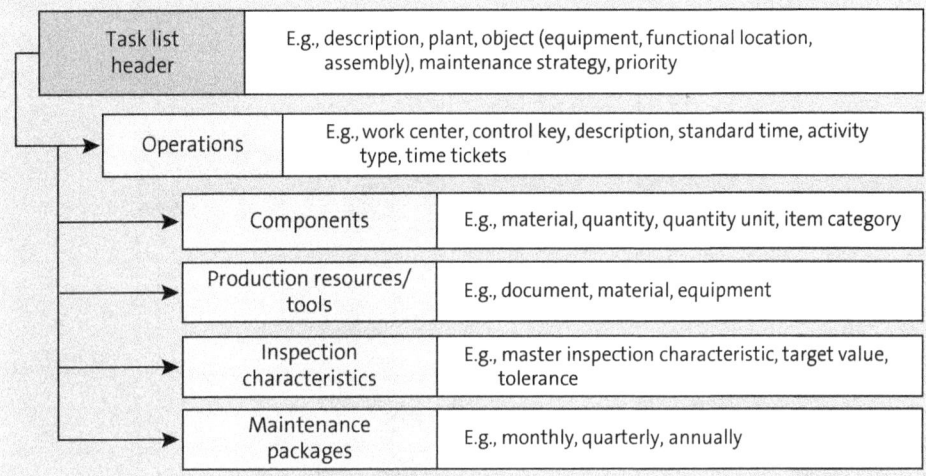

Figure 6.2 Task List: Structure and Content

The following Customizing functions are available for adjusting the management and use of task lists to your requirements.

Maintain Task List Status

You use this Customizing function to define which task list status you want to manage, and you then use the task list status to control whether the task list is still locked or to control the functions for which it's released.

Prerequisites
There are no special prerequisites.

Customizing Path
Plant Maintenance and Customer Service • Maintenance Plans, Work Centers, Task Lists, and PRTs • Task Lists • General Data • Maintain Task List Status

Transaction
OIL1

Settings

The standard SAP delivery contains the following statuses, among others (see Figure 6.3):

- **1 Created**
  Checking the box for this status doesn't release the task lists for orders (**RelInd**) or costing (**Cstng**).
- **4 Released (General)**
  Checking the box for this status releases the task lists for orders and costing.

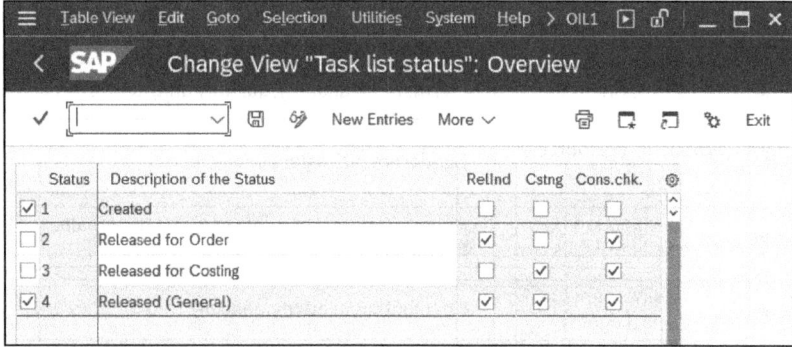

Figure 6.3 Task List: Status

> [+] **Task List Statuses 1 and 4 Usually Suffice**
>
> If you nevertheless want to define your own statuses, make sure that you check the box to define at least one status that releases the task lists for both costing and orders.

Define Task List Usage

In the SAP system, task lists are used not only in SAP S/4HANA Asset Management but also in the following other areas of SAP S/4HANA:

- Production planning for discrete industries, where they are used as routing or as reference operation sets
- Production planning for process industries, where they are used as master recipes
- Project systems, where they are used as standard networks
- Quality management, where they are used as inspection plans

You use this Customizing function to define task list usages.

Prerequisites

There are no special prerequisites.

Customizing Path

Plant Maintenance and Customer Service • Maintenance Plans, Work Centers, Task Lists, and PRTs • Task Lists • General Data • Define Task List Usage

Settings

The standard SAP delivery provides task list **Usage 4** for **Plant Maintenance** (see Figure 6.4).

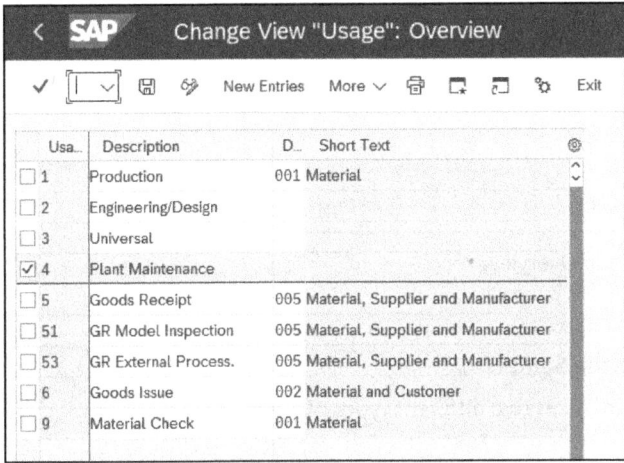

Figure 6.4 Task List: Usage

> **Task List Usage 4 Usually Suffices**
>
> If you nevertheless want to define your own task list usages, make sure to make the right selection in all follow-up actions (e.g., when assigning the task list usage in the work center).

Configure Planner Group

You use this Customizing function to define the groups or persons responsible for maintaining task lists for each plant.

Prerequisites

You must maintain the plants beforehand.

Customizing Path

Plant Maintenance and Customer Service • Maintenance Plans, Work Centers, Task Lists, and PRTs • Task Lists • General Data • Configure Planner Group

6 Configuring Preventive Maintenance

Settings

At the plant level, you can define descriptive or numeric planner groups that contain a maximum of three numbers or characters (see Figure 6.5).

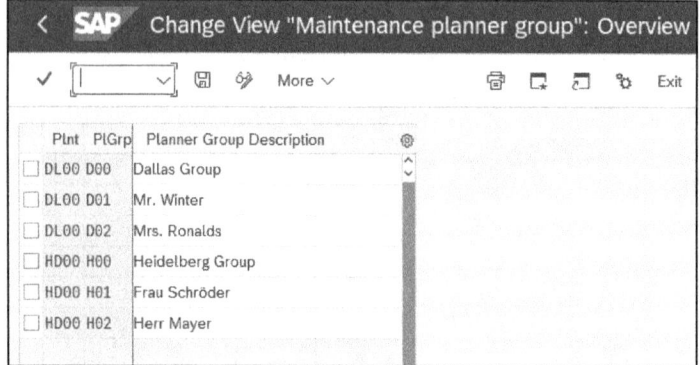

Figure 6.5 Task List: Planner Group Overview

> **Not All Planner Groups Are the Same**
>
> Note that the SAP system recognizes two different types of planner groups:
>
> - **Planner group**
>   You assign this group as the responsible planner group to a task list at the header level.
> - **Maintenance planner group**
>   You assign this group as the responsible planner group to a piece of equipment, functional location, notification, order, or maintenance item.

Define Profiles with Default Values

You use this Customizing function to maintain default values such as increment or unit. When you create a task list, you can call one of the profiles and transfer its values to the new task list as default values.

Prerequisites

If necessary, you must maintain the following values as default values beforehand:

- Graphic profile (see the **Network • Define Graphic Profiles** Customizing function)
- Field key for user fields (see the **Define User Fields** Customizing function)
- Purchasing organization
- Purchasing group
- Vendor
- Cost element

- Material group
- Task list usage (see the **Define Task List Usage** Customizing function)
- Task list status (see the **Maintain Task List Status** Customizing function)

Customizing Path

Plant Maintenance and Customer Service • Maintenance Plans, Work Centers, Task Lists, and PRTs • Task Lists • Control Data • Define Profiles with Default Values

Transaction

OIL6

Settings

You can assign the following default values to a profile (see Figure 6.6):

- Oper./act. increment

  Use this option to determine the number assignment for operations (generally, 0010).

- Group and Name

  Use these two fields to assign a previously defined graphic profile so that you can create a network within a task list.

- Field key

  Use this field to assign additional fields to the task list that are relatively freely definable and don't exist in the standard system.

- Calculation Key

  Use this field to control the relationship between the **Work** and **Duration** fields. Here, you have the following options:
  - 0 = Maintain manually
  - 1 = Calculate duration
  - 2 = Calculate work
  - 3 = Calculate no. of capacities

- Unit for work

  Use this field to determine the units to be defined as default values for the **Work** and **Duration** fields.

- External procurement

  In this field group, define default values for the purchasing organization, purchasing group, currency, vendor, cost element, and material group if an operation is processed externally.

- Usage and Status

  Use these two fields to assign default values to the task list for task list usage and the task list status, respectively.

6 Configuring Preventive Maintenance

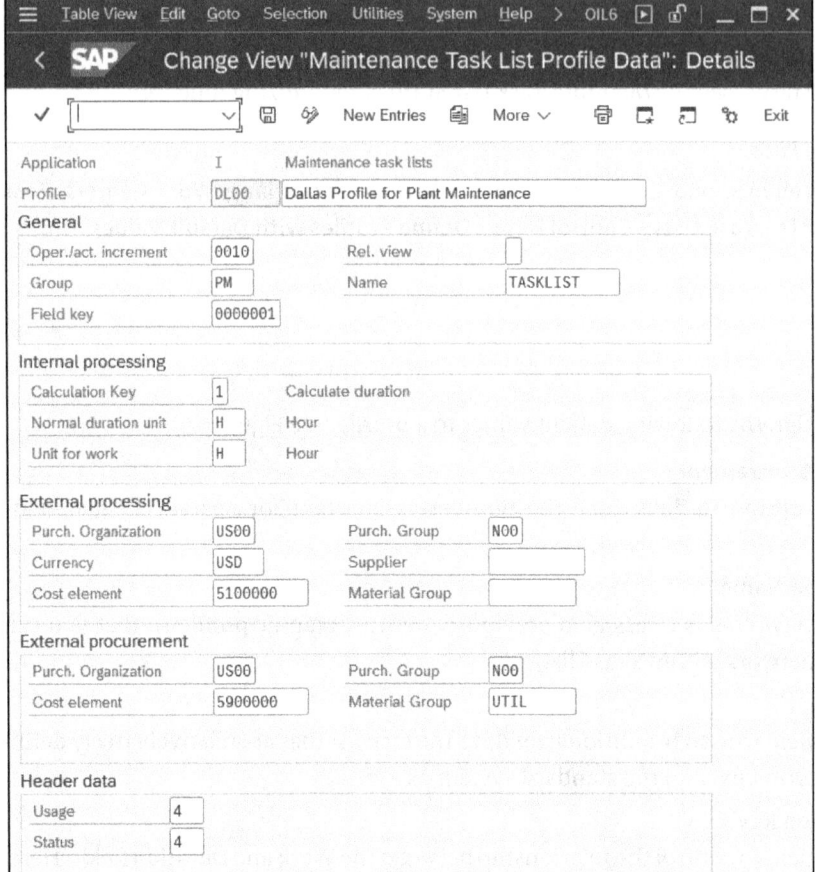

Figure 6.6 Task List: Profile with Default Values

> [+] **Define and Use Default Value Profiles**
> Create different default value profiles so that your users don't have to repeatedly enter the same data manually.

Define Presetting for Free Assignment of Material

In the maintenance task lists, you can use *free material assignment*, which means you don't have to create a BOM beforehand. Instead, you can assign the materials directly to the task list. Purely for technical reasons, the system automatically generates a BOM in the background for the materials and the task list. Don't use the normal dialog transactions (e.g., Transactions CS02, IB02, IB12) to maintain this BOM.

Every BOM requires a *BOM usage*, and you use this Customizing function to define the BOM usage, which is assigned to those BOMs that are automatically generated by the system when you use free material assignment.

6.1 Task Lists

Prerequisites

You must define the BOM usage beforehand (see the **Define BOM Usages** Customizing function).

Customizing Path

Plant Maintenance and Customer Service • Maintenance Plans, Work Centers, Task Lists, and PRTs • Task Lists • Control Data • Define Presetting for Free Assignment of Material

Settings

In the standard SAP delivery, the maintenance BOMs are assigned **BOM Usg 4** (see Figure 6.7).

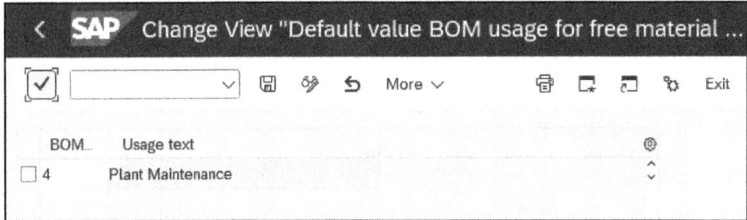

Figure 6.7 Task List: Default Value for BOM Usage

> **BOM Usage 4 Is Generally Used**
> The only time you must enter a different BOM usage is when you define your own usages for maintenance BOMs.

Define User Fields

You can use this Customizing function to define fields that aren't contained in the standard system but are required in your maintenance task lists. You can then choose **Operation • User Data** to call or assign these fields in the task list, and then, you can use the **Define Profile with Default Values** Customizing function to propose field keys.

Prerequisites

There are no special prerequisites.

Customizing Path

Plant Maintenance and Customer Service • Maintenance Plans, Work Centers, Task Lists, and PRTs • Task Lists • Operation Data • Define User Fields

Transaction

OILJ

Settings

In particular, the following field types are available (see Figure 6.8):

- Two 20-digit **Text** fields
- Two 10-digit **Text** fields
- Two **Quantity** fields
- Two **Value** fields
- Two **Date** fields
- Two **Checkboxes** fields

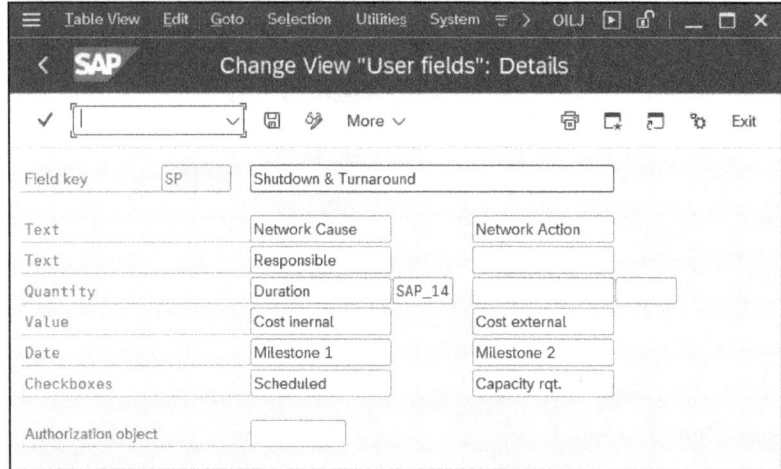

Figure 6.8 Task List: User Fields

Maintain Suitabilities

You use this Customizing function to define, for each plant, the requisite *suitability* for performing a particular activity. Then, on the **Default Values in Work Center** master data screen, you decide whether you want to define the suitability as a reference or as a default value in the task list. Otherwise, you can assign a suitability for each operation manually.

Prerequisites

You must define the plants beforehand.

Customizing Path

Plant Maintenance and Customer Service • Maintenance Plans, Work Centers, Task Lists, and PRTs • Task Lists • Operation Data • Maintain Suitabilities

Settings

Figure 6.9 shows a sample list of potential suitabilities. In each case, you check the box to assign a two-digit **Suitability** number, and you enter a description (under **Text for suitability**) that contains a maximum of 30 characters.

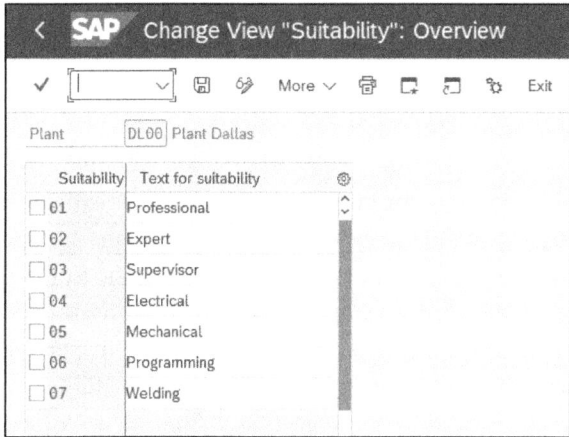

Figure 6.9 Task List: Suitabilities

6.2 Maintenance Plans

SAP S/4HANA Asset Management provides you with maintenance plans that support the business processes associated with preventive maintenance. The following maintenance plan types are available for selection (see Figure 6.10):

- **Single cycle plans**
  You create single cycle plans when you have to perform the same maintenance action tasks in full at regular intervals (time based or performance based). In this case, including a task list is optional.

- **Maintenance strategies and strategy plans**
  You create maintenance strategies and strategy plans when you have to perform maintenance action tasks that are interrelated or supersede each other—either as a time-based strategy (e.g., every three, six, or twelve months) or as a performance-based strategy (e.g., every 10,000, 20,000, or 40,000 kilometers). In this case, you must include a task list, specifically one that pursues the same strategy as the maintenance plan.

- **Multiple counter plans with one cycle**
  You create multiple counter plans if determination of the maintenance date is dependent on several influencing factors (e.g., every 6 months, 10,000 kilometers, or 1,000 operation hours). Including a task list here is optional.

- **Multiple counter plans with multiple cycles**
  Multiple counter plans with multiple cycles enable you to define several cycles that are based on one another as the following examples show:
  - Cycle set 1: every 1,000 operation hours, 5,000 miles, or 12 months
  - Cycle set 2: every 3,000 operation hours, 15,000 miles, or 36 months

  Including several task lists is necessary here.

- **Inspection rounds plans**
  How do the functions for inspection rounds differ from those for maintenance planning described in the previous sections? Maintenance planning usually involves a single object for which a series of activities, some rather complex, are to be executed. When it comes to inspection rounds, this occurs in the reverse direction. With an inspection round, you'll process a large number of objects, executing the same or similar activities on each object. These activities aren't usually overly complex and require the same tools, spare parts, and qualifications. You can map inspection rounds in the SAP system in two different ways: by using an *object list* (different objects but the same activity) or a *task list* (different objects and different activities).

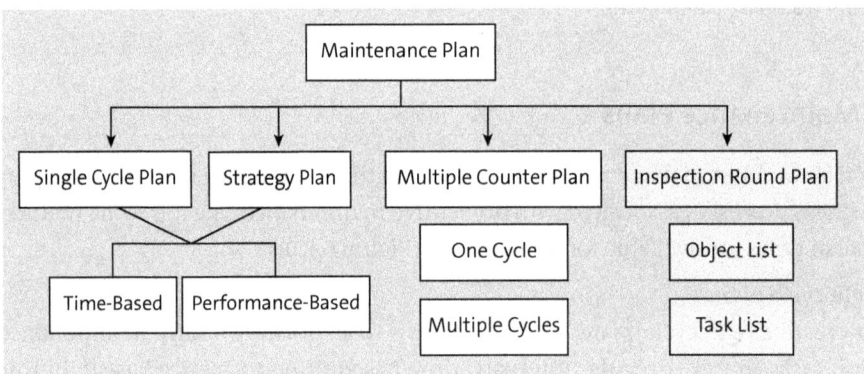

Figure 6.10 Maintenance Plans Overview

The following Customizing functions are available for adapting the management and use of maintenance plans to your needs.

Set Maintenance Plan Categories

You use this Customizing function to define all the control parameters for the maintenance plans (the maintenance call object or reference object, among others).

Prerequisites

If you want to assign reasons and causes to maintenance items, you must activate business function LOG_EAM_CI_13.

6.2 Maintenance Plans

Customizing Path

Plant Maintenance and Customer Service · Maintenance Plans, Work Centers, Task Lists, and PRTs · Maintenance Plans · Set Maintenance Plan Categories

Settings

Figure 6.11 shows the potential call objects. You can use this setting (see Figure 6.12) to define which of the following maintenance call objects are to be called from the maintenance plan when they are due:

- **<Blank> = Maintenance or service order**
  Select this if you generate an order and want to execute the maintenance plan specifications unchanged, without any further planning.

- **1 = Service entry sheet**
  Select this if you have an outline agreement on regular services with an external company and you want the system to automatically generate, at regular intervals, the service entry sheets required for acceptance.

- **2 = Notification**
  Select this if you want to perform other detailed planning when it's due (e.g., if you want to combine several tasks into an order on the basis of the current capacity usage). In such cases, several task lists are then copied into a single order as an operations list.

- **3 = Inspection lot**
  Select this if you conduct stability studies in quality management and therefore require corresponding inspection lots at regular intervals.

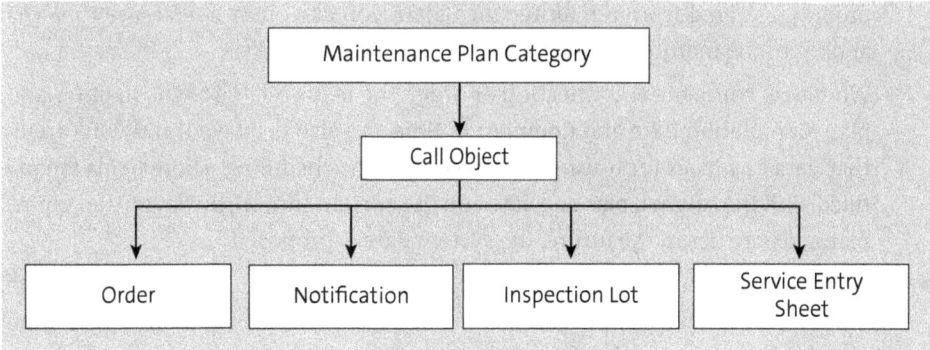

Figure 6.11 Maintenance Plan: Call Objects

In addition to the maintenance call object, you can use the maintenance plan category to specify the following (see Figure 6.12):

- **Change document**
  Check the box to set this indicator if you want to activate change documents. The change documents will later contain information for this maintenance plan category,

such as who changed the maintenance plan, what change was made, and when the change was made.

- **Document release**
  Check the box to set this indicator if you want to activate change documents when a maintenance call is released. The change documents will later contain information for this maintenance plan category, such as who released the maintenance call from the maintenance plan, which maintenance call was released, and when the maintenance call was released.

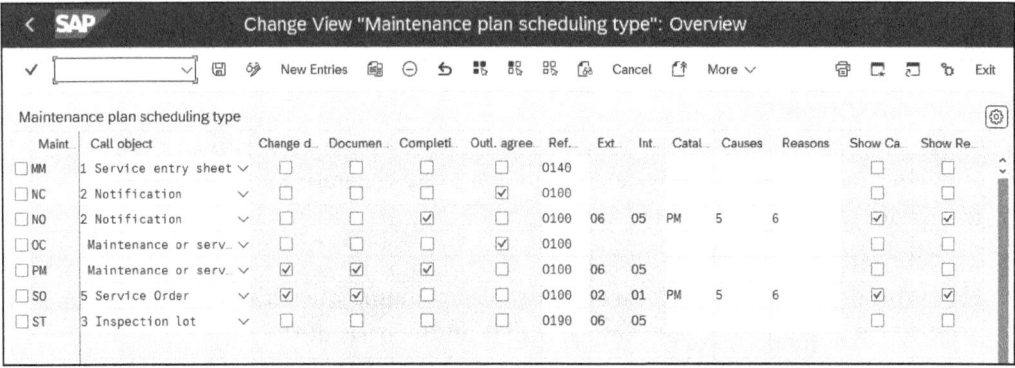

Figure 6.12 Maintenance Plan Categories

- **Completion data**
  Check the box to set this indicator so you can do the following:
  - When you complete one of the maintenance call objects (e.g., notification or order) generated from a maintenance plan, you can enter a *reference time* that applies to the maintenance plan only.
  - When you complete a notification or order, the fields related to the maintenance plan, **Completion date** and **Completion time**, are also displayed, and you can use that data as a basis for future maintenance plan scheduling. These fields are prefilled with the original planned date and the current time. If, however, the original planned date lies in the future, the system date is proposed.

- **Outl. agreement**
  Check the box to set this indicator to establish a reference to an outline agreement within the maintenance plan (e.g., if you call service entry sheets).

- **Ref. object**
  Set this to define the technical object that you want to assign to the maintenance item:
  - **O100** = Functional location + piece of equipment + assembly
  - **O110** = Equipment + assembly
  - **O120** = Functional location (30) + equipment + assembly

- O130 = Serial number + material number + device data
- O140 = Without a reference object
- O150 = Equipment only
- O160 = Functional location only
- O170 = Equipment + serial number + material number
- O180 = Functional location 1:1 + equipment + assembly
- O190 = Physical sample

A new functionality comes up with business function LOG_EAM_CI_13: each maintenance plan is created for some *reasons* and addresses a certain *cause* that has an impact on the deterioration mechanism of equipment or functional locations. If cause and reason aren't captured and documented well, it is quite challenging to evaluate the effectiveness of preventive tasks. Inefficient preventive measures will affect too much or too little maintenance intensity, which may result in unnecessary costs.

To enable this feature, you have to configure the maintenance plan category indicating which default **Catalog profile** would be used for a maintenance plan, which catalog type would be used for documenting reasons, and which would be used for documenting causes.

In the example shown in Figure 6.12, there is a maintenance plan category **NO** that allows users to capture reasons and causes. Here, the default catalog profile is **PM**, the default catalog type for causes is **5**, and the default catalog type for reasons is **6**.

> **Recommendations Concerning the Maintenance Plan Category**
>
> The following settings are used most often:
> - **Maintenance or service order** call object
> - Change documents (**Change document** and **Document release**)
> - **Completion data**
> - **Ref. object** (functional location and/or equipment)

Define Sort Fields for Maintenance Plan

You use this Customizing function to define the potential content of the sort field for the maintenance plan. You assign the sort field to a maintenance plan on the **Maintenance Plan: Additional Data** tab (see the next section). The sort field then becomes an important grouping and selection criterion during maintenance plan scheduling (Transaction IP30H [Mass Scheduling Maintenance Plans] or program RISTRA20H).

Prerequisites

There are no special prerequisites.

6 Configuring Preventive Maintenance

Customizing Path

Plant Maintenance and Customer Service • Maintenance Plans, Work Centers, Task Lists, and PRTs • Maintenance Plans • Define Sort Fields for Maintenance Plan

Settings

You can freely determine the sort field (see Figure 6.13) via the following categorization criteria:

- Object type (e.g., PC, pump, engine, fleet object)
- Scheduling frequency (daily, weekly, monthly)
- Object location (e.g., country, plant, building)

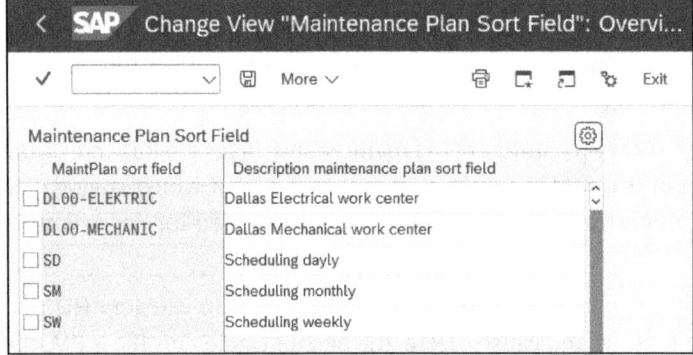

Figure 6.13 Maintenance Plan: Sort Field

Configure Special Functions for Maintenance Planning

This Customizing function offers you two options:

- First, you can activate the *enhanced multiple counter plan*, which enables you to define several interrelated cycles, as illustrated previously in Figure 6.10. If you don't activate this enhancement function, you can define only one cycle in the multiple counter plan (Transaction IP43 [Add Multiple Counter Plan]).

- Second, you can activate the nontemplate copying of maintenance plans. If you don't activate this enhancement function, you can only copy maintenance plans that you've characterized as **Reference Maintenance Plan** (Transactions IP41, IP42, and IP43 or menu path **Create Maintenance Plan • Functions • Copy Template • Allow**).

Prerequisites

If you want to copy maintenance plans, you must activate business function LOG_EAM_CI_10.

6.3 Summary

Customizing Path

Plant Maintenance and Customer Service • Maintenance Plans, Work Centers, Task Lists, and PRTs • Maintenance Plans • Configure Special Functions for Maintenance Planning

Settings

If you want to activate enhancement functions, check the boxes to set the **Enh. MultipleCounterPlan** indicator for the enhanced multiple counter plan and the **Enable Non-template-based Copying** indicator for the opportunity to copy all maintenance plans (see Figure 6.14).

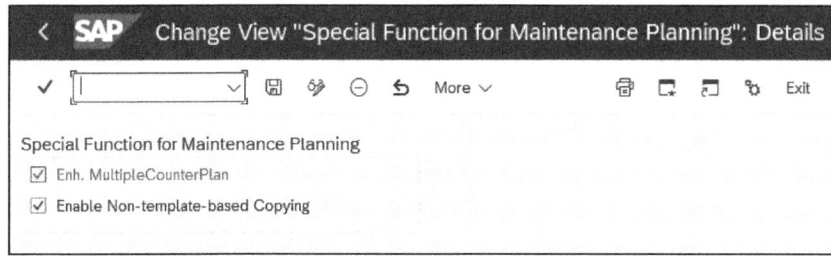

Figure 6.14 Maintenance Plan: Special Functions

Take Care When Enhancing the Multiple Counter Plan

After you check the box to set the **Enh. MultipleCounterPlan** indicator, you cannot uncheck it. However, there are no negative consequences if you set the indicator and don't create any enhanced multiple counter plans.

Enable Non-Template-Based Copying

You should enable this option because it's much easier to allow all maintenance plans as references than to characterize maintenance plans manually for reference plans.

6.3 Summary

SAP S/4HANA Asset Management offers only a few Customizing functions for preventive maintenance. In this chapter, the focus was on the options available for adjusting task lists and maintenance plans to your requirements.

The key takeaways from this chapter regarding preventive maintenance are as follows:

- For task lists, you'll configure task list status and task list usage. Normally you can stay with SAP's standard values.

- You'll define maintenance plan categories that determine, for example, reference call objects, number assignments, completion information, causes and reasons assignments, and creation of change documents.
- You can activate the enhanced multiple counter plan.
- You can set all maintenance plans to be available as reference plans.

Chapter 7
Configuring Additional Business Processes

This chapter describes the Customizing functions for those business processes that are built on the basic work order cycle and that give you enhanced options in relation to your plant maintenance (e.g., external processing, checklists, refurbishment, subcontracting, calibration processes).

In Chapter 5, you learned about the Customizing functions that can influence the basic work order cycle: notifications, orders, and completion confirmations. This chapter introduces you to Customizing functions for those business processes that are built on the basic work order cycle and that provide you with additional business processes for your plant maintenance. These processes include the following:

- External processing of maintenance work, either as an individual purchase order or as a service specification
- Processing of refurbishment tasks
- Subcontracting
- Calibrating test/measurement equipment
- Inspection checklists
- Processing pool asset management
- Using shift notes and shift reports
- Processing maintenance projects using the Maintenance Event Builder
- Phase model-based work order cycles

Because the following explanations are based on the information provided in Chapter 5, not all of the Customizing functions will be repeated here. Instead, I'll discuss only the following:

- Customizing functions that you now require *in addition* to those described in Chapter 5
- *Special attributes, features, or changes* concerning the Customizing functions discussed earlier and now used in one of the aforementioned processes

7 Configuring Additional Business Processes

7.1 External Processing

External processing plays a major role in plant maintenance, considerably more than in production, for example. A nonrepresentative short survey of SAP user companies in the German-speaking SAP user group showed that, on average, approximately half of their maintenance costs result from external processing. Some companies don't have their own maintenance workshops. Instead, they have coordination points (e.g., work scheduling or planners), which are responsible for planning, monitoring, and approving external processing.

You use a control key to initiate external processing. Furthermore, you use different control keys, depending on what type of external processing is to take place.

The control key attributes control the type of external processing that is to take place (see Figure 7.1), as follows:

- **Work center for external company**
  If you've configured a work center for an external company and you want to execute external processing using an internal order in the same way as you do for your own work centers, you should initiate this type of external processing using a control key (PM01 or similar) for which the **Ext. Processing** option is set to **Internally processed operation** and the **Service** indicator isn't set. Remember that in Chapter 2, Section 2.4, you learned how to define control keys (see also the **Maintain Control Keys** Customizing function).

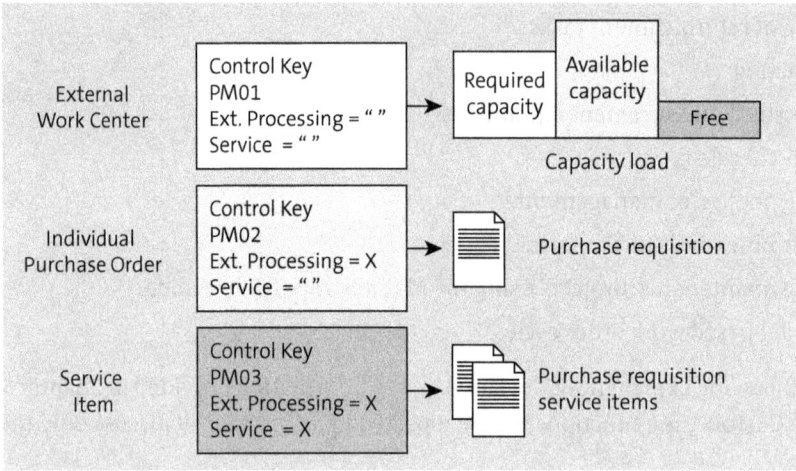

Figure 7.1 External Processing: Control Keys

- **Individual purchase order**
  If you want to execute external processing using a purchase requisition and an individual standard purchase order, you should initiate this type of external processing using a control key (PM02 or similar) for which the **Ext. Processing** option is set to

Externally processed operation and the **Service** indicator isn't set. The additional Customizing functions required for this business process and the Customizing settings that need to be changed will be discussed in Section 7.1.1.

- **Service specifications**
  If you want to execute external processing using service items or service specifications and a subsequent service entry sheet, you should initiate this type of external processing using a control key (PM03 or similar) for which the **Ext. Processing** option is set to **Externally processed operation** and the **Service** indicator is set. The additional Customizing functions required for this business process and the Customizing settings that need to be changed will be discussed in Section 7.1.2.

7.1.1 External Processing as an Individual Purchase Order

If you want to commission external processing as an individual purchase order, the process is as follows (see Figure 7.2):

1. When you plan external processing in an order, a purchase requisition is automatically triggered in the background.
2. The purchasing department (or maintenance planner) converts the purchase requisition into a purchase order.

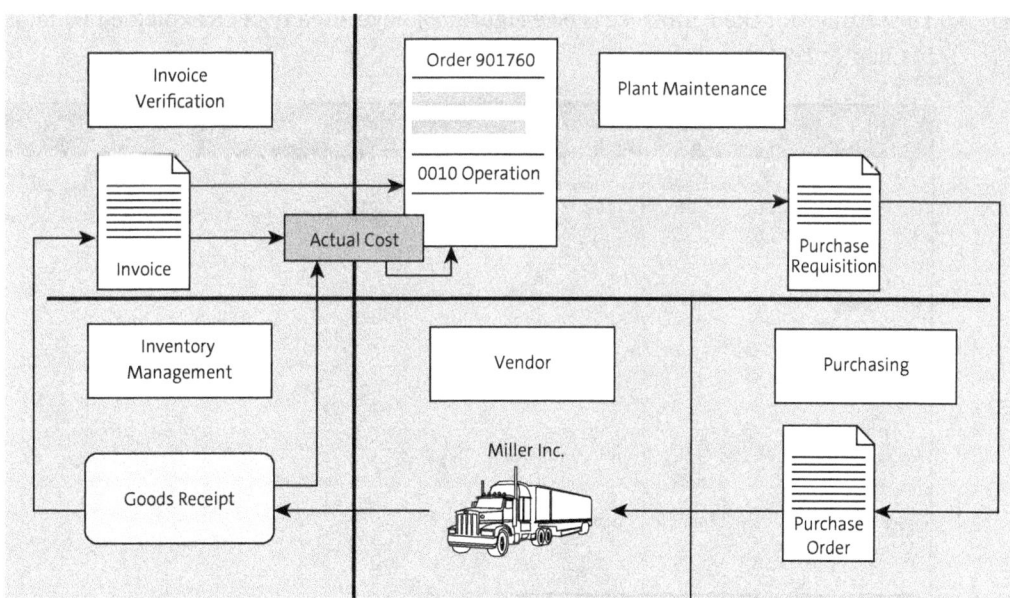

Figure 7.2 External Processing: Individual Purchase Order

3. After the external company has rendered the services, you enter them into the system. However, you don't confirm external processing in the same way as normal time confirmations. Instead, you enter a service confirmation as a goods receipt for

7 Configuring Additional Business Processes

the purchase order. If the goods receipt is valuated (an option available for the service item in the purchase order), actual costs are posted to the order at this point in time.

4. The invoice receipt concludes this process. If the invoice amount differs from the amount on the purchase order, the costs in the order are automatically corrected and the order shows the net costs from the invoice.

The settings required for this business process are described in the following sections.

Maintain Control Keys

To initiate external processing as an individual purchase order, you assign a specific control key to an operation associated with a task list (but more importantly here, to an operation associated with an order).

Prerequisites

There are no specific prerequisites.

Customizing Path

Plant Maintenance and Customer Service • Maintenance Plans, Work Centers, Task Lists, and PRTs • Work Centers • Task List Data • Maintain Control Keys

Settings

The settings for the control keys (see Figure 7.3) have already been explained in detail in Chapter 2, Section 2.4.

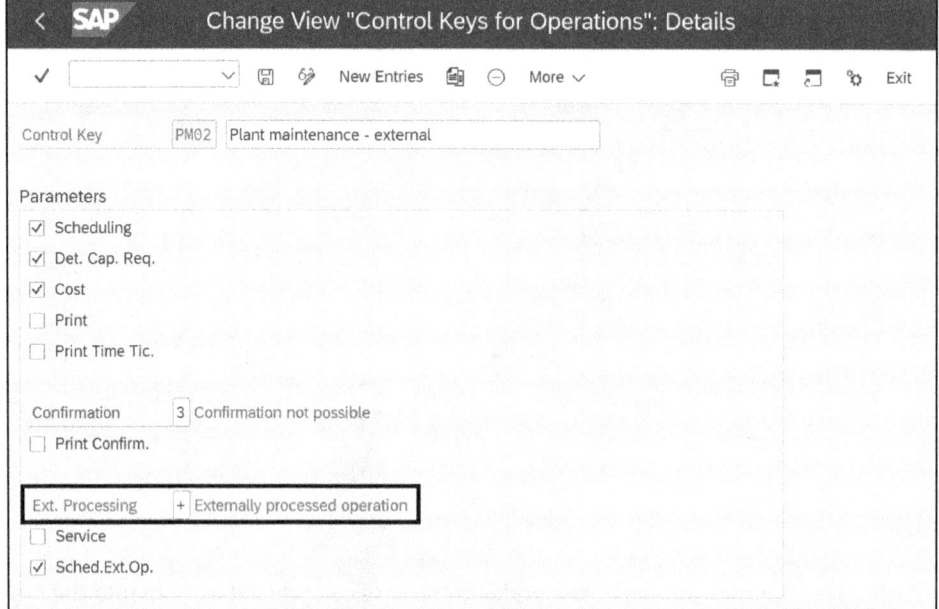

Figure 7.3 External Processing: Control Keys for Individual Purchase Orders

For external processing as an individual purchase order, configure a special control key that has the following properties:

- The **Ext. Processing** option *must* be set to **Externally processed operation**.
- The **Cost** indicator *should* be checked so that the external processing costs are included in the costing.
- The **Confirmation** option *should* be set to **Confirmation not possible** so that none of the internal employees can write their hours to the same operation.
- To prevent the system from creating a service specification, you're *not permitted* to check the box to set the **Service** indicator.
- You *can* check the box to set the **Sched.Ext.Op.** (schedule external operation) indicator if you order external processing on an hourly basis and you want to include it in order scheduling.

Create Default Value Profiles for External Procurement

You use this Customizing function to define profiles for the external procurement of services and materials. In the profile, you can define default values for generating purchase requisitions from orders, and in each profile, you can define default values for the following:

- Cost element
- Purchasing organization
- Purchasing group
- Material group

In the case of external procurement, it's always necessary to differentiate between the external procurement of materials and the external procurement of services.

Prerequisites

You must define the requisite values for the cost element, purchasing organization, purchasing group, and material group beforehand.

Customizing Path

Plant Maintenance and Customer Service • Maintenance and Service Processing • Maintenance and Service Orders • Functions and Settings for Order Types • Procurement • Create Default Value Profiles for External Procurement

Settings

Make all the entries that will be copied to the purchase requisitions at a later stage (see Figure 7.4). Then, use the **Default Values for Task List Data and Profile Assignments** Customizing function to make the relevant assignment for each order type and plant.

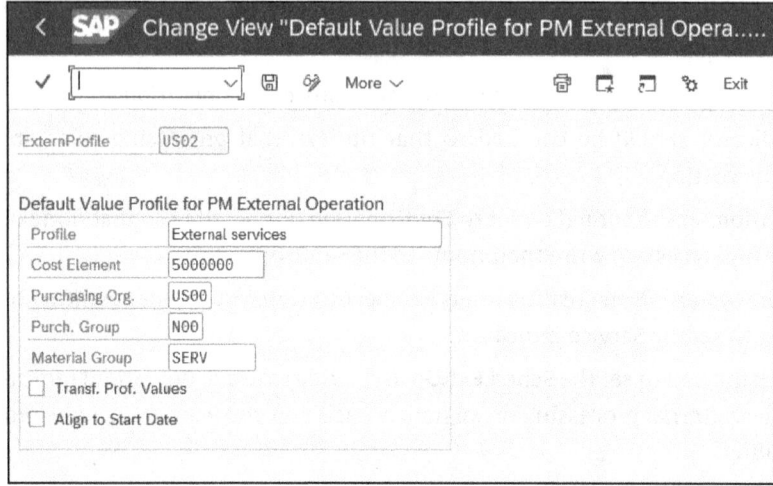

Figure 7.4 External Processing: Default Value Profile for External Operation Data

Default Values for Task List Data and Profile Assignments

You use this Customizing function to define, for each order type and plant, the profiles for external procurement as well as other default values.

Prerequisites

You must define the order types, plants, default value profiles, and external material and external processing profiles beforehand.

Customizing Path

Plant Maintenance and Customer Service • **Maintenance and Service Processing** • **Maintenance and Service Orders** • **Functions and Settings for Order Types** • **Default Values for Task List Data and Profile Assignments**

Settings

Make the assignment for each order type and plant. You can also mask the order type with the * entry (see Figure 7.5).

Use **ExternProfile** to assign the profile with default values for the cost element, purchasing organization, material group, and purchasing group for external processing.

[+] **One Profile for All Types of External Processing**
The profile for external processing applies not only to external processing with an individual purchase order but also to external processing with service specifications. The SAP system doesn't make any distinction here.

7.1 External Processing

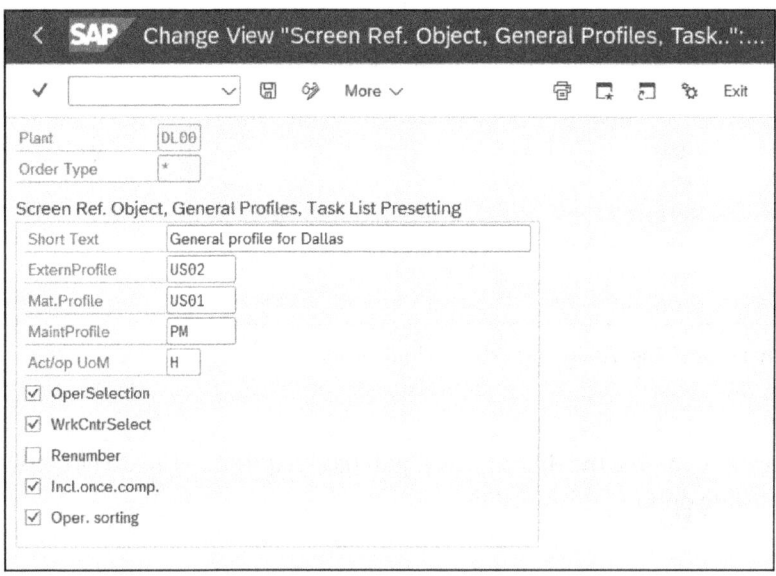

Figure 7.5 External Processing: General Profile

Define Change Documents, Collective Purchase Requisition, and MRP Relevance

You use this Customizing function to define various order parameters, including parameters that you can use to control external processing, for each order type and plant.

Prerequisites

You must define the plants and order types beforehand.

Customizing Path

Plant Maintenance and Customer Service • Maintenance and Service Processing • Maintenance and Service Orders • Functions and Settings for Order Types • Procurement • Define Collective Purchase Requisition and MRP Relevance

Settings

The table shown in Figure 7.6 contains the following parameters for controlling external processing:

- CollReqstn (collective purchase requisition)
  If you check the box to set this indicator, all the purchase requisition items that result from an order number are grouped together under a single purchase requisition number. If you don't set this indicator, the system generates a separate purchase requisition number for each purchase requisition item.

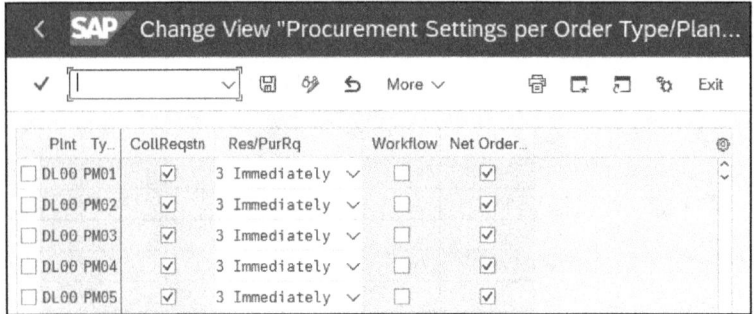

Figure 7.6 External Processing: Assigning External Indicators

- Res/PurRq
  Use this indicator to define the time at which purchase requisitions are to be generated. The following options are available:
  - 1: Never
  - 2: From Release
  - 3: Immediately
- Net Order Price
  Check the box to set this indicator to specify that the net price from the purchase requisition must be copied unchanged to the purchase order and can't be changed there.

[+] **One Setting for All Types of External Processing**
The settings for the **Collective Purchase Requisition, MRP Relevance** Customizing function apply not only to external processing with an individual purchase order but also to external processing with service specifications. Once again, the SAP system doesn't make any distinction here.

7.1.2 External Processing with Service Specifications

The *external processing with service specifications* business process differs from the *external processing with individual purchase order* business process in that the services to be rendered by the external company aren't broadly outlined in the short and long text of the purchase order item; instead, they are specifically listed individually in a service specification. This gives rise to the following differences in the process flow (see Figure 7.7):

- You use a *service specification* to execute order planning.
- Within order planning, you can define limits for planned and unplanned services.
- The document that you generate here is called a *service entry sheet*, and you execute it rather than entering a goods receipt. This can also be carried out by the vendor.

- Unlike the goods receipt, the service entry sheet enables you to add unplanned items.
- You must use a service acceptance to release the services entered (under the principle of dual control). Only then are the services integrated into the materials management process so an invoice can be issued.

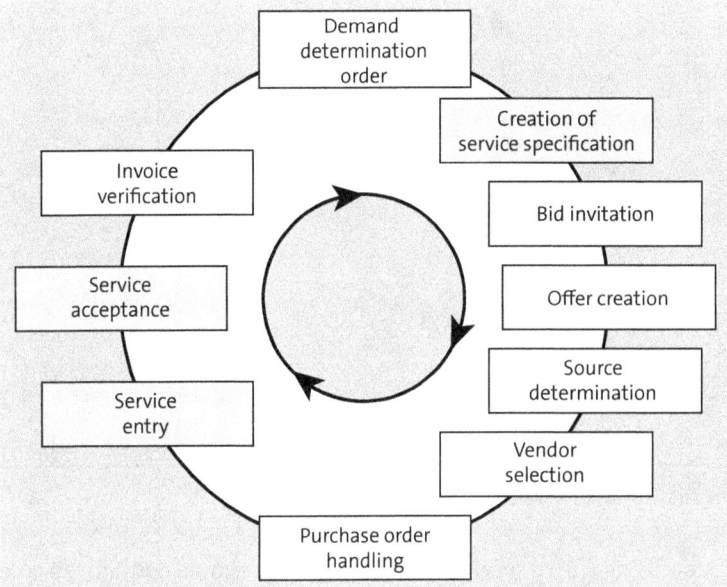

Figure 7.7 External Processing with Service Specifications

Some of the settings required for this business process are described in the following sections.

Maintain Control Keys

To initiate external processing with a service specification, you once again assign a specific control key to an operation associated with a task list or an operation associated with an order.

Prerequisites
There are no specific prerequisites.

Customizing Path
Plant Maintenance and Customer Service • Maintenance Plans, Work Centers, Task Lists, and PRTs • Work Centers • Task List Data • Maintain Control Keys

Settings
Figure 7.8 shows the settings for the control key for external processing with service specifications.

7   Configuring Additional Business Processes

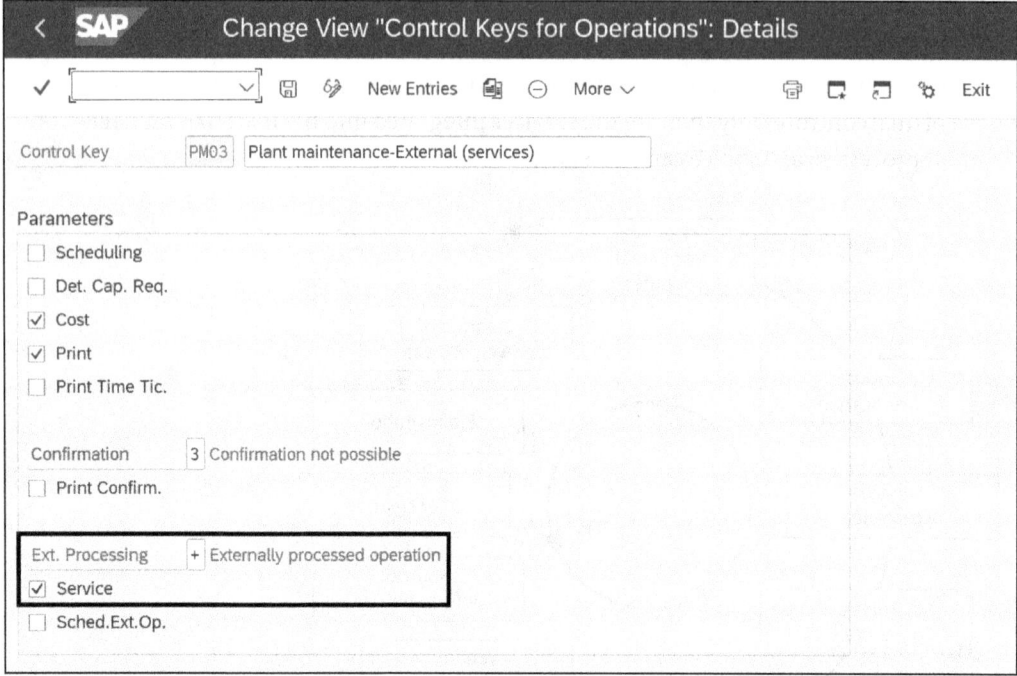

Figure 7.8 External Processing with Service Specifications: Control Keys

To initiate external processing with a service specification, you should configure a control key that contains the following indicators:

- The **Ext. Processing** indicator *must* be set to **Externally processed operation**.
- The **Service** indicator *must* be checked so that a service specification can be created.
- The **Cost** indicator *should* be checked so that the external processing costs are included in the costing.
- The **Print** indicator *should* be checked so that the service entry sheet can be printed.
- The **Confirmation** option *should* be set to **Confirmation not possible** so that none of the internal employees can write their hours to the same operation.
- You *can* check the box to set the **Sched.Ext.Op.** (schedule external operation) indicator if you've converted the services to an hourly basis and want to include them in order scheduling.

Define Service Category

The *service master* is the master record that describes a service, and it's the counterpart to the *material master*, which is a master record that describes a material. In the same way as a material type must be assigned to each material master record, a service category must be assigned to a service master record. The service category is the most important grouping criterion for service masters, and it affects the screen sequence, field selection, and number assignment, among other things. You use this Customizing function to define your service categories.

Prerequisites

You must define the account category references and the organizational statuses beforehand.

Customizing Path

Materials Management • External Services Management • Service Master • Define Service Category

Settings

The following service categories are available (see Figure 7.9):

- AS01

    Master records that describe services only with basic data

- AS02

    Master records that describe services with basic and controlling data

- AS03

    Master records that describe services with basic, controlling, and purchasing data

Figure 7.9 External Processing with Service Specifications: Service Category

Use the **Define Organizational Status for Service Categories** Customizing function to maintain the organizational status along with the assignment of the relevant functions.

Our example uses organizational status **AS03** for services to be procured externally. The service category also contains the assignment to an *account category reference* (**ARef**) that influences the automatic account determination for SAP S/4HANA Finance.

Define Attributes of System Messages

You use this Customizing function to define the manner in which a system message concerning service entry sheets is to be output (e.g., in an error message or a warning).

7 Configuring Additional Business Processes

Prerequisites

There are no specific prerequisites.

Customizing Path

Materials Management • External Services Management • Define Attributes of System Messages

Settings

Define how each message is to be output. The following options are available to you (see Figure 7.10):

- No message
- **W**: Warning
- **E**: Error message
- **S**: Result message
- **P**: Dialog box
- **I**: Information
- **X**: Program termination (a little joke among developers)

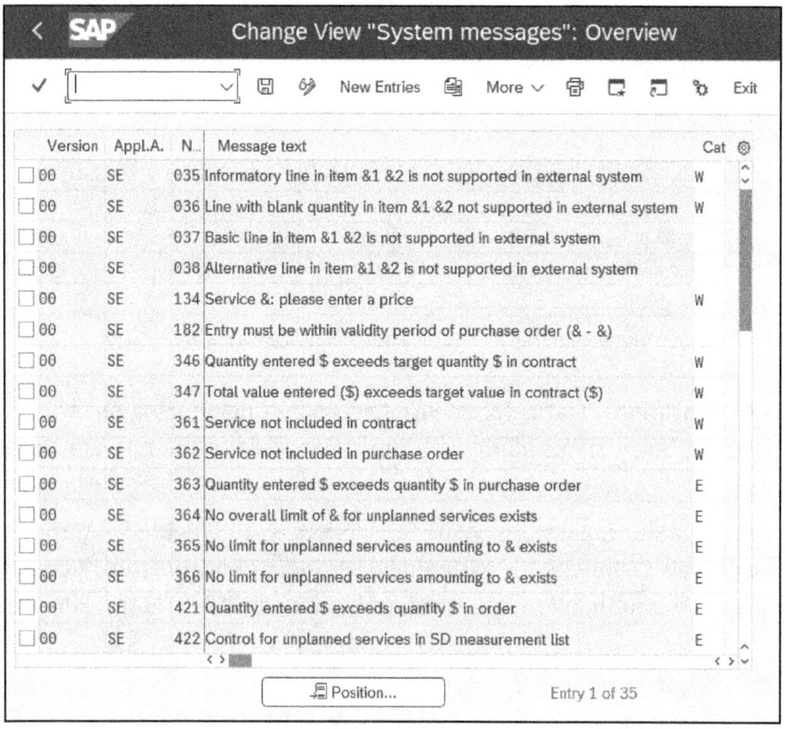

Version	Appl.A.	N...	Message text	Cat
00	SE	035	Informatory line in item &1 &2 is not supported in external system	W
00	SE	036	Line with blank quantity in item &1 &2 not supported in external system	W
00	SE	037	Basic line in item &1 &2 is not supported in external system	
00	SE	038	Alternative line in item &1 &2 is not supported in external system	
00	SE	134	Service &: please enter a price	W
00	SE	182	Entry must be within validity period of purchase order (& - &)	
00	SE	346	Quantity entered $ exceeds target quantity $ in contract	W
00	SE	347	Total value entered ($) exceeds target value in contract ($)	W
00	SE	361	Service not included in contract	W
00	SE	362	Service not included in purchase order	W
00	SE	363	Quantity entered $ exceeds quantity $ in purchase order	E
00	SE	364	No overall limit of & for unplanned services exists	E
00	SE	365	No limit for unplanned services amounting to & exists	E
00	SE	366	No limit for unplanned services amounting to & exists	E
00	SE	421	Quantity entered $ exceeds quantity $ in order	E
00	SE	422	Control for unplanned services in SD measurement list	E

Figure 7.10 External Processing with Service Specifications: Messages

Source Determination and Default Values

You can use this Customizing function to configure source determination at the client or purchasing organization level and to define the following values:

- The unit of measure of the service at the item level
- The material group of the service at the item level
- The line number increment for the service specification
- The update of conditions in the service master record with the conditions in the purchasing documents

The settings at the purchasing organization level have priority over the settings at the client level.

Prerequisites

You must maintain the purchasing organizations and the default values (e.g., the material group or unit of measure) beforehand.

Customizing Paths

There are two important Customizing paths for source determination, as follows:

- **Materials Management • External Services Management • Source Determination and Default Values • For Client**
- **Materials Management • External Services Management • Source Determination and Default Values • For the Purchasing Organizations**

Settings

The Customizing functions contain the following settings (see Figure 7.11):

- Define your default values for **Line No. Increment in Serv. Specifications**, **Default Material Group at Item Level**, and **Unit of Measure at Item Level**.
- If you check the box to set the **Automatic PO Generation for Service Requisitions** indicator, the system can automatically generate purchase orders from those purchase requisitions to which the vendor is assigned as a source of supply (e.g., in Transaction ME59N).
- The **Set Condition Update Indicator as Default in Quotation** and **Set Condition Update Indicator as Default in Purchase Order** options allow the system to update the conditions in the service master record with the service conditions in the purchasing document (i.e., the purchase order or quotation) by default. The user can accept or reject this update on a line-by-line basis.
- The settings in the **Source Determination** screen area support you in assigning contracts to the purchase order (e.g., when using the material group).

7 Configuring Additional Business Processes

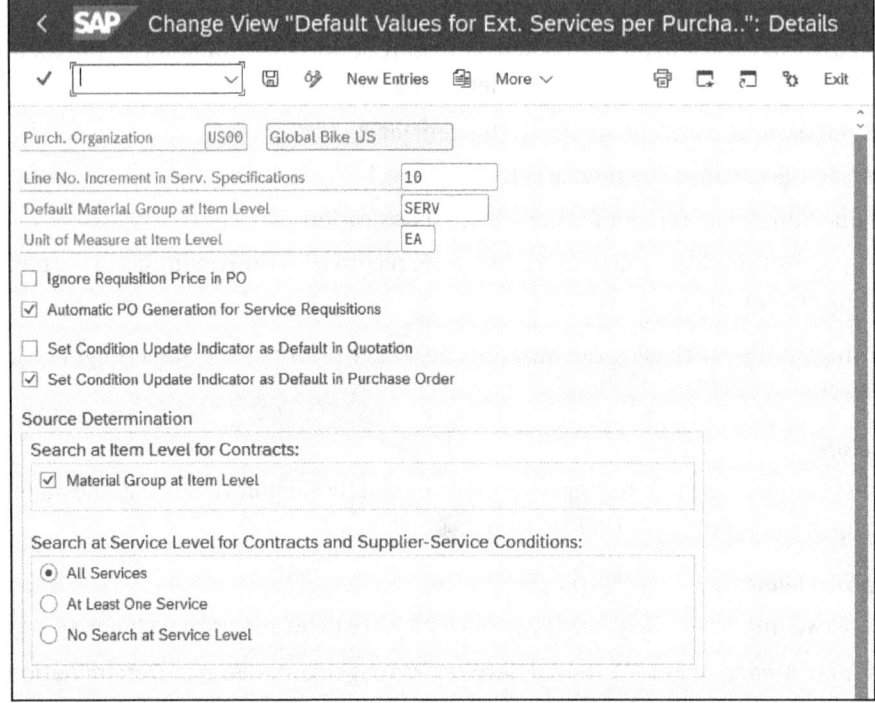

Figure 7.11 External Processing with Service Specifications: Default Values

Define Release Procedure for Service Entry Sheet

The service entry sheet is subject to a release procedure, and you use this Customizing function to define the release procedure.

Prerequisites

You must create a class with class type 032 (e.g., REL_SES [Release Strategy for Service Entry Sheets]) and assign the relevant release characteristics (e.g., purchasing organization and value).

Customizing Path

Materials Management • External Services Management • Define Release Procedure for Service Entry Sheet

Transaction

CUST_VV_T16FS_3

Settings or Recommended Settings

The **Define Release Procedure for Service Entry Sheet** Customizing function contains several subfunctions that you use as follows (see Figure 7.12):

- In the first subfunction, create a **Release Group** for object category **03 (Release Entry Sheet)**. Assign the class to this release group.
- Use the next subfunction to create your **Release Codes** for the release group. You can use these codes to define those positions involved in the release procedure (e.g., **01** = accounting release, **02** = technical release).
- Use the **Release Indicators** subfunction to define those statuses that flag a service entry sheet as released or not released (e.g., **N** = not released, **B** = **Business released**, and **T** = **Technical released**).
- The **Release Strategy** subfunction comprises several other subfunctions. You use the subfunction as follows:
  - Define which **Release Strategy** applies to which **Release Group**. Here, specify which positions (release codes) are involved in a strategy.
  - For **Release Prerequisites**, define the order in which the individual positions must be released.
  - For **Release Statuses**, define which status a service entry sheet has after certain positions have been released.

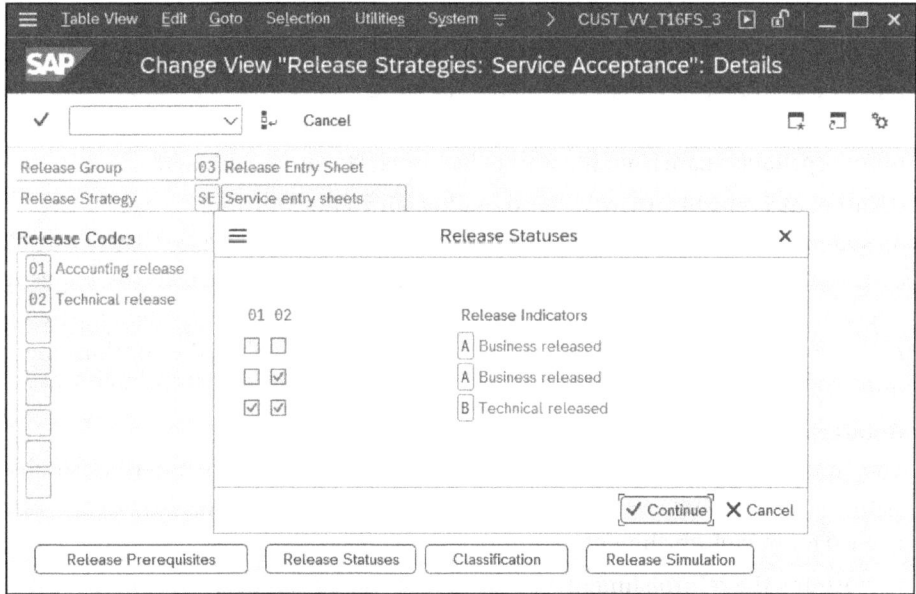

Figure 7.12 External Processing with Service Specifications: Release Strategy

The release procedure for purchasing documents isn't discussed further here. Instead, refer to SAP Note 207490, which provides a detailed description of the release procedure and the necessary Customizing settings.

7.2 Refurbishment

The *refurbishment* business process is characterized by the fact that repairable spares are held in stock (e.g., to ensure the availability of a particular asset). There are different statuses for repairable spares (e.g., **New**, **Operational Again**, **Faulty**). Faulty parts are *refurbished*—meaning they are restored to an operational status—by your own staff or external staff.

As a prerequisite, you must manage repairable spares with different accounting values in the warehouse. When you refurbish a repairable spare, it has a higher value than it would have had in a faulty status. The repairable spares are either managed as material only or as individual units (material serial numbers).

You use the following order refurbishment processes to process repairable spares (see Figure 7.13):

- **Procurement of repairable spares**
  You store repairable spares for certain critical and high-value components used in an asset so that you can replace the components immediately if a breakdown occurs.

- **Goods issue of intact repairable spares and goods receipt of faulty repairable spares**
  If a material (single unit) managed as a repairable spare is faulty in an asset, you replace it with an intact repairable spare. To do this, you remove the faulty repairable spare from the asset and return the spare to the warehouse. You then withdraw an intact repairable spare from the warehouse and install it on the asset.

- **Creation of a refurbishment order (possibly with a notification)**
  As soon as the number of faulty repairable spares in the warehouse has reached a certain amount, you create a refurbishment order. You schedule all the requisite operations, materials, tools, and so on, for the refurbishment of repairable spares. If necessary, you can create a refurbishment notification before the refurbishment order, either to trigger the refurbishment process or to ensure a complete history.

- **Goods issue from the warehouse**
  The employees responsible for the refurbishment withdraw the faulty repairable spares from the warehouse, including all other materials scheduled in the order and needed for the refurbishment.

- **Performing the refurbishment**
  You carry out the refurbishment. You can post completion confirmations for internal services, goods receipts, or service entry sheets for external material or external processing.

- **Goods receipt to warehouse**
  You use a goods receipt to return the refurbished repairable spares to the warehouse. In the case of repairable spares that can't be refurbished, you cancel the reservation and post a scrapping.

- **Scrapping**
  If faulty repairable spares can no longer be refurbished, you scrap them. Don't forget to also post a goods issue for this situation.

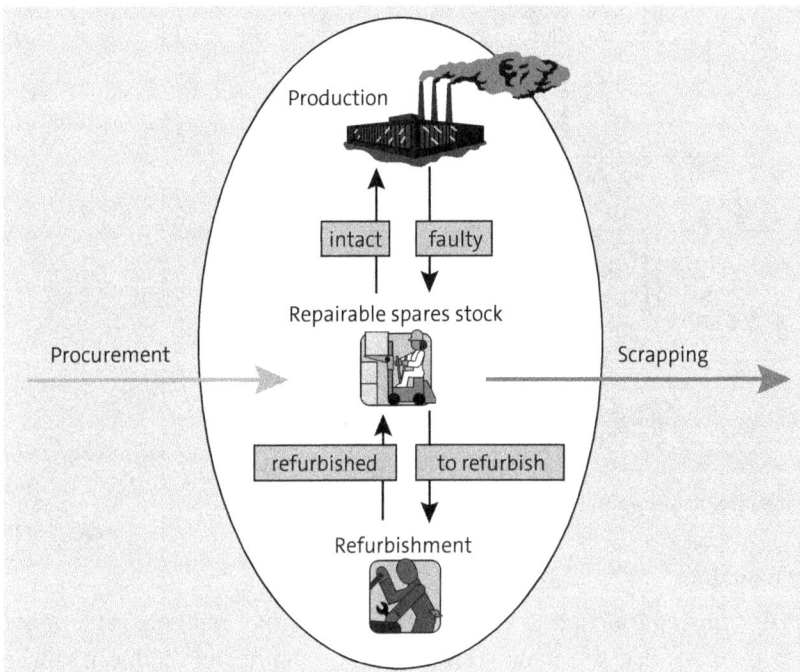

Figure 7.13 Refurbishment Process

To ensure that the process works as described here, you must make some Customizing settings.

7.2.1 Material Master

To use the material master of the repairable spare for the refurbishment, you define the *valuation category* in the accounting data at the plant level and then create several *valuation types* for the valuation category (see Figure 7.14).

In the following example, the valuation types are to be defined for valuation category C (status):

- **C1**: Like new
- **C2**: Refurbished and operational again
- **C3**: Faulty and no longer operational

To perform a split valuation for a material master, you must first activate split valuation in Customizing and then configure it (i.e., define valuation categories and valuation types).

7   Configuring Additional Business Processes

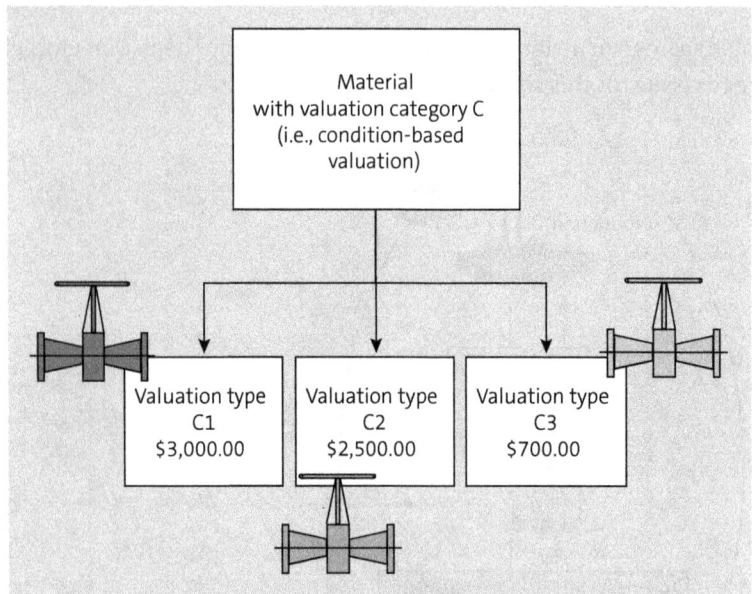

Figure 7.14 Material Master: Valuation Category and Valuation Type

Activate Split Valuation

You use this Customizing function to configure whether a split material valuation is permitted in your enterprise. If you allow split valuation in principle, it doesn't mean that you need to perform split valuation for every single material. On the contrary, when you create the material master record, you decide whether you want to perform split valuation for the material by entering a valuation category in the material master and then creating several valuation types.

Prerequisites

There are no specific prerequisites.

Customizing Path

Materials Management • Valuation and Account Assignment • Split Valuation • Activate Split Valuation

Transaction

OMWO

Settings

Activate split valuation by clicking the radio button to set the **Split material valuation active** indicator (see Figure 7.15).

7.2 Refurbishment

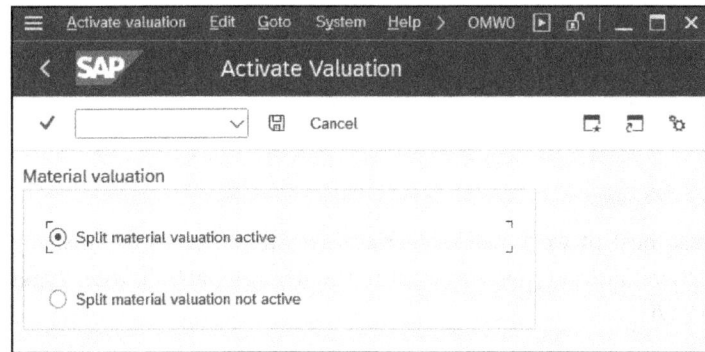

Figure 7.15 Material Master: Activate Split Valuation

Configure Split Valuation

You use this Customizing function to define split valuation in detail. Here, you define the following (see Figure 7.16):

- Valuation types you want to differentiate from one another (*global valuation types*)
- Valuation categories you want to differentiate from one another (*global valuation categories*)
- Global valuation types you want to activate for specific global valuation categories
- Global valuation categories you want to use locally (i.e., for each plant), which automatically gives rise to *local valuation types*

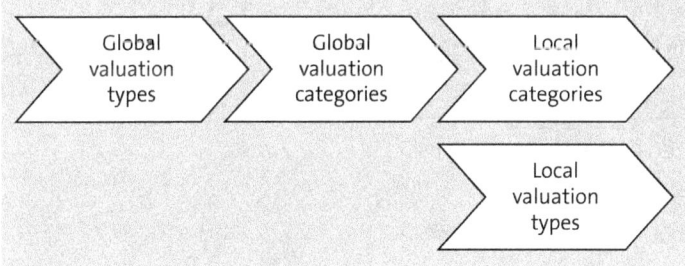

Figure 7.16 Material Master: Valuation

Prerequisites

You must define the account category references and plants beforehand. You must also activate split valuation.

Customizing Path

Materials Management • Valuation and Account Assignment • Split Valuation • Configure Split Valuation

Transaction

OMWC

451

7 Configuring Additional Business Processes

Settings

When you call the Customizing function, you obtain the following subfunctions:

- Global Types
- Global Categories
- Local Definitions

In this sequence, it's also best to work with split material valuation. To do this, you should first define which valuation types you want to use (e.g., global **Valuation Types** **C1**, **C2**, and **C3** in Figure 7.17).

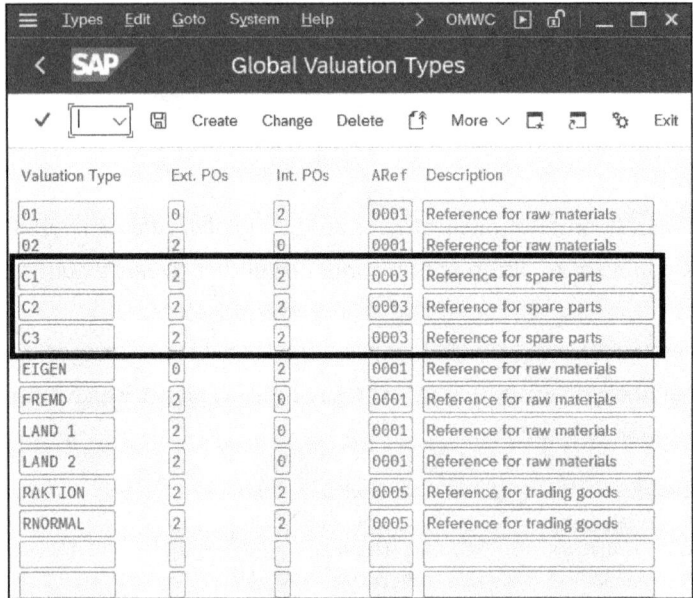

Figure 7.17 Material Master: Global Valuation Types

Define the following:

- **Ext. POs**
  The valuation type is permitted for external purchase orders.

- **Int. POs**
  The valuation type is permitted for refurbishment orders.

> **Valuation Type for Internal Purchase Orders**
>
> To enable you to use a valuation type for refurbishment orders, you must set the **Int. POs** indicator to **2** (**Allowed**).

Then, define which global valuation categories you want to use (e.g., global valuation category **C** = **Condition** in Figure 7.18).

452

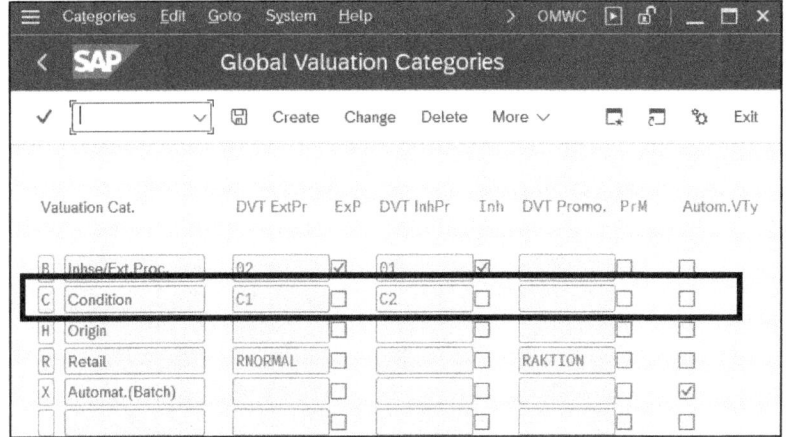

Figure 7.18 Material Master: Global Valuation Categories

Use the **Types · Cat.** function to make an assignment in relation to which global valuation types are permitted for which global valuation categories. You do this by activating the valuation types for each valuation category (see Figure 7.19).

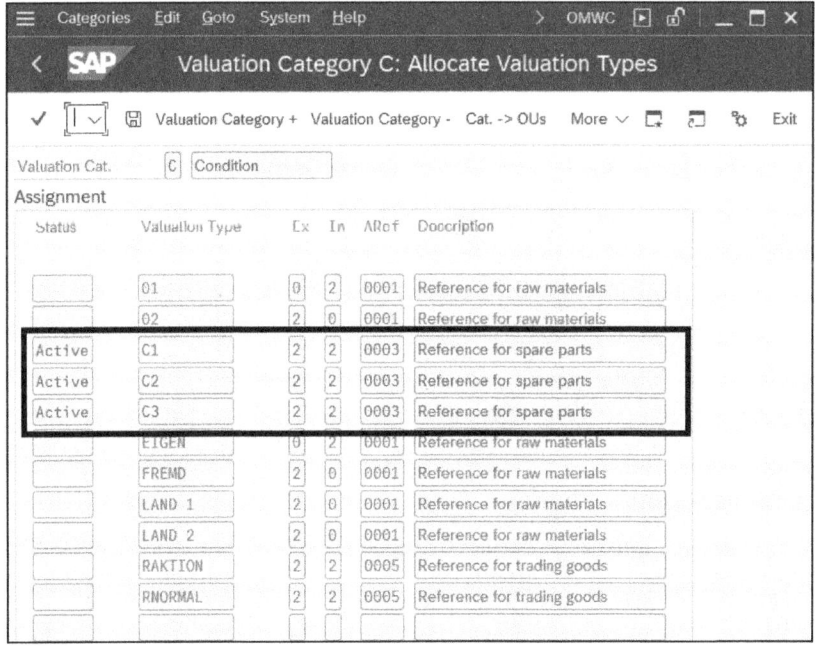

Figure 7.19 Material Master: Global Valuation Category and Global Valuation Type

Finally, use the **Local Definitions** option to define the global valuation categories as local valuation categories. To do this, activate the relevant valuation type for the plant by clicking **Cats → OU** (see Figure 7.20).

7 Configuring Additional Business Processes

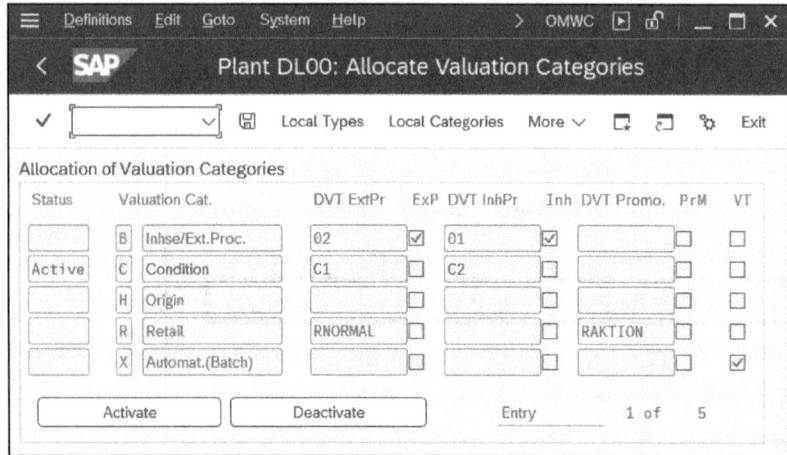

Figure 7.20 Material Master: Activate Local Valuation Categories

Activating the global valuation category for the plant and assigning the global valuation types to the global valuation categories cause the system to automatically create the local valuation types permitted for the plant (see Figure 7.21). Therefore, you don't need to activate or assign these separately.

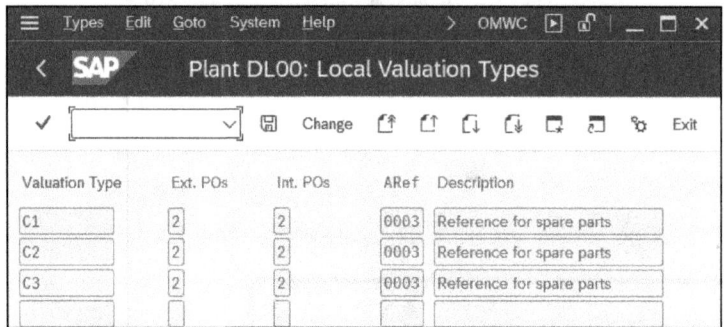

Figure 7.21 Material Master: Local Valuation Types

7.2.2 Refurbishment Notifications

You can use a *refurbishment notification* to trigger the refurbishment process. The refurbishment notification is optional, however. You use refurbishment notifications in the following scenarios:

- You attach great importance to having complete documentation in the history, so you want to use notifications and orders to document the actions of the plant maintenance department.
- You always want to use a notification as a first step and, therefore, as a trigger for your maintenance processes.

7.2 Refurbishment

If you use refurbishment notifications, we recommend that you use a separate notification type to distinguish refurbishment notifications from other maintenance notifications and to control them separately. However, the Customizing functions and settings are essentially the same as those for other notifications (see Chapter 5, Section 5.1), and the only exception is the screen type object (**ScrnType Object**).

Overview of Notification Type

Within this Customizing function, you use the **Screen Areas in Notification Header** subfunction to define the notification header and reference objects you want to use for each notification type.

Prerequisites

You must define the notification types themselves and the screen layout for these notification types beforehand.

Customizing Path

Plant Maintenance and Customer Service • Maintenance and Service Processing • Maintenance and Service Notifications • Overview of Notification Types • Subfunction Screen Areas in Notification Header

Settings

In the screen type header (**Screen Type Hdr**), always enter "H100" (**Header maintenance notification**) unless you've defined your own headers (see Figure 7.22). The **Header maintenance notification** contains information such as the notification type, notification number, short text, status, and assigned order number.

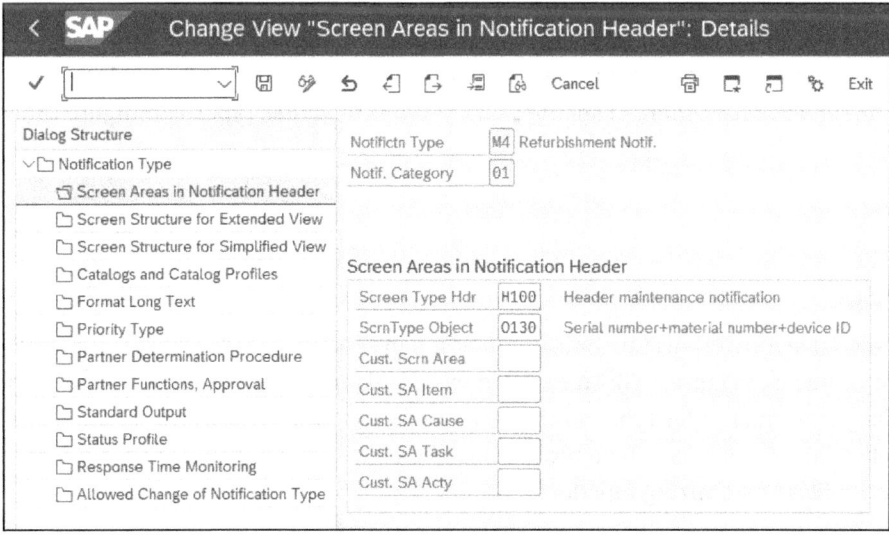

Figure 7.22 Refurbishment: Notification Type

For **ScrnType Object**, enter a setting that contains the material number because the refurbishment process always refers to a material number. Enter screen type "O130" (**serial number + material number + device ID**) for this purpose.

7.2.3 Refurbishment Orders

Next, we'll describe some additional Customizing functions as well as some special features within the Customizing settings in relation to an order type for the refurbishment process. All other Customizing functions and settings are the same as those for other orders (see Chapter 5, Section 5.2).

Configure Order Types

You use this Customizing function to define the order type and some basic parameters.

> **Separate Order Type for Refurbishment**
>
> You need a separate order type for the refurbishment process. You can't use this order type for regular maintenance processes, and you can't use a regular order type for a refurbishment process.

Prerequisites

Because order types are used by plant maintenance in SAP S/4HANA Asset Management as well as by other user departments and their corresponding areas in SAP S/4HANA, you must consult with your colleagues from the other user departments on the naming convention and number ranges for order types.

Customizing Path

Plant Maintenance and Customer Service • Maintenance and Service Processing • Maintenance and Service Orders • Functions and Settings for Order Types • Configure Order Types

Transaction

OIDA

Settings

The special feature of this Customizing function in relation to order types is the settlement profile (see Figure 7.23). This settlement profile *must* differ from the settlement profile for normal order types so that the orders can be settled based on a material number. Further details will be provided in the next section on the **Maintain Settlement Profiles** Customizing function.

7.2 Refurbishment

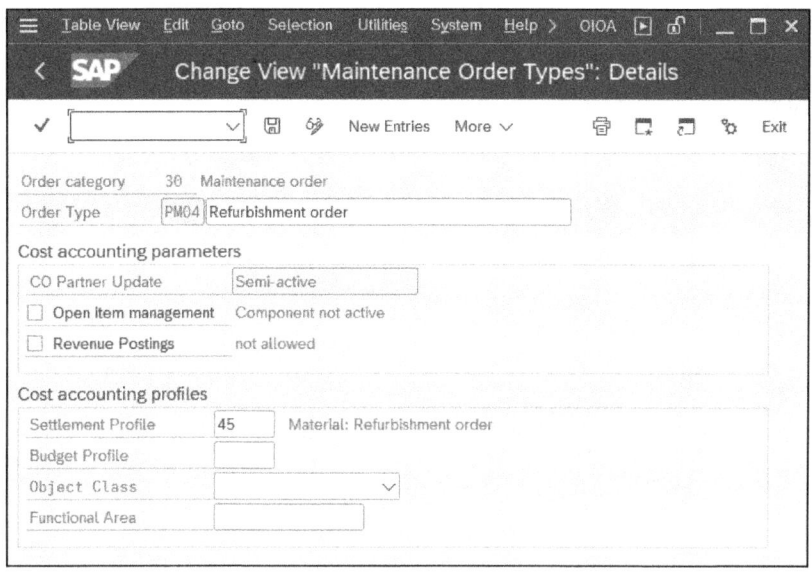

Figure 7.23 Refurbishment: Order Type

Maintain Settlement Profiles

You use this Customizing function to define the allocation structure, permitted account assignment objects, and other settlement parameters.

Prerequisites

You must define the allocation structures and document types for the settlement documents beforehand.

Customizing Path

Plant Maintenance and Customer Service • Maintenance and Service Processing • Basic Settings • General Order Settlement • Maintain Settlement Profiles

Settings

Figure 7.24 shows our recommendations for the settlement profile for refurbishment orders.

> **Configure a Separate Settlement Profile**
> Because refurbishment orders have settlement requirements that are different from those of standard maintenance orders, you need to create your own settlement profile.

In contrast to the settlement profile for normal order types, the settlement profile for refurbishment orders has two special features:

7 Configuring Additional Business Processes

- In the **Valid Receivers** section, you *must* set the account assignment object **Material** to **1 Settlement Optional**.
- You should enter "MAT" (**Material**) as the **Default Object Type**.

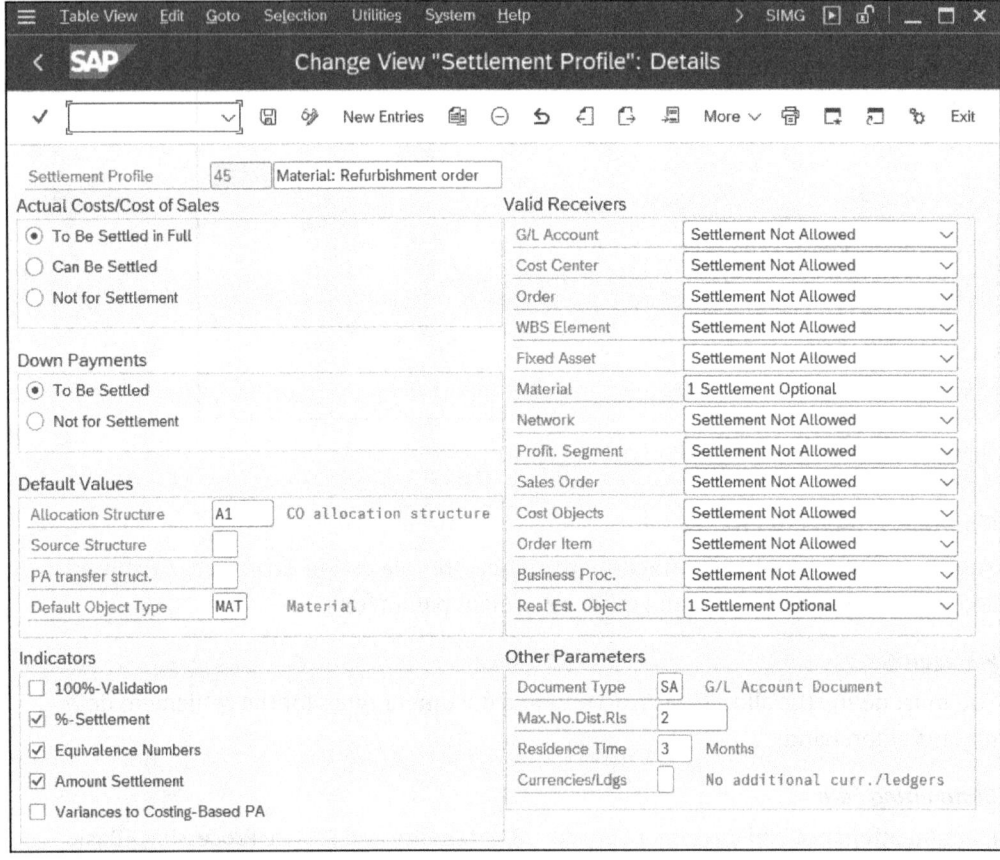

Figure 7.24 Refurbishment: Settlement Rule

Indicate Order Types for Refurbishment Processing

You use this Customizing function to identify an order type as a refurbishment order type. You're then unable to use this order type for normal maintenance processes.

Prerequisites

You must define the order type beforehand.

Customizing Path

Plant Maintenance and Customer Service • Maintenance and Service Processing • Maintenance and Service Orders • Functions and Settings for Order Types • Indicate Order Types for Refurbishment Processing

7.2 Refurbishment

Settings

Check the boxes to select the order types that you want to use for the refurbishment (see Figure 7.25).

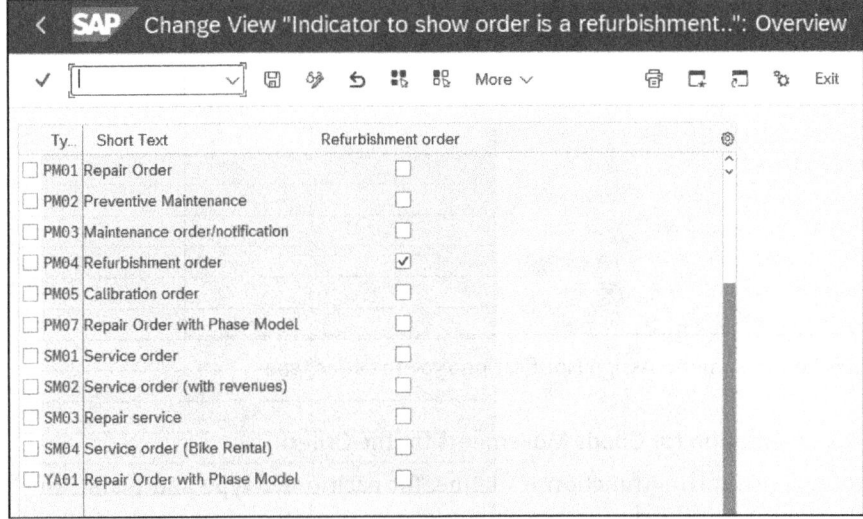

Figure 7.25 Refurbishment: Indicate Refurbishment Order Types

Indicate Order Type for Refurbishment
To use a given order type for refurbishments, you must identify the order type as a refurbishment order type. Note, however, that it then won't be possible to use the order type for regular maintenance processes.

Assign Notification Types to Order Types

If you've also configured a notification type for the refurbishment, you should use this Customizing function to specify that the order type for the refurbishment is proposed from the notification type for the refurbishment if an order is generated from the notification.

Prerequisites

You must define the notification types and order types beforehand.

Customizing Path

Plant Maintenance and Customer Service • Maintenance and Service Processing • Maintenance and Service Notifications • Notification Creation • Notification Types • Assign Notification Types to Order Types

7 Configuring Additional Business Processes

Settings

For the notification type, enter the order type that you've earmarked for the refurbishment (see Figure 7.26).

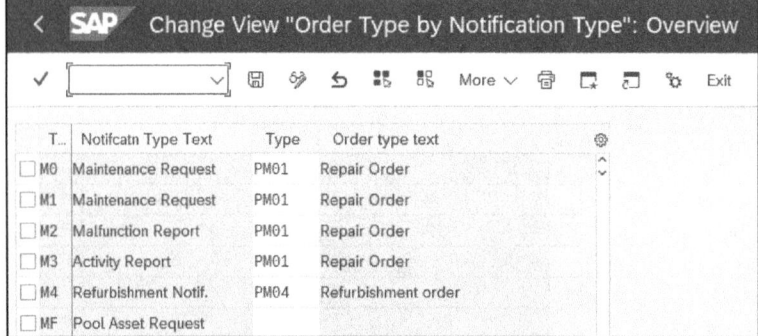

Figure 7.26 Refurbishment: Assign Notification Type to Order Type

Define Documentation for Goods Movements for the Order

You use this Customizing function to define, for each order type and plant, which material movements are to be recorded. You can then view these in the order (either in the document flow or in the documented goods movements).

Prerequisites

You must define the order type and plant beforehand.

Customizing Path

Plant Maintenance and Customer Service • Maintenance and Service Processing • Maintenance and Service Orders • Functions and Settings for Order Types • Goods Movements for Order • Define Documentation for Goods Movements for the Order

Settings

You can document the following goods movements for the maintenance orders separately (see Figure 7.27):

- **PurchOrder**
  Goods receipts for the purchase order
- **PlGoodsIss**
  Planned goods issues (i.e., goods issues for a reservation)
- **UnplGdsIss**
  Unplanned goods issues (i.e., goods issues without a reservation)
- **GR refrbshmnt**
  Goods receipts for refurbishment orders

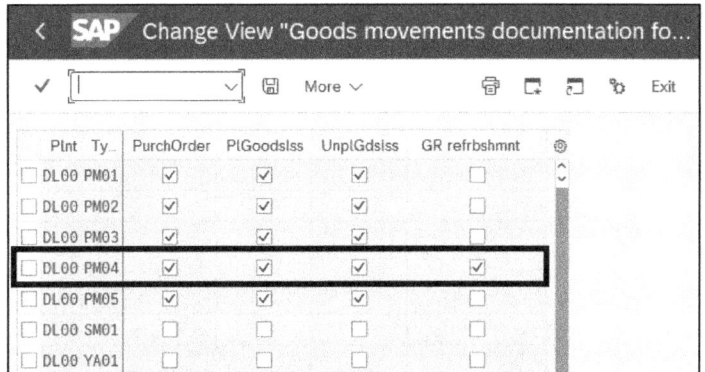

Figure 7.27 Refurbishment: Goods Movement Documentation

You can only set **GR refrbshmnt** for those order types that are identified as an order type for the refurbishment.

You also require further settings for this order type. However, they aren't discussed here because they are the same settings associated with normal order types. Instead, refer to Chapter 5, Section 5.2, for order-specific settings; Section 5.3 for completion confirmations; and Chapter 3 for generic functions.

7.2.4 Material Requirements Planning

You can now also generate refurbishment orders in MRP, specifically in the stock/requirements list (Transaction MD04).

The business background is as follows: Whenever the quantity of operational parts falls below the reorder point but nonoperational parts are in stock, MRP automatically generates planned orders. Previously, it was only possible to convert planned orders to purchase requisitions for purchasing, production orders in discrete manufacturing, or process orders in process manufacturing. Now, however, you can also convert the automatically generated planned orders to refurbishment orders to ensure that you always have the required quantity of operational parts.

> **Business Function**
>
> Before you can convert a planned order to a refurbishment order, make sure that business functions LOG_EAM_ROTSUB, LOG_EAM_ROTSUB_2, and LOG_MM_SERNO are activated within S/4H_ALWAYS_ON_FUNCTIONS.

Material Master

To convert planned orders to refurbishment orders, you must assign a spare part class code (in the field of the same name) on the **Basic Data** screen in the material master.

You must define it as a repairable spare part with a Component Maintenance Manual (CMM) (**Spare Part Class Code 2**), as shown in Figure 7.28, or as a repairable spare part without a CMM (**Spare Part Class Code 6**).

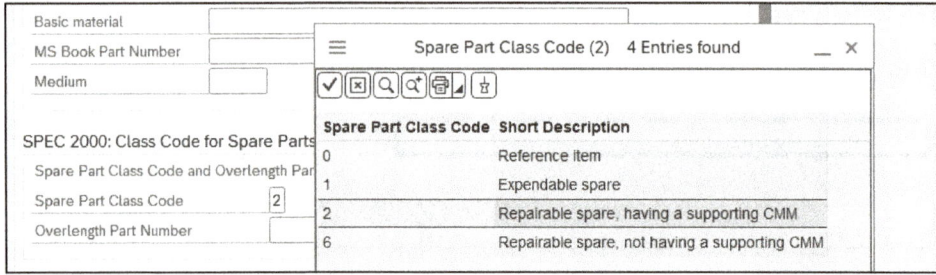

Figure 7.28 Material Master: Spare Part Class Code

If this subscreen doesn't exist, you must configure the layout of the data screens for the material master as described next.

Define Structure of Data Screens for Each Screen Sequence

You use this Customizing function to specify a screen sequence and to assign the necessary main screens and their subscreens.

Prerequisites
There are no specific prerequisites.

Customizing Path
Logistics - General • Material Master • Configuring the Material Master • Define Structure of Data Screens for Each Screen Sequence

Transaction
OMT3B

Settings
To specify a screen sequence, follow these steps:

1. Select the screen sequence that you want to use for your refurbishment materials.
2. Select the data screen on which you want to position the spare part class code.
3. Choose **Subscreens** to position the subscreen for the spare part class code there. To do this, enter the **Program** "SAPLADRT21" and **Scr.** (screen number) "2000" (see Figure 7.29).

Then, in the material requirements/stock list (Transaction MD04), you can convert a planned order into a refurbishment order.

These Customizing settings should now enable you to execute the refurbishment process.

7.3 Subcontracting

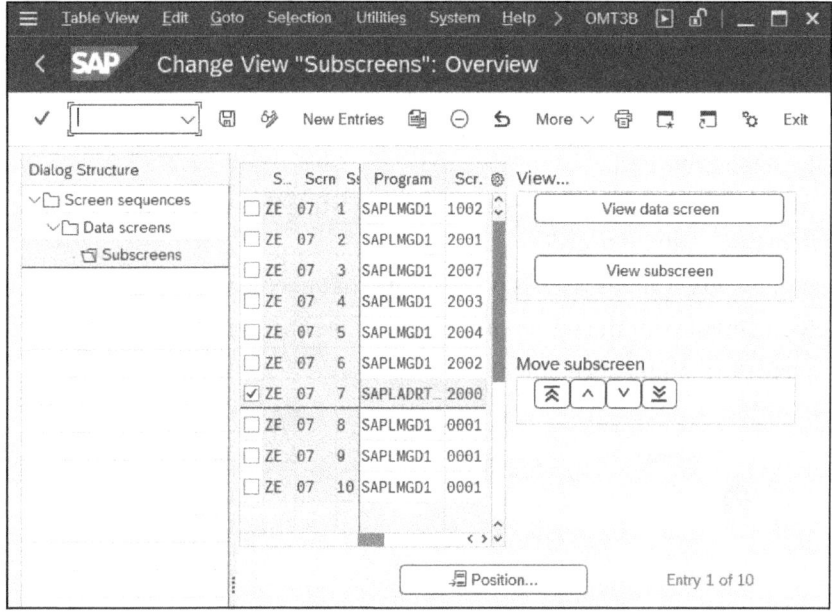

Figure 7.29 Material Master: Screen Layout for Refurbishment

7.3 Subcontracting

The *subcontracting* business process, also known as subcontracting for maintenance, repair, and overhaul (MRO), describes the process associated with having a piece of equipment (i.e., a material serial number) repaired by a service provider. In contrast to the processes used for external processing, the object to be repaired or maintained is sent to the service provider, processed there, and then returned.

The subcontracting process comprises the following steps (see Figure 7.30):

❶ You have a faulty part that requires repair and is to be processed by a service provider, so you create a maintenance order with a material serial number and a refurbishment order type. In this order, you enter a subcontracting operation (for which the **Subcontracting** indicator is set) and assign a material (for which the **Refurbishment** indicator is set) to this operation.

❷ The SAP S/4HANA system now uses the maintenance order to create a purchase requisition for the external repair or maintenance service with a subcontracting item (item category L) and the material serial number.

❸ You convert the purchase requisition to a purchase order for the external repair or maintenance service. The purchase order item is identified as a subcontracting item (item category L) and is assigned the part that is to be returned following repair (material serial number).

❹ You use an outbound delivery to send the part to be repaired to the subcontractor.

463

7   Configuring Additional Business Processes

❺ The parts provided are managed as stock of material provided to the vendor, which is also known as subcontracting stock. The provision represents a transfer posting from unrestricted-use stock to subcontracting stock.

❻ The subcontractor repairs, modifies, replaces, or exchanges the faulty part and returns the serviceable part.

❼ For the part delivered, you create a goods receipt posting that refers to the subcontracting item in the purchase order.

❽ For the components, a goods issue is posted from the subcontracting stock.

❾ You receive an invoice for the service provided by the subcontractor.

❿ You complete the maintenance order.

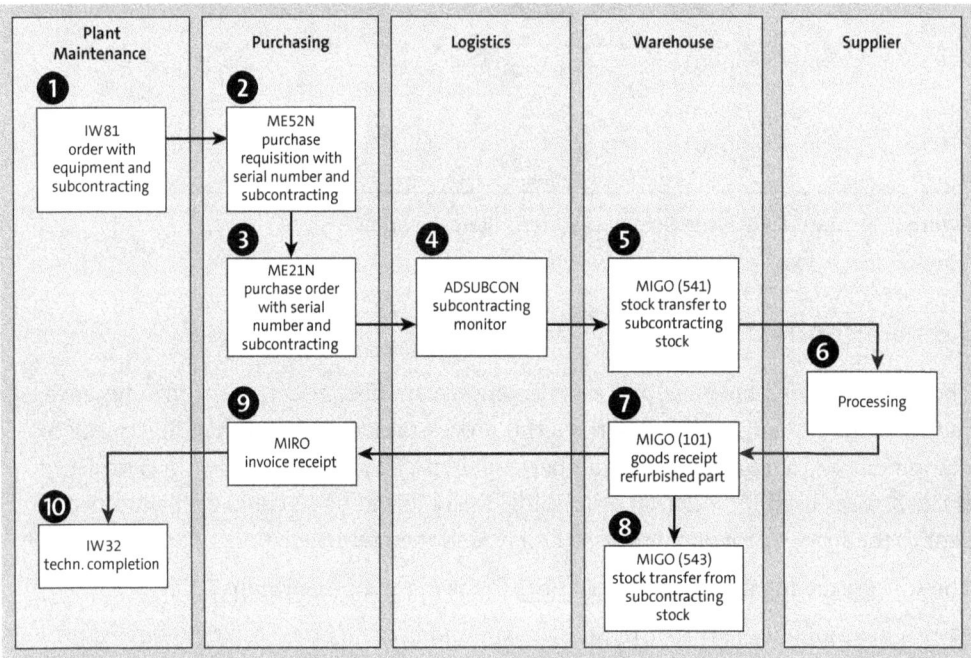

Figure 7.30 Subcontracting Process

Business Functions
Make sure that business functions LOG_EAM_ROTSUB, LOG_EAM_ROTSUB_2, and LOG_MM_SERNO are activated within S/4H_ALWAYS_ON_FUNCTIONS.

You must then configure the Customizing settings discussed in the following sections.

Define Serial Number Profiles

You use this Customizing function to define serial number profiles, and you must assign a material serial number to each piece of equipment that you want to refurbish by means of a subcontracting process. In the material master record, you must assign a serial number profile to this material at the plant level.

Prerequisites
There are no special prerequisites.

Customizing Path
Plant Maintenance and Customer Service • Master Data in Plant Maintenance and Customer Service • Technical Objects • Serial Number Management • Define Serial Number Profiles

Transaction
OIS2

Settings
You must assign at least the following three serializing procedures to each of the serial number profiles that you want to use for the subcontracting process (see Figure 7.31):

- MMSL
  You can enter a goods receipt and goods issue for the material serial number (see steps ❹, ❺, ❼, and ❽ in Figure 7.30).

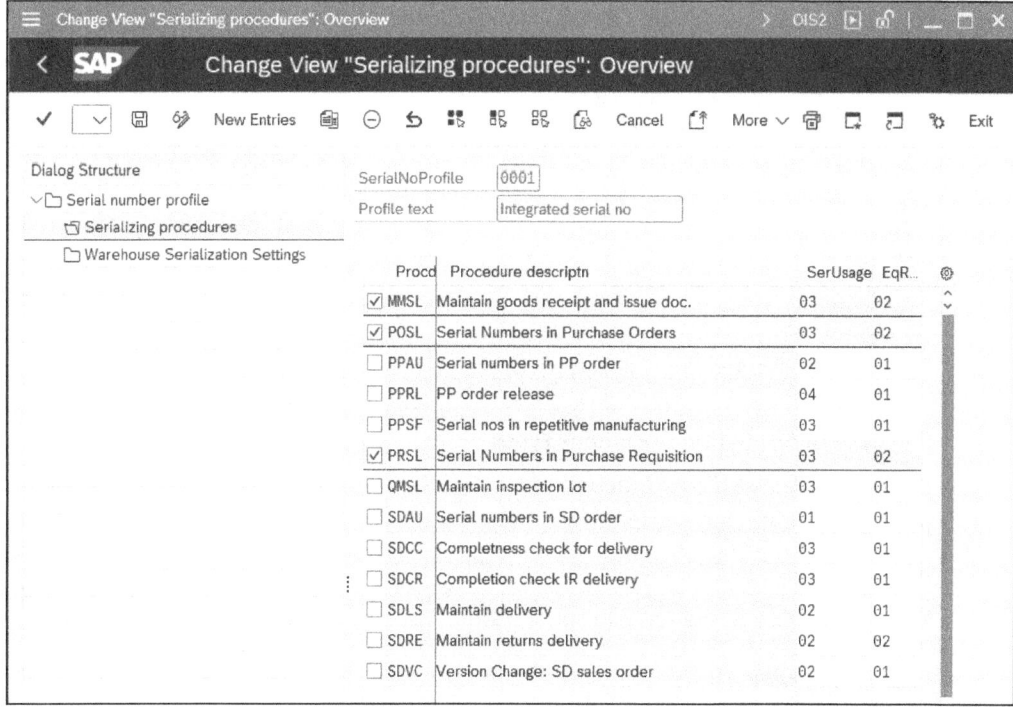

Figure 7.31 Subcontracting: Serial Number Profile

7 Configuring Additional Business Processes

- **POSL**
  You can manage serial numbers in purchase orders (see step ❸).
- **PRSL**
  You can manage serial numbers in purchase requisitions (see step ❹).

Purchase Requisition: Define Document Types

You use this Customizing function to manage the document types and item categories permitted for purchase requisitions.

Prerequisites

You must maintain the serial number profiles beforehand.

Customizing Path

Materials Management • Purchasing • Purchase Requisition • Define Document Types

Settings

For the document type that you want to use for the subcontracting process (e.g., **NB = Standard Purchase Requisition**), you should call item category **L** (**Subcontracting**) and define the serial number profile that you want to use (see Figure 7.32). You can then assign a material serial number to a purchase requisition item (see step ❷ in Figure 7.30).

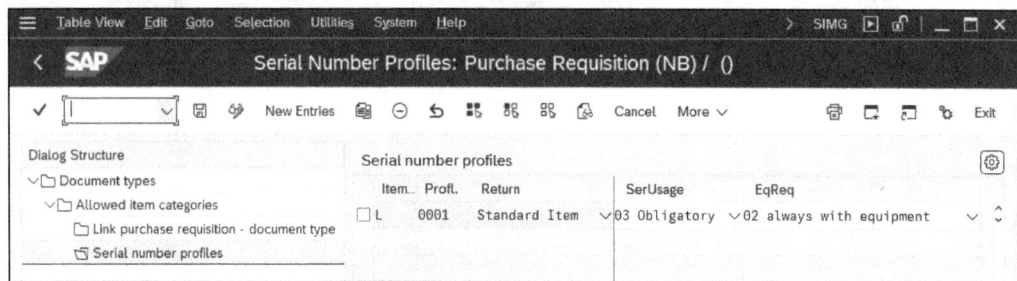

Figure 7.32 Subcontracting: Purchase Requisition Item Category

Purchase Order: Define Document Types

You use this Customizing function to manage the document types and item categories permitted for purchase orders.

Prerequisites

You must maintain the serial number profiles beforehand.

Customizing Path

Materials Management • Purchasing • Purchase Order • Define Document Types

Settings

For the document type that you want to use for the subcontracting process (e.g., **NB** = **Standard Purchase Order**), you should call item category **L** (**Subcontracting**) and define the serial number profile that you want to use (see Figure 7.33). You can then assign a material serial number to a purchase order item (see step ❸ in Figure 7.30).

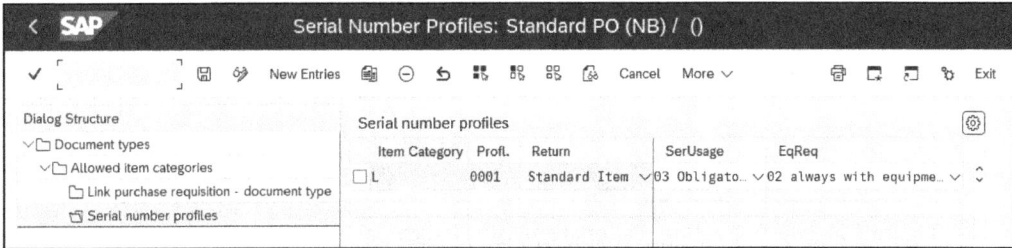

Figure 7.33 Subcontracting: Purchase Order Item Category

Define Material Provision Indicators

To provide a material for subcontracting, you must assign a material provision indicator to the item in the order. This indicator then flags the item as a refurbishment material (see step ❶ in Figure 7.30).

You use this Customizing function to define material provision indicators.

Prerequisites
There are no specific prerequisites.

Customizing Path
Plant Maintenance and Customer Service • Master Data in Plant Maintenance and Customer Service • Bills of Material • Item Data • Define Material Provision Indicators

Transaction
OICO

Settings
Define a material provision indicator for which you check the box to set the **AD MP** (aerospace and defense material provision) indicator (see Figure 7.34).

> **Subcontracting Is Available for All Customers**
> Subcontracting was originally developed for the SAP for Aerospace & Defense industry solution but was later made available for general use. It's part of S/4H_ALWAYS_ON_FUNCTIONS and is automatically activated.

These Customizing settings should now enable you to execute the subcontracting process.

7 Configuring Additional Business Processes

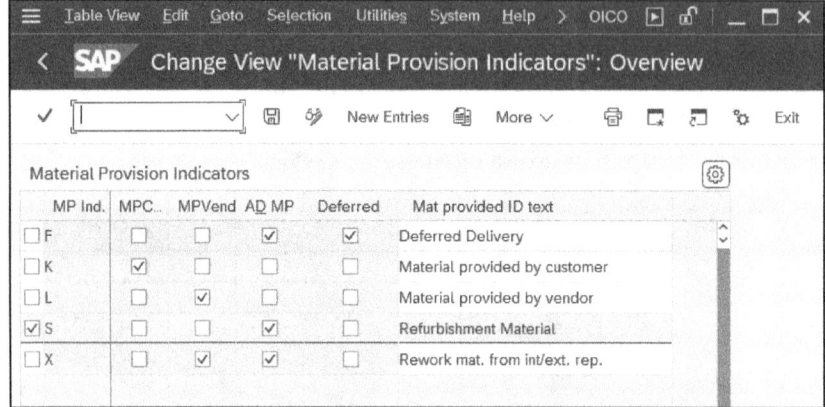

Figure 7.34 Subcontracting: Material Provision Indicator

7.4 Calibration of Test/Measurement Equipment

In many companies, test/measurement equipment—such as scales, gauges, and calipers—is used for quality inspections in the intermediate and final checking of products and equipment. To ensure that the test equipment being used always meets the specified performance criteria, it's regularly checked and calibrated. You can use the functions of test/measurement equipment management to perform the following activities:

- Manage equipment
- Plan and schedule inspections
- Execute orders and inspection lots for processing calibration inspections on equipment

Figure 7.35 provides an overview of the objects and processes associated with test/measurement equipment management:

- Test/measurement equipment is managed as equipment.
- A general task list is created for the test/measurement equipment.
- The master inspection characteristics describe the properties that must be measured (e.g., visual inspections and linear measurements). They are assigned to task list operations.
- The sampling procedures describe the type and scope of the sample.
- Equipment and a general task list are incorporated into a maintenance plan.
- The maintenance plan generates an order and an inspection lot.
- Each inspection involves processes from inspection lot management (e.g., results recording and the usage decision) and processes from order management (e.g., time confirmation and technical completion confirmation).

7.4 Calibration of Test/Measurement Equipment

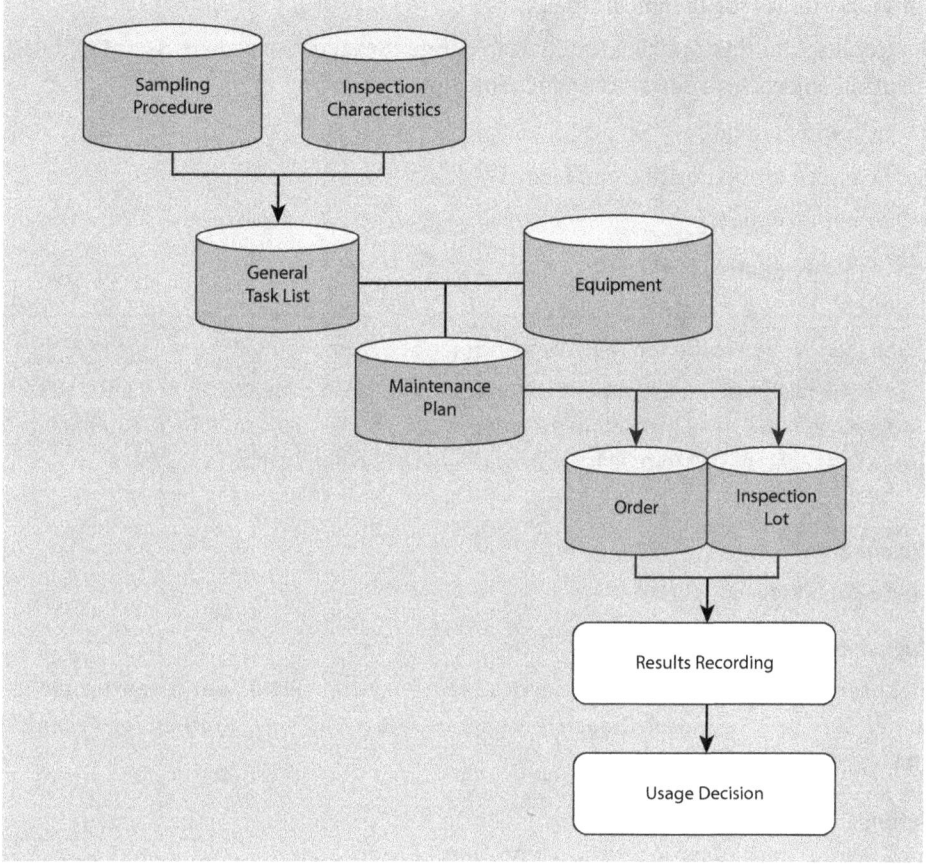

Figure 7.35 Calibration Process

In the next section, we'll cover which Customizing settings are required or recommended to implement the test/measurement equipment management process, separated into plant maintenance and quality management.

> **Plant Maintenance/Quality Management Coupling**
>
> Because Customizing functions from plant maintenance and quality management are required to calibrate test/measurement equipment, these functions are called *PM/QM coupling* as well.

7.4.1 Customizing Functions for Plant Maintenance

The Customizing functions for plant maintenance include specific settings for the following objects:

- Equipment
- Notifications
- Orders
- Maintenance plans

7 Configuring Additional Business Processes

Set View Profiles for Technical Objects

You can use this Customizing function to define the layout of screens associated with technical objects. You determine the following:

- The number of tabs
- The screen group positions on tabs
- The name of the tab
- The icons assigned to the tab

> **Separate View Profile for Test/Measurement Equipment**
>
> To distinguish test/measurement equipment from other pieces of equipment and to therefore make test/measurement equipment independent of other equipment, define a separate view profile for test/measurement equipment.

Prerequisites

There are no special prerequisites.

Customizing Path

Plant Maintenance and Customer Service • Master Data in Plant Maintenance and Customer Service • Technical Objects • General Data • Set View Profiles for Technical Objects

Settings

First, define a name for the view profile and make it valid for equipment (see Figure 7.36).

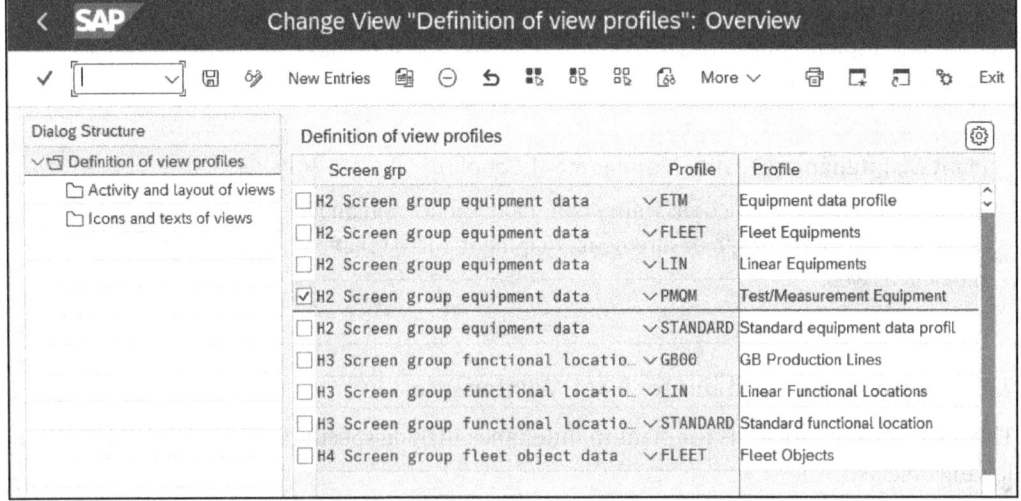

Figure 7.36 Test/Measurement Equipment: View Profile Overview

470

Then, use the **Activity and layout of views** subfunction to assign the relevant screen areas to the tabs (see Figure 7.37).

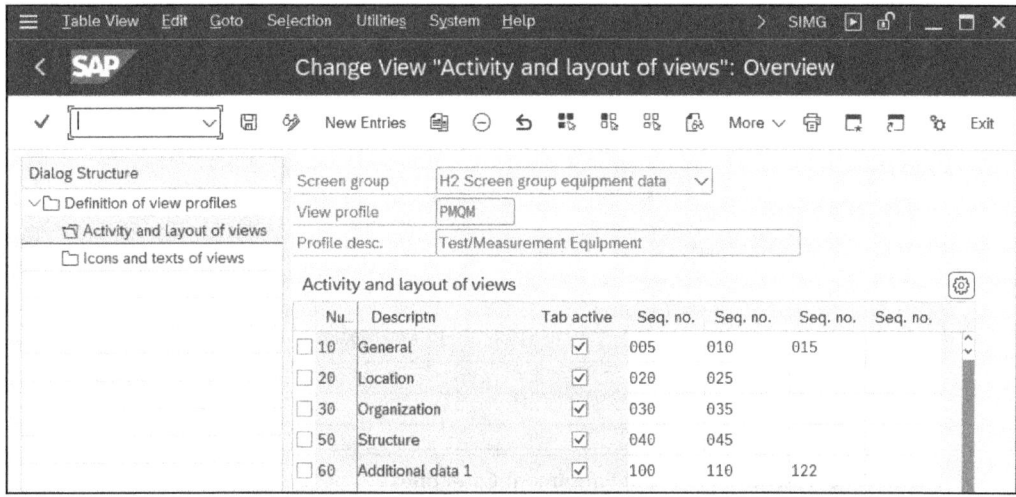

Figure 7.37 Test/Measurement Equipment: View Profile Detail

Finally, you can use the **Icons and texts of views** subfunction to define your own titles for the tabs and to assign an icon.

Maintain Equipment Category

You use this Customizing function to define equipment categories. This enables you to control which basic properties you want a corresponding equipment master record to have. You can use control properties to define the number of equipment categories you require.

> **Separate Equipment Category for Test/Measurement Equipment**
>
> To ensure that you can execute the business process for calibrating test/measurement equipment, you require a separate equipment category, which you must assign to the **Production Resource/Tool Reference Category**.

Prerequisites

If you want to assign object information keys and a view profile to the equipment categories, you must define them beforehand.

Customizing Path

Plant Maintenance and Customer Service • Master Data in Plant Maintenance and Customer Service • Technical Objects • Equipment • Equipment Categories • Maintain Equipment Category

Settings

Figure 7.38 shows an equipment category for test/measurement equipment management.

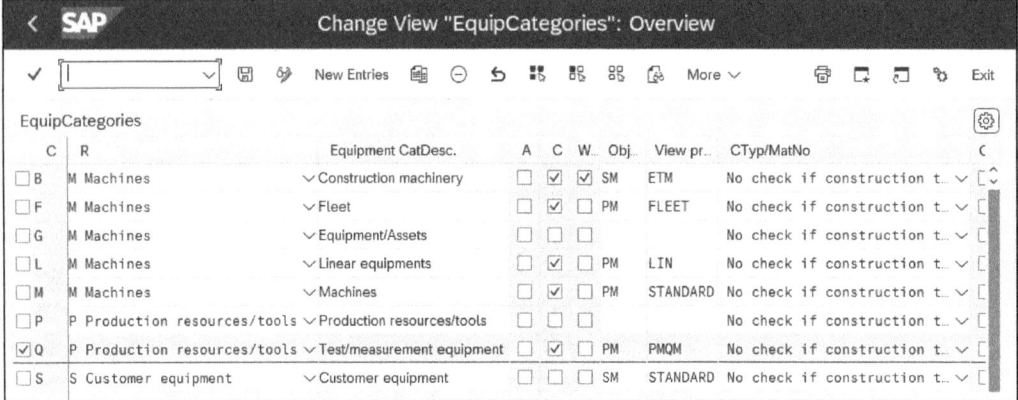

Figure 7.38 Test/Measurement Equipment Categories

From a test/measurement management perspective, you should make the following settings in the equipment category:

- Assign a reference category (in the **R** column), which is used to define history-relevant fields in an equipment master record. The reference category must be **P Production resources/tools** for equipment categories associated with test/measurement management.
- Assign the **View profile** that you've defined for test/measurement equipment.

Define Additional Business Views for Equipment Categories

In this work step, you can define additional views for each equipment category. The views represent additional tabs within the piece of equipment.

Prerequisites

You must define the equipment categories beforehand.

Customizing Path

Plant Maintenance and Customer Service • Master Data in Plant Maintenance and Customer Service • Technical Objects • Equipment • Equipment Categories • Define Additional Business Views for Equipment Categories

Settings

Figure 7.39 shows you which additional views you can activate.

To perform the business process for calibrating test/measurement equipment, you must set the **Production Resources/Tools** view to **Selection** because this business process doesn't work for equipment categories for which the **Production Resources/Tools** view is **Blank**.

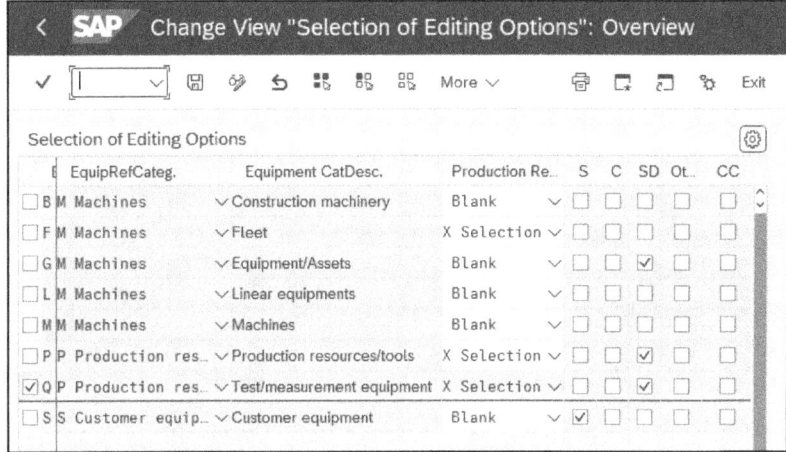

Figure 7.39 Test/Measurement Equipment: Additional Views

If you want to place your test/measurement equipment in storage, it must also have a material number and a serial number. You must therefore activate the **SD** (serial data) view, in which case, you still need to assign a serial number profile to the material number.

Define Serial Number Profiles

You use this Customizing function to define serial number profiles. In the material master, you must assign a serial number profile at the plant level to each *material* that you want to serialize. This means that for each plant, you can assign a separate serial number profile to a material. A material may also require serial numbers in one plant but not in another.

Prerequisites
There are no special prerequisites.

Customizing Path
Plant Maintenance and Customer Service • Master Data in Plant Maintenance and Customer Service • Technical Objects • Serial Number Management • Define Serial Number Profiles

Transaction
Transaction OIS2

Settings
On the overview screen, define a profile name and a profile description (see Figure 7.40).

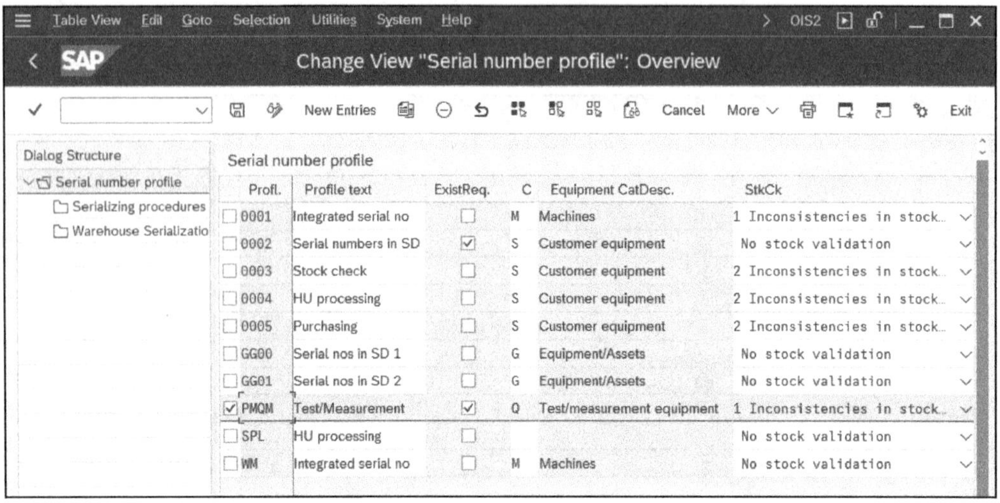

7   Configuring Additional Business Processes

![Serial number profile overview screen]

Figure 7.40 Test/Measurement Equipment: Serial Number Profile Overview

If, for your test/measurement equipment, you want the system to automatically create serial numbers for a business transaction such as the goods receipt, you should specify the equipment category that you've earmarked for your test/measurement equipment in the **C** (category) column. Then, you should define the **Serializing procedures** that are permitted for each of the serial number profiles (see Figure 7.41) on the detail screen.

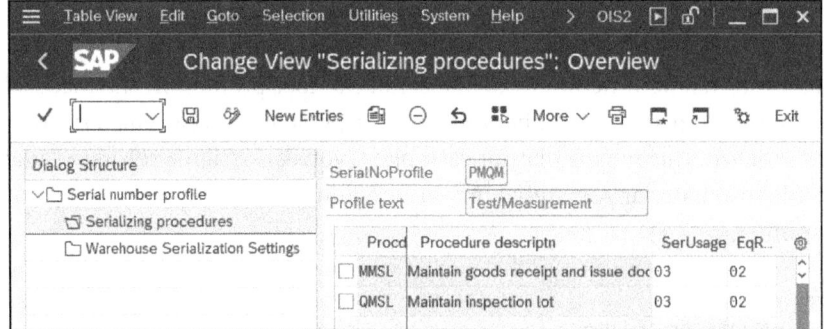

Figure 7.41 Test/Measurement Equipment: Serial Number Profile Details

You must assign the following serializing procedures to an equipment category that you want to use for the business process for calibrating test/measurement equipment:

- **MMSL: Maintain goods receipt and issue doc.**
- **QMSL: Maintain inspection lot**

You also require other settings for this equipment category. However, they are the same as the settings associated with other equipment categories, so we won't discuss them here. Instead, refer to Chapter 4, Section 4.3, for information about equipment-specific settings, and refer to Chapter 3 for information about generic functions.

Configure Order Types

You use this Customizing function to define the order type and some basic parameters.

> **Separate Order Type for Test/Measurement Equipment Management**
> You require a separate order type for the business process for calibrating test/measurement equipment. You can't use this order type for normal maintenance processes, and you also can't use a normal order type for test/measurement equipment management.

Prerequisites

Because order types are by plant maintenance as well as by other user departments and their areas in SAP S/4HANA, consult with your colleagues from the other user departments on the naming convention and number ranges for order types.

Customizing Path

Plant Maintenance and Customer Service • Maintenance and Service Processing • Maintenance and Service Orders • Functions and Settings for Order Types • Configure Order Types

Transaction

OIDA

Settings

When you use this Customizing function, make sure that you use the correct reference object. If you manage your test/measurement equipment as equipment, use **Screen RefObject O160** (equipment). If, however, you also manage the equipment as material and serial numbers, you should choose **Screen RefObject O170** (equipment, material, and serial number; see Figure 7.42).

You also require further settings for this order type. However, they are the same as the settings associated with normal order types, so they aren't covered here. Instead, refer to Chapter 5, Section 5.2, for order-specific settings; Section 5.3 for completion confirmations; and Chapter 3 for generic functions.

> **Calibration Notification Is Optional**
> You don't necessarily require a notification for the business process for calibrating test/measurement equipment, but you can use one.

If you want to use a notification with the business process for calibrating test/measurement equipment, note the following special features.

7   Configuring Additional Business Processes

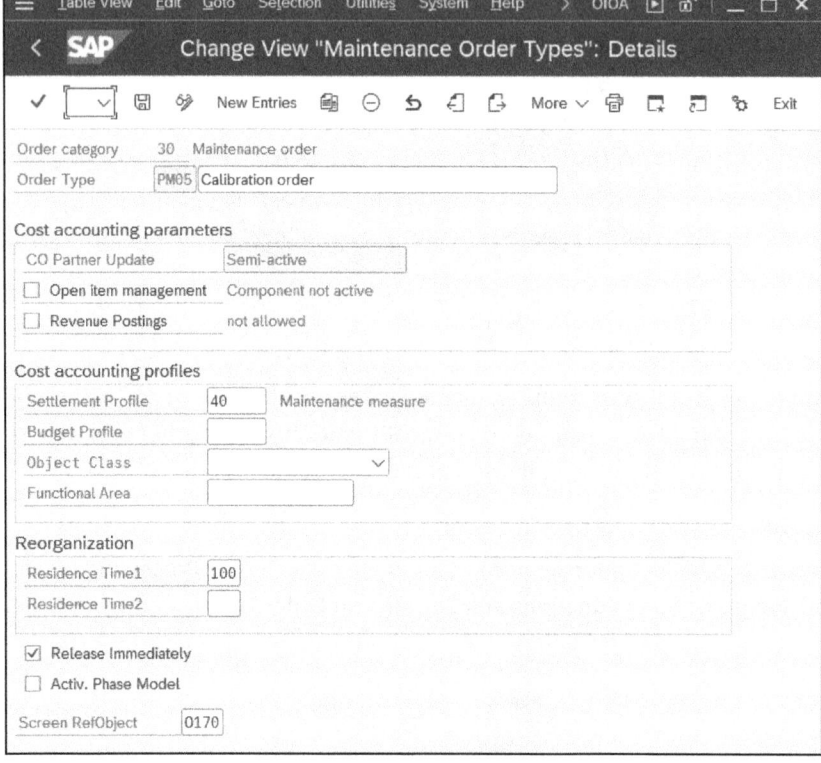

Figure 7.42 Test/Measurement Equipment: Order Type

Define Notification Types

You use this Customizing function to define the notification types that you require. The functions described previously should be taken into consideration when defining the number of notification types you need, as well as which particular notification types you want to use.

> **Separate Notification Type for Calibration**
>
> If you decide to use a notification with the business process for calibrating test/measurement equipment, you should define a separate notification type to distinguish it from other notification types and to control it independently.

Prerequisites
You must define the catalog profile beforehand.

Customizing Path
Plant Maintenance and Customer Service • Maintenance and Service Processing • Maintenance and Service Notifications • Notification Creation • Notification Types • Define Notification Types

7.4 Calibration of Test/Measurement Equipment

Settings

A special feature of this notification type is the assignment of a **Catalog Profile**, which is tailored to the interests of the business process for calibrating test/measurement equipment (see Figure 7.43).

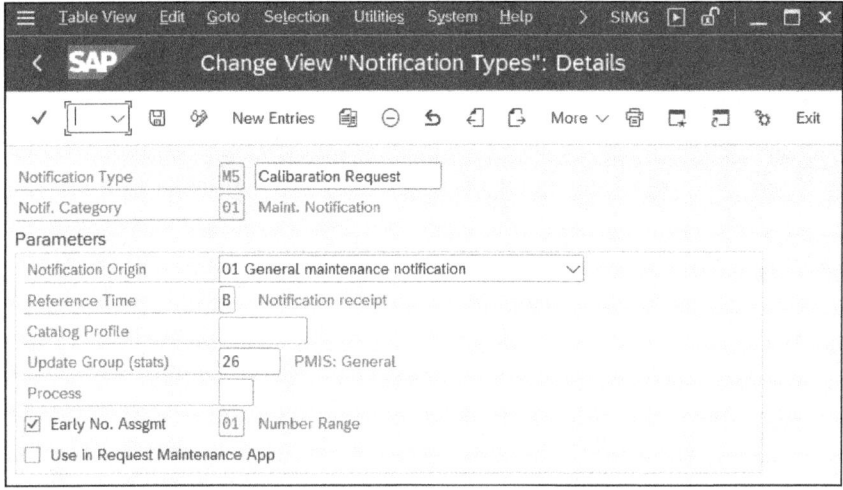

Figure 7.43 Test/Measurement Equipment: Notification Type

Maintain Catalogs

You use this Customizing function to maintain the catalogs, code groups, and codes that you want to use.

> **Catalog 9 for Calibration Notification**
> When you configure codes and code groups for test/measurement equipment management, I recommend that you use **Catalog 9** (**Defect Types**) because this can be addressed directly in Customizing for quality management (see Figure 7.44).

Prerequisites

There are no special prerequisites. However, you must consult with your colleagues from quality management, claims management, and customer service on the naming convention for catalogs because they also use catalogs.

Customizing Path

Plant Maintenance and Customer Service • Maintenance and Service Processing • Maintenance and Service Notifications • Notification Creation • Notification Content • Maintain Catalogs

Transaction

QS41

Settings

In the catalog for **Defect Types**, configure a code group and codes that you can use to document exceeding upper limits and falling short of lower limits for target values (see Figure 7.44).

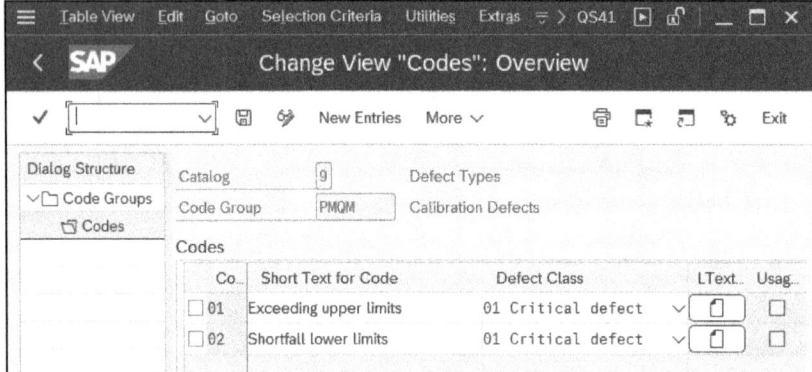

Figure 7.44 Test/Measurement Equipment: Defect Types

Define Catalog Profile

To call the defect codes in the calibration notification in a useful manner, I recommend that you configure a catalog profile.

Prerequisites

You must define the catalog types, code groups, and codes beforehand. If you also want to assign a class, you must use Transaction CT04 to define the necessary characteristics and then Transaction CL02 to create a class of class type 015.

Customizing Path

Plant Maintenance and Customer Service • Maintenance and Service Processing • Maintenance and Service Notifications • Notification Creation • Notification Content • Define Catalog Profile

Transaction

OQN6

Settings

Figure 7.45 shows the assignment of a code group from **Catalog** type **9** to a **Catalog Profile**.

7.4 Calibration of Test/Measurement Equipment

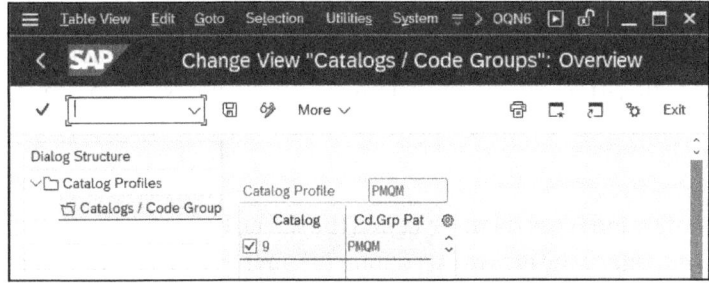

Figure 7.45 Test/Measurement Equipment: Catalog Profile

Change Catalogs and Catalog Profiles for Notification Type

You use this Customizing function to assign your chosen catalogs to your notification type for calibrating test/measurement equipment.

Prerequisites

You must define the catalog types, catalog profiles, and notification types beforehand.

Customizing Path

Plant Maintenance and Customer Service • Maintenance and Service Processing • Maintenance and Service Notifications • Notification Creation • Notification Content • Change Catalogs and Catalog Profile for Notification Type

Settings

Assign the catalog profile and enter catalog "9" in the **Problems** category (see Figure 7.46).

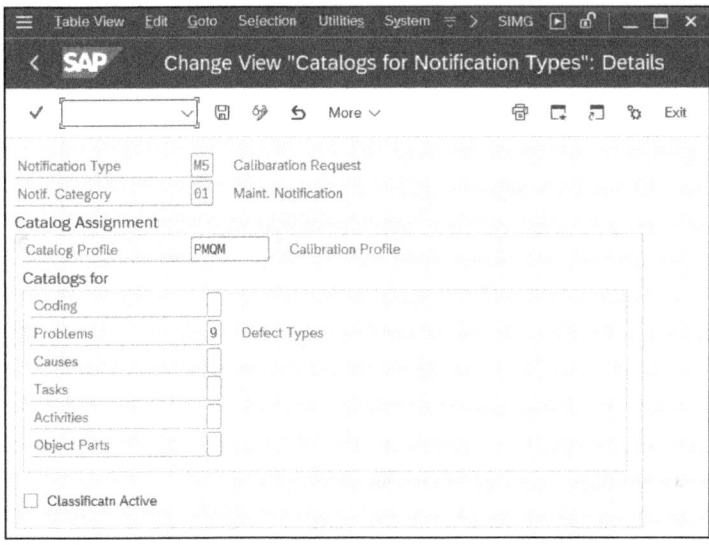

Figure 7.46 Test/Measurement Equipment: Catalog Profile and Notification Type

479

7 Configuring Additional Business Processes

Overview of Notification Type

Within this Customizing function, you use the **Screen Areas in Notification Header** subfunction to define the notification header and reference objects you want to use for each notification type.

Prerequisites

You must define the notification types themselves and the screen layout for these notification types beforehand. One of the tabs must contain the **Subscreen 005 = Reference Object** entry.

Customizing Path

Plant Maintenance and Customer Service • Maintenance and Service Processing • Maintenance and Service Notifications • Overview of Notification Types • Subfunction Screen Areas in Notification Header

Settings

In **Screen Type Hdr** (screen type header), always enter "H100" (**Header maintenance notification**) unless you've defined your own headers. The **Header maintenance notification** contains information such as the notification type, notification number, short text, status, and assigned order number.

In **ScrnType Object**, assign either "O150" (**Equipment Only**) or "O170" (**Equipment + Serial No. + Material No.**), as shown in Figure 7.47.

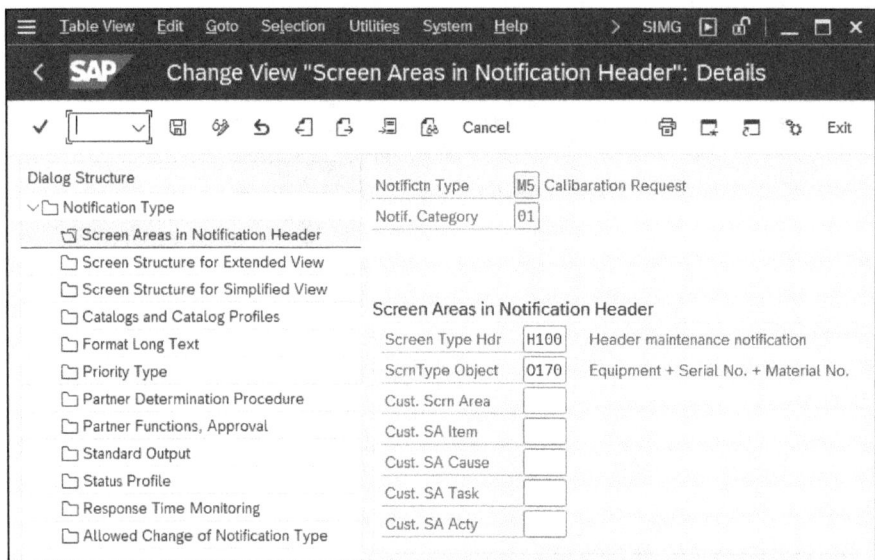

Figure 7.47 Test/Measurement Equipment: Notification Screen Areas

7.4 Calibration of Test/Measurement Equipment

Define Notification and Order Integration

Generally, you don't manually create orders for calibrating test/measurement equipment. Instead, the orders are automatically generated as a call in relation to maintenance plans. If you also want to use a notification for an order, it makes sense for the system to automatically create the notification for the order. You use this Customizing function to achieve this.

Prerequisites
You must define the order types and notification types beforehand.

Customizing Path
Plant Maintenance and Customer Service • Maintenance and Service Processing • Maintenance and Service Orders • Functions and Settings for Order Types • Define Notification and Order Integration

Settings
To enable the system to automatically create a calibration notification for a calibration order, check the box to set the indicator in the **Notifctn** column and enter the notification type that you want to create in the **Notif.Type** column (see Figure 7.48).

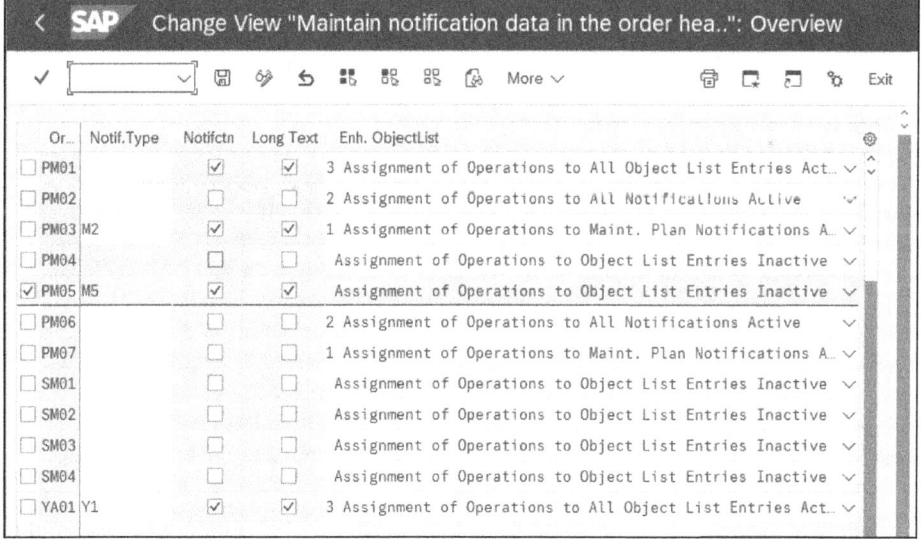

Figure 7.48 Test/Measurement Equipment: Order and Notification Integration

You also require further settings for this notification type, but they aren't discussed here because they are the same as the settings associated with normal notification types. Instead, refer to Chapter 5, Section 5.1, for information about notification-specific settings and refer to Chapter 3 for information about generic functions.

Set Maintenance Plan Categories

You use this Customizing function to define the control parameters for the maintenance plans (the maintenance call object or reference object, among others).

> **Separate Maintenance Plan Category for Calibration**
> You should define a separate maintenance plan category for calibrating test/measurement equipment to distinguish it from other maintenance plans and to control it independently.

Prerequisites

There are no special prerequisites.

Customizing Path

Plant Maintenance and Customer Service • Maintenance Plans, Work Centers, Task Lists, and PRTs • Maintenance Plans • Set Maintenance Plan Categories

Settings

Figure 7.49 shows the settings for a maintenance plan category for test/measurement equipment management.

Maint. plan cat.	Name	Call object	Change d...	Documen...	Completi...	Outl. agree...	Ref. object	Ext...	IntNoR...
☑ CA	Calibration Order	Maintenance or serv...	☑	☑	☑	☐	0170	06	05
☐ MM	Service Procurement	1 Service entry sheet	☐	☐	☐	☐	0140		
☐ NC	Service Notification with Outli...	2 Notification	☐	☐	☐	☑	0100		
☐ NO	Inspection (Only Notification)	2 Notification	☐	☐	☑	☐	0100	06	05
☐ OC	Service Order with Outline A...	Maintenance or serv...	☐	☐	☐	☑	0100		
☐ PM	Maintenance Order	Maintenance or serv...	☑	☑	☑	☐	0100	06	05
☐ SO	Service Order	5 Service Order	☑	☑	☐	☐	0100	02	01
☐ ST	Stability Study	3 Inspection lot	☐	☐	☐	☐	0190	06	05

Figure 7.49 Test/Measurement Equipment: Maintenance Plan Category

The following settings are important for test/measurement equipment management:

- The most important setting is the **Call object**. For the business process for calibrating test/measurement equipment, choose the **Maintenance or service order** call object (and not the **Inspection lot** call object, which you may think makes more sense here).
- In the **Ref. object** column, assign "O150" (**Equipment Only**) or "O170" (**Equipment + Serial No. + Material No.**).

You also require further settings for this maintenance plan category, but we won't discuss them here because they are the same as the settings for other maintenance plan

7.4.2 Customizing Functions for Quality Management

categories. Instead, refer to Chapter 6 for maintenance plan-specific settings and Chapter 3 for information about generic functions.

7.4.2 Customizing Functions for Quality Management

The Customizing functions for test/measurement equipment management comprise the following phases:

- Quality planning
- Quality inspection
- Quality notification

Define Statistical Process Control Criteria

Generally, in quality planning, you use this Customizing function to define the criteria according to which inspection results for control charts are grouped together and which processes are assigned. The statistical process control (SPC) criteria define whether individual or global control charts are managed for different orders, work centers, materials, manufacturers, or customers.

For quality planning within test/measurement equipment management, in particular, you use this Customizing function to control the inspection characteristics that you assign to the operation in an inspection plan. Here, it's necessary to distinguish between two different types of inspection characteristics:

- **Master inspection characteristics**
  These are inspection characteristics whose properties you initially assign as a master record and then assign to the operation in the inspection plan

- **Task list characteristics**
  These are inspection characteristics that you assign directly to the operation and whose properties you define directly in the inspection plan.

Prerequisites
You must define the task list usages (see Chapter 6, Section 6.1) beforehand.

Customizing Path
Quality Management • Quality Planning • Basic Data • Sampling, SPC • Statistical Process Control • Define SPC Criteria

Settings
You require two SPC criteria:

- You use one SPC criterion (such as **SPC Criterion 410** in Figure 7.50) to control the inspection characteristics that you define directly in the inspection plan. You must assign the QRKS_CHARACTERISTIC_PM_OBJECT function module to this plan.

- You use the other SPC criterion (such as **SPC Criterion 420** in Figure 7.51) to control the inspection characteristics that you initially create as master inspection characteristics and then assign to the inspection plan. You must assign the QRKS_MASTER_ CHAR_PM_OBJECT function module to this plan.

In addition, both SPC criteria must be assigned task list **Usage 004** (maintenance task lists) or **009** (all task list types) so that you can use them in general maintenance task lists or equipment task lists.

Furthermore, the **New assignment in each insp. point** radio button must be set for both SPC criteria because it's used to control the quality control charts for technical objects.

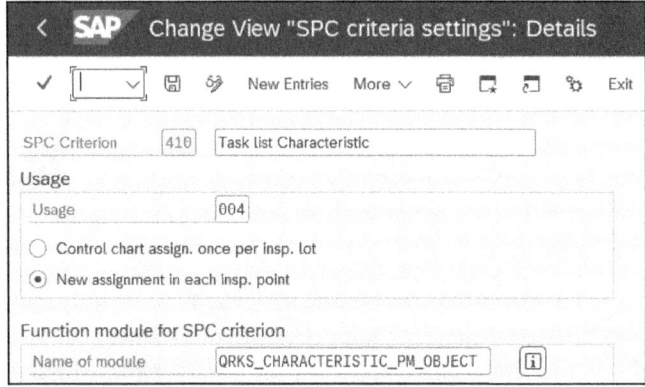

Figure 7.50 SPC Criteria: Task List Inspection Characteristic

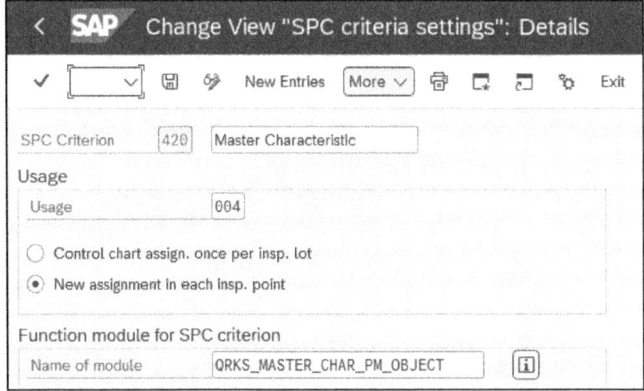

Figure 7.51 SPC Criteria: Master Inspection Characteristic

Define Identifier for Inspection Points

You use this Customizing function to define the inspection points and their attributes. You can use inspection points in several application areas, and the following inspection point types are available for selection:

7.4 Calibration of Test/Measurement Equipment

- **Freely defined inspection points**
  You use this inspection point type, for example, if you create inspection lots for a production order and want to carry out several inspections for each operation during the production process, or if you want to identify partial quantities in production.

- **Inspection points to identify samples**
  You use this inspection point type if you want to manage samples and record inspection results for physical samples.

- **Inspection points to identify maintenance objects**
  You use this inspection point type if you want to carry out calibration inspections and record their results.

Prerequisites

You must define the order types and usage decisions beforehand.

Customizing Path

Quality Management • Quality Planning • Inspection Planning • General • Define Inspection Points

Settings

Define an inspection point and assign **InsPt type 1 (Inspection Point for Equipment)** to this inspection point as shown in Figure 7.52.

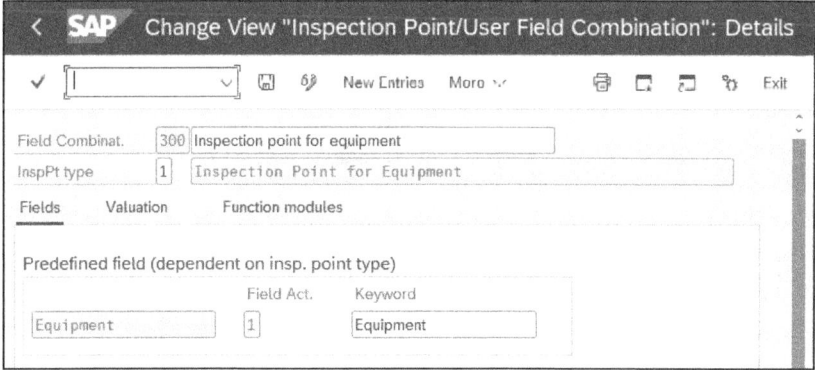

Figure 7.52 Inspection Point for Equipment

For this inspection point, define how you want to deal with the usage decision (see Figure 7.53):

- **Default for valuating inspection point**
  For the **Plant**, use the **Selected Set** to define the code group that is to be used for the equipment usage decision.

485

- **Default when all chars. are accepted**
  Define the **Code** that you want to use if the inspection lot for the piece of equipment is to be accepted.
- **Default when at least one char. is rejected**
  Define the **Code** that you want to use if the inspection lot for the piece of equipment is to be rejected.

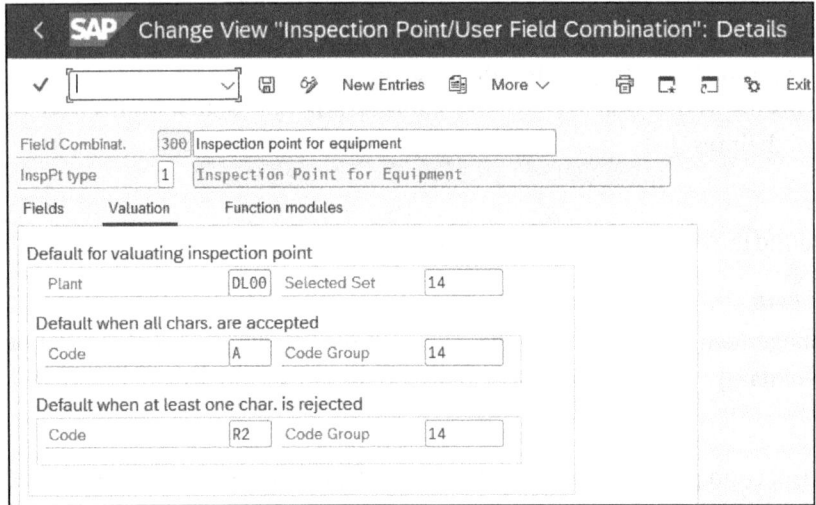

Figure 7.53 Inspection Point for Equipment: Valuation

To define the catalog for usage decisions, along with its code groups and codes, see the **Maintain Catalogs for Usage Decisions** Customizing function.

> **One Inspection Point for Each Plant**
>
> Because the default values for the usage decision are selected on a plant-specific basis, I recommend that you create a separate inspection point for each plant in which you calibrate test/measurement equipment.

Define Control Keys for Inspection Operations

You can use the control key of an operation to define which business functions you want to execute or how you want to handle the operation with regard to costing, printing, completion confirmations, scheduling, and so on. The function and use of control keys was discussed in Chapter 2, Section 2.4.

> **Control Keys with Inspection Characteristics**
>
> From a calibration of test/measurement equipment perspective, the control key has the following important property: **Insp. Char. Required** (inspection characteristic

7.4 Calibration of Test/Measurement Equipment

required). You must therefore assign a control key with this particular property to all operations that are to trigger a calibration inspection in the form of an inspection lot.

Prerequisites

There are no technical prerequisites. Because, however, the control key is used not only in plant maintenance but also in production, quality management, and project processing, you should consult with your colleagues from these areas on the naming convention for the control key.

Customizing Path

Quality Management • Quality Planning • Inspection Planning • Operation • Define Control Keys for Inspection Operations

Settings

Configure one or more control keys for which the **Insp. Char. Required** indicator is set (see Figure 7.54).

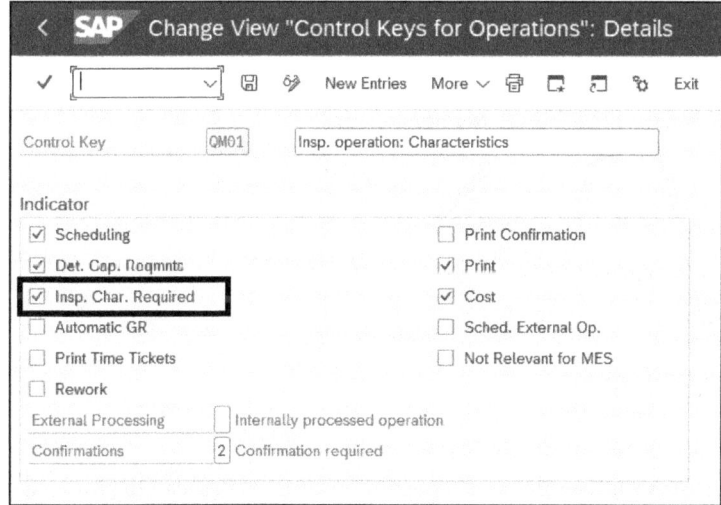

Figure 7.54 Test/Measurement Equipment: Control Key

> **[+] Separate Control Key for Test/Measurement Equipment Management**
> To distinguish test/measurement equipment management not only from the other control keys in plant maintenance but also from all other control keys that generate inspection lots, I recommend that you define a separate control key solely for use with the inspection lots associated with test/measurement equipment management.

7 Configuring Additional Business Processes

Maintain Inspection Types

You use this Customizing function to define your inspection types, and it is the basis for many other Customizing functions. Just as the order type is at the heart of maintenance task processing and can be used to significantly influence maintenance processes, the inspection type is at the heart of inspection task processing and can be used to significantly influence inspection processes (see Figure 7.55). For example, you can control which order type in plant maintenance is to be linked with which inspection type in quality management.

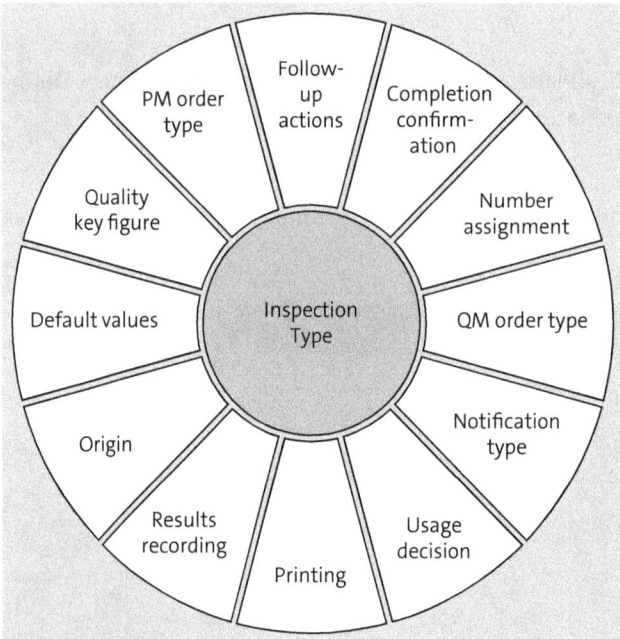

Figure 7.55 Inspection Type: Overview

Prerequisites

You must define the status profile, quality management order types, catalog for usage decisions, and notification types beforehand.

Customizing Path

Quality Management • Quality Inspection • Inspection Lot Creation • Maintain Inspection Types

Settings

Figure 7.56 shows you the basic settings for an inspection type.

You require a separate inspection type for calibrating test/measurement equipment. You can't use other inspection types (e.g., incoming inspections or in-process inspections in production).

488

7.4 Calibration of Test/Measurement Equipment

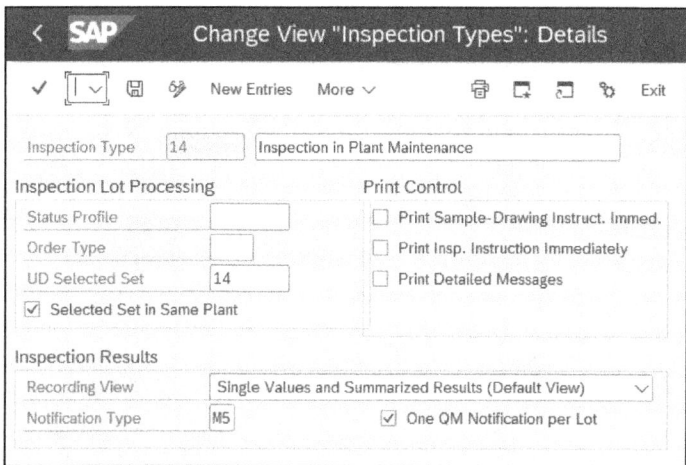

Figure 7.56 Inspection Type

The following settings are important from a test/measurement equipment management perspective:

- **UD Selected Set**
  This defines the usage decisions in relation to accepting or rejecting the inspection lots.

- **Notification Type**
  This is the notification type created if an inspection lot is rejected.

For information about the notification type, see the **Define Notification Types** Customizing function in Chapter 5, Section 5.1.1. For information about defining the catalog for usage decisions, along with its code groups and codes, see the **Maintain Catalogs for Usage Decisions** Customizing function in this section.

Maintain Inspection Lot Origins and Assign Inspection Types

You use this Customizing function to assign your inspection type to an inspection lot origin, which, for plant maintenance, is **14**. The inspection lot origin is predefined by SAP and can't be changed, but you can define inspection types yourself. Each inspection type must refer to an inspection lot origin, and several inspection types can be assigned to the same origin.

Prerequisites
You must define the inspection types beforehand.

Customizing Path
Quality Management · Quality Inspection · Inspection Lot Creation · Maintain Inspection Lot Origins and Assign Inspection Types

489

7 Configuring Additional Business Processes

Settings

In the dialog structure below the **Inspection Lot Origin** subfunction, the standard SAP delivery contains entry **14 Plant Maintenance** (see Figure 7.57). In the dialog structure below the **Inspection Types for the Origin** subfunction, you must assign your inspection types for calibrating test/measurement equipment to entry **14** (see Figure 7.58).

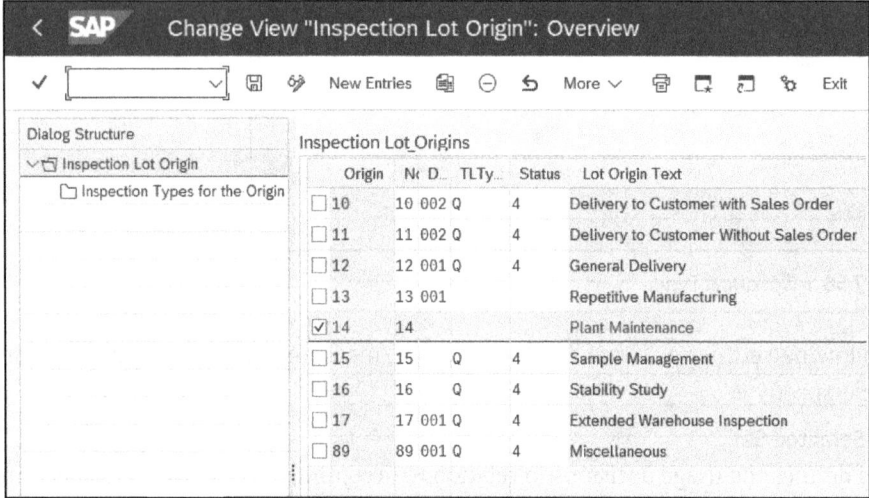

Figure 7.57 Inspection Lot Origins

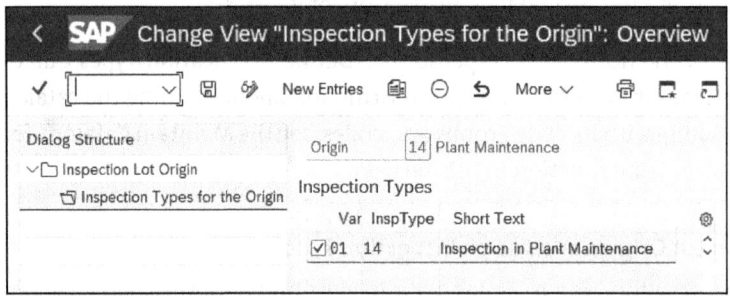

Figure 7.58 Inspection Lot Origins: Detail

> [!] **Assign Origin 14**
>
> You must assign origin **14** to all your inspection types for calibrating test/measurement equipment management, or you won't be able to use them for test/measurement equipment management.

Define Default Values for Inspection Type

You can use this Customizing function to define default values for the inspection types.

7.4 Calibration of Test/Measurement Equipment

Prerequisites
You must define the inspection types beforehand.

Customizing Path
Quality Management • Quality Inspection • Inspection Lot Creation • Define Default Values for Inspection Type

Settings
The following settings make sense for inspection types for test/measurement equipment management (see Figure 7.59):

- **Inspect by Task List**
  Because the only way to inspect test/measurement equipment is by means of a task list (a general maintenance task list or an equipment task list), this indicator is preselected and can't be changed.

- **Record Characteristic Results**
  Check the box to set this indicator because you generally want to record the results for test/measurement equipment management. You can then perform characteristic-based results recording for the inspection lot.

- **Automatic Usage Decision**
  If you want to make manual usage decisions only, don't check the box to set this indicator. However, you should set this indicator if you want to permit automatic usage decisions in addition to manual usage decisions.

- **Q-Score Procedure**
  Use this dropdown list to define how the quality code for the inspection lot is to be determined. In the case of test/measurement equipment management, the correct selection is usually **06 From usage decision code**.

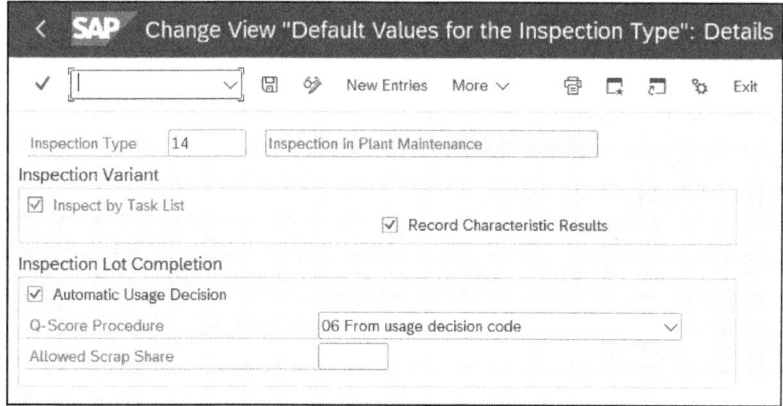

Figure 7.59 Inspection Type: Default Values

Define Inspections in Plant Maintenance

This Customizing function is *the* central element of PM/QM coupling. Here, you link the order types that you've defined for test/measurement equipment management to the corresponding inspection types for each plant that you've defined. Afterward, if you create an order in this plant and for this order type (in general, this is automatically created from the maintenance plans), the system automatically generates an inspection lot with the specified inspection type because of the linking of the order types.

Prerequisites

You must define the plants, order types, and inspection types beforehand.

Customizing Path

Quality Management • Quality Inspection • Inspection Lot Creation • Define Inspections in Plant Maintenance

Settings

Figure 7.60 shows you how to assign, for each plant, an inspection type to an order type. For information about configuring the order type, see the **Configure Order Types** Customizing function in Chapter 5, Section 5.2.1.

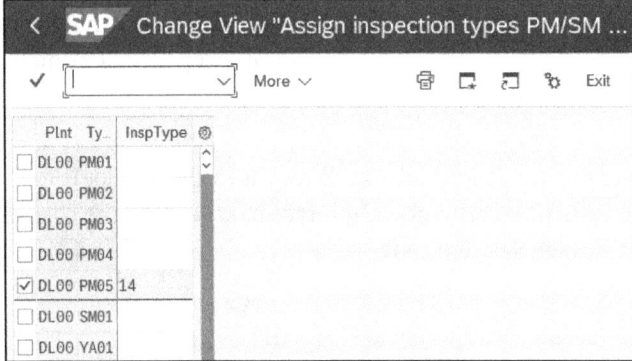

Figure 7.60 Order Types and Inspection Type

> [!] **Assign Inspection Types to Order Types**
>
> Make sure that you assign your inspection types fully and correctly to the order types for each plant. If you don't ensure this assignment, there is no PM/QM coupling, the system doesn't create any inspection lots for the orders, and you can't calibrate the test/measurement equipment.

7.4 Calibration of Test/Measurement Equipment

Maintain Catalogs for Usage Decisions

You use this Customizing function to define the usage decisions for the inspection lots for test/measurement equipment management (in particular, acceptance and rejection).

Prerequisites

You must define the follow-up actions and maintenance plants beforehand.

Customizing Paths

The following are the Customizing paths available for maintaining catalogs for usage decisions:

- **Quality Management** • **Quality Notifications** • **Notification Creation** • **Notification Content** • **Maintain Catalogs**
- **Quality Management** • **Quality Inspection** • **Inspection Lot Completion** • **Edit Code Groups and Codes for Usage Decisions**
- **Quality Management** • **Quality Inspection** • **Inspection Lot Completion** • **Edit Selected Sets for Usage Decisions**

Settings

These Customizing functions comprise three interrelated subfunctions:

1. Use the first menu path and then the **Define Catalogs** subfunction to define a number and description for the usage decision catalog. In the standard SAP delivery, this is **Catalog 3** (see Figure 7.61), which you can also use in general.

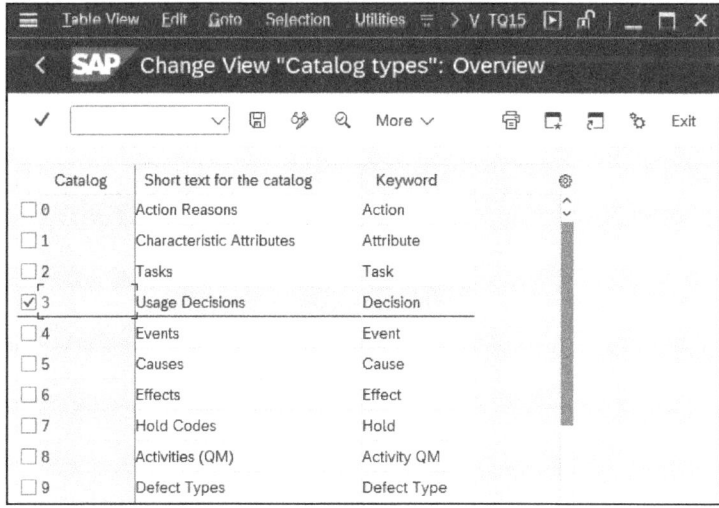

Figure 7.61 Catalog for Usage Decisions

2. Use the second menu path to define, for the usage decision catalog, one or more code groups for test/measurement equipment management (see Figure 7.62).

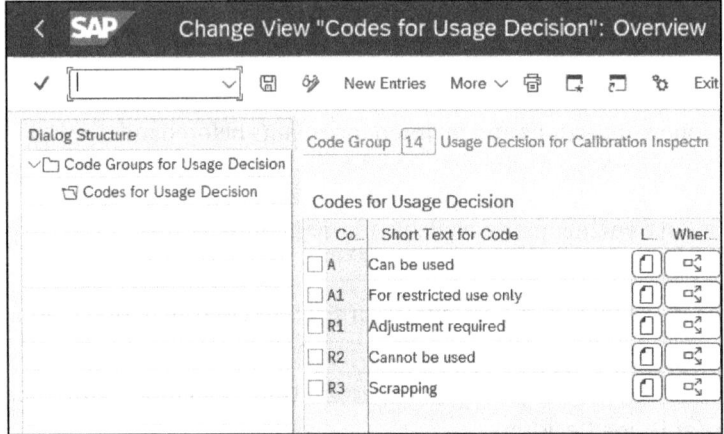

Figure 7.62 Code Group for Usage Decisions

3. Use the third menu path to define the codes in question and their consequences for each plant (see Figure 7.63):

- **Code Valuation**: In this column, define whether the code will result in the inspection lot being accepted or rejected.
- **Qualit...**: In this column, refine this yes/no decision further on a scale of 0 to 100 (e.g., **0 = not OK, 100 = OK, 66 = OK with Restrictions**).
- **FolUpAct**: In this column, define whether and, if so, which follow-up actions are to be triggered (e.g., lock the piece of equipment, technically complete the order). For more information, see the **Define Follow-up Action** Customizing function in this section.

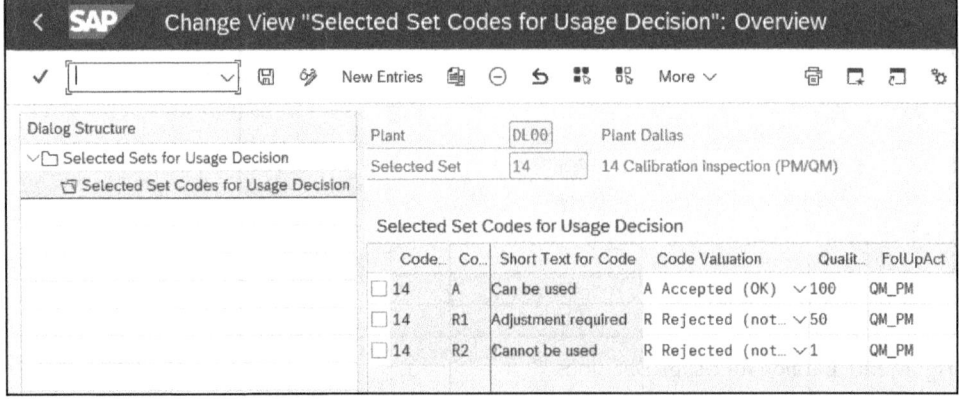

Figure 7.63 Selected Sets for Usage Decision

7.4 Calibration of Test/Measurement Equipment

> **At Least One Code for Acceptance and One Code for Rejection** [+]
>
> To ensure that the inspection lots for test/measurement equipment management are completed with a useful usage decision, you need at least one code for accepting the inspection lot and one code for rejecting the inspection lot.
>
> You must also assign these codes to each plant in which you want to implement test/measurement equipment management.

Define Follow-Up Action

You use this Customizing function to define the follow-up actions that the system will automatically trigger when a usage decision is made.

Prerequisites

There are no specific prerequisites.

Customizing Path

Quality Management • Quality Inspection • Inspection Lot Completion • Define Follow-up Action

Transaction

V_TQ07

Settings

In the dialog structure, create a group of follow-up actions (in the **FolUpAct** column) for test/measurement equipment management (e.g., **QM_PM** in Figure 7.64).

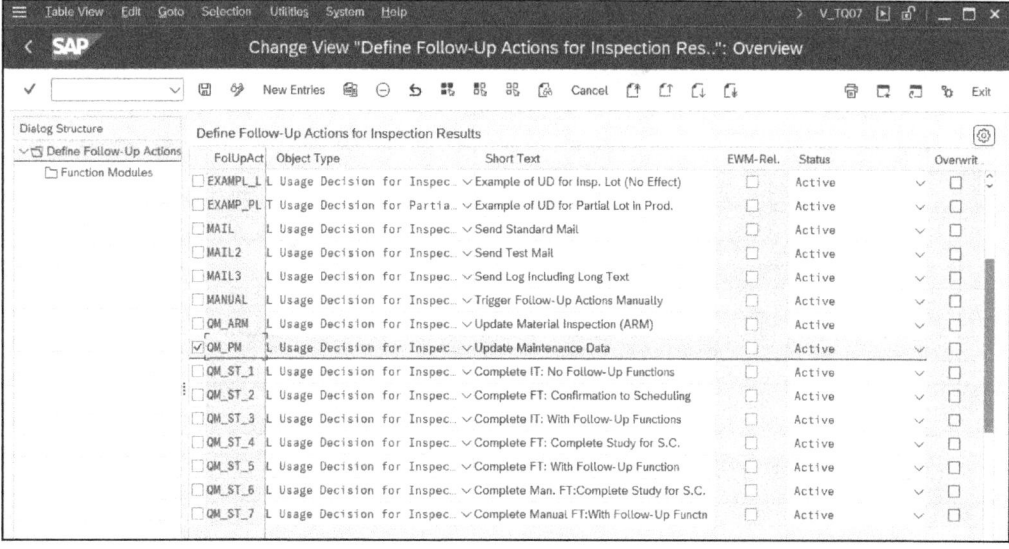

Figure 7.64 Follow-Up Actions: Overview

495

7 Configuring Additional Business Processes

> **L = Usage Decision for Inspection Lot**
> It's important to assign function L to the test/measurement equipment group so that the follow-up actions are controlled based on the usage decision for the inspection lot.

In the next step, use the **Function Modules** subfunction to define which follow-up actions are to be triggered within the group. From a test/measurement equipment management perspective, the following three follow-up actions are of interest (see Figure 7.65):

- The QFOA_QM_CHAR_TO_PM_E_POINTS function module activates the **Transfer of QM Characteristic Results to PM Measuring Points** follow-up action. You can use this follow-up action to transfer inspection results to equipment as measurement documents. Quantitative inspection characteristics (measurement readings) and qualitative inspection characteristics (characteristics with catalog attributes) are taken into consideration here. A class characteristic is used to establish a link between the master inspection characteristics and the measuring points.

- The QFOA_OBJECT_STATUS_SET function module activates the **Set Equipment Status/Change Cycle Modification Factor** follow-up action. When you make the usage decision, the system displays a dialog box in which you can change the status of the piece of equipment from **ReadyForUse** to **Not ReadyForUse** or vice versa, and you can change the **Cycle ModificFactor** (cycle modification factor) of the maintenance plan.

- The QFOA_ORDER_TECHNICAL_COMPLETE function module activates the **Technically Complete PM Order** follow-up action. When you make the usage decision, the system technically completes the order automatically (in the background).

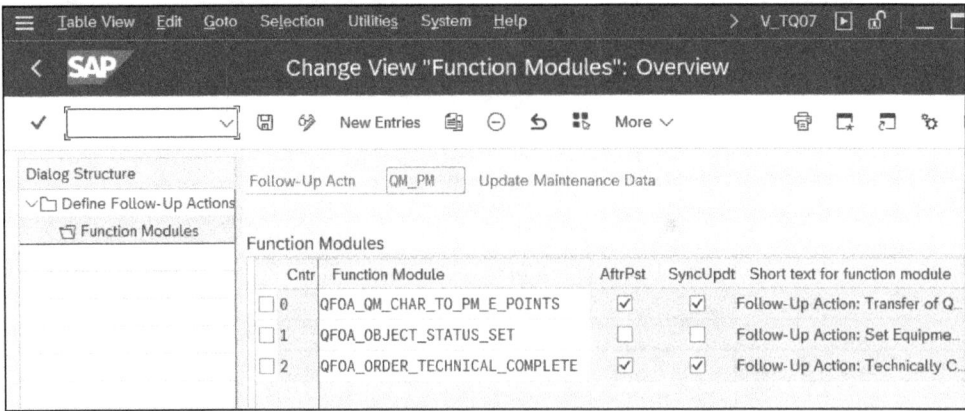

Figure 7.65 Follow-Up Actions: Details

7.4 Calibration of Test/Measurement Equipment

> **Activate Follow-Up Actions**
>
> Don't neglect to use the option to automatically trigger follow-up actions in the background. In particular, the following options are very helpful:
>
> - The option to declare a piece of equipment as ready for use or not ready for use
> - The option to technically complete orders automatically

Plan Automatic Usage Decision

You use this Customizing function to plan the process for the automatic usage decision for inspection lots. The inspection lots must satisfy the following conditions:

- All characteristics have been completed.
- None of the characteristics have been rejected.
- No defects have been recorded, and no quality notifications have been created.

The system then waits until the waiting time defined in this Customizing function has elapsed. During this period, you can still intervene in the automatic process. The system records the automatic usage decision in the long text.

Prerequisites

You must have the **Automatic Usage Decision** indicator set in the **Define Default Values for Inspection Type** Customizing function.

Furthermore, you must define the plants, recording configurations, and usage decisions beforehand.

Customizing Path

Quality Management • Quality Inspection • Inspection Lot Completion • Plan Automatic Usage Decision

Settings

Many of the settings available in this Customizing function don't apply to the inspection lots for test/measurement equipment management. Only the following settings are relevant:

- On the **Results Recording** tab (see Figure 7.66), define the **Recording Configuration** by setting it to **1 Automatic close**. For more information, refer to the **Define Recording Configuration** Customizing function.
- The same tab contains the **Proposal for inspection point valuation** field group (not shown). Here, you define which code the system is to be automatically set to for the usage decision, if the aforementioned prerequisites are fulfilled. Usually, this is the **Acceptance** usage decision.

7 Configuring Additional Business Processes

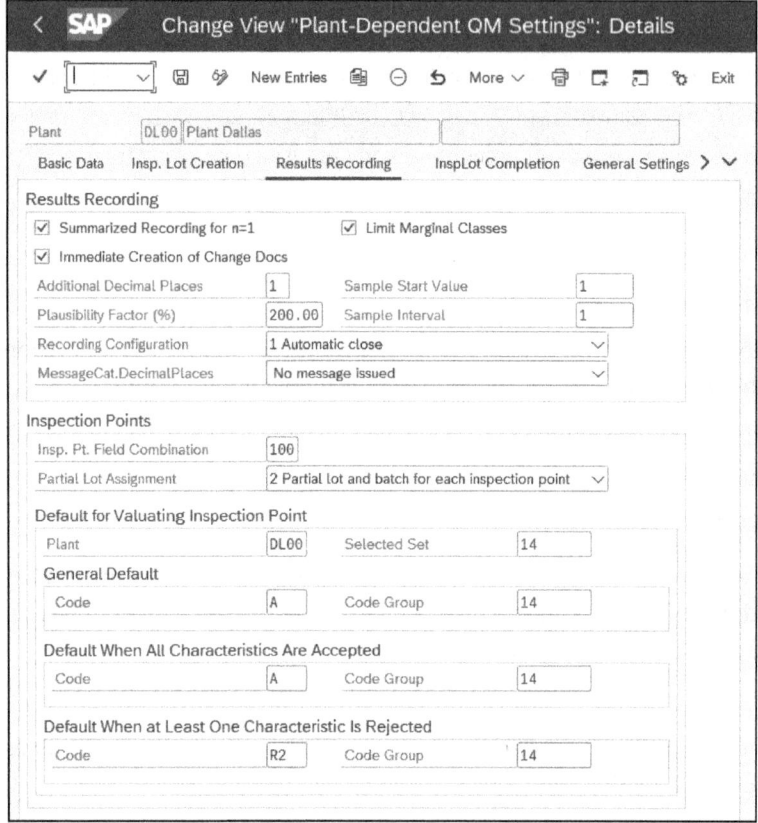

Figure 7.66 Automatic Usage Decision: Results Recording

- The **InspLot Completion** tab (see Figure 7.67) contains the **Waiting Time** in hours and minutes. Here, you specify how much time must have elapsed since the inspection lot was created before the system can make the automatic usage decision.

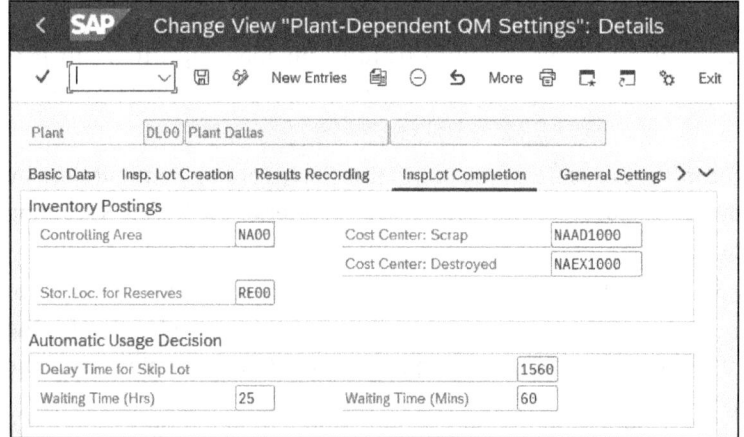

Figure 7.67 Automatic Usage Decision: Inspection Lot Completion

7.4 Calibration of Test/Measurement Equipment

> **Recommendations for the Automatic Usage Decision** [+]
> - Set the **Recording Configuration** option to **Automatic close**.
> - Enter a usage decision for acceptance.
> - Schedule sufficient waiting time since the creation of the inspection lot.

Define Recording Configuration

You use this Customizing function to define different background functions for recording characteristic inspection results. If certain prerequisites are fulfilled, you can prompt the system to complete certain functions in the background when it records inspection results.

Prerequisites

There are no specific prerequisites.

Customizing Path

Quality Management • Quality Inspection • Results Recording • Define Recording Configuration

Settings or Recommended Settings

Figure 7.68 shows the recording configurations contained in the standard SAP delivery.

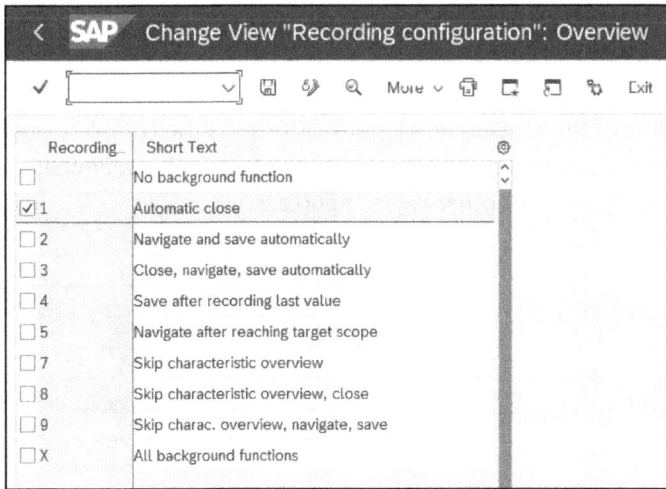

Figure 7.68 Recording Configuration: Overview

Then, on the detail screen, define which functions are triggered. For the recording configuration recommended in the automatic usage decision (1 = **Automatic close**), these are the following:

7 Configuring Additional Business Processes

- Close When Choosing Continue
- Close When Navigating
- Close When Saving

> [+] **Recommendation for Recording Configuration**
> Retain the standard SAP settings. If, however, you want to change them, press F1 and carefully read the help text for the individual functions.

These Customizing settings should now enable you to execute the business process for calibrating test/measurement equipment.

7.5 Pool Asset Management

You can use *pool asset management* to manage objects that are contained in a pool from which they can be borrowed for a certain length of time. These can include the following objects:

- Fleet objects
- IT equipment (notebooks, projectors, etc.)
- Cell phones
- Tools
- Other objects

These objects are returned to the pool after they have been borrowed, and the service is charged to the cost object (e.g., the cost center). Figure 7.69 shows the complete pool asset management process, using the example of a pool of fleet objects. Pool asset management works in the same way for other types of pools.

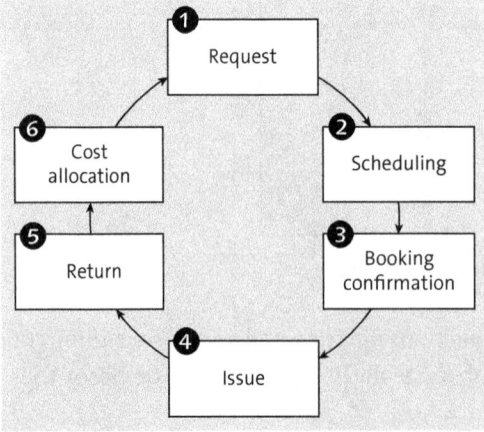

Figure 7.69 Pool Asset Management Process

Here, pool asset management is composed of the following steps:

❶ Request
An employee enters a request for a fleet object in the system, and he creates a pool asset management notification with information about the trip, driver, and settlement. For this, you configure a separate notification type, a separate partner determination procedure, and the equipment characteristics of the object being requested in Customizing.

❷ Vehicle scheduling
A vehicle scheduler assigns a fleet object to the request, and you configure the content and layout of the planning board in Customizing.

❸ Booking confirmation
The employee who requested the fleet object receives an automatic booking confirmation by email, the content and layout of which is determined in Customizing.

❹ Fleet object issue
A mileage reading is taken from the fleet object, the time at which it's required is recorded, and the fleet object is then issued.

❺ Fleet object return
The employee returns the fleet object, and the mileage reading and date are recorded.

❻ Cost allocation
The costs associated with the employee's use of the fleet object are determined and allocated to the account assignment object. The system creates a pool asset management order for the allocation in the background. This order contains the piece of equipment as the reference object and the settlement cost center for the pool asset management notification as the settlement rule. For this, you configure the order type required for the settlement as well as the activity types that you require in Customizing.

> **Business Function**
>
> Before you can use pool asset management, you must activate the LOG_EAM_PAM business function. Some of the following Customizing functions for configuring the business process for pool asset management don't become available until this business function is activated.

Overview of Notification Type

You use a pool asset management requirement notification to trigger the business process for pool asset management. You need a separate notification type here.

This Customizing function provides central access to a wide range of settings for a notification type. For the pool asset management notification, you need the following:

7   Configuring Additional Business Processes

- The **Screen Areas in Notification Header** subfunction to define the notification header and reference object
- The **Screen Structure for Extended View** subfunction to define the layout
- The **Partner Determination Procedure** subfunction to define the partners

Prerequisites

You must define the notification types beforehand.

Customizing Path

Plant Maintenance and Customer Service • Maintenance and Service Processing • Maintenance and Service Notifications • Overview of Notification Type

Settings

Use the **Screen Areas in Notification Header** subfunction to determine the header and reference object for the notification and make the following settings (see Figure 7.70):

- In **Screen Type Hdr**, always enter "H100" (**Header maintenance notification**), even for the pool asset management requirement notification.
- In **ScrnType Object**, enter "O150" (**Equipment only**) for the pool asset management notification because the objects in an asset management pool must be pieces of equipment.

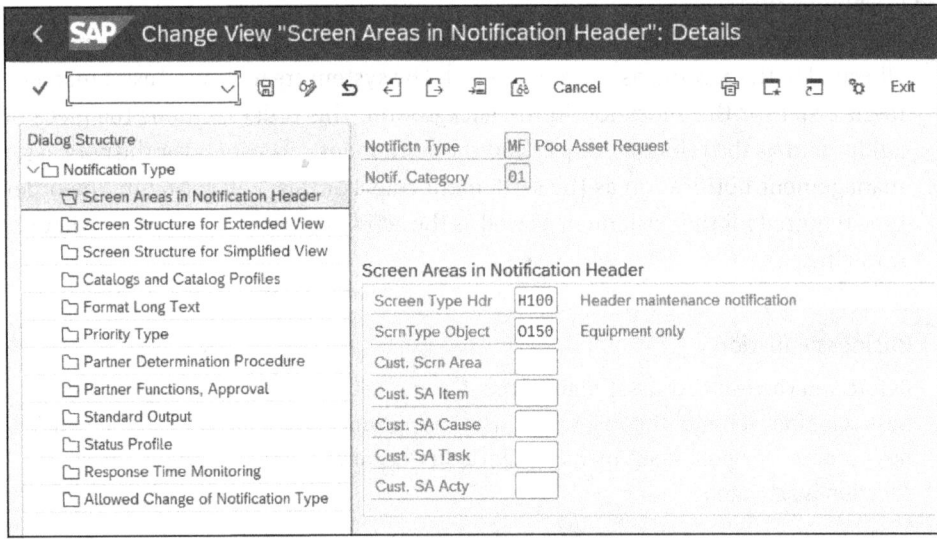

Figure 7.70 Pool Asset Management: Notification Header

Use the **Screen Structure for Extended View** subfunction to define the layout of the pool asset management notification. You require at least two tabs, as follows (see Figure 7.71):

- **10\TAB23**

  This must be one of the tabs. SAP has predefined this tab for a pool asset management notification, and it contains information about the trip data, equipment characteristics, driver, and settlement.

- **065 Partner Overview (Table Control)**

  A second tab should then contain this subscreen. Use the **Partner** subfunction to define the partners involved in the pool asset management notification. The **10\TAB01**, **10\TAB02**, and **10\TAB03** tabs are suitable here.

You can include additional tabs (e.g., **Location**, **Date Overview**) if necessary.

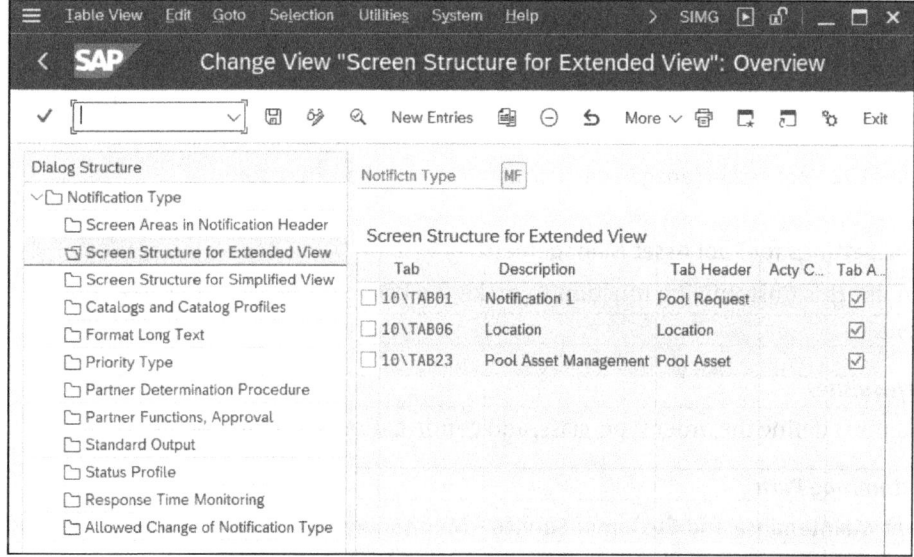

Figure 7.71 Pool Asset Management: Notification Screens

Use the **Partner Determination Procedure** subfunction to assign the partner determination procedure to the notification type (see Figure 7.72). The partner determination procedure that you use for the pool asset management notification should contain at least one partner function for the object user and one partner function for the requester. The partner functions should have either partner type **PE** (personnel) or **US** (SAP user). Later, these partners will receive emails containing the reservation confirmation, among other things. For more detailed information about these terms and about configuring a partner determination procedure, see Chapter 3, Section 3.7.

> **Recommendations for the Pool Asset Management Notification Type**
>
> One tab must be **10\TAB23** so that you can report the pool asset management requirement data. The assigned partner types should be either **PE** or **US** so that an email containing the reservation confirmation can be sent later.

7 Configuring Additional Business Processes

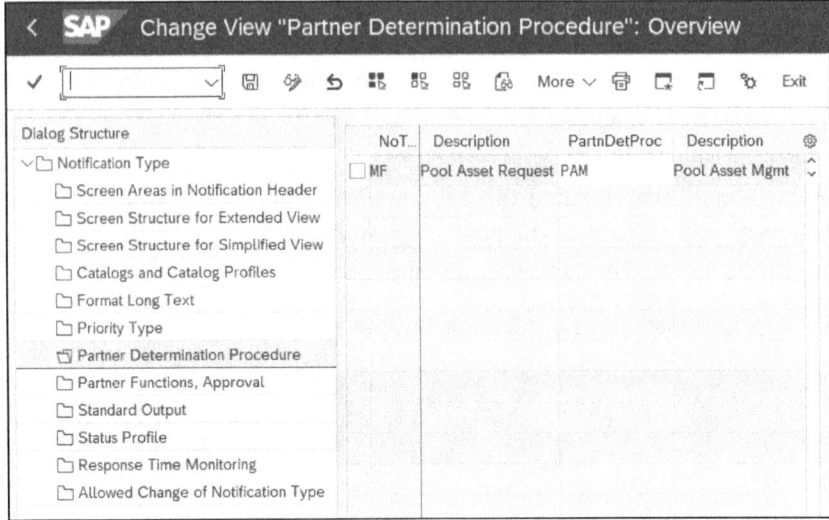

Figure 7.72 Pool Asset Management: Partner Determination Procedure

Basic Settings for Pool Asset Management

You use this Customizing function to make the basic settings for pool asset management.

Prerequisites

You must define the order type, class, and confirmation text beforehand.

Customizing Path

Plant Maintenance and Customer Service • Maintenance and Service Processing • Pool Asset Management • Basic Settings for Pool Asset Management

Settings

In this Customizing function, you can make the following settings (see Figure 7.73):

- **Application**
  For this field, the fixed value **FL Pool Asset Management** is always preselected.

- **Reservation Txt**
  For this field, enter the SAPscript form that is to be used to send the notification email to the requester in the event of a reservation or cancellation. Use Transaction SE71 to maintain the SAPscript form itself and to incorporate this form into print control (see the next Customizing function for print control).

- **OrdTy. IntAlloc**
  For this field, enter the order type for the internal allocation. This order type determines the internal allocations for object usage and will be used when an internal pool usage is settled.

7.5 Pool Asset Management

- **OrdTy. ExtAlloc**
  For this field, enter the order type for the external allocation if you have customers renting assets from your pool. This order type determines the external allocations for object usage and will be used when an external pool usage is settled.

- **Class Type and Class**
  For **Class Type** (fixed value: **PAM**) and **Class**, specify the class. This refers to the pool asset management-relevant characteristics of the object (piece of equipment) and the request (maintenance notification).

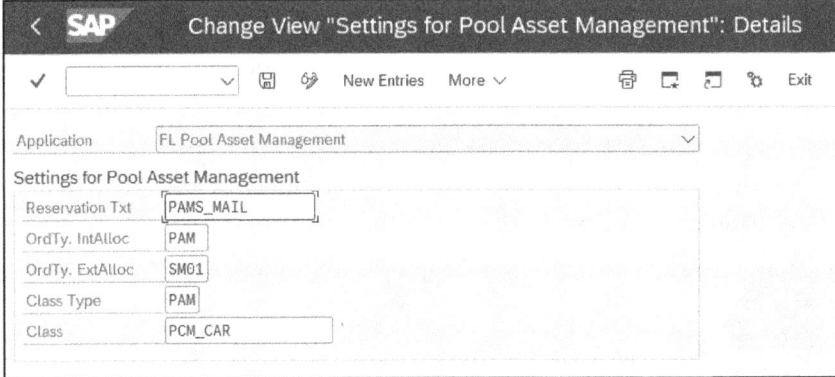

Figure 7.73 Pool Asset Management: Basic Settings

> **You Can Assign Only One Class**
>
> One weakness of pool asset management is the fact that you can assign only one class (e.g., for fleet objects). Even though other pools (e.g., tools, cranes, notebooks) require other classes with other characteristics, at present, pool asset management can only be used for one pool and its full scope of functions (i.e., including the equipment characteristics).

Define Shop Papers, Forms, and Output Programs

To enable the output or sending of an email containing the reservation confirmation, you must incorporate this form into normal print control.

Prerequisites

You must develop the output program and the form itself beforehand.

Customizing Path

Plant Maintenance and Customer Service • Maintenance and Service Processing • Maintenance and Service Notifications • Notification Processing • Notification Print Control • Define Shop Papers, Forms, and Output Programs

7 Configuring Additional Business Processes

Transaction

OIDA

Settings

This Customizing function has two subfunctions: **Define Shop Papers** and **Define Shop Papers for Notification Type**. However, you need only the first subfunction. In this subfunction, define the following (see Figure 7.74):

- **Shop paper**
  A four-digit number (maximum) for the shop paper.
- **Program name and ABAP form**
  The report name and the form routine used for print control.
- **Form**
  The SAPscript form used as a basis for designing the pages. The form contains different elements that are used to control the layout of the individual pages and to show design information for the texts to be output on the individual pages.

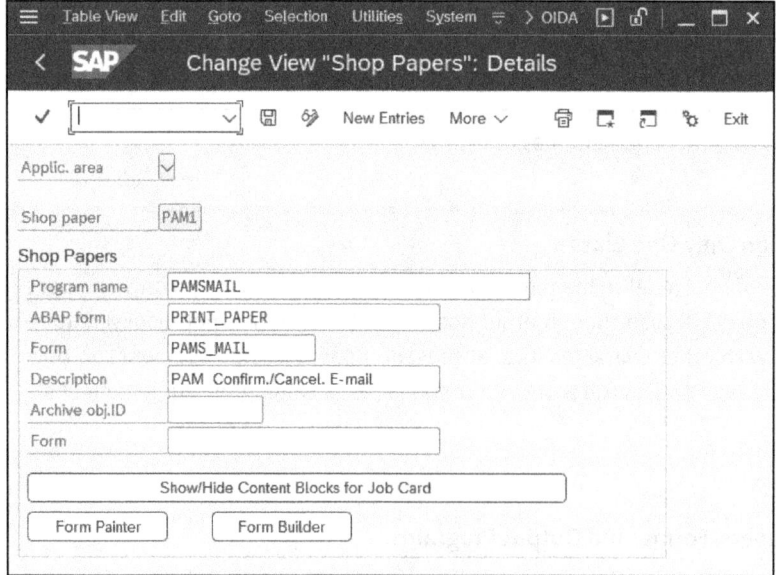

Figure 7.74 Pool Asset Management: Define Shop Paper

Create Classes and Characteristics

In the **Basic Settings for Pool Asset Management** Customizing function, you assign a class to the pool asset management notification type. You use this class so that you can specify the equipment characteristics of the equipment to be borrowed in the pool asset management notification. For a transport fleet, this might include the following characteristics:

7.5 Pool Asset Management

- Fleet object category
- Number of persons
- Transmission type
- Winter equipment

Here, use Transaction CT04 to create the requisite characteristics. Then, use Transaction CL02 to integrate them into a class of the **Class type PAM** (see Figure 7.75).

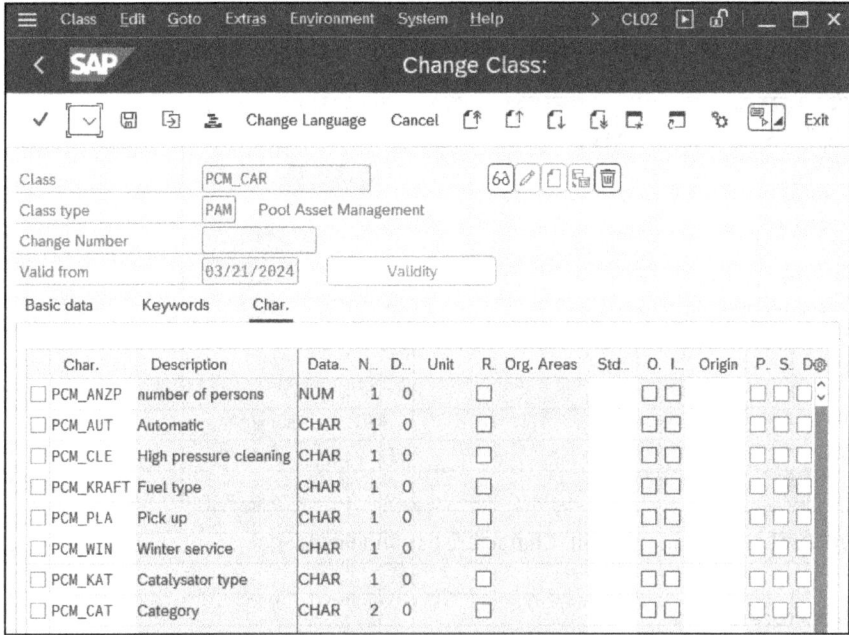

Figure 7.75 Pool Asset Management: Class and Characteristics

Edit Relationships for Characteristics and Fields

You can use this Customizing function to edit the relationships for characteristics and fields. You can define up to twelve equipment characteristics, the object category (used for settlement determination), and the number of persons (in the case of a fleet object) that are to be displayed in the planning board.

Prerequisites
You must define the class with the characteristics beforehand and then assign this class to pool asset management in the **Basic Settings for Pool Asset Management** Customizing function.

Customizing Path
Plant Maintenance and Customer Service • Maintenance and Service Processing • Pool Asset Management • Edit Relationships for Characteristics and Fields

7 Configuring Additional Business Processes

Settings

Specify the properties of the respective characteristics (see Figure 7.76) as follows:

- **Equipment**
  This determines whether the characteristic concerns a normal equipment characteristic.

- **No.Persons**
  This determines whether the characteristic represents the number of persons.

- **PAM Type**
  This determines whether the characteristic defines the fleet object category.

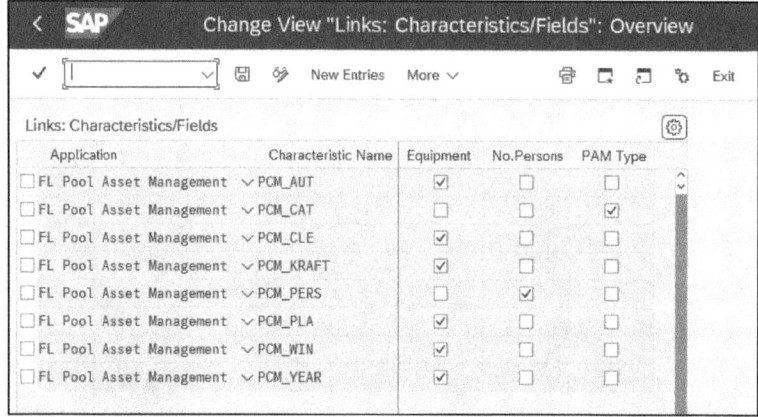

Figure 7.76 Pool Asset Management: Characteristics and Fields

Edit Object Processing

You can use this Customizing function to define which objects are to be considered in the planning board. Even though the Customizing function already knows the order type, the current version of pool asset management only takes notifications—not orders—into account.

Prerequisites

You must define the notification types beforehand.

Customizing Path

Plant Maintenance and Customer Service • Maintenance and Service Processing • Pool Asset Management • Edit Object Processing

Settings

You can make the following settings for each notification type (see Figure 7.77):

- **Typ**
  You need at least two notification types: one (here, **MF**) for requesting the pool asset and another (e.g. **M1**) for repairs to the pool asset.

- **Req. Role**
  Use the **Requester** partner function to specify the partner function for the request. The user obtains information about the piece of equipment that has been reserved or the cancellation of a request.
- **User Role**
  Use the **User** partner function to specify the partner function to be used (e.g., a driver in the case of a fleet object).
- **Dir. Sched**
  Use the direct scheduling option to control the direct transfer of the notification to the planning board without the need for manual scheduling (e.g., for notifications from maintenance plans or for malfunction reports).

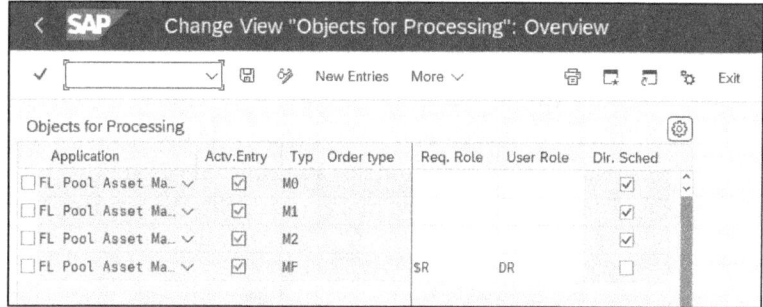

Figure 7.77 Pool Asset Management: Objects for Processing

Define Settings for Planning Board

You use this Customizing function to make some basic settings for the graphical planning board.

Prerequisites

You must define the calendar and graphic profiles beforehand.

Customizing Path

Plant Maintenance and Customer Service • Maintenance and Service Processing • Pool Asset Management • Define Settings for Planning Board

Settings

The following settings are available to you (see Figure 7.78):

- **Factory Calendar ID and Holiday Cal. ID**
  In these fields, specify which factory calendar and holiday calendar you want to use.
- **Group, Name, and Index**
  In these columns, specify which graphic profile you want to use. For information about the settings for the graphic profile, see Chapter 5, Section 5.2.5.

- **Number and Char01 fields**
  According to SAP, these fields contain system settings that can't be changed (and therefore could have been omitted from this screen).

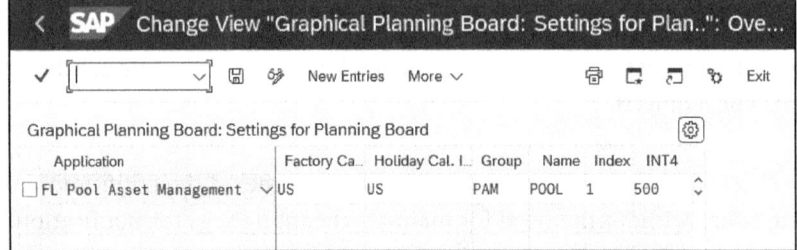

Figure 7.78 Pool Asset Management: Settings for Planning Board

Edit Display Variants of Planning Board

You use this Customizing function to define display variants for the graphical planning board (Transaction PAM03), that is, which fields are to be displayed in the upper section of the planning board (fields relating to the equipment) and which fields are to be displayed in the lower section of the planning board (fields relating to the requests; see Figure 7.79).

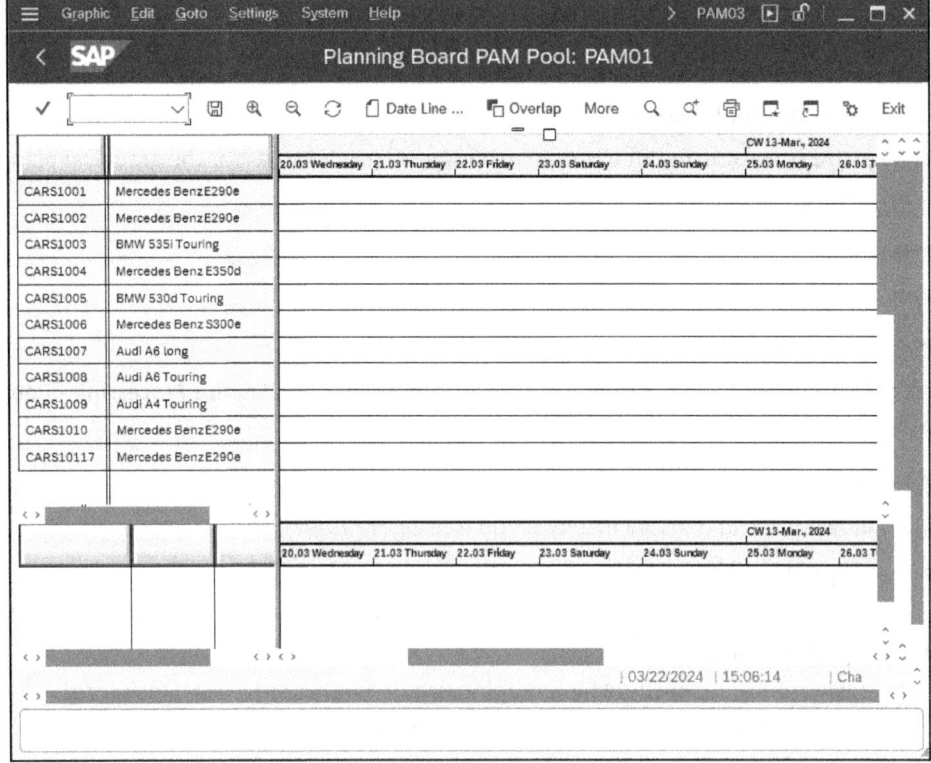

Figure 7.79 Pool Asset Management: Graphical Board

7.5 Pool Asset Management

Prerequisites
There are no special prerequisites.

Customizing Path
Plant Maintenance and Customer Service • Maintenance and Service Processing • Pool Asset Management • Edit Display Variants of Planning Board

Settings
For the FL Pool Asset Management app, you need to specify the following (see Figure 7.80):

- **Display Variant**
  Enter a name for the display variant.

- **ChartID 1**
  Use this field to define the fields in the upper section of the planning board (i.e., the fields relating to the equipment).

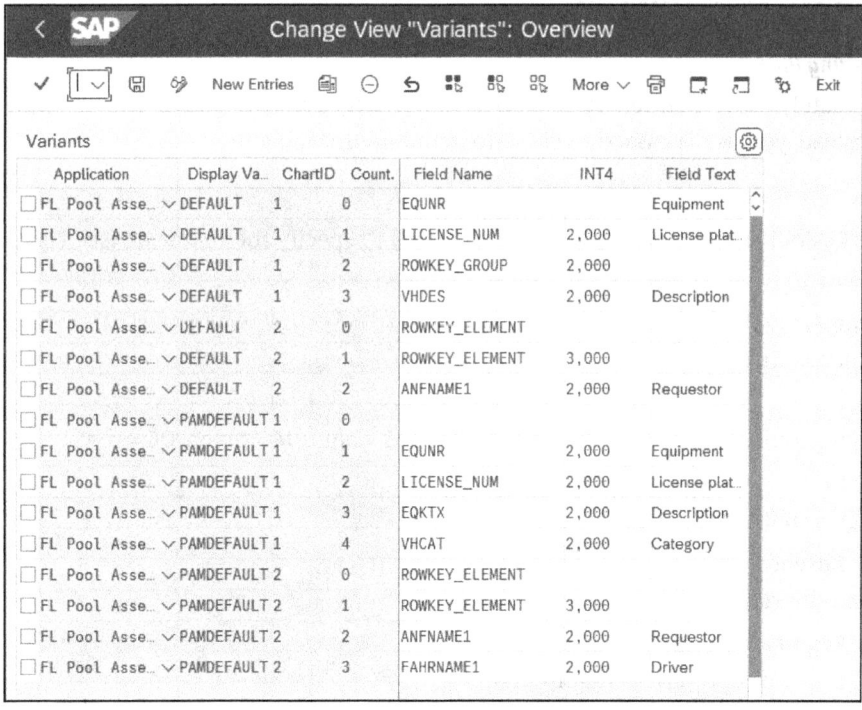

Figure 7.80 Pool Asset Management: Display Variant of Planning Board

- **ChartID 2**
  Use this field to define the fields in the lower section of the planning board (i.e., the fields relating to the requests).

- **Field Name**
  Choose from 33 fields.

511

- **Field Text**
  You can define this yourself.

> **Note**
>
> In the upper section, the equipment number is displayed for the **ROWKEY_ELEMENT** field; in the lower section, the notification number is displayed for this field. The sequence numbers aren't displayed.

Maintain Status

You use this function to define how the bars that represent the requested elements are to be displayed on the planning board (in particular, which color is to be assigned to each status).

Prerequisites

There are no specific prerequisites.

Customizing Path

Unfortunately, table maintenance for PAMS_VSTAI hasn't been integrated into Customizing. Instead, you use Transaction SM30 to maintain this table.

Settings

For the FL Pool Asset Management app, you need to specify the following (see Figure 7.81):

- **Status**
  Define a status.
- **Color**
  Assign a color to your status (e.g., **01** = yellow, **03** = blue, **04** = green, **06** = red).
- **Split Ty.**
  Assign a split type to your status:
  - **A: Absence**
    The object is unavailable due to other tasks.
  - **B: Reqmt**
    This status represents a requirement.
  - **E: Completed Split**
    The requirement is completed.
  - **S: Productive Split**
    The object is currently in use.
- **SysSt**
  Link your status to a system status. The following system statuses are useful:

- I3901: NDIS (Requested)
- I3902: DISP (Allocated)
- I3903: GELD (Deleted by planner)
- I3904: STOR (Reserved request canceled)
- I3905: GELA (Deleted by requester)
- I3906: STAN (Canceled)
- I3907: RESM (Reserved)
- I3908: ABGH (Picked)
- I3909: RGBE (Returned)
- I3910: AUFE (Settled)

- **Status Text**
  Assign a status text to your status.

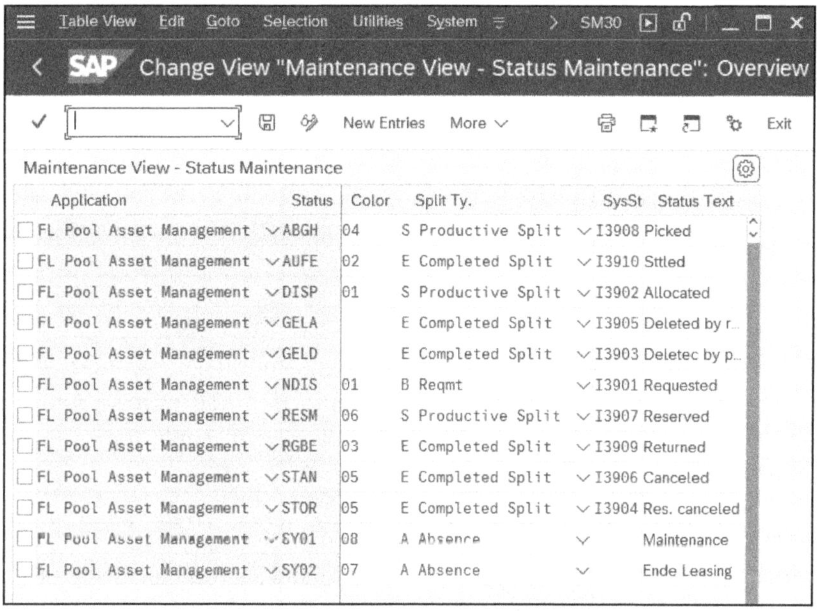

Application	Status	Color	Split Ty.	SysSt Status Text
FL Pool Asset Management	ABGH	04	S Productive Split	I3908 Picked
FL Pool Asset Management	AUFE	02	E Completed Split	I3910 Sttled
FL Pool Asset Management	DISP	01	S Productive Split	I3902 Allocated
FL Pool Asset Management	GELA		E Completed Split	I3905 Deleted by r...
FL Pool Asset Management	GELD		E Completed Split	I3903 Deletec by p...
FL Pool Asset Management	NDIS	01	B Reqmt	I3901 Requested
FL Pool Asset Management	RESM	06	S Productive Split	I3907 Reserved
FL Pool Asset Management	RGBE	03	E Completed Split	I3909 Returned
FL Pool Asset Management	STAN	05	E Completed Split	I3906 Canceled
FL Pool Asset Management	STOR	05	E Completed Split	I3904 Res. canceled
FL Pool Asset Management	SY01	08	A Absence	Maintenance
FL Pool Asset Management	SY02	07	A Absence	Ende Leasing

Figure 7.81 Pool Asset Management: Status

Define Pool Categories

You use this Customizing function to define pool categories. You'll need these categories to allocate the scheduled requirements later.

Prerequisites

There are no specific prerequisites.

Customizing Path

Plant Maintenance and Customer Service • Maintenance and Service Processing • Pool Asset Management • Define Pool Categories

Settings

Define the pool categories and their descriptions (see Figure 7.82).

7 Configuring Additional Business Processes

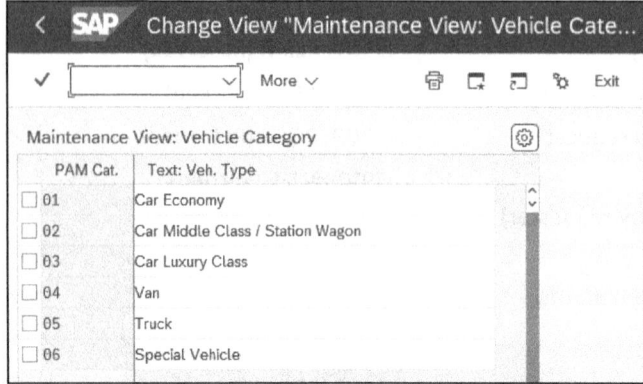

Figure 7.82 Pool Asset Management: Vehicle Category

Define Activity Types for Pool Categories

You use this Customizing function to define, for each controlling area and pool category, which activity types are to be used for the various allocation frequencies (e.g., for a day rate).

Prerequisites

You must define the controlling areas, pool asset management categories, and activity types beforehand. You create activity types with Transaction KL01 (Create Activity Type). You define rates for activity types per controlling area, cost center, and year with Transaction KP26 (Change Plan Data for Activity Types).

Customizing Path

Plant Maintenance and Customer Service • Maintenance and Service Processing • Pool Asset Management • Define Service Types for Pool Categories

Settings

The following cost units are available to you (see Figure 7.83):

- **RentalPr.**
  For the rental price, the costs associated with the borrowing time are calculated based on the unit of measurement for the activity type (hours, days).

- **PrAdd.Unit**
  For the price/unit, the activity-dependent costs (e.g., mileage when using a fleet object pool) are determined based on the unit of measurement for the activity type.

- **Flat Rate1 and Flat Rate2**
  These are used to allocate flat-rate costs (e.g., cleaning costs, refueling costs).

- **Units Incl**
  This is used to specify whether there are free miles for the cost unit **PrAdd.Unit** (e.g., free miles when using a fleet object pool).

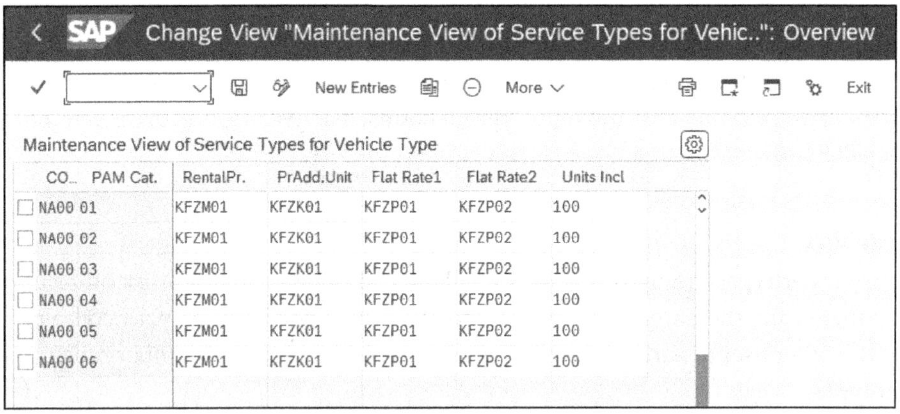

7.5 Pool Asset Management

[Screenshot: Change View "Maintenance View of Service Types for Vehic..": Overview, showing a table with columns CO.., PAM Cat., RentalPr., PrAdd.Unit, Flat Rate1, Flat Rate2, Units Incl, with rows NA00 01 through NA00 06, all containing KFZM01, KFZK01, KFZP01, KFZP02, 100.]

Figure 7.83 Pool Asset Management: Activity Types per Category

Use Transaction KP26 to maintain the prices for each fiscal year (see Figure 7.84).

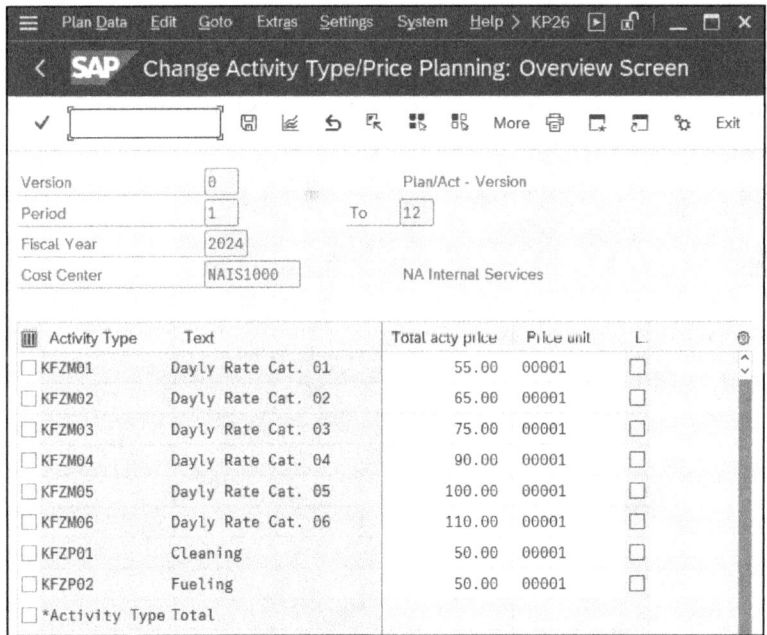

Figure 7.84 Pool Asset Management: Activity Type

Configure Order Types

If you settle a pool asset management requirement, the system creates an order in the background, which is used to settle the costs to the account assignment object in the pool asset management notification (order and cost center). You use this Customizing function to define the order type to be used for this purpose and to define some basic parameters.

515

7 Configuring Additional Business Processes

Define a Separate Order Type for Pool Asset Management
You should define a separate order type for the business process for pool asset management. You shouldn't use this order type for normal maintenance processes, and you shouldn't use a normal order type for this business process.

Prerequisites

Because order types are used not only by plant maintenance but also by other user departments and their areas of SAP S/4HANA, you must consult with your colleagues from the other user departments on the naming conventions and number ranges for order types.

Customizing Path

Plant Maintenance and Customer Service • Maintenance and Service Processing • Maintenance and Service Orders • Functions and Settings for Order Types • Configure Order Types

Settings

The settings for the pool asset management order type are the same as the settings for normal order types (see Figure 7.85).

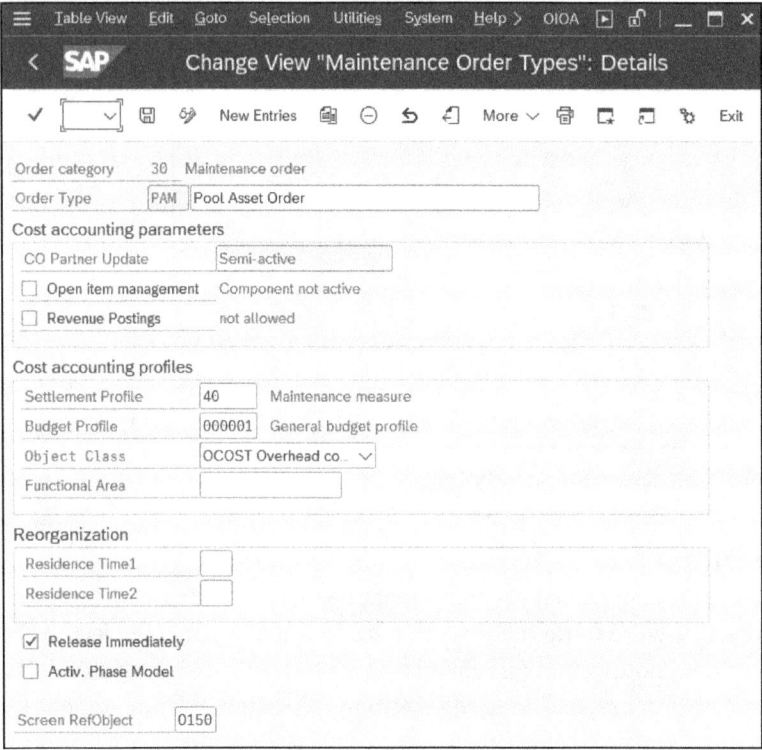

Figure 7.85 Pool Asset Management: Order Type Details

These Customizing settings should now enable you to execute the pool asset management business process.

7.6 Maintenance Event Builder

You can use the *Maintenance Event Builder* via Transaction WPS1 to plan smaller maintenance projects in the form of individual work packages. The Maintenance Event Builder is technically a workbench that supports you when you perform the following tasks:

- Check the worklist of notifications (i.e., the backlog).
- Bundle the notifications into revisions.
- Create orders from the notifications.
- Assign the tasks.
- Display different information such as open work requirements, due dates, orders, and so on.
- Check the availability situation of resources.

If you want to process maintenance tasks within the Maintenance Event Builder, you need to use revisions, which have start and end dates and use status management (e.g., **Created**, **Released**, **Assignments Exist**). You create revisions either explicitly in Transaction IWR1 (Create/Change Revision) or implicitly in the Maintenance Event Builder.

> **Business Function**
> Before you can use the Maintenance Event Builder, make sure that the business functions LOG_EAM_POM and LOG_EAM_POM_2 are activated. They are part of S/4H_ALWAYS_ON_FUNCTIONS and should be activated automatically.

At the heart of the Maintenance Event Builder is the *revision* to which all notifications and orders that need to be executed are assigned (see Figure 7.86).

Maintain Revision Type

Each revision must be assigned to a revision type, and you use this Customizing function to define revision types. You determine whether you want to use the revision type in the Maintenance Event Builder or in normal maintenance processing.

Prerequisites
There are no specific prerequisites.

7 Configuring Additional Business Processes

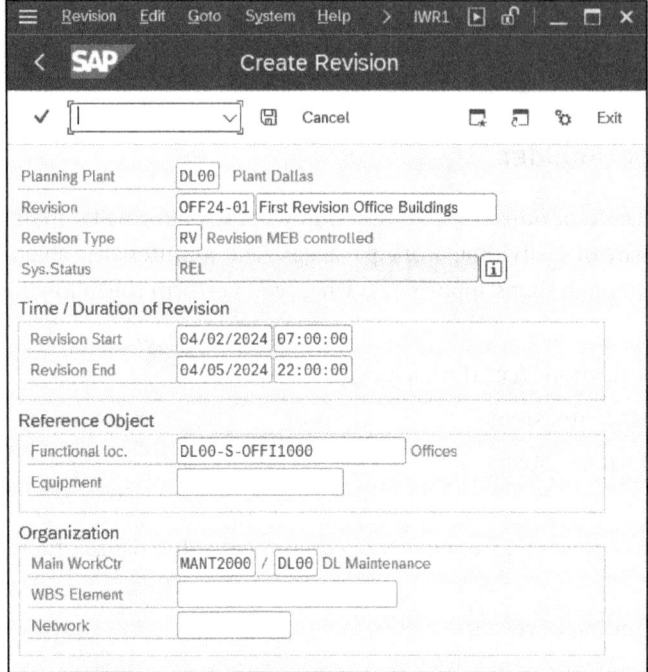

Figure 7.86 Maintenance Event Builder: Define Revision

Customizing Path
Plant Maintenance and Customer Service • Maintenance and Service Processing • Maintenance Event Builder • Maintain Revision Type

Transaction
DIWPSC4

Settings or Recommended Settings
Assign revision type **Not WPS-Controlled** to a revision that you want to use in normal maintenance processing (see Figure 7.87). You can then enter these revisions in notifications and orders, but you can't use these revisions in the Maintenance Event Builder.

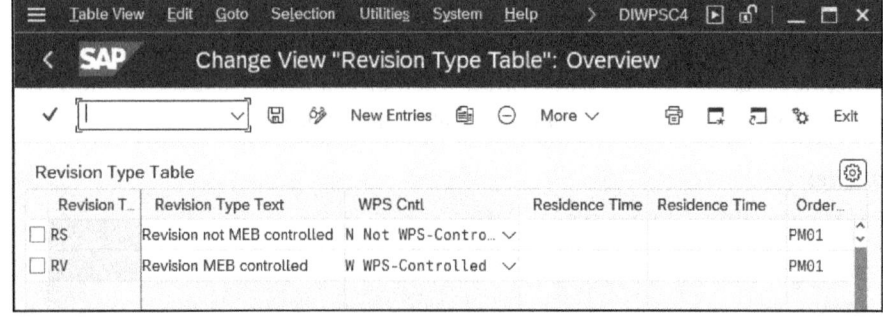

Figure 7.87 Maintenance Event Builder: Revision Type Overview

The opposite behavior is associated with **WPS-Controlled** revision types; that is, you can only use them in the Maintenance Event Builder (not in normal maintenance processing).

This Customizing setting should now enable you to use the Maintenance Event Builder.

7.7 Shift Reports and Shift Notes

You use shift notes and shift reports to document events that occur during a shift. In a *shift note*, you enter information about an event (e.g., comments, times, objects). Shift notes can contain the following information:

- General notes (e.g., shift interruptions, power outages)
- Breakdown documentation (e.g., a turning machine breakdown from/to)
- Suggestions for improvement (e.g., "Reduce machine speed by 10% because...")
- Notes on employees (e.g., "Mr. Henry finished one hour early because...")
- Notes on material usage (e.g., "Casing should be clamped with a maximum pressure of 10 bars.")
- Comments on the use of tools (e.g., "Hand brace 9700 isn't suitable for material T-B400.")

A *shift report* is a PDF document generated at the end of a shift by the person responsible for the shift. It comprises the shift notes that have been recorded as well as other documents. If necessary, a digital signature can be used to sign a shift report. A shift report may include the following elements:

- Shift notes
- Production activities
- Completion confirmations
- Goods movements
- Maintenance notifications
- Quality notifications
- Maintenance orders
- Measurement documents
- Graphical analyses

> **Business Functions** [+]
> Before you can use shift notes and shift reports, you need activated business functions LOG_PP_SRN_CONF and LOG_PP_SRN_02. They are part of S/4H_ALWAYS_ON_FUNCTIONS and should be activated automatically.

The following Customizing functions are available for adjusting shift notes and shift reports to meet your requirements.

Overview of Notification Type

From a technical data perspective, shift notes are notifications that *must* be created with a separate notification type. This Customizing function provides central access to a wide range of settings for a notification type. For a shift note, you require the following:

- The **Screen Areas in Notification Header** subfunction to define the notification header and reference object
- The **Screen Structure for Extended View** subfunction to define the layout

Prerequisites

You must define the notification types beforehand. The notification type for shift notes *must* be assigned to notification category **05** (general notification).

Customizing Path

Plant Maintenance and Customer Service • Maintenance and Service Processing • Maintenance and Service Notifications • Overview of Notification Type

Settings

Use the **Screen Areas in Notification Header** subfunction to determine the notification header. In **Screen Type Hdr**, enter "H700" (**Header general notification**) for the shift note type (see Figure 7.88).

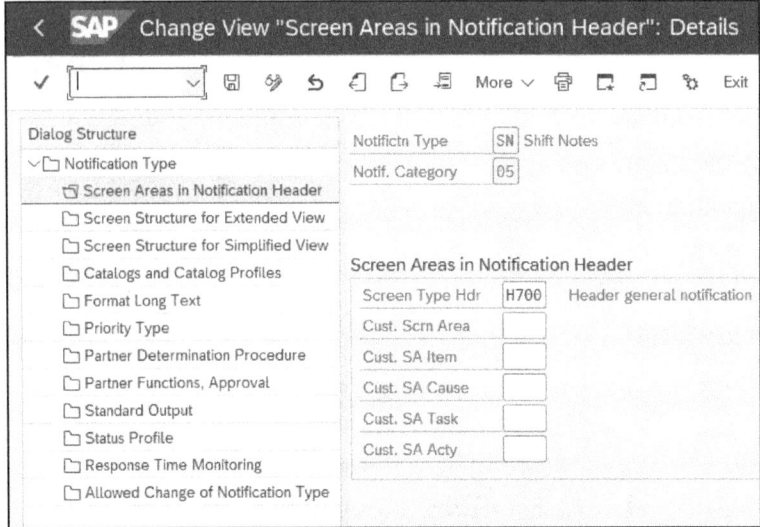

Figure 7.88 Shift Note: Screen Area Header Details

7.7 Shift Reports and Shift Notes

Use the **Screen Structure for Extended View** subfunction to define the layout of the shift note. You require a single tab (see Figure 7.89), and on this tab, enter "130" (**Shift Note**) in **Screen Area 1**.

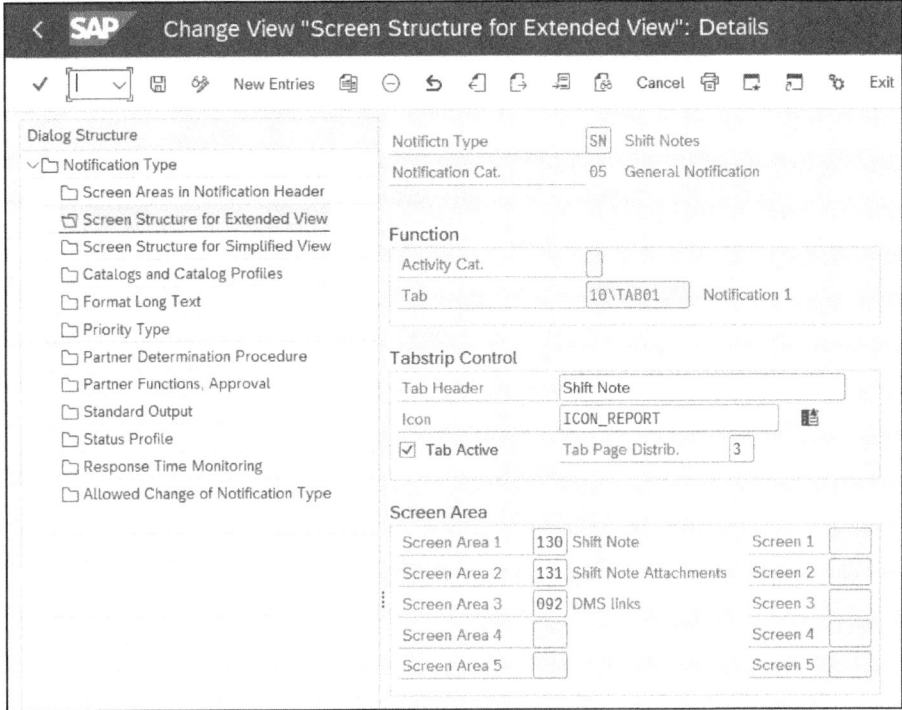

Figure 7.89 Shift Note: Screen Area Tab

Make Settings for Shift Note Type

This Customizing function is the central function for configuring shift notes. Here, you define the main properties of your shift notes (e.g., whether you want to assign reference objects or categories to the shift note).

Prerequisites

You must define the notification types and the catalog for the categories beforehand.

Customizing Path

Plant Maintenance and Customer Service • Maintenance and Service Processing • Shift Reports/Notes • Settings for Shift Notes • Make Settings for Shift Note Type

Settings

The following settings are available to you on the screen shown in Figure 7.90:

- **Note Number**
  Check the box to set this indicator if you want the system to always display the shift note number in the processing transactions associated with the shift note.

7 Configuring Additional Business Processes

- **Origin**
  Check the box to set this indicator if you want the system to always display the origin in the processing transactions associated with the shift note. If the origin is **Work Center/Hierarchy,** the system displays the number and short text of the work center as well as the hierarchy indicator and plant. If the origin is **Functional Location/Equipment**, the system displays the number and short text of the functional location as well as the number and short text of the piece of equipment.

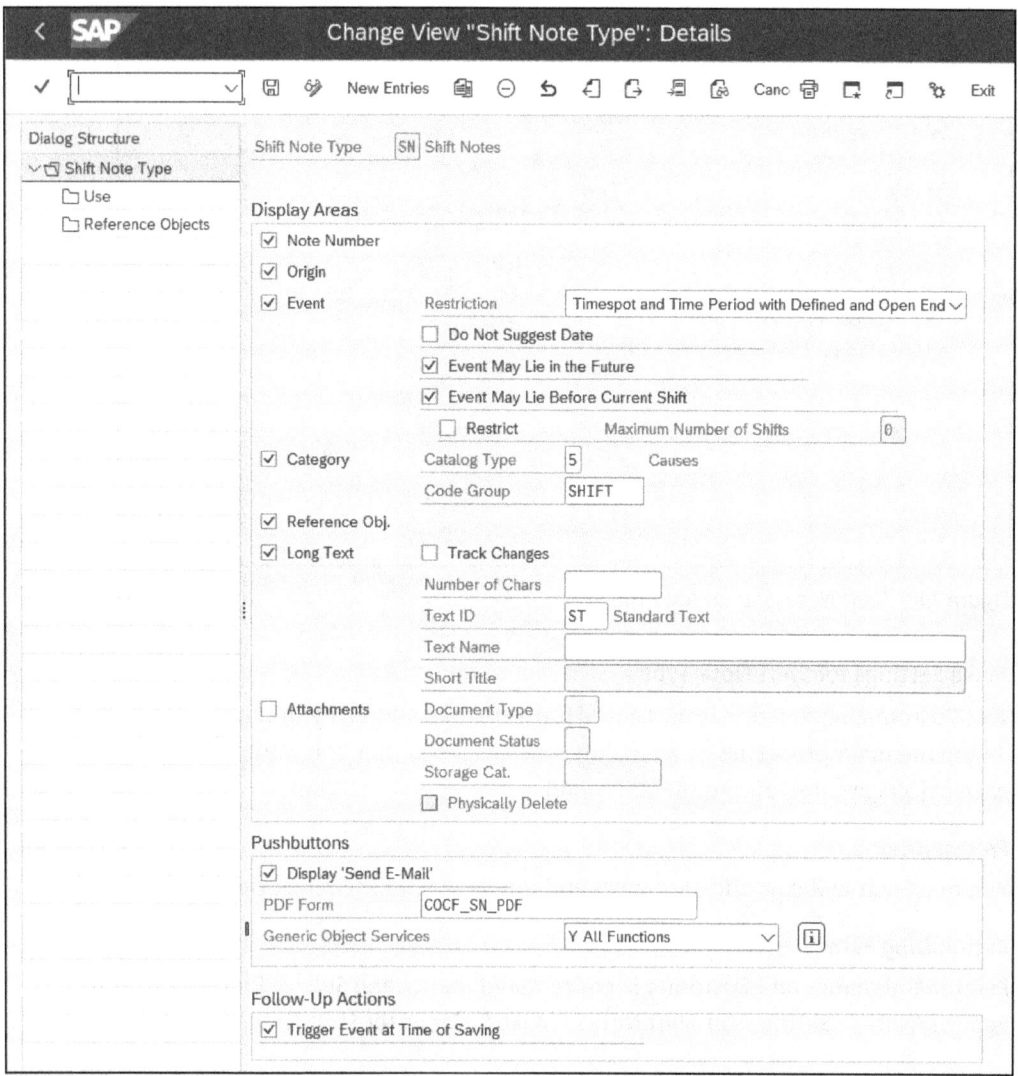

Figure 7.90 Shift Note: Basic Settings

- **Event**
  Check the box to set this indicator if you want to record the time or period in which the event took place in the shift note.
- **Category**
  Check the box to set this indicator if you want to assign a category to the shift note. In the **Catalog Type** and **Code Group** fields, specify the categories that you want to access. Use the **Maintain Catalogs** Customizing function to maintain the categories themselves.
- **Reference Obj.**
  Check the box to set this indicator if you want to enter a reference object (e.g., a piece of equipment or functional location) in the shift note. Use the **Reference Objects** subfunction to define which reference objects you want to use.
- **Long Text**
  Check the box to set this indicator and an OCX window for the long text will be displayed in the shift note.
- **Track Changes**
  Check the box to set this indicator and the system will log who made which changes to the long text and when these changes were made.
- **Attachments**
  Check the box to set this indicator and the system will display the **Attachments** area in the lower half of the shift note. You can then add documents to the shift note as attachments, and then, you must specify a **Document Type** and storage category (**Storage Cat.**) for the documents. In the **Screen Structure for Extended View** Customizing function, you must assign subscreen "131" in addition to subscreen "130."
- **Display 'Send E-Mail'**
  Check the box to set this indicator if you want the system to display the **Send** function key in the processing transactions associated with the shift note. Use the **PDF Form** to define the layout of the email.
- **Generic Object Services**
  Check the box to set this indicator and the system will display the icon and drop-down list for generic object services in the processing transactions associated with the shift note.

Use the **Use** subfunction to define whether you want to create independent shift notes or whether the shift notes are to be displayed as a screen area within a normal notification (see Figure 7.91). If you want to display them in a notification, you can define whether the **Origin** is to be **Work Center/Hierarchy** or **Functional Location/Equipment**.

7 Configuring Additional Business Processes

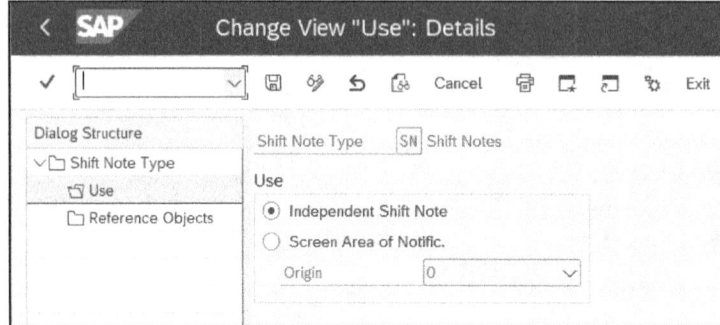

Figure 7.91 Shift Note: Use

Use the **Reference Objects** subfunction to define what you consider to be important reference objects for your shift note (see Figure 7.92).

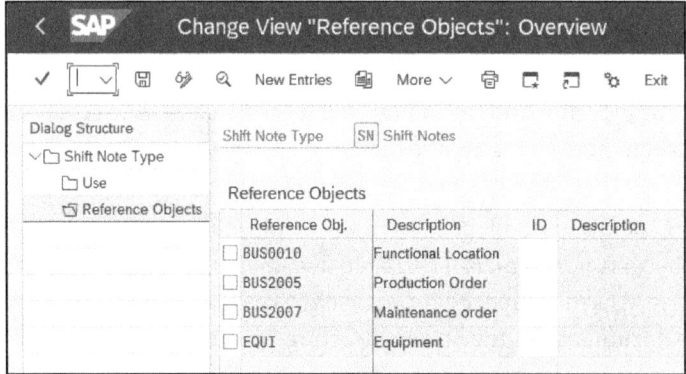

Figure 7.92 Shift Note: Reference Objects

You can choose from the following reference objects:

- BUS1001002: Batch
- EQUI: Equipment
- BUS2005: Production Order
- BUS2116: Production order confirmation
- BUS1038: Production resource/tool
- BUS2007: Maintenance order
- BUS2128: Maintenance order confirmation
- BUS2038: Maintenance notification
- BUS1001: Material
- BUS2092: Measurement document
- SAP_65106: Measuring point
- BUS2004: Planned order

7.7 Shift Reports and Shift Notes

- BUS2016: Process order confirmation
- BUS0001: Process Order
- BUS2045: Inspection lot
- BUS2078: Quality notification
- COCF_SR: Shift report
- COCF_SN: Shift note
- BUS0010: Functional Location

Maintain Catalogs

The shift note categories are defined by a catalog's code group. You use this Customizing function to maintain the catalogs and their associated code groups and codes.

Prerequisites

There are no special prerequisites.

Customizing Path

Plant Maintenance and Customer Service • Maintenance and Service Processing • Maintenance and Service Notifications • Notification Creation • Notification Content • Maintain Catalogs

Transaction

QS41

Settings

You decide which catalog to use for the shift note categories. You can define your own catalog, or you can use an existing one. Figure 7.93 shows the code group categories that were defined in the cause of damage catalog.

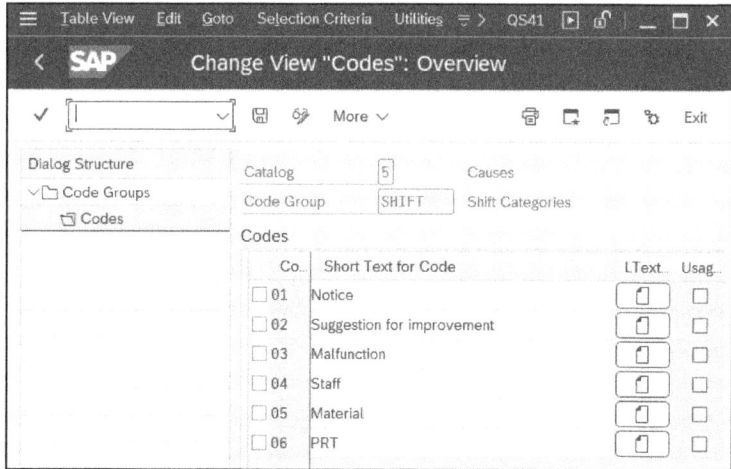

Figure 7.93 Shift Note: Categories

Define Shift Report Types

This Customizing function is *the* central function for configuring shift reports. Here, you define how your shift reports are to look after they are generated and which data they should contain.

Prerequisites

If you want to use digital signatures, you must establish a signature strategy beforehand.

Customizing Path

Plant Maintenance and Customer Service • Maintenance and Service Processing • Shift Reports/Notes • Settings for Shift Reports • Define Shift Report Types

Transaction

ORPS7

Settings

Use the **Shift Report Type** subfunction (see Figure 7.94) to define the shift report types you require. If you require different shift reports that are to contain different structures and content, because of the different requirements of the technical objects or work centers, you must create several shift report types. In this subfunction, also define the following properties for each shift report type:

- Usage
  Use this indicator to control whether the shift report type is to be generated later for **Work Center/Hierarchy** or for **Functional Location/Equipment**.

- Gap-Free Reports
  Use this indicator to control whether a shift report must start at the time when the preceding shift report ends.

- Protect Documents
  If you check the box to set this indicator, the system uses encryption to protect the PDF document against changes.

- Check Documents
  If you check the box to set this indicator, the system checks whether the report has been changed since the last time it was saved, each time you call the shift report.

- Use Signature
  Use this indicator to define whether digital signatures are required when saving and changing the shift report. If they are required, specify the relevant digital signature strategy in the **Signature Strategy** field.

- Complete
  Use this indicator to specify whether a new shift report can only be created if all preceding shift reports have been completed. Shift reports that require a digital signature aren't completed until they have been signed, meaning the next shift report

can't be created until the preceding shift report has been signed. Shift reports that don't require a signature are completed immediately after they are generated.

- **Display 'Send E-Mail' Function Key**
  Check the box to set this indicator if you want the system to display the **Send** function key in the processing transactions associated with the shift report.

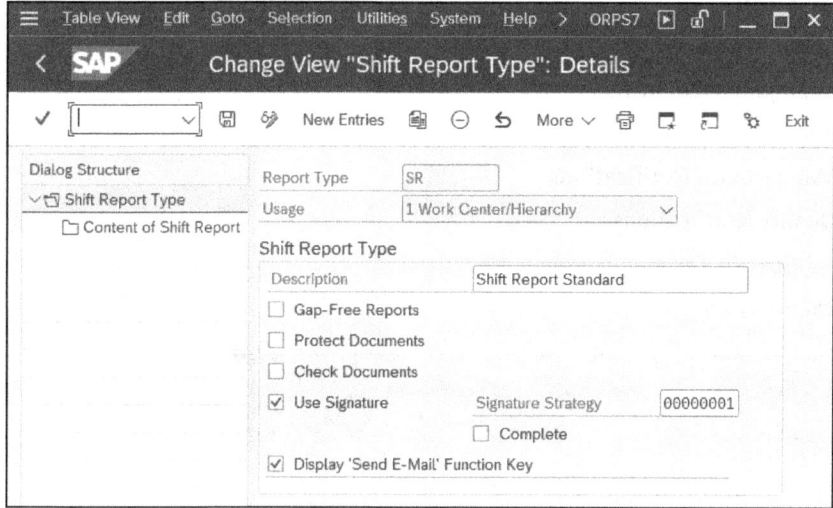

Figure 7.94 Shift Report: Type Details

Use the **Content of Shift Report** subfunction to assign a class and PDF form (see Figure 7.95):

- **Class/Interface**
  SAP has provided the standard class CL_COCF_SR_PDF for this purpose.
- **Form**
  The standard PDF form COCF_SR_PDF_LAYOUT is assigned to this standard class.

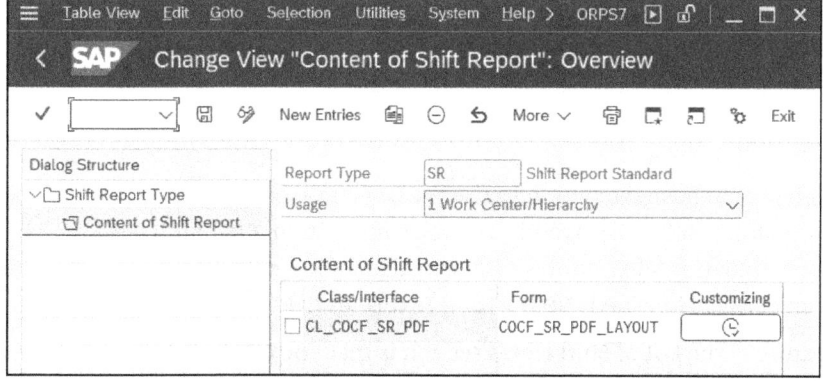

Figure 7.95 Shift Report: Content

A separate Customizing, which you can call via the **Customizing** function key, is predefined for the standard class. Here, specify those sections of the shift report that are to comprise the PDF document. The following sections are available to you (see Figure 7.96):

- Cover Page
- List of Shift Notes
- List of Completion Confirmations
- List of Aggregated Quantities
- List of Aggregated Activities
- List of Maintenance Notifications
- List of Quality Notifications
- Customer-Specific Data
- Graphical Analyses

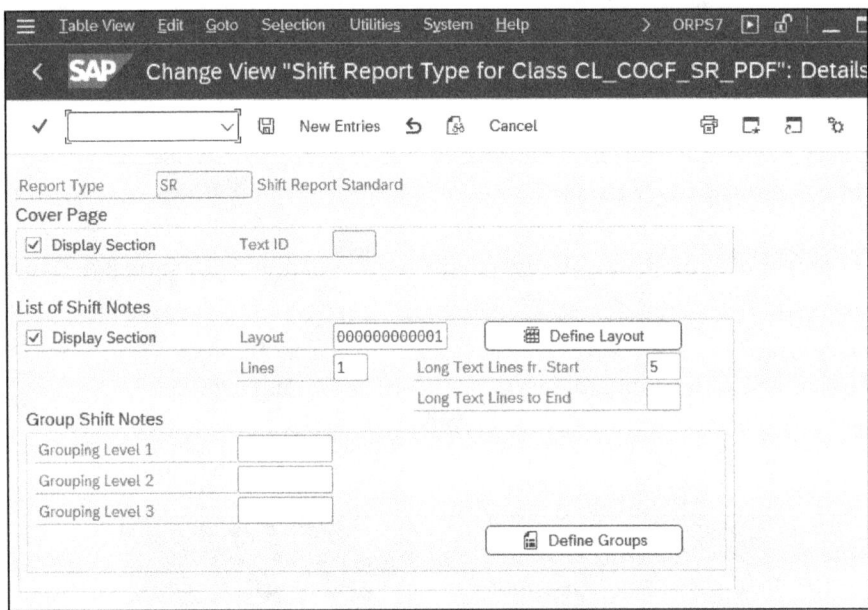

Figure 7.96 Shift Report: Layout

For each section, you have the option to use the **Define Layout** function key to define those fields that are to be displayed in the relevant section. Chapter 3, Section 3.10, described how to define such layouts. You can also define multiline displays (in the **Lines** field) and display long texts (in the **Long Text Lines** fields).

A special feature of the **List of Shift Notes** section is that you can group the shift notes in the shift report according to various criteria. These grouping levels enable you to obtain sorted shift notes in the shift report. For example, you want to sort the list of shift notes according to the functional location and then, within a functional location, according to the equipment number. To do this, use the **Define Groups** function key to call the **Shift Note Group** subfunction, which you then use to define the following two classification criteria: **Functional Location** and **Equipment** (see Figure 7.97).

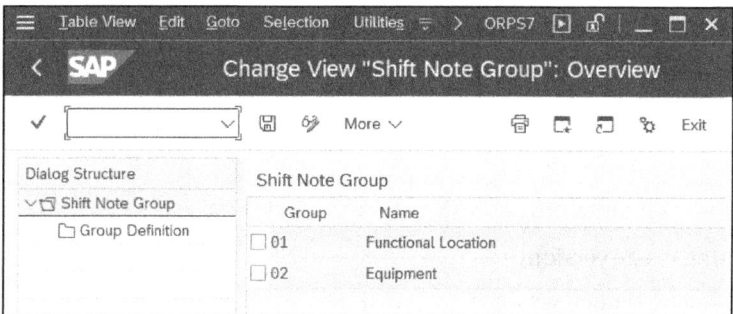

Figure 7.97 Shift Report: Groups Overview

For each classification criterion, define which field is to determine the grouping and sorting on the detail screen (see Figure 7.98). Here, you can press F4 and choose from almost 200 fields in the help text. You also have the option to define a layout, which overrides the general layout for shift notes.

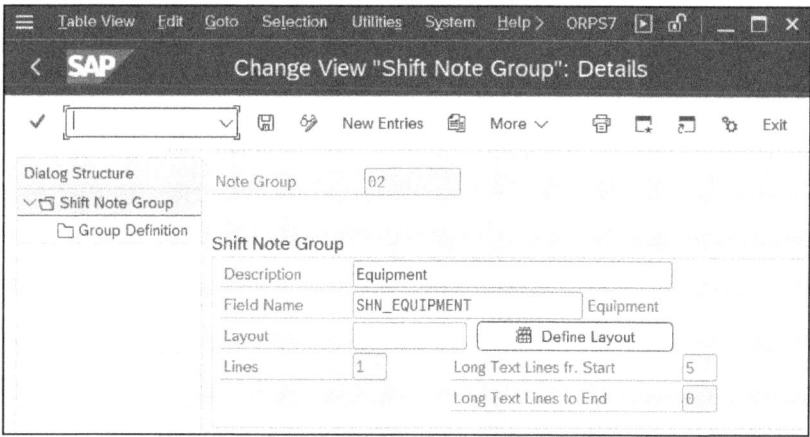

Figure 7.98 Shift Report: Group Details

This Customizing setting should now enable you to use shift notes and shift reports.

7.8 Checklists

Checklists are needed in plant maintenance for performing inspections and checks of assets, and they serve as legally binding documentation agendas.

7 Configuring Additional Business Processes

The inspection checklist process is an end-to-end process starting from the creation of inspection plans to recording results and follow-up actions in plant maintenance.

The checks carried out produce one of the following results:

- Simply an "OK" or a "not OK"
- A qualitative description of the condition (e.g., degree of corrosion)
- A quantitative measurement (e.g., temperature)

Inspections are typically carried out on a regular basis within which the maintenance orders, including inspection lots, are generated based on maintenance plans.

There are two options for maintaining checklists in the SAP system:

- Checklist processing (basic version)
- Checklist processing (extended version)

7.8.1 Checklist Processing (Basic Version)

Figure 7.99 provides an overview of the objects and processes associated with the basic version of the checklist process:

- The master inspection characteristics describe the properties that must be measured (e.g., OK/not OK, visual inspections, or measurements). They are assigned to the general task list on the operation level.
- The technical objects to be checked are managed as equipment or functional location.
- A general task list is created for checking tasks.
- The sampling procedures describe the type and scope of the sample.
- Functional location, equipment and general task list are incorporated into a maintenance plan.
- The maintenance plan generates an order and an inspection lot.
- Each check involves processes from inspection lot management (e.g., results recording and the usage decision) and processes from order management (e.g., time confirmation and technical completion confirmation).

As you can see from Figure 7.99, this process is very similar to the process for calibrating equipment described in Section 7.4. There are only a few differences:

- During the calibration process, the technical object is always a piece of equipment. The technical object of the checklist process can be a functional location and/or a piece of equipment.
- You should define an order type.
- In the follow-up actions, a piece of equipment can be blocked, but a functional location cannot.

7.8 Checklists

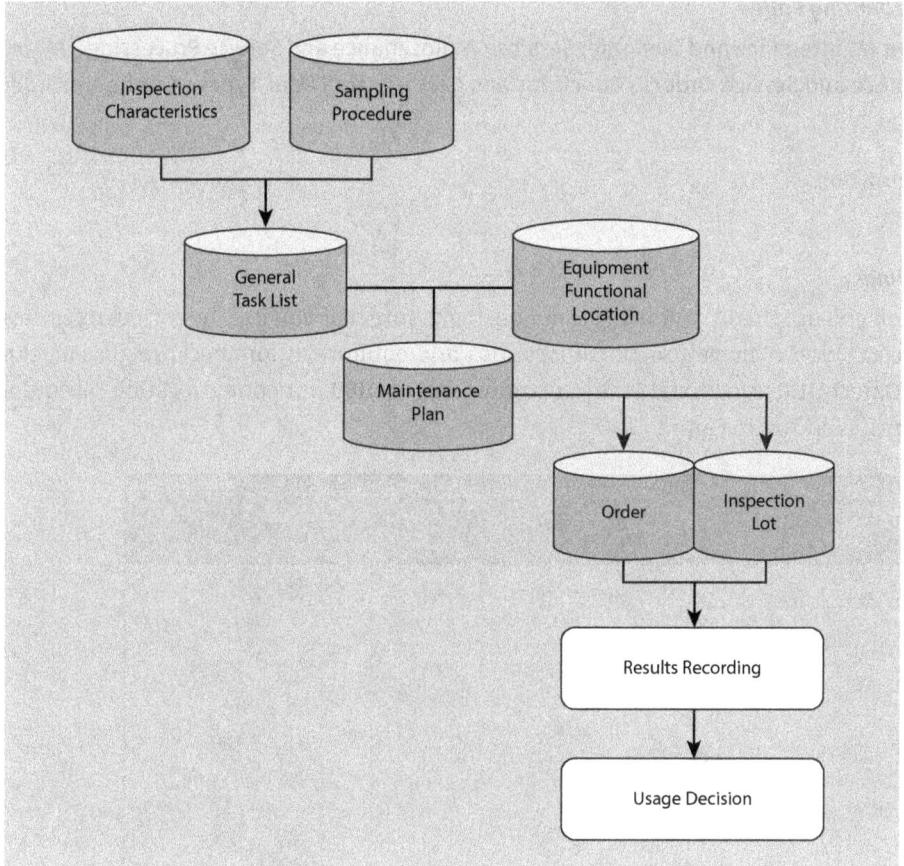

Figure 7.99 Basic Version of Checklist Processing

The following sections describe only those changes in customizing that result from the calibration process (compare with Section 7.4).

Configure Order Types

You use this Customizing function to define the order type and some basic parameters.

> **[+] Separate Order Type for Checklist Processing**
>
> You should define a separate order type for the business process for checklist processing. You can't use this order type for normal maintenance processes, and you can't use a normal order type for test/measurement equipment management.

Prerequisites

Because order types are by plant maintenance as well as by other user departments and their areas in SAP S/4HANA, consult with your colleagues from the other user departments on the naming conventions and number ranges for order types.

7 Configuring Additional Business Processes

Customizing Path
Plant Maintenance and Customer Service • Maintenance and Service Processing • Maintenance and Service Orders • Functions and Settings for Order Types • Configure Order Types

Transaction
OIDA

Settings
When you use this Customizing function, make sure that you use the correct reference object. If you manage functional locations and equipment for checklists, use **Screen RefObject O110** (functional location, equipment) or **O180** (functional location 1:1, equipment), as shown in Figure 7.100.

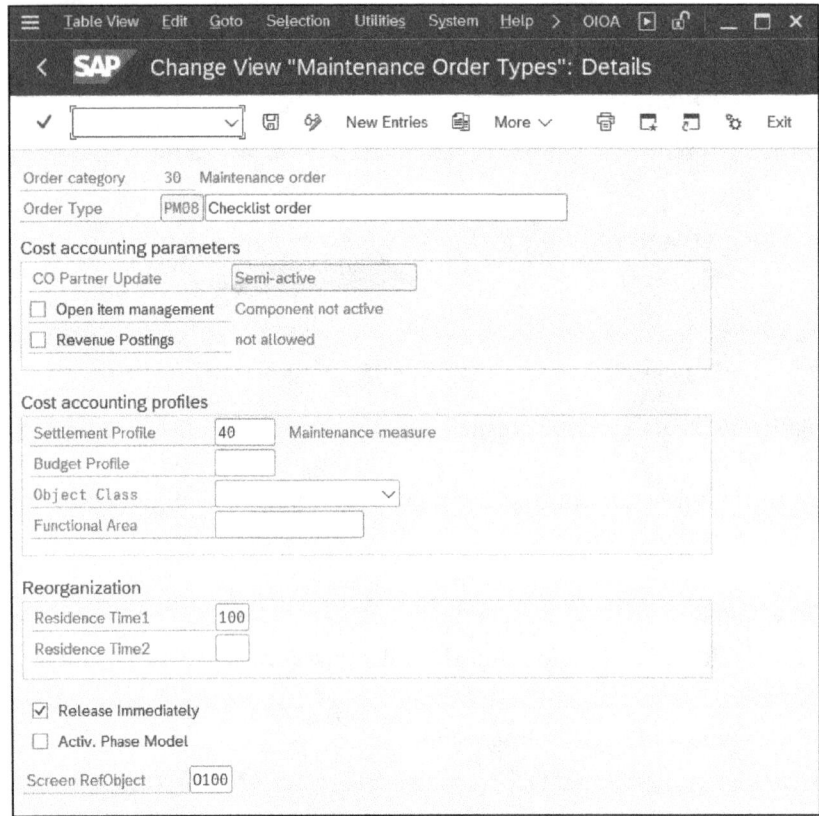

Figure 7.100 Order Type: Checklist Processing

[+] **Checklist Notification Is Optional**
You don't necessarily require a notification for the checklist processing, but you can use one.

If you want to edit notifications by the checklist process, the following Customizing functions have to be carried out:

- **Define Notification Types**
  Plant Maintenance and Customer Service • Maintenance and Service Processing • Maintenance and Service Notifications • Notification Creation • Notification Types • Define Notification Types
- **Change Catalogs and Catalog Profile for Notification Type**
  Plant Maintenance and Customer Service • Maintenance and Service Processing • Maintenance and Service Notifications • Notification Creation • Notification Content • Change Catalogs and Catalog Profile for Notification Type
- **Overview of Notification Type**
  Plant Maintenance and Customer Service • Maintenance and Service Processing • Maintenance and Service Notifications • Overview of Notification Types • Subfunction Screen Areas in Notification Header
- **Define Notification and Order Integration**
  Plant Maintenance and Customer Service • Maintenance and Service Processing • Maintenance and Service Orders • Functions and Settings for Order Types • Define Notification and Order Integration

Define Identifier for Inspection Points

This Customizing function is to be used to define inspection points and their attributes. Inspection points could be installed in several application areas, and for equipment as well as for a functional location, an inspection point has to be defined.

Prerequisites

Order types and usage decisions need to be defined beforehand.

Customizing Path

Quality Management • Quality Planning • Inspection Planning • General • Define Inspection Points

Settings

Define an inspection point and assign inspection point type **2** (**Inspection Point for Funct. Location**) as shown in Figure 7.101 or inspection point type **1** (**Inspection Point for equipment**).

For both inspection points, you must define how you want to deal with the usage decision (see Figure 7.102):

- **Default for valuating inspection point**
  For the plant setting, the equipment usage decision is to be met by using the **Selected Set** to define the code group.

7 Configuring Additional Business Processes

- **Default when all chars. are accepted**
  Define the code you want to use if the inspection lot for the piece of equipment is to be accepted.

- **Default when at least one char. is rejected**
  Define the code you want to use if the inspection lot for the piece of equipment is to be rejected.

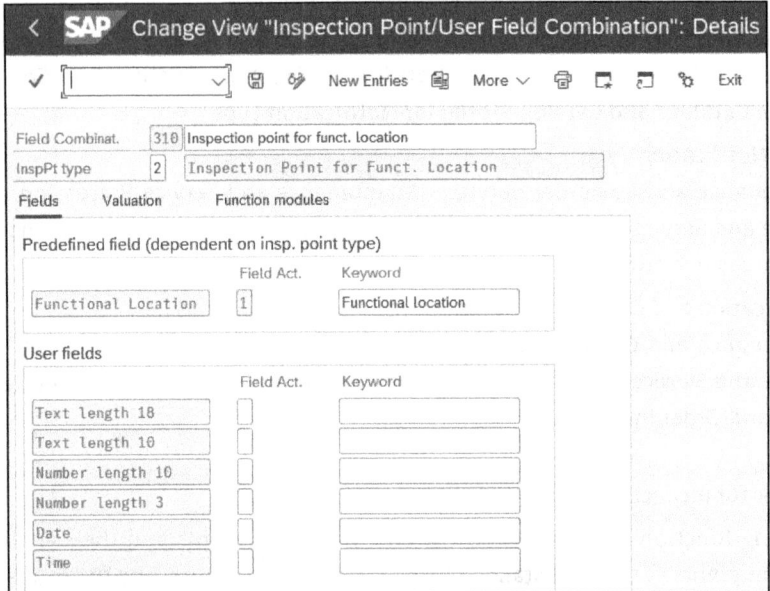

Figure 7.101 Inspection Point

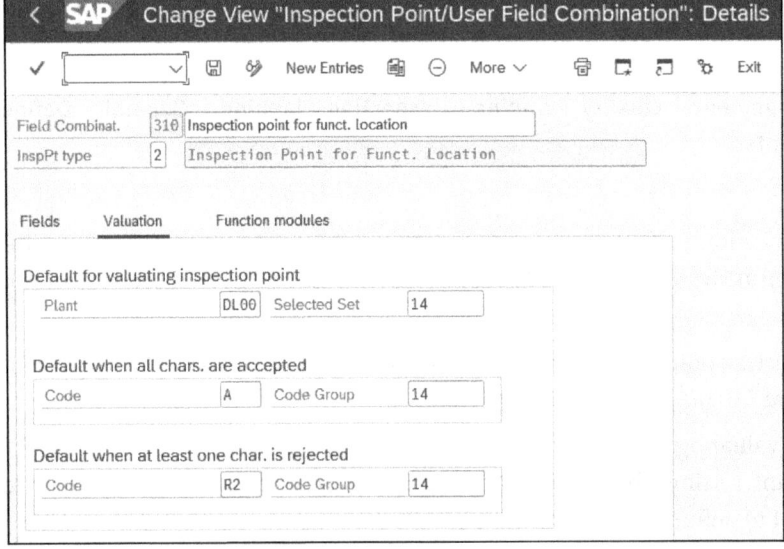

Figure 7.102 Inspection Point: Valuation

534

To define the catalog for usage decisions along with its code groups and codes, see the **Maintain Catalogs for Usage Decisions** Customizing function.

> **Inspection Points for Each Plant**
> Because the default values for the usage decision are selected on a plant-specific basis, I recommend creating separate inspection points for each plant.

Define Inspections in Plant Maintenance

This Customizing function is *the* central element of PM/QM coupling. First, you link the order types you've defined for the checklist process to the corresponding inspection types for each plant defined. Then, if you create an order in this plant and for this order type (in general, it is automatically created by the maintenance plans), the system generates an inspection lot with the specified inspection type because it links the order types automatically.

Prerequisites

Plants, order types, and inspection types are to be defined beforehand.

Customizing Path

Quality Management • Quality Inspection • Inspection Lot Creation • Define Inspections in Plant Maintenance

Settings

Figure 7.103 shows how to assign an inspection type to an order type for each plant. For information about configuring the order type, see the **Configure Order Types** Customizing function in Chapter 5, Section 5.2.1.

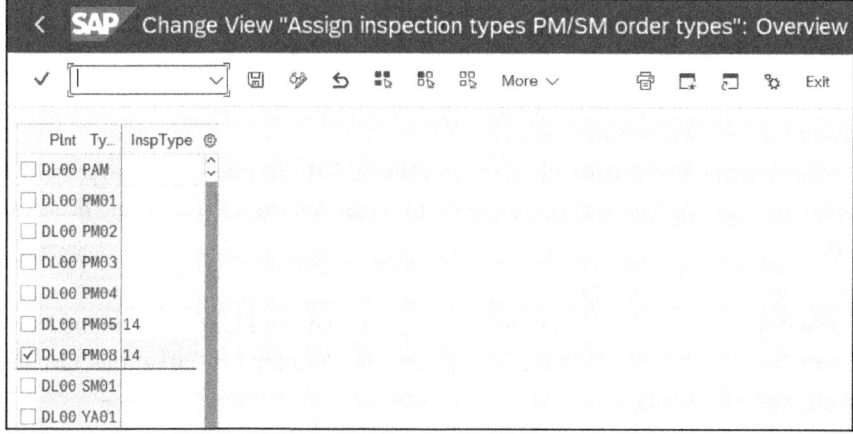

Figure 7.103 Order Type and Inspection Type

> **[!] Assign Inspection Types to Order Types**
>
> Make sure that you assign your inspection types fully and correctly to the order types for each plant. If you don't ensure this assignment, then there is no PM/QM coupling, the system doesn't create any inspection lots for the orders, and you can't calibrate the test/measurement equipment.

Define Follow-Up Action

A third and final difference to the calibration process is defining follow-up actions.

Prerequisites

There are no specific prerequisites.

Customizing Path

Quality Management • Quality Inspection • Inspection Lot Completion • Define Follow-Up Action

Transaction

V_TQ07

Settings

In Section 7.4, I said that one of the automatic follow-up actions to the usage decision can be the automatic locking of a piece of equipment. This also applies to the checklist process, but only if the technical object is a piece of equipment. This is done by the QFOA_OBJECT_STATUS_SET function module.

However, if the technical object is a functional location, there is no automatic follow-up action to be used to lock it.

Other Customizing Settings

All other Customizing settings are the same as in the calibrating process:

- Set View Profiles for Technical Objects
  Plant Maintenance and Customer Service • Master Data in Plant Maintenance and Customer Service • Technical Objects • General Data • Set View Profiles for Technical Objects
- Maintain Equipment Category
  Plant Maintenance and Customer Service • Master Data in Plant Maintenance and Customer Service • Technical Objects • Equipment • Equipment Categories • Maintain Equipment Category

- **Define Additional Business Views for Equipment Categories**
  Plant Maintenance and Customer Service · Master Data in Plant Maintenance and Customer Service · Technical Objects · Equipment · Equipment Categories · Define Additional Business Views for Equipment Categories
- **Define Serial Number Profiles**
  Plant Maintenance and Customer Service · Master Data in Plant Maintenance and Customer Service · Technical Objects · Serial Number Management · Define Serial Number Profiles
- **Maintain Catalogs**
  Plant Maintenance and Customer Service · Maintenance and Service Processing · Maintenance and Service Notifications · Notification Creation · Notification Content · Maintain Catalogs
- **Define Catalog Profile**
  Plant Maintenance and Customer Service · Maintenance and Service Processing · Maintenance and Service Notifications · Notification Creation · Notification Content · Define Catalog Profile
- **Set Maintenance Plan Categories**
  Plant Maintenance and Customer Service · Maintenance Plans, Work Centers, Task Lists, and PRTs · Maintenance Plans · Set Maintenance Plan Categories
- **Maintain Settings at Plant Level**
  Quality Management · Basic Settings · Maintain Settings at Plant Level
- **Define SPC Criteria**
  Quality Management · Quality Planning · Basic Data · Sampling, SPC · Statistical Process Control · Define SPC Criteria
- **Define Control Keys for Inspection Operations**
  Quality Management · Quality Planning · Inspection Planning · Operation · Define Control Keys for Inspection Operations
- **Maintain Inspection Types**
  Quality Management · Quality Inspection · Inspection Lot Creation · Maintain Inspection Types
- **Maintain Inspection Lot Origins and Assign Inspection Types**
  Quality Management · Quality Inspection · Inspection Lot Creation · Maintain Inspection Lot Origins and Assign Inspection Types
- **Define Default Values for Inspection Type**
  Quality Management · Quality Inspection · Inspection Lot Creation · Define Default Values for Inspection Type
- **Maintain Catalogs for Usage Decisions**
  - Quality Management · Quality Notifications · Notification Creation · Notification Content · Maintain Catalogs

- Quality Management • Quality Inspection • Inspection Lot Completion • Edit Code Groups and Codes for Usage Decisions
- Quality Management • Quality Inspection • Inspection Lot Completion • Edit Selected Sets for Usage Decisions

- Plan Automatic Usage Decision
  Quality Management • Quality Inspection • Inspection Lot Completion • Plan Automatic Usage Decision
- Define Recording Configuration
  Quality Management • Quality Inspection • Results Recording • Define Recording Configuration

7.8.2 Checklist Processing (Extended Version)

Careful consideration must be given to whether you need the extended version of checklist processing, as it has significantly more requirements than basic checklist processing. You only need extended checklist processing in cases where you want to generate several checklists per order. Therefore, in most cases, basic checklist processing is sufficient.

In cases where you need to use the extended version of checklist processing, business function LOG_EAM_CHECKLIST has to be activated.

Figure 7.104 provides an overview of the objects and processes associated with the extended version of checklist processing:

- The master inspection characteristics describe the properties to be measured (e.g., OK/not OK, visual inspections, measurements). Each inspection characteristic must contain a checklist type, and inspection characteristics are assigned to the quality management inspection plan on the operation level.
- The technical objects to be checked are managed as equipment or functional locations. Technical objects should be classified.
- A quality management inspection plan is created for checking tasks. The inspection plan can be classified, and the classification must then correspond to the classification of the technical object.
- A plant maintenance task list must be created with a checklist type on the operation level. The checklist type must correspond to the checklist type of the inspection characteristic.
- A plant maintenance task list is incorporated into a maintenance plan.
- The maintenance plan generates an order with an object list.
- If there are matches between a technical object within this object list and a quality management inspection plan, the system automatically creates several inspection lots at order release or you manually create inspection lots using the new Transaction IW92.

- The inspection lot processing (e.g., results recording, usage decision) and order processing (e.g., time confirmation, technical completion) are the same as in basic checklist processing.

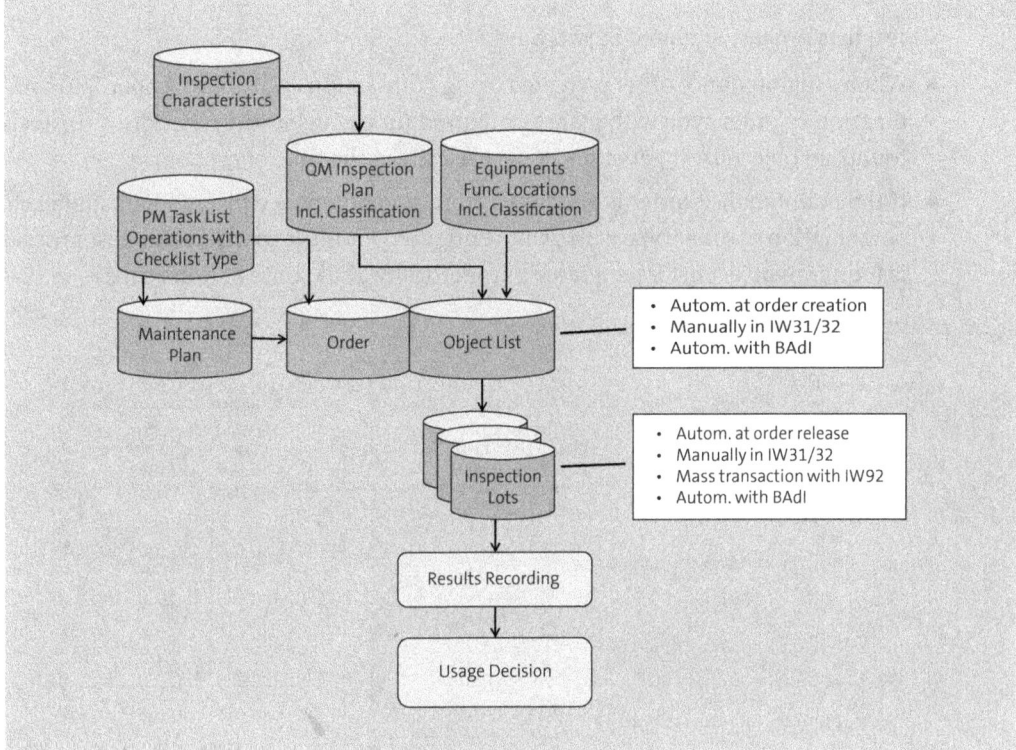

Figure 7.104 Checklist Processing Extended Version

The exact relationship among the objects, which is the prerequisite for the extended checklist process, is shown in Figure 7.105:

- Technical objects (class type 002 for equipment, class type 003 for functional locations) have to be classified. The example shows class PUMP_CHK with characteristic CHK_TYPE for checklist type.
- Quality management inspection plans (class type 018 for task lists) must be classified. The example shows class PUMP_CHK with characteristic CHK_TYPE for checklist type.
- By using Transaction IW98, you produce matching pairs of classification characteristics in order to generate inspection checklists. The characteristic name of a quality management inspection plan has to be entered, and the value must be CHK_TYPE as the characteristic of the technical object. The value EAM_CL_TYPE determines that the characteristic must match the checklist type defined in the order operation for the generation of the inspection checklist.

- You define checklist types with language-dependent short descriptions. Using Transaction IW99, checklist types with language-dependent short descriptions are defined. The example shows checklist type PUMPS and the assignment to plant maintenance task list operations.

Object lists can be generated in two ways:

- When a maintenance order is created using Transaction IW31 and an applicable combination of order type with plant configured for the inspection checklist process is found, an object list is generated automatically at the time of order save.
- If the maintenance order is generated from a maintenance plan and an applicable combination of order type with plant configured for the inspection checklist process is found, an object list is generated automatically at the time of order save.

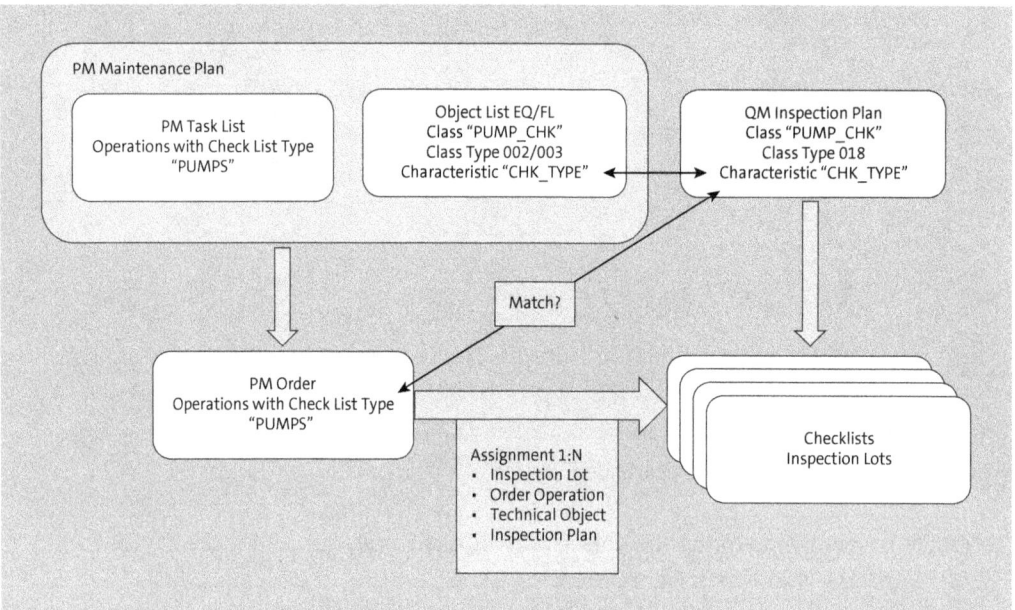

Figure 7.105 Objects in Extended Version of Checklist Processing

Inspection lots can be created in three ways:

- You can use the new **Generate** button in Transactions IW31 or IW32 to generate an inspection checklist.
- You can use Customizing feature **Configure Checklists for Maintenance Order Types and Planning Plants to** enable automatic checklist generation at maintenance order release.
- You can use Transaction IW92 for mass generation of inspection checklists.

As you can see in the examples of Figure 7.104 and Figure 7.105, the master data requirements and the steps of the extended checklist process differ widely from those of the basic checklist process. However, there are only a few differences in the Customizing requirements:

- There is a new Customizing function to configure checklists for order types.
- You have to define your own specific order type.
- A new inspection type is to be defined and assigned to the new order type.
- A new notification type is to be determined for checklist processing.

The following sections only describe the changes in Customizing resulting from the checklist processing basic version from Section 7.8.1.

Configure Order Types

Use the Customizing function to define the order type and some basic parameters.

> **Separate Order Type for Checklist Processing** [+]
>
> You must define a separate order type. For the business process for checklist processing extended version, a separate order type is to be defined. This order type can't be used for normal maintenance processes, and you can't use a normal order type for test/measurement equipment management.

Prerequisites

Order types are made up by plant maintenance as well as by other user departments and their areas in SAP S/4HANA. Consultation with colleagues on the naming convention and number ranges for order types is advisable.

Customizing Path

Plant Maintenance and Customer Service • Maintenance and Service Processing • Maintenance and Service Orders • Functions and Settings for Order Types • Configure Order Types

Transaction

OIDA

Settings

While using this Customizing function, make sure that you use the correct reference object. If you manage functional locations and equipment for checklists, use screen reference object **O100** (functional location, equipment) or **O180** (functional location 1:1, equipment), as shown in Figure 7.106.

7 Configuring Additional Business Processes

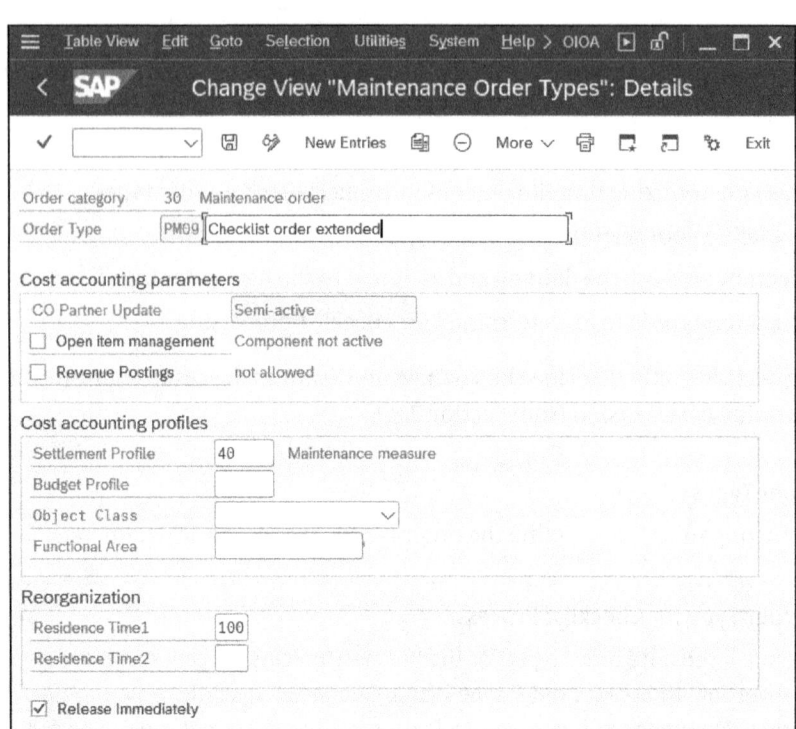

Figure 7.106 Order Type: Extended Checklist Processing

Configure Checklists for Maintenance Order Types and Planning Plants

This is a new Customizing function that you find by activating business function LOG_ EAM_CHECKLIST.

In this Customizing activity, inspection checklists for combinations of maintenance order types and planning plants can be configured. With this functionality in place, you can carry out quality inspections for technical objects in the object list at the operation level in maintenance orders, based on inspection lots.

Prerequisites

You must have maintained a maintenance planning plant as well as an order type, and thus, you must have assigned it to a maintenance planning plant.

You must have entered a suitable inspection type (see the Customizing function).

To be able to print your checklists, you must set the document creation parameters and customize a shop paper (see the Customizing function).

A document type and a storage category must be chosen.

7.8 Checklists

Customizing Path

Plant Maintenance and Customer Service • Maintenance and Service Processing • Maintenance and Service Orders • Functions and Settings for Order Types • Configure Checklists for Maintenance Order Types and Planning Plants

Settings

You must determine the following (see Figure 7.107):

- The inspection type that is assigned to the automatically generated inspection lots
- Whether an object list should be automatically generated at order creation
- Whether checklists are to be automatically generated at order release
- Parameters for document creation (document type and storage category)
- Notification type, if notifications should be created for checklists

Also note that the Customizing setting defines whether all usage decisions have to be made for all checklists before an order can be completed technically.

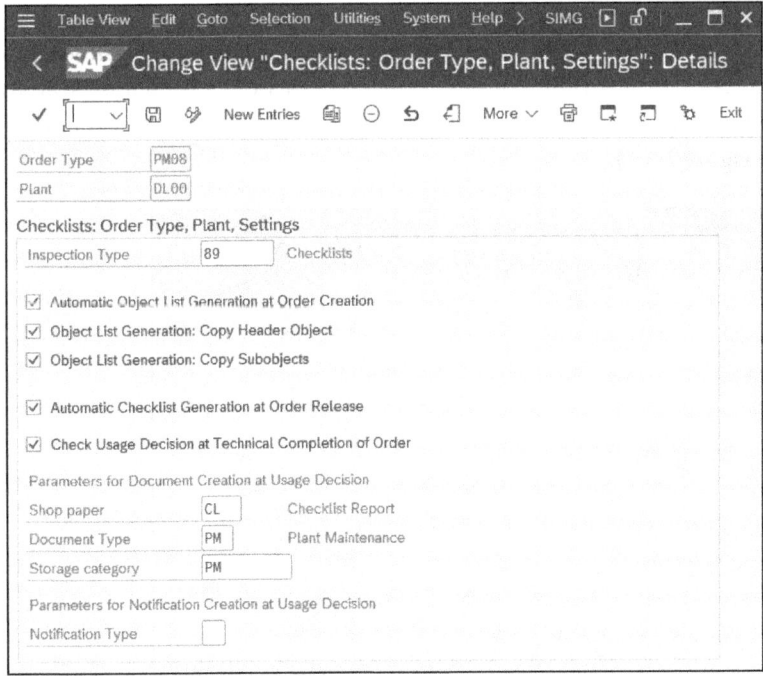

Figure 7.107 Checklists: Maintenance Order Types and Planning Plants

Define Shop Papers, Forms, and Output Programs

This Customizing function may be used to define your shop papers and their print layout. It could also be determined which shop paper is to be printed for which order type.

When you activate business function EAM_LOG_CHECKLIST, it makes a new paper available for printing orders with checklists.

7 Configuring Additional Business Processes

Prerequisites

You must program the print report as well as the layout defined in an SAPscript form beforehand. The same applies to the order type.

If you want to print papers before an order is released or after an order is technically completed, you should activate business function LOG_EAM_CI_7.

Customizing Path

Plant Maintenance and Customer Service • Maintenance and Service Processing • Maintenance and Service Orders • Print Control • Define Shop Papers, Forms, and Output Programs

Transactions

The following transactions are available for defining and assigning shop papers:

- OIDF (PM Shop Papers for Orders)
- OIDG (PM Shop Papers by Order Type)

Settings

This Customizing function has two subfunctions: **Define Shop Papers** and **Define Shop Papers for Order Type**. Use the **Define Shop Papers** subfunction (see Figure 7.108) to define the shop paper and assign it to an order type with the **Shop Papers by Order Type** subfunction (see Figure 7.109).

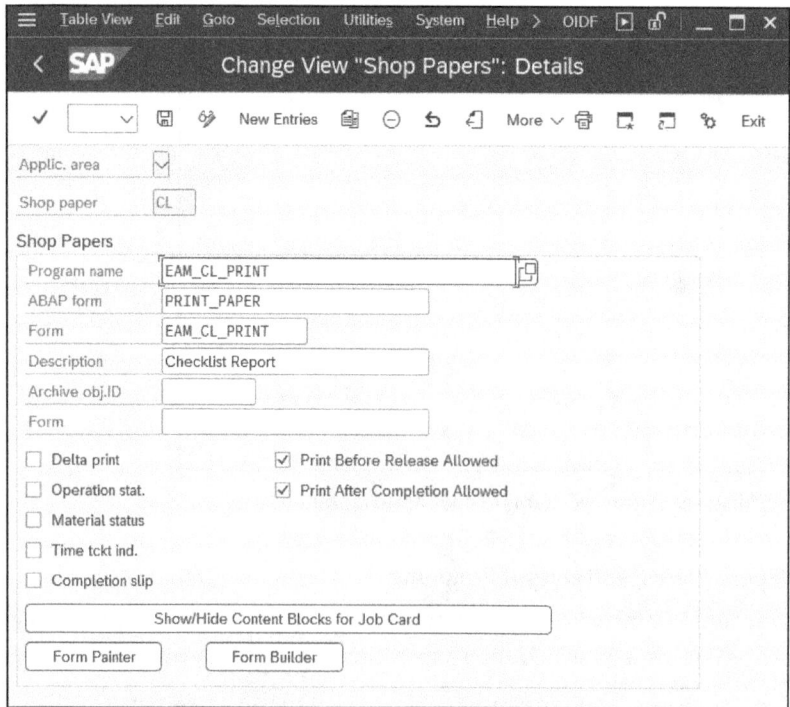

Figure 7.108 Shop Papers: Checklist Processing

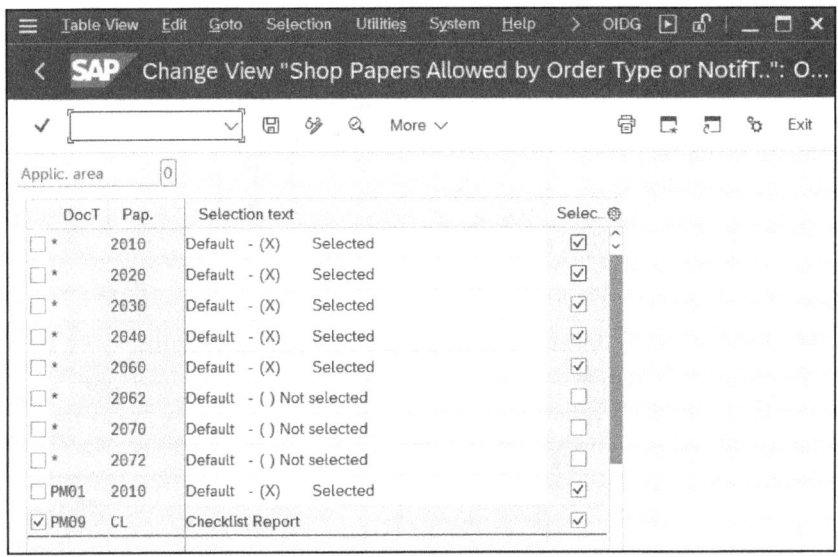

Figure 7.109 Shop Papers: Assign to Order Type

> **Checklist Notification Is Optional**
> You don't necessarily require a notification for the checklist processing, but you may use one.

If you want to use a notification with checklist processing, the following Customizing functions have to be carried out:

- Define Notification Types
  Plant Maintenance and Customer Service • Maintenance and Service Processing • Maintenance and Service Notifications • Notification Creation • Notification Types • Define Notification Types

- Change Catalogs and Catalog Profile for Notification Type
  Plant Maintenance and Customer Service • Maintenance and Service Processing • Maintenance and Service Notifications • Notification Creation • Notification Content • Change Catalogs and Catalog Profile for Notification Type

- Overview of Notification Type
  Plant Maintenance and Customer Service • Maintenance and Service Processing • Maintenance and Service Notifications • Overview of Notification Types • Subfunction Screen Areas in Notification Header

- Define Notification and Order Integration
  Plant Maintenance and Customer Service • Maintenance and Service Processing • Maintenance and Service Orders • Functions and Settings for Order Types • Define Notification and Order Integration

7 Configuring Additional Business Processes

Define Inspections in Plant Maintenance

This Customizing function is *the* central element of PM/QM coupling. Here, you link the order types defined for checklist processing to the corresponding inspection types for each previously defined plant. In the subsequent creation of an order in this plant and for this order type (in general, this is automatically created from the maintenance plans), the system automatically generates an inspection lot with the specified inspection type because it links the order types.

Prerequisites

You must define the plants, order types, and inspection types beforehand.

Customizing Path

Quality Management • Quality Inspection • Inspection Lot Creation • Define Inspections in Plant Maintenance

Settings

Figure 7.110 shows how to assign, for each plant, an inspection type to an order type for the extended version of checklist processing.

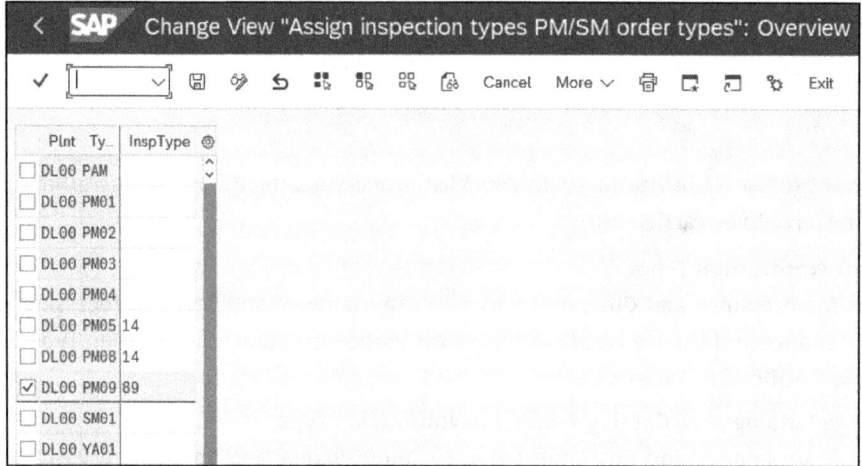

Figure 7.110 Assign Inspection Type to Order Type

> [!] **Assign Inspection Types to Order Types**
>
> Make sure that you assign your inspection types fully and correctly to the order types for each plant. If you don't ensure this assignment, there will be no PM/QM coupling, the system won't create any inspection lots for the orders, and you won't be able to calibrate the test/measurement equipment.

546

7.8 Checklists

Maintain Inspection Types

Just as the order type is at the heart of maintenance task processing and can be used to significantly influence maintenance processes, the inspection type is at the heart of inspection task processing and could also be used to considerably influence inspection processes.

If you activate business function LOG_EAM_CHECKLIST, a new inspection lot origin **89** becomes available for the extended version of checklist processing.

Prerequisites

You must define the status profile, quality management order types, catalog for usage decisions, and notification types beforehand.

Customizing Path

Quality Management • Quality Inspection • Inspection Lot Creation • Maintain Inspection Types

Settings

Figure 7.111 shows you the basic settings for an inspection type.

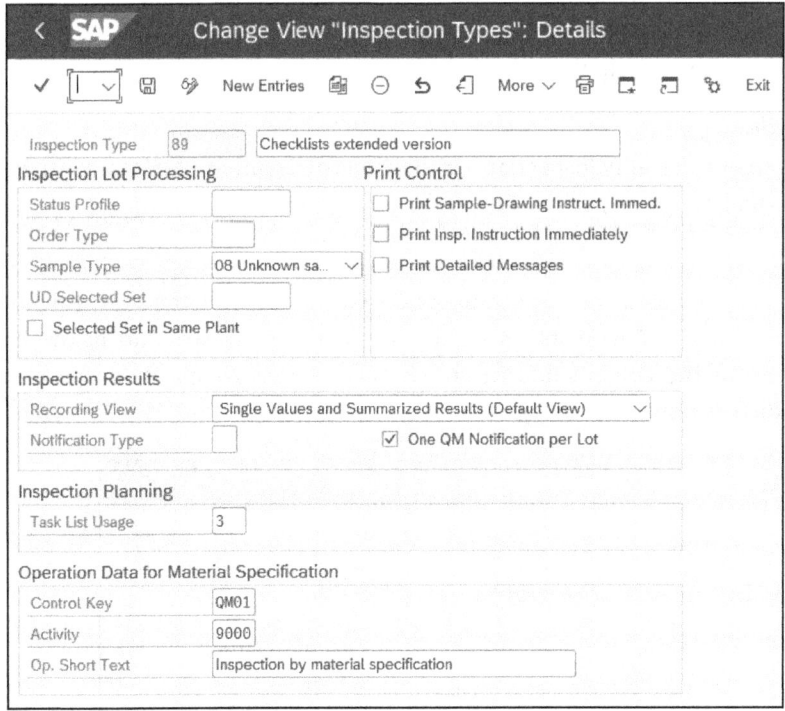

Figure 7.111 Inspection Type for Checklist Processing: Extended Version

7 Configuring Additional Business Processes

For the extended version of checklist processing, a separate inspection type is required. Other inspection types (e.g., incoming inspections, in-process inspections in production) are to no avail.

From a test/measurement equipment management perspective, the following settings are important:

- **UD Selected Set**
  This defines the usage decisions in relation to accepting or rejecting the inspection lots.
- **Notification Type**
  This notification type is created in case an inspection lot is rejected.

Maintain Inspection Lot Origins and Assign Inspection Types

This Customizing function is used to assign your inspection type to an inspection lot origin, which, for the extended version of checklist processing, is **89**. The inspection lot origin is predefined by SAP and can't be changed. You can define inspection types yourself. Each inspection type must refer to an inspection lot origin, and several inspection types may be assigned to the same origin.

Prerequisites

Inspection types are to be designed beforehand.

Customizing Path

Quality Management • Quality Inspection • Inspection Lot Creation • Maintain Inspection Lot Origins and Assign Inspection Types

Settings

In the dialog structure below the **Inspection Lot Origin** subfunction, the standard SAP delivery contains entry **89 Checklists** (see Figure 7.112). In the dialog structure below the **Inspection Types for the Origin** subfunction, you need to assign specific inspection types for checklists to entry **89** (see Figure 7.113).

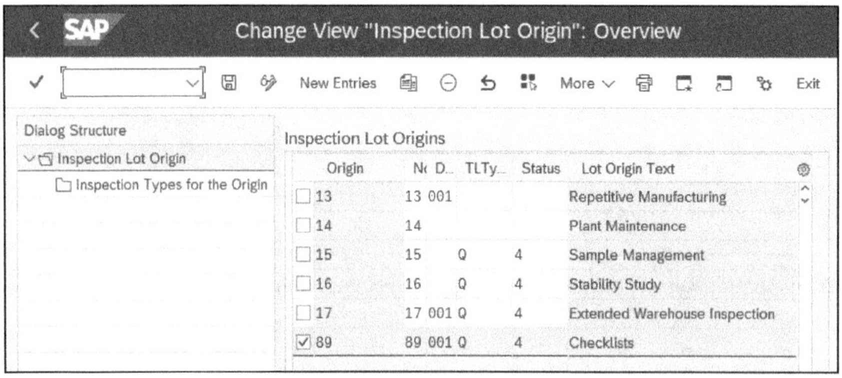

Figure 7.112 Inspection Types: Overview

7.8 Checklists

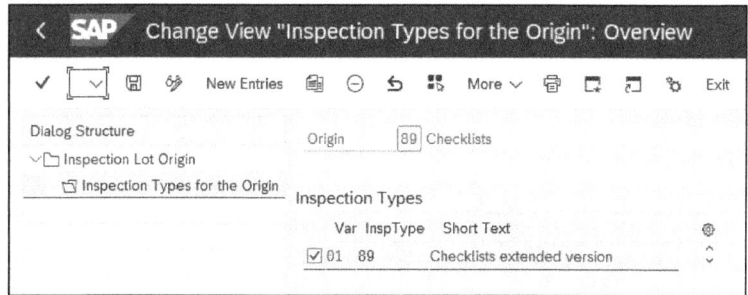

Figure 7.113 Inspection Types: Inspection Origin

> **[!] Assign Origin 89**
> It's imperative to assign origin **89** to all inspection types for the extended version of checklist processing. Otherwise, use of these inspection types for the extended version of checklist processing is not possible.

Other Customizing Settings

All other Customizing settings are the same as in the checklist process basic version:

- Set View Profiles for Technical Objects
  Plant Maintenance and Customer Service • Master Data in Plant Maintenance and Customer Service • Technical Objects • General Data • Set View Profiles for Technical Objects
- Maintain Equipment Category
  Plant Maintenance and Customer Service • Master Data in Plant Maintenance and Customer Service • Technical Objects • Equipment • Equipment Categories • Maintain Equipment Category
- Define Additional Business Views for Equipment Categories
  Plant Maintenance and Customer Service • Master Data in Plant Maintenance and Customer Service • Technical Objects • Equipment • Equipment Categories • Define Additional Business Views for Equipment Categories
- Define Serial Number Profiles
  Plant Maintenance and Customer Service • Master Data in Plant Maintenance and Customer Service • Technical Objects • Serial Number Management • Define Serial Number Profiles
- Maintain Catalogs
  Plant Maintenance and Customer Service • Maintenance and Service Processing • Maintenance and Service Notifications • Notification Creation • Notification Content • Maintain Catalogs

549

- **Define Catalog Profile**
  Plant Maintenance and Customer Service • Maintenance and Service Processing • Maintenance and Service Notifications • Notification Creation • Notification Content • Define Catalog Profile

- **Set Maintenance Plan Categories**
  Plant Maintenance and Customer Service • Maintenance Plans, Work Centers, Task Lists, and PRTs • Maintenance Plans • Set Maintenance Plan Categories

- **Maintain Settings at Plant Level**
  Quality Management • Basic Settings • Maintain Settings at Plant Level

- **Define SPC Criteria**
  Quality Management • Quality Planning • Basic Data • Sampling, SPC • Statistical Process Control • Define SPC Criteria

- **Define Control Keys for Inspection Operations**
  Quality Management • Quality Planning • Inspection Planning • Operation • Define Control Keys for Inspection Operations

- **Define Identifier for Inspection Points**
  Quality Management • Quality Planning • Inspection Planning • General • Define Inspection Points

- **Define Default Values for Inspection Type**
  Quality Management • Quality Inspection • Inspection Lot Creation • Define Default Values for Inspection Type

- **Maintain Catalogs for Usage Decisions**
  - Quality Management • Quality Notifications • Notification Creation • Notification Content • Maintain Catalogs
  - Quality Management • Quality Inspection • Inspection Lot Completion • Edit Code Groups and Codes for Usage Decisions
  - Quality Management • Quality Inspection • Inspection Lot Completion • Edit Selected Sets for Usage Decisions

- **Define Follow-Up Action**
  Quality Management • Quality Inspection • Inspection Lot Completion • Define Follow-Up Action

- **Plan Automatic Usage Decision**
  Quality Management • Quality Inspection • Inspection Lot Completion • Plan Automatic Usage Decision

- **Define Recording Configuration**
  Quality Management • Quality Inspection • Results Recording • Define Recording Configuration

7.9 Work Order Cycle Using the Phase Model

SAP has introduced the *phase model* for notification and order processing since SAP S/4HANA version 2021. In contrast to the standard work order cycle, which consists of six phases (see Chapter 5), the phase-based work order cycle has the following nine phases (see Figure 7.114):

1. **Initiation**
   During the *initiation phase*, you can create maintenance requests for a technical object such as an item of equipment or a functional location using the Create Maintenance Request (F1511A) app. You can enter all necessary details to help in screening, processing, planning, and execution of the request, and you can also include attachments and links. When you submit a maintenance request, the new request is available in the My Maintenance Requests (F4513) app. Only your requests are displayed.

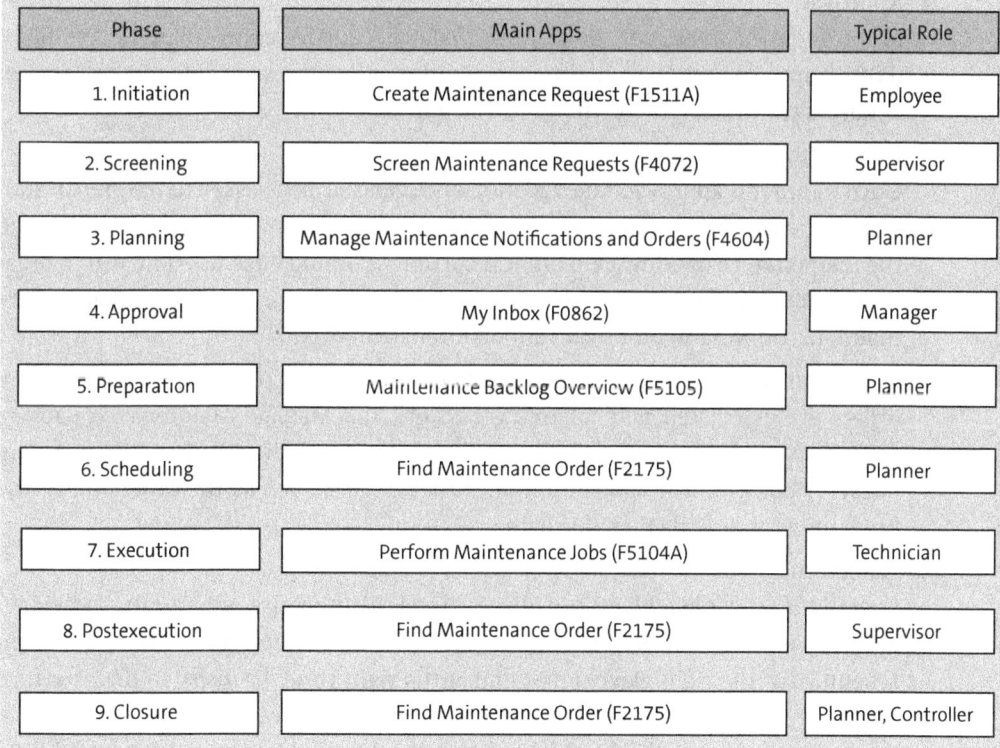

Phase	Main Apps	Typical Role
1. Initiation	Create Maintenance Request (F1511A)	Employee
2. Screening	Screen Maintenance Requests (F4072)	Supervisor
3. Planning	Manage Maintenance Notifications and Orders (F4604)	Planner
4. Approval	My Inbox (F0862)	Manager
5. Preparation	Maintenance Backlog Overview (F5105)	Planner
6. Scheduling	Find Maintenance Order (F2175)	Planner
7. Execution	Perform Maintenance Jobs (F5104A)	Technician
8. Postexecution	Find Maintenance Order (F2175)	Supervisor
9. Closure	Find Maintenance Order (F2175)	Planner, Controller

Figure 7.114 Work Order Cycle Using Phase Model: Overview

2. **Screening**
   The submitted maintenance request moves to the *screening phase*, when maintenance requests are screened and accepted. As a supervisor, you can review all open maintenance requests in the Screen Maintenance Requests (F4072) app. If information is insufficient, you can send the request back to the initiator, and when the

initiator provides information and resubmits the request, you can review the request again.

3. **Planning**
The *planning phase* begins when a maintenance request is accepted. Being the maintenance planner, you can now create and plan orders. In the Manage Maintenance Notifications and Orders (F4604) app, maintenance orders can be created for one or more maintenance notifications. Alternatively, an existing maintenance order can be assigned to a maintenance notification, and information from the leading maintenance notification is copied into the maintenance order.

Orders can also be created in the Process Maintenance Order (W0017) app, where you can add information related to operation, material, labor, services and cost to plan the order. If the maintenance order type is configured for cost approval, the planner will submit the order for cost approval.

4. **Approval**
In the *approval phase*, you view orders that are ready for approval in the My Inbox (F0862) app. To approve a maintenance order, select the workflow request, review details of the order, enter comments, and approve.

5. **Preparation**
Once an order is approved for execution and released, it is passed to the *preparation phase*. The Manage Maintenance Planning Buckets (F3888) app helps you manage the maintenance backlog. One-off or recurrent planning buckets allow you to organize major maintenance events (such as a planned shutdown) and define weekly maintenance windows for recurrent maintenance work.

When the maintenance planning buckets have been created and maintenance orders are associated with planning buckets, the Manage Maintenance Backlog (F5105) app provides a list of all the maintenance orders in a specific planning bucket. Here, you can view important order details, such as the order status, the order priority, and the final due date.

6. **Scheduling**
During the *scheduling phase*, you dispatch the maintenance order or the individual order operations and suboperations and thereby confirm that they have been scheduled at the right work center and at the right time. To dispatch maintenance orders, you can either select several maintenance orders in the Find Maintenance Order (F2175) app or call individual orders in the Process Maintenance Order (W0017) app.

7. **Execution**
In the *execution phase*, a maintenance technician receives all the orders and operations that need to be executed. The technician then performs the required maintenance tasks according to the recommendations provided in the order operation. With the Perform Maintenance Jobs (F5104A) app, the maintenance technician can

review, execute, and report the findings for the jobs dispatched for execution. This app allows you to start, pause, and mark the job as done, and it also allows you to record the time for the job, failure data, and measurement readings.

8. **Postexecution**
In the *postexecution phase*, the supervisor reviews the failure data, confirmed times, and issued materials, using the Find Maintenance Order (F2175) app. The supervisor can also review the actual order costs using the Maintenance Order Costs (F4603) app. The status of the order is set to technically completed, and the supervisor then reviews the notification status and approves it as completed if the order has the associated notification.

9. **Completion**
During the *completion phase*, the maintenance planner or controller reviews maintenance orders that are technically completed, using the Find Maintenance Order (F2175) app. The maintenance planner or controller then settles the orders and performs the business completion.

Difficult Phase Model Configuration

The phase-based order cycle requires extensive and difficult Customizing. SAP itself says the following on page 21 its manual *S43400 Exploring Advanced Functions in Maintenance Processing*:

"Please note that the best method to implement the phase model is the use of the Best Practices and the pre-configured content of scope items 4HH, 4HI, 4VT and 4WM. This will allow an automatic Customizing. If you cannot use the Best Practices scope items, you can configure the phase model manually. However, this will result in a much bigger effort."

After having manually configured the phase model in our own SAP S/4HANA system, I can do nothing but confirm this statement. However, let me now describe the configurations you need to make if you want to use the phase model.

Activate Business Feature

This Customizing function is the starting point for Customizing. All further Customizing settings are only available if you activate the phase model.

Prerequisites
There are no specific prerequisites.

Customizing Path
ABAP Platform • Application Server • Business Management • SAP Business Feature • Activate Business Feature

7 Configuring Additional Business Processes

Settings

You have to add the business feature **EAM_PHASE_MODEL_PROCESSING** and to set it to 3 = **Active** (see Figure 7.115).

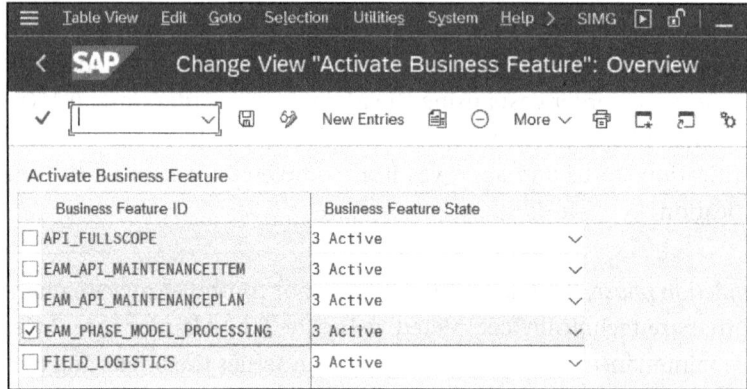

Figure 7.115 Phase Model: Activate

Define Notification Types

You use this Customizing function to define the notification types that you require. The standard notification type for phase model processing is **Y1**.

Prerequisites

There are no technical prerequisites, but you must consult with colleagues from quality management, service management, and claims management on the naming convention for notification types because they also need to use notifications.

Customizing Path

Plant Maintenance and Customer Service • Maintenance and Service Processing • Maintenance and Service Notifications • Notification Creation • Notification Types • Define Notification Types

Settings

Specifically, you can or should configure the settings for notification type **Y1** (see Figure 7.116).

You can find an explanation of the settings in Chapter 5, Section 5.1.1.

Assign Notification Types to Order Types

You use this Customizing function to define which order type is to be proposed for which notification type if an order is generated from a notification. The standard order type for phase model processing is **YA01**.

7.9 Work Order Cycle Using the Phase Model

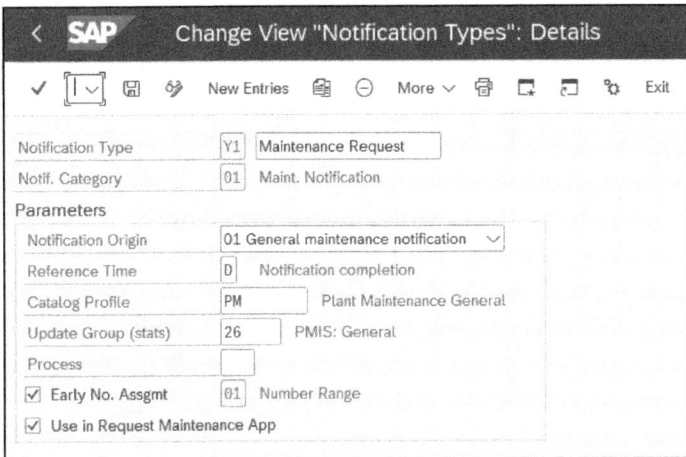

Figure 7.116 Notification Type Y1

Prerequisites
You must define the notification type **Y1** and order type **YA01** beforehand.

Customizing Path
Plant Maintenance and Customer Service • Maintenance and Service Processing • Maintenance and Service Notifications • Notification Creation • Notification Types • Assign Notification Types to Order Types

Settings
Enter order type **YA01** to be proposed for notification type **Y1** (see Figure 7.117).

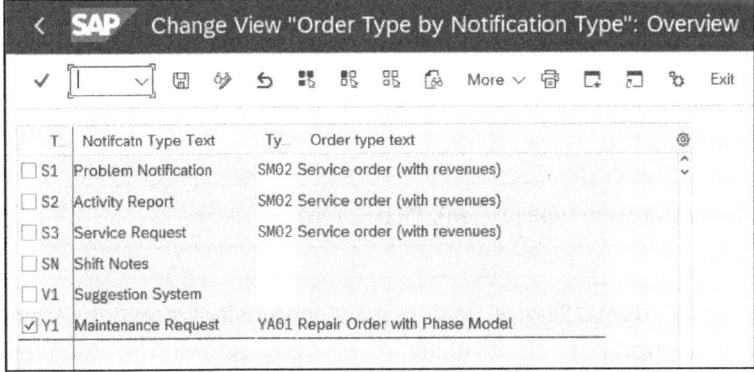

Figure 7.117 Notification Type Y1 Assigned to Order Type YA01

Event Prioritization
Event prioritization enables you to assess priorities based on consequence categories, consequences, and likelihoods, but only if you create maintenance notifications using

7 Configuring Additional Business Processes

the Create Maintenance Request (F1511A) app. On selecting consequences and likelihoods, the system calculates and suggests the applicable priority for the maintenance request (see Figure 7.118).

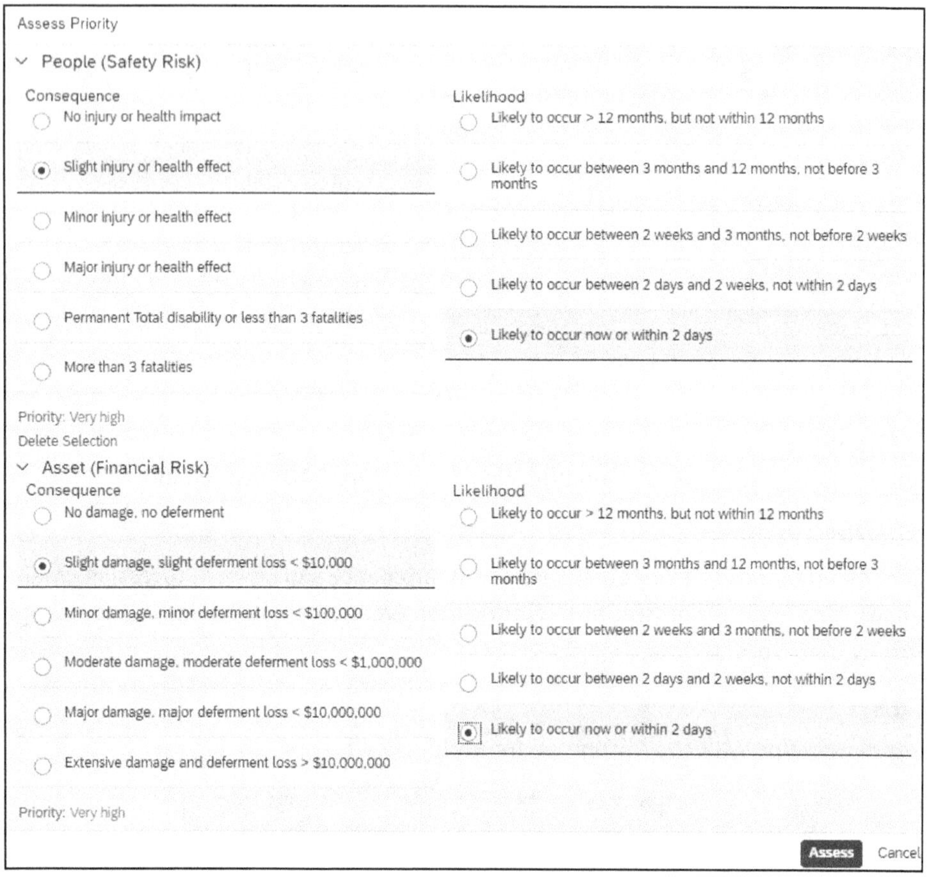

Figure 7.118 Priority Calculated by Consequences and Likelihoods

Prerequisites

You must define the notification type (e.g., **M1**, **Y1**) beforehand.

Customizing Path

Plant Maintenance and Customer Service • Maintenance and Service Processing • Maintenance and Service Notifications • Notification Processing • Response Time Monitoring • Event Prioritization • Define Consequence Categories, Consequences, and Likelihoods and Define Prioritization Profiles

Settings

First, you should **Define Consequence Categories** (e.g., people, assets, environment, reputation) in the **Define Consequence Categories, Consequences, and Likelihoods** Customizing function (see Figure 7.119).

7.9 Work Order Cycle Using the Phase Model

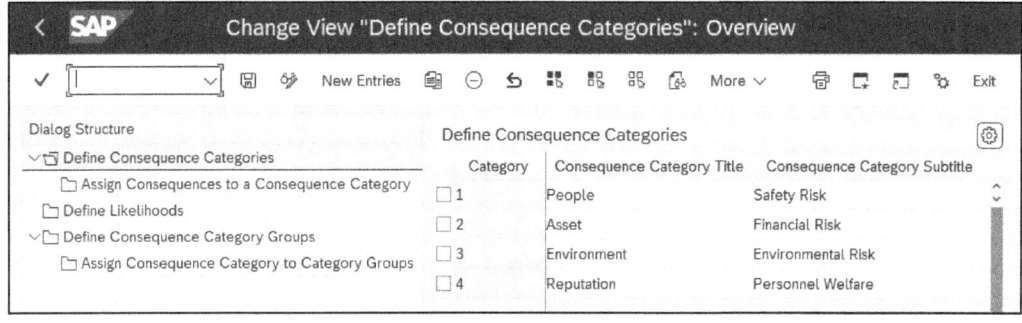

Figure 7.119 Event Prioritization: Consequence Categories

Next, assign consequences to a consequence category. For example, to the **People** category, assign **No injury or health impact** or **Major injury or health effect** (see Figure 7.120).

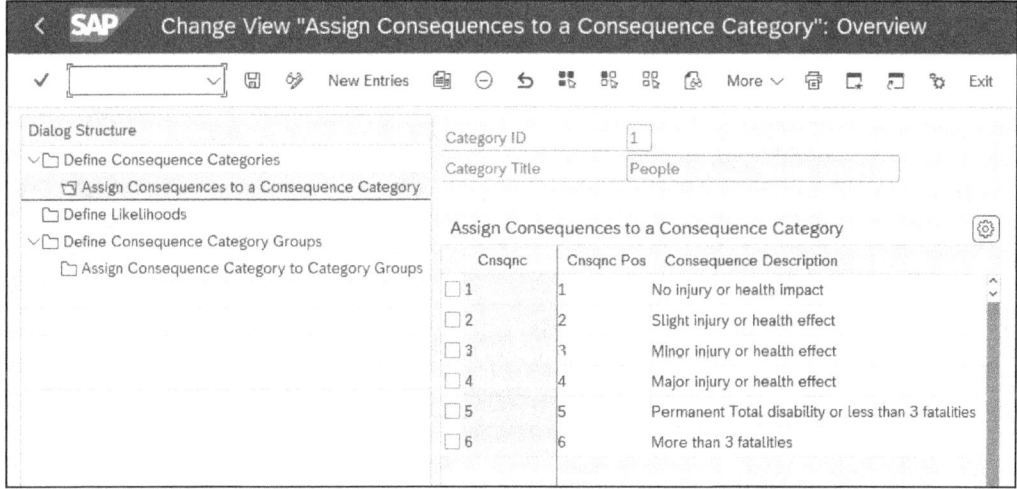

Figure 7.120 Event Prioritization: Assign Consequences to Consequence Category

Next, you define the probabilities of occurrence within a certain period (e.g., **Likely to occur now or within 2 days**; **Likely to occur > 12 months, but not within 12 months**; see Figure 7.121).

Finally, you define one or more category groups and assign consequence categories (see Figure 7.122).

The Customizing function **Define Prioritization Profiles** allows you to maintain prioritization of one or more prioritization profiles and to assign them to a notification type within a plant (see Figure 7.123).

557

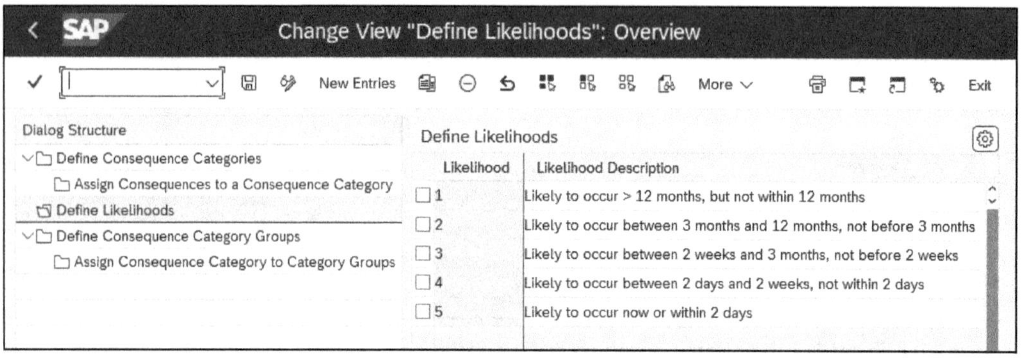

7 Configuring Additional Business Processes

Figure 7.121 Event Prioritization: Define Likelihoods

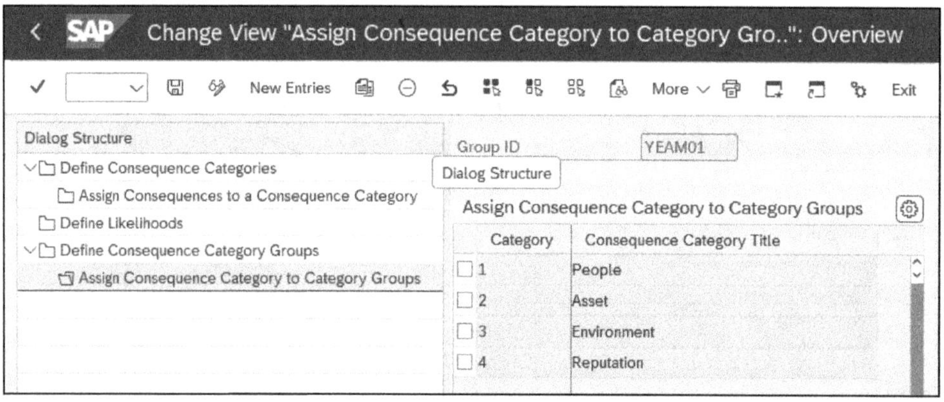

Figure 7.122 Event Prioritization: Consequence Category Groups

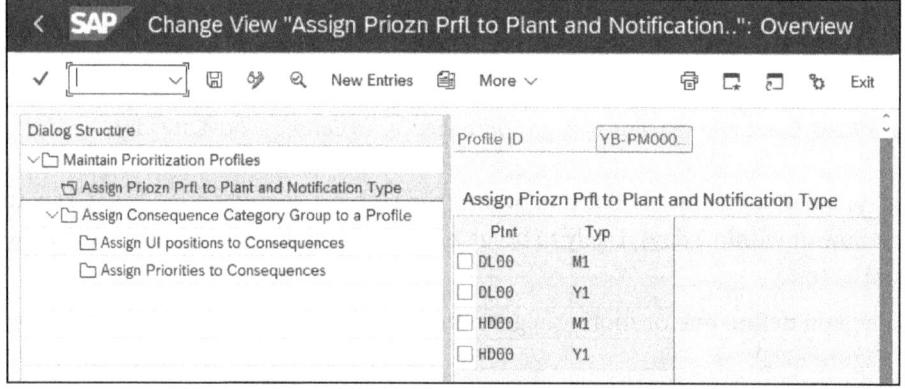

Figure 7.123 Event Prioritization: Prioritization Profiles

Next, you must assign a consequence category group to a prioritization profile (see Figure 7.124). Each profile can be assigned to only one group, but you can assign the same group to one or more profiles.

7.9 Work Order Cycle Using the Phase Model

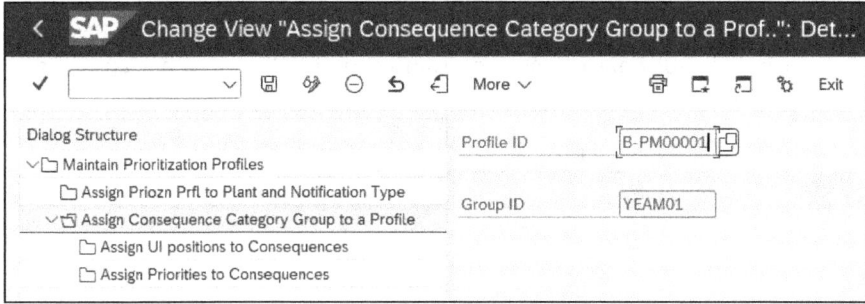

Figure 7.124 Event Prioritization: Assign Consequence Category Group to Profile

The penultimate step now is to assign the user interface position to a consequence category (see Figure 7.125). Within a group, you can choose to arrange the consequence categories in a particular sequence, and when you assess priority, you'll see the consequence categories in this sequence.

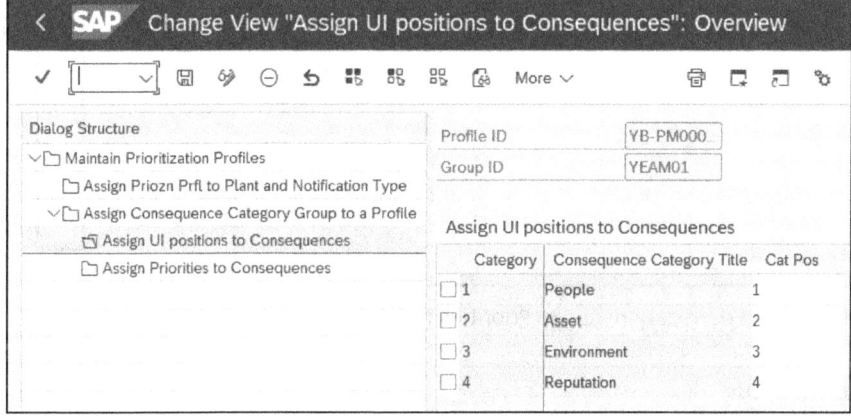

Figure 7.125 Event Prioritization: Consequence Category UI Positions

The last step is to assign an appropriate priority to a combination of consequence categories, consequences, and likelihoods (see Figure 7.126). It is advisable to use great caution when selecting priorities to be assigned to the combinations, as the priorities have significant impacts on other aspects of maintenance requests such as the final due date, required start date, and required end date.

> **Prioritization Is Useful but Only Available within the Create Maintenance Request App**
>
> Prioritization is a very useful function. It doesn't allow a priority to be assigned arbitrarily—in which case, over 90% of the notifications would have priority 1. Instead, the priority is calculated based on categories and likelihoods.

559

7 Configuring Additional Business Processes

> Unfortunately, this functionality is only available in the Create Maintenance Request app and not in SAP GUI transactions (e.g., Transaction IW21) or in other apps (e.g., Request Maintenance).

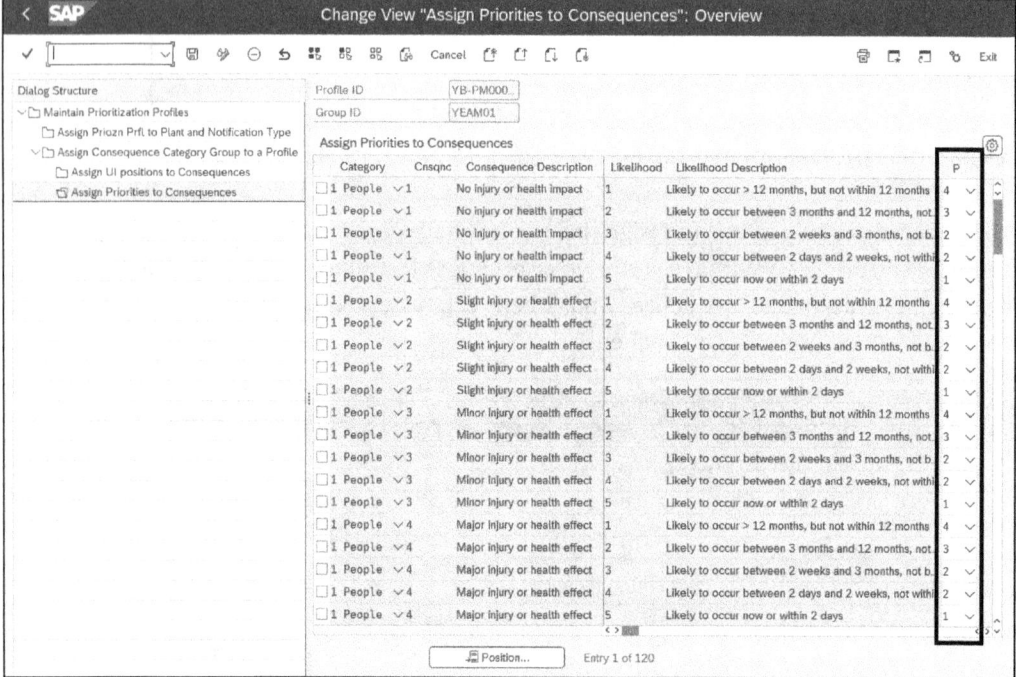

Figure 7.126 Event Prioritization: Assign Priorities to Categories and Likelihoods

Detection Methods

The *detection methods* list shows possible occasions on which an error was identified. You can use it only if you create maintenance notifications via the Create Maintenance Request (F1511A) app (see Figure 7.127).

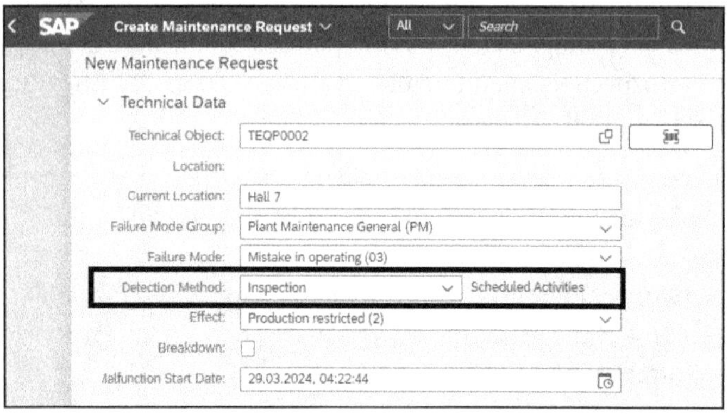

Figure 7.127 Detection Method in Create Maintenance Request App

560

7.9 Work Order Cycle Using the Phase Model

Prerequisites

You must define the notification type (e.g., **M1, Y1**) beforehand.

Customizing Path

Plant Maintenance and Customer Service • Maintenance and Service Processing • Maintenance and Service Notifications • Notification Creation • Notification Content • Additional Functions for Notification Types • Define Detection Methods and Maintain Additional Functions

Settings

First, define within the **Define Detection Methods** Customizing function the detection method itself (e.g., **Periodic Maintenance, Inspection, Continuous Condition Monitoring**; see Figure 7.128).

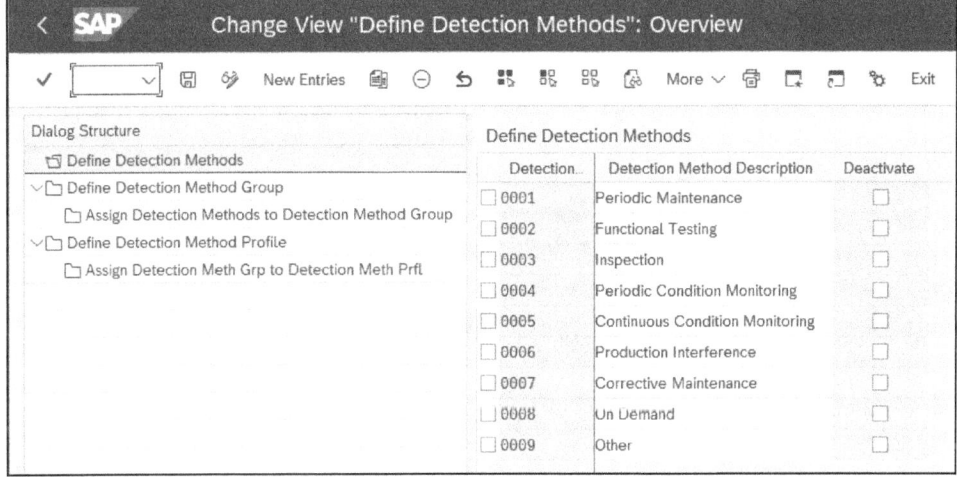

Figure 7.128 Detection Methods: Overview

Furthermore, detection method groups like **Scheduled Activities** (e.g., continuous monitoring) are to be defined, and detection methods (e.g., **Periodic Maintenance**) are to be assigned to detection method groups (see Figure 7.129). It's possible to assign the same detection method to several groups.

Next, you define a detection method profile and assign the appropriate detection method groups (**Dtctn Grp**) (see Figure 7.130).

Finally, with the Customizing function **Maintain Additional Functions**, you assign a detection method profile to a notification type (see Figure 7.131).

7 Configuring Additional Business Processes

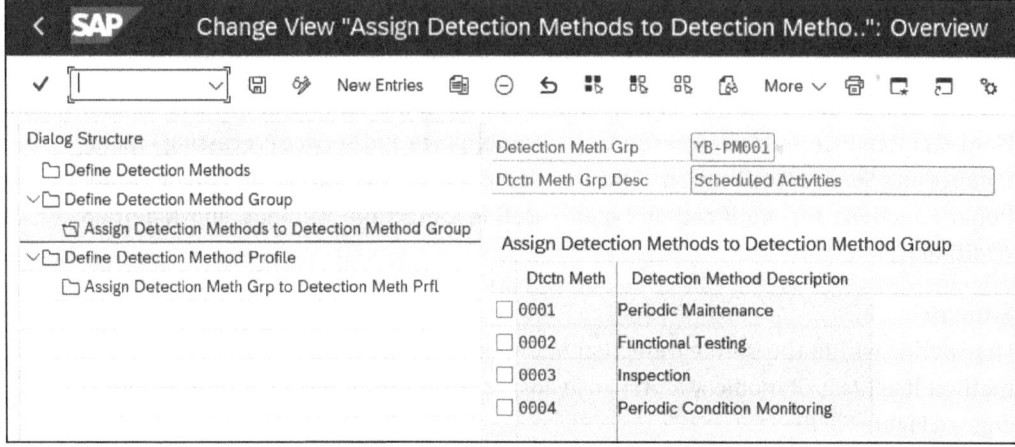

Figure 7.129 Detection Methods: Group with Detection Methods

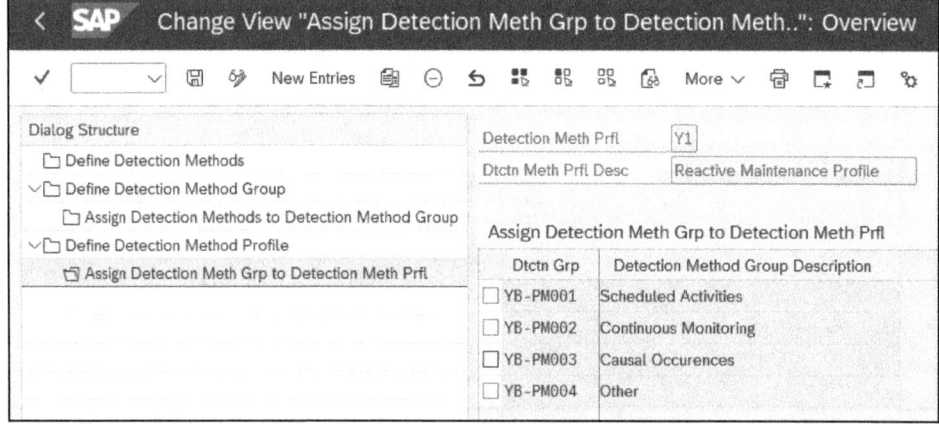

Figure 7.130 Detection Methods Profile: Detection Method Groups

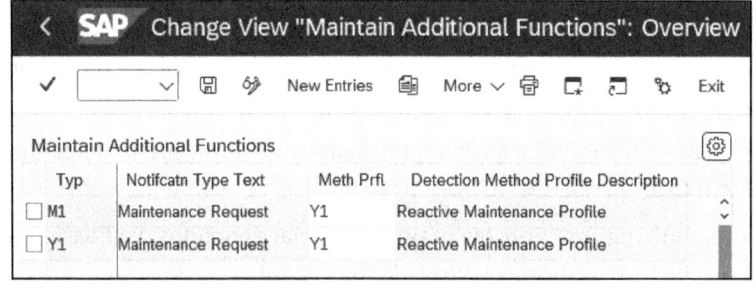

Figure 7.131 Notification Type and Detection Method Profile

> **Detection Methods Are Useful but Only Available within the Create Maintenance Request App**
>
> Detection methods are very useful functions that enable you to define the source of a notification.
>
> Unfortunately, this functionality is only available in the Create Maintenance Request app and not in the SAP GUI transactions (e.g., Transaction IW21) or in other apps (e.g., Request Maintenance).

Configure Order Types

You use this Customizing function to activate the phase model-based work order cycle within the order type.

Customizing Path

Plant Maintenance and Customer Service • Maintenance and Service Processing • Maintenance and Service Orders • Functions and Settings for Order Types • Configure Order Types

Transaction

OIOA

Prerequisites

Because order types are used not only in SAP S/4HANA Asset Management but also in other user departments and their corresponding areas in SAP S/4HANA, you must consult with your colleagues from the other user departments (production planning, quality management, project management, controlling, customer service, and materials management) on the naming convention and number ranges for order types.

A settlement profile and a budget profile have to be defined beforehand, and business feature **EAM_PHASE_MODEL_PROCESSING** has to be activated.

Settings

Check the box to set the **Activ. Phase Model** parameter (see Figure 7.132).

Job Template for Material Availability Check

You use this Customizing function to define the batch jobs for the automatic material availability check.

You can use the Schedule Material Availability Check app to execute material availability checks for several maintenance orders, and you can also schedule these checks as a recurrent batch job. The **Material Availability Check for Maintenance Orders** template enables you to reassign committed stock from one maintenance order to another of higher priority and urgency.

7 Configuring Additional Business Processes

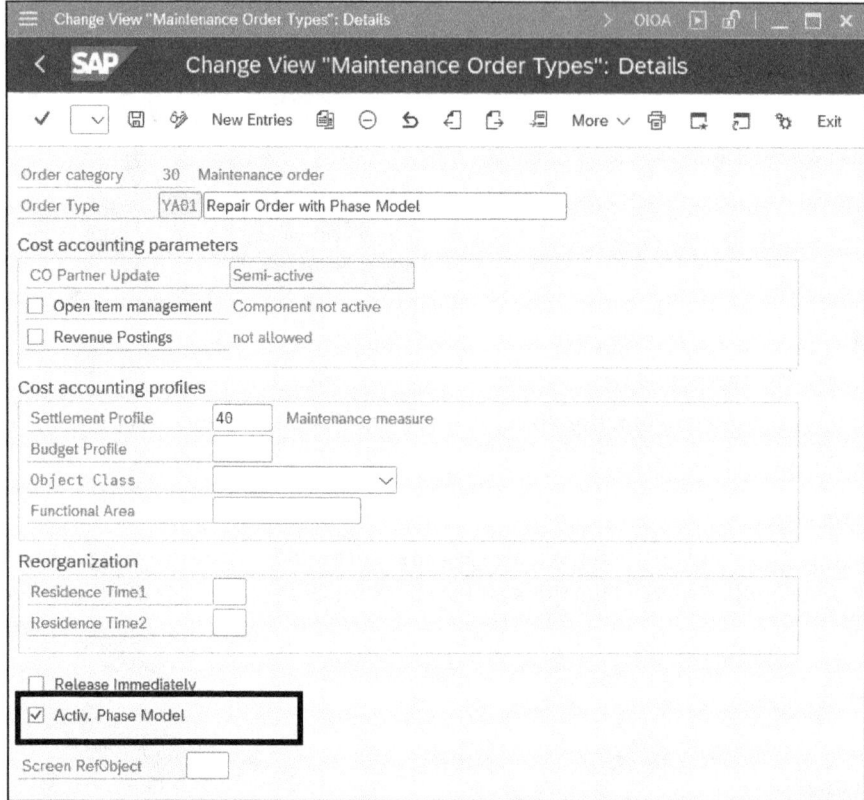

Figure 7.132 Order Type: Details

Customizing Path
ABAP Platform • Application Server • System Administration • Activation of Scope-Dependent Application Job Catalog Entries

Prerequisites
There is no specific prerequisite.

Settings
Add the two templates **SAP_PM_EAM_ATP_RIAUFK20** and **SAP_PM_EAM_MATL_AVAI-LY_CHECK** (see Figure 7.133).

7.9 Work Order Cycle Using the Phase Model

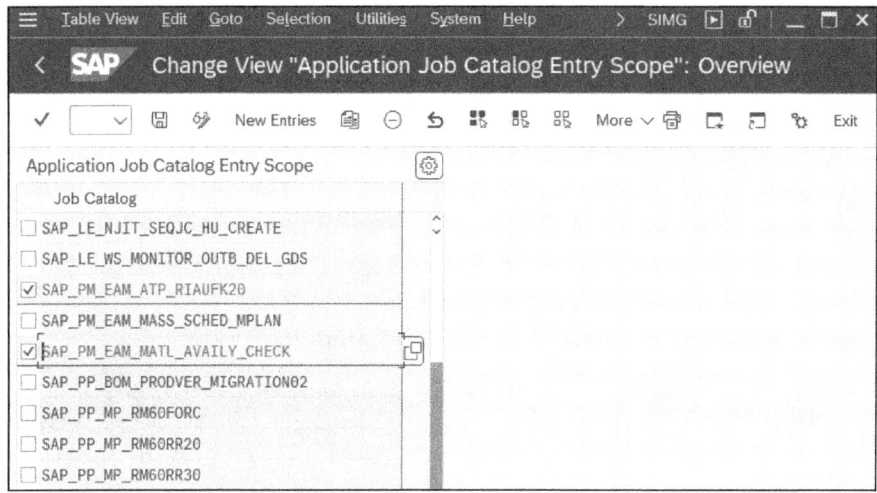

Figure 7.133 Jobs for Material Availability Check

Monitoring Procurement Milestones

As a maintenance planner, you can monitor the procurement of nonstock components and external services in the SAP Fiori Maintenance Backlog Overview app. For this purpose, the system tracks the completion of several milestones in the procurement process, such as the following:

- Purchase Requisition Created
- Purchase Requisition Released
- Purchase Order Sent to Vendor
- Order Confirmed by Vendor
- Order Shipped
- Order Received in Full

Customizing Path

Plant Maintenance and Customer Service • Maintenance and Service Processing • Maintenance and Service Orders • Functions and Settings for Order Types • Procurement • Activate Event Type Linkage for Procurement Milestones and Define Assignment Rules for Procurement Milestones

Prerequisites

There is no specific prerequisite.

Settings

The completion of the milestones **Purchase Order Sent to Vendor**, **Order Confirmed by Vendor**, **Order Shipped**, and **Order Received in Full** is triggered by certain changes in the purchase order and the posting of the goods receipt. To enable the system to react to

565

7　Configuring Additional Business Processes

these changes and update the corresponding milestones automatically in the Manage Maintenance Backlog app and the Maintenance Backlog Overview app, you need to activate event type linkages in the **Activate Event Type Linkage for Procurement Milestones** Customizing function for the following object types (see Figure 7.134):

- **BUS2012** for purchase orders
- **CL_APOC_OUTPUT_STATUS_EVENTS** for purchase orders
- **CL_MMIM_MATDOC_EVENT** for goods receipts
- **CL_MM_PUR_WF_OBJECT_PO** for purchase orders
- **CL_MM_PUR_WF_OBJECT_PR** for purchase requisitions

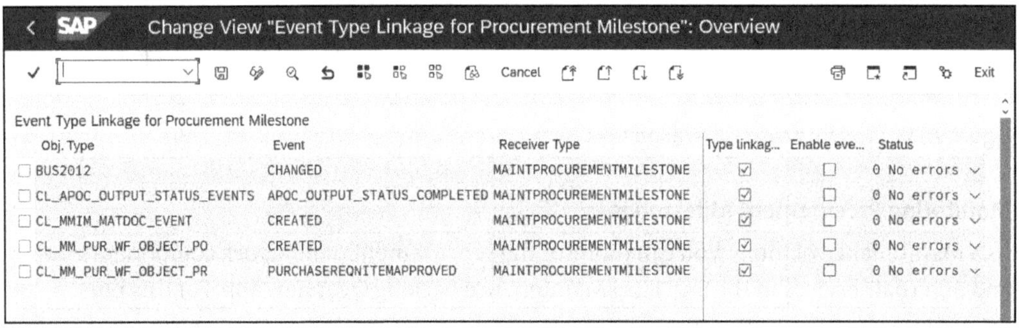

Figure 7.134 Event Type Linkage for Procurement Milestones

In the **Define Assignment Rules for Procurement Milestones** Customizing function, you can change the default settings that determine how the system monitors procurement milestones in the Manage Maintenance Backlog app and the Maintenance Backlog Overview app (see Figure 7.135).

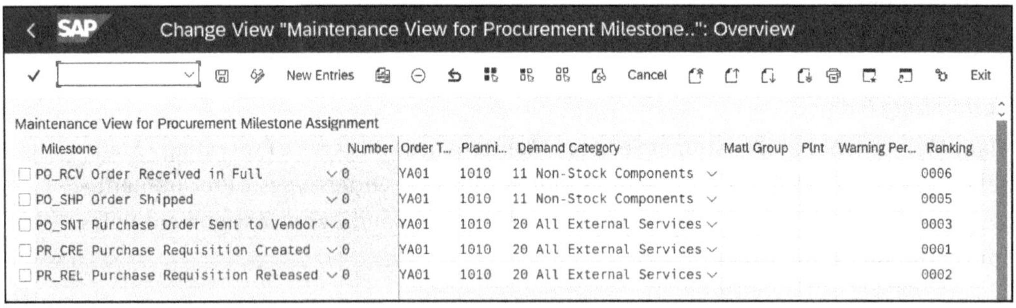

Figure 7.135 Assignment Rules for Procurement Milestones

Custom rules should be defined to determine which procurement milestones are monitored for a combination of order type, procurement demand category, maintenance planning plant, maintenance plant, and material group.

The following procurement milestones and demand categories are available (see Table 7.1).

Milestone	Demand Category
Purchase Requisition Created	Nonstock components, external operations, lean services, all external services
Purchase Requisition Released	Nonstock components, external operations, lean services, all external services
Purchase Order Sent to Vendor	Nonstock components, external operations, lean services, all external services
Order Confirmed by Vendor	Nonstock components, external operations, lean services, all external services
Order Shipped	Nonstock components only
Order Received in Full	Nonstock components only

Table 7.1 Procurement Milestones and Demand Categories

The *warning period* specifies the time between the date on which a milestone becomes due and the date on which the system starts to display the procurement status as a warning. By default, the warning period is two days before the milestone due date, but the default warning period can be changed.

Define Notification and Order Integration

This Customizing function is to be used to define the following:

- Whether you want to maintain notification and order data on one screen
- Whether you want the system to automatically create a notification when you enter an order
- Whether the system is to automatically copy the long text of the notification header to the order header if an order is generated from a notification
- Under which condition and to what extent you'll permit the assignment of operations to object list entries in an order

Prerequisites

Order types and notification types are to be defined beforehand.

For enhanced object list assignment, business function LOG_EAM_CI_4 has to be activated.

Customizing Path
Plant Maintenance and Customer Service • Maintenance and Service Processing • Maintenance and Service Orders • Functions and Settings for Order Types • Define Notification and Order Integration

7 Configuring Additional Business Processes

Settings

The order types for the phase model (e.g., **YA01**) must be linked to the notification type (e.g., **Y1**). In addition, the **Assignment of Operations to All Object List Entries** must be activated for the order type (e.g. **YA01**; see Figure 7.136).

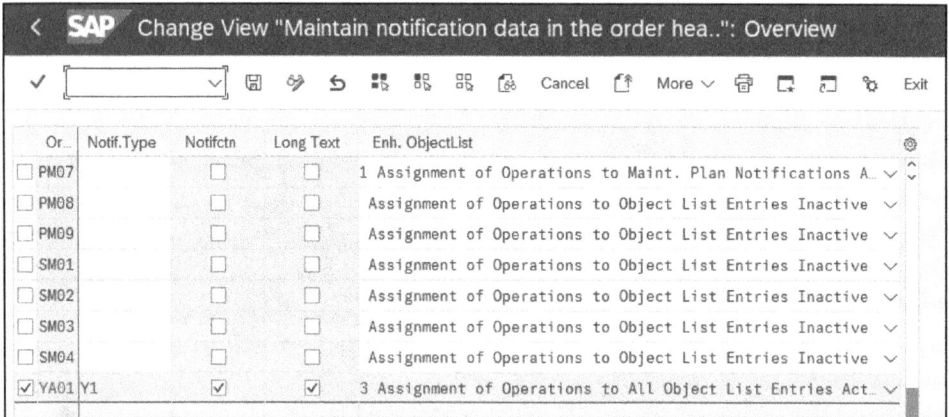

Figure 7.136 Order Type and Notification Type Integration

Teams and Responsibilities

Within these Customizing functions, the profiles, roles, and other functions necessary for the approving process of a maintenance order are defined.

Prerequisites

There are no specific perquisites.

Customizing Paths

Cross-Application Components • General Application Functions • Responsibility Management • Functions • Define Functions

Cross-Application Components • General Application Functions • Responsibility Management • Functions • Define Function Profiles

Cross-Application Components • General Application Functions • Responsibility Management • Teams and Responsibilities • Team Categories • Validate Team Members • Assign Functions to Template Roles

Cross-Application Components • General Application Functions • Responsibility Management • Teams and Responsibilities • Team Categories • Assign Function Profiles to Team Category

Settings

Within the Customizing function **Define Functions**, you define the tasks within a workflow that you use to determine who is responsible for approving them (see Figure 7.137).

7.9 Work Order Cycle Using the Phase Model

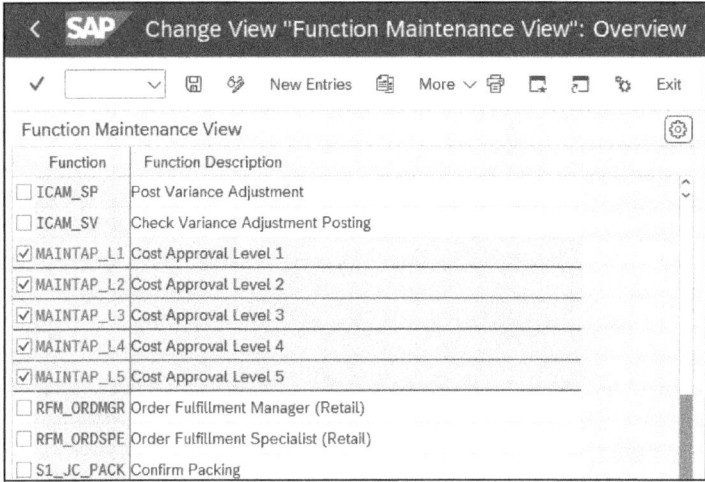

Figure 7.137 Teams and Responsibilities: Define Function

Use the Customizing function **Define Function Profiles** to group several functions (see Figure 7.138).

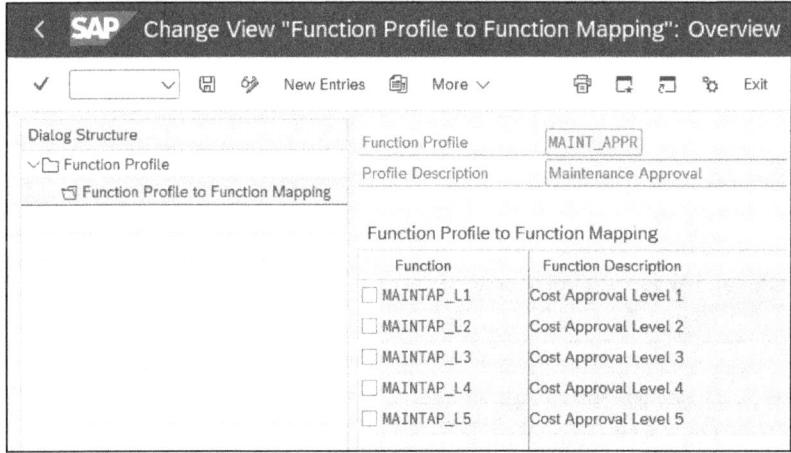

Figure 7.138 Teams and Responsibilities: Function Profile

Function profiles need to be assigned to a team category using the Customizing function **Assign Function Profiles to Team Category**. The standard team category for Asset Management **Maintenance Approval** is **MAINT_P_E** (see Figure 7.139).

The team category and the functions must be linked to a corresponding role for the user in the backend system using Customizing function **Assign Functions to Template Roles** (see Figure 7.140).

569

7   Configuring Additional Business Processes

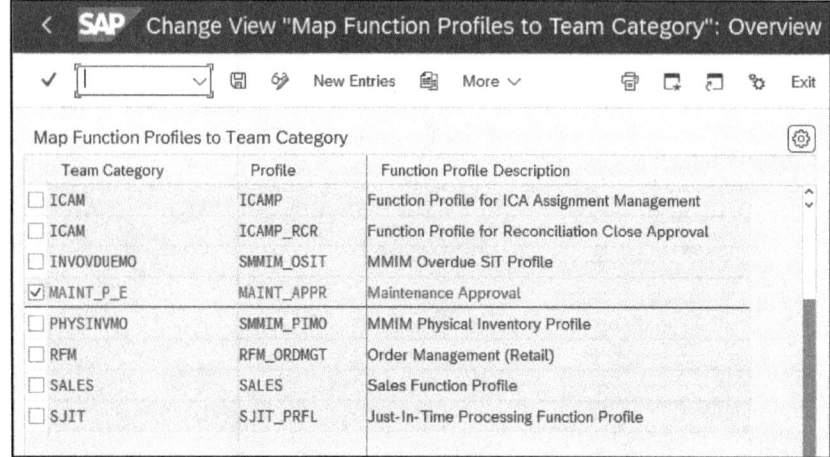

Figure 7.139 Teams and Responsibilities: Assign Function Profiles to Team Category

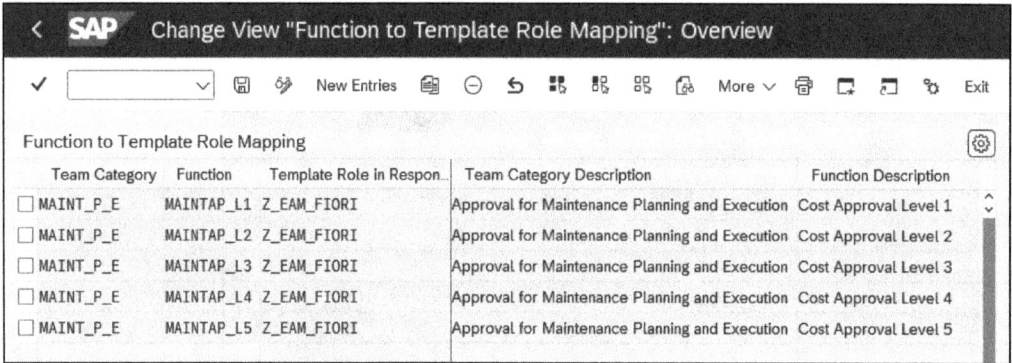

Figure 7.140 Teams and Responsibilities: Assign Functions to Template Roles

Workflows

The next step is to determine the workflow settings to set up the workflows according to the settings in **Teams and Responsibilities**.

Prerequisites

There are no specific prerequisites.

Customizing Paths

ABAP Platform • Application Server • Business Management • SAP Business Workflow • Maintain Standard Settings

ABAP Platform • Application Server • Business Management • SAP Business Workflow • Flexible Workflow • Scenario Activation

7.9 Work Order Cycle Using the Phase Model

ABAP Platform • SAP Gateway Service Enablement • Content • Task Gateway • Task Gateway Service • Scenario Definition

ABAP Platform • SAP Gateway Service Enablement • Content • Workflow Settings • Maintain Task Names and Decision Options

Settings

The scenario for maintenance orders **WS02000019 - WFL_EAM_MO Workflow for Maintenance Order** needs to be activated with the Customizing function **Scenario Activation** (see Figure 7.141).

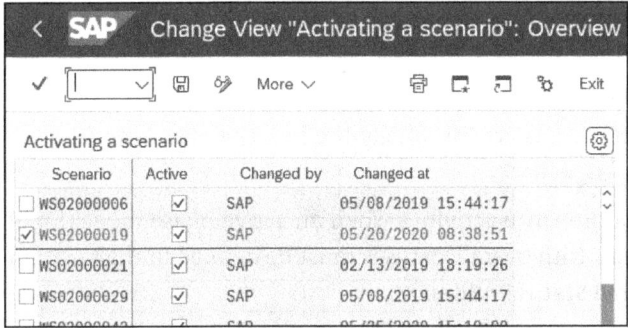

Figure 7.141 Workflow: Activating Scenario for Maintenance Orders

Use Customizing function **Scenario Definition** to assign scenario **WFL_EAM_MO** to consumer types **DESKTOP**, **MOBILE**, and **TABLET** (see Figure 7.142). Also assign scenario **WFL_EAM_MO** to task type **TS02000040**.

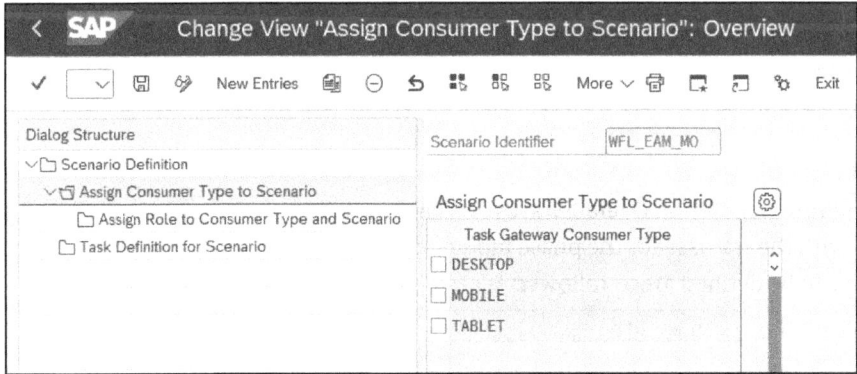

Figure 7.142 Workflow: Assign Scenario to Consumer Types

So that work items reach the right recipient along with the right contents in the My Inbox app, the **Maintain Task Names and Decision Options** Customizing function need to be included in the task filter along with their names and decision keys (see Figure 7.143).

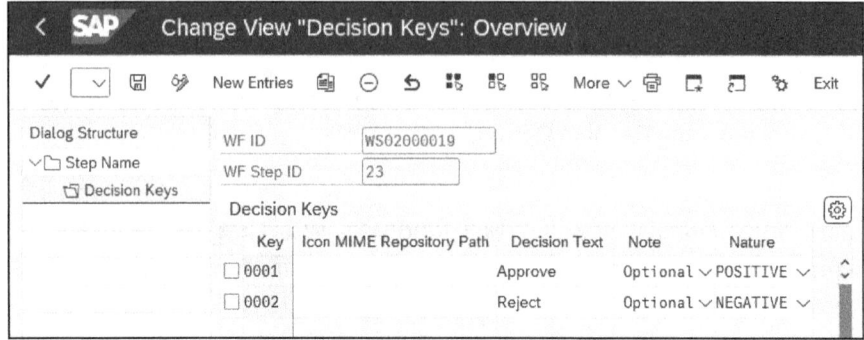

Figure 7.143 Workflow: Task Names and Decision Options

Overall Status Profile and Phases

The following Customizing functions are essentially the heart of phase model-based work order cycle processing.

To record the progress of maintenance activities when an active phase model for an order type is set, status profiles with overall statuses must be defined in the Customizing function **Configure Overall Status Profiles**.

Prerequisites

The order type and notification type must be defined beforehand.

Customizing Path

Plant Maintenance and Customer Service · Maintenance and Service Processing · SAP Fiori Apps for Maintenance Processing · General Settings · Configure Overall Status Profiles

Transaction

VC_EAM_OVRL_STS

Settings

In the subfunction **Define Overall Status Profile**, you define one or more overall status profiles with the statuses for the phase model-based work order cycle (see Figure 7.144). The items to be defined are as follows:

- **Profile** name
- **Status** number
- **Sequence** (defines the step in the process)
- **Entity** (to which object the step belongs: **QMI** = **Notification Header**, **ORI** = **Order Header**, **OVG** = **Order Operation**)
- Description

7.9 Work Order Cycle Using the Phase Model

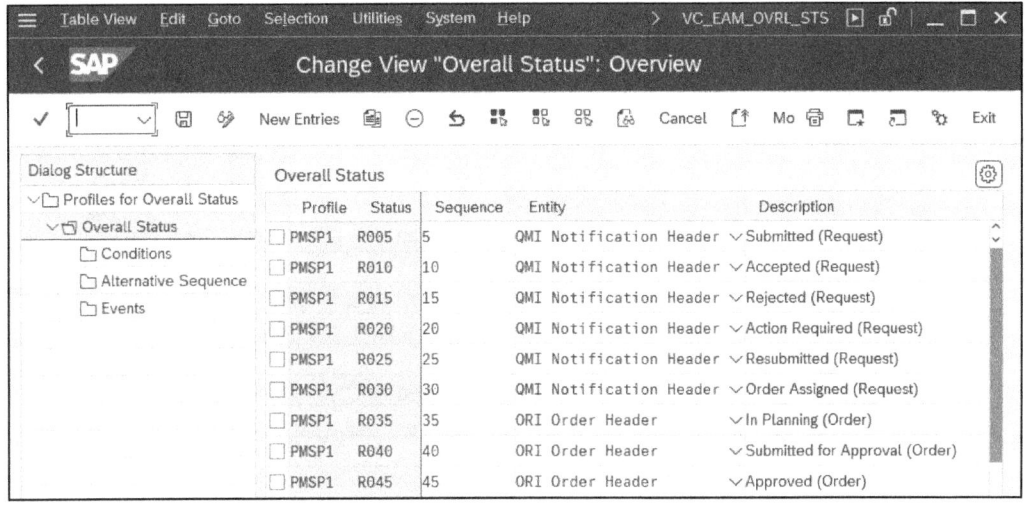

Figure 7.144 Overall Status Profile

Table 7.2 shows all statuses that are available when you implement SAP Best Practices 4HH or 4HI.

Profile	Status	Sequence	Entity	Description
PMSP1	R005	5	QMI Notification Header	Submitted (Request)
PMSP1	R010	10	QMI Notification Header	Accepted (Request)
PMSP1	R015	15	QMI Notification Header	Rejected (Request)
PMSP1	R020	20	QMI Notification Header	Action Required (Request)
PMSP1	R025	25	QMI Notification Header	Resubmitted (Request)
PMSP1	R030	30	QMI Notification Header	Order Assigned (Request)
PMSP1	R035	35	ORI Order Header	In Planning (Order)
PMSP1	R040	40	ORI Order Header	Submitted for Approval (Order)
PMSP1	R045	45	ORI Order Header	Approved (Order)
PMSP1	R050	50	ORI Order Header	Rejected (Order)
PMSP1	R055	55	ORI Order Header	In Preparation (Order)
PMSP1	R060	60	ORI Order Header	Ready to Schedule (Order)
PMSP1	R065	65	ORI Order Header	Ready for Execution (Order)
PMSP1	R070	70	ORI Order Header	Main Work Started (Order)

Table 7.2 Overall Status Profile: Standard Statuses

7   Configuring Additional Business Processes

Profile	Status	Sequence	Entity	Description
PMSP1	R075	75	ORI Order Header	Main Work Completed (Order)
PMSP1	R080	80	ORI Order Header	Work Done (Order)
PMSP1	R085	85	ORI Order Header	Technically Complete (Order)
PMSP1	R090	90	QMI Notification Header	Completed (Request)
PMSP1	R092	92	ORI Order Header	Work Not Performed (Order)
PMSP1	R095	95	ORI Order Header	Closed (Order)
PMSP1	R100	100	QMI Notification Header	Deletion Flag (Request)
PMSP1	R105	105	ORI Order Header	Deletion Flag (Order)
PMSP1	R110	110	OVG Order Operation	In Planning
PMSP1	R115	115	OVG Order Operation	In Preparation
PMSP1	R120	120	OVG Order Operation	Ready to Schedule
PMSP1	R125	125	OVG Order Operation	Ready for Execution
PMSP1	R130	130	OVG Order Operation	Work in Execution
PMSP1	R135	135	OVG Order Operation	Work Paused
PMSP1	R140	140	OVG Order Operation	Work Finished
PMSP1	R145	145	OVG Order Operation	Technically Complete
PMSP1	R150	150	OVG Order Operation	Closed

Table 7.2 Overall Status Profile: Standard Statuses (Cont.)

Within the subfunction **Conditions**, you define how each overall status depends on a user status or system status set for a maintenance order or notification and whether this status needs to be applied to one or more operations or components. For example, Figure 7.145 shows the conditions for status **R080** = **Work Done**. You must specify the following:

- Link
  The link between two conditions (**And** or **Or**)
- Entity
  QMI = Notification Header, ORI = Order Header, OVG = Order Operation
- Entity Qualifier
  01 = Main Order Operations

7.9 Work Order Cycle Using the Phase Model

- **Status Profile**
  Whether the status belongs to a user status within the status profile
- **Condition Qualifier**
  -I = Is Not Set, I = Is Set, 0 = Was Never Set, ~ = Was Set
- **Condition Qualifier**
  0.* = Optional, 1.* = Applies to at least one element, M.M = Applies to all elements

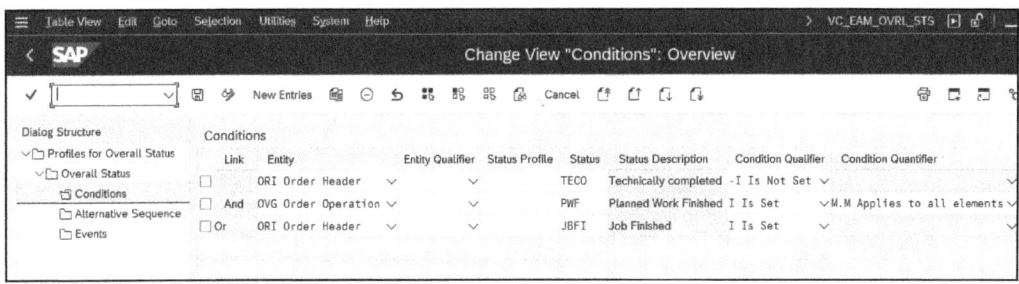

Figure 7.145 Overall Status Profile: Conditions

Table 7.3 shows all conditions available when you implement SAP Best Practices 4HH or 4HI.

Status	Link	Entity	Entity Qualifier	Status	Condition Qualifier	Condition Quantifier
R005		QMI		APRQ	I Is set	
R010		QMI		NOPR	I Is set	
	AND	QMI		APOK	I Is set	
R015		QMI		NOCO	I Is set	
	AND	QMI		APRF	I Is set	
R020		QMI		OSNO	I Is set	
	AND	QMI		AIRQ	I Is set	
	AND	QMI		APRQ	~ Was set	
R025		QMI		OSNO	I Is set	
	AND	QMI		APRQ	I Is set	
	AND	QMI		AIRQ	~ Was set	
R030		QMI		NOPR	I Is set	
	AND	QMI		ORAS	I Is set	

Table 7.3 Overall Status Profile: Standard Conditions

Status	Link	Entity	Entity Qualifier	Status	Condition Qualifier	Condition Quantifier
R035		ORI		CRTD	I Is set	
	AND	ORI		ORAI	-I Is not set	
	AND	ORI		ORAP	-I Is not set	
	AND	ORI		ORRJ	-I Is not set	
R040		ORI		ORAI	I Is set	
R045		ORI		ORAP	I Is set	
	AND	ORI		REL	-I Is not set	
	AND	ORI		TECO	-I Is not set	
	AND	ORI		CLSD	-I Is not set	
	AND	ORI		DLFL	-I Is not set	
R050		ORI		ORRJ	I Is set	
R055		ORI		REL	I Is set	
	AND	ORI		ORSC	-I Is not set	
	AND	ORI		OMWC	-I Is not set	
	AND	ORI		JBFI	-I Is not set	
	AND	OVG		DSPT	-I Is not set	M.M
R060		ORI		ORSC	I Is set	
	AND	OVG		DSPT	-I Is not set	1.*
R065		OVG	01	DSPT	I Is set	1.*
	AND	OVG	01	JBFI	-I Is not set	1.*
R070		OVG	01	PCNF	I Is set	1.*
	OR	OVG	01	JIPR	I Is set	1.*
	OR	OVG	01	CNF	I Is set	1.*
	AND	OVG	01	JBFI	-I Is not set	1.*
	OR	OVG	01	JBFI	I Is set	1.*
	OR	OVG	01	PWS	I Is set	1.*

Table 7.3 Overall Status Profile: Standard Conditions (Cont.)

7.9 Work Order Cycle Using the Phase Model

Status	Link	Entity	Entity Qualifier	Status	Condition Qualifier	Condition Quantifier
R075		ORI		OMWC	I Is set	
	AND	ORI		CNF	-I Is not set	
	AND	ORI		JBFI	-I Is not set	
	AND	ORI		TECO	-I Is not set	
R080		ORI		TECO	-I Is not set	
	AND	OVG		PWF	I Is set	M.M
	OR	ORI		JBFI	I Is set	
R085		ORI		TECO	I Is set	
R090		QMI		NOCO	I Is set	
	AND	QMI		APRF	-I Is not set	
R092		ORI		NCMP	I Is set	
	AND	ORI		CLSD	I Is set	
R095		ORI		CLSD	I Is set	
	AND	ORI		NCMP	-I Is not set	
R100		QMI		DLFL	I Is set	
R105		ORI		DLFL	I Is set	
R110		OVG		CRTD	I Is set	0.*
R115		OVG		REL	I Is set	0.*
	AND	OVG		ORSC	-I Is not set	0.*
	AND	OVG		DSPT	-I Is not set	0.*
	AND	OVG		JIPR	-I Is not set	0.*
	AND	OVG		JBFI	-I Is not set	0.*
	AND	OVG		PCNF	-I Is not set	0.*
	AND	OVG		CNF	-I Is not set	0.*
R120		OVG		ORSC	I Is set	0.*

Table 7.3 Overall Status Profile: Standard Conditions (Cont.)

7 Configuring Additional Business Processes

Status	Link	Entity	Entity Qualifier	Status	Condition Qualifier	Condition Quantifier
R125		OVG		DSPT	I Is set	0.*
	AND	OVG		JIPR	-I Is not set	0.*
	AND	OVG		PCNF	-I Is not set	0.*
	AND	OVG		CNF	-I Is not set	0.*
R130		OVG		JIPR	I Is set	0.*
	OR	OVG		PCNF	I Is set	0.*
	OR	OVG		PWS	I Is set	0.*
	OR	OVG		SECF	I Is set	0.*
	OR	OVG		SEPC	I Is set	0.*
	OR	OVG		EOPD	I Is set	0.*
	OR	OVG		EODL	I Is set	0.*
R135		OVG		JIPR	~ Was set	0.*
R140		OVG		TECO	-I Is not set	0.*
	AND	OVG		PWF	I Is set	0.*
	OR	OVG		OMWC	I Is set	0.*
R145		OVG		TECO	I Is set	0.*
R150		OVG		CLSD	I Is set	0.*

Table 7.3 Overall Status Profile: Standard Conditions (Cont.)

You may be surprised that a large table like this is included in the book. The reason for this lies in the complexity of configuring the status logic manually. This table gives a hint of how statuses and conditions could be defined, and with a template like this, it will certainly be easier for you to configure your own status profile.

> **Single Condition = OR, Several Conditions = AND**
>
> If you configure the statuses and conditions manually, here is a little a little tip for you regarding the links. You can't manually set the link to OR or AND, so do the following instead:
>
> - If you need AND links, click on new entries and enter several conditions in a list at once.
> - If you need an OR link, click on a new entry, enter exactly one condition, and save it.

> **There Are Three Reasons Against, One Reason For**
>
> What is a summary of the phase model-based work order cycle?
>
> First, the configuration is complex because many Customizing settings must be made in addition to the general Customizing settings of the simple work order cycle described in Chapter 5.
>
> Second, some configurations are very complicated (e.g., configuring the overall status profile manually).
>
> Third, the business process of the nine-phase work order cycle is more rigid and more time consuming to handle in day-to-day business.
>
> However, some SAP Fiori apps can only be used to their full extent if they are integrated into the phase model-based work order cycle.

7.10 Summary

This chapter described the Customizing functions for business processes built on a basic work order cycle (which we discussed in Chapter 5) and provided enhanced options for your plant maintenance. The key takeaways from this chapter are as follows:

- Setting an appropriate control key for individual purchase processing and another one for service specification processing is crucial for external processes.
- The following configurations are necessary for the refurbishment process:
  - Split valuation for material master records must be activated and configured.
  - A separate order type and a separate settlement profile must be set up.
- For the subcontracting process, it's necessary to configure document types for purchase requisitions and purchase orders as well as to set up an appropriate material provision indicator.
- For the calibration of test equipment processes, some plant maintenance settings and some quality management settings have to be configured:
  - In plant maintenance, you set up an appropriate equipment type and a separate order type.
  - In quality management, you set up control keys, inspection points, a separate inspection type, codes, and catalogs for usage decisions and follow-up actions.
- For the pool asset management process, you need a separate notification type for requesting, basic settings for initiating, and a separate order type for settling.
- If you are to use the Maintenance Event Builder, a revision type has to be defined.

- If you want to use shift notes, a separate notification type and Customizing of the basic settings (e.g., origin indicator, categories) must be configured. If you want to create shift reports, you must set up a shift report type.
- The basic version of checklist processing, where one inspection lot per order is generated, is very similar to the process for calibration of test equipment and should be sufficient in most cases.
- The extended version of checklist processing can create several inspection lots for one order, but it also places significantly higher demands on the configuration and master data.
- The phase model-based work order cycle configuration is complex, so use your best judgement on whether its additional features are convincing enough to be worthwhile.

Chapter 8
Configuring the SAP Fiori Launchpad for Plant Maintenance

The SAP Fiori launchpad can be the single point of entry for accessing the SAP S/4HANA system. This chapter explains the basics of SAP Fiori, SAP's newest interface technology, and the configuration of the SAP Fiori launchpad with apps that use SAPUI5 technology and apps that don't.

As SAP's latest user experience (UX) product, SAP Fiori provides access to SAP systems on all devices—not only via desktop or notebook but also on mobile devices such as tablets and smartphones. Therefore, SAP Fiori represents an alternative to SAP GUI, SAP Business Client, and SAP's mobile solutions.

This chapter contains basic information on SAP Fiori as well as detailed information on how to configure your own SAP Fiori launchpad with different types of SAP Fiori apps, including SAP Fiori apps with and without SAPUI5 technology.

8.1 Basics of SAP Fiori

The number of SAP Fiori apps offered by SAP is growing rapidly, and the current volume can be looked up in the SAP Fiori apps reference library (*https://fioriappslibrary.hana.ondemand.com*). As of the time of writing, it shows about 15,000 SAP Fiori apps for SAP S/4HANA. In the Asset Management lines of business, there are approximately 250 SAP Fiori apps, including many apps for project management, materials management, and quality management, which all have points of contact with plant maintenance (e.g., apps for inventory in embedded extended warehouse management [EWM] in SAP S/4HANA and for usage decisions in an inspection lot). There are about 50 SAP Fiori apps currently available for actual maintenance topics.

8.1.1 SAP Fiori Design

SAP Fiori offers a frontend design with *tiles* that allow users to select functions and apps via the SAP Fiori launchpad (see Figure 8.1). Users may create their own **My Home** page out of tiles they are entitled to use like favorites in SAP GUI.

8 Configuring the SAP Fiori Launchpad for Plant Maintenance

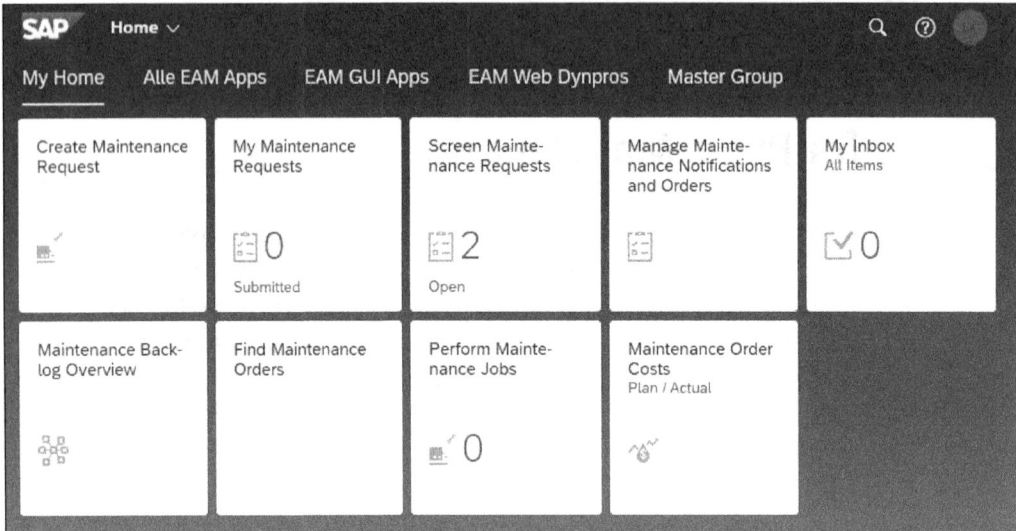

Figure 8.1 SAP Fiori Launchpad: My Home

If these functions or apps are clicked, the browser shows the screens of an SAPUI5 front-end. (Note that SAPUI5 is based on HTML5.)

8.1.2 Types of SAP Fiori Apps

SAP Fiori apps can be classified three different ways (see Figure 8.2):

- **According to their content (the classification method preferred by SAP)**
  Transactional apps, analytical apps, and fact sheet apps
- **According to their dynamism**
  Static apps and dynamic apps (only listing apps)
- **According to their results**
  Real apps with added value, real apps without added value, and pseudo apps

Let's now look at how we can classify SAP Fiori apps within each of these categories.

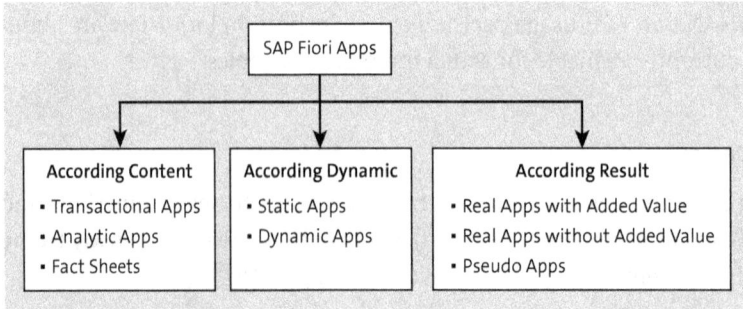

Figure 8.2 Types of SAP Fiori Apps

8.1 Basics of SAP Fiori

Classifying SAP Fiori Apps Based on Their Content

SAP prefers to classify apps according to their content. In this classification method, there are three different kinds of SAP Fiori apps (see Figure 8.3):

- ❶ **Transactional apps**

  These apps execute certain business processes (e.g., creating a malfunction report, recording a confirmation), and they conform 1:1 to a transaction in SAP GUI. In addition, a couple of functions can be combined in one app.

- ❷ **Fact sheet apps**

  These apps provide the most important information about one object (e.g., supplier, material master, customer) within one screen display.

- ❸ **Analytical apps**

  These apps can be used to produce evaluations, key figures, statistics, and diagrams for your business processes.

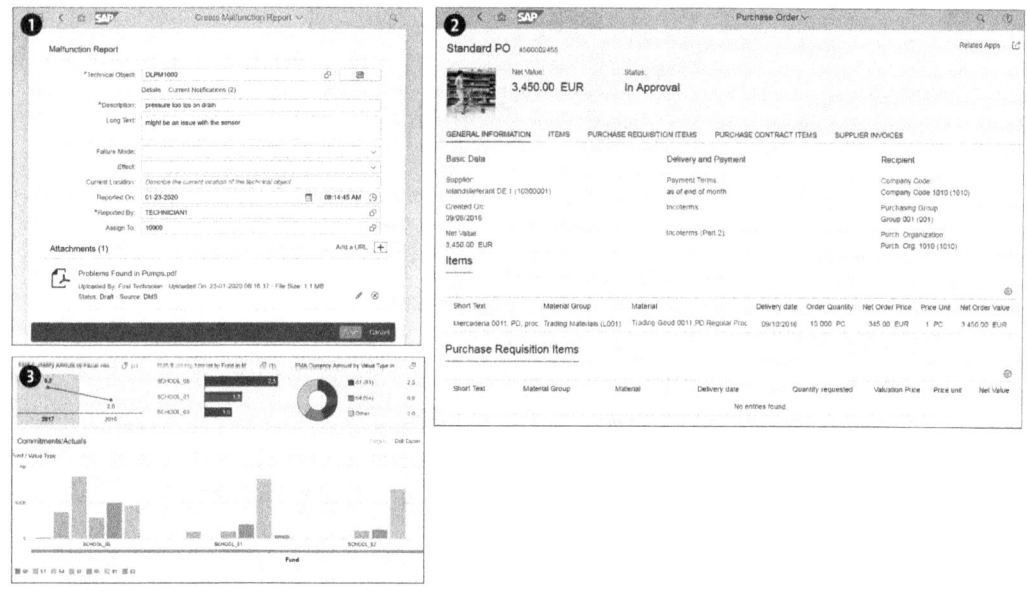

Figure 8.3 SAP Fiori App Types: Content Based

Classifying SAP Fiori Apps Based on Their Dynamism

We can also classify SAP Fiori apps that issue lists by dividing them into static and dynamic apps, as follows:

- **Static apps**

  These apps operate like list transactions in SAP GUI. You enter your selection criterion (e.g., a plant) in an upstream selection screen, and the list will show the relevant findings. If you want to change the selection criterion (e.g., to another plant or by adding a plant), you return to the selection screen, widen your selection, and subsequently see the new (static) findings.

8   Configuring the SAP Fiori Launchpad for Plant Maintenance

- **Dynamic apps**
  These apps show the filter area and findings in a list. If you change the filter, the findings are adjusted dynamically. Figure 8.4 shows the Find Technical Object app as an example of a dynamic app.

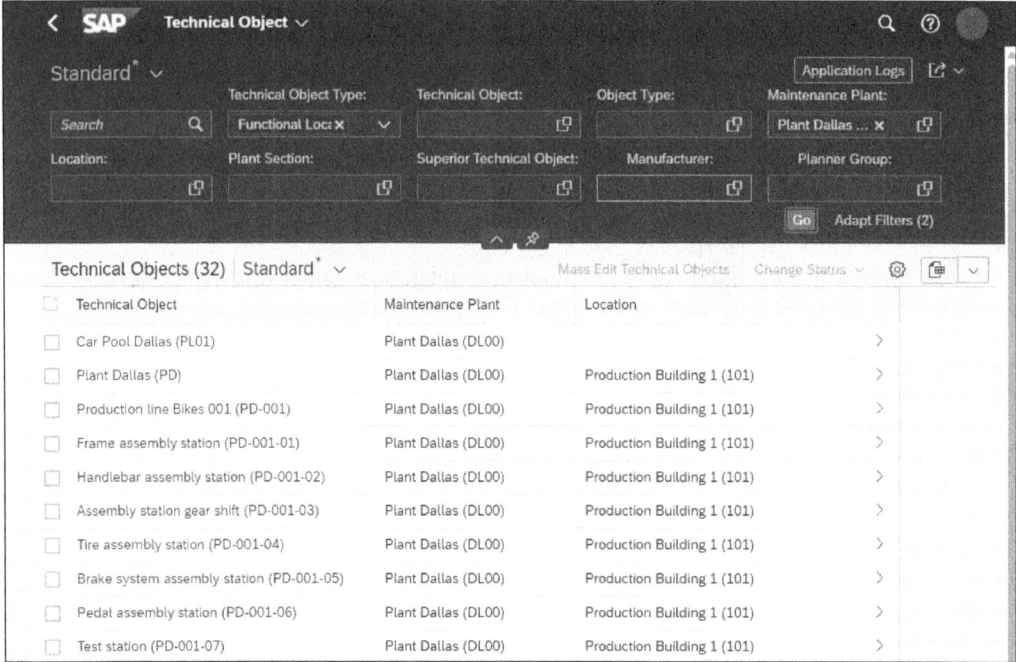

Figure 8.4 SAP Fiori App: Find Technical Object

Classifying SAP Fiori Apps Based on Their Results

Finally, SAP Fiori apps can be classified according to their results, such as real SAP Fiori apps with added value, real SAP Fiori apps without added value, and pseudo-SAP Fiori apps.

Let us first define *real SAP Fiori apps with added value*. In essence, these apps use SAPUI5 technology and add value relative to SAP GUI. Adding value can mean several things, including the following:

- Extracting a single function out of a complex function. The Release Maintenance Orders app and the Approve Purchase Orders app do this.
- Pooling a couple of transactions into just one. The Post Incoming Invoices app does this.
- Providing functions that don't exist in SAP GUI. Dynamic apps do this.
- Preparing and consolidating statistics for graphic representation.
- Showing key figures on the SAP Fiori tile (e.g., the number of maintenance notifications).

By contrast, real SAP Fiori apps without added value use SAPUI5 technology but don't provide any additional value by using it. See, for example, the Create Purchase Requisition app, which uses SAP Fiori technology but doesn't offer any added value compared to Transaction ME51N in SAP GUI.

Pseudo SAP Fiori apps, however, are SAP Fiori apps that don't use SAPUI5 and don't deliver any additional value. See for example, the Overall Completion app, which doesn't use SAPUI5 and only displays Transaction IW42 with a SAP GUI surface. Therefore, there is no additional user value.

8.1.3 Characteristics of SAP Fiori

Value-adding apps basing on SAPUI5 facilitate users' workload in the following ways:

- The surface is responsive, which means it identifies the terminal device and accessing data and aligns itself to the device.
- The operation is coherent. In all apps, all functions and all devices of the operation are unified, and all processes share and are controlled by the same data.
- Just like in SAP GUI, there is a role-based allocation. This is why each user will only find the specific functions personally assigned to him on his SAP Fiori launchpad.
- Just like in SAP GUI, numerous possibilities for personalization are available (e.g., selection variants, field selection).
- Each SAP Fiori app may be used on mobile devices, which enables users to access their data or tasks from anywhere at any time.
- Provision is made for developing specific customer requirements, which means users may integrate their own self-developed apps within the SAP-provided development workbench (e.g., the SAP Business Application Studio, formerly known as the SAP Web IDE toolkit).

Moreover, these characteristics of SAP Fiori are useful in day-to-day-business:

- If it's necessary or relevant, important key figures will be shown on the SAP Fiori launchpad (see Figure 8.5).
- You can send emails (e.g., to send a list or a link to a document) from within each app.
- All lists may be exported to Microsoft Office.

Figure 8.5 SAP Fiori Launchpad: Tiles with Key Figures

On the other hand, in everyday business, some aspects of SAP Fiori can be rather off-putting:

- SAPUI5 surfaces contain a lot of empty space, so they aren't compact, which is also why screen pictures are needlessly big and blind.
- Because SAP GUI uses tabstrips, information is assigned according to relevant headlines and then allotted to tabstrips. This tabstrip technique isn't available at HTML5, so in SAP Fiori, items of information are shown one below the other instead of in tabstrips. In many cases, this creates very long screens requiring a lot of scrolling to get the desired information.
- Positioning of functional keys isn't standardized yet.

8.2 How to Configure an SAP Fiori Launchpad with SAPUI5 Apps

This section explains the configuration of the SAP Fiori launchpad, which could be the single point of entry for users of the SAP S/4HANA system. This section supports your understanding of the role-based definition of users and the underlying SAP functionalities. You'll be shown how to define roles for users and how to integrate the standard Create Maintenance Request app for plant maintenance (see Figure 8.6).

The key features of the SAP Fiori launchpad are as follows:

- Maintenance requests for technical objects can be created, and the technical object may be a functional location or a piece of any equipment.
- Barcodes can be scanned to enter the ID of the technical object.
- A failure mode of the technical object can be assigned, and a maintenance request has to be prepared for it. Failure modes are maintained in Customizing function **Maintain Catalogs** (see Chapter 5, Section 5.1.2).
- To indicate how a failure was discovered, a detection method can be assigned. Detection methods are maintained in Customizing function **Detection Methods** (see Chapter 7, Section 7.9).
- Effects for production can be attributed and are maintained in Customizing function **Define Operational Effects** (see Chapter 5, Section 5.1.3).
- Malfunction information (e.g., malfunction start date and time) can be entered.
- A template can be used to deliver a detailed description of the maintenance request.
- Priorities can be allocated and are maintained in Customizing function **Define Priorities** (see Chapter 5, Section 5.1.3).

8.2 How to Configure an SAP Fiori Launchpad with SAPUI5 Apps

- A priority can be determined by using risk assessment and selecting a combination of consequence categories. These are consequences and likelihoods that use the **Assess Priority** functionality. Prioritization profiles are defined by using the Customizing function **Event Prioritization** (see Chapter 7, Section 7.9).
- It is possible to upload documents related to the maintenance request, and a link can be established to documents related to the maintenance request.

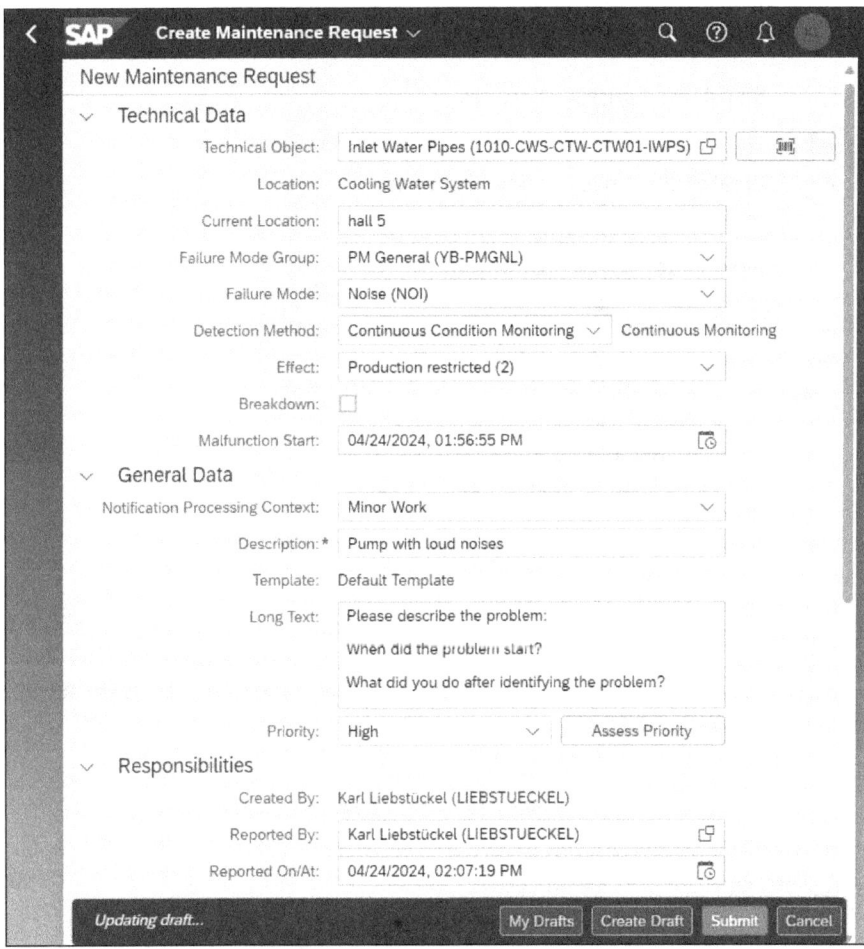

Figure 8.6 SAP Fiori Create Maintenance Request App

The process of configuring the SAP Fiori launchpad consists of eight steps, as shown in Figure 8.7.

8 Configuring the SAP Fiori Launchpad for Plant Maintenance

Figure 8.7 Eight Steps to Configure an SAP Fiori Launchpad with SAPUI5 Apps

Transactions for the SAP Fiori Launchpad

Before we start with the configuration, here's an overview of the transactions you need for the SAP Fiori launchpad:

- Transaction /UI2/FLP (Start SAP Fiori Launchpad)
- Transaction /IWFND/MAINT_SERVICE (Activate OData Service)
- Transaction /IWFND/V4_ADMIN (Activate OData Service Groups)
- Transaction SICF (Activate ICF Services)
- Transaction /UI2/FLPCM_CUST (Maintain Client-Specific Catalogs)
- Transaction /UI2/FLPCM_CONF (Maintain Cross-Client Catalogs)
- Transaction /UI2/FLPD_CUST (Start Launchpad Designer)
- Transaction /UI2/FLPAM (SAP Fiori Launchpad App Manager)
- Transaction PFCG (Maintain Roles)
- Transaction SU01 (Maintain Users)

8.2.1 Step 1: Check SAP System Data

The relevant SAP system data can be found by choosing **System • Status**. A popup window appears that shows an overview of your SAP system information (see Figure 8.8).

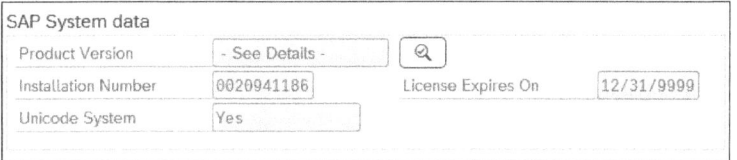

Figure 8.8 SAP System Data

Choose the icon in the **SAP System data** area to see **Installed Software Component Versions** (see Figure 8.9).

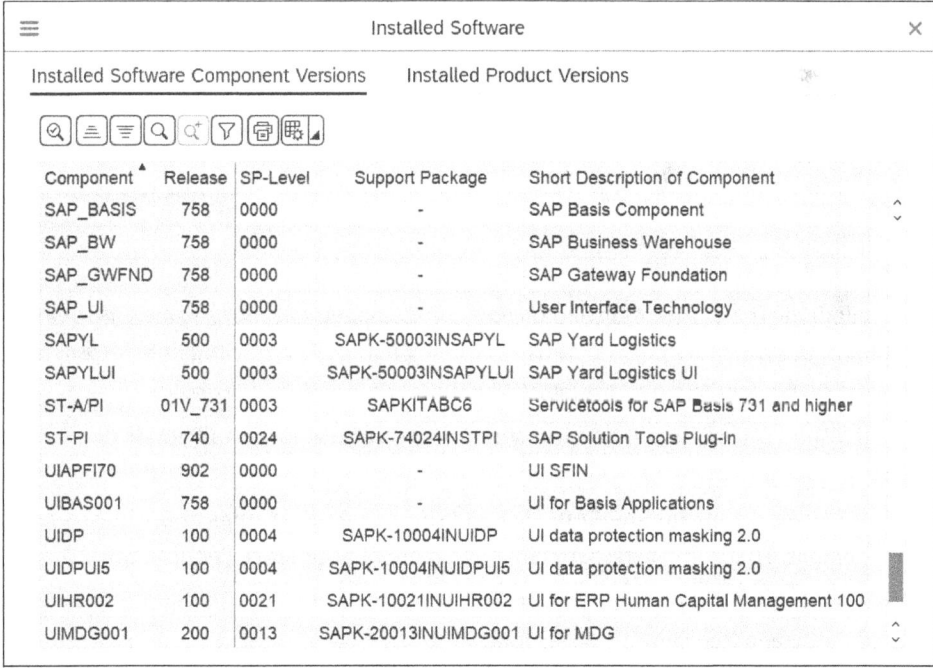

Figure 8.9 Installed Software Component Versions

You'll have to find versions of the following components:

- SAP_BASIS
- SAP_UI
- S4CORE
- UIS4HOP1
- UIAPFI70 (only if you want to implement finance apps)

8 Configuring the SAP Fiori Launchpad for Plant Maintenance

Furthermore, you'll have to look up these installed product versions (see Figure 8.10):

- On-premise SAP S/4HANA (in this case, the 2023 initial shipment stack)
- SAP Fiori for SAP S/4HANA (in this case, the 2023 initial shipment stack)

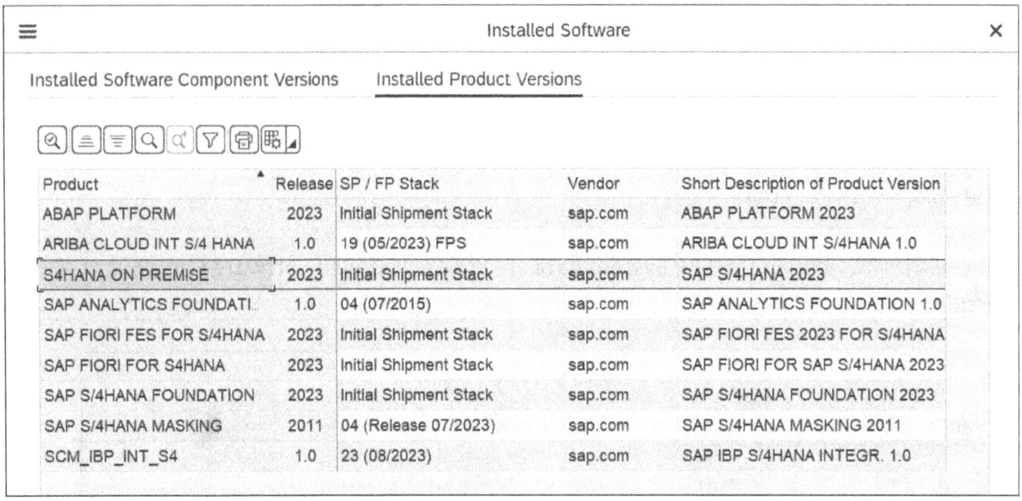

Product	Release	SP / FP Stack	Vendor	Short Description of Product Version
ABAP PLATFORM	2023	Initial Shipment Stack	sap.com	ABAP PLATFORM 2023
ARIBA CLOUD INT S/4 HANA	1.0	19 (05/2023) FPS	sap.com	ARIBA CLOUD INT S/4HANA 1.0
S4HANA ON PREMISE	2023	Initial Shipment Stack	sap.com	SAP S/4HANA 2023
SAP ANALYTICS FOUNDATI...	1.0	04 (07/2015)	sap.com	SAP ANALYTICS FOUNDATION 1.0
SAP FIORI FES FOR S/4HANA	2023	Initial Shipment Stack	sap.com	SAP FIORI FES 2023 FOR S/4HANA
SAP FIORI FOR S4HANA	2023	Initial Shipment Stack	sap.com	SAP FIORI FOR SAP S/4HANA 2023
SAP S/4HANA FOUNDATION	2023	Initial Shipment Stack	sap.com	SAP S/4HANA FOUNDATION 2023
SAP S/4HANA MASKING	2011	04 (Release 07/2023)	sap.com	SAP S/4HANA MASKING 2011
SCM_IBP_INT_S4	1.0	23 (08/2023)	sap.com	SAP IBP S/4HANA INTEGR. 1.0

Figure 8.10 Installed Product Versions

8.2.2 Step 2: Explore the SAP Fiori Apps Reference Library

To implement the Create Maintenance Request app, you retrieve some information about it in the SAP Fiori apps reference library by using the following link: *https://fiori-appslibrary.hana.ondemand.com/sap/fix/externalViewer/#/home*.

On the resulting page, you'll find information about the library, and you'll also see the different app categories on the left (see Figure 8.11).

Choose **SAP Fiori apps for SAP S/4HANA** and then **All Apps** or **Lines of Business • Asset Management**. There, you can search for the Create Maintenance Request app.

After you locate the app, some important information will show up on the right-hand side of the screen (see Figure 8.12), as follows:

- **Required Back-End Product** (SAP S/4HANA or SAP S/4HANA Cloud)
- **Application Type Transactional**
- **Database HANA DB** (exclusive)
- **Device Type(s) Desktop, Smartphone, Tablet** (i.e., appropriate devices)
- **App ID F1511A**

8.2 How to Configure an SAP Fiori Launchpad with SAPUI5 Apps

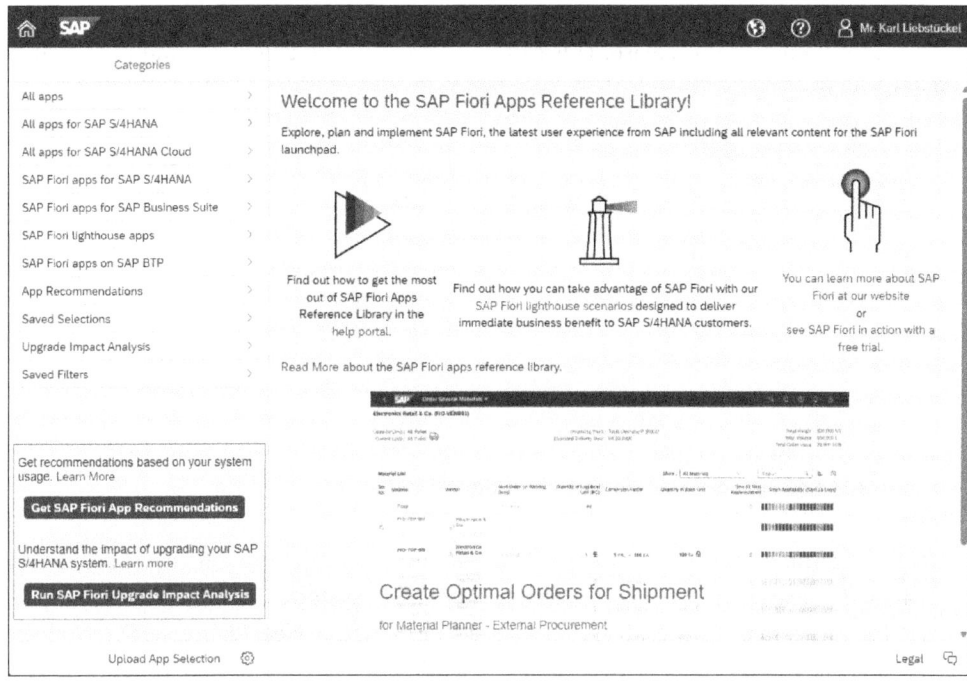

Figure 8.11 SAP Fiori Apps Reference Library

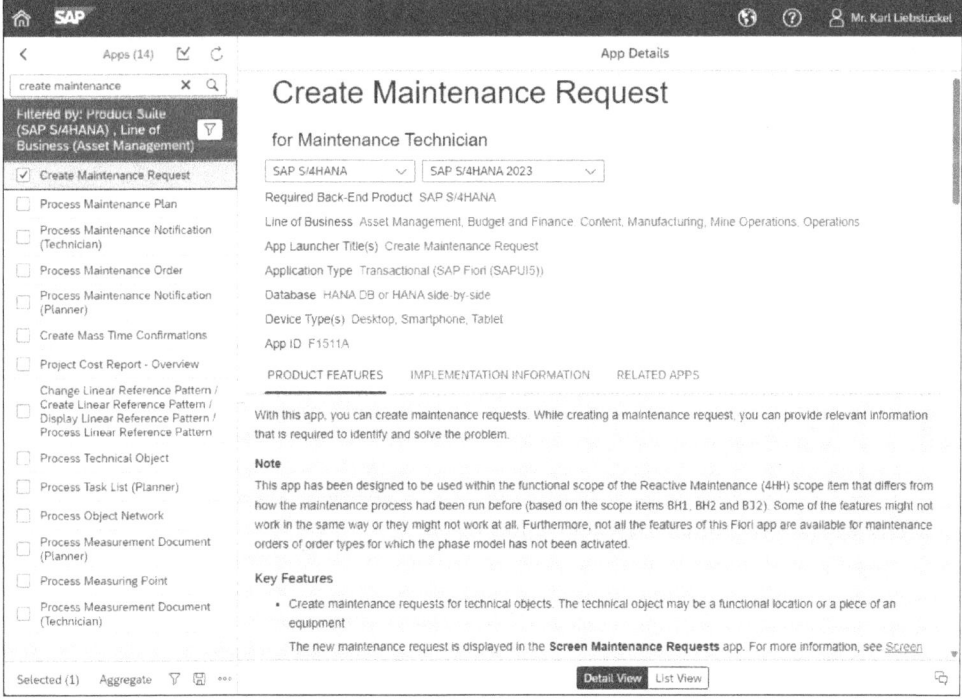

Figure 8.12 SAP Fiori App: Product Features

Navigate to the **IMPLEMENTATION INFORMATION** tab and ensure that the app is appropriate for your system release. This can be checked in the dropdown menu shown in Figure 8.13.

Figure 8.13 SAP Fiori App: Version Information

In the example shown, the Create Maintenance Request app can be used in SAP S/4HANA releases from 2021 on, which fits the target system. Under **Installation**, you'll see the necessary frontend and backend components (see Figure 8.14), so check whether the listed versions are compatible with your system and remember the information from step 1, where you checked the components.

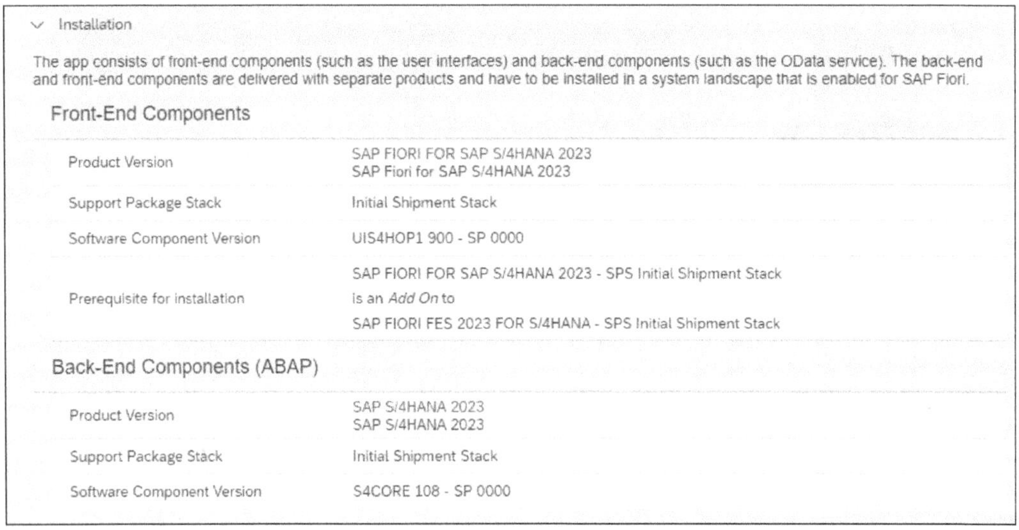

Figure 8.14 SAP Fiori App: Installation Information

8.2 How to Configure an SAP Fiori Launchpad with SAPUI5 Apps

In the upper part of the **Configuration** screen, you'll find the technical requirements that must be fulfilled by the backend system (see Figure 8.15). In this case, you must activate the following:

- SAPUI5 Application EAM_WREQ_CRTS1 using Transaction SICF with **Path to ICF Node** /sap/bc/ui5_ui5/sap/eam_wreq_crts1
- OData Service UI_MAINTWORKREQUESTOVW_V2 using Transaction /IFWND/MAINT_SERVICE
- OData V4 Service Group using Transaction /IWFND/V4_ADMIN

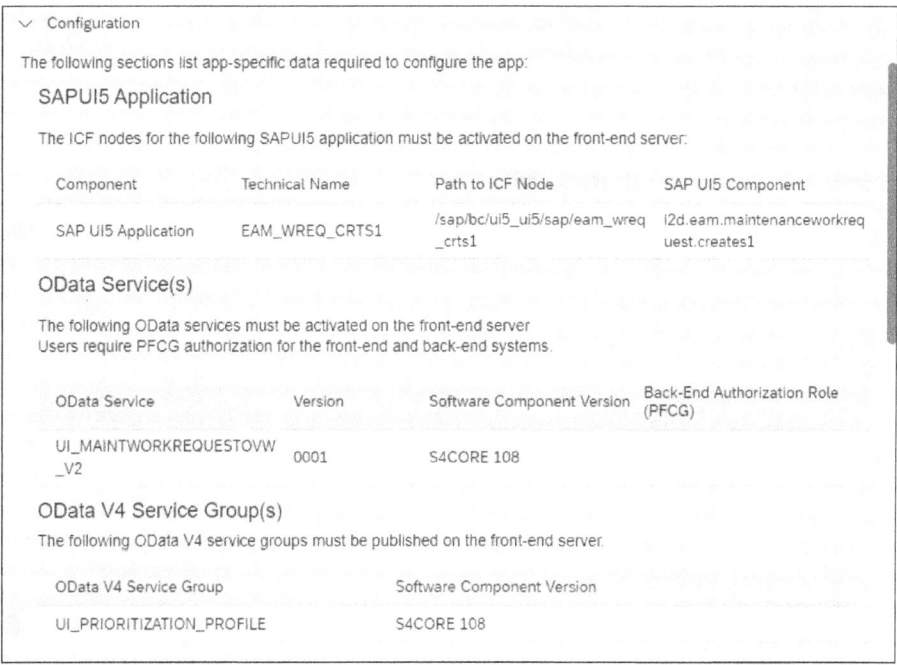

Figure 8.15 SAP Fiori App: Configuration (1 of 2)

In the lower part of the **Configuration** screen, you'll find the catalogs and groups, which you can use to create your SAP Fiori launchpad (see Figure 8.16). In this case, you must activate the following:

- **SAP_TC_COMMON** as the **Technical Catalog**
- **SAP_EAM_BC_MREQ_MNG** as the **Business Catalog**
- **SAP_EAM_BCG_MREQ_MNG** as the **Business Group**

8 Configuring the SAP Fiori Launchpad for Plant Maintenance

Technical Catalog(s)	
Technical Catalog	Technical Catalog Description
SAP_TC_EAM_COMMON	EAM - Technical Catalog

Business Catalog(s)	Extend Apps Selection
Business Catalog	Business Catalog Description
☐ SAP_EAM_BC_MREQ_MNG	EAM - Maintenance Request

Business Group(s)	
Business Group	Business Group Description
SAP_EAM_BCG_MREQ_MNG	Maintenance Request

Figure 8.16 SAP Fiori App: Configuration (2 of 2)

8.2.3 Step 3: Configure OData Services

Return to the backend system and activate the necessary OData services. Use Transaction /IFWND/MAINT_SERVICE to check if OData service **UI_MAINTWORKREQUEST-OVW_V2** is already activated. If the OData service is activated (as shown in Figure 8.17), you don't need to do anything else.

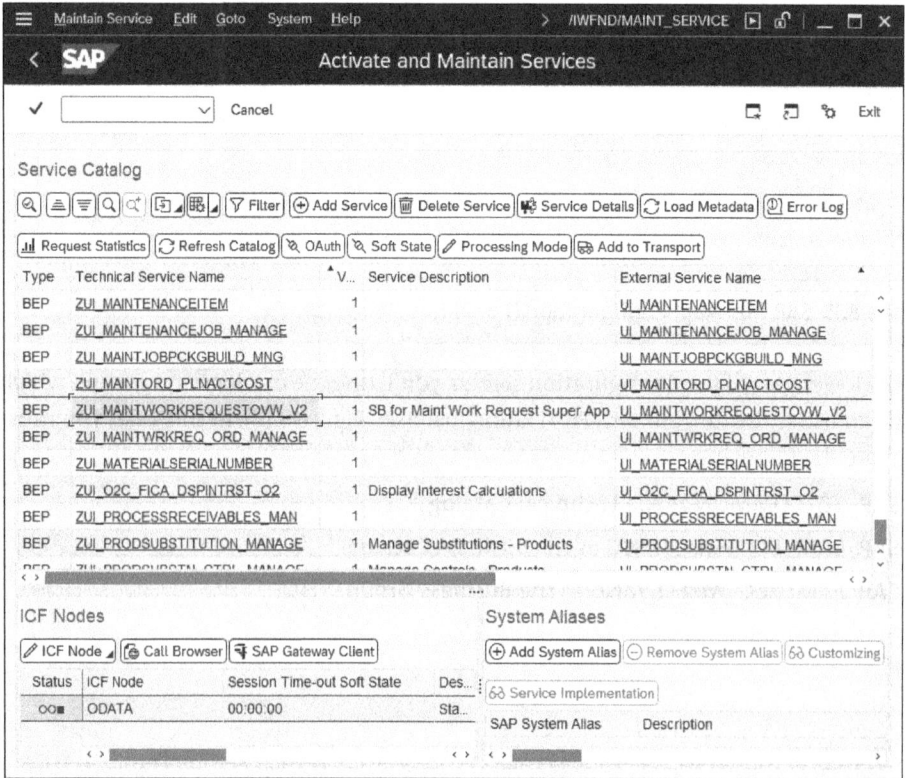

Figure 8.17 OData Services

8.2 How to Configure an SAP Fiori Launchpad with SAPUI5 Apps

If the OData service hasn't yet been activated, you must import the OData service using the **+ Add Service** button. On the next screen, enter the requested OData service name, and type in the **System Alias** as "LOCAL" (see Figure 8.18).

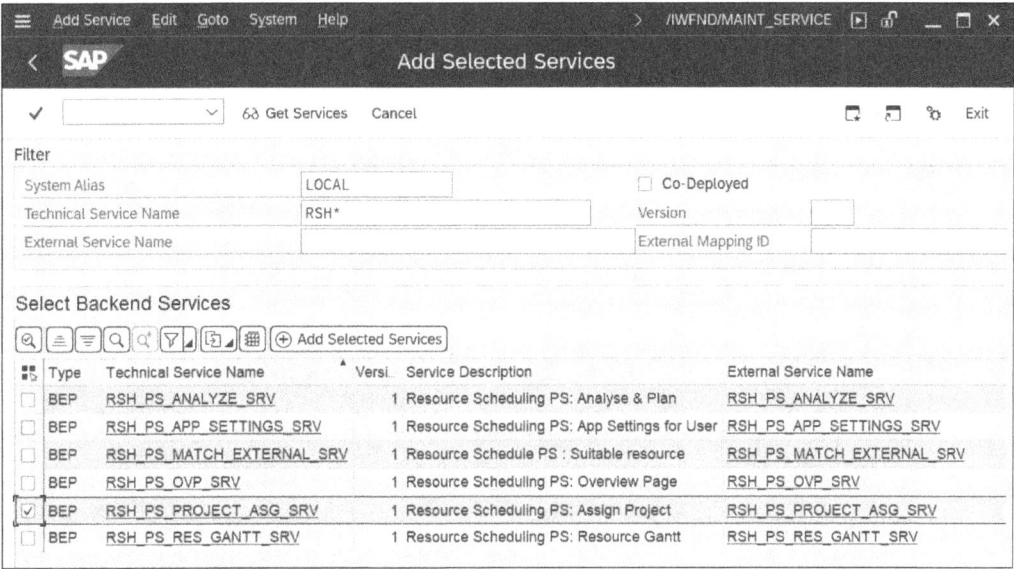

Figure 8.18 Adding New OData Services

Click the **Get Services** button to retrieve the requested OData services and mark the service that should be installed using the **+ Add Selected Services** button. Here are the possible outcomes:

- If you try to install an OData service that is only available in a later release not yet implemented in your backend system, you'll receive the **No Backend Services Found** error message.
- If you try to install an OData service that is already implemented, you'll receive the **Backend Services Are Already Registered** error message.
- In all other cases, you should get the **Service was created and its metadata was loaded successfully** message (see Figure 8.19).

Use Transaction /IFWND/V4_ADMIN to check whether the necessary OData **Service Group UI_PRIORITIZATION_PROFILE** is already activated. If the OData service is activated, you don't need to do anything else (see Figure 8.20).

If the OData service group hasn't yet been activated, you must import the OData service group using the **Publish Service Groups** button. The rest of the procedure is identical to that for importing a single OData service.

8   Configuring the SAP Fiori Launchpad for Plant Maintenance

Figure 8.19 OData Service Installed

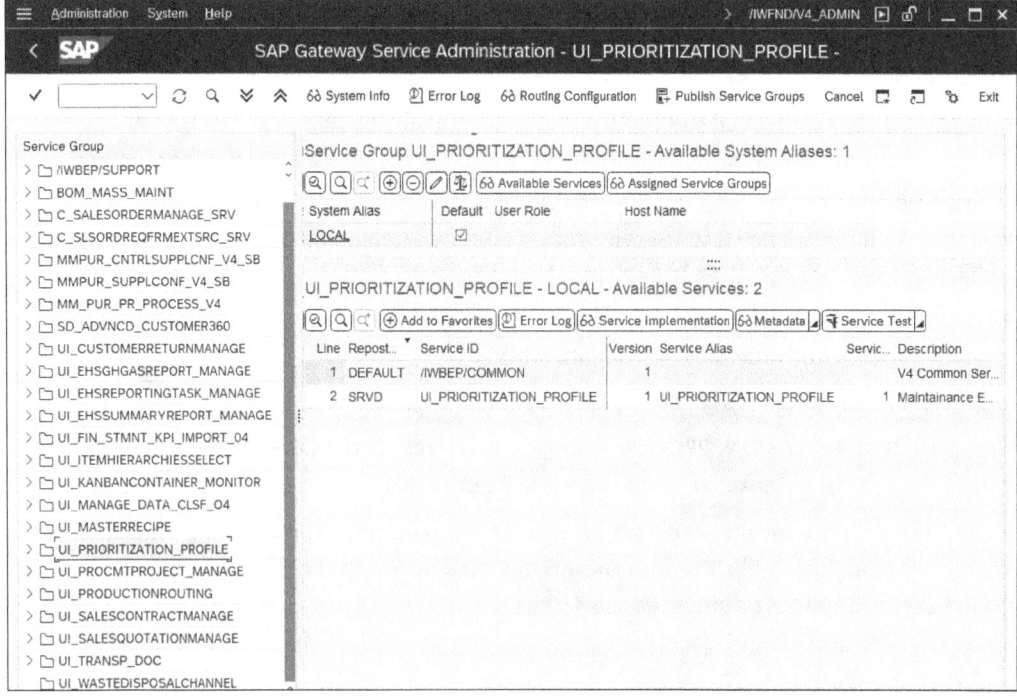

Figure 8.20 OData Service Groups

8.2.4 Step 4: Configure ICF Services

Start Transaction SICF, which will take you to the **Define Services** screen. In this transaction, services are maintained (see Figure 8.21).

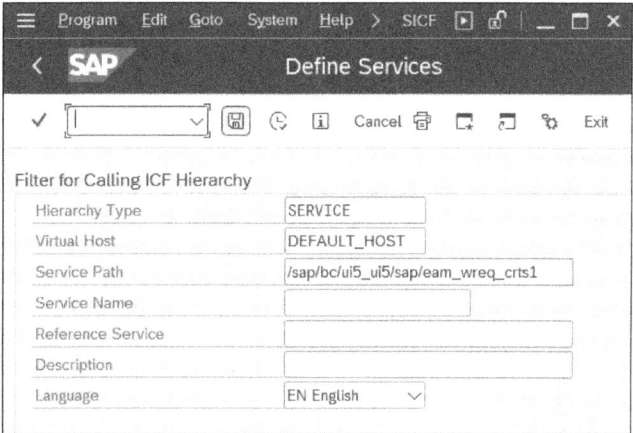

Figure 8.21 Define ICF Services

After installation of the SAP NetWeaver Application Server for ABAP (SAP NetWeaver AS for ABAP), all Internet Communication Framework (ICF) services are delivered inactive for security reasons. After the installation, you must decide which services should be activated manually.

Because several services can be executed when a URL is entered, all service nodes must be activated in the Transaction SICF tree.

To create and configure an ICF service, you usually perform the following steps:

1. Create the service.
2. Maintain the logon procedure.
3. Maintain service options.
4. Determine security requirements.
5. Set up the error pages.
6. Include request handlers.
7. Activate/deactivate service.

In our case, we'll create and activate the service for the Create Maintenance Request app by choosing the menu path **/sap/bc/ui5_ui5/sap/eam_wreq_crts1** (see Figure 8.22).

8   Configuring the SAP Fiori Launchpad for Plant Maintenance

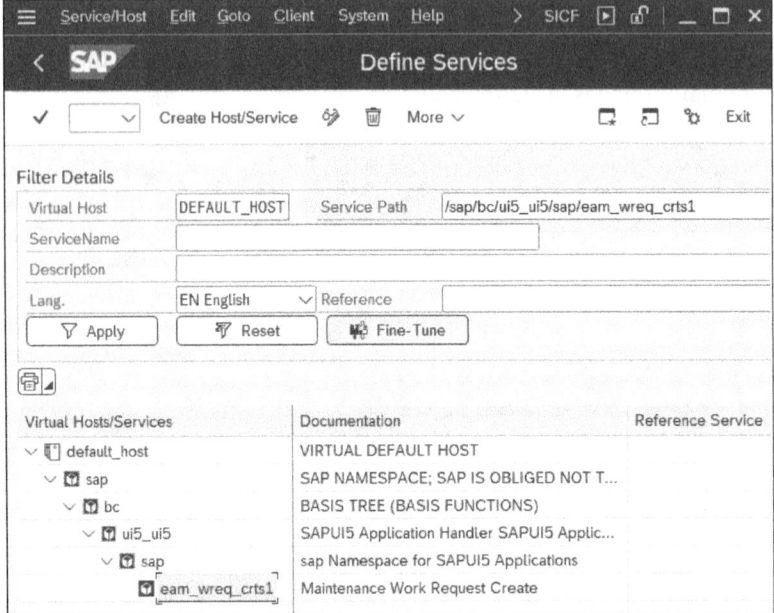

Figure 8.22 Service for Create Maintenance Request App

8.2.5 Step 5: Create a Launchpad Catalog for the SAP Fiori Launchpad

Creating a new launchpad catalog is not a must; it's an optional step. You can also use the technical catalog or the business catalogs.

A *catalog* is a set of apps that you can make available to a role. Depending on the role and the catalogs associated with the role, users can browse the catalogs and select the apps they want to view on the SAP Fiori launchpad homepage.

A *group* is a subset of apps from one or more catalogs, and the tiles displayed on a user's homepage depend on the groups associated with the user role. In addition, the user can personalize the homepage by predefining apps and adding or removing self-defined groups. All the tiles added are displayed in the group overview.

There are two different transactions you can use to create a new business catalog:

- Transaction /UI2/FLPCM_CUST (SAP Fiori Launchpad Content Manager Client-Specific)
- Transaction /UI2/FLPCM_CONF (SAP Fiori Launchpad Content Manager Cross-Client)

In our case, we want to create a new catalog for a client, so we use Transaction /UI2/FLPCM_CUST. If you start the transaction, you get an overview of all existing catalogs (see Figure 8.23).

8.2 How to Configure an SAP Fiori Launchpad with SAPUI5 Apps

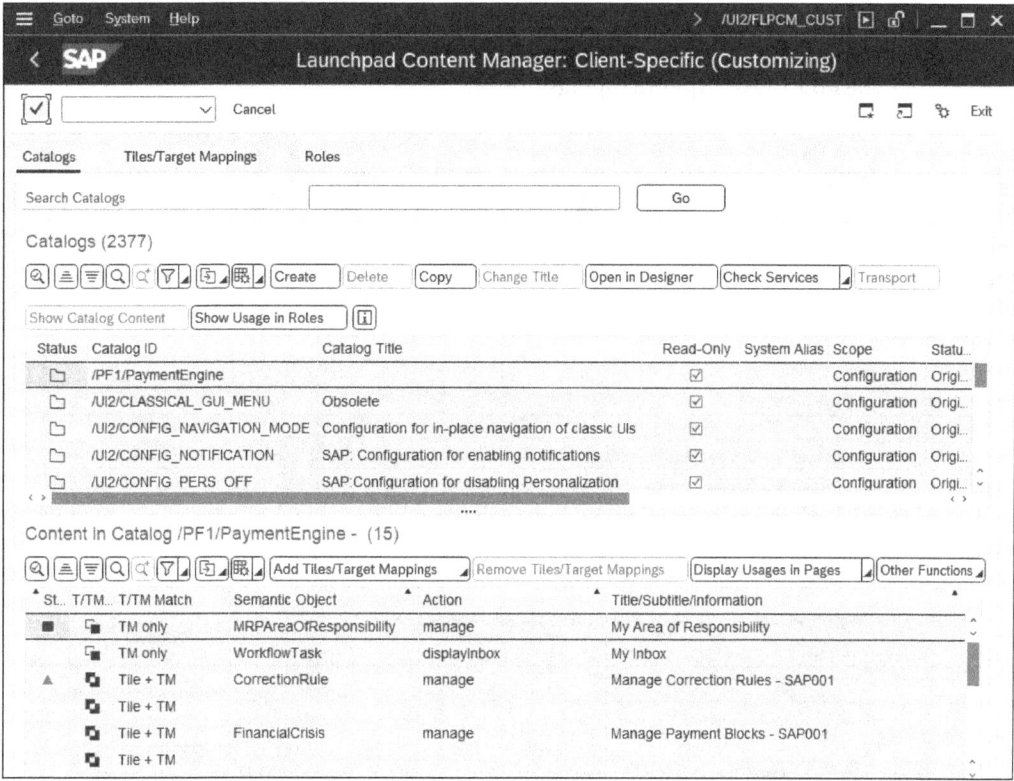

Figure 8.23 SAP Launchpad Content Manager: Overview

In our example, we want to create a new catalog for maintenance technicians in the Dallas plant. Therefore, we must click the **Create** button and enter the necessary details (see Figure 8.24).

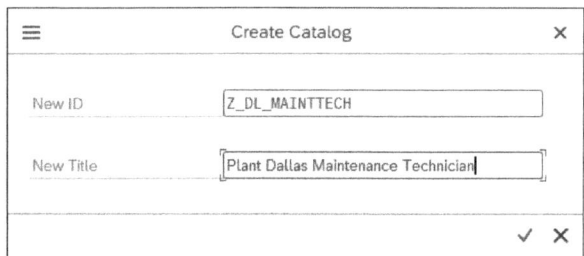

Figure 8.24 SAP Launchpad Content Manager: Create New Catalog

Next, we must search for the technical catalog because the easiest way to assign content to a new business group is to add tiles from the technical catalog. Therefore, we search for the **SAP_TC_EAM_COMMON** technical catalog (see Figure 8.25).

8 Configuring the SAP Fiori Launchpad for Plant Maintenance

On the bottom, you see all tiles that are assigned to the technical catalog. If you want to assign tiles from the technical catalog to your new business catalog, mark the tiles and click the **Add Tiles/Target Mappings** button.

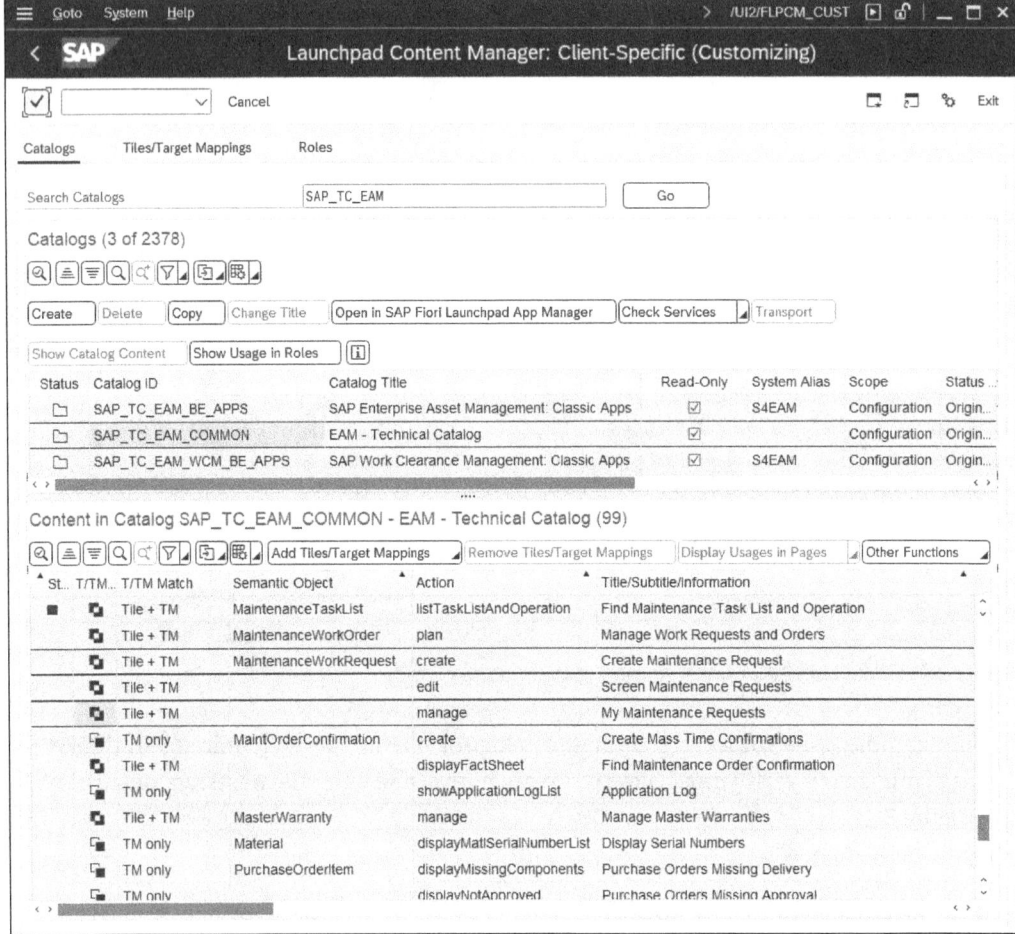

Figure 8.25 SAP Launchpad Content Manager: Catalog with Tiles

On the next screen, you search for your new business catalog (see Figure 8.26), and with the **Add Tile/TM Reference** button, you assign a tile reference to your new business catalog.

8.2 How to Configure an SAP Fiori Launchpad with SAPUI5 Apps

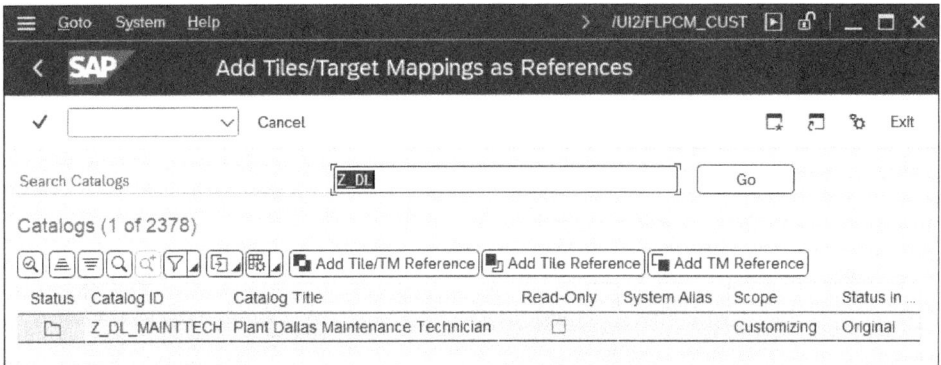

Figure 8.26 SAP Launchpad Content Manager: Assigning Tiles to New Catalog

8.2.6 Step 6: Create a Business Group for the SAP Fiori Launchpad

The next step is to create a new business group, so you must start the SAP Fiori launchpad designer using Transaction /UI2/FLPD_CUST or browser link *https://myserver/sap/bc/ui5_ui5/sap/arsrvc_upb_admn/main.html*. Replace *myserver* with your own launchpad server.

The SAP Fiori launchpad designer is a client-specific SAP system, and the client (e.g., **Client 201)** is shown in the upper right-hand corner.

The changes made in the SAP Fiori launchpad designer are distributed to other systems via transports, so a transport order must be assigned at the beginning of the configuration. In our case, the configuration shouldn't be distributed to other systems because we only want to make local changes.

To add a new group, you click on **Groups** and then click ⊕.

We'll now create a new group with apps for maintenance technicians in the Dallas plant (see Figure 8.27). By checking the **Enable users to personalize their group** box, you allow users to move apps to the group or delete them on their launchpad. You'll get an empty screen like the one shown in Figure 8.28.

Figure 8.27 SAP Fiori Launchpad Designer: New Group

601

8 Configuring the SAP Fiori Launchpad for Plant Maintenance

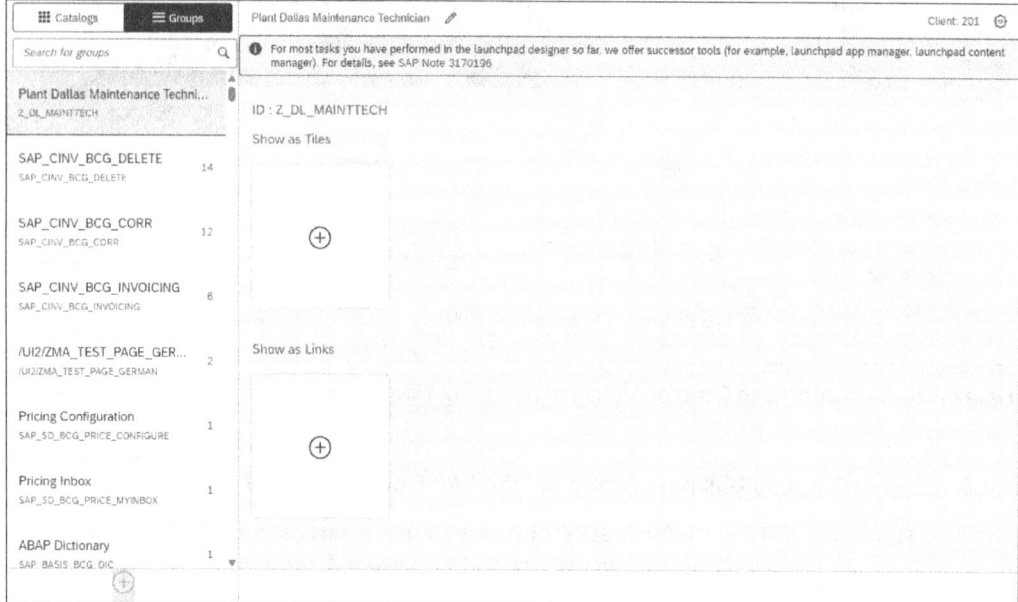

Figure 8.28 SAP Fiori Launchpad Designer: New Group Initial Screen

You add new tiles by clicking ⊕.

On the next screen, you must search for your new business catalog (see Figure 8.29).

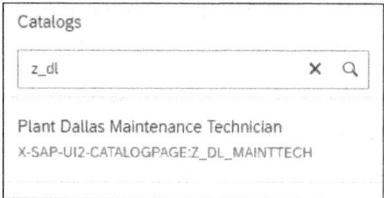

Figure 8.29 SAP Fiori Launchpad Designer: Searching Catalog

There, you can assign tiles to your new group by using the ✓ button (see Figure 8.30).

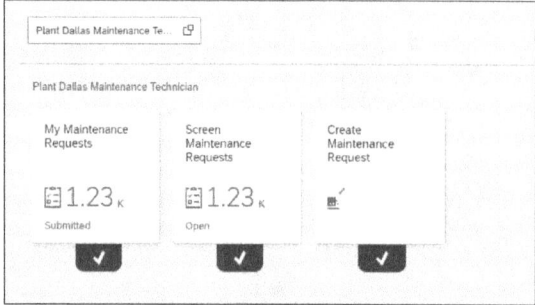

Figure 8.30 SAP Fiori Launchpad Designer: Assigning Tiles to New Group

602

8.2.7 Step 7: Create a Role and Assign Authorizations

The authorization concept of SAP S/4HANA is role based. *Roles* are groups of related transactions, reports, web links, and apps that are used for the execution of certain tasks. You create a role and then assign it to a user or group of users who need the embedded functions to perform their daily business tasks. In our example, we are going to create a role for a maintenance technician in the Dallas plant.

Roles are based on your company's organizational plan, and they link the user with the appropriate permissions. Roles contain the structure and layout of a user's menu and are defined as single roles with Transaction PFCG (Role Maintenance).

Authorization profiles are generated from single roles and contain the individual authorization objects combined with authorization fields.

Appendix A, Section A.2 contains a summary of authorization objects available in SAP S/4HANA Asset Management, as well as the organizational units, functions and fields that are checked in each use case.

For our example, we'll create a single role for a maintenance technician in the Dallas plant and assign the role to user KARL.

The first step is to assign the **Launchpad Catalog** and the **Launchpad Group** on the **Menu** tab (see Figure 8.31).

On the **Authorizations** tab, you configure the authorizations (see Figure 8.32).

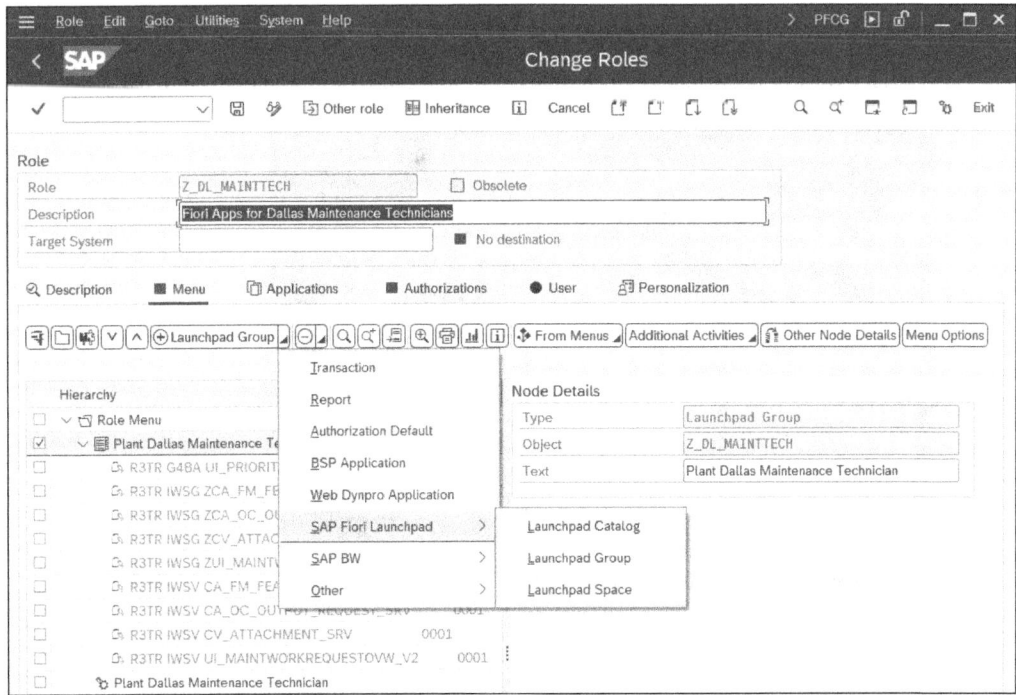

Figure 8.31 Single Role: Defining Launchpad Catalog and Launchpad Group

603

8 Configuring the SAP Fiori Launchpad for Plant Maintenance

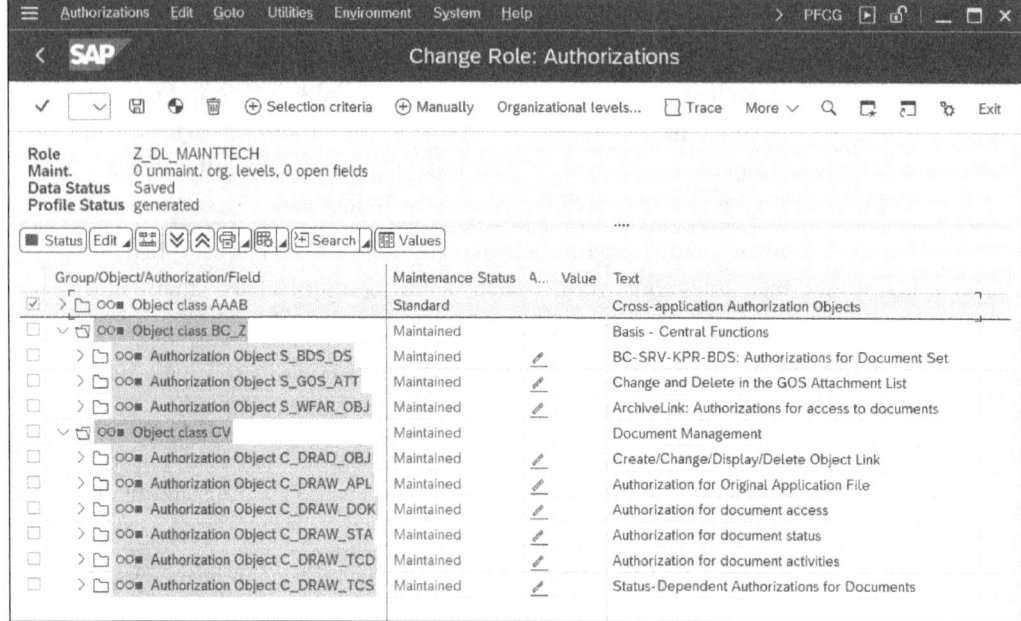

Figure 8.32 Single Role: Defining Authorizations

On the **User** tab, you assign the role to one or more users.

8.2.8 Step 8: Initialize the App

If a user starts the SAP Fiori launchpad, the new group appears (see Figure 8.33).

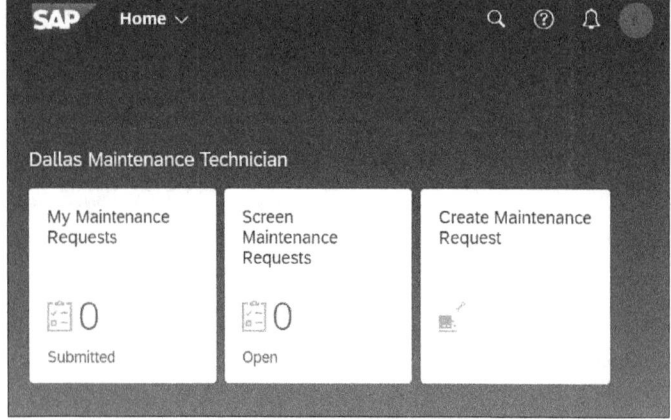

Figure 8.33 SAP Fiori Launchpad with New Group

8.3 Tabular Overview of All SAP Fiori Apps

In Table 8.1 you'll see a list of all available SAPUI5 apps for SAP S/4HANA Asset Management. To configure these for the SAP Fiori launchpad, you need information on the OData and ICF services. The list therefore contains the following information:

- App ID and short text
- Description
- ICF nodes
- OData services

App	Description	ICF Nodes	OData Services
F1511 Request Maintenance	Request repairs to a technical object.	/sap/bc/ui5_ui5/sap/eam_ntf_cres1 /sap/bc/ui5_ui5/sap/eam_ntf_reuses1	EAM_NTF_CREATE
F2072 Find Technical Object	View a list of technical objects in your system.	/sap/bc/ui5_ui5/sap/eam_to_mans1	EAM_OBJPG_TECHNICALOBJECT_SRV
F2021 Find Maintenance Notification	Display a list of notifications, mass modification of notifications.	/sap/bc/ui5_ui5/sap/eam_ntf_mans1	EAM_OBJPG_MAINTNOTIFICATION_SRV
F2175 Find Maintenance Order	Display a list of orders, mass modification of messages.	/sap/bc/ui5_ui5/sap/eam_ord_mans1	EAM_OBJPG_MAINTENANCEORDER_SRV
F2173 Find Maintenance Orders and Operations	View a list of orders and related tasks, add time-by-time confirmations.	/sap/bc/ui5_ul5/sap/eam_ordop_mons1	EAM_OBJPG_MAINTORDANDOPER_SRV
F2174 Find Maintenance Order Confirmation	View a list of order confirmations.	/sap/bc/ui5_ui5/sap/eam_oconf_mans1	EAM_OBJPG_ORDERCONFIRMATION_SRV
F2222 Manage Work Center Utilization	Analyze and manage the utilization of your work centers.	/sap/bc/ui5_ui5/sap/rsh_eam_dets1 /sap/bc/ui5_ui5/sap/rsh_eam_libs1 /sap/bc/ui5_ui5/sap/rsh_eam_oplvls1 /sap/bc/ui5_ui5/sap/rsh_eam_overvs1	RSH_EAM_ANALYZE_SRV

Table 8.1 SAP Fiori Apps

App	Description	ICF Nodes	OData Services
F2227 Resource Scheduling	Monitor important key performance indicators (KPIs) for your work centers.	/sap/bc/ui5_ui5/sap/rsh_eam_overvs1 /sap/bc/ui5_ui5/sap/rsh_eam_libs1	RSH_EAM_ANALYZE_SRV
F2774 Mass Schedule Maintenance Plans	Schedule all maintenance plans due within a certain period.	/sap/bc/ui5_ui5/sap/nw_aps_apj /sap/bc/ui5_ui5/sap/nw_aps_apj_lib	APJ_JOB_MANAGEMENT_SRV
F2812 Analytical List Page for Technical Object Breakdown Analysis	Analyze failures and calculate the duration of breakdowns and repairs and time between repairs.	/sap/bc/ui5_ui5/sap/ast-brkdwn_mons1	C_MAINTOBJBREAKDOWNQUERY_CDS
F2023 Report and Repair Malfunction	Report a fault in a technical object, plan the necessary repair work, and document and carried-out actual times.	/sap/bc/ui5_ui5/sap/eam_malf_mans1	EAM_MALFUNCTION_MANAGE
F2465 Schedule Material Availability Check	Perform material availability checks for various maintenance orders at runtime or schedule them as recurring batch jobs.	/sap/bc/ui5_ui5/sap/nw_aps_apj /sap/bc/ui5_ui5/sap/nw_aps_apj_lib	APJ_JOB_MANAGEMENT_SRV
F2412 Manage Teams and Responsibilities	Manage teams and team members who take on specific tasks.	/sap/bc/ui5_ui5/sap/rsm_team_mans1	APS_CHANGE_DOCUMENTS_SRV CA_RSM_TEAM_ACTION_SRV CA_RSM_TEAM_SRV C_RESPYMGMTTEAMHIERARCHY_CDS
F2603 Maintenance Scheduling Board	Visualize orders, tasks, and subtasks in your workstations on a timeline.	/sap/bc/ui5_ui5/sap/rsh_eam_ordgts1 /sap/bc/ui5_ui5/sap/rsh_eam_libs1	RSH_EAM_ORDER_GANTT_SRV RSH_SB_MAINTENANCE_ORDER UI_RSHPERSON

Table 8.1 SAP Fiori Apps (Cont.)

8.3 Tabular Overview of All SAP Fiori Apps

App	Description	ICF Nodes	OData Services
F2953 My Inbox - Maintenance Management	Process your workflow tasks via mobile or desktop devices.	/sap/bc/ui5_ui5/sap/ca_fiori_inbox	/IWPGW/TASKPROCESSING
F3075 Technical Object Damages	Display common damage and identify the parts of the technical object that cause this damage, the number of causes, actions, and parts of the object.	/sap/bc/ui5_ui5/sap/astdmg_mons1	C_DAMAGEANALYSIS-QUERY_CDS
F2828 Maintenance Planning Overview	Display a list of critical factors, such as missing spare parts or overdue orders; various cards illustrate the results in numbers and colored charts.	/sap/bc/ui5_ui5/sap/eam_ord_mons1 /sap/bc/ui5_ui5/sap/eam_po_mons1 /sap/bc/ui5_ui5/sap/eam_procmts1	EAM_ORDER_MONITOR
F3065 Procurement for Maintenance Planners (Purchase Requisition)	Display a list of all purchase requisitions that have been created for your maintenance orders.	/sap/bc/ui5_ui5/sap/eam_procmts /sap/bc/ui5_ui5/sap/eam_ord_mons1 /sap/bc/ui5_ui5/sap/eam_po_mons1	EAM_OBJPG_PURCH_SRV
F2827 Procurement for Maintenance Planners (Purchase Orders)	Display a list of all purchase orders that have been created for your maintenance orders.	/sap/bc/ui5_ui5/sap/eam_po_mons1 /sap/bc/ui5_ui5/sap/eam_ord_mons1 /sap/bc/ui5_ui5/sap/eam_procmts1	EAM_OBJPG_PURCHORDER_SRV
F2660 Find Maintenance Task List	Find and view maintenance routing schedules.	/sap/bc/ui5_ui5/sap/eam_tl_mans1	EAM_OBJPG_TASKLIST_SRV
F2661 Find Maintenance Task List and Operation	Find and view maintenance routing and operations.	/sap/bc/ui5_ui5/sap/eam_tlop_mons1	EAM_OBJPG_TASKLISTANDOP_SRV

Table 8.1 SAP Fiori Apps (Cont.)

App	Description	ICF Nodes	OData Services
F3326 Manage Schedules	Set up and monitor schedules for maintenance order operations.	/sap/bc/ui5_ui5/sap/rsh_eam_oplvls1 /sap/bc/ui5_ui5/sap/rsh_eam_dets1 /sap/bc/ui5_ui5/sap/rsh_eam_libs1 /sap/bc/ui5_ui5/sap/rsh_eam_overvs1	RSH_EAM_MAINT_SCHED_SIMULATION_SRV
F3622 Find Maintenance Plans	Find and view maintenance schedules.	/sap/bc/ui5_ui5/sap/eam_mplan_mans1	EAM_OBJPG_MAINTPLAN_SRV
F3567 Actual Maintenance Cost Analysis	Analyze actual costs incurred by current orders from different perspectives.	/sap/bc/ui5_ui5/sap/eam_acost_mons1	EAM_ORDER_ACTUAL-COST_MONITOR
F3925 Create Mass Time Confirmations	View status of time confirmations and post them.	/sap/bc/ui5_ui5/sap/eam_oconf_crts1 /sap/bc/ui5_ui5/sap/eam_ordop_mons1	EAM_ORD_MASS_CONFIRMATION_SRV
F5008 Find Maintenance Items	Search for and display maintenance items.	/sap/bc/ui5_ui5/sap/eam_mitem_mans1	EAM_OBJPG_MAINTPLAN_SRV
F4577 Manage Maintenance Schedule for Assets	Visualize maintenance windows and planned shutdowns.	/sap/bc/ui5_ui5/sap/rsh_eam_astbds1 /sap/bc/ui5_ui5/sap/rsh_eam_libs1	RSH_EAM_MAINT_ASSET_BOARD_SRV
F5104A Perform Maintenance Jobs	Review and execute all tasks scheduled for execution and report the actual data.	/sap/bc/ui5_ui5/sap/eam_pmntj_mans1	API_MAINTNOTIFICATION API_MAINTORDERCONFIRMATION UI_MAINTENANCE-JOB_MANAGE
F5147 Display Serial Numbers	Display a list of serial numbers assigned to a material.	/sap/bc/ui5_ui5/sap/eam_serno_diss1	UI_MATERIALSERIALNUMBER

Table 8.1 SAP Fiori Apps (Cont.)

8.3 Tabular Overview of All SAP Fiori Apps

App	Description	ICF Nodes	OData Services
F5105 Overview of Maintenance Backlog Overview	Display information about maintenance worklists (requisitions and orders) for a specific planning book.	/sap/bc/ui5_ui5/sap/eam_bcklg_mons1	EAM_MAINT_PLANNER_OVERVIEW_SRV
F1511A Create Maintenance Request	Create requirements.	/sap/bc/ui5_ui5/sap/eam_wreq_crts1	UI_MAINTWORKREQUESTOVW_V2
F4513 My Maintenance Requests	View requests you've submitted.	/sap/bc/ui5_ui5/sap/eam_wreq_ovws1	UI_MAINTWORKREQUESTOVW_V2
F4072 Screen Maintenance Requests	Review and accept or reject requests.	/sap/bc/ui5_ui5/sap/eam_wreq_ovws1	UI_MAINTWORKREQUESTOVW_V2
F4604 Manage Maintenance Notifications and Orders	Manage notifications and orders that are processed by stages.	/sap/bc/ui5_ui5/sap/eam_wrord_mans1	EAM_OBJPG_MAINTENANCEORDER_SRV EAM_OBJPG_MAINTNOTIFICATION_SRV UI_MAINTWORKREQUESTOVW_V2 UI_MAINTWRKREQ_ORD_MANAGE
F3888 Manage Maintenance Planning Buckets	Create planning folders to manage your worklist.	/sap/bc/ui5_ui5/sap/eam_bkt_mans1	EAM_PLNGBUCKET_MANAGE
F4073 Manage Maintenance Backlog	Manage orders that have been automatically assigned to a specific planning book.	/sap/bc/ui5_ui5/sap/eam_bl_mans1 /sap/bc/ui5_ui5/sap/eam_bkt_mans1	EAM_BACKLOG_MANAGE
F4603 Maintenance Order Costs	Evaluate estimated, baseline, plan, and actual costs from current orders.	/sap/bc/ui5_ui5/sap/eam_pacst_mons1	UI_MAINTORD_PLNACTCOST

Table 8.1 SAP Fiori Apps (Cont.)

App	Description	ICF Nodes	OData Services
F4691 Create Work Permit Request	Create a new work permit either independently or with reference to an existing maintenance order or a predesigned work permit template.	/sap/bc/ui5_ui5/sap/eam_wp_mans1	UI_WORKPERMIT
F6065 Manage Work Packs	Find existing work packs and send them to the output immediately or schedule the mass output.	/sap/bc/ui5_ui5/sap/eam_wrkpk_mans1	UI_MAINTJOBPCKG-BUILD_MNG
F5325 Manage Maintenance Plans	Create, change, and schedule maintenance plans.	/sap/bc/ui5_ui5/sap/eam_mplanmans1	UI_MAINTENANCE_PLAN
F3556 Manage Maintenance Items	Create or change maintenance items.	/sap/bc/ui5_ui5/sap/EAM_MITEMMANS1	UI_MAINTENANCEITEM

Table 8.1 SAP Fiori Apps (Cont.)

8.4 How to Configure an SAP Fiori Launchpad with Non-SAPUI5 Apps

As previously explained, you can use the apps predefined by SAP for the launchpad configuration. In addition to SAP Fiori SAPUI5 apps, you can add the following:

- Web Dynpro apps
- SAP GUI transactions

As an example for a Web Dynpro app, we use Confirm Unplanned Job (app ID EAMS_WDA_JOBUC_OIF; see Figure 8.34).

Key features are as follows:

- The app can be used for after-event recording. This means that the technician carries out the repair without an order and records the work carried out with actual data afterward.
- Actual times can be confirmed.

8.4 How to Configure an SAP Fiori Launchpad with Non-SAPUI5 Apps

- Materials issued can be confirmed.
- Measurement or counter readings can be entered.
- Malfunction data, damage codes, causes, and activity codes can be entered.
- You end the entry by clicking **Save and Complete**. In the background, the system creates a technically completed order with actual times, materials used, actual costs, and a completed notification.

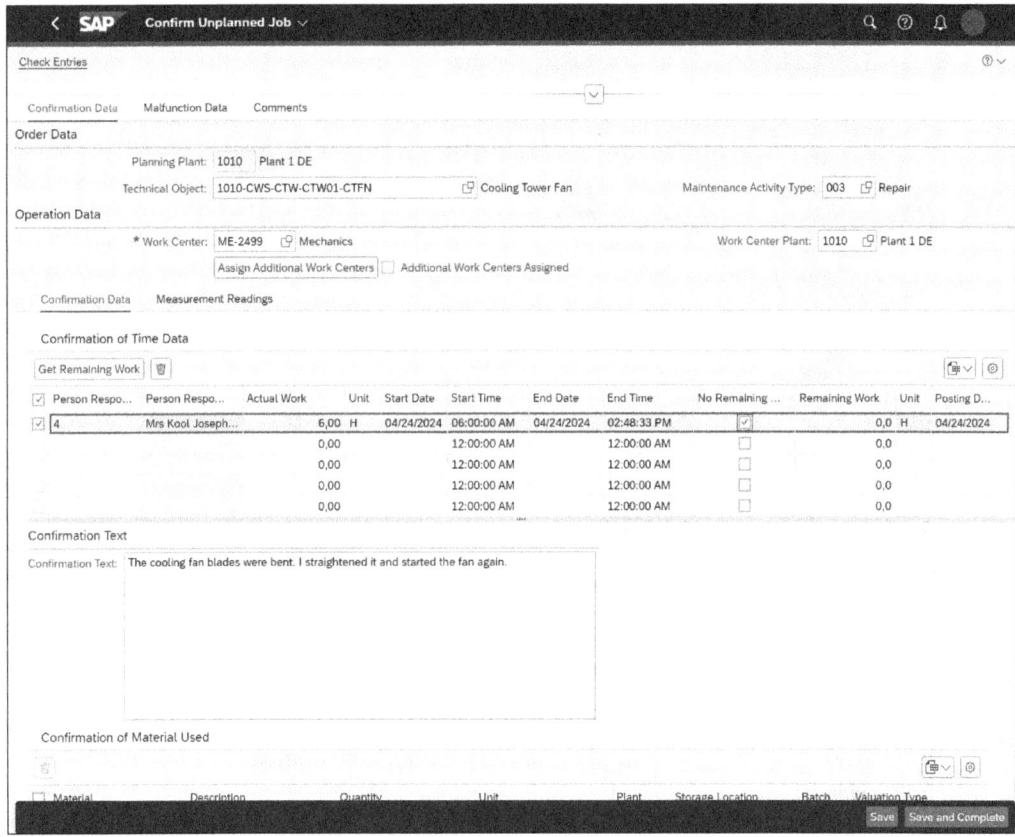

Figure 8.34 Web Dynpro Confirm Unplanned Job App

As an example for an SAP GUI transaction, we use Transaction IH01 (Functional Location Structure List). The app shows the same as the original SAP GUI Transaction IH01 (app ID IH01), not with the locally installed SAP GUI but as an HTML GUI (see Figure 8.35).

Configuring the SAP Fiori launchpad with non-SAPUI5 apps consists of the same eight steps as configuring the SAP Fiori launchpad with SAPUI5 apps shown in Figure 8.7. Therefore, in the following sections, we only describe the differences you'll find when configuring the launchpad with non-SAPUI5 apps.

611

8   Configuring the SAP Fiori Launchpad for Plant Maintenance

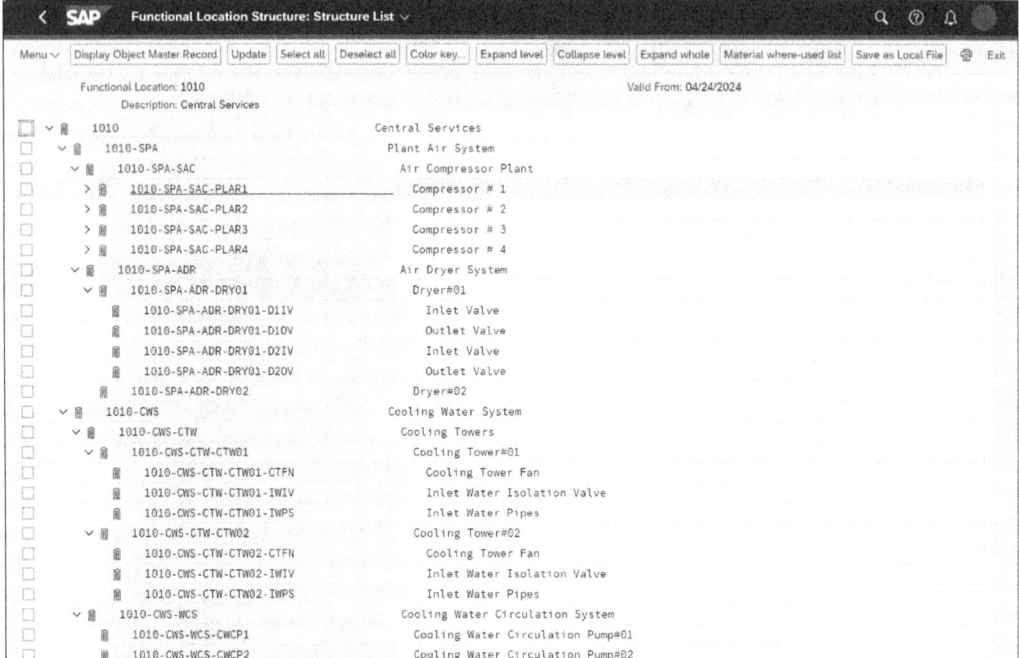

Figure 8.35 SAP GUI Transaction IH01 Functional Location Structure List

8.4.1 Step 1: Check SAP System Data

This is the same as described in Section 8.2.1.

8.4.2 Step 2: Browse the SAP Fiori Apps Reference Library

Now, you'll retrieve information about the apps in the SAP Fiori apps reference library by using the following link: *https://fioriappslibrary.hana.ondemand.com/sap/fix/externalViewer/#/home*.

Search for "Confirm Unplanned Job." You see that this app has **Application Type Web Dynpro** (see Figure 8.36).

Search for term "IH01." This app has **Application Type SAP GUI** (see Figure 8.37).

Navigate to the **IMPLEMENTATION INFORMATION** tab. First, you'll have to ensure that this app exists for your system release:

- The Confirm Unplanned Job app is available for all SAP S/4HANA releases starting with 1809 FPS01.

- The Functional Location Structure app is available for all SAP S/4HANA releases starting with 1610. Under **Installation**, you'll find the necessary frontend and backend components. Find out whether or not the listed versions are compatible with your system.

8.4 How to Configure an SAP Fiori Launchpad with Non-SAPUI5 Apps

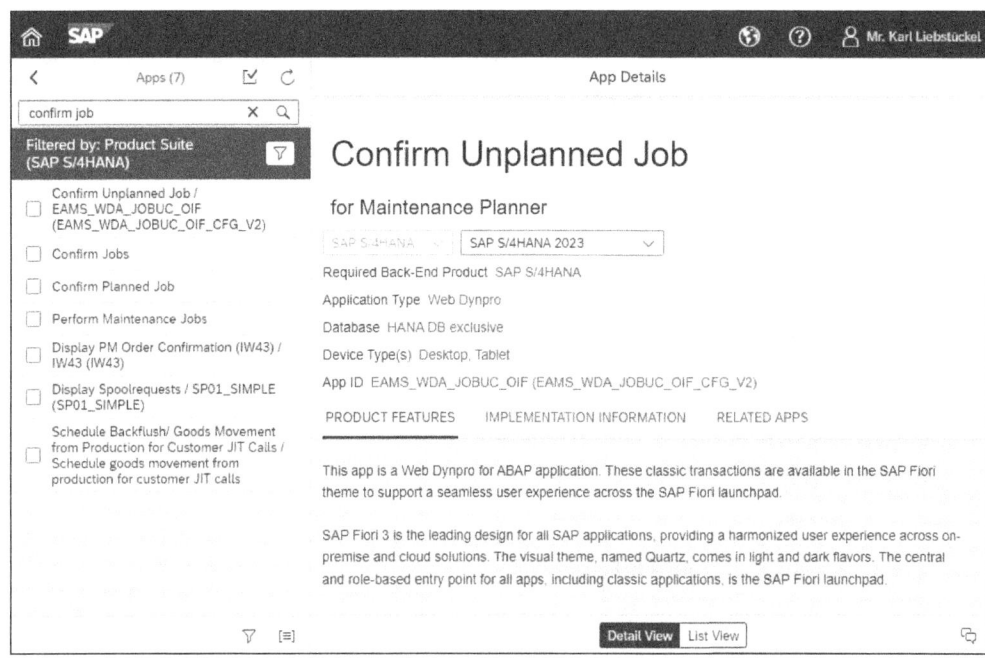

Figure 8.36 SAP Fiori Apps Reference Library: Web Dynpro App

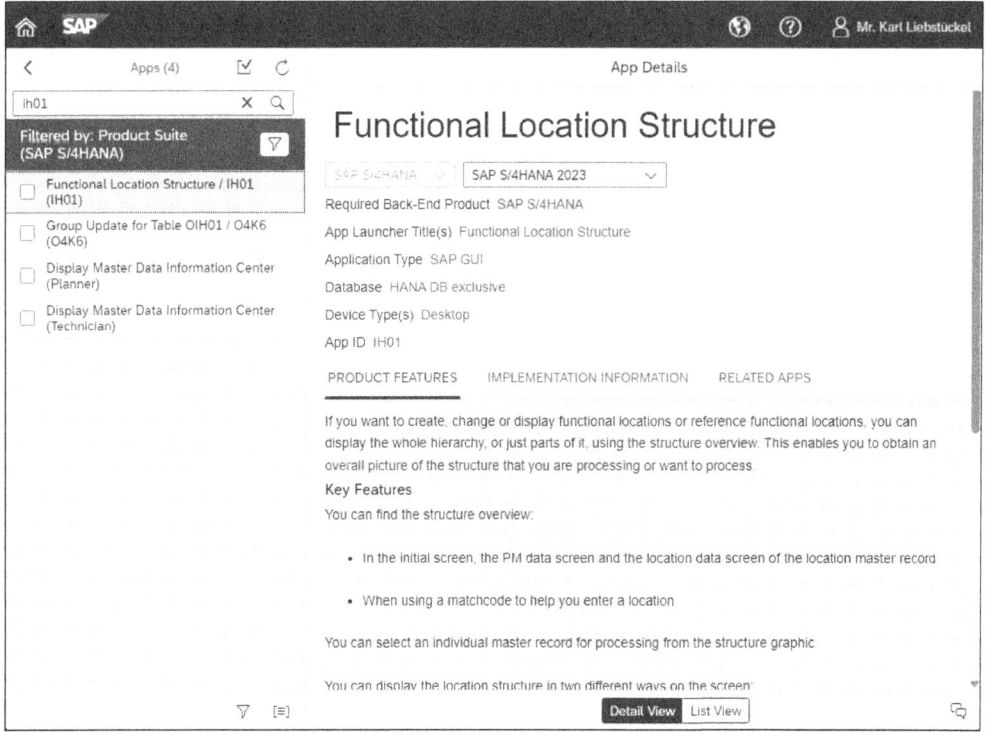

Figure 8.37 SAP Fiori Apps Reference Library: SAP GUI Transaction

8 Configuring the SAP Fiori Launchpad for Plant Maintenance

Under the **Configuration** paragraph, you'll see all necessary information to set up the app. All Web Dynpro and SAP GUI apps for SAP S/4HANA Asset Management are listed in **Technical Catalog SAP_TC_EAM_BE_APPS** (see Figure 8.38).

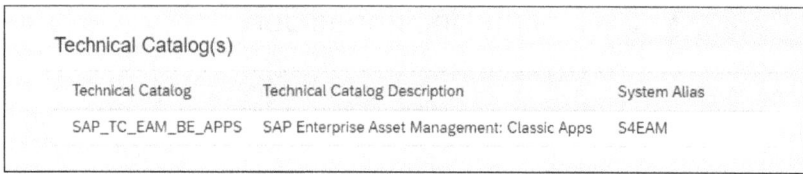

Figure 8.38 SAP Fiori Non-SAPUI5 App Configuration

8.4.3 Step 3: Configure OData Services

No OData services must be activated for the apps described.

8.4.4 Step 4: Configure ICF Services

For these apps, the following ICF services must be activated:

- Web Dynpro Application Confirm Unplanned Job: EAMS_WDA_JOBUC_OIF
- SAP GUI Transaction IH01 (Functional Location Structure)

An ICF service doesn't need to be activated for any SAP GUI transaction.

Start Transaction SICF, which will take you to the **Define Services** screen. In this transaction, services are maintained (see Figure 8.39).

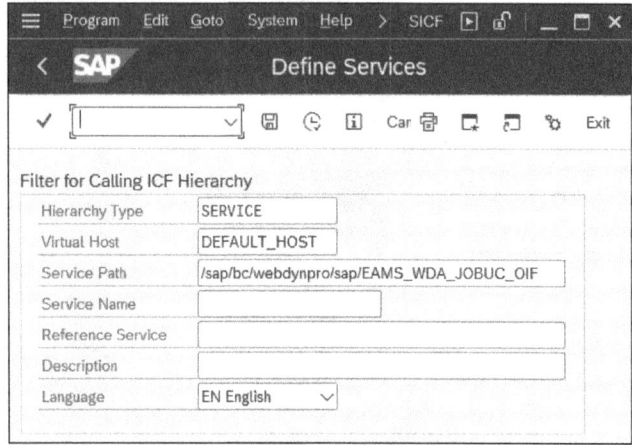

Figure 8.39 Define Services

8.4 How to Configure an SAP Fiori Launchpad with Non-SAPUI5 Apps

In our case, we'll create and activate the service for the Confirm Unplanned Job app by choosing the menu path **/sap/bc/webdynpro/sap/EAMS_WDA_JOBUC_OIF** (see Figure 8.40).

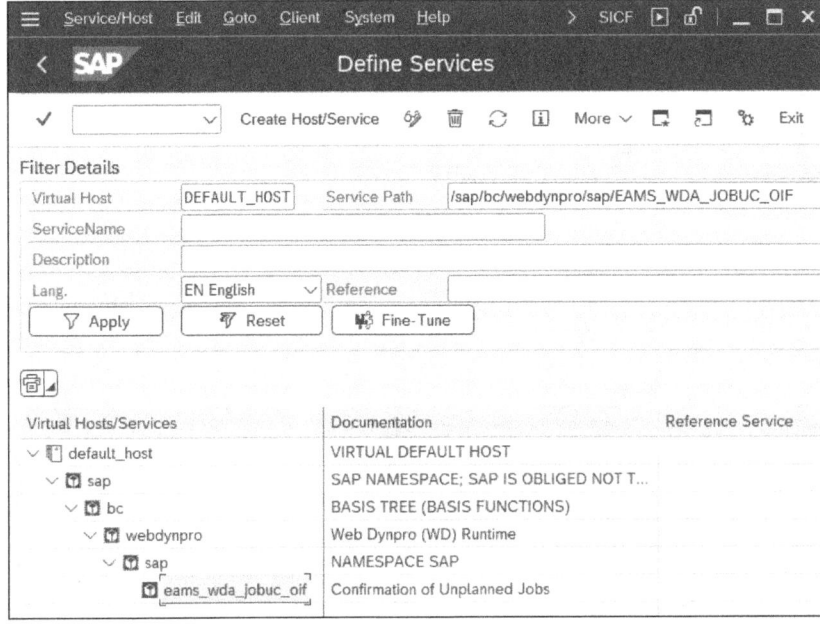

Figure 8.40 Creating and Activating Service for Confirm Unplanned Job App

8.4.5 Step 5: Create a Business Catalog for the SAP Fiori Launchpad

Creating a new launchpad catalog is an optional step, but you may as well make use of the technical catalog or one of SAP's business catalogs to create your own.

To achieve this, two different transactions have to be made:

- Transaction /UI2/FLPCM_CUST (SAP Fiori Launchpad Content Manager Client-Specific)
- Transaction /UI2/FLPCM_CONF (SAP Fiori Launchpad Content Manager Cross-Client)

In our case, we want to create a new catalog for a client. If you start Transaction /UI2/FLPCM_CUST, you get an overview of all existing catalogs (see Figure 8.23).

To extend our example further, we don't want to create a new business catalog, but we want to expand the business catalog from Section 8.2.5 by including the two non-SAPUI5 apps. That is why we are looking for **Technical Catalog SAP_TC_EAM_BE_APPS**, which has the necessary two apps (see Figure 8.41).

615

8 Configuring the SAP Fiori Launchpad for Plant Maintenance

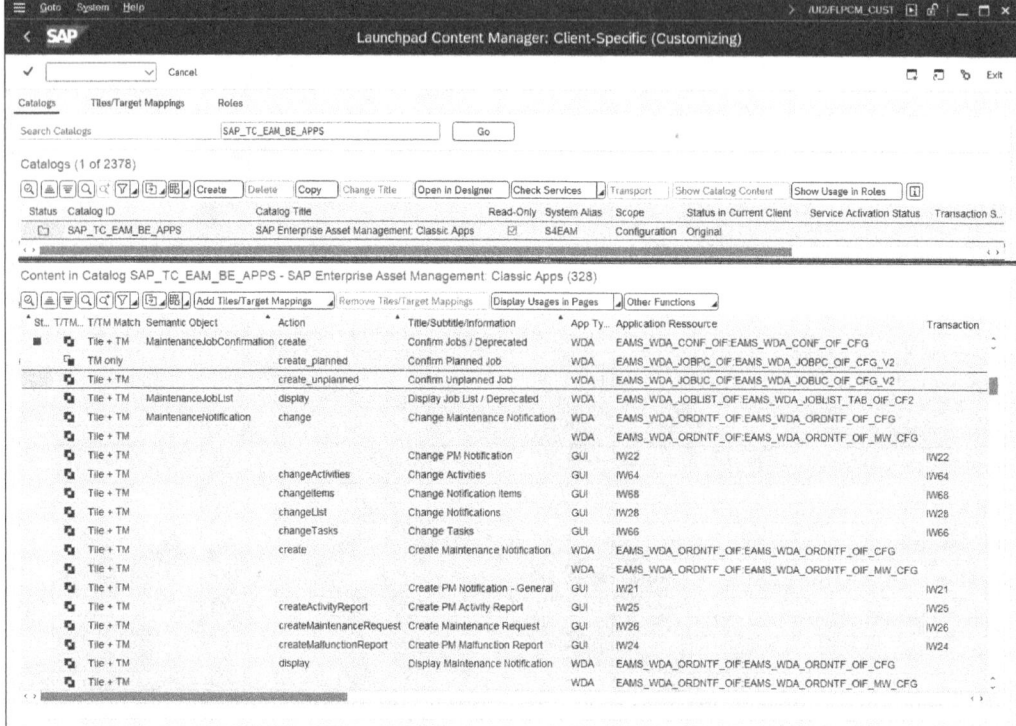

Figure 8.41 SAP Launchpad Content Manager Technical Catalog

At the bottom, you see all tiles assigned to the technical catalog. There are 328 non-SAPUI5 apps in **Technical Catalog SAP_TC_EAM_BE_APPS**, and there, we search for Transaction IH01 and Confirm Unplanned Job. If you want to attribute tiles from the technical catalog to your business catalog, mark the tiles and click the **Add Tiles/Target Mappings** button.

On the next screen, you search for the business catalog (see Figure 8.42). By clicking the **Add Tile/TM Reference** button, you assign a tile reference to your new business catalog.

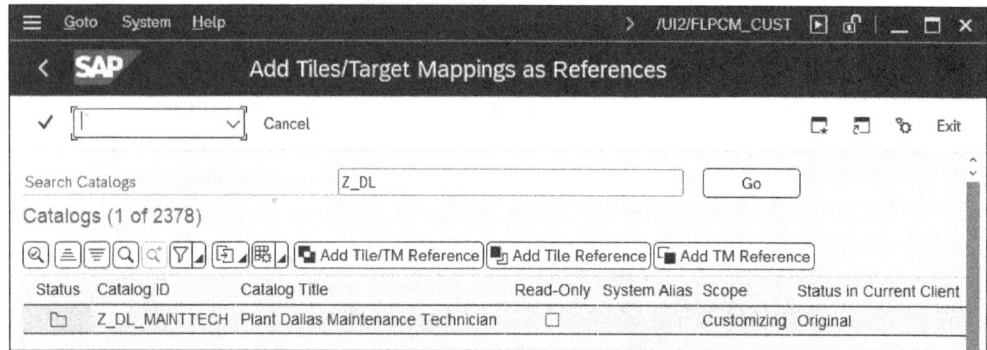

Figure 8.42 SAP Launchpad Content Manager: Assigning Tiles to Catalog

616

8.4 How to Configure an SAP Fiori Launchpad with Non-SAPUI5 Apps

8.4.6 Step 6: Create a Business Group for the SAP Fiori Launchpad

The next step brings us to creating a new business group. In our case, we want to extend the existing business group with the two non-SAPUI5 apps. Therefore, you start the SAP Fiori launchpad designer using *https://myserver/sap/bc/ui5_ui5/sap/arsrvc_upb_admn/main.html*. Replace *myserver* with your own launchpad server.

The SAP Fiori launchpad designer is a client-specific SAP system, and the client (**Client 201**) is shown in the upper right-hand corner.

We look for our business group **Z_DL_MAINTTECH** (see Figure 8.43), and we add new tiles by clicking ⊕.

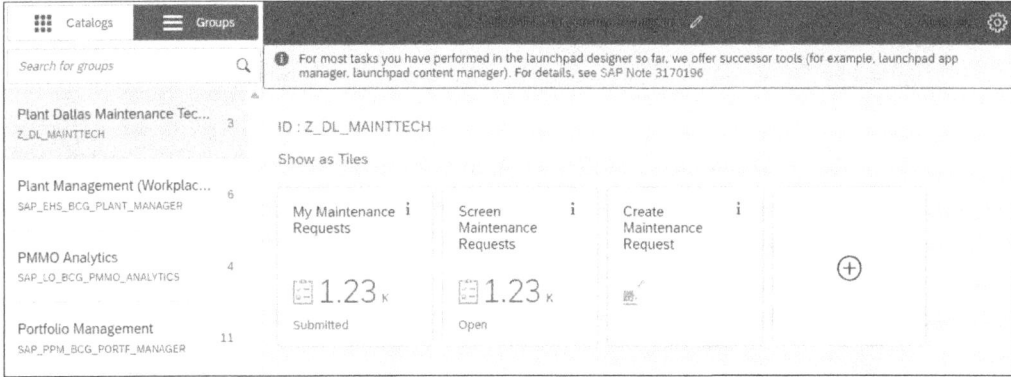

Figure 8.43 SAP Fiori Launchpad Designer: Business Group

On the next screen, you must find the relevant business catalog (see Figure 8.44).

Figure 8.44 SAP Fiori Launchpad Designer: Searching Catalog

There, you can assign tiles to your new group by using the ✓ button (see Figure 8.45).

617

8   Configuring the SAP Fiori Launchpad for Plant Maintenance

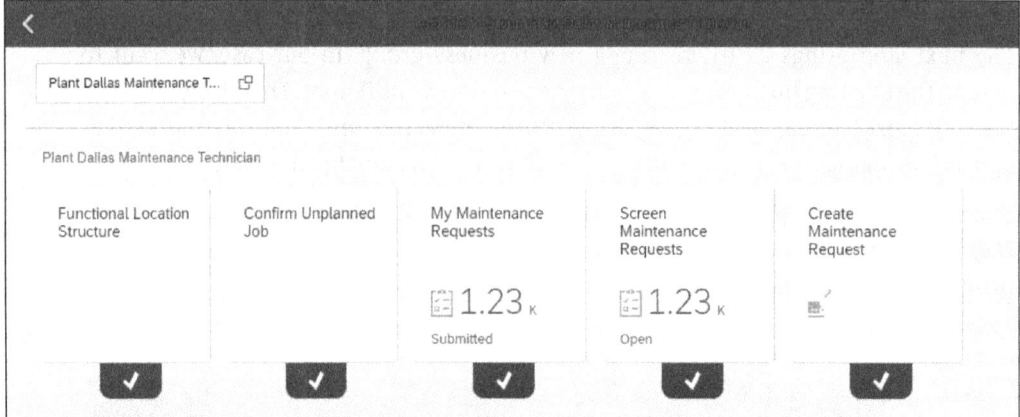

Figure 8.45 SAP Fiori Launchpad Designer: Assigning Tiles to Business Group

8.4.7   Step 7: Create or Change a Role and Assign Authorizations

Keeping our example, we must reassign the **Launchpad Catalog** and the **Launchpad Group** on the **Menu** tab (see Figure 8.46).

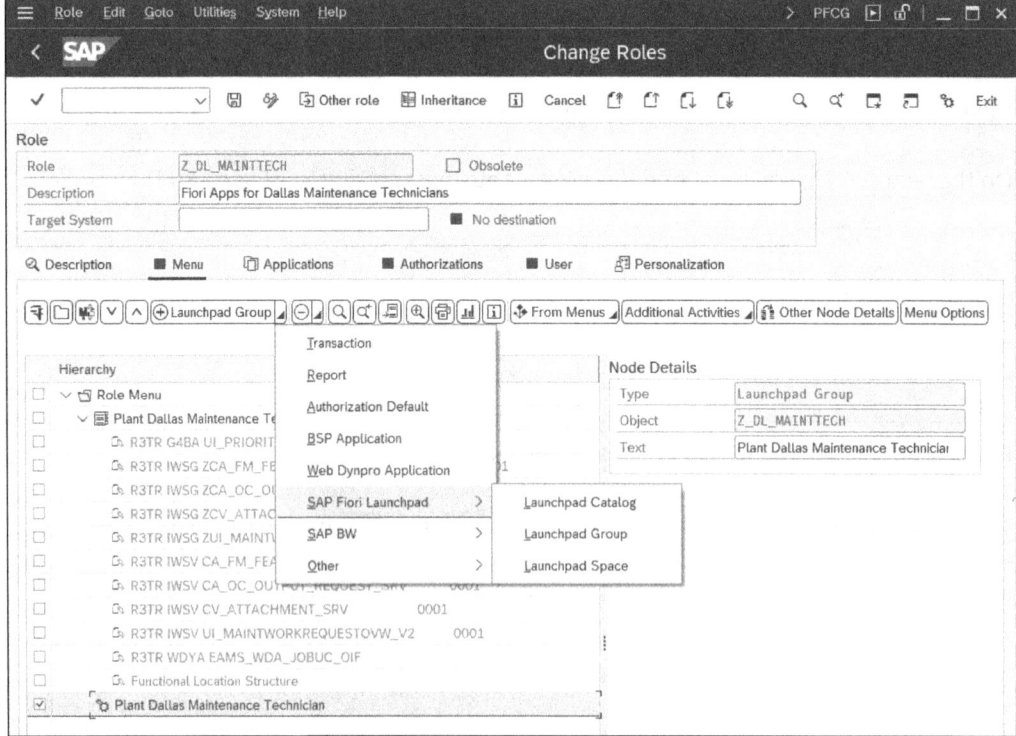

Figure 8.46 Single Role: Defining Catalog and Group

On the **Authorizations** tab, we configure the authorizations (see Figure 8.47).

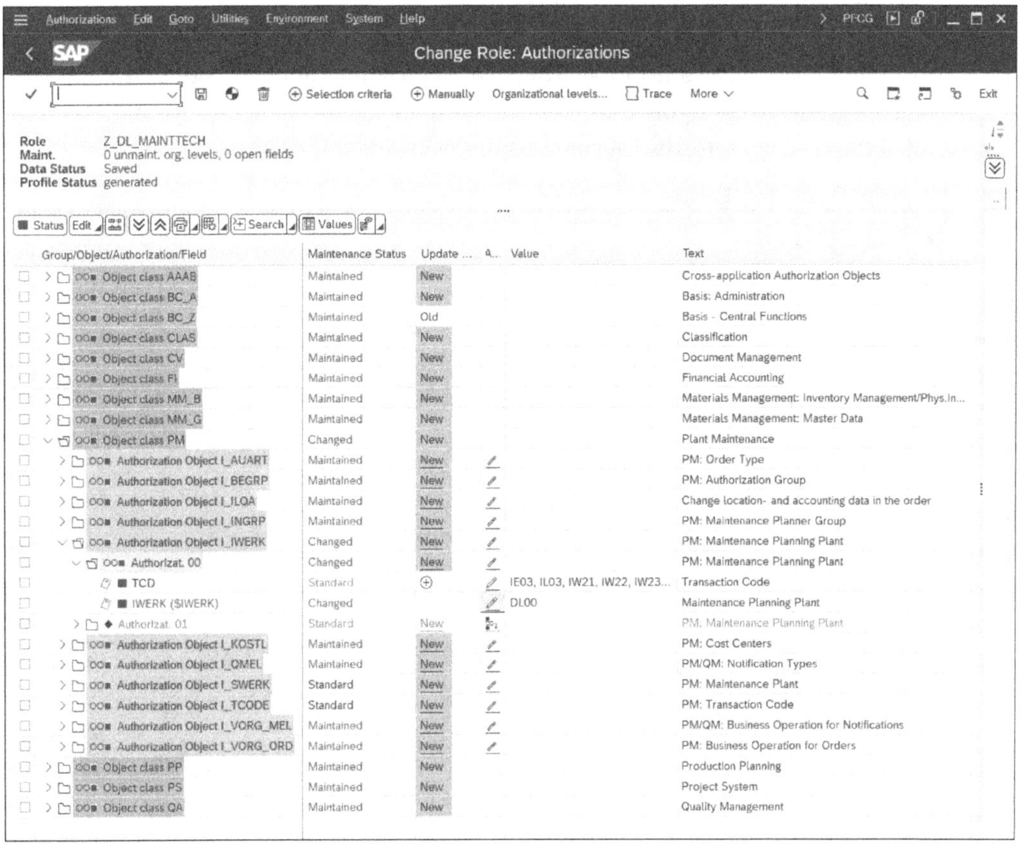

8.4 How to Configure an SAP Fiori Launchpad with Non-SAPUI5 Apps

Figure 8.47 Single Role: Defining Authorizations

8.4.8 Step 8: Initialize the App

If we start our SAP Fiori launchpad, the newly added apps appear (see Figure 8.48).

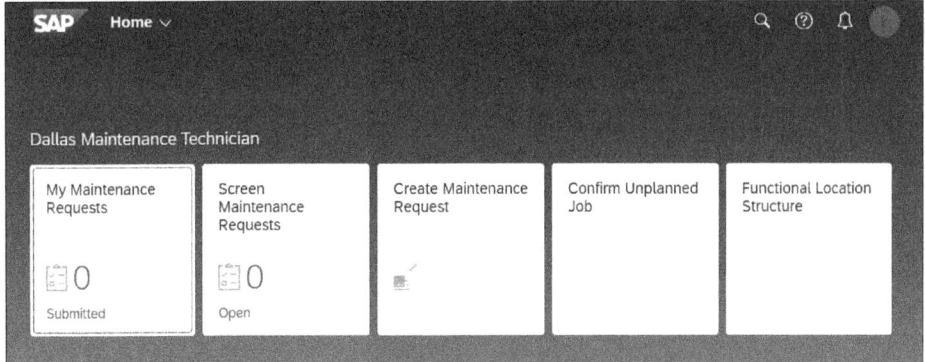

Figure 8.48 SAP Fiori Launchpad with New Apps

Now, your SAP Fiori launchpad, including SAPUI5 and non-SAPUI5 apps, is ready to use.

619

8.5 Tabular Overview of All Web Dynpro Apps

In Table 8.2 you'll see a list of all available Web Dynpro apps for SAP S/4HANA Asset Management. To configure these for the SAP launchpad, you need information on the ICF services. Therefore, the list contains the following information:

- App ID and short text
- Description
- ICF nodes

Short Text	Description	ICF Nodes /sap/bc/webdynpro/sap/...
Asset Viewer	Display complex object relationships (e.g., technical objects, materials, measurement documents), navigation between objects, hierarchical structure display, and navigation.	EAMS_WDA_NAV_OIF
Confirm Unplanned Job	After-event recording of actual times, consumed materials, and counter readings.	EAMS_WDA_JOBUC_OIF
Confirm Planned Job	Recording of actual times, consumed materials, and counter readings.	EAMS_WDA_JOBPC_OIF
W0017 Process Maintenance Order	Create, change, and display a maintenance order.	EAMS_WDA_ORDNTF_OIF
W0029 Process Technical Object	Create, change, and display a technical object.	EAMS_WDA_TECHOBJ_OIF
W0003 Process Maintenance Notification	Create, change, and display a maintenance notification.	EAMS_WDA_ORDNTF_OIF
W0023 Process Linear Reference Pattern	Create, change, and display linear patterns.	EAML_WDA_LRP_OIF
W0020 Confirm Jobs	Create order confirmations.	EAMS_WDA_CONF_OIF
W0016 Display Job List	Display a list of notifications, orders and operations, and printing job cards.	EAMS_WDA_JOBLIST_OIF

Table 8.2 Web Dynpro Apps

Short Text	Description	ICF Nodes /sap/bc/webdynpro/sap/...
W0024 Display Maintenance Item	Display a maintenance item and corresponding data (technical object, orders, etc.).	EAMS_WDA_MPOS_OIF
W0009 Process Maintenance Plan	Create, change, and display maintenance plans.	EAMS_WDA_MPLAN_OIF_V2
W0031 Process Measuring Point	Create, change, and display measuring points.	EAMS_WDA_MP_OIF
W0013 Process Measuring Documents	Create, change, and display measuring documents.	EAMS_WDA_MD_OIF
W0021 Process Task List	Create, change, and display maintenance task lists.	EAMS_WDA_TL_OIF_V2
W0192 Maintenance Plan Scheduling Overview	View maintenance plan scheduling details.	FPM_BICS_OVP

Table 8.2 Web Dynpro Apps (Cont.)

8.6 Summary

This chapter described various features of SAP Fiori apps and how to make use of them in an SAP Fiori launchpad. It covered these particular areas:

- Some basic aspects of SAP Fiori, mainly different types of SAP Fiori apps (e.g., transactional apps that are categorized based on their content, SAPUI5 apps with added value that are categorized based on their results), as well as special characteristics of SAP Fiori apps (e.g., dynamic design, static design)
- Step-by-step instructions on how to configure the SAP Fiori launchpad to allocate SAP Fiori apps based on SAPUI5 technology
- Step-by-step instructions on how to configure the SAP Fiori launchpad to allocate SAP Fiori apps that are not based on SAPUI5 technology, namely SAP GUI apps and Web Dynpro apps
- A tabular summary of all SAPUI5 apps, including information about how to configure them in the SAP Fiori launchpad
- A tabular summary of all Web Dynpro apps, including information about how to configure them in the SAP Fiori launchpad
- An explanation of the fact that SAP GUI apps don't need any configuring prerequisites to include them in the SAP Fiori launchpad

Chapter 9
Usability

Although there are enduring prejudices in terms of usability not being a major feature of the SAP system, this chapter shows a variety of customer options for how to substantially improve SAP's usability.

This chapter presents the tools and options have to improve SAP's usability, and you'll learn how to define each of these options. The options are subdivided into the following three categories (listed in the same order in which you should check their use in your enterprise):

- **Category 1**
  The user's own options
- **Category 2**
  The IT department's nonprogramming options
- **Category 3**
  The IT department's programming options

Table 9.1 provides an overview of the categories, who uses them, and what options fall into each of the categories.

Category	Who?	What?
1	User	<ul><li>General user parameters and default values</li><li>Maintenance-specific user parameters</li><li>Role menus and favorites menus</li><li>List variants and dynamic lists</li><li>Personalized input help</li><li>Buttons and key combinations</li><li>Table controls</li></ul>
2	IT department (nonprogramming options)	<ul><li>Transaction variants</li><li>Customizing</li><li>Action boxes</li><li>SAP Service and Asset Manager</li><li>SAP Fiori</li><li>GuiXT</li><li>SAP Screen Personas</li></ul>

Table 9.1 Options for Improving Usability

Category	Who?	What?
3	IT department (programming options)	■ Upstream transactions ■ Business Application Programming Interfaces (BAPIs) ■ Web interfaces ■ Customer exits ■ Business add-ins (BAdIs) and enhancement points ■ SAP Business Workflow

Table 9.1 Options for Improving Usability (Cont.)

Next, let's take a look at these options. You'll find out not only *what* you can do but also *how* you should do it.

9.1 Options at the User's Disposal

Let's start with the easiest options to implement, that is, those parameters and settings the user can change.

9.1.1 General User Parameters and Default Values

From a user's point of view, many entries required by the SAP system remain unchanged over a period of time. This is due to the following reasons:

- The user belongs to a specific work center.
- The work center is assigned to a specific cost center.
- The cost center belongs to a specific company code and controlling area.
- The user works in a particular planning and maintenance plant.
- A particular purchasing organization works for the plant maintenance department.
- The user is responsible for a particular functional location.
- The spare parts are stored in a specific storage location.

The use of *user parameters* to prepopulate fields contributes greatly to the improvement of usability, and access to these defaults is to be found in Transaction SU3 (Maintain User Profile) or via the menu path **System • User Profile • User Data** (see Figure 9.1). Once defined, these fields are automatically prepopulated with the assigned values when business processes are processed.

9.1 Options at the User's Disposal

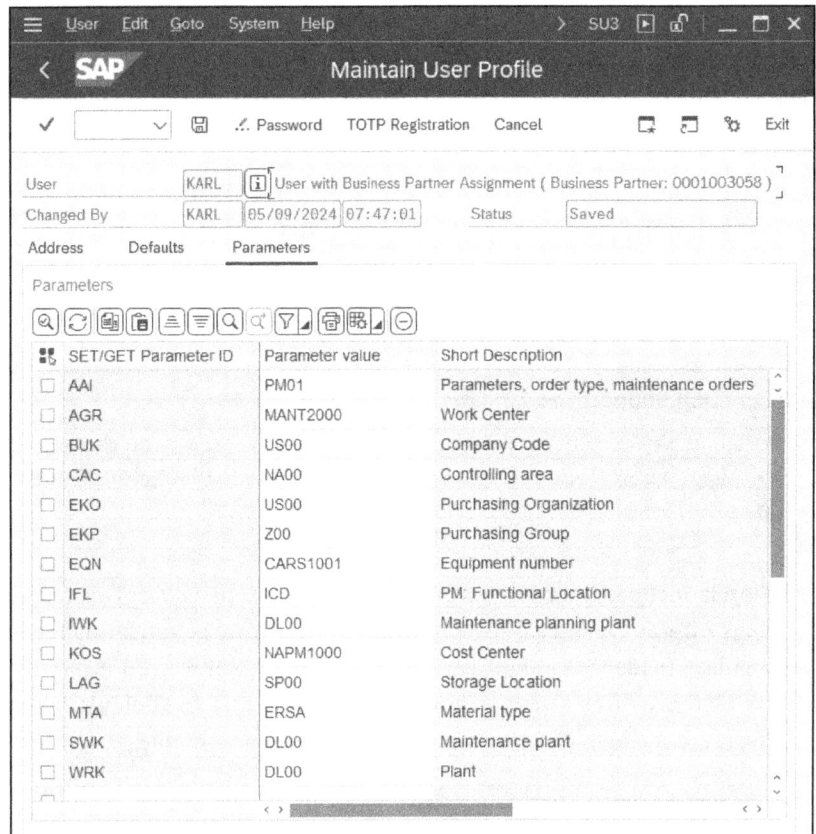

Figure 9.1 General User Parameters in SAP GUI

Viewed from the perspective of maintenance, the following parameters are in frequent use:

- AAI: Order Type
- AAR: Work Center Type
- AGR: Work Center
- BUK: Company Code
- CSA: BOM Application
- CSV: BOM Usage
- EKG: Purchasing Group
- EKO: Purchasing Organization
- EQN: Equipment Number
- EQT: Equipment Category
- IFL: Functional Location
- IMD: Measurement Document
- IME: Measurement Recording List
- IP1: Measuring Point Category
- IPT: Measuring Point
- IRL: Reference Functional Location
- ISR: Structure Indicator for Functional Location
- IWK: Maintenance Planning Plant
- KOK: Controlling Area
- KOS: Cost Center
- LAG: Storage Location
- LGP: Storage Bin
- LIS: CATS Profile for List Display

9 Usability

- MAT: Material Number
- MPL: Maintenance Plan
- MPS: Maintenance Item
- MTA: Material Type
- PER: Personnel Number
- PIN: PP Profile/Default Values
- PLN: Task List Group

- Q_ALV_GRID_INACTIVE: Deactivates the ALV Grid
- QMR: Notification Type
- SWK: Maintenance Plant
- VAP: Main Work Center
- WAT: Maintenance Plan Category
- WGR: Material Group
- WRK: Plant

> **[+] The (F1) Help Function Supports You in Finding Parameter IDs**
> Processing time can be saved by using parameters to prepopulate fields. You can generally view the relevant parameter ID by pressing (F1) and looking in the **Technical Info** of the help field.

Within the SAP Fiori launchpad, there is also an option to preassign values. This can be reached via **Settings** • **Default Values** (see Figure 9.2), but there are significantly fewer options to prepopulate fields that might be interesting and helpful from a maintenance perspective. Here, you don't need to know the parameter ID; the default values are directly on display, but a help function via the (F4) key is not available.

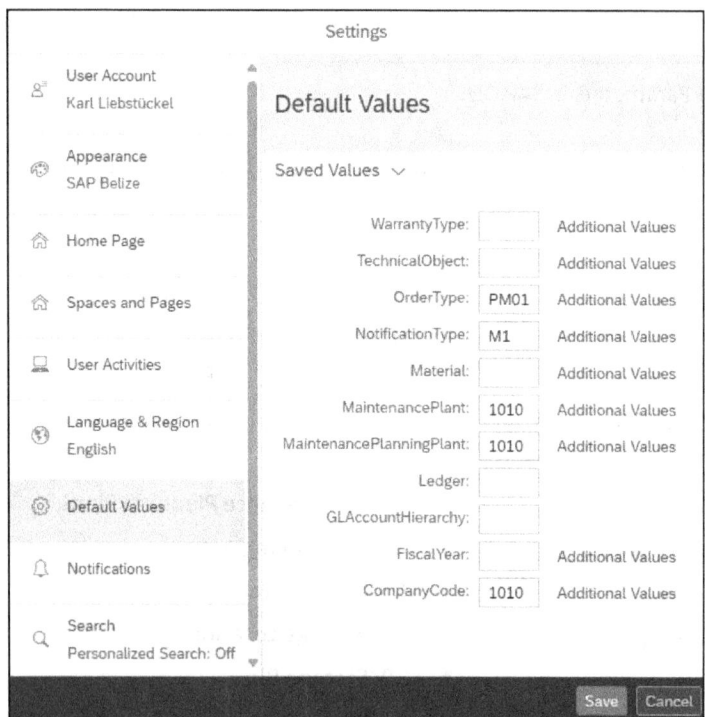

Figure 9.2 Default Values in SAP Fiori Launchpad

9.1.2 Maintenance-Specific User Parameters

In addition to the general user parameters, there are maintenance-specific user parameters available in the notification via **Extras** • **Setting** • **Control/Default Values** and in the order via **Extras** • **Settings** • **Default Values**.

A total of six tabs are available there. On the **General** tab, you can define personal default values for the order type, notification type, planning plant, planner group, and main work center (see Figure 9.3). These parameters have priority over the general user parameters shown in Figure 9.1 and the default values shown in Figure 9.2. This means that if values are entered for the general user parameters and default values, these values are overwritten by the maintenance-specific user parameters.

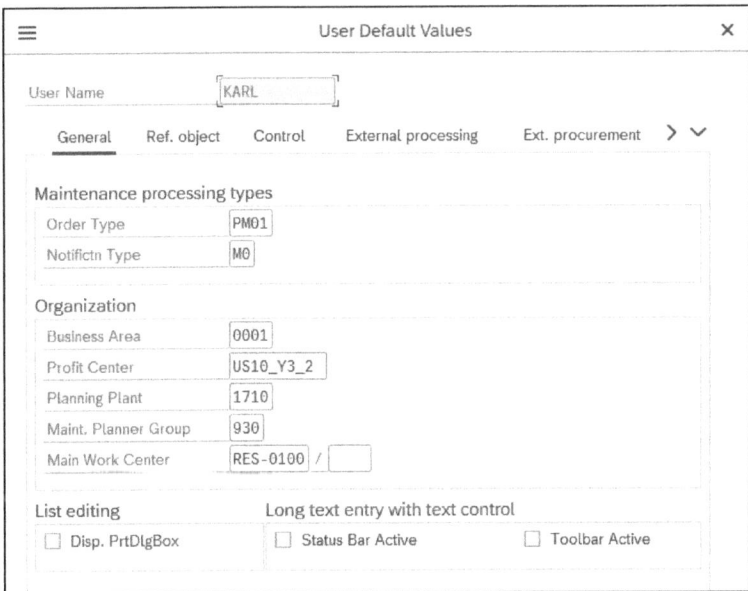

Figure 9.3 Maintenance-Specific Default Values: General

On the **Ref. object** tab (see Figure 9.4), the scenario for the reference object is to be defined.

The following scenarios are available:

- O100: Functional location + equipment + assembly
- O110: Equipment + assembly
- O120: Functional location (30-digit) + equipment + assembly
- O130: Serial number + material number + device ID
- O140: Without a reference object
- O150: Equipment only
- O160: Functional location only

9 Usability

- O170: Equipment + serial number + material number
- O180: Functional location 1:1 + equipment + assembly
- O190: Material sample

≡	User Default Values	×

User Name: KARL

General | Ref. object | Control | External processing | Ext. procurement | > ∨

View

Scenario	0120	Functional location(30)+equipment+assbly

Objects

Functional Location	1710-CWS
Equipment	
Assembly	
Serial Number	
Material	

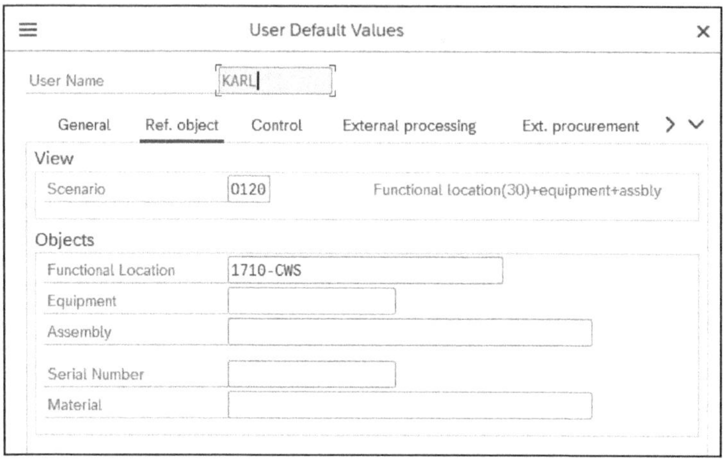

Figure 9.4 Maintenance-Specific Default Values: Reference Object

If you use the **Overview of Notification Type** Customizing function to assign a scenario to the notification type (see Chapter 5, Section 5.1.1) and the **Configure Order Types** Customizing function to assign a scenario to the order type (see Chapter 5, Section 5.2.1), the maintenance-specific user parameters override these Customizing settings for the notification type and order type. If, however, you use the **Set Maintenance Plan Categories** Customizing function to assign a scenario to a maintenance plan category (see Chapter 6, Section 6.2), they won't be overridden. Here, you can also assign a specific technical object, which then is proposed in the notification and order as well.

The **Control** tab is available in the order only. There, you have the option to influence different dialog boxes (see Figure 9.5):

- **Function: Put order in process**
  In this part of the screen, you can set the terms for the **Selection dialog** and therefore override the default **Display dialog** setting.

- **Print control**
  In this part of the screen, you can override the default **Print with dialog** setting.

- **Dialog: Maintain settlement rule**
  In this part of the screen, you can override the default setting from the **Settlement Rule: Define Time and Creation of Distribution Rule** Customizing function (see Chapter 5, Section 5.2.9).

- **No dialog completion**
  If you check the box to set this indicator, you can suppress the dialog box for

technical completion of the order, which means the order will always be completed using the default values (e.g., the current date and time).

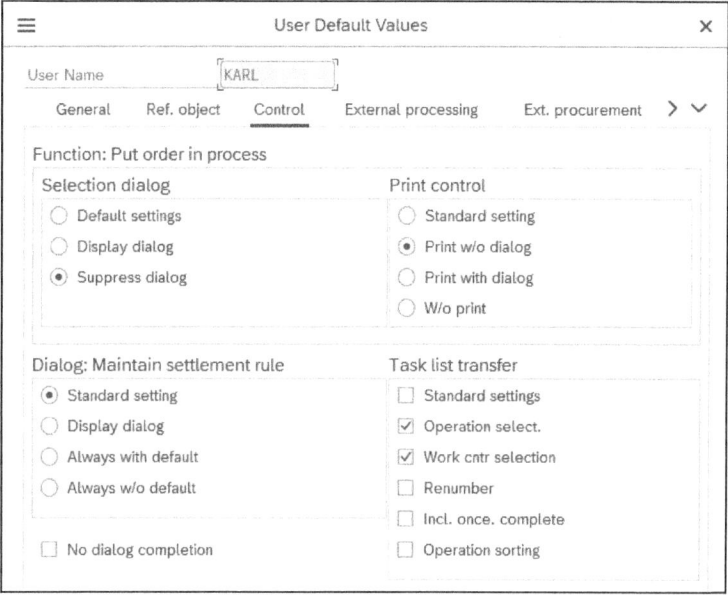

Figure 9.5 Maintenance-Specific Default Values: Control

In addition to these default values, this tab can be used to define in which way task lists are to be transferred to orders:

- **Operation select.**
  You can check this box to indicate a dialog box for selecting operations is to appear.
- **Work cntr selection**
  You can check this box to indicate a dialog box for selecting work centers is to appear.
- **Renumber**
  You can check this box to indicate the operations selected are to be renumbered consecutively in intervals of ten.
- **Incl. once. complete**
  You can check this box to indicate that a task list should be fully integrated only once (maximum).
- **Operation sorting**
  You can check this box to indicate that the sequence of operations is the same as in the task list.

You can use these settings to override the default settings from the **Default Values for Task List Data and Profile Assignments** Customizing function (see Chapter 5, Section 5.2.11).

The **External processing** (see Figure 9.6) and **Ext. procurement** (see Figure 9.7) tabs are available in the order only. There, you can define personal default values for the procurement of services and materials (e.g., for the purchasing group or material group).

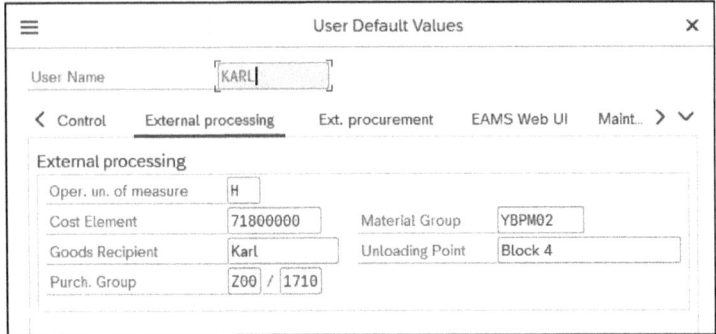

Figure 9.6 Maintenance-Specific Default Values: External Processing

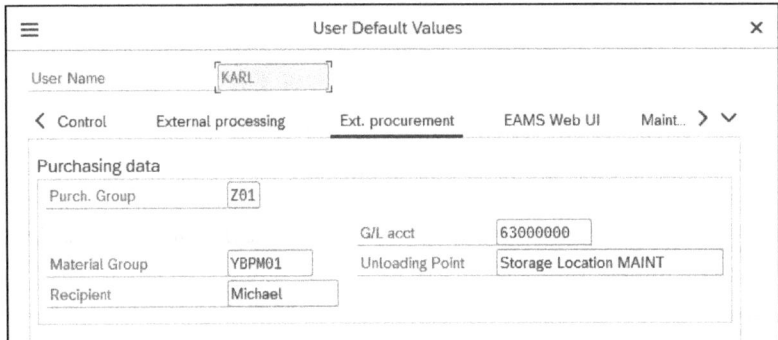

Figure 9.7 Maintenance-Specific Default Values: External Procurement

These personal default values override the default values from the **Create Default Value Profiles for External Procurement** Customizing function (see Chapter 5, Section 5.2.4), which are assigned to the plant and order type.

On the **EAMS Web UI** tab (see Figure 9.8), you define some default settings for different purposes:

- You can determine how the system should react to messages when you perform a specific function in the order (e.g., if you complete orders technically, you want to complete notifications as well).
- If you check the box for **Hide SF Relationship**, a start-finish relationship can't be selected.
- You can predefine the order type if you use the Web Dynpro app **Confirm Unplanned Job** function (Chapter 8, Section 8.4).
- You can predefine the plant and storage location if you post good issues within the Report and Repair Malfunction app.

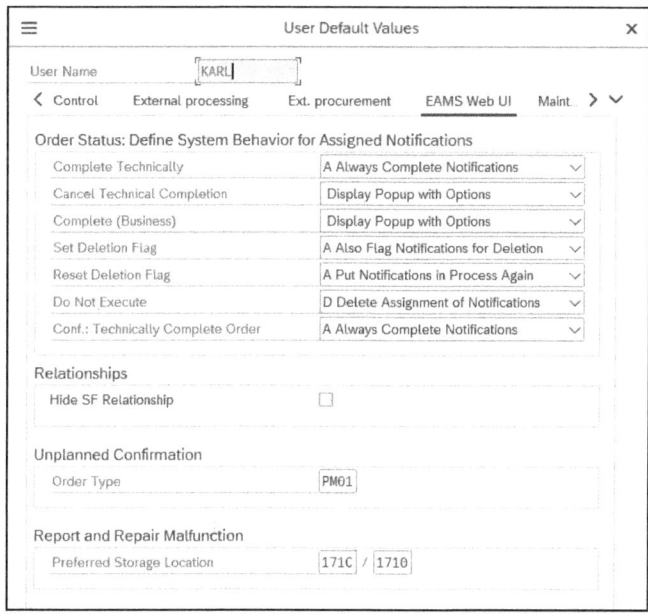

Figure 9.8 Maintenance-Specific Default Values: EAMS Web UI

On the one hand, the **Maintenance Plan** tab (see Figure 9.9) offers the option of preassigning certain scheduling parameters like shift factors, scheduling indicators, or call horizons. On the other hand, you can make preassignments (e.g., the objects to be copied, the initialization of the start) when you copy a maintenance plan.

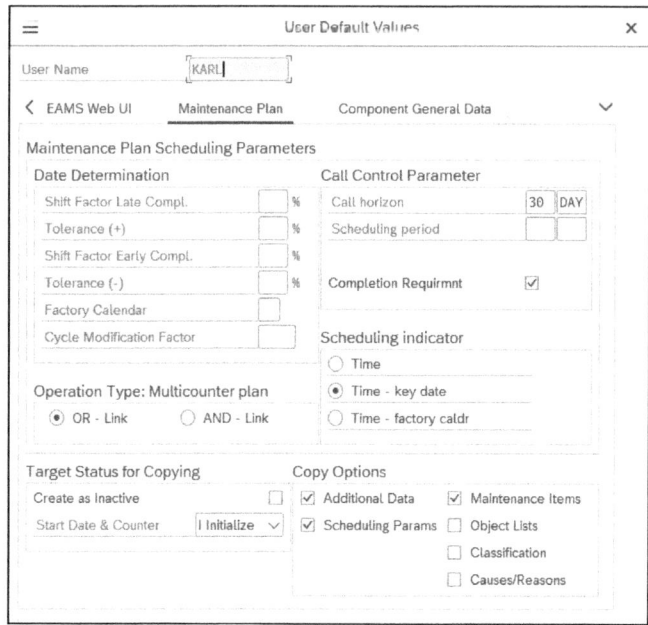

Figure 9.9 Maintenance-Specific Default Values: Maintenance Plan

The **Component General Data** tab offers the option of preassigning certain default values when you handle spare parts as stock components like storage location, unloading point, or goods recipient (see Figure 9.10).

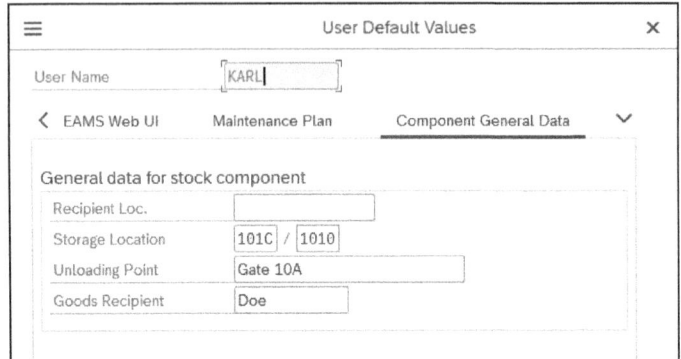

Figure 9.10 Maintenance-Specific Default Values: Component General Data

The **Action box** tab (see Figure 9.11) is available in the notification only.

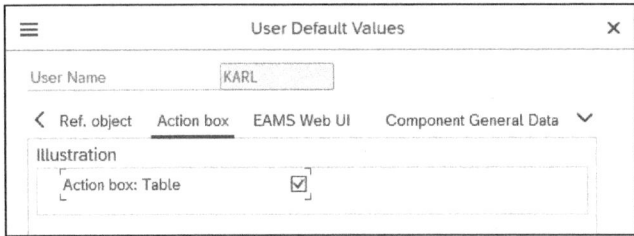

Figure 9.11 Maintenance-Specific Default Values: Action Box

If you check the box to set the **Action box: Table** indicator, the actions that you assigned to the notification type (in the **Define Action Box** Customizing function; see Chapter 5, Section 5.1.3) aren't displayed as a sidebar but as a table in the notification header (see Figure 9.12).

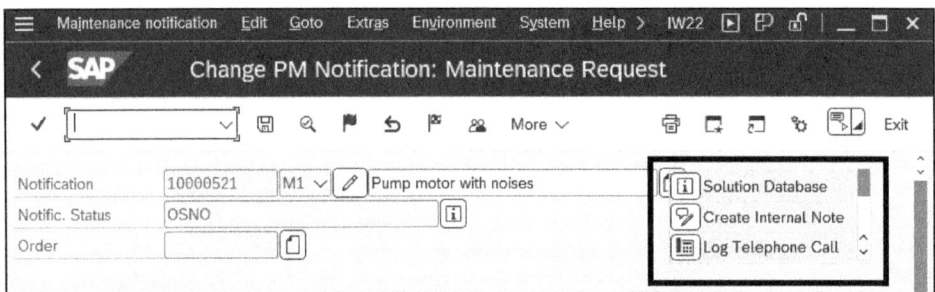

Figure 9.12 Notification: Action Box as Table

The maintenance-specific user parameters are stored in table TPMUS. With Transaction SE16N, it is possible to display and manage this table (see Figure 9.13).

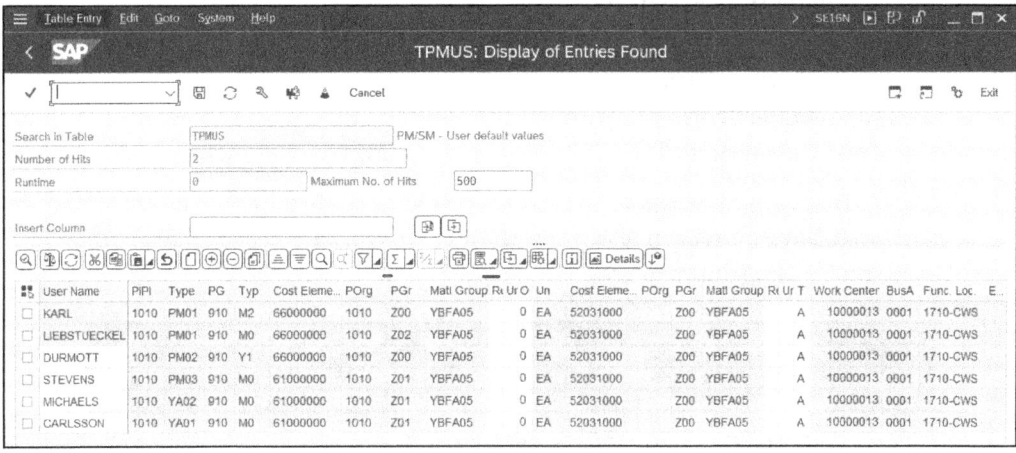

Figure 9.13 Maintenance-Specific Default Values: Table TPMUS

Here, it's possible to do the following:

- Create a new entry.
- Change an existing entry.
- Copy an entry from one user to another.
- Delete an entry.

Central Administration via Transaction SE16N for Table TPMUS [+]

You can also centrally manage the maintenance-specific user parameters for your end users. Windows key combinations Ctrl+Y, Ctrl+C, and Ctrl+V can be used to copy entries from one user to another.

Use Maintenance-Specific Parameters [+]

Using maintenance-specific parameters to prepopulate fields reduces processing time and enables control of processing steps.

9.1.3 Roles, Favorites, and My Home

To create role-based menus, start Transaction PFCG. Role-based menus have a much simpler structure than the standard SAP menu, which, in plant maintenance, can consist of up to seven levels. Starting a transaction from a role-based menu is, therefore, much faster than starting it from a standard SAP menu. If a role-based menu is assigned to you, the system automatically displays this menu on the initial screen (see

Figure 9.14). A further simplification is provided by favorites, with which the user adds only those transactions he requires (see Figure 9.14). Favorites menus can be single level or multilevel, and starting a transaction from a favorites menu is generally much faster than starting it from a standard SAP menu.

Note the following restriction: Each user creates his own favorites menu, which means a user can't directly use the favorites menu of his colleagues. However, you can use **Favorites • Download to PC** or **Upload from PC** to copy favorites menus to another user.

The **My Home** page (see Figure 9.15) in the SAP Fiori launchpad is similar to the favorites in SAP GUI. Every user can place the SAP Fiori apps that are needed most in their daily business by using drag and drop.

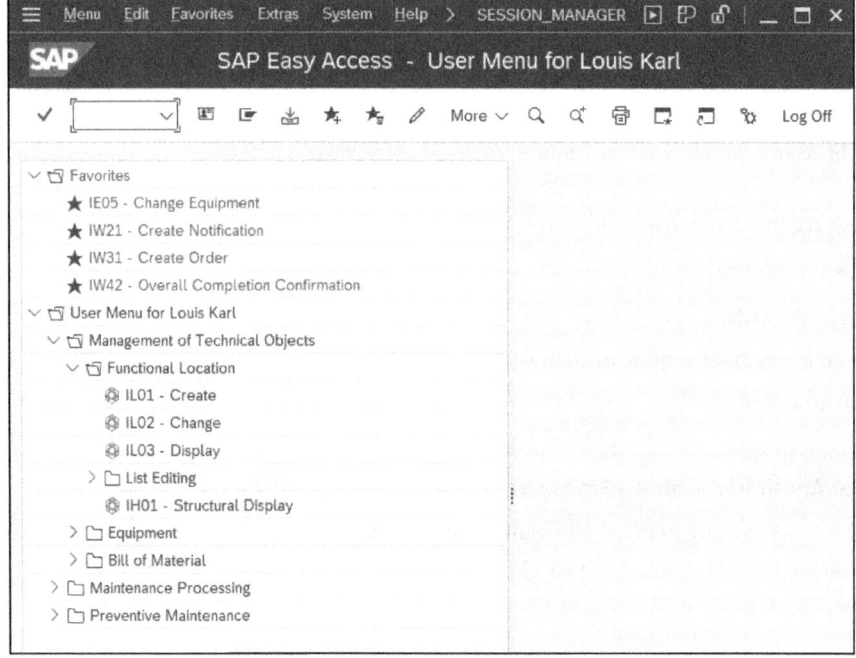

Figure 9.14 User Menu and Favorites in SAP GUI

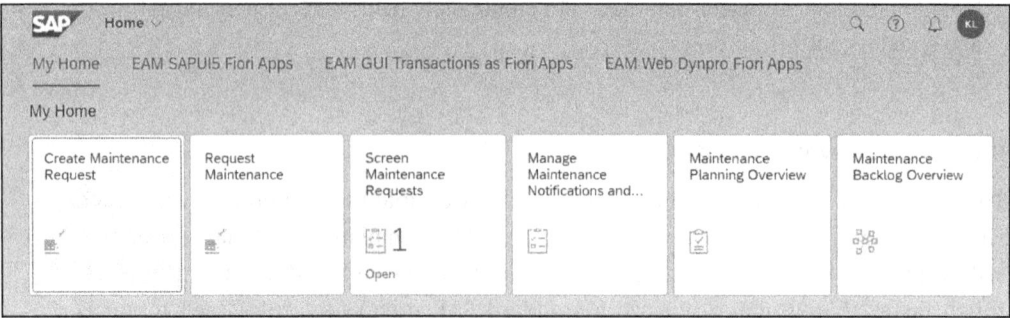

Figure 9.15 My Home in SAP Fiori Launchpad

9.1.4 List Variants and Dynamic Lists

Configuring selection and display variants in SAP GUI's SAP List Viewer and configuring dynamic lists in the SAP Fiori launchpad are different ways to make a considerable contribution toward improving SAP's usability and increasing user acceptance.

SAP List Viewer doesn't present information to you as a rigid list. Instead, you can flexibly adapt the list to your own information requirements.

The following lists are available to you in SAP S/4HANA Asset Management using SAP GUI. In each case, the relevant transactions are provided in parentheses:

- List of functional locations (Transactions IL05 and IH06)
- List of reference functional locations (Transactions IL15 and IH07)
- List of equipment (Transactions IE05 and IH08)
- List of fleet objects (Transactions IE36 and IE37)
- List of object links and object networks (Transactions IN15, IN16, IN18, and IN19)
- List of measurement documents (Transactions IK17 and IK18)
- List of material serial numbers (Transactions IQ08 and IQ09)
- List of materials (Transaction IH09)
- List of measuring points (Transactions IK07 and IK08)
- List of reference measuring points (Transactions IK07R and IK08R)
- List of notifications (Transactions IW28 and IW29)
- List of tasks (Transactions IW66, IW67)
- List of activities (Transactions IW64 and IW65)
- List of notification items (Transactions IW68 and IW69)
- List of orders (Transactions IW38 and IW39)
- List of order operations (Transactions IW37 and IW49)
- Combined orders/operations list (Transactions IW37N and IW49N)
- List of components (Transactions IW3K and IW3L)
- List of permits (Transactions IPM2 and IPM3)
- List of completion confirmations (Transaction IW47)
- List of goods movements (Transaction IW3M)
- List of maintenance plans (Transactions IP15 and IP16)
- List of maintenance items (Transactions IP17 and IP18)
- List of maintenance dates (Transaction IP24)
- List of task lists (Transactions IA08 and IA09)
- List of task lists and operations (Transaction IA38)
- List of shift notes (Transactions SHN4 and ISHN4)
- List of shift reports (Transactions SHR4 and ISHR4)

9 Usability

In the following sections, I'll use the example of the order list (Transaction IW38) to introduce SAP List Viewer. However, the information provided can be applied to any of the lists.

Selection

When you start a list, a selection screen is shown containing all the selection options. This extends over two to four pages, depending on the list and the screen resolution. Experience shows that it is advisable to reduce the options to one screen, so the first step should be to create a selection variant.

> **[+] Create a Selection Variant**
>
> Define a selection variant whose selection conditions span a maximum of one page. All selection variants that you'll create in the future should also follow this basic rule.

To create a selection variant, choose 💾 to save the selection screen. Then, on the next screen, you should make extensive and purposeful use of the option to hide the selection conditions (see Figure 9.16).

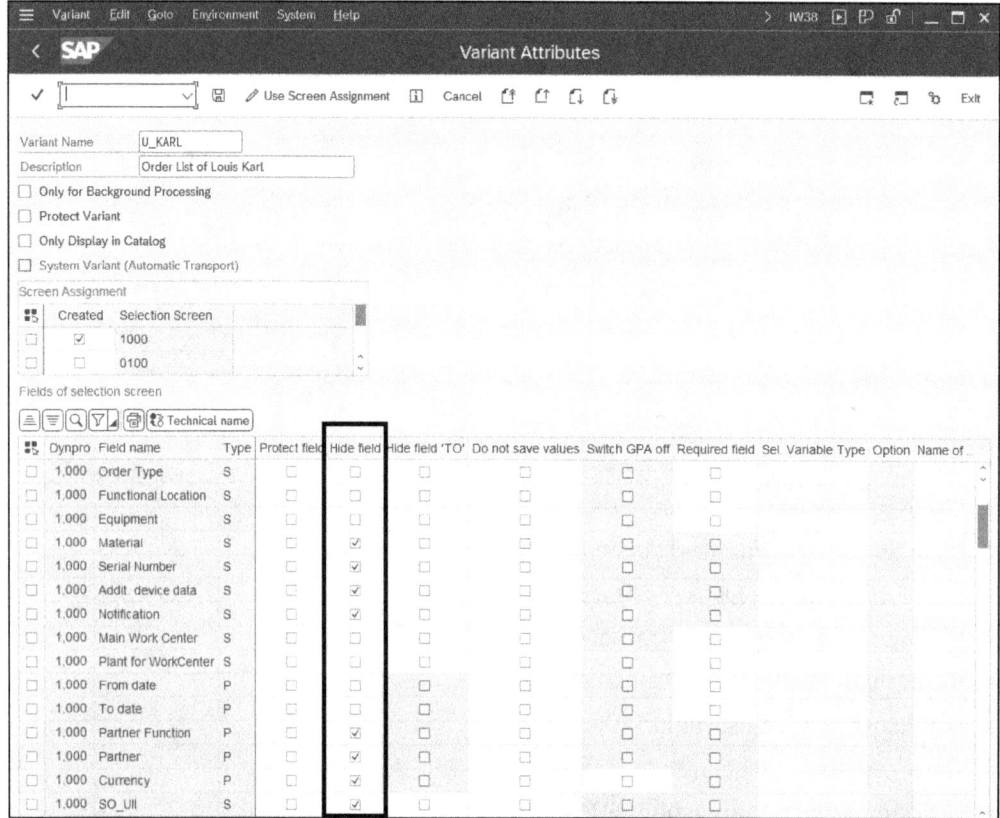

Figure 9.16 List Variants: Hiding Selection Fields

Then, name the list variants, which are displayed on the initial screen of every list. In this way, you can display a list of selection variants and then choose the list variant that you need. This list can become quite extensive over time, however, so follow this next practical tip.

> **Default Variant U_<USERNAME>**
>
> The selection variant that you use most often should be named U_ followed by your SAP username (see Figure 9.17). This selection variant is then automatically displayed when you start the list.
>
>
>
> **Figure 9.17** List Variant: Default Name

The following selection options are available for each selection field:

- **Single value**
  This searches for a single value (e.g., order type PM01) or multiple single values (e.g., order types PM01, PM05, and PM10).

- **Interval**
  This searches for an interval (e.g., order types PM01–PM05) or multiple intervals.

- **Wildcard search**
  This performs a wildcard search (e.g., order type PM*, which selects all order types that begin with PM).

- **Exclusion of elements**
  This excludes values and intervals (e.g., not order types PM03 and PM05–PM08).

- **Clipboard**
  This pastes in search values from the Windows clipboard.

Of particular help is the definition of selection variables for the *dynamic date calculation*—whereby, depending on the relevant current date, the from/to date selection is calculated dynamically based on the option selected (see Figure 9.18).

> **Dynamic Date Calculation**
>
> You can use the **Dynamic Date Calculation** selection variables to determine the selection date of the list dynamically.

Let's look at the following example: if the list is configured with the selection **Start date equals current date − 180 days and + 60 days** on September 1, then the system selects data from March 4 to October 31.

9   Usability

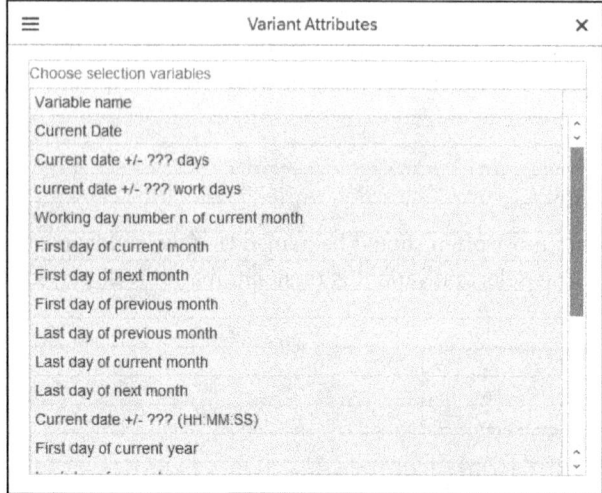

Figure 9.18 List Variant: Dynamic Date Calculation

[+] **Activate the Monitor**
Some lists (but not all) allow you to activate a monitor. Depending on the parameters selected (e.g., basic date), the list entries are red (for overdue orders), yellow (if the start date has been reached but the finish date hasn't been reached yet), or green (for future orders).

List Display

If you now start the list, you see an initial basic list based on the selections and settings made (see Figure 9.19).

The following options for adjusting the layout of the list are available:

- **Show and hide fields**
  This shows additional fields or hides the fields already displayed. Almost all the fields of the object to be analyzed are available for selection, and classification data regarding functional locations and equipment can be displayed.

- **Sort**
  This sorts by a single criterion (e.g., by date) or by several criteria (e.g., by cost center and then, within the cost center, by date).

- **Calculate totals**
  You can use this to calculate totals and display subtotals (e.g., for each order type) for value and quantity fields (e.g., actual costs).

- **Count**
  This lets the system show the total number of entries in total and for each sort criterion.

9.1 Options at the User's Disposal

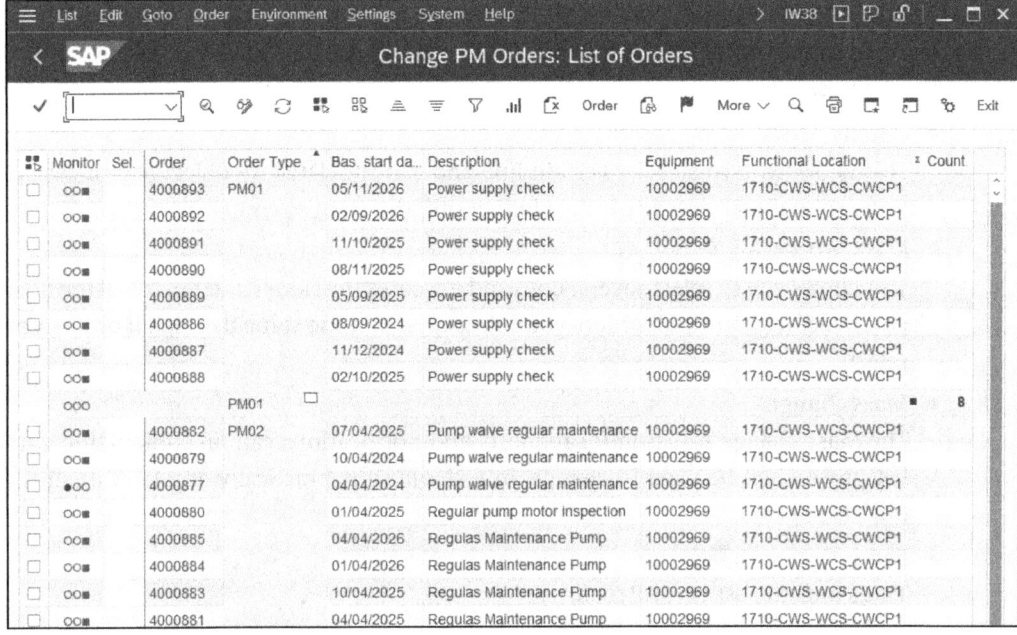

Figure 9.19 List Variant: Display Variant

- **Change column width**
  This optimizes the column width.
- **Search**
  This searches a list for a certain term (e.g., *leaks*), and it is an especially useful function for searching long lists.
- **Filter**
  This filters a displayed list (e.g., it displays only entries with the system status **REL**).
- **Perform ABC analysis**
  This uses a key figure to perform an activity-based costing (ABC) analysis (e.g., an ABC analysis of orders relative to actual costs).
- **Graphical display**
  This generates a graphical display (e.g., a bar chart with the number of orders for each order type).
- **Use display variants**
  This saves your settings as a display variant.

Presetting for a Display Variant
The most frequently used display variant should be marked as a presetting.

Further Processing

If you now have the list in the required form, the following options are available for further processing:

- **Details**
  This allows you to select a certain line and call the database object (e.g., a certain order so that you change the date in that order).

- **Mass processing**
  This allows you to select several lines and executes the same function for all the lines you select (e.g., print all orders you select or assign the same date to all orders you select).

- **Mass change**
  The **Mass change** function for requests and notifications can (in conjunction with list processing) be used to change, in a single step, practically any field in all the objects selected.

- **Send**
  This sends the list via SAP mail.

> [+] **No Email Button?**
> If you can't find a button for sending a list (e.g., **Send**), choose **List • Save • Office**.

- **Download**
  This saves the list in any current Microsoft Office format and processes it further there (e.g., it downloads the order list to Excel to display it there using pivot functions).

> [+] **Schedule a List as a Periodic Job**
> You can also schedule a list as a periodic job, which then runs automatically at regular intervals and performs certain follow-up functions (e.g., it can send an email containing the maintenance plan orders for the coming week to the production managers).

Dynamic Lists

Dynamic lists in the SAP Fiori launchpad show a filter area in the top part of the screen, whereas the findings are in a list at the bottom. If you change the filter, the findings are adjusted dynamically.

The following lists are available in SAP S/4HANA Asset Management using the SAP Fiori launchpad:

- F2021 Find Maintenance Notification
- F2072 Find Technical Object

9.1 Options at the User's Disposal

- F5147 Display Serial Numbers
- F1511 Monitor Maintenance Requests
- F1511A Screen Maintenance Requests
- F2175 Find Maintenance Order
- F4604 Manage Maintenance Notifications and Orders
- F2173 Find Maintenance Orders and Operations
- F2174 Find Maintenance Order Confirmation
- F2660 Find Maintenance Task List
- F2661 Find Maintenance Task List and Operation
- F3622 Find Maintenance Plans
- F5325 Manage Maintenance Plans
- F5008 Find Maintenance Items
- F3556 Manage Maintenance Items

The following sections look at an example of the Find Maintenance Order app (F2175, Figure 9.20) to introduce dynamic lists. However, the information provided can be applied to any of the lists. The following possibilities are available:

- Use the filter bar to select the findings.
- Use the **Adapt Filters (2)** button to define your personal selection criteria.

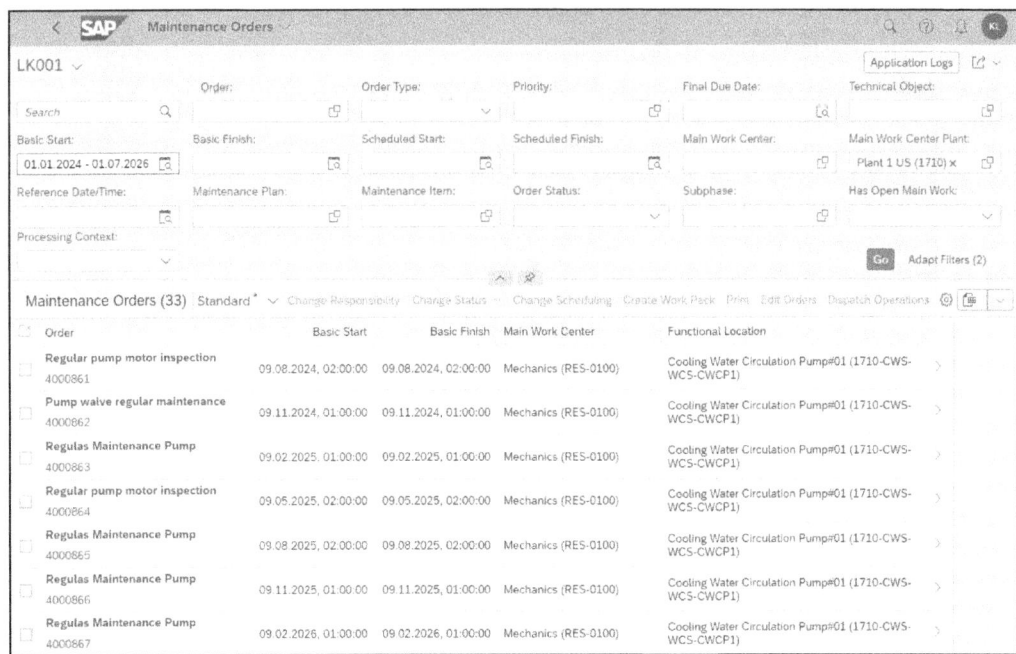

Figure 9.20 Dynamic List: Find Maintenance Orders

9 Usability

- Use date options for basic start dates like from/to, year to date, this month, and last quarter.
- Save the filter as your private filter or make it public.
- Define one of the filters as a default.
- Use the ⚙ button to determine the fields as well as the sort filter and group criteria.
- Use the [🖶|∨] button to download a list (e.g., to Microsoft Excel).
- Use the 📧 button to send the list (e.g., to your colleague).

9.1.5 Personalizing Input Help

Pressing [F4] to bring up a help box usually lets you display all possible entries. Often, however, this list of all entries (e.g., material groups, object types, causes of damage) is very long even though the individual user generally requires only a certain number of entries from this list.

To solve this problem, there is an option to define a *personal value list*. To do this, press [F4] to call up the help menu in the corresponding field (e.g., **Material Group**, **G/L Accounts**), select the entries you require, and click ★ to assign them to your personal value list.

If you now press [F4] to call up the help for this field, the system automatically displays your personal value list instead of the complete list (see Figure 9.21). You can click 🌐 to return to the complete value list at any time. To delete an entry from your personal value list, click ★.

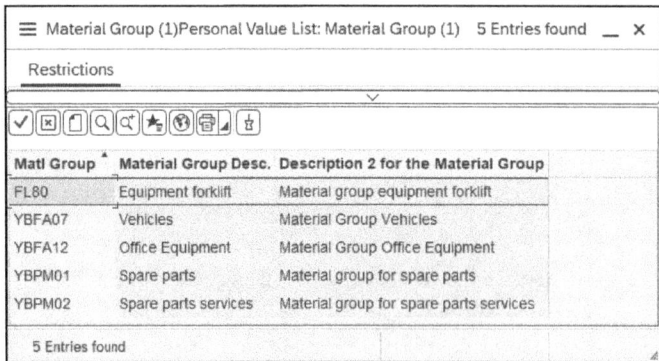

Figure 9.21 Personal Input Help

[+] **Use Personal Value Lists in SAP GUI**

Pressing [F4] to call up a help menu and create a personal value list saves time in relation to searching for the correct entry in the complete list.

However, personal value lists are not available in SAP Fiori apps.

9.1.6 Buttons and Key Combinations

Since the mouse was introduced as a control element for computers, opinions have been divided as to whether operation is faster with or without a mouse. If you prefer to use the keyboard instead of the mouse, key combinations and buttons can be helpful.

> **Create Your Own Key Combinations** [+]
>
> You can create important and general valid key combinations for users who prefer the keyboard.

Examples, not only in the area of plant maintenance, include the following:

- [F11]
  Use this key to save a document. This function corresponds to the disk icon in the system toolbar.

- [F4]+[Enter] **key combination**
  Use this key combination in date and time fields to copy the current date or time into those fields.

- [F4]
  Use this key in general to call up a value list.

- **Clipboard history**
  The commands [Ctrl]+[C] and [Ctrl]+[V] are known and most probably used to copy the last entries. You can also call up the complete clipboard history by pressing [Windows]+[V] and selecting the desired entry.

- **Share desktop**
  If you no longer see the desktop with the windows open, you don't have to minimize the windows individually to reveal the view of the surface. With the keyboard shortcut [Windows]+[D], you can minimize them in one visit to the taskbar. If you want to restore the old state, press the same combination again and all windows will be enlarged again.

- **Virtual desktops**
  Using smaller windows is a way of not losing track of things on a small display. However, if many programs or windows are to be open at the same time, virtual desktops are the better option. They allow you to use several screens at the same time and switch back and forth among them without any problems. You open the overview with [Windows]+[Tab]. There, it's possible to add further desktops using the + sign. You can also press [Windows]+[Ctrl]+[D] to create a new virtual desktop and switch directly there. You can assign programs in the task view by dragging them onto the miniature of a desktop. If you want to remove a desktop, press [Windows]+[Ctrl]+[F4] while it's displayed.

9.1.7 Table Controls

Table controls are largely unknown and unused options that end users can use to adjust the SAP system to their requirements. In the past, SAP GUI screens were preprogrammed, which meant that each field occupied a permanently fixed position on the screen. Since then, however, many screens have been provided with table controls, which you can use to define your own screen layout and field sequence.

Examples of table controls in plant maintenance and related areas are as follows:

- Operation lists (see Figure 9.22), component lists, service specifications, and object lists in the order
- All partner overviews
- Item, cause, measure, and activity overviews in the notification
- Time confirmation, measurement recording, goods movements, and activities in the overall completion confirmation
- Item overviews in the maintenance plan
- Operation overviews, component overviews, maintenance packages, and service specifications in the task list
- Collective confirmation
- Item overviews for material withdrawal
- Item overviews in purchase requisitions
- Item overviews in purchase orders

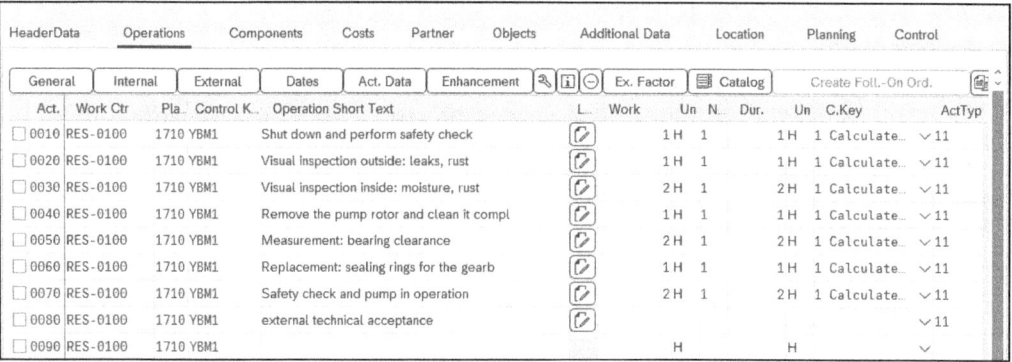

Figure 9.22 Table Controls in SAP GUI

The following settings are available for all table controls:

- You can use drag and drop to arrange the columns in the order in which you require them most often.
- You can make each column wider or narrower so that it accommodates all of the entries you usually make there.
- You can hide fields you don't need by setting the width of the column to zero.

9.1 Options at the User's Disposal

The most important setting of all is the option to save these settings as a user-specific variant. The upper-right corner of each table control contains a ⚙ icon for accessing a detailed screen for creating settings as a variant, and it can therefore let you override the default settings (see Figure 9.23). If you activate business function LOG_EAM_CI_5, SAP S/4HAHA Asset Management makes the following special feature available: the ability for all purchasing data (e.g., vendor, material group, ship-to party, unloading point, or purchasing group) to be maintained in both of the table controls. Previously, the table controls for the **Operations** overview and the **Components** overview consisted of a selection of fields predefined by SAP. The purchasing data was sorely missed, so this has been changed.

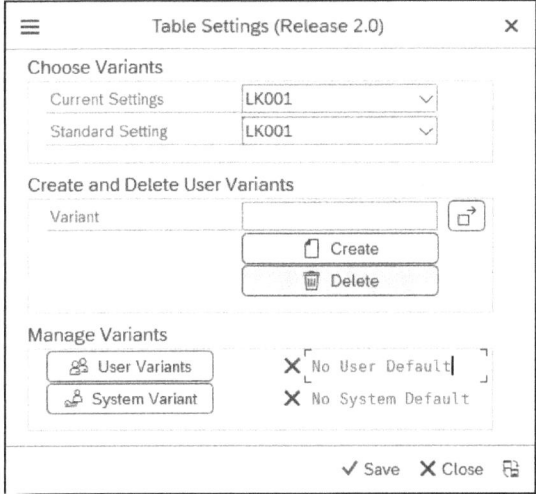

Figure 9.23 Table Controls: Save Table Settings

> **[+] Make Extensive Use of Table Controls in SAP GUI**
>
> There are individualized table controls and table controls stored as a variant to considerably reduce the time spent making entries and searching for entries in the screen templates. This makes a significant contribution to improving SAP's usability.
>
> However, table controls are not available in SAPUI5 apps. Only Web Dynpro apps or SAPUI5 apps that use Web Dynpro contain the table control technology.

Here is a way to find an example of a SAPUI5 app that uses Web Dynpro. Start the Find Maintenance Orders SAPUI5 app, select a specific order for details, and switch to edit mode. The system then switches from SAPUI5 surface to Web Dynpro surface (see Figure 9.24), and there, the table control option is available.

9 Usability

General Data		Location Data	Organizational Data	Operation Data	Object List	Costs	Documents	Permits		Material Overview		
Operations												
Standard * ∨		New		Assign Task List	Create Task List							
☐ Op...	S	Description		Lo...	Work Center	Plant	Technical Object	Contr...	W...	Unit	N...	Unit
☑ 0010		Shut down and perform safety check		⊕	RES-0100	1710		YBM1	1	H	1	
☐ 0020		Visual inspection outside: leaks, rust		⊕	RES-0100	1710		YBM1	1	H	1	
☐ 0030		Visual inspection inside: moisture, rust		⊕	RES-0100	1710		YBM1	2	H	1	
☐ 0040		Remove the pump rotor and clean it compl		⊕	RES-0100	1710		YBM1	1	H	1	
☐ 0050		Measurement: bearing clearance		⊕	RES-0100	1710		YBM1	2	H	1	
☐ 0060		Replacement: sealing rings for the gearb		⊕	RES-0100	1710		YBM1	1	H	1	
☐ 0070		Safety check and pump in operation		⊕	RES-0100	1710		YBM1	2	H	1	
☐ 0080		external technical acceptance		⊕	RES-0100	1710		YBM1	0,0		0	

Figure 9.24 Table Control in Web Dynpro App

This concludes the description of all the options available within Category 1, whereby each user can configure their own individual settings.

Let's now take a look at the options offered by Category 2. In most enterprises, these options are part of the responsibility of the IT department, but options contained in this category don't require any programming.

9.2 The IT Department's Nonprogramming Options

First, we want to introduce the approach of creating any number of adjusted variants (known as *transaction variants*) before moving on to the rest of the nonprogramming options.

9.2.1 Transaction Variants

If a variety of user groups use a transaction to process different business transactions, it often makes sense to adjust the transaction process to the relevant business transaction or user group. The process of creating an order varies, depending on the user group or groups creating it, examples of which are as follows:

- Internal or external parties
- Electricians or mechanics
- Planners or technicians

Transaction variants are suitable in these cases.

[+] **Use Transaction Variants to Adjust Each Transaction**

You can create any number of transaction variants for an original transaction. In a transaction variant, you can do the following:

9.2 The IT Department's Nonprogramming Options

- Hide entire screens.
- Hide individual tabs.
- Disable menu functions.
- Disable buttons.
- Set the field selection control for individual fields (display, required, or hide).
- Prepopulate the field content.
- Modify the column sequence, change the column width, and hide columns for table controls.
- Assign a custom name to the transaction.

Using Transaction IW31 (Create Order), this section will demonstrate how to create an adjusted transaction variant. First, define a new transaction variant in Transaction SHD0 (Transaction and Screen Variants). On the initial screen (see Figure 9.25), specify the original transaction by entering "IW31" in the **Transaction Code** field, and then enter the name of the **Transaction Variant** (e.g., "Z_IW31_M" for an adjusted Transaction IW31 for the mechanical workshop).

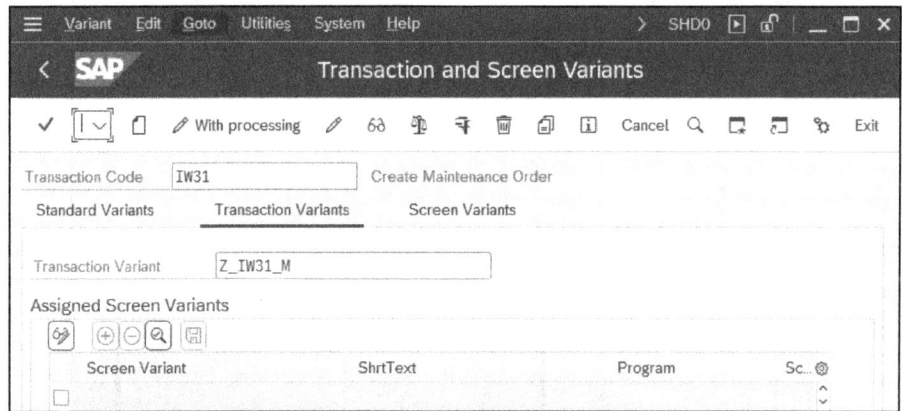

Figure 9.25 Transaction SHD0: Create Transaction Variant

Now, click 🗋 to start the recording. The system then guides you through the screens in the original transaction. Figure 9.26 shows the first screen that you see when you start Transaction IW31.

For each screen, you specify whether or not you want it to be displayed (see Figure 9.27). Check the box to set the **Do not display screen** indicator if the screen should be hidden.

9   Usability

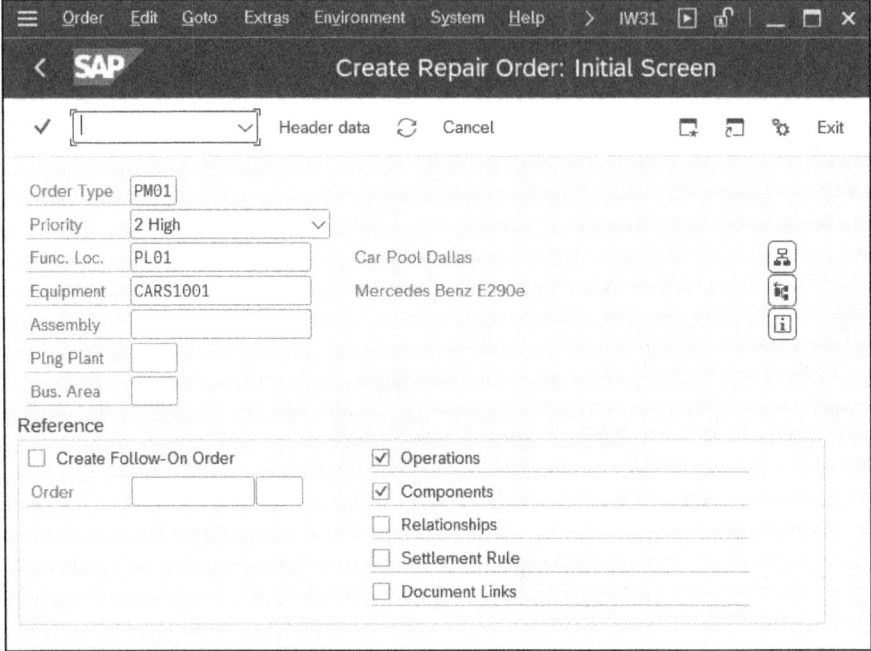

Figure 9.26 Transaction Variant: Initial Screen

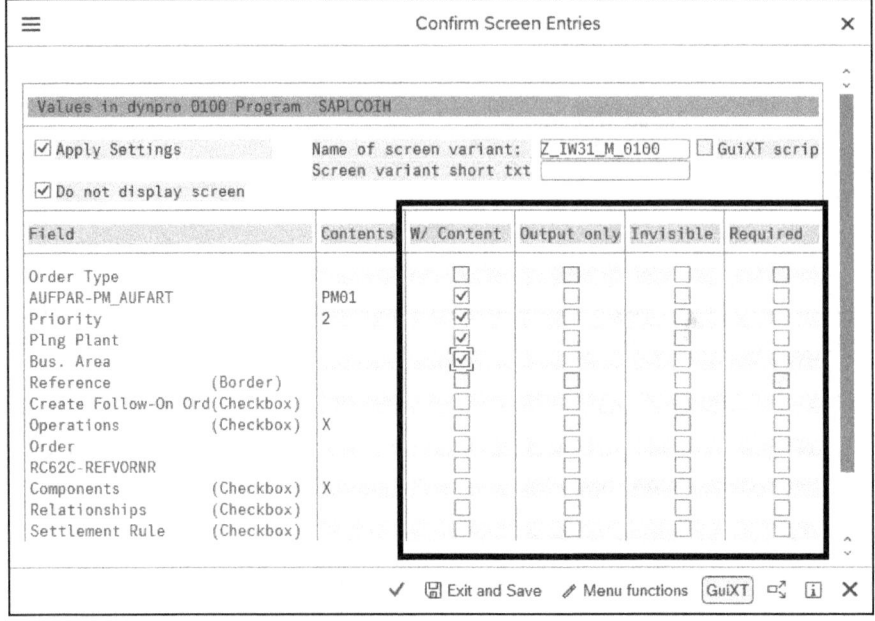

Figure 9.27 Transaction Variant: Field Selection

Use each additional field on the screen to do the following:

- **W/ Content**
  Check this box if you want to transfer the entered content to the field.
- **Output only**
  Check this box if you want it to be a display-only field and not a ready-for-input field.
- **Invisible**
  Check this box if you want to hide the field.
- **Required**
  Check this box if you want to make it mandatory for a user to make an entry in the field.

If the transaction has several tabs, the system will ask which of these tabs are to be displayed in the transaction variant (see Figure 9.28). If the **Invisible** indicator is set, the relevant tab should be completely hidden.

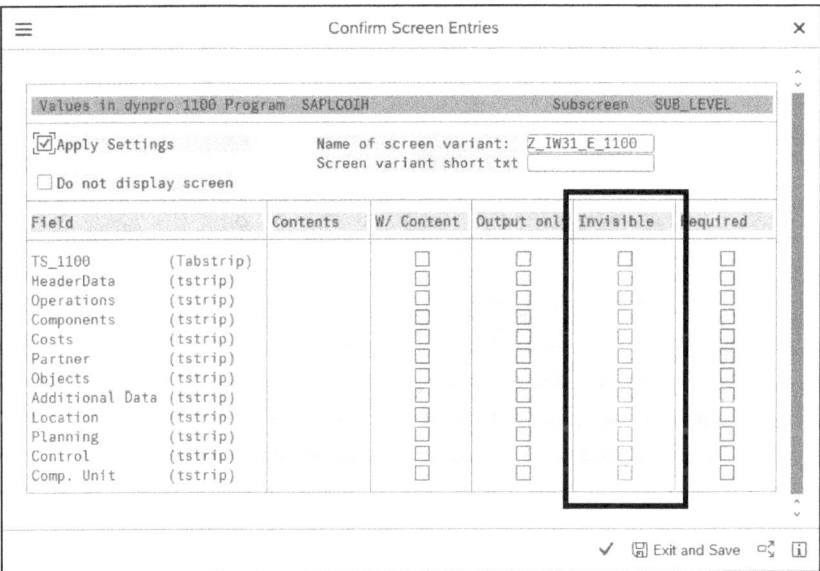

Figure 9.28 Transaction Variant Tab Selection

If the transaction menu and the function keys should be manipulated, choose the **Menu functions** button.

Place the cursor on a menu option in the **Menu Bar** area and click 🔲 to enable or disable this option. All menu options that have been disabled are highlighted in color (see Figure 9.29). They are subsequently displayed on the menu of the transaction variant, but they are grayed out and therefore inactive.

If you set the cursor on a function key in the **Function Key Setting** area, you can click 🔲 to enable or disable this function. All function keys that have been disabled are highlighted in color (see Figure 9.30) and are subsequently not displayed at all in the transaction variant.

649

9 Usability

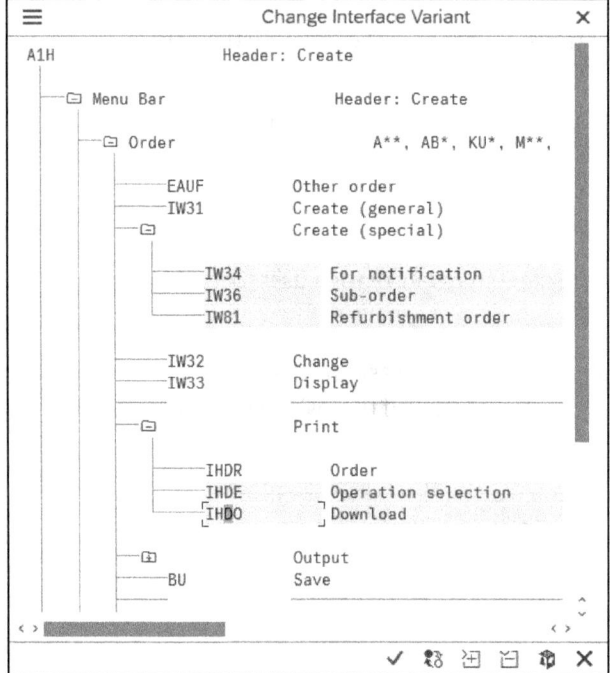

Figure 9.29 Transaction Variant: Enable or Disable Menu Bar Functions

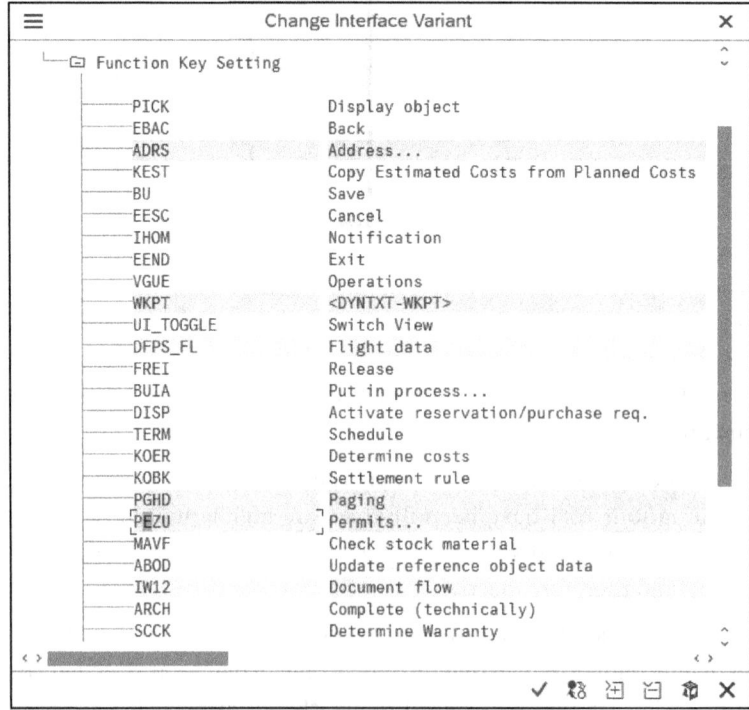

Figure 9.30 Transaction Variant: Enable or Disable Function Keys

9.2 The IT Department's Nonprogramming Options

Use Transaction SE93 (Maintain Transaction Codes) to assign a transaction name to the transaction variant (see Figure 9.31).

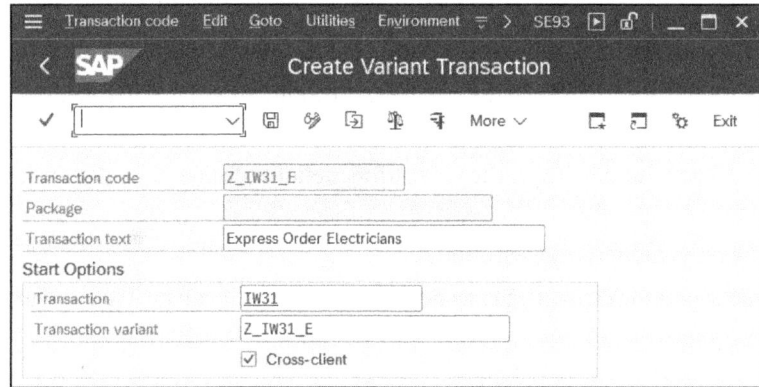

Figure 9.31 Transaction SE93 to Create New Transaction

If you now start this transaction, the original transaction is displayed in a new screen. Figure 9.32 shows the new Transaction Z_IW31_E.

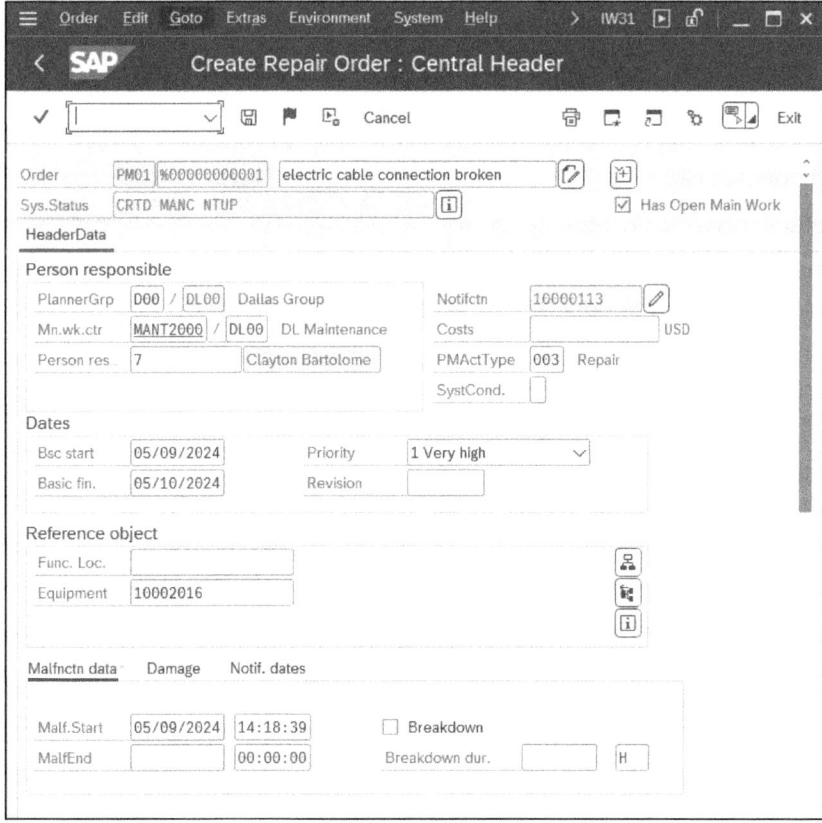

Figure 9.32 Transaction Variant: Result

651

To simplify the transaction, the following measures have been taken:

- The initial screen for Transaction IW31 has been skipped.
- The content of certain fields has been prepopulated (e.g., **Order, Priority**).
- Apart from one tab, all other tabs have been hidden (e.g., **Operations, Material**).
- Menu options and buttons have been disabled (e.g., **Settlement Rule, Schedule, Paging**).
- Subscreens have been hidden (e.g., the operation detail subscreen in the order header).
- Some fields have been hidden (e.g., **Assembly**).
- A unique transaction name (Z_IW31_E) has been assigned, and it can be used to start the transaction variant.

The result is a transaction whose content has been significantly reduced compared to the original transaction and contains only those screens, fields, and functions that the user needs.

9.2.2 Customizing

Customizing also provides many options in terms of improving usability. The most important of these Customizing functions are listed here, along with a description of each transaction's function if not immediately apparent from its name. The order in which these Customizing functions are listed is the order in which they occur in the SAP Implementation Guide.

- **Generic functions (see Chapter 3)**
  - **Documents • Define Profiles**
    This assigns applications and storage categories automatically.
- **Technical objects (see Chapter 4)**
  - **Configuring the Material Master**
    This adjusts the screen layout for material master data.
  - **Material Master Field Selection**
    This defines required entry fields and hides fields.
  - **Set View Profiles For Technical Objects**
    This adjusts the screen layout for equipment and functional locations.
  - **Define Field Selection for Functional Locations**
    This defines required entry fields and hides fields.
  - **Define Field Selection for Equipment Master Record**
    This defines required entry fields and hides fields.
  - **Assign View Profile and Equipment Categories to Fleet Object Types**
    This adjusts the screen layout for fleet equipment.

- Set Field Selection for Specific Fields in Fleet Management
  This defines required entry fields and hides fields.
- Define Transaction-Based Default Values for Object Types
  This prepopulates the equipment category or functional location category.
- Maintenance plans, task lists, and work centers (see Chapter 6)
  - Work Centers • Define Field Selection
    This defines required entry fields and hides fields.
  - Work Centers • Configure Screen Sequence for Work Center
    This adjusts the screen layout for work centers.
  - Task Lists • Define Profiles with Default Values
    This defines default values for task lists.
  - Task Lists • Define Field Selection
    This defines required entry fields and hides fields.
  - Configure Special Functions for Maintenance Planning
    This enables nontemplate-based copying.
- Work order cycle (see Chapter 5)
  - Define Printer
    This prepopulates the values for print output.
  - Send Shop Papers by Email
    This sends shop floor papers using SAP Mail or email instead or printing.
  - Maintain Settlement Profile
    This defines default values for account assignment.
  - Overview of Notification Type • Screen Areas in Notification Header
    This defines a reference object for the notification.
  - Set Screen Templates for the Notification Type
    This adjusts the screen layout for notifications.
  - Set Field Selection for Notifications
    This defines required entry fields and hides fields.
  - Define Transaction Start Values
    This prepopulates the notification type and skips the initial screen.
  - Define Allowed Changes of Notification Type
    This allows you to switch between notification types.
  - Assign Notification Types to Order Types
    This prepopulates the order types for the notification types.
  - Event Prioritization
    This determines priority by consequences and likelihoods.
  - Define Response Monitoring
    This proposes the service profile and response profile for notifications.

- **Configure Order Types**
  This proposes the settlement profile, budget profile, and reference object for the order and integrates the order into the phase model.
- **Create Default Profiles for External Procurement**
  This proposes values for material and service procurement.
- **Activate Event Type Linkage for Procurement Milestones**
  This monitors the procurement of nonstock components and external services in the SAP Fiori Maintenance Backlog Overview app.
- **Text Types For Purchase Requisitions • Define Text Types**
  This defines up to five target text types for purchase requisitions.
- **Create Default Value Profiles for General Order Data**
  This proposes values for graphics and scheduling.
- **Default Values for Maintenance Task List Data and Profile Assignments**
  This proposes values for external data profiles, graphic profiles, and task list transfers.
- **Define Default Order Types for Maintenance Items**
  This proposes the order types for maintenance orders.
- **Define Notification and Order Integration**
  This allows you to enter the notification and order data into one screen and automatically transfer the long text from the notification to the order. Each entry in the object list automatically creates an order operations list.
- **Maintain Default Values for Control Keys for Order Types**
  This proposes a value for the control key.
- **Default Values for Maintenance Activity Type for Each Order Type**
  This proposes a value for the maintenance activity type.
- **Basic Order View • Define View Profiles**
  This adjusts the screen layout for orders.
- **Define Default Values for Component Item Categories**
  This proposes a value for the item category for each material type.
- **Activate Default Value for Current Date as Basic Date**
  This defines a value for prepopulating the date.
- **Set Scheduling Parameters**
  This activates automatic scheduling.
- **Message Control**
  This controls whether a warning, an error message, or no message is to be issued.
- **Define Object Information Keys**
  This displays object information keys automatically.
- **Settlement Rule: Define Time and Creation of Distribution Rule**
  This creates settlement rules automatically.

9.2 The IT Department's Nonprogramming Options

- Define Transfer of Project or Investment Program
  This transfers project numbers or investment programs automatically.
- Define Field Selection for Order Header Data (PM)
  This defines required entry fields and hides fields.
- Define Field Selection for Order Operations (PM and CS)
  This defines required entry fields and hides fields.
- Define Field Selection for Components (PM and CS)
  This defines required entry fields and hides fields.
- Set Screen Templates for Completion Confirmation
  This adjusts the screen layout for the overall completion confirmation.
- Set Field Selection for Completion Confirmation
  This defines required entry fields and hides fields.
- CATS: Maintain Data Entry Profiles and Choose Fields
  This defines appropriate profiles for maintenance confirmations.

- Additional business processes (see Chapter 7)
  - Create Default Value Profiles for External Procurement
    This proposes values for service procurement.
  - Text Types For Purchase Requisitions • Define Text Types
    This defines up to five target text types for purchase requisitions.
  - Source Determination and Default Values • for Client or for the Purchasing Organizations
    This sets default values and source determination process for external services.
  - Configure Split Valuation
    This defines valuation categories and valuation types to enable the refurbishment process.
  - Define Serial Number Profiles
    This defines serial profile procedures to enable the subcontracting process.
  - Set View Profiles for Technical Objects
    This defines screen templates to enable calibration of the test/measurement equipment process.
  - Define Default Values for Inspection Type
    This defines default values for calibration of the test/measurement equipment process in quality management.
  - Maintain Catalogs for Usage Decisions
    This sets default values for usage decisions for calibration of the test/measurement equipment process.
  - Define Follow-Up Action
    This automates follow-up actions such as equipment status changes or order technical completion.

- **Maintain Revision Type**
  This defines a special type of revision that lets you process the Maintenance Event Builder.
- **Define Settings for Shift Note Screen Templates**
  This adjusts the screen layout and field selection for shift notes.
- **Define Settings for Shift Note Type**
  This proposes values for categories.
- **Basic Settings for Pool Asset Management**
  This proposes values for order type, class, and text to enable the pool asset management process.

> **[+] Purposeful Customizing Improves Usability**
>
> Customizing provides many options to enhance the usability of SAP S/4HANA Asset Management, especially options for adjusting the screen layout in notifications, orders, and overall completion confirmations.
>
> In addition, you should make specific use of field selection controls, particularly for hiding information.

9.2.3 Action Box

When you process notifications, you can use the *action box* to execute follow-up actions that may make notification processing easier for you. After you've executed a follow-up action, it's documented as an activity or measure.

You can start the functions directly from the notification. Then, when you complete the function, the system automatically returns you to the notification. Examples of actions within the action box include the following:

- Entering an internal note
- Documenting a telephone call
- Initiating a telephone call via SAPphone
- Sending an SAP mail or email
- Posting a goods movement
- Creating a repair order
- Searching the solution database for a problem or solution
- Creating a quality notification
- Generating an 8D report
- Creating a maintenance plan
- Assigning a BOM

9.2 The IT Department's Nonprogramming Options

The action box concerns functions that would normally require you to exit notification processing, so it makes a significant contribution to simplifying a business process.

Use the **Define Action Box** Customizing function to define functions in addition to those shown in Figure 9.33.

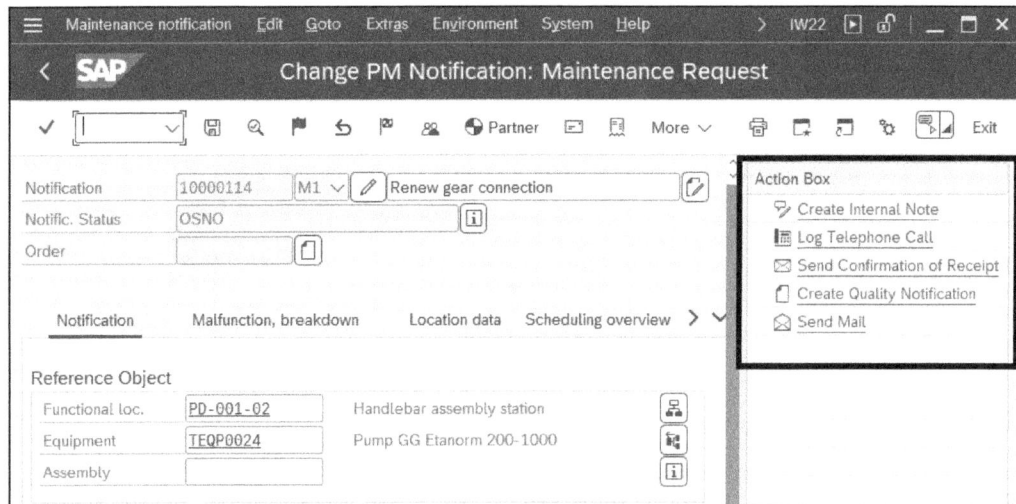

Figure 9.33 Notification with Action Box

In Chapter 5, Section 5.1.3, you learned how to define an action box and how to assign it to a notification type.

9.2.4 SAP Service and Asset Manager

SAP Asset Manager is the newest SAP solution for mobile maintenance. This was developed to implement a more user-friendly native interface for the frontend. SAP Service and Asset Manager was also designed from the outset to integrate modern technologies such as the Internet of Things (IoT) and predictive analytics.

Local Layouts

Depending on whether you're using a tablet or a smartphone as a local device, different layouts are available that are adapted to the respective hardware conditions.

Figure 9.34 shows three screenshots from SAP Service and Asset Manager on an Android smartphone. The one on the left has the entry with object map and first orders, the one in the middle has the order list, and the one on the right has the order details.

In comparison, Figure 9.35 shows the layout of SAP Service and Asset Manager on a iOS tablet. Due to the size, you can of course view more information here, but it's not possible to display an order list on the left and details of an order to the right. Unfortunately, it's necessary to jump to the order detail; on doing so, the list vanishes.

9   Usability

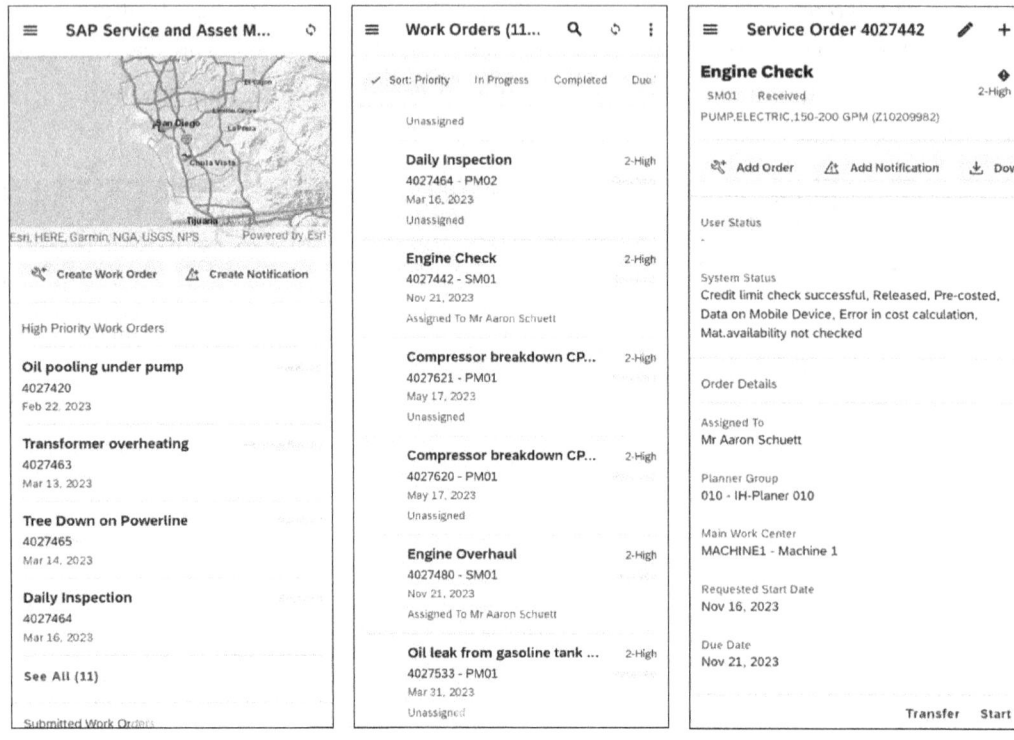

Figure 9.34 SAP Service and Asset Manager: Smartphone Layout

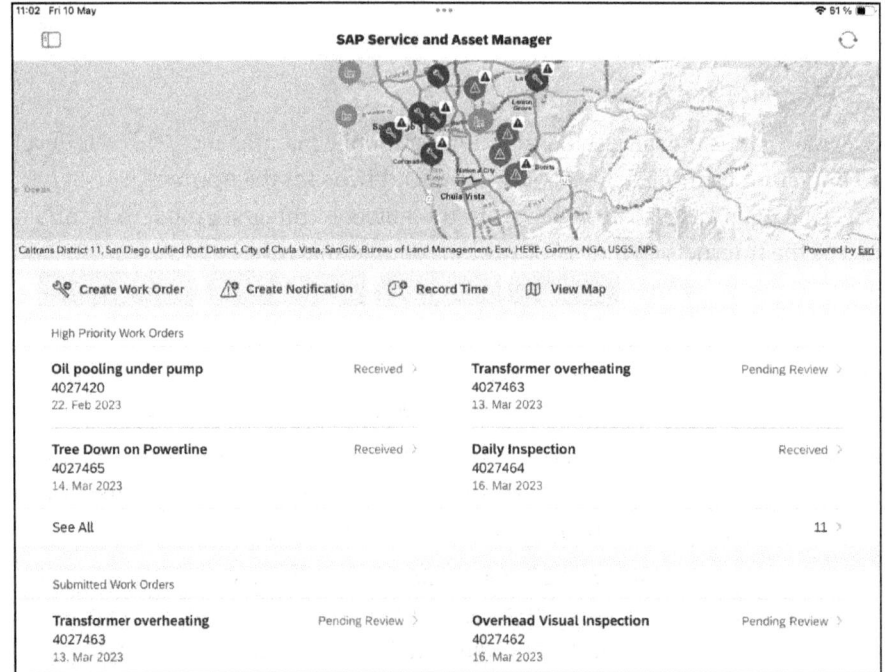

Figure 9.35 SAP Service and Asset Manager: Tablet Layout

SAP Service and Asset Manager includes functions for four roles (so-called *personas*):

- Maintenance technician
- Field service technician
- Safety technician
- Inventory clerk

You switch between the personas in the **Profile Settings**.

Maintenance Technician

SAP Service and Asset Manager currently offers the following functionalities for the maintenance technician persona (version 2310 from November 2023):

- **Order processing**
  - Show high-priority work orders.
  - Show the order list.
  - Show order details.
  - Change orders.
  - Change order status.
  - Create follow-on orders.
  - Create follow-on notifications
  - Create new orders.
  - Display operations for the order.
  - Create new operations for the order.
  - Display spare parts for the order.
  - Create new spare parts for the order.
  - Issue spare parts from the storage location or return them.
  - Manage PRTs.
  - Display notifications about the order.
  - Display documents for the order.
  - Using object lists.
  - Add electronic signatures.
  - Delete orders that haven't yet been transferred.
- **Checklists**
  - Record characteristic results.
  - Record defects.
- **Calibration orders**
  - Show order and inspection lot overviews.
  - Display, change, or create a quality management notification.

- Record results for inspection characteristics.
- Display inspection methods.
- Valuate inspection lots.
- Assign usage decisions to inspection lots.

- **Notification processing**
    - Show maintenance request overviews.
    - Show maintenance request details.
    - Show related maintenance requests
    - Create new maintenance requests.
    - Change maintenance requests.
    - Add notification items.
    - Add, change, and display tasks.
    - Add, change, and display activities (actions).
    - Create an order for the notification.
    - Delete notifications that haven't yet been transferred.
    - Complete the notification.

- **Time sheets**
    - Show existing time sheets.
    - Add a time confirmation to a time sheet.
    - Work with overtime labor entries.
    - Work with clock-in and clock-out.

- **Measuring points and counters**
    - Display overviews.
    - Install, replace and dismantle counters.
    - Change counters.
    - Take a counter reading.

- **Equipment and functional locations**
    - Display overviews.
    - Display hierarchies (functional location, equipment, equipment hierarchy, and BOM).
    - View details.
    - Install and dismantle equipment in functional locations.
    - Create new equipment and functional locations.
    - Display classifications.
    - Display documents.
    - Display measuring points and measurement documents.

9.2 The IT Department's Nonprogramming Options

- Display guarantees.
- Display business partners.
- Display checklists.
- Display, change, or add linear asset management (LAM) data.

Field Service Technician

SAP Service and Asset Manager currently offers specific functionality for the field service technician persona (version 2310 from November 2023). Although most of the maintenance technician functions are also included within the field service technician role, there are some useful special functions for a field service technician:

- Working with service orders
- Working with service requests
- Working with vehicle stock
- Working with travel expenses and mileage
- Working with smart forms (e.g., for activity reports)
- Working with maps and routes, including orders and technical objects (see Figure 9.36)
- Working with business partners, especially customers
- Working with attachments like files and photos
- Working with notes and reminders

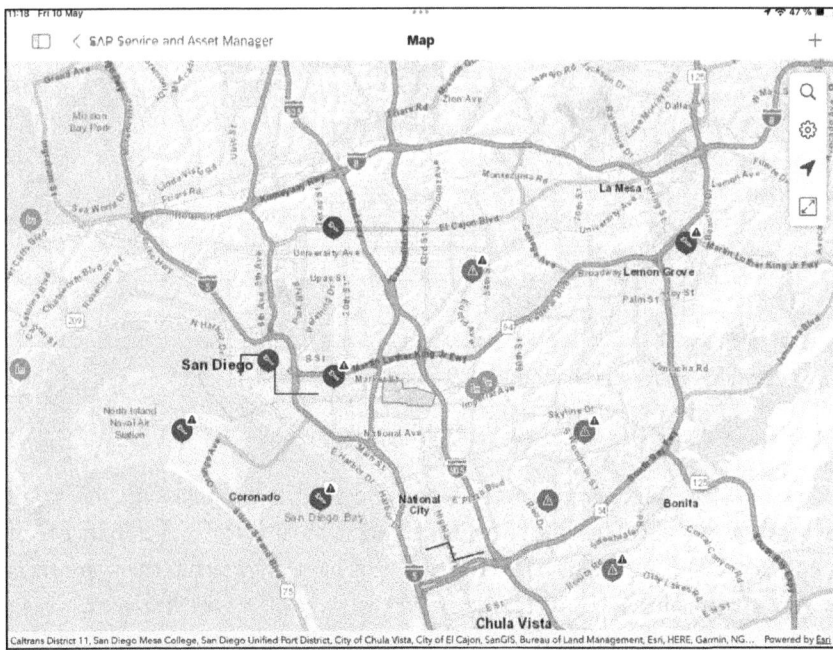

Figure 9.36 SAP Service and Asset Manager: Maps and Routes

661

Inventory Clerk

SAP Service and Asset Manager currently offers specific functionality for the inventory clerk persona (version 2310 from November 2023), because a maintenance technician must also be able to manage spare parts. The most important functions are as follows:

- Working with goods issues (see Figure 9.37, left side)
- Working with goods receipt
- Working with stock transfers and transfer postings
- Performing stock lookups (see Figure 9.37, middle)
- Performing physical inventory
- Creating purchase requisitions (see Figure 9.37, right side)

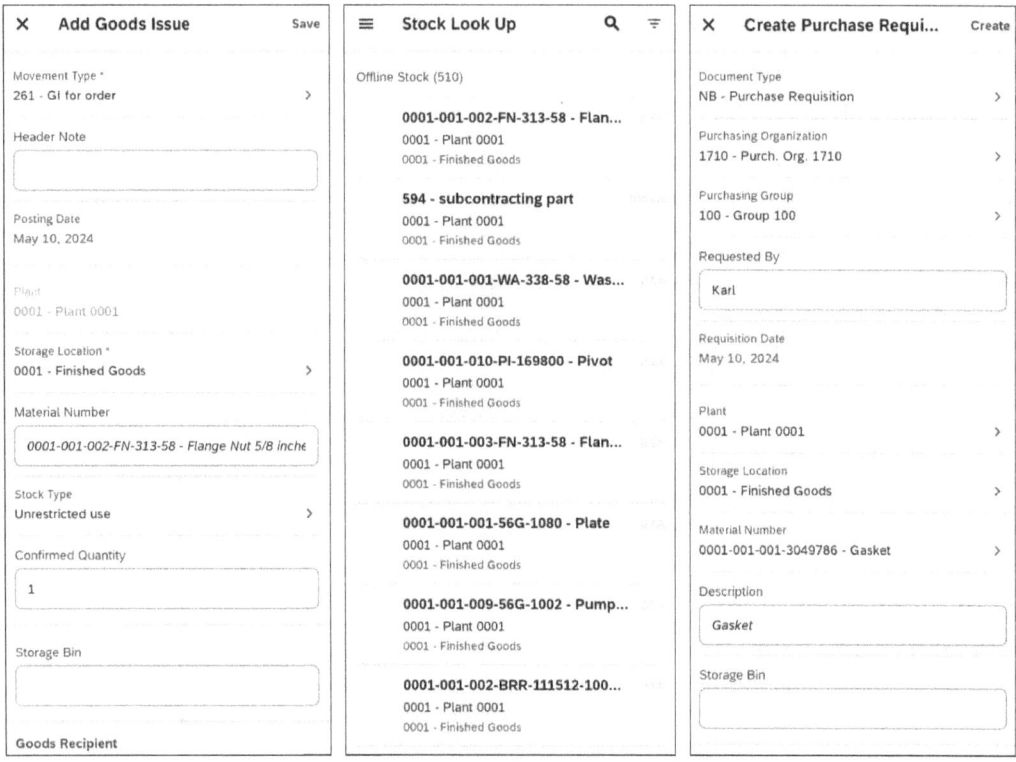

Figure 9.37 SAP Service and Asset Manager: Warehouse Tasks

Safety Technician

SAP Service and Asset Manager currently offers specific functionality for the safety technician persona (version 2310 from November 2023). Most of the maintenance technician functions are also included within the safety technician role, but there are some useful special functions for a safety technician:

- Working with work permits (e.g., to display permit requirements of the technical object)
- Working with work approvals (e.g., to issue an approval)
- Working with isolation certificates for lockout and tagout procedures

> **Additional Information**
>
> For more information on SAP Asset Manager, see the *SAP Service and Asset Manager User Guide* for each of the personas at *https://help.sap.com/docs/service-asset-manager?locale=en-US*.

Usability Advantages of Mobile Work

Regardless of the detailed requirements of your company, the general advantages associated with mobile work in terms of usability are evident:

- Less manual effort is required for data entry; you can directly enter data electronically, rather than handwritten and then electronically.
- The direct electronic recordings on-site and the electronic transmission to the SAP S/4HANA system reduce the risk of transmission errors and thus the error rate.
- The overall data quality is higher, which should result in fewer complaints from the customer due to incorrect order settlements.
- You can use a device (a smartphone or tablet) that is best suited to your day-to-day business (based on its weight, size, and display, the data to be recorded, etc.).
- The elimination of manual paper transports reduces the lead time; the technician doesn't have to pick up or return the order papers.
- Another important advantage results from access to electronic documents: from the tablet, the on-site technician has access to electronic documents (e.g., drawings, procedural instructions) at any time when performing the maintenance order. These could be documents from SAP 3D Visual Enterprise Viewer, a file server, or a document management system (DMS). The technician can consult the documents on a case-by-case basis without going back to the planning office to get the relevant documents from physical folders.

9.2.5 SAP Fiori

In Chapter 8, you were introduced to SAP Fiori, including the following:

- The SAP Fiori UI
- SAP Fiori properties
- Types of SAP Fiori apps
- How to configure the SAP Fiori launchpad with SAPUI5 apps

- How to configure the SAP Fiori launchpad with non-SAPUI5 apps
- How to configure the backend system to use SAP Fiori apps

The question that arises from this information is this: What kind of influence does the SAP Fiori UI and working with SAP Fiori apps have on usability? Here are some answers:

- Most SAP Fiori apps only require a reduced amount of data—both data displayed and data to be entered—compared to SAP GUI transactions.
- Some SAP Fiori apps combine functions on one screen, which in SAP GUI requires a sequence of several transactions.
- With the modern UI, the user has the feeling of working with a modern system.
- The SAP Fiori UI can be used with any browser (e.g., Edge, Chrome, Firefox).
- The SAP Fiori UI is independent of devices, so the user can use the device that is best suited for the task at hand: desktop, notebook, tablet, or smartphone. With SAP GUI, only a desktop or notebook can be used.
- Because of the preceding two characteristics (browser based and usable on any device), the user can also do mobile work and isn't tied to their workspace.
- The mobile orientation brings many of the same advantages that were mentioned for the mobile apps:
  - Less manual effort for data entry
  - Reduced risk of transmission errors and reduced error rate
  - Higher data quality
  - Fewer complaints from customers
  - Reduced lead time by eliminating manual paper transports
  - Access to existing electronic documents

You also can create new electronic documents (e.g., photos) and assign them directly to notifications or orders.

9.2.6 GuiXT

GuiXT is an SAP GUI component that enables you to design SAP transactions according to your daily requirements without having to modify the SAP programs or screens. The following options are available:

- Prepopulate fields with values.
- Hide fields and field groups.
- Move fields.
- Add and change text.
- Add field help.
- Add new screen elements (e.g., checkboxes, buttons, graphics, documentation).

9.2 The IT Department's Nonprogramming Options

- Adjust table controls.
- Change menu entries.
- Change field labels throughout the system.

GuiXT works with a script language that is stored as a TXT file. In a script, you can define the relevant layout in the form of individual statements for each SAP screen.

The GuiXT window is the central switching point for GuiXT and all components (see Figure 9.38). By default, the display window for screen elements and scripts is opened initially.

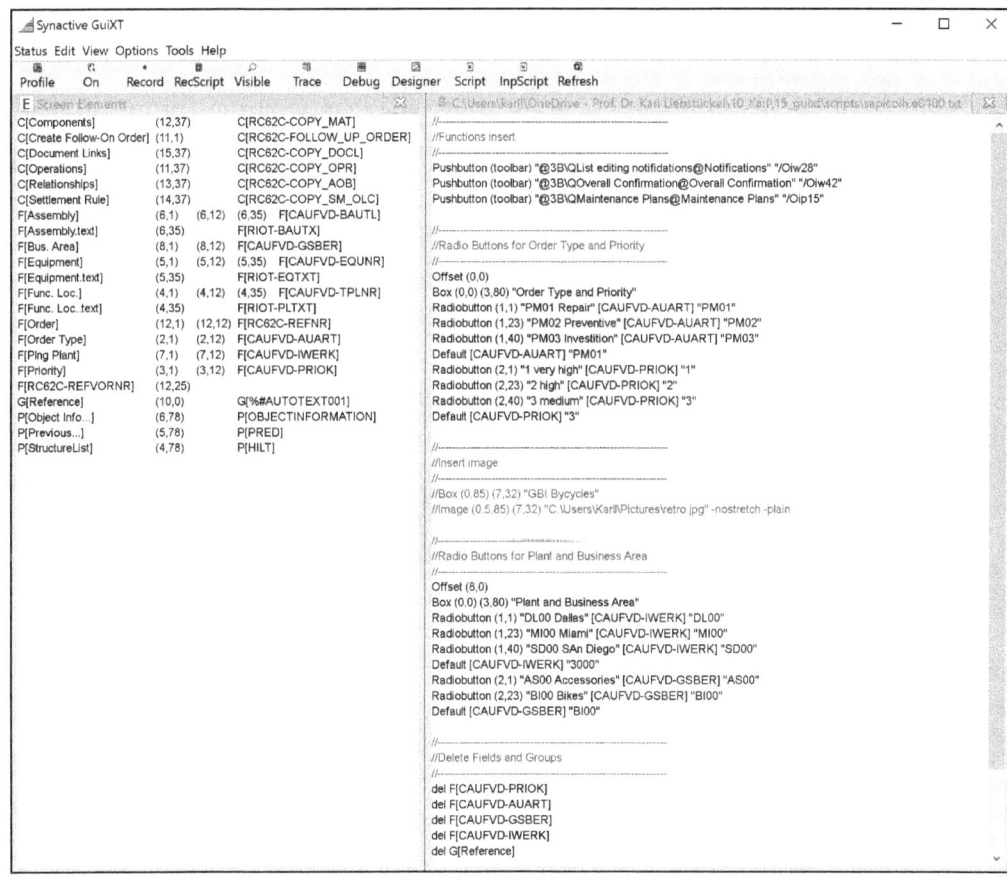

Figure 9.38 GuiXT Control Window

Click **Script** to access your custom-defined editor for creating scripts. In the default setting, this is the *notepad*, but you can use another editor. You can store the script files at different storage locations:

- As local files (e.g., at *C:\GuiXT\Scripts*)
- On central file servers (e.g., *\\demo\GuiXT\Scripts*)

9 Usability

- On HTTP servers (e.g., *www.demo.com/demo/scripts*)
- On FTP servers (e.g., *ftp://demo.com/demo/scripts*)
- In the SAP Web Repository (SAP database) (e.g., *sapwdf11.guixt.demo*)

Each of these storage options has advantages and disadvantages. For example, it's very quick and easy to save the script files as local files, but they are then available on the local PC only.

Figure 9.39 shows an example to outline the procedure and options associated with using GuiXT. The following changes should be made to the layout of the standard initial screen in Transaction IW31:

❶ In the toolbar, the user should be provided with additional options for calling a list of notifications, a list of orders, and the overall completion confirmation.

❷ Due to being error prone, the user shouldn't be able to select the order type as a normal [F4] selection list. Instead, the user should make use of radio buttons to choose from the order types PM01, PM02, and PM03.

❸ For priorities, the user shouldn't be able to choose the priority from a dropdown list. Otherwise, the user will always choose priority 1. Instead, the screen should contain radio buttons for priorities 1–3. Priority 3 is recommended to be the default setting, meaning the user must make the conscious decision to choose a higher priority.

❹ Because maintenance processing occurs in planning plants DL00, MI00, and SD00 only, the system is made to display these three options only (as radio buttons).

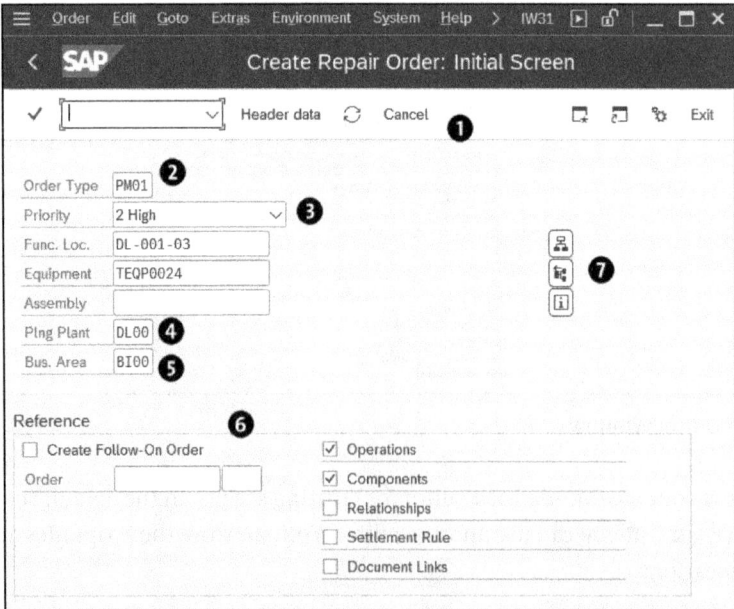

Figure 9.39 GuiXT: Standard Screen

666

9.2 The IT Department's Nonprogramming Options

❺ Because maintenance processing occurs in business BI00 and AS00 only, the system is made to display these two options only (as radio buttons).

❻ Because the user doesn't require a **Reference** subscreen, it should be hidden. This can also be done using the normal field selection. The disadvantage of this, however, is that the empty line is retained. Therefore, all lower fields should be moved up.

❼ On the right-hand side of the screen, an image should be displayed that represents your company. Because the example is a bicycle manufacturer, an image depicting a bicycle should be displayed on the screen.

Listing 9.1 shows a script that satisfies the preceding requirements.

```
//----------------------------------------------------------------
//Functions insert
//----------------------------------------------------------------
Pushbutton (toolbar) "@3B\QList editing notifidations@Notifications" "/Oiw28"
Pushbutton (toolbar) "@3B\QOverall Confirmation@Overall Confirmation" "/Oiw42"
Pushbutton (toolbar) "@3B\QMaintenance Plans@Maintenance Plans" "/Oip15"


//----------------------------------------------------------------
//Radio Buttons for Order Type and Priority
//----------------------------------------------------------------
Offset (0,0)
Box (0,0) (3,80) "Order Type and Priority"
Radiobutton (1,1) "PM01 Repair" [CAUFVD-AUART] "PM01"
Radiobutton (1,23) "PM02 Preventive" [CAUFVD-AUART] "PM02"
Radiobutton (1,40) "PM03 Investment" [CAUFVD-AUART] "PM03"
Default [CAUFVD-AUART] "PM01"
Radiobutton (2,1) "1 very high" [CAUFVD-PRIOK] "1"
Radiobutton (2,23) "2 high" [CAUFVD-PRIOK] "2"
Radiobutton (2,40) "3 medium" [CAUFVD-PRIOK] "3"
Default [CAUFVD-PRIOK] "3"


//----------------------------------------------------------------
//Insert image
//----------------------------------------------------------------
Image (1,85) "C:\bike.jpg" -nostretch -nobuffer


//----------------------------------------------------------------
//Radio Buttons for Plant and Business Area
//----------------------------------------------------------------
Offset (8,0)
Box (0,0) (3,80) "Plant and Business Area"
Radiobutton (1,1) " DL00 Dallas" [CAUFVD-IWERK] "DL00"
Radiobutton (1,23) "MI00 Miami" [CAUFVD-IWERK] "MI00"
```

9 Usability

```
Radiobutton (1,40) " SD00 San Diego" [CAUFVD-IWERK] "SD00"
Default [CAUFVD-IWERK] "DL00"
Radiobutton (2,1) "BI00 Bikes" [CAUFVD-GSBER] "BI00"
Radiobutton (2,23) "AS00 Accessories" [CAUFVD-GSBER] "AS00"
Default [CAUFVD-GSBER] "BI00"

//----------------------------------------------------------------
//Delete Fields and Groups
//----------------------------------------------------------------
del F[CAUFVD-PRIOK]
del F[CAUFVD-AUART]
del F[CAUFVD-GSBER]
del F[CAUFVD-IWERK]
del G[Reference]

//----------------------------------------------------------------
//Positioning fields
//----------------------------------------------------------------
pos F[CAUFVD-IWERK] F[CAUFVD-BAUTL]+(1,0)
```

Listing 9.1 Script for Adjusting the Standard Screen for Transaction IW31

To activate the GuiXT script, use menu path **SAPGUI Settings and Actions • Activate GuiXT**. The next time you start Transaction IW31, it will no longer have its standard layout. Instead, it will have the GuiXT-manipulated layout (see Figure 9.40).

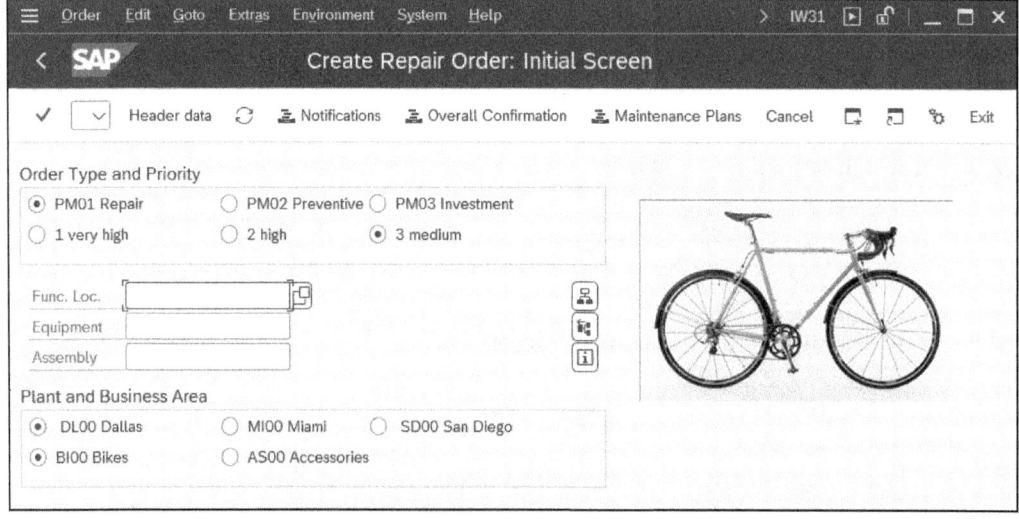

Figure 9.40 GuiXT: Manipulated Screen

9.2 The IT Department's Nonprogramming Options

Figure 9.41 shows an example of how you can turn a standard SAP transaction (Transaction IW28) into a maintenance cockpit.

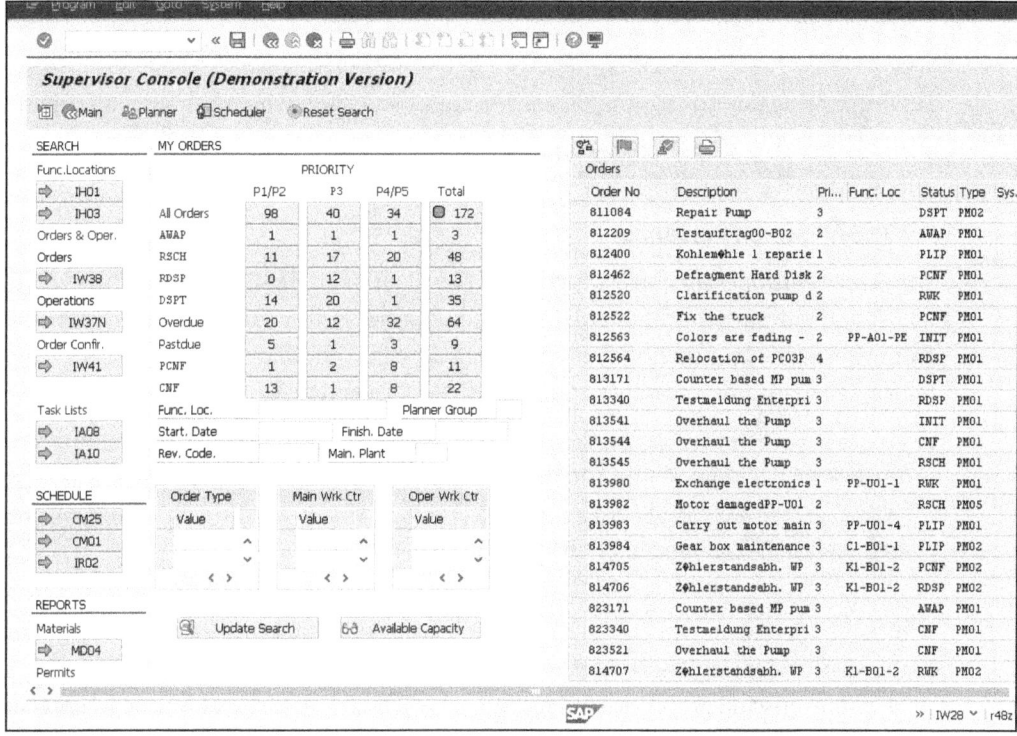

Figure 9.41 GuiXT Maintenance Cockpit

The **SEARCH** section contains buttons for starting the most important transactions needed for your daily work (e.g., Transaction IH01, IW38, IW41).

The **MY ORDERS** section provides an overview of your current order situation. It displays key figures and the number of orders in each status (e.g., overdue or confirmed [**CNF**]) and in each priority, both of which were assigned to the order when it was created.

If you choose one of these buttons, a list of corresponding orders is displayed in the **Orders** section within the cockpit (e.g., which orders have priority **1** or **2** and the **Overdue** status assigned to them). You can use selection criteria such as the order type or functional location to restrict this list further.

> **Note**
>
> The Synactive website (*www.synactive.com*) provides further extensive information about GuiXT.

669

9 Usability

> [+] **You Can Use GuiXT to Adjust Your Screens**
>
> GuiXT can also help you adjust business transaction processing in plant maintenance to your requirements and simplify and accelerate processing.

This now brings us to the SAP Screen Personas software, which can also be used to change the layout of screens.

9.2.7 SAP Screen Personas

SAP Screen Personas version 3.0 is currently available, and like GuiXT, it enables you to adjust and redesign the standard SAP screen templates to your own requirements. It comprises various functions for changing screens, thus supporting you in your endeavors to improve usability and increase user acceptance.

> **SAP Screen Personas Works for SAP GUI and the SAP Fiori Launchpad**
>
> SAP Screen Personas can be used for SAP GUI, and as of SAP S/4HANA 2021, SAP Screen Personas can be configured for the SAP Fiori launchpad.

A modified initial screen for notification creation that contains the user's company logo and provides only those functions that the user requires for daily work in the form of large buttons and graphics (see Figure 9.42) will enjoy much greater user acceptance than the standard initial screen of the SAP access menu.

Figure 9.42 SAP Screen Personas Modified Initial Screen

Such modified screens, whereby a specific SAP transaction is personalized in a particular way, are known as *flavors*. Note that a flavor is always linked to one transaction (e.g., Transaction IW31), but you can create any number of flavors for each transaction. After you finish creating a flavor, you can release it for certain users or for all users of your SAP system.

You use a *User Interface Editor* (UI Editor) to edit the screens themselves. Here, you can edit all the control elements of a screen:

- Titles
- Radio buttons
- Checkboxes
- Function keys
- Input fields
- Dropdown lists
- Field groups
- Texts

Then, depending on the control element, you have the following options (note, however, that not all these options apply to all control elements):

- Completely show or hide individual control elements or entire groups of control elements.
- Define new control elements (e.g., assign a URL to a button).
- Use drag and drop to move individual control elements to the positions that you want them to occupy.
- Change the size of the control elements.
- Assign a new quick info to the control element.
- Change the default value of a control element (e.g., by prepopulating an input field with a particular value).
- Adjust the title and text of a control element.
- Use sticky notes to add information to the relevant control element. Unlike quick infos, sticky notes are only displayed when the control element is selected.

Figure 9.43 shows a **Confirmation Cockpit** screen adjusted this way. This confirmation cockpit provides the following functions:

- Single confirmation (Transaction IW41)
- Overall completion confirmation (Transaction IW42)
- Display confirmation (Transaction IW43)
- List of confirmations (Transaction IW47)
- Counter readings and measurement documents (Transaction IK11)

9 Usability

- Material withdrawal for order (Transaction MIGO)
- Entry of technical data for notification (Transaction IW22)
- List of notification items (Transaction IW69)
- List of goods movements (Transaction MB51)
- Technically complete order (Transaction IW32)

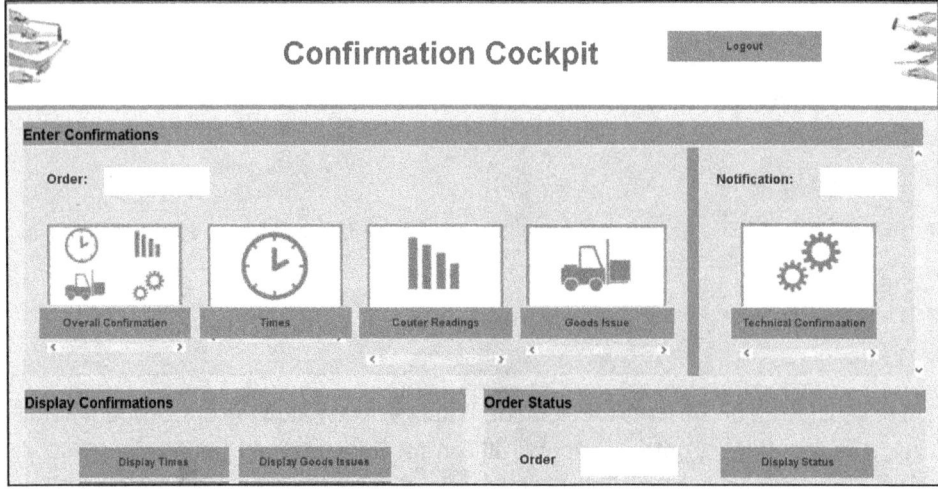

Figure 9.43 SAP Screen Personas Confirmation Cockpit

Additional Resources

Thanks to Charlotte Papenburg for her bachelor's thesis, "Confirmation Cockpit for Plant Maintenance: Concept and Realization Using SAP Screen Personas."

The **Storage Location Cockpit** screen (see Figure 9.44) follows a similar approach and can help you perform various detailed tasks, including the following:

- Enter a goods receipt for purchase orders or production orders.
- Enter a goods issue for orders or cost centers.
- Display a material stocks overview.
- Display a delivery status.
- Display a material master.
- Cancel material documents.

Additional Resources

Thanks to Daniela Loos for her bachelor's thesis, "Storage Location Cockpit: Concept and Realization Using SAP Screen Personas."

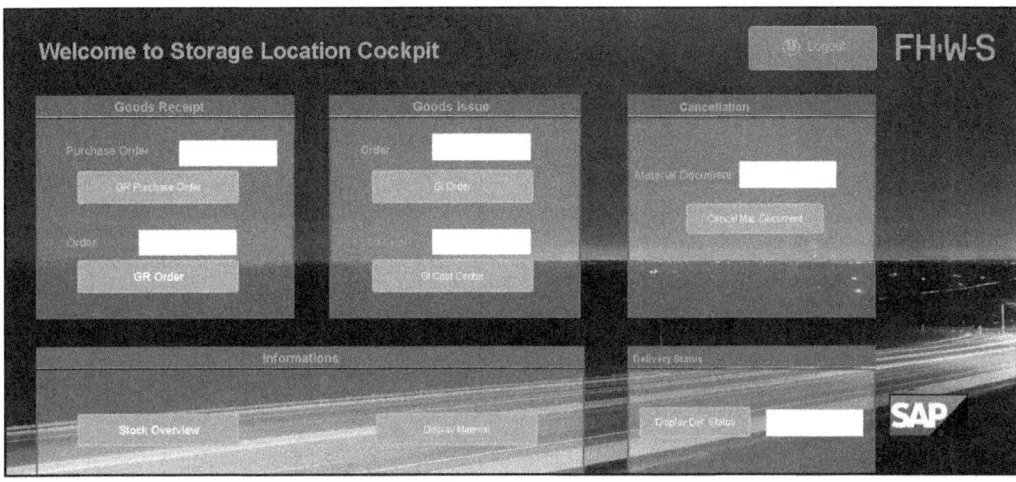

Figure 9.44 SAP Screen Personas Storage Location Cockpit

In addition to individual control elements, you can change the complete *theme* of the flavor in the following ways:

- Define your own background screens (e.g., insert logos, display context-dependent images, define borders and color schemes).
- Adjust the following standard SAP themes: Horizon, Quartz, Belize, Corbu, and Blue Crystal.
- Change the color scheme for Quartz, Belize, Corbu, and Blue Crystal.

> **More Information**
>
> Further information is available at *www.sapscreenpersonas.com* and *http://help.sap.com/personas*.

Let's now look at the options available to IT for improving usability with programming.

9.3 The IT Department's Programming Options

In this section, we'll discuss IT's options for improving usability using programming, including BAPIs, BAdIs, SAP Business Workflow, and others.

9.3.1 Upstream Transactions

When processing business transactions, it's usually necessary to start several transactions in succession. Furthermore, within each transaction, the fields to be entered are distributed across several screens.

The basic idea of upstream transactions is to create a custom transaction containing only one or a few screens. This transaction starts the original SAP transactions in the background and transfers the data. Alternatively, the custom transaction can be used as a cockpit for controlling the original transactions and thus simplifying the process.

Let's explain this further using a real-world example: The business process for calibrating test/measurement equipment is highly complex, requiring several SAP transactions to be started in succession. In the case of an automotive supplier who needed to manage more than 20,000 pieces of measurement equipment per plant, the standard processes were unmanageable. Therefore, an upstream Transaction ZMV01 (see Figure 9.45) was created to do the following:

- Issue the measurement equipment from the warehouse and install it in a functional location (standard Transaction IE4N).
- Dismantle the measurement equipment in the functional location and return it to the warehouse (standard Transaction IE4N).
- Generate a maintenance order, including an inspection lot, from the maintenance plan (standard Transaction IP10).
- Record the measurement results (standard Transaction QE17).
- Make the usage decision (standard Transaction QA11), including technical completion of the maintenance order using follow-up actions.
- Confirm the maintenance order with actual times (standard Transaction IW41).

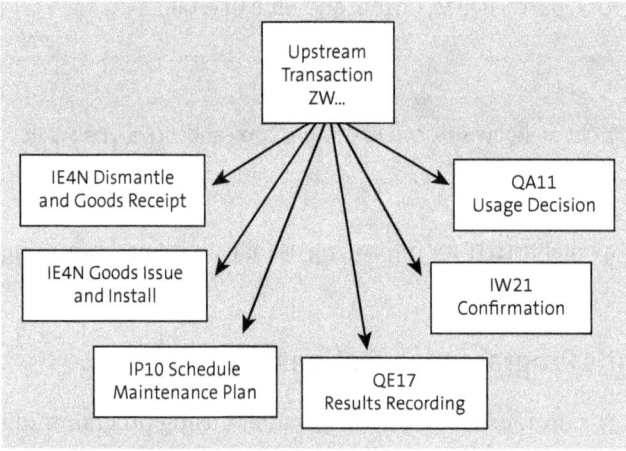

Figure 9.45 Upstream Transaction

[+] **The Programming Effort Associated with Upstream Transactions Is Worthwhile**
An upstream transaction should substantially accelerate business transaction processing. Otherwise, it has failed in its purpose.

BAPI technology is frequently used to develop upstream transactions and other custom developments.

9.3.2 Business Application Programming Interfaces

To simplify the development effort associated with upstream transactions and other custom developments, SAP provides a range of *Business Application Programming Interfaces* (BAPIs), which can be used to do the following:

- Connect SAP systems to the internet.
- Enable SAP components to communicate with one another.
- Develop upstream transactions.
- Connect third-party software and legacy systems to SAP systems.
- Enable PC programs to be used as a frontend for SAP systems.
- Develop web applications that will communicate with the SAP system.
- Enable workflow applications to communicate across system boundaries.
- Enable SAP Business Workflow applications to communicate using the internet.

You can start the BAPI Explorer in Transaction BAPI (see Figure 9.46). Navigating through the hierarchy is a relatively simple way to see an overview of all available BAPIs.

SAP provides the following BAPIs for plant maintenance:

- **For equipment**
  The PieceOfEquipment BAPI, with methods such as Create to create equipment master records or Dismantle to dismantle equipment from functional locations or equipment hierarchies; and the Equipment BAPI for EA-APPL extension enhancements (e.g., installing and dismantling with goods movements)

- **For functional locations**
  The FunctionalLocation BAPI, with methods such as InheritChange to change the inheritance parameters or StrucAssign to replace the assignment of superior functional locations

- **For notifications**
  The MaintNotificBAPIs BAPI, with methods such as NotifCreate to create notifications or NotifListPlangroup to create notification lists for each planner group

- **For orders**
  The MaintenanceOrderBAPI BAPI, with methods such as OrderMaintain to create and change orders

- **For order confirmations**
  The MaintOrdConfirmation BAPI, with methods such as ConfCreate to create completion confirmations

9   Usability

- **For material master records**
  The Material BAPI, with methods such as Edit to change material master data or GetMRPList to generate the MRP list for a material
- **For BOMs**
  The MaterialBOM BAPI, with methods such as CreateBomGroup to general material BOMs

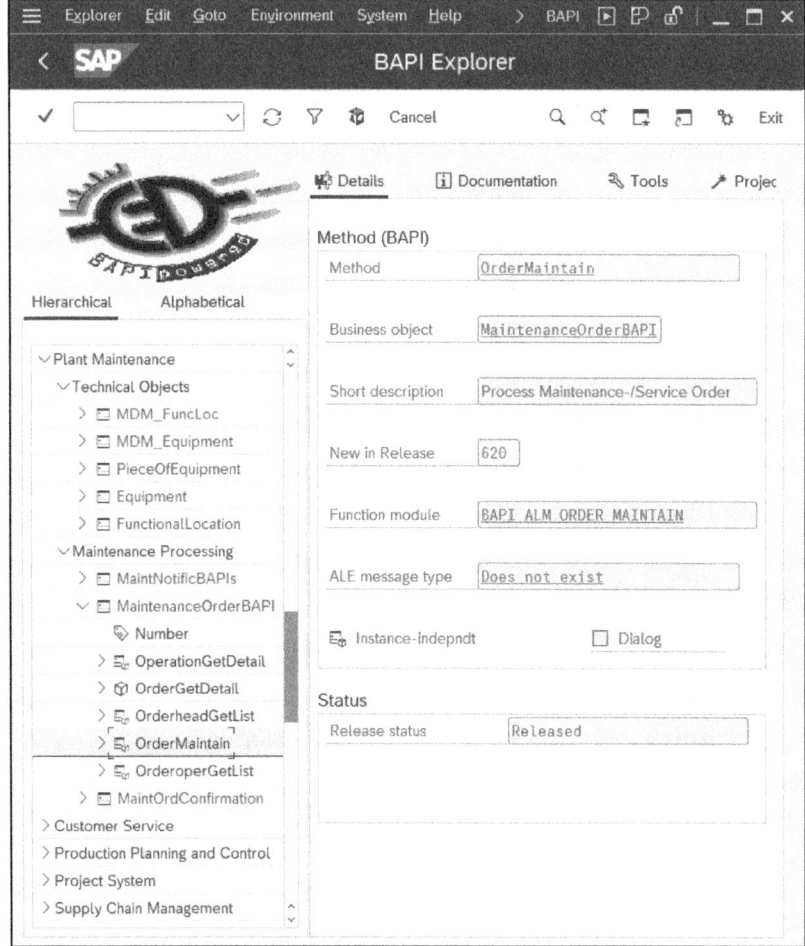

Figure 9.46 BAPI Explorer

A complete list of methods is provided in Appendix A, Section A.3. Unfortunately, there are no BAPIs for other objects such as maintenance task lists or maintenance plans.

In general, you also use BAPIs to develop web applications that will communicate with SAP systems.

9.3.3 Web Interface

A popular way to collect data for the SAP system or to display data from the SAP system without coming into direct contact with the SAP system is to create a web interface.

As an example of a web application, you'll now be introduced to the *EAM app*, which was developed at the Institute for Design and Information Systems (IDIS) at Technical University Würzburg-Schweinfurt for a well-known customer of the automotive industry.

The main functions of the EAM app are displayed in the main menu immediately after you log in to the app (see Figure 9.47):

- **Express Confirmation**
  This function allows you to create completion confirmations for one or more employees involved in an existing order or order operation. All such confirmations can be done on just one screen. Here, you can also perform planned and unplanned material withdrawals as well as enter downtimes, completion confirmation texts, and causes of damage.

- **Overall Confirmation**
  This function enhances the **Express Confirmation** function to include the recording of measurement readings and counter readings. To provide you with a better overview, the individual functions are spread across several tabs.

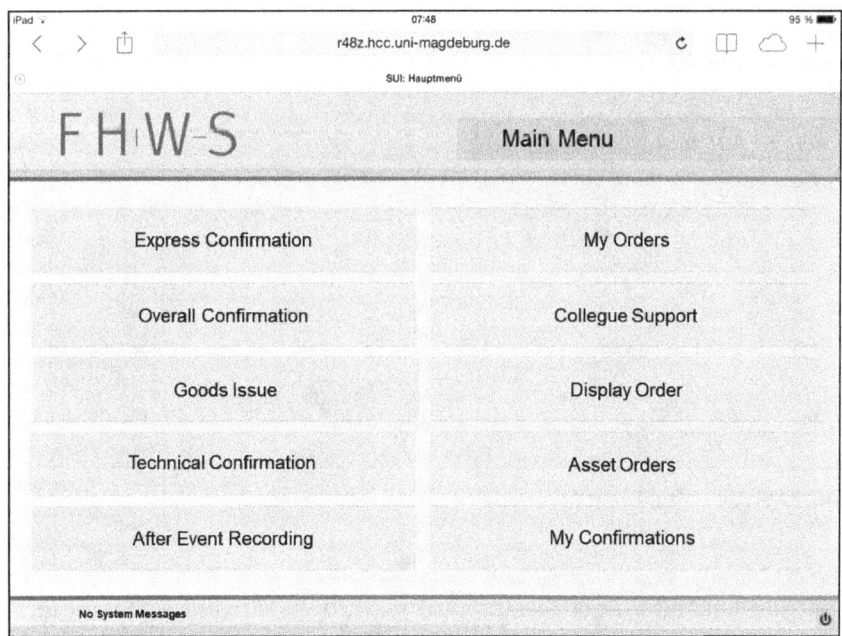

Figure 9.47 EAM App: Main Menu

9   Usability

- **Technical Confirmation**
  This function contains only the technical aspects of the completion confirmation (e.g., downtimes or causes of damage). It doesn't contain any time or material confirmations.
- **After Event Recording**
  This function (see Figure 9.48) is used to record work that has already been carried out. For example, you can enter an order for a technical object (a functional location or piece of equipment) and perform a time confirmation for one or more employees. You can also perform planned and unplanned material withdrawals as well as specify downtimes, completion confirmation texts, and causes of damage. The following process is associated with transferring such data to the SAP system: Notification created → Order created → Order released → Material withdrawal posted → Time confirmation posted → Technical completion confirmation performed → Order and notification completed. This EAM app function is therefore comparable to the Web Dynpro Confirm Unplanned Job app (see Chapter 8, Section 8.4).

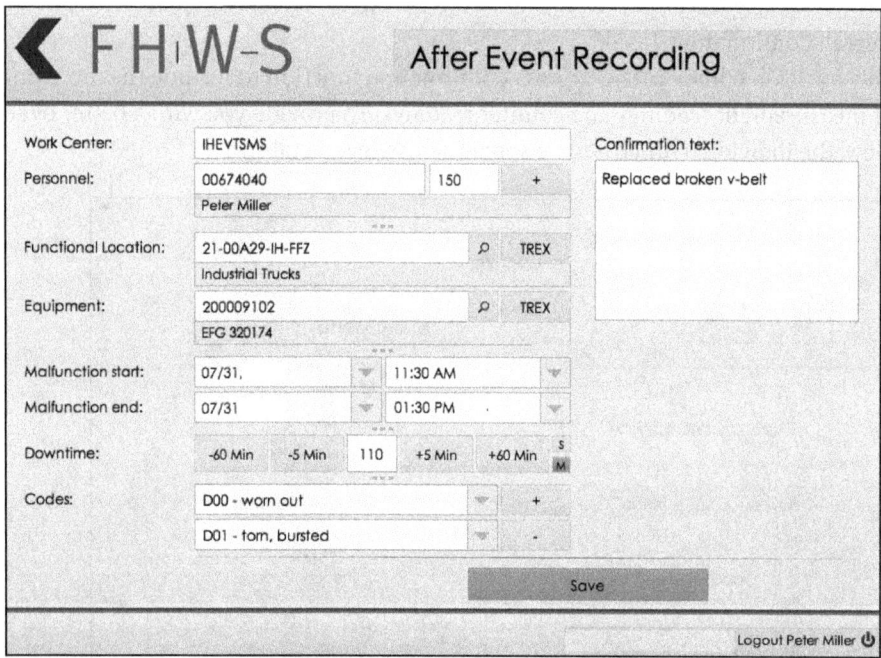

Figure 9.48 EAM App: After Event Recording

- **My Orders**
  This function contains a list of operations currently open or in process for the technician.
- **Colleague Support**
  This function contains a list of operations currently open for the technician's colleagues (i.e., the operations of those personnel numbers assigned to the same work

center as the technician). The technician can then assign their colleagues' orders (e.g., if a colleague is ill or hindered in their daily work for some other reason).

- **Display Order**
  This function makes it possible to display a particular order according to its order number.
- **Asset Orders**
  This function enables you to view a list display of all open and completed operations for a functional location or a piece of equipment.
- **My Confirmations**
  This function contains the technician's completion confirmations for a predefined period (e.g., this week) or for any time interval of your choice.

In addition to pure data entry, these main functions contain additional subfunctions such as the following:

- Display details
- Print
- Create follow-up order
- Technically complete order

In addition to these functional properties, the following technical properties of the EAM app are worth noting:

- The EAM app was completely developed as an online scenario, so it has a permanent connection via local area network (LAN), wireless LAN (WLAN), or Global System for Mobile (GSM) communications to the backend system and can immediately detect wrongly entered data and output an error message to that effect.
- A BAPI is used to check the validity of the data entered (e.g., whether a valid equipment number or order number was entered).
- A BAPI (e.g., the `OrderMaintain` BAPI) is used to transfer the data to the SAP system and post it there.
- Because the EAM app was developed for HTML5, you can use it on an iPad, iPhone, desktop PC, industry PC, Android smartphone, or Android tablet.
- The EAM app can be fully operated using a mouse and keyboard or touchscreens (e.g., in the case of an iPad).

> **Web Applications Enjoy High User Acceptance**
> Almost all users are familiar with web applications or apps for their private iPhone, iPad, or smartphone. For this reason, apps are also very easy to learn and use in a professional capacity and are therefore very popular among users.

9.3.4 Customer Exits

You can use *customer exits* to add your own functions to standard SAP applications without modifying the original SAP programs. In standard applications, SAP creates customer exits for certain programs, screens, and menus. Initially, these exits don't have any function. Instead, they serve as predefined entry and exit points for adding your own additional functions to the SAP system.

You use Transaction SMOD (SAP Enhancement Management) to manage customer exits, and you can then view them by entering the following:

- Technical objects: ITOB*, IEQM*, ILOM*, IHCL*, CCM*
- Measuring points/counters: IMRC*
- Warranties: BG*
- Task lists: IAIH*
- Maintenance plans: IPRM*, CI*
- Notifications: QQMA*
- Orders: IWO*, CNEX*, COZF*, IREV*
- Capacity planning: COI*, CYPP*
- Completion confirmations CMFU*, CONFPM*
- Cross-application time sheets (CATS): CATS*
- Information systems: MCI*
- Data transfers: IBIP*
- Graphic modules: IMSM*

By way of example, the following are five applications in which data entry was simplified and user acceptance increased significantly as a result:

- **Transfer characteristics**
  Customer exit IHCL0001 allows you to transfer the characteristic valuation from a material master record to the characteristic valuation of a piece of equipment. If you often add new equipment with reference to a material number, you can simplify the essential maintenance of the classification data by storing information such as performance, type specification, or other technical data once as a classification in the material master and then transferring these characteristics as default values to your new piece of equipment. Ideally, you'll then no longer have to maintain the classification.

- **Right plant, right storage location**
  Customer exit CNEX0027, which, strictly speaking, is part of the project system but can also be used for plant maintenance, supports you in specifying the right plant and the right storage location during material planning. This may be useful, for example, when a spare part is stored in only one plant that all other plants can access. If,

within this customer exit, you then program a smart search strategy for your enterprise, you no longer have to worry about having the right plant in material planning for the order.

- **Right task list**
  A significant improvement is also provided by customer exits IWO10021 and IWO20001. The combination of these two customer exits, for example, ensures that the right task list is derived from the damage or cause code in a malfunction report. When the order is created from the malfunction report, the correct operations are already listed in the order. This eliminates the need to find the right task list and include it in the order and therefore accelerates order processing for standardized processes in particular.

- **Right data in notification and order**
  Even though they don't make data entry any faster, customer exits that can run customized data checks when saving orders (IWO10009) or notifications (QQMA0014) help to eliminate the need for data corrections. In the notification, for example, you can immediately check whether certain combinations of breakdown and cause codes make sense. These checks help inexperienced users allay their fears about operating the system.

- **Right data in purchase requisitions**
  Customer exits COZF0001 and COZF0002 allow you to automatically prepopulate certain fields for purchase requisitions generated within the order according to your own specifications. For example, you can prepopulate the **Requester** field with the name of the user logged on to the system, or you can automatically enter your department acronym in the **Requirement Number** field. This may not sound like much progress, but if a user has to enter the same department acronym thirty-five times for thirty-five nonstock items, it's easy to see why optimization becomes a priority.

The following are some more examples from professional experiences with customer projects:

- **Checks at technical completion of orders**
  You can use customer exit IWO10004 to implement order completion checks. For example, one customer wanted to prevent orders from being technically completed if any open purchase requisitions or unconfirmed operations exist or if only some of the objects in the object list have a processing indicator.

- **Defaults for time confirmation**
  You can use customer exit CONFPM01 to determine your own default values when confirming an order. For example, a customer wanted to prevent the activities of certain employees from being allocated on a cost basis (an indicator in the HR master record). Now, if one of those employees enters a completion confirmation, the system proposes an activity type without a valuation.

- **Creation or manipulation of settlement rules**
  You can use customer exit IW010027 to generate your own settlement rule. If you use object lists, you frequently want to proportionally allocate the resulting order costs to the cost centers involved (see Figure 9.49). You can program this customer exit in such a way that the cost centers associated with the equipment are determined from the object list and the corresponding entries are configured in the settlement rule and assigned percentages.

- **Right priority for an order**
  You can use customer exit IW010012 to control priority handling within the order. For example, one customer wanted the system to use the ABC indicator for equipment to automatically determine the priority of an order (e.g., priority 1 is automatically set for "A" equipment).

- **Additional fields in notification**
  You can use customer exit QQMA0001 to define additional fields. For example, one customer wanted to connect their own solution database and wanted notifications to contain, where applicable, a reference to the solution number, so this customer exit was used to define additional fields in the notification header.

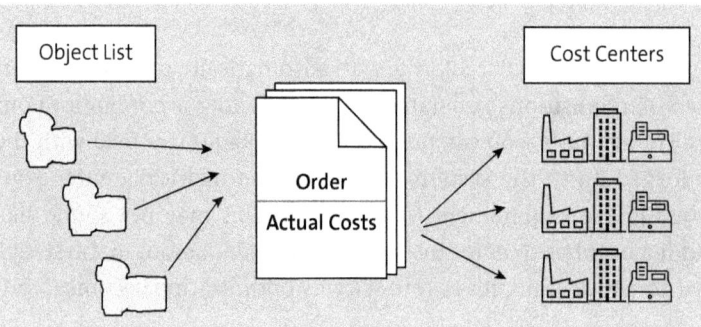

Figure 9.49 Customer Exit: Settlement Rule

These examples clearly show the various ways in which customer exits can be used, and ultimately, there are no limits to your imagination. More than a hundred customer exits are provided for SAP S/4HANA Asset Management. A complete list of customer exits for plant maintenance is provided in Appendix A, Section A.4.

Customer Exits Simplify the Standard System and Make It More Secure
Customer exits enable you to adjust business transaction processing in plant maintenance to your requirements and to simplify and accelerate processing. However, you need to give considerable thought to whether custom programming really has the desired effect and increases user acceptance.

9.3 The IT Department's Programming Options

For the sake of completeness, let's now look at other programming techniques for enhancing or changing the standard SAP system without having to implement a modification.

9.3.5 Business Add-Ins and Enhancement Points

Business add-ins (BAdIs) are used to create predefined enhancement options in SAP components. They therefore pursue the same goal as customer exits: adding more functions to the standard SAP system without having to implement a modification. Unlike customer exits, however, BAdIs are based on ABAP objects.

In contrast to customer exits, which support a two-tier system landscape (SAP and the customer), BAdIs support a multilevel system landscape (SAP, country versions, industry solutions, partners, customers, etc.). You can create BAdI definitions and implementations at each of these levels within the system landscape, meaning that you can implement BAdIs several times. On the other hand, customer exits can be applied only once.

You use Transaction SE19 (BAdI Builder) to create BAdIs. There are two types of BAdIs (see Figure 9.50):

- Classic BAdIs
- New BAdIs (also known as *enhancement points*)

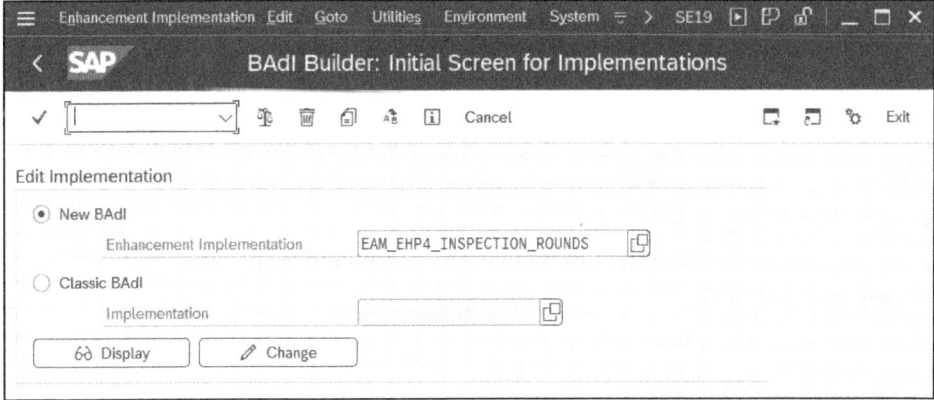

Figure 9.50 BAdI Builder

The following is a list of examples of new BAdIs (enhancement points) that can be used in SAP S/4HANA Asset Management (more than two hundred are available for plant maintenance purposes) and how they can be used:

683

- `EAM_EHP4_CI_SFWS_SC_LIST_ENH`
  To enhance the lists used in plant maintenance (e.g., to display maintenance plans, maintenance items, and maintenance packages in multilevel lists)
- `EAM_EHP4_CI_SFWS_SC_INSP_ROUND`
  To enhance inspection rounds (e.g., to activate the PRT as a measuring point)
- `EAM_EHP4_INSPECTION_ROUNDS`
  To perform Inspections for measuring point as a PRT
- `BADI_EAM_ITOBBAPI_SHIFTNOTE`
  To add new fields to the BAPIs for the technical object
- `EAM_WKCTR_SYNC_IMPL`
  To perform synchronization of work centers for maintenance orders to notifications
- `EAM_WS_ORDER_RELEASE_IMPL`
  For security-relevant enhancements when releasing orders
- `EAM_EHP4_CI_SFWS_SC_AUTH_ENH`
  To perform additional authorization checks for orders
- `EAML_BADI_EAM_STRUCTURE_LIST`
  To add linear data to structure lists
- `EAMWS_ORDER_CHECKLIST`
  To perform status management for the list of relevant risks in SAP S/4HANA Asset Management
- `COCF_ES_SN_LIST`
  To define your own selection options in the list of shift notes

The following is a list of examples that demonstrate how classic BAdIs can be used in plant maintenance (more than twenty are available in total for plant maintenance purposes):

- `EAML_PM_NOTIFICATION`
  To send notifications with linear data at the header and item levels
- `IMPL_EAM_DOC_PMAUFK`
  With order documents
- `IMPL_EAM_DOC_PMPLKO`
  With task list documents
- `IQS_MASS_CHANGE`
  To make mass changes to notifications
- `IWO_MASS_CHANGE`
  To make mass changes to orders
- `PLM_CATALOG_OCI`
  To call the catalog selection via the Open Catalog Interface (OCI)

9.3 The IT Department's Programming Options

> **[+] BAdIs and Enhancement Points Simplify the Standard System and Make It More Secure**
>
> As is the case with customer exits, you can program BAdIs and enhancement points in such a way that you can adjust business transaction processing in plant maintenance to your requirements and thereby simplify and accelerate processing. However, you need to give considerable thought to whether custom programming really has the desired effect and increases user acceptance.

9.3.6 SAP Business Workflow

SAP Business Workflow enables you to define business processes that aren't yet mapped in the system. These can range from simple release and approval procedures to more complex business processes such as creating a material master and coordinating the various departments involved. SAP Business Workflow is at its most efficient when a large number of workflows need to be processed repeatedly or the business process requires a large number of processors in a precisely defined sequence.

Workflows are called and configured in the SAP menu under **Tools • Business Workflow**. Figure 9.51 shows, for example, Transaction SWDD (Workflow Builder).

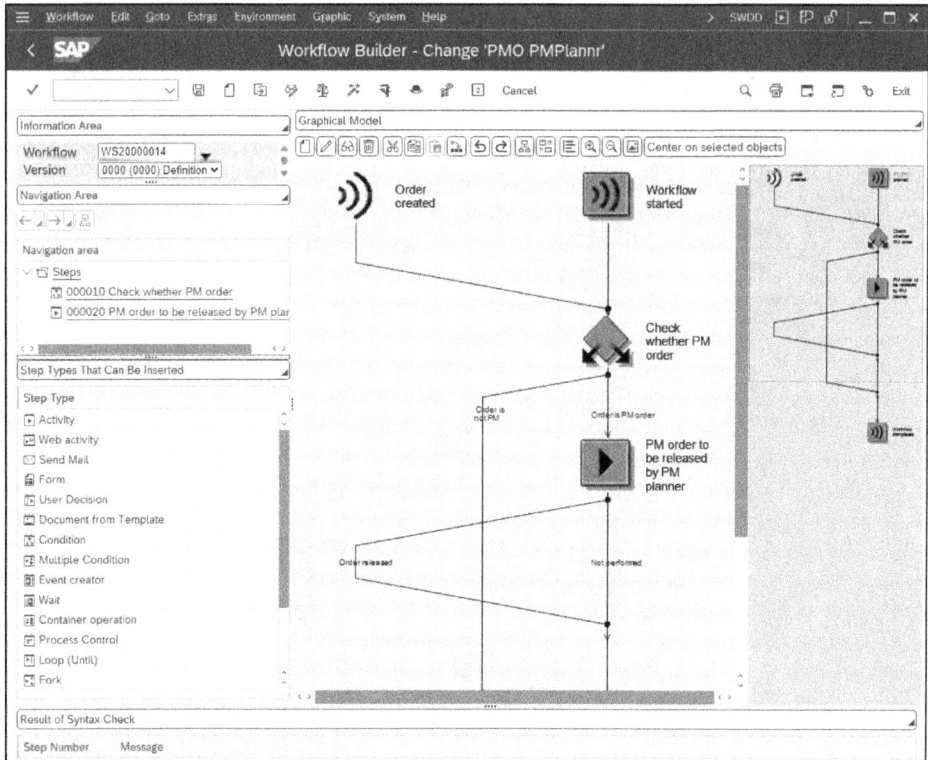

Figure 9.51 SAP Business Workflow: Workflow Builder

9 Usability

From a maintenance processing perspective, the following workflows are particularly relevant:

- WS 00200065 PM: Process Notification
- WS 00200095 PM: Put Notification in Process
- WS 00200075 PM: Complete Notification
- WS 00200074 PM: Complete Task
- WS 20000014 PMO (Maintenance Planner): Release Order, Technically Complete Order, among others
- WS 20000021 PMO (Order Creator): Release Order, Technically Complete Order, among others
- WS 20000031 PMO: Confirm Operation

The purpose of these scenarios is to support you in processing, monitoring, and completing newly created notifications and orders. For example, you can inform the person who created the order (the order creator) when their order has been released or technically completed.

If you're entered in the workflow definition as a processor, you receive a *work item* on your **Business Workplace** screen (Transaction SWBP, see Figure 9.52). You can directly execute the work item (**Process Notification**, in this case).

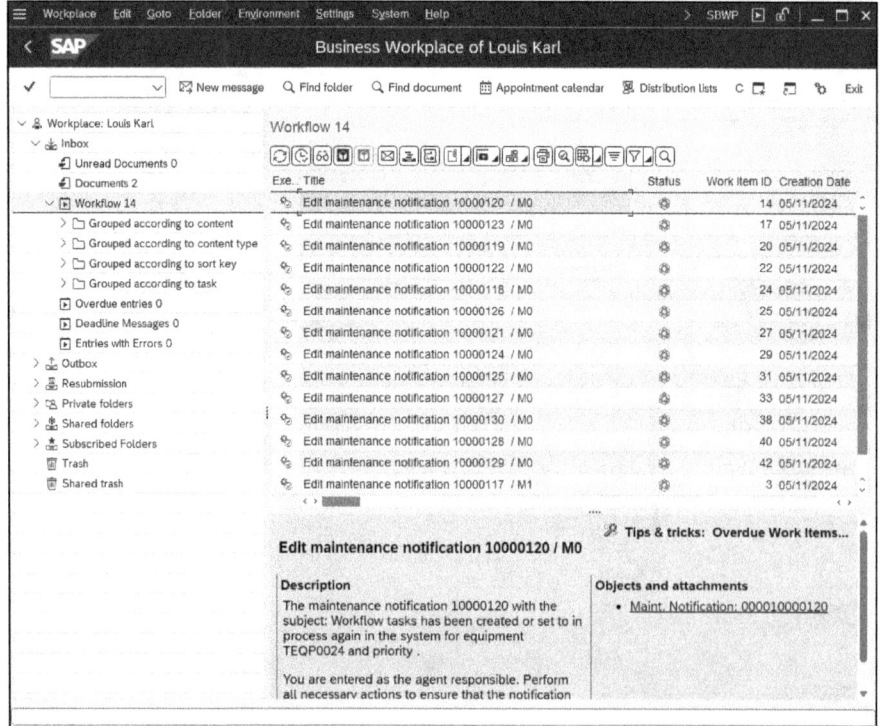

Figure 9.52 SAP Business Workplace Work Items

Use the following Customizing paths to maintain the processor of maintenance notifications and maintenance orders:

- Plant Maintenance and Customer Service • Maintenance and Service Processing • Maintenance and Service Notifications • Notification Processing • Set Workflow for Maintenance Notifications
- Plant Maintenance and Customer Service • Maintenance and Service Processing • Maintenance and Service Orders • Functions and Settings for Order Types • Set Workflow for Orders

> **Recommendations for Notifications and Order Workflows** [+]
>
> When you set up and activate workflows, your SAP system becomes an active system with a push mechanism that enables you to address all users who are to execute actions in notification and order processing (e.g., release an order or complete a task) but who otherwise have very little to do with the SAP system.
>
> In particular, you can simplify task processing considerably by forwarding work items to the usual email client of the recipient (e.g., Outlook), thereby giving recipients the option to use a smartphone to process a task.

From a maintenance perspective, there is one more workflow—WS00400037—that you can use to synchronize equipment master data in SAP S/4HANA Asset Management with the asset data in asset accounting. For example, you can configure the system to automatically create an asset master record in the background when you create an equipment master record, and vice versa. The requisite workflow already exists in the system, and all you need to do is activate some Customizing settings.

Specify Conditions for Synchronization of Master Data

You use this Customizing function to define the framework conditions in relation to whether and, if so, how equipment and assets are to be synchronized.

Prerequisites

You must define the asset class, equipment category, and object type beforehand.

Customizing Path

Financial Accounting • Asset Accounting • Master Data • Automatic Creation of Equipment Master Records • Specify Conditions for Synchronization of Master Data

Settings

Use this Customizing function to establish a link between the **Asset Class** and the **EquipCategory** or **Object Type** (see Figure 9.53). You define which equipment category or order type is to be assigned when an asset master record is created with a specific asset class. Then, you can use **Edit asset master record** and **Edit equipment master** record to regulate how the master records are to be synchronized in create and change modes. You can use the dropdown lists to choose among these options:

- **No synchronization**
  When you create an asset master record, the system doesn't propose any equipment. If you manually create the related pieces of equipment at a later stage, the system doesn't synchronize the master data.

- **Direct synchronization after saving**
  When you create an asset master record, the system proposes a piece of equipment, and, when you save the asset, the system creates the corresponding equipment master record (and the assignment between both master records) immediately. Subsequent changes to the asset master record can be synchronized in the equipment master record (and vice versa).

- **Saving triggers a workflow**
  When you create an asset master record, the system proposes a piece of equipment, and, when you save the asset, the system triggers a specific workflow event. Depending on the attributes of the assigned workflow (e.g., a workflow to notify the person responsible), the system generates the associated equipment master record when you execute this workflow. Subsequent changes to the asset master record can be synchronized in the equipment master record (and vice versa).

- **Direct synchronization plus workflow**
  When you create an asset master record, the system proposes a piece of equipment, and, when you save the asset, the system generates the corresponding equipment master record (and assigns it to the asset) immediately. It also triggers a specific workflow event.

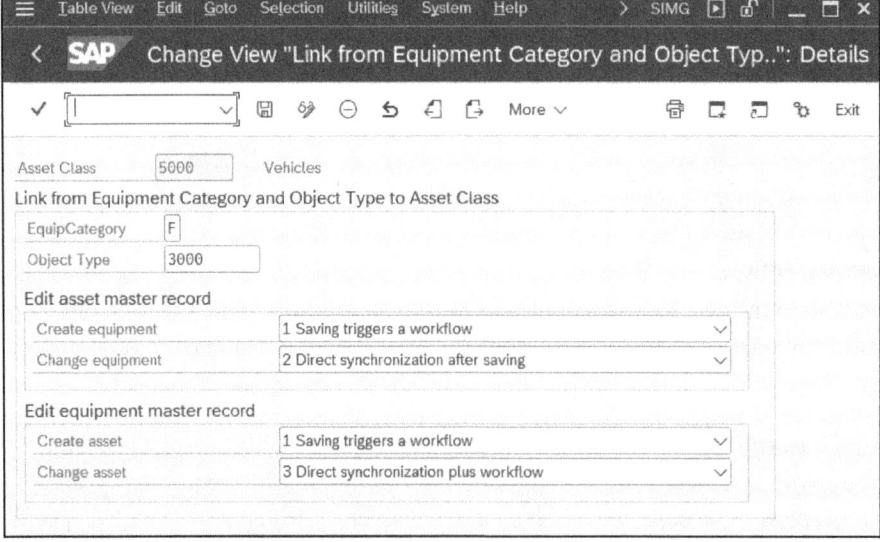

Figure 9.53 SAP Business Workflow: Synchronize Equipment and Asset Master

Assign Master Data Fields of Assets and Equipment

Use this Customizing function to assign to the fields in the equipment master record those fields in the asset master record that are to be compared when the asset master records and equipment master records are synchronized.

Prerequisites
You must activate and define the synchronization of asset master data and equipment master data beforehand.

Customizing Path
Financial Accounting • Asset Accounting • Master Data • Automatic Creation of Equipment Master Records • Assign Master Data Fields of Assets and Equipment

Settings
All the usual fields are contained (see Figure 9.54).

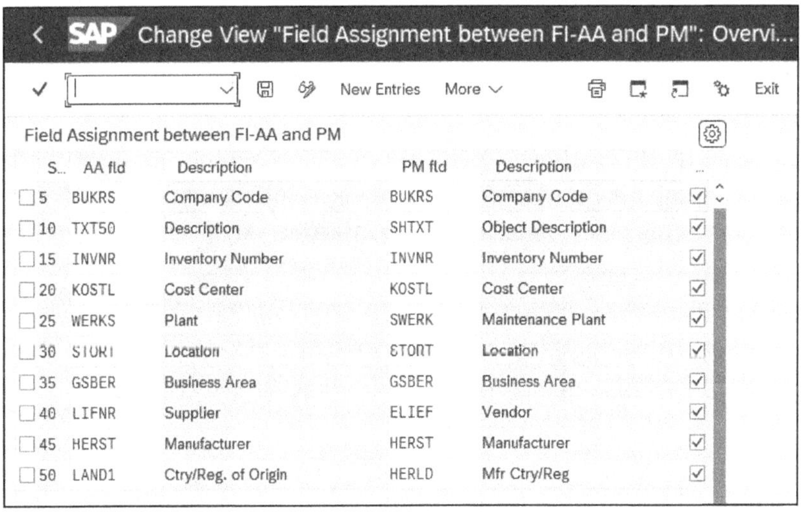

S..	AA fld	Description	PM fld	Description	
5	BUKRS	Company Code	BUKRS	Company Code	✓
10	TXT50	Description	SHTXT	Object Description	✓
15	INVNR	Inventory Number	INVNR	Inventory Number	✓
20	KOSTL	Cost Center	KOSTL	Cost Center	✓
25	WERKS	Plant	SWERK	Maintenance Plant	✓
30	STORT	Location	STORT	Location	✓
35	GSBER	Business Area	GSBER	Business Area	✓
40	LIFNR	Supplier	ELIEF	Vendor	✓
45	HERST	Manufacturer	HERST	Manufacturer	✓
50	LAND1	Ctry/Reg. of Origin	HERLD	Mfr Ctry/Reg	✓

Figure 9.54 SAP Business Workflow: Synchronized Fields for Equipment and Asset Master

It's possible to make other assignments by selecting additional fields in Asset Accounting and then assigning an additional field in plant maintenance to these fields. The fields available in Asset Accounting are contained in the Data Dictionary (DDIC) structure SAAPM1, while the fields available in plant maintenance are contained in the DDIC structure SAAPM2.

If it's necessary to make a more extensive field assignment, use customer exit AAPM0001. You can also use this customer exit to derive different asset classes from the fields in the equipment master record and different equipment categories from the fields in the asset master record.

> **Recommendations for the Asset Workflow**
>
> The workflow for synchronizing equipment and assets must be viewed somewhat critically because maintenance workers and accountants have a completely different point of view on assets. In general, maintenance workers prefer a "narrow" structure in which the technical aspects of the assets are central, while accountants prefer a "broad" structure in which the taxation aspects of the assets are crucial.
>
> Therefore, achieving a 1:1 relationship between a piece of equipment and an asset is rather unlikely.

9.4 Summary

This chapter explained that there is a wide variety of options to both improve usability and have a powerful instrument for daily work. These options and tools can be differentiated into three categories:

- All the options a user has to improve usability, such as setting up list variants or configuring his table controls
- Options and tools for IT departments to improve usability without the need of programming (for example, by implementing a mobile solution, setting up SAP Fiori launchpads, or creating new frontends using transaction variants or screen personas)
- Options for IT departments to improve usability based on the need for programming (for example, using z-programs for upstream transactions, creating web interfaces with integrating BAPIs, or programming customer exits and BAdIs)

Appendix A
Project Plans and Overviews

To support your project work, this appendix contains three tables and a list, as briefly described here:

- A table of available functions in SAP S/4HANA Asset Management to prioritize your project scope
- A table of available authorization objects in SAP S/4HANA Asset Management for setting up an authorization concept and the plant maintenance roles
- A table of available Business Application Programming Interfaces (BAPIs) for SAP S/4HANA Asset Management and their methods to set up interfaces (e.g., to develop a web application and transfer data to SAP)
- A list of available customer exits for SAP S/4HANA Asset Management to set up additional functionality in your system

A.1 Functional Scope of SAP S/4HANA Asset Management and Assignment of Priorities

This section contains tables that you can fill with your own dates and comments for your project planning.

You can also download this file from *www.sap-press.com/5900*. From there, navigate to the **Product supplements** area on this book's individual page.

Table A.1 is a table that you can use to prioritize your project scope. The priority levels are as follows:

- A: Must be implemented in the first phase
- B: Will be implemented in a later expansion phase
- C: Not required, according to current knowledge

Area	Function	A	B	C
Asset management	Functional locations			
	Reference functional locations			

Table A.1 Priorities in SAP S/4HANA Asset Management Projects

Area	Function	A	B	C
	Alternative labeling			
	Linear objects			
	Equipment			
	Serial numbers			
	Classification			
	BOMs			
	Change management			
	Equipment hierarchies			
	Variant configuration			
	Documents			
	Permits			
	Warranties			
	PRTs			
	Partners			
	Counters			
	Measuring points			
	Equipment and asset synchronization			
Notification	User status			
	Reference objects			
	Priorities			
	Partners			
	Telephone integration			
	Paging			
	Addresses			
	Object parts			
	Damage			
	Causes of damage			

Table A.1 Priorities in SAP S/4HANA Asset Management Projects (Cont.)

A.1 Functional Scope of SAP S/4HANA Asset Management and Assignment of Priorities

Area	Function	A	B	C
	Tasks			
	Activities			
	Notification items			
	Classification			
	Print			
	Fax			
	Download			
	Breakdowns			
	Permits			
	Response time monitoring			
	Revisions			
	Solution database			
Order	Reference objects			
	User status			
	Priorities			
	Partners			
	Telephone integration			
	Paging			
	Addresses			
	Print			
	Fax			
	Download			
	Permits			
	Operations			
	Scheduling			
	Relationships			
	Capacity planning			

Table A.1 Priorities in SAP S/4HANA Asset Management Projects (Cont.)

Area	Function	A	B	C
	Capacity availability check			
	Reservation of stock material			
	Stock material availability check			
	Ordering of nonstock material			
	Catalog integration (internet and intranet catalogs, vendor catalogs)			
	Estimated costs			
	Planned/actual costing			
	Order budgets			
	Object list			
	PRTs			
	PRTs availability check			
	Suborders			
	Utilization of production capacities			
Completion confirmation	Time confirmations			
	Technical completion confirmations			
	Goods receipts			
	Material withdrawals			
	Overhead costing			
	Order settlement			
External processing	External processing via service specifications			
	External processing as a purchase order			
	External processing via work centers			
	Revisions			
Preventive maintenance	General maintenance task lists			
	Equipment task lists			
	Functional location task lists			

Table A.1 Priorities in SAP S/4HANA Asset Management Projects (Cont.)

A.1 Functional Scope of SAP S/4HANA Asset Management and Assignment of Priorities

Area	Function	A	B	C
	Maintenance strategies			
	Time-based maintenance plans			
	Performance-based maintenance plans			
	Single cycle plans			
	Strategy plans			
	Basic multiple counter plans			
	Enhanced multiple counter plans			
	Order maintenance call objects			
	Notification maintenance call objects			
	Inspection lot maintenance call objects			
	Service entry sheet maintenance call objects			
	Simulation of planned costs			
	Automatic deadline monitoring			
Condition-based maintenance	PM-PCS interface			
Refurbishment	Refurbishment of serial numbers			
	Refurbishment of materials			
	Settlement based on standard price			
	Settlement based on moving average price			
Subcontracting	Subcontracting of serial numbers			
	Subcontracting of materials			
	Settlement based on standard price			
	Settlement based on moving average price			
Pool asset management	Pools			
	Planning table			

Table A.1 Priorities in SAP S/4HANA Asset Management Projects (Cont.)

Area	Function	A	B	C
Calibration of test/measurement equipment	Test/measurement equipment as equipment			
	General task list, including inspection characteristics			
	Inspection maintenance plans			
	Results recording			
	Completion confirmation			
	Usage decision			
Project-based maintenance	Work breakdown structure elements			
	Networks			
	Manual assignment			
	Automatic assignment			
	Maintenance Event Builder			
Checklist processing	Equipment and/or functional locations			
	Inspection plans or general task list, including inspection characteristics			
	Inspection maintenance plans			
	Results recording			
	Completion confirmation			
	Usage decision			
Work order cycle using phase model	Notification types			
	Order types			
	Event prioritization			
	Teams and responsibilities			
	Status profile and phases			

Table A.1 Priorities in SAP S/4HANA Asset Management Projects (Cont.)

A.2 Authorization Objects in SAP S/4HANA Asset Management

Table A.2 shows all available authorization objects in SAP S/4HANA Asset Management that are relevant for setting up your configuration concept and roles.

A.2 Authorization Objects in SAP S/4HANA Asset Management

Authorization Object	Description	Checked Fields
I_AER	Follow-up order creation	- Maintenance planning plant - Order type
I_AUART	PM: Order type	- Maintenance planning plant - Order type
I_AUFK_AFV	Creating order operations	- Work center - Plant - Control key
I_AUSWK	Effect on operation	- Effect on operation
I_BEGRP	PM: Authorization group	- Transaction code - Authorization group for technical object
I_BETRVORG	PM: Business transaction	- Business transaction
I_CL_LOT	Process inspection lots for checklists	- Activity
I_CONFSTOR	PM: Mass confirmation reversal	- Confirmation reversal
I_EAM_OM	Output management	- Plant - Planner group - Work center
I_EQART	Technical object type	- TO type
I_EQTYP	Equipment category	- EQ category
I_FLTYP	Functional location category	- FL category
I_GAART	Warranty type	- Warranty type - Activity
I_ILART	Maintenance activity type	- Maintenance activity type
I_ILOA	Changing location and account assignment data in orders	- Maintenance planning plant - Order type
I_INGRP	PM: Maintenance planner group	- Transaction code - Maintenance planning plant - Planner group for customer service and plant maintenance

Table A.2 Authorization Objects in SAP S/4HANA Asset Management

Authorization Object	Description	Checked Fields
I_IWERK	PM: Maintenance planning plant	- Transaction code - Maintenance planning plant
I_KOSTL	PM: Cost centers	- Transaction code - Controlling area - Cost center
I_MASS	PM: Mass data change	- Mass data change for object types
I_MPTYP	Maintenance plan category	- MP category
I_OST_PROF	Overall status profile	- Overall status
I_PHSE_CTL	Maintenance order phase control codes	- Order type - Plant - Activity
I_QMEL	PM/QM: Notification types	- Transaction code - Notification type
I_ROUT	PM: Task list	- Action
I_ROUT1	PM: Task lists by maintenance planning plant, work scheduler, and status	- Transaction code - Maintenance planning plant - Responsible planner group/department - Status of task list
I_SOGEN	PM: Permit	- Maintenance plant - Permit key
I_SWERK	PM: Maintenance plant	- Transaction code - Maintenance plant
I_TCODE	PM: Transaction code	- Transaction code
I_VORG_MEL	PM/QM: Business transaction for notifications	- Notification type - Business transaction
I_VORG_MP	PM: Business transaction for maintenance planning	- Maintenance plan category - Business transaction
I_VORG_ORD	PM: Business transaction for orders	- Order type - Business transaction
I_WPS_MEB	Revision planning: Maintenance Event Builder	- Maintenance Event Builder screen areas

Table A.2 Authorization Objects in SAP S/4HANA Asset Management (Cont.)

Authorization Object	Description	Checked Fields
I_WPS_REV	Revision planning: Authorization object revision	- Revision type - Work center - Plant - Revision operations for the work packaging and sequencing process

Table A.2 Authorization Objects in SAP S/4HANA Asset Management (Cont.)

A.3 BAPIs for SAP S/4HANA Asset Management

Table A.3 shows all available BAPIs for SAP S/4HANA Asset Management and their methods for setting up your interfaces (e.g., to develop a web application and transfer data to SAP).

Object	Method	Content
Equipment	Change	You can use this method to change an existing piece of equipment.
	Create	You can use this method to create a new piece of equipment.
	CreateByReference	You can use this method to create a new piece of equipment with reference to an existing piece of template equipment.
	Dismantle	You can use this method to dismantle a piece of equipment from a functional location or a superior piece of equipment.
	Dismantle_IE4N	You can use this method to dismantle a piece of equipment from its superior piece of equipment or functional location. The piece of equipment must have a serial number. During dismantling, the material is transferred to the warehouse (in a process known as a goods movement), and a maintenance notification is created to document the dismantling of the piece of equipment.

Table A.3 BAPIs in SAP S/4HANA Asset Management

Object	Method	Content
	Exchange	You use this method to exchange two pieces of equipment at their installation locations. The pieces of equipment must have serial numbers. Each time a piece of equipment is dismantled or installed, a maintenance notification is created to document the dismantling or installation of the piece of equipment. Consequently, four maintenance notifications are created.
	GetCatalogProfile	You can use this method to determine the catalog profile for a piece of equipment on the current date.
	GetDetail1	You can use this method to read an existing piece of equipment.
	GetList	You can use this method to select a list of equipment according to predefined selection criteria.
	GetStatus	You can use this method to read the system status and user status of an existing piece of equipment.
	Install	You can use this method to install a piece of equipment in a functional location or to install a superior piece of equipment.
	InstallNonSerializedMaterial	You can use this method to install a nonserialized material in the BOM header of a piece of equipment without requiring a user action in Transaction IE4N. A maintenance notification is created to document the installation.
	Install_IE4N	You can use this method to install a piece of equipment on another piece of equipment or in a functional location. The piece of equipment must have a serial number. During installation, the material is transferred from the warehouse (in a process known as a goods movement). A maintenance notification is created to document the installation of the piece of equipment.
	RemoveNonSerializedMaterial	You can use this method to remove a nonserialized material from the BOM header of a piece of equipment without requiring a user action in Transaction IE4N. During dismantling, the material is transferred to the warehouse (in a process known as a goods movement).

Table A.3 BAPIs in SAP S/4HANA Asset Management (Cont.)

Object	Method	Content
	Update	You use this method to change a piece of equipment.
Functional Location	Change	You can use this method to change the master data of a single functional location.
	Create	You can use this method to create a single functional location.
	GetDetail	You can use this method to read the master data of a single functional location.
	GetList	You can use this method to read all functional locations that satisfy certain selection conditions.
	GetStatus	You can use this method to read the system status and user status of an existing functional location.
	InheritChange	You can use this method to change the inheritance indicator of a functional location.
	StrucAssign	You can use this method to assign a superior functional location to a functional location.
	StrucReplace	You can use this method to assign a new superior functional location to a functional location. This replaces the existing assignment to a superior functional location.
	StrucUnassign	You can use this method to unassign a superior functional location from a functional location.
Maint Notific BAPIs	Notif Changeusrstat	You can use this method to change the user status of a notification.
	NotifClose	You can use this method to close a notification. The notification is assigned the system status **Notification Completed**.
	NotifCreate	You can use this method to create a new notification. You can create certain fields in the notification header as well as items, causes, tasks, activities, and long texts for the notification. You can also maintain partner data.
	NotifDataAdd	You can use this method to add items, causes, activities, tasks, and long texts to a notification.

Table A.3 BAPIs in SAP S/4HANA Asset Management (Cont.)

Object	Method	Content
	`NotifDataDelete`	You can use this method to delete items, causes, activities, tasks, and partners from a notification.
	`NotifDataModify`	You can use this method to change the header, items, causes, activities, tasks, and partners in a notification.
	`NotifGetDetail`	You can use this method to display all detailed information about a notification.
	`NotifListEqui`	You can use this method to select notifications that are assigned to a piece of equipment.
	`NotifListFuncloc`	You can use this method to select notifications that are assigned to a functional location.
	`NotifListPartner`	You can use this method to select notifications that are assigned to a certain combination of partner functions and partners.
	`NotifListPlangroup`	You can use this method to select notifications that are assigned to a certain combination of planner groups and planning plants.
	`NotifListSortfield`	You can use this method to select notifications according to the specified sort field.
	`NotifPostpone`	You can use this method to postpone the notification. The notification is assigned the system status **Notification Postponed**.
	`NotifPutinprogress`	You can use this method to put the notification in process. The notification is assigned the system status **Notification in Process**.
	`NotifTaskComplete`	You can use this method to complete a notification task. The task is assigned the system status **Task Completed**.
	`NotifTaskRelease`	You can use this method to release a notification task. The task is assigned the system status **Task Released**.
	`NotifTaskSuccess`	You can use this method to indicate that a notification task has succeeded. The task is assigned the system status **Task Successful**.

Table A.3 BAPIs in SAP S/4HANA Asset Management (Cont.)

A.3 BAPIs for SAP S/4HANA Asset Management

Object	Method	Content
Maintenance OrderBAPI	ComponentGetDetail	You can use this method to display detailed information about an order component. Long texts are also returned.
	OperationGetDetail	You can use this method to read detailed information about an order component. Long texts are also returned.
	OrderGetDetail	You can use this method to read information about an order.
	OrderheadGetList	You can use this method to determine a list of orders in accordance with a list of selection conditions.
	OrderMaintain	You can use this method to change orders and subobjects. You can edit the following order objects: - Order header - Partners - User status - Order operations - Relationships - Components - Long texts for the order header, operations, and components - Settlement rules - Object lists - Link to object lists - PRTs - Classifications of a service package - Service lines in a service package - Limits of a service package - Contract limits of a service package
	OrderoperGetList	You can use this method to determine a list of operations in accordance with a list of selection conditions.

Table A.3 BAPIs in SAP S/4HANA Asset Management (Cont.)

Object	Method	Content
MaintOrd Confirmation	Cancel	You can use this method to cancel an order confirmation. Any goods movements that were posted together with the completion confirmation (backflushes) are automatically canceled with the completion confirmation.
	ConfCreate	You can use this method to enter time confirmations for order operations. Goods movements for a completion confirmation are determined in accordance with the standard logic for backflushes and an automatic goods receipt for the completion confirmation.
	GetDetail	You use this method to provide confirmation data for an order. Any goods movements that were posted together with the completion confirmation are also provided.
	GetList	This method delivers a list of all completion confirmations that satisfy the predefined selection criteria (order, operation, suboperation, or confirmation number).
	GetProp	You can use this method to provide default data for time confirmations for order operations.
Material	Availability	You can use this method to determine, in accordance with the available-to-promise (ATP) logic, the quantities still available for a specific material in a specific plant.
	Display	You can use this method to display a material.
	Edit	You can use this method to change a material.
	GetBatches	You can use this method to generate a list of batches for a material, and you can then use input parameters to restrict this list.
	GetDetail	This method delivers a selection of client, plant, valuation area, and valuation type-specific detailed information about a predefined material.
	GetList	This method delivers a list of materials with a short text for the specified selection parameters.
	GetMRPList	You can use this method to select the MRP list for a material.

Table A.3 BAPIs in SAP S/4HANA Asset Management (Cont.)

Object	Method	Content
	GetStockRequirementsList	You can use this method to select the current stock/requirements list for a material.
MaterialBOM	CreateBomGroup	You can use this method to create a material BOM and all of its assigned objects.
	ExistenceCheckBomGroup	You can use this method to check whether a BOM exists for a specified material, plant, and usage.

Table A.3 BAPIs in SAP S/4HANA Asset Management (Cont.)

A.4 Customer Exits for SAP S/4HANA Asset Management

The following is a list of the most important customer exits available for SAP S/4HANA Asset Management:

- **Master data**
  - IEQM0001: Additional checks when installing equipment in a functional location
  - IEQM0002: Additional checks when defining equipment hierarchies
  - IEQM0003: Additional checks prior to posting equipment
  - IEQM0004: Object is permitted for contract partner (order maintenance contract)
  - IEQM0005: Object is permitted for a sales contract
  - IEQM0006: Object is permitted for a sales contract (maintain maintenance contract)
  - IEQM0007: Check/change vendor field in equipment master
  - IHCL0001: Create equipment with material: classes/characteristics template
  - ILOM0001: Additional checks prior to posting a functional location
  - ILOM0002: Structural check of location numbers
  - ITOB0001: Customer includes subscreen for functional location master data
  - ITOB0002: Field changes in template
  - ITOB0003: Customer includes subscreen for fleet object data
  - ITOB0004: Fleet identification data: checks
- **Measuring points**
  - IMRC0001: Define field content for measuring points and measurement documents
  - IMRC0002: Measuring point: menu exit for custom function
  - IMRC0003: Measurement document: menu exit for custom function
  - IMRC0004: Checks for new measurement document
  - IMRC0005: Measuring point: exit in AUTHORITY_CHECK_IMPT

- **Warranties**
  - BG000001: Warranty check
  - BG000002: Warranty check dialog box
  - BG000003: Change warranty check result
- **Task lists**
  - IAIH0001: Enhance task list header with customer-specific fields
- **Maintenance plans**
  - IPRM0002: Determine planned dates for maintenance plan
  - IPRM0003: User fields: maintenance plan
  - IPRM0004: Maintenance plan/item: customer check during **Save**
  - IPRM0005: Determine offset for service-based strategy plans
- **Notifications**
  - QQMA0001: User subscreen for notification header
  - QQMA0008: User subscreen for notification item
  - QQMA0010: User subscreen for additional data on cause
  - QQMA0011: User subscreen for additional data on task
  - QQMA0012: User subscreen for additional data on activity
  - QQMA0014: Checks prior to saving a notification
  - QQMA0015: Exit prior to calling [F4] help for catalog
  - QQMA0016: **User Data** function in **Goto** • **Task** menu
  - QQMA0017: **User Data** function in **Goto** • **Activities** menu
  - QQMA0018: Date assignment after entry of priorities
  - QQMA0019: Default partner when a notification is added
  - QQMA0021: **User Data** function in **Goto** menu
  - QQMA0022: **User Data** function in **Goto** • **Cause** menu
  - QQMA0023: QM/PM/SM: **User Data** function in **Goto** • **Item** menu
  - QQMA0024: Deactivation of function codes in menu
  - QQMA0025: Default values when a notification is added
  - QQMA0026: Authorization check when starting a notification transaction
  - QQMA0027: Default values when a task is added
  - QQMA0029: Changing the notification type
  - QQMA0030: Checking a status change for permissibility
- **Orders**
  - COZF0001: Changing a purchase requisition for an external operation
  - COZF0002: Changing a purchase requisition for an external component

A.4 Customer Exits for SAP S/4HANA Asset Management

- IW010001: Creating a plant maintenance suborder
- IW010002: Customer check for order release
- IW010004: Customer check for order completion
- IW010005: Customer-specific determination of profit center
- IW010006: Exclusion of function codes
- IW010007: Customer enhancement: permits in order
- IW010008: Customer enhancement: determination of tax jurisdiction code
- IW010009: Customer check during **Save**
- IW010010: Customer enhancement for determining the WBS element
- IW010011: Customer enhancement for component selection
- IW010012: Priority handling in central header
- IW010015: [F4] help for user fields in operation
- IW010016: Customer enhancement for checking operation user fields
- IW010017: Determine external order number by customer logic
- IW010018: User fields in order header
- IW010020: Automatic task list integration
- IW010021: Automatic task list transfer during order creation from notification
- IW010022: Determine calendar
- IW010025: Determine responsible cost center
- IW010026: Set **Do not execute** status
- IW010027: Generate custom-specific settlement rule
- IW010029: Include BOM in order
- IW010030: Prepopulate fields for results object
- IW010031: Hide personnel number in order
- IW010033: Customer-specific authorization check
- IW010034: Operation status is based on capacity requirements status
- IW020001: Task list transfer to order
- IWOC0001: Create notification: determination of reference object
- IWOC0002: Check status change permissibility
- IWOC0003: Reference object and planner group authorization check
- IWOC0004: Change single-level list processing in ALV settings
- CNEX0026: Customer enhancement for general material checks
- CNEX0027: Customer enhancement: determine plant and storage location for a component

- **Completion confirmations**
  - CMFU0001: Define customer-specific screen layout
  - CMFU0002: Set parameters for time confirmation and goods movements
  - CONFPM01: Order confirmation: determine customer-specific default values
  - CONFPM02: PM/SM order confirmation: customer-specific input checks 1
  - CONFPM03: Order confirmation: customer-specific check after operation selection
  - CONFPM04: Order confirmation: customer-specific input checks 2
  - CONFPM05: Order confirmation: customer-specific amendments during save process
- **Information system**
  - MCI10001: MCI1: PMIS/QMIS update

Appendix B
The Author

Prof. Dr. Karl Liebstückel is a professor of information management and business software at the Technical University Würzburg-Schweinfurt in Germany. He has more than 35 years of experience in SAP plant maintenance. Over the course of his career, Karl Liebstückel has gained extensive practical experience in close to 100 plant maintenance projects. He was a member of the executive board of the German SAP User Group (DSAG) from 2003 to 2012 and chairman from 2007 to 2012. Furthermore, he presided over DSAG's Plant Management and Service Management work group from 2001 to 2008. Previously, he worked at SAP AG for 13 years in the areas of development, consulting, and training for plant maintenance and service management. The last two positions he held prior to leaving SAP AG were platinum consultant and global product manager.

Prof. Dr. Liebstückel also owns a consulting firm, and he has written several books on the topic of logistics.

You are welcome to contact the author by email at *karl@liebstueckel.consulting*.

Appendix

The Archive

Index

A

ABC analysis 639
Access sequence 326
Account assignment category 323
Account category reference 443
Action box 278, 279, 632, 656
   define 657
Activity category 259
Activity types 297
   number 298
Actual costs 381
Address data 325
After event recording 678
Allocation structure 378
   assignments 379
   criteria 380
Allowed property value 208
Alternative BOM 246
Alternative determination 246
Alternative labeling 180, 181
Analytical apps 583
Approval phase 552
As-is analysis 56, 67
Asset master record 687
Asset under construction (AuC) 390
Assignment keys 385
Assignments 379
   source 380
Authorization concept 32, 67
Authorization object 79, 696
Authorizations 603, 618
Automatic account assignment 216
Auxiliary materials 213
Availability checks 304, 307, 563
   advanced 309
   inspection control 306
   scope of check 311
Available capacity 346
Available-to-promise (ATP) quantity 304

B

BAPI Explorer 675
Bar charts 342
Bill of materials (BOM) 43, 171, 239, 422
   alternative 246
   equipment 171, 239

Bill of materials (BOM) (Cont.)
   for a functional location 171
   header 242
   item 244
   item categories 242
   material BOM 171
   restrict 242
   status 240
   usage 241, 422
   usage priorities 245
Booking confirmation 501
Breakdown notification 67
Budget profile 291
Buffer 333
Business add-ins (BAdIs) 683
Business Application Programming
   Interface (BAPI) 675, 699
Business blueprint 31
Business case 56
Business catalog 600, 615
   add tile 600
Business feature 553
Business function 23, 464, 501, 517, 519
Business group 601, 617
Business process master list 56
Business Process Model and
   Notation (BPMN) 63
Business process modeling 60
Business processes 655

C

CAD drawings 151
CAD system 69
Calculation key 421
Calibration 468
   catalog 9 477
   maintenance plan category 482
   notification 475, 481, 532, 545
   orders 659
Capacity category 347
   labor 347
Capacity evaluation 346, 350
Capacity leveling 346, 351
   profiles 353
Capacity planning 345
   prerequisites 345
   shift sequences 349

Capacity planning (Cont.)
  stages 346
Capacity requirements 346
Capatog profiles 255, 265, 270, 429, 477–479, 533, 537, 545, 550
  define 269
Catalog type 429
Catalogs 265, 477, 537, 549, 598
  assign to order types 315
  categorizaion 266
  change 271, 479, 533, 545
  define 313, 493
  maintain 266, 525, 586
  types 266
  usage decisions 493, 537, 550
Category groups 557
Change document 165, 186, 427, 439
  activate 167
  Customizing paths 166
Change management 49
Change requests 37
Characteristics 508
  linear data 204
Chart of accounts 83
Checking groups 305, 308, 310
Checking rules 305
  define 305
Checklist 529, 659
  configure for order types 542
  configure for plants 542
Checklist processing 530, 696
  basic 530
  extended 538
  follow-up actions 536
  inspection points 533
  order types 531, 541
  shop papers 544
Class type 205, 505
Classification system 201
Code group 478
Codes 268
Collective purchase requisition 327, 439
Company code 81
  assigning controlling area 86
  edit 81
  global parameters 83
  parameters 81
  tax 84
Compatability mode 332
Competency 51
Completion confirmation 401, 694, 708
  control parameters 402

Completion confirmation (Cont.)
  screen areas 406
  screen templates 405
Completion phase 553
Component item categories 321
Condition-based maintenance 695
Conditions 574
Confirmation 677
Consequence categories 556
  assign 557
Consistency checks 368
Construction type 416
Consumable type 193
Content repository 143, 147
Continuous improvement (CI) 24
Control key 104, 105, 434, 436, 441, 486
  indicators 442
  properties 437
Controlling area 84
  assigning to company code 86
  maintain 84
  one per enterprise 86
  plant 93
Conversion modules 319, 320
  use 320
Copy assignment 386
Cost allocation 501
Cost data 283
Cost elements 363
  assign 365
Cost settings 376
Costing 362
  missing order costs 377
  parameters 368, 375
  planned vs. actual 376
Costing sheets 368, 369
  baseline 370
  overhead rate 371
  rows 369
Costing types 368
Costing variants 374
Counter 128, 196, 197, 660
Create Maintenance Request app 586, 590
Create Purchase Requisition app 585
Create Supplier Invoice app 585
Cross-application time sheet (CATS) 401
  confirmations 407
  data entry profile 409
  field selection 410
  hide fields 411
Cross-plant maintenance 93
Customer exit 680, 705

Index

Customizing 652
Cutover plan 57, 76
Cycle modification factor 496

D

Damage catalogs 265
Data entry 72
   profiles 407
Debit costs 380
Defect types 478
Demand categories 566
Detection methods 560, 586
   groups 561
   profile 561
Dialog box 628
Direct scheduling 509
Display variant 639
Distribution rules 383
   options 383
Document 141
   assignment 153
   create 146
   number range 144
   object 142
   object link 145
   release 428
   status 144
   version number 143
Document types 142, 323, 466
   define 143
   subcontracting 467
Dynamic apps 584
Dynamic date calculation 637
Dynamic list 640
Dynamic segmentation 201

E

EAM app 677
   technical properties 679
Earliest scheduled start 337
Early number assignment indicator 255
Electronic parts catalogs 312
End-user training 76
Engine type 194
Enhanced multiple counter plan 430
Enhanced object list 295
Enhanced procurement mode 332
Enhancement point 683
Enterprise asset management 153

Equipment 170, 182, 660
   Customizing 183
   linear data 202
   master 232
   reference category 185
   task list 416
Equipment category 185, 474, 536, 549
   additional views 186, 472
   installation 190
   maintain 185
   serial number 234
   test/measurement 471
Estimated costs 363
Evaluation profiles 351
Event prioritization 555, 587
Event type linkage 566
Event-driven process chain (EPC) 61
Execution phase 552
Expendable supplies 213
External number assignment 119
External processing 421, 434, 435, 694
   external indicators 440
   profiles 438
   release strategy 447
   service messages 444
   with service specifications 440
External procurement 324
   default value profile 437

F

Fact sheet apps 583
Favorites 633, 634
Feasibility study 41
   organizational structures 42
   plant maintenance processes 44
   purchase and warehouse processes 46
   reporting and key figures 47
   results 48
   system background 42
   technical asset structuring 43
Field selection 100, 154, 180, 227
   CATS 410
   configuration 155
   data screen 230
   functional location 156
   group 227
   industry sector 228
   multilevel lists 163
   object 164
   position 165
   priority rules 158, 231

Index

File type	149
Final confirmation	402
Find Maintenance Order app	641
Find Technical Object app	584
Fiscal year variant	84
Flavor	671
theme	673
Fleet consumption	197
Fleet management	191
consumption values	197
object type	192
Fleet object	192
consumable type	193
consumption	198
consumption values	197
counter	196
engine type	194
fuel consumption	195
issue/return	501
special measurement position	195
type	192
usage type	193
Follow-up actions	278, 495, 536, 550
Formula definitions	333
configure	335
Formula key	337
Formula parameters	333
define	334
Fuel consumption	195
units of measurement	195
Fuel counter	197
Function modules	279
Function profile	569
Functional location	169, 171, 660
alternative labeling	180
category	155, 157, 177, 178
installation	189
linear data	201
numbers	181
structural display	179
structure indicator	172
task list	416
Functional test	75

G

Gap list	57, 73
General maintenance task list	416
General order data	398
Generic functions	109
Generic notifications	254
Global valuation categories	451, 454
Global valuation types	451
Go-live	76
Goods movements	328, 460
documenting	329
refurbishment	460
Graphic profiles	343
define	344
Groups	120, 598
GuiXT	664
activate script	668
control window	665
options	666
script	667

H

Header data	416
HTML fields	315
target fields	317
values	318

I

ICF service	597, 614
create	597
Icon	177, 259
Immediate orders	293
Implementation project	27
buy-in	51
decision-making	52
documentation	56
effort estimation	54
goal	36, 50
go-live	76
information acquisition	47
input planning	54
legacy system	54
marketing	59
priorities	691
project preparation	49
risk factors	35
stages	28
strategy	27
success factors	38
team	52
testing	75
tips	39
training	58
users	66
Indirect assignment	171
Individual purchase order	434
Industry sector	228

Index

Influencing field 154, 158
Information system 708
Initiation phase 551
Input help 642
Inspection characteristics 416
Inspection control 306
Inspection lots 485
   generate 540
   maintenance call object 427
   origins 489, 537, 548
Inspection points 484, 533, 550
   usage decision 485, 533
Inspection results transfer 496
Inspection rounds plan 426
Inspection types 484
   assign 489, 537, 548
   assign to order types 492, 536, 546
   default values 490, 537, 550
   maintain 488, 537, 547
   settings 488, 547
Inspections 492, 535, 546
Integration test 75
Internal number assignment 119
Interval document 129
Investment management 384
   assignment key 387
Investment measures 391
Investment profiles 390, 392
   define 390
Investment programs 385
   automatic assignment 386
   fields 386
Item category 242
   default values 322

K

Key combination 643
Knowledge Provider 143

L

Latest scheduled start 337
Latest staging date 340
Launchpad catalog 598, 618
Lean services 332
Legacy data transfer 31, 69
Legacy System Migration
   Workbench (LSMW) 69, 71
Linear asset 131, 170, 201, 207
Linear asset management 201
   field selection 209

Linear asset management (Cont.)
   offset type 203
Linear reference patterns 203
   type 204
List display 209
List variant 158, 635, 637
   Customizing paths 160
   layout 162, 638
   periodic job 640
   selection fields 161
Local valuation types 451
Location 95
Log line 262

M

Main screen 224
   order 224
Maintenance assembly 170, 211, 213
Maintenance call object 427
   completion data 428
Maintenance cockpit 669
Maintenance Event Builder 517, 518
Maintenance item 428
Maintenance objects 485
Maintenance order 687
   goods movement 460
Maintenance package 417
Maintenance plan 425, 631, 653, 706
   category 427, 429, 482, 537, 550, 628
   multiple counter plan 425
   set categories 426
   single cycle plan 425
   sort fields 429
   special functions 430
   strategy plan 425
   type 425
Maintenance planner group 94, 420
Maintenance plant 91
Maintenance status 220
Maintenance, repair, and
   overhaul (MRO) 463
Masking 358
Mass change 640
Master data 705
   assign fields 689
   maintenance 69
   synchronization 687
Master inspection characteristics 483
Material master 211, 442, 449, 461
   checking group 310
   field selection 227

Index

Material master (Cont.)
　global valuation types 452
　layout 217
　screen sequence 217, 462
　valuation category 453
Material planning 321
　parameters 327
Material provision indicators 467
Material requirements
　planning (MRP) 305, 461
　relevance 327, 439
Material type 212, 214, 225
　attributes 215
　BOM header 242
　BOM item 244
　spare parts 215
Materials 170, 211
　availability check 563
　free assignment 422
　list 283, 416
　number 232
　release 308
　reservations 322
　status 215, 356
　type 212
Materials management 46
Measurement document 128
　linear data 201
Measurement position 130, 196
Measuring point 127, 660, 705
　category 130
　linear data 201
　system settings 128
Message control 394
Microsoft Access 71
MIME technology 151
Movement type 237, 311, 322
　assign group 238
　group 237
Multilevel list 163
　Customizing paths 164
　field selection 163
Multilingual text maintenance 190
Multiple counter plan 425
My Inbox app 571

N

Net price 440
Network 198

Network attribute category 206, 207
　define 207
　network type 209
Network attribute properties 208
Network graphic profiles 345
Network group 206
Network type 206
　network attribute categories 209
Nonstock material 213
Notepad 665
Notification 249, 250, 517, 632, 681, 687,
　692, 706
　action box 278, 656
　activity details 278
　calibration 475
　catalog profile 270
　catalog types 266
　category 254
　code groups 267
　Customizing 251
　integration with orders 294
　maintenance call object 427
　order integration 481, 533, 545
　origin 254
　pool asset management 501
　priorities 274
　priority type 273
　processing 656, 660
　refurbishment 454
　required start/end 274
　service profile 277
　structure 251
Notification types 252, 254, 455, 489, 501,
　533, 545, 628
　allow change 263
　assign priorities 275
　assign to order types 264
　assing to order types 554
　calibration 476
　catalog profiles 271
　define 253, 554
　header 260
　long text control 261
　order types 459
　overview 260, 480, 533, 545
　response time monitoring 277
　screen areas 258
　screen templates 255
　screen types 261
　settings 508
　shift notes 520
　tabs 256

Number assignment 119
    transactions 262
    Customizing paths 120
Number range 119
    define 122

O

Object information 110
    assign key 113
    display automatically 111
    keys 110
    notification 112
Object link 170, 198
    category 199
    define media 200
Object list 283, 295, 426, 540
Object network 170, 205
Object processing 508
Object type 199
    define 199
OData service 594
    add service 595
    groups 595
Offset type 202
Open Catalog Interface (OCI) 316
Operating conditions 392
Operating resources 213
Operational effects 282, 586
Operations 283, 416
Options profile 344
Order 517, 693, 706
    default values 681
    value categories 367
Order types 287, 456, 475, 505, 531, 541, 628
    assign catalogs 315
    assign to plant 292
    assigning catalogs 315
    checklist 542
    configure 289, 515, 563
    control keys 296
    Customizing 288
    default activity types 299
    default values 293
    investment management 387
    notification integration 295
    notification type integration 568
    notification types 264, 459, 554
    phase model 563
    planning indicator 293
    reference time 397
    refurbishment processing 458

Order types (Cont.)
    shop papers 356
    valid activity types 299
Orders 249, 283, 334, 482, 669, 679, 681
    allocation structure 379
    bar charts 343
    catalog fields 318
    catalogs 314
    categories 289
    commitments 291
    completion check 681
    costing activity valuation 372
    create 284
    Customizing 285
    default current date 341
    electronic parts catalog 312
    formula definitions 336
    goods movement 328
    header 283
    investment profiles 391
    maintenance call object 427
    missing order costs 377
    network graphic profile 344
    notification integration 294, 567
    operation-level costs 376
    order sheets 369
    print diversion 359
    processing 659
    PRT control keys 393
    refurbishment 456
    settlement 377
    settlement heirarchy 383
    shop papers 355
    text labels 331
    value categories 364
    view profiles 300
Organizational area 204
Organizational structure 79, 80
Organizational unit 42, 79, 80
    maintenance-specific 94
    overview 90
    system-wide 80
Overall profile 350
    capacity leveling 351
    subprofiles 351
Overhead rate 370

P

Partner 134
    assign determination procedure 138
    determination procedure 135, 137, 503

Index

Partner (Cont.)
  editing 136
  field selection 139
  functions 135
  partner function 135
  type 135, 137
Performance-based maintenance 45
Permit 132
  categories 133
Personal value list 642
Phase model 292, 551, 696
  activate 563
  order types 568
  teams and responsibilities 568
Physical samples 485
Planned orders 293
Planner group 94, 419
Planning board 509
  display variants 510
  objects 508
Planning indicator 294
Planning phase 552
Planning plant 90
Plant 87, 107, 680
  address 88
  assign company code 89
  assign order types 292
  checklist 542
  define sections 96
  define suitability 424
  define, delete, check 87
  field reference 229
  inspection types 492, 535, 546
  maintain 90
  maintenance perspective 89
  maintenance planning plant 90
  maintenance plant 91
  screen selection 229
Plant maintenance batch input 69, 71
Plant maintenance information
  system (PMIS) 196, 283
Plant-specific maintenance 92
PM/PS reference element 385, 389
  fields 388
PM/QM coupling 469, 492, 535, 546
Pool asset management 500, 695
  assign class to notification type ... 506
  basic settings 504
  characteristics 506
  display variant 511

Pool asset management (Cont.)
  fields 507
  notification types 501, 503
  notifications 502
  objects 508
  order types 515
  planning board 509
  pool categories 513
  single class 505
  status 512
Pool categories 513
  activity type 514
Postexecution phase 553
Preparation phase 552
Preventive maintenance 415, 694
Price control 216
Primary counter 197
Print diversion 359
Printing 353
  control options 355
  define printer 357
  documents 353
  output 353
  user-specific 358
Priorities 273, 586
  type 273
Prioritization 559
  profile 557
Priority handling 682
Priority types 273
Procurement default values 630
Procurement milestones 565, 566
  assignment rules 566
  event type linkage 566
Procurement type 216
Production resources/tools (PRTs) 187,
    283, 416
  control keys 393
Profile 151, 420
  assignments 400, 438
  budget 291
  capacity evaluation 351
  catalog 255, 265, 269, 429, 479
  data entry 407
  default values 420
  define 151
  detection method 561
  function 569
  graphic 342
  investment 390
  options 344
  overall 350

Profile (Cont.)
prioritization 557
response 275, 276
serial number 233, 464
service 275
settlement 291, 381, 456
status 572
value 437
view 175, 201, 300, 470
workstation applications 152
Profile assignments 400, 438
Project management 36, 53, 384
Project plan 56, 691
Project-based maintenance 696
Prototype 72
Purchase order 434, 435, 440, 463
define document types 466
Purchase requisition 329, 435, 440, 463, 681
collective 439
document types 466
text types 330

Q

Q score 491, 494
Quality management 483
inspection types 488

R

Recording configuration 499, 500, 538, 550
Reducing downtime 51
Reference functional location 170, 171
Reference group 66
Reference objects 524, 627
Reference time 397, 428
Refurbishment 448, 695
notifications 454
order 456
processing 458
settlement profile 457
Refurbishment order 448
Release material 308
Release procedure 446
Release strategy 447
Rental price 514
Repairable spare 448
Replenishment lead time 311
Report and Repair Malfunction app 630
Request Maintenance app 255
Response monitoring 275
Response profiles 275, 276
tasks 276

Results analysis keys 375
Results recording 497
Revision 517
Revision type 517
Roadmap Viewer 33
Role-based menu 633
Roles 603, 618, 633

S

SAP Activate 29
accelerator 58
deploy phase 32, 75
discover phase 30, 39
explore phase 31, 65
phases overview 30
prepare phase 30, 48
realize phase 32, 74
run phase 33, 77
SAP Business Workflow 685
SAP Extended Warehouse
 Management (SAP EWM) 237
SAP Fiori 581, 663
app implementation 592
app search 590
app types 582
apps 605
basics 581
benefits 585
characteristics 585
design 581
downsides 586
initialize app 619
real vs. pseudo apps 584
reference library 590, 612
SAP Fiori launchpad 581
business groups 601
configure 586
default values 626
designer 601, 617
dynamic lists 640
features 586
launchpad catalog 598
my home 634
non-SAPUI5 apps 610
roles and authorization 603, 618
SAPUI5 apps 586
transactions 588
SAP GUI 585, 586
transactions 611
SAP List Viewer 635
further processing 640

Index

SAP List Viewer (Cont.)
   list display 638
   monitor 638
   selection option 637
   selection variant 636
SAP NetWeaver Application Server
   for ABAP 597
SAP S/4HANA Asset Management 22
   roadmap 33
SAP S/4HANA Finance 81
SAP S/4HANA trial 40
SAP Screen Personas 670
   confirmation cockpit 671
   control element 671
   storage location cockpit 672
SAP Service and Asset Manager 657
   basic functions 659, 661, 662
   local layout 657
   tablet 657
SAPscript form 355, 504, 506
SAPUI5 .... 582, 585, 586
   apps 605
Scenario test 75
Scheduled repair 249
Scheduling 332
   current date 341
   dates 340
   parameters 339
   phase 552
   type 337
Scope of check 305
   define 310
Scrapping 449
Screen area 220
   accounting 222
   basic data 220
   MRP 222
   plant/storage 222
   purchasing 221
Screen group 100
Screen layout
   available screens 176
   technical object 174
Screen reference 225
Screen sequence 218, 462
   assign 225
   control 226
   data screen 219
   refurbishment 462
Screen structure 219
Screening phase 551

Secondary screen 223
   order 224
Secure storage area 142
Selection variant 636
Serial data 187
Serial number 170, 232, 448
   goods issue 238
   previously created 234
   profile 233
   serializing procedures 235
Serial number profiles 464, 473, 537, 549
Serialization attributes 237
Serializing procedure 235, 236, 474
   level 237
   maintenance processing 236
   subcontracting 465
Service category 442, 443
Service entry sheets 440
   maintenance call object 427
   release procedure 446
Service master 442
Service profiles 275, 277
Service specification 435, 440
Settlement 377, 391
   cost elements 380
Settlement profiles 291, 456, 457
   maintain 381
Settlement rules 283, 381, 682
   distribution rules 383
   time 383
Shift definitions 349
Shift notes 45, 519
   categories 525
   independent 523
   notifications 520
   origin 522
   settings for type 521
   type 520
Shift reports 519
   classification 529
   grouping 528
   layout 528
   PDFs 527
   type 526
Shift sequences 348, 349
Shop papers 354, 505, 543
   define 354, 544
   email 360
   status control 362
Single cycle plan 425
Single roles 603
Software component versions 589

Sort field 430
Source determination 445
Spare parts 46, 213, 215
   assignment 239
   class code 461
Special period 84
Split valuation 450
   configure 451
Standard value key 102
Static apps 583
Statistical process control (SPC) 483
Status control
   shop papers 362
Status management 113
Status number 116
Status profile 116, 572
   assign to object 118
   conditions 574
   object types 117
Stock check indicator 234
Stock types 311
Storage category 148
Storage location 680
   cockpit 672
Storage system 147
Strategy plan 425
Structure description 239
Structure indicator 172
   details 173
   label 174
Structuring of technical systems 169
Subcontracting 374, 463
   document types 467
   process 463
Subnetworks 389
Suitability 424
Summarization levels 363
System conditions 281
System data 589
System landscape 64
System messages 395, 443
System status 113, 307, 512

T

Table control 644
   operations/components 645
   settings 644
Task list 98, 103, 400, 416, 426, 538, 653, 681, 706
   characteristics 483
   data 438

Task list (Cont.)
   status 417
   transfer to orders 629
   usage 418
Tasks 278
Tasks catalog type 275
Team category 569
Teams and responsibilities 568
   define functions 568
   workflows 570
Technical catalog 600, 614
Technical completion 397
Technical objects 174, 652
Technical systems 169
Test plan 57
Test/measurement equipment 468, 472, 674, 696
   code groups 494
   control key 487
   equipment category 472
   follow-up action 495
   notification screen 480
   order type 475
   serial numbers 474
   storage 473
   task list 491
   view profiles 470
Time sheet 660
Tolerance period 131
Training course 59
Transaction 611
   /IFWND/MAINT_SERVICE 593, 594
   /IFWND/V4_ADMIN 595
   /IWFND/MAINT_SERVICE 588
   /IWFND/V4_ADMIN 588, 593
   /UI2/FLP 588
   /UI2/FLPAM 588
   /UI2/FLPCM_CONF 588, 615
   /UI2/FLPCM_CUST 588, 598, 615
   /UI2/FLPD_CUST 588, 601
   ADPMPS 388
   BAPI 675
   BG20 122
   CAC2 410
   CL02 .... 269, 478, 507
   CM01 .... 346, 350, 351
   CM33 .... 346, 351
   CM34 346
   CT04 .... 269, 478, 507
   CUST_VV_T16FS_3 446
   CV01N–CV04N 141
   DC10 143

Transaction (Cont.)
- DIWPSC4 518
- GM01 124
- GM04 126
- I_GRAPH_MONITOR 342
- IA08 635
- IA09 635
- IA10 163
- IBIP 69
- IE05 635
- IE07 163
- IE36 635
- IE37 635
- IE4N 674
- IH01 .... 179, 210, 246
- IH03 246
- IH04 246
- IH05 246
- IH07 635
- IH08 635
- IH09 635
- IH12 246
- IH18 163
- IK07 635
- IK07R 635
- IK08 635
- IK08R 635
- IK09 122
- IK11 671
- IK17 635
- IK18 635
- IK19 122
- IL05 635
- IL07 163
- IL15 635
- IN15 635
- IN16 635
- IN18 635
- IN19 635
- IN20 122
- IN21 205
- IN24 205
- IN25 205
- IP10 674
- IP15 635
- IP16 635
- IP17 635
- IP18 635
- IP20 122
- IP21 122
- IP24 635
- IP30H 429

Transaction (Cont.)
- IP43 430
- IPM2 635
- IPM3 635
- IPMD 132
- IQ08 635
- IQS21 263
- IQS22 263
- ISHN4 635
- ISHR4 635
- IW20 122
- IW21 .... 263, 560
- IW22 672
- IW26 263
- IW28 .... 635, 669
- IW29 635
- IW30 163
- IW31 .... 540, 647, 666, 671
- IW32 .... 344, 672
- IW37 635
- IW37N 635
- IW38 635
- IW39 635
- IW3K 635
- IW3L 635
- IW3M 635
- IW40 163
- IW41 .... 671, 674
- IW42 .... 405, 671
- IW43 671
- IW47 .... 635, 671
- IW49 635
- IW49N 635
- IW64 635
- IW65 635
- IW66 635
- IW67 635
- IW69 672
- IW92 .... 538, 540
- IW99 540
- IWR1 517
- KL01 514
- KP26 .... 377, 514, 515
- LSMW 69
- MB51 672
- MCIZ 196, 197
- MD04 .... 461, 462
- ME59N 445
- MIGO 672
- OAC0 148
- OACT 149
- OIAB 96

Index

Transaction (Cont.)

OIAD	156
OIAE	156
OIAF	156
OIAL	156
OIAN	156
OIBS	116
OICD	241
OICG	242
OICH	240
OICJ	246
OICK	243
OICMPD	156
OICO	467
OICP	244
OICQ	247
OIDA	456, 475, 506, 532, 541
OIDF	354, 544
OIDG	354, 544
OIDH	357
OIDI	357
OIDJ	357
OIEN	122
OIEV	138
OIEZ	189
OIK1	364
OIK2	365
OIL0	122
OIL1	417
OIL4	122
OIL5	122
OIL6	421
OILJ	423
OIMD	111
OIMF	276
OINM	200
OIO4	298
OIO5	298
OIO6	296
OIO9	342
OIOA	290, 563
OIOD	292
OIOF	375
OIOI	306
OIOM	138
OION	122
OIOPD	156
OIOR	402
OIOS	293
OIPK	173
OIPU	181
OIPV	181

Transaction (Cont.)

OIR1–OIR8	140
OIS2	234, 465, 473
OITA	390
OIVC	367
OIW3	161
OIW5	161
OIW6	161
OIWI	161
OIWL	161
OIWO	161
OIWP	179
OIWU	161
OIWW	161
OIWY	161
OIYC	161
OIYH	161
OIZ2	102
OIZA	99
OIZD	103
OIZN	156
OIZU	106
OKP6	374
OKP8	372
OLI5N	377
OMS3	229
OMS9	230
OMSA	229
OMSR	228
OMT3B	219, 462
OMT3E	225
OMT3R	224
OMT3Z	223
OMW0	450
OMWC	451
OP4A	348
OP5A–OP5H	156
OP7B	334
OPA6	350
OPD0	351
OPFA	101, 156
OPJJ	311
OPJN	338
OPK3	336
OPTP	389
OPU7	339
OQN6	270, 478
ORPS7	526
OVZ2	309
OX06	84
OX18	89
OX19	81, 86

Transaction (Cont.)
    PAM03 510
    PFCG 588, 603, 633
    QA11 674
    QE17 674
    QS41 .... 266, 478, 525
    QS51 .... 538, 550
    S_ALR_87000248_2 273
    S_ALR_87000310_1 238
    S_ALR_87000310_2 238
    S_ALR_87000811_2 138
    S_ALR_87000921_1 276
    S_ALR_87000995_1 278
    S_ALR_87001010_1 266
    S_ALR_87001148_1 273
    S_ALR_87001148_3 273
    SBWP 361
    SCOT 361
    SE16N 633
    SE19 683
    SE93 651
    SFW5 .... 22, 313
    SHD0 647
    SHN4 635
    SHR4 635
    SICF 588, 593, 597, 614
    SMOD 680
    SPAD 361
    start values 262
    SU01 588
    SU3 624
    SWBP 686
    SWDD 685
    upstream 673
    V_T001 87
    V_T355E 276
    V_TQ07 495
    VC_EAM_OVRL_STS 572
    VOP2 136
    WPS1 517
    Z_IW31_E 651
    ZMV01 674
Transaction control 117
Transaction variant 646
    field selection 647
    function key 649
    menu bar 649
    tab selection 649
    uses 646
Transaction/event chain diagram (ECD) 62
Transactional apps 583
Transfer characteristics 680

U

UI Editor 671
Units for operation 342
Unplanned orders 293
Upstream transaction 673
Usability 66, 623
    mobile 663
    SAP Fiori 664
Usage 242
    BOM 423
    priorities 245
    task list 418
Usage decision 485, 491, 493, 533, 537, 550
    automatic 497, 499, 538, 550
    inspection lot 496
Usage history 187
    reference category 188
User acceptance 66
User department 215
User field 423
    type 424
User parameter 624, 627
    commonly used 625
    maintenance-specific 627, 633
User status 114
    define 114
    object 114
    without status number 116

V

Valid receivers 381
Valuation category 449, 453
Valuation code 494
Valuation types 449, 452
Valuation variants 368, 372
Value categories 363, 364
    cost elements 365
    default values 366
    retroactive changes 366
Value chain diagram (VCD) 60
Value list 643
Value profiles 398, 437
    default 398
Value updating 217
Variances 404
Vehicle scheduling 501
Vendor field selection 140

Index

View profiles 175, 192, 201, 300, 470, 536, 549
   assign 303
   buttons 303
   define 300
   navigation 303
   screen areas 301
Virtual desktop 643

W

Warning period 567
Warranty 44, 124, 706
   category 124
   counter 126
   delete 125
   type 125
   vendor vs. manufacturer 125
Web Dynpro 154, 610, 645
   apps 620

Web interface 677
Work break schedules 348
Work center 94, 97, 103, 108, 333, 434, 653
   category 98, 99
   define 98
   person as a work center 97, 108
   screen sequence 106
   standard value key 102
   type 101
Work item 686
Work order cycle 249, 653, 696
   phase model 551
Work scheduler group 420
Workflow 570, 685
   activate scenario 571
Workflow Builder 685
Workshop 44, 97
Workstation application 149, 152
WRICEF list 57

- Run preventive maintenance, repairs, refurbishment, and more in SAP S/4HANA
- Manage routine plant operations with step-by-step instructions
- Improve your usability and workflows with mobile apps, SAP Fiori, and SAP Intelligent Asset Management

Karl Liebstückel

Plant Maintenance with SAP S/4HANA

Business User Guide

Your company is now on SAP S/4HANA—so how do you run your plant maintenance operations on this new system? Between these pages, you'll find the detailed, step-by-step instructions you need for your routine (and non-routine) functions. Ample screenshots walk you through scheduling repairs, planning maintenance cycles, completing inspections, and all the tasks you perform to keep your assets in shape. With information on new UIs and mobile apps, this guide is the only one you need!

665 pages, pub. 10/2020
E-Book: $74.99 | **Print:** $79.95 | **Bundle:** $89.99

www.sap-press.com/5180

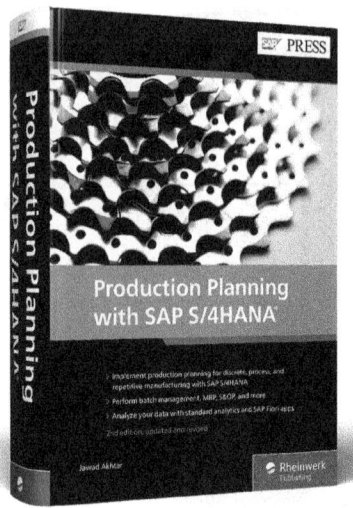

- Implement production planning for discrete, process, and repetitive manufacturing with SAP S/4HANA
- Perform batch management, MRP, S&OP, and more
- Analyze your data with standard analytics and SAP Fiori apps

Jawad Akhtar

Production Planning with SAP S/4HANA

Streamline your production planning process with SAP S/4HANA! Get step-by-step instructions for configuring and using SAP S/4HANA for discrete, process, and repetitive manufacturing. Then dive into production tools and functionalities like batch management, S&OP, predictive MRP, DDMRP, and the Early Warning System. This foundational guide is full of industry examples to help you maximize your production planning!

1092 pages, 2nd edition, pub. 08/2021
E-Book: $84.99 | **Print:** $89.95 | **Bundle:** $99.99

www.sap-press.com/5373

Interested in reading more?

Please visit our website for all new book
and e-book releases from SAP PRESS.

www.sap-press.com